Judith E. Brown

Nutrition

Through the Life Cycle

third edition

Dietary Reference Intakes (DRI)

The Dietary Reference Intakes (DRI) include two sets of values that serve as goals for nutrient intake—Recommended Dietary Allowances (RDA) and Adequate Intakes (AI). The RDA reflect the average daily amount of a nutrient considered adequate to meet the needs of most healthy people. If there is insufficient evidence to determine an RDA, an AI is set. AI are more tentative than RDA, but both may be used as goals for nutrient intakes. (Chapter 1 provides more details.)

In addition to the values that serve as goals for nutrient intakes (presented in the tables on these two pages), the DRI include a set of values called Tolerable Upper Intake Levels (UL). The UL represent the maximum amount of a nutrient that appears safe for most healthy people to consume on a regular basis. Turn the page for a listing of the UL for selected vitamins and minerals.

Estimated Energy Requirements (EER), Recommended Dietary Allowances (RDA), and Adequate Intakes (AI) for Water, Energy, and the Energy Nutrients

Age(yr)	Reference BMI (kg/m²)	Reference height, cm (in)	Reference weight, kg (lb)	Water[a] AI (L/day)	Energy EER[b] (kcal/day)	Carbohydrate RDA (g/day)	Total fiber AI (g/day)	Total fat AI (g/day)	Linoleic acid AI (g/day)	Linolenic acid[c] AI (g/day)	Protein RDA (g/day)[d]	Protein RDA (g/kg/day)
Males												
0–0.5	—	62 (24)	6 (13)	0.7[e]	570	60	—	31	4.4	0.5	9.1	1.52
0.5–1	—	71 (28)	9 (20)	0.8[f]	743	95	—	30	4.6	0.5	13.5	1.5
1–3[g]	—	86 (34)	12 (27)	1.3	1046	130	19	—	7	0.7	13	1.1
4–8[g]	15.3	115 (45)	20 (44)	1.7	1742	130	25	—	10	0.9	19	0.95
9–13	17.2	144 (57)	36 (79)	2.4	2279	130	31	—	12	1.2	34	0.95
14–18	20.5	174 (68)	61 (134)	3.3	3152[h]	130	38	—	16	1.6	52	0.85
19–30	22.5	177 (70)	70 (154)	3.7	3067[h]	130	38	—	17	1.6	56	0.8
31–50				3.7	3067[h]	130	38	—	17	1.6	56	0.8
>50				3.7	3067[h]	130	30	—	14	1.6	56	0.8
Females												
0–0.5	—	62 (24)	6 (13)	0.7[e]	520	60	—	31	4.4	0.5	9.1	1.52
0.5–1	—	71 (28)	9 (20)	0.8[f]	676	95	—	30	4.6	0.5	13.5	1.5
1–3[g]	—	86 (34)	12 (27)	1.3	992	130	19	—	7	0.7	13	1.1
4–8[g]	15.3	115 (45)	20 (44)	1.7	1642	130	25	—	10	0.9	19	0.95
9–13	17.4	144 (57)	37 (81)	2.1	2071	130	26	—	10	1.0	34	0.95
14–18	20.4	163 (64)	54 (119)	2.3	2368	130	26	—	11	1.1	46	0.85
19–30	21.5	163 (64)	57 (126)	2.7	2403[i]	130	25	—	12	1.1	46	0.8
31–50				2.7	2403[i]	130	21	—	12	1.1	46	0.8
>50				2.7	2403[i]	130	21	—	11	1.1	46	0.8
Pregnancy												
1st trimester				3.0	+0	175	28	—	13	1.4	+25	1.1
2nd trimester				3.0	+340	175	28	—	13	1.4	+25	1.1
3rd trimester				3.0	+452	175	28	—	13	1.4	+25	1.1
Lactation												
1st 6 months				3.8	+330	210	29	—	13	1.3	+25	1.1
2nd 6 months				3.8	+400	210	29	—	13	1.3	+25	1.1

NOTE: For all nutrients, values for infants are AI. Dashes indicate that values have not been determined.

[a] The water AI includes drinking water, water in beverages, and water in foods; in general, drinking water and other beverages contribute about 70 to 80 percent, and foods, the remainder. Conversion factors: 1 L = 33.8 fluid oz; 1 L = 1.06 qt; 1 cup = 8 fluid oz.

[b] The Estimated Energy Requirement (EER) represents the average dietary energy intake that will maintain energy balance in a healthy person of a given gender, age, weight, height, and physical activity level. The values listed are based on an "active" person at the reference height and weight and at the midpoint ages for each group until age 19. Chapter 8 and Appendix F provide equations and tables to determine estimated energy requirements.

[c] The linolenic acid referred to in this table and text is the omega-3 fatty acid known as alpha-linolenic acid.

[d] The values listed are based on reference body weights.

[e] Assumed to be from human milk.

[f] Assumed to be from human milk and complementary foods and beverages. This includes approximately 0.6 L (~3 cups) as total fluid including formula, juices, and drinking water.

[g] For energy, the age groups for young children are 1–2 years and 3–8 years.

[h] For males, subtract 10 kcalories per day for each year of age above 19.

[i] For females, subtract 7 kcalories per day for each year of age above 19.

SOURCE: Adapted from the *Dietary Reference Intakes* series, National Academies Press. Copyright 1997, 1998, 2000, 2001, 2002, 2004, by the National Academies of Sciences.

Recommended Dietary Allowances (RDA) and Adequate Intakes (AI) for Vitamins

Age (yr)	Thiamin RDA (mg/day)	Riboflavin RDA (mg/day)	Niacin RDA (mg/day)[a]	Biotin AI (μg/day)	Pantothenic acid AI (mg/day)	Vitamin B_6 RDA (mg/day)	Folate RDA (μg/day)[b]	Vitamin B_{12} RDA (μg/day)	Choline AI (mg/day)	Vitamin C RDA (mg/day)	Vitamin A RDA (μg/day)[c]	Vitamin D AI (μg/day)[d]	Vitamin E RDA (mg/day)[e]	Vitamin K AI (μg/day)
Infants														
0–0.5	0.2	0.3	2	5	1.7	0.1	65	0.4	125	40	400	5	4	2.0
0.5–1	0.3	0.4	4	6	1.8	0.3	80	0.5	150	50	500	5	5	2.5
Children														
1–3	0.5	0.5	6	8	2	0.5	150	0.9	200	15	300	5	6	30
4–8	0.6	0.6	8	12	3	0.6	200	1.2	250	25	400	5	7	55
Males														
9–13	0.9	0.9	12	20	4	1.0	300	1.8	375	45	600	5	11	60
14–18	1.2	1.3	16	25	5	1.3	400	2.4	550	75	900	5	15	75
19–30	1.2	1.3	16	30	5	1.3	400	2.4	550	90	900	5	15	120
31–50	1.2	1.3	16	30	5	1.3	400	2.4	550	90	900	5	15	120
51–70	1.2	1.3	16	30	5	1.7	400	2.4	550	90	900	10	15	120
>70	1.2	1.3	16	30	5	1.7	400	2.4	550	90	900	15	15	120
Females														
9–13	0.9	0.9	12	20	4	1.0	300	1.8	375	45	600	5	11	60
14–18	1.0	1.0	14	25	5	1.2	400	2.4	400	65	700	5	15	75
19–30	1.1	1.1	14	30	5	1.3	400	2.4	425	75	700	5	15	90
31–50	1.1	1.1	14	30	5	1.3	400	2.4	425	75	700	5	15	90
51–70	1.1	1.1	14	30	5	1.5	400	2.4	425	75	700	10	15	90
>70	1.1	1.1	14	30	5	1.5	400	2.4	425	75	700	15	15	90
Pregnancy														
≤18	1.4	1.4	18	30	6	1.9	600	2.6	450	80	750	5	15	75
19–30	1.4	1.4	18	30	6	1.9	600	2.6	450	85	770	5	15	90
31–50	1.4	1.4	18	30	6	1.9	600	2.6	450	85	770	5	15	90
Lactation														
≤18	1.4	1.6	17	35	7	2.0	500	2.8	550	115	1200	5	19	75
19–30	1.4	1.6	17	35	7	2.0	500	2.8	550	120	1300	5	19	90
31–50	1.4	1.6	17	35	7	2.0	500	2.8	550	120	1300	5	19	90

NOTE: For all nutrients, values for infants are AI.
[a]Niacin recommendations are expressed as niacin equivalents (NE), except for recommendations for infants younger than 6 months, which are expressed as preformed niacin.
[b]Folate recommendations are expressed as dietary folate equivalents (DFE).
[c]Vitamin A recommendations are expressed as retinol activity equivalents (RAE).
[d]Vitamin D recommendations are expressed as cholecalciferol and assume an absence of adequate exposure to sunlight.
[e]Vitamin E recommendations are expressed as α-tocopherol.

Recommended Dietary Allowances (RDA) and Adequate Intakes (AI) for Minerals

Age (yr)	Sodium AI (mg/day)	Chloride AI (mg/day)	Potassium AI (mg/day)	Calcium AI (mg/day)	Phosphorus RDA (mg/day)	Magnesium RDA (mg/day)	Iron RDA (mg/day)	Zinc RDA (mg/day)	Iodine RDA (μg/day)	Selenium RDA (μg/day)	Copper RDA (μg/day)	Manganese AI (mg/day)	Fluoride AI (mg/day)	Chromium AI (μg/day)	Molybdenum RDA (μg/day)
Infants															
0–0.5	120	180	400	210	100	30	0.27	2	110	15	200	0.003	0.01	0.2	2
0.5–1	370	570	700	270	275	75	11	3	130	20	220	0.6	0.5	5.5	3
Children															
1–3	1000	1500	3000	500	460	80	7	3	90	20	340	1.2	0.7	11	17
4–8	1200	1900	3800	800	500	130	10	5	90	30	440	1.5	1.0	15	22
Males															
9–13	1500	2300	4500	1300	1250	240	8	8	120	40	700	1.9	2	25	34
14–18	1500	2300	4700	1300	1250	410	11	11	150	55	890	2.2	3	35	43
19–30	1500	2300	4700	1000	700	400	8	11	150	55	900	2.3	4	35	45
31–50	1500	2300	4700	1000	700	420	8	11	150	55	900	2.3	4	35	45
51–70	1300	2000	4700	1200	700	420	8	11	150	55	900	2.3	4	30	45
>70	1200	1800	4700	1200	700	420	8	11	150	55	900	2.3	4	30	45
Females															
9–13	1500	2300	4500	1300	1250	240	8	8	120	40	700	1.6	2	21	34
14–18	1500	2300	4700	1300	1250	360	15	9	150	55	890	1.6	3	24	43
19–30	1500	2300	4700	1000	700	310	18	8	150	55	900	1.8	3	25	45
31–50	1500	2300	4700	1000	700	320	18	8	150	55	900	1.8	3	25	45
51–70	1300	2000	4700	1200	700	320	8	8	150	55	900	1.8	3	20	45
>70	1200	1800	4700	1200	700	320	8	8	150	55	900	1.8	3	20	45
Pregnancy															
≤18	1500	2300	4700	1300	1250	400	27	12	220	60	1000	2.0	3	29	50
19–30	1500	2300	4700	1000	700	350	27	11	220	60	1000	2.0	3	30	50
31–50	1500	2300	4700	1000	700	360	27	11	220	60	1000	2.0	3	30	50
Lactation															
≤18	1500	2300	5100	1300	1250	360	10	14	290	70	1300	2.6	3	44	50
19–30	1500	2300	5100	1000	700	310	9	12	290	70	1300	2.6	3	45	50
31–50	1500	2300	5100	1000	700	320	9	12	290	70	1300	2.6	3	45	50

Tolerable Upper Intake Levels (UL) for Vitamins*

Age (yr)	Niacin (mg/day)[a]	Vitamin B_6 (mg/day)	Folate (µg/day)[a]	Choline (mg/day)	Vitamin C (mg/day)	Vitamin A (µg/day)[b]	Vitamin D (µg/day)	Vitamin E (mg/day)[c]
Infants								
0–0.5	—	—	—	—	—	600	25	—
0.5–1	—	—	—	—	—	600	25	—
Children								
1–3	10	30	300	1000	400	600	50	200
4–8	15	40	400	1000	650	900	50	300
9–13	20	60	600	2000	1200	1700	50	600
Adolescents								
14–18	30	80	800	3000	1800	2800	50	800
Adults								
19–70	35	100	1000	3500	2000	3000	50	1000
>70	35	100	1000	3500	2000	3000	50	1000
Pregnancy								
≤18	30	80	800	3000	1800	2800	50	800
19–50	35	100	1000	3500	2000	3000	50	1000
Lactation								
≤18	30	80	800	3000	1800	2800	50	800
19–50	35	100	1000	3500	2000	3000	50	1000

* UL = The maximum level of daily nutrient intake that is likely to pose no risk of adverse effects. Unless otherwise specified, the UL represents total intake from food, water, and supplements. Due to lack of suitable data, ULs could not be established for vitamin K, thiamin, riboflavin. vitamin B_{12}, pantothenic acid, biotin, or carotenoids. In the absence of ULs, extra caution may be warranted in consuming levels above recommended intakes.

[a] The UL for niacin and folate apply to synthetic forms obtained from supplements, fortified foods, or a combination of the two.
[b] The UL for vitamin A applies to preformed vitamin A only.
[c] The UL for vitamin E applies to any form of supplemental α-tocopherol, fortified foods, or a combination of the two.

Tolerable Upper Intake Levels (UL) for Minerals*

Age (yr)	Sodium (mg/day)	Chloride (mg/day)	Calcium (mg/day)	Phosphorus (mg/day)	Magnesium (mg/day)[d]	Iron (mg/day)
Infants						
0–0.5	—[e]	—[e]	—	—	—	40
0.5–1	—[e]	—[e]	—	—	—	40
Children						
1–3	1500	2300	2500	3000	65	40
4–8	1900	2900	2500	3000	110	40
9–13	2200	3400	2500	4000	350	40
Adolescents						
14–18	2300	3600	2500	4000	350	45
Adults						
19–70	2300	3600	2500	4000	350	45
>70	2300	3600	2500	3000	350	45
Pregnancy						
≤18	2300	3600	2500	3500	350	45
19–50	2300	3600	2500	3500	350	45
Lactation						
≤18	2300	3600	2500	4000	350	45
19–50	2300	3600	2500	4000	350	45

* UL = The maximum level of daily nutrient intake that is likely to pose no risk of adverse effects. Unless otherwise specified, the UL represents total intake from food, water, and supplements. Due to lack of suitable data, ULs could not be established for arsenic, chromium, and silicon. In the absence of ULs, extra caution may be warranted in consuming levels above recommended intakes.
[d] The UL for magnesium applies to synthetic forms obtained from supplements or drugs only.
[e] Source of intake should be from human milk (or formula) and food only.

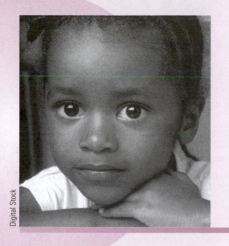

Nutrition
Through the Life Cycle
THIRD EDITION

Judith E. Brown
Ph.D., M.P.H., R.D.
University of Minnesota

with

Janet S. Isaacs, Ph.D., R.D.
Johns Hopkins Hospital

U. Beate Krinke, Ph.D., M.P.H., R.D.
University of Minnesota

Maureen A. Murtaugh, Ph.D., R.D.
University of Utah School of Medicine

Carolyn Sharbaugh, M.S., R.D.
Nutrition Consultant

Jamie Stang, Ph.D., M.P.H., R.D.
University of Minnesota

Nancy H. Wooldridge, M.S., R.D., L.D.
University of Alabama at Birmingham

THOMSON
WADSWORTH

Australia • Brazil • Canada • Mexico • Singapore
Spain • United Kingdom • United States

AQUISITIONS EDITOR: Peter Adams
DEVELOPMENT EDITOR: Nedah Rose
ASSISTANT EDITOR: Elesha Feldman
EDITORIAL ASSISTANT: Kate Franco/Elizabeth Downs
TECHNOLOGY PROJECT MANAGER: Ericka Yeoman-Saler
MARKETING MANAGER: Jennifer Somerville
MARKETING ASSISTANT: Jessica Capizano
MARKETING COMMUNICATIONS MANAGER: Shemika Britt
CONTENT PROJECT MANAGER, EDITORIAL
    PRODUCTION: Teresa Trego
CREATIVE DIRECTOR: Rob Hugel
ART DIRECTOR: John Walker
PRINT BUYER: Rebecca Cross
PERMISSIONS EDITOR: Roberta Broyer
PRODUCTION SERVICE: PreMedia
TEXT DESIGNER: Ann Borman
PHOTO RESEARCHER: Pre-PressPMG
COPY EDITOR: Pre-PressPMG
ILLUSTRATOR: David Farr, Pre-PressPMG
COVER DESIGNER: Cuttris & Hambleton, Berkeley
COVER IMAGES: Getty Images and iStock
    International Inc.
COVER PRINTER: Edwards Brothers
COMPOSITOR: Pre-PressPMG
PRINTER: Edwards Brothers

Printed in the United States of America
1 2 3 4 5 6 7 11 10 09 08 07

THOMSON HIGHER EDUCATION
10 Davis Drive
Belmont, CA 94002–3098
USA

ASIA
THOMSON LEARNING
5 Shenton Way
#01–01 UIC Building
Singapore 068808

AUSTRALIA/NEW ZEALAND
Thomson Learning Australia
102 Dodds Street
Southbank, Victoria 3006
Australia

CANADA
Thomson Nelson
1120 Birchmount Road
Toronto, Ontario M1K 5G4
Canada

UK/EUROPE/MIDDLE EAST/AFRICA
Thomson Learning
High Holborn House
50/51 Bedford Row
London WC1R 4LR
United Kingdom

LATIN AMERICA
Thomson Learning
Seneca, 53
Colonia Polanco
11560 Mexico D.F.
Mexico

SPAIN/PORTUGAL
Thomson Paraninfo
Calle Magallanes, 25
28015 Madrid, Spain

Library of Congress Control Number: 2007903321
Student Edition:
ISBN-13: 978-0-495-11637-0
ISBN-10: 0-495-11637-8

Contents in Brief

Preface   xix

Chapter 1
Nutrition Basics   1

Chapter 2
Preconception Nutrition   47

Chapter 3
Preconception Nutrition   67
CONDITIONS AND INTERVENTIONS

Chapter 4
Nutrition during Pregnancy   83

Chapter 5
Nutrition during Pregnancy   129
CONDITIONS AND INTERVENTIONS

Chapter 6
Nutrition during Lactation   155

Chapter 7
Nutrition during Lactation   189
CONDITIONS AND INTERVENTIONS

Chapter 8
Infant Nutrition   219

Chapter 9
Infant Nutrition   245
CONDITIONS AND INTERVENTIONS

Chapter 10
Toddler and
Preschooler Nutrition   265

Chapter 11
Toddler and
Preschooler Nutrition   293
CONDITIONS AND INTERVENTIONS

Chapter 12
Child and
Preadolescent Nutrition   307

Chapter 13
Child and
Preadolescent Nutrition   335
CONDITIONS AND INTERVENTIONS

Chapter 14
Adolescent Nutrition   353

Chapter 15
Adolescent Nutrition   383
CONDITIONS AND INTERVENTIONS

Chapter 16
Adult Nutrition   407

Chapter 17
Adult Nutrition   437
CONDITIONS AND INTERVENTIONS

Chapter 18
Nutrition and Older Adults   455

Chapter 19
Nutrition and Older Adults   487
CONDITIONS AND INTERVENTIONS

Appendix A
CDC Growth Charts   A-1

Appendix B
Nutrient Intakes of Adults
Aged 70 and Older   A-9

Appendix C
Measurement Abbreviations
and Equivalents   A-11

Appendix D
Body Mass Index (BMI)   A-13

Glossary   G-1

Index   I-1

Contents

Preface   xix

All chapters include Resources and References

Chapter 1
Nutrition Basics   1

Introduction   2

Principles of the Science of Nutrition   2
Essential and Nonessential Nutrients   3
Dietary Intake Standards   3
Carbohydrates   4
Protein   5
Fats (Lipids)   6
Vitamins   12
Other Substances in Food   13
Minerals   13
Water   21

Nutritional Labeling   32
Nutrition Facts Panel   32
Ingredient Label   33
Dietary Supplement Labeling   33
Herbal Remedies   33
Functional Foods   34

**Meeting Nutritional Needs across
the Life Cycle   35**
Dietary Considerations Based on Ethnicity   35
Dietary Considerations Based on Religion   35

Nutritional Assessment   36
Community-Level Assessment   36
Individual-Level Nutritional Assessment   36
Dietary Assessment   36
Anthropometric Assessment   38
Biochemical Assessment   38

Monitoring the Nation's Nutritional Health   38

Public Food and Nutrition Programs   39
WIC   39

**Nationwide Priorities for Improvements
in Nutritional Health   39**
U.S. Nutrition and Health Guidelines   40

Chapter 2
Preconception Nutrition   47

Introduction   48

Preconception Overview   48
2010 Nutrition Objectives for the Nation Related
to the Preconceptional Period   48

Reproductive Physiology   49
Female Reproductive System   49
Male Reproductive System   52

Sources of Disruptions in Fertility   52

Nutrition-Related Disruptions in Fertility   52
Undernutrition and Fertility   53
Body Fat and Fertility   54

Nutrition Time Line ⟶

1621
First Thanksgiving
feast at Plymouth
colony

Photo Disc

1702
First coffeehouse
in America opens
in Philadelphia

Photo Disc

1734
Scurvy recognized

Exercise and Infertility   55
Diet and Fertility   55
Other Factors Contributing to Infertility in Males   57

Nutrition and Contraceptives   58
Oral Contraceptives and Nutritional Status   59
Contraceptive Injections   59
Contraceptive Implants   59
Contraceptive Patches   59
Emerging Forms of Contraceptives   60

Other Preconceptual Nutrition Concerns   60
Very Early Pregnancy Nutrition Exposures   60
Recommended Dietary Intakes for Preconceptional
    Women   61

Model Preconceptional Nutrition Programs   61
Preconceptional Benefits of WIC   61
Decreasing Iron Deficiency in Preconceptional Women
    in Indonesia   62
Preconception Care: Preparing for Pregnancy   62

Chapter 3
Preconception Nutrition   67
Conditions and Interventions

Introduction   68

Premenstrual Syndrome   68
Caffeine Intake and PMS   68
Exercise and Stress Reduction   68
Magnesium, Calcium, Vitamin D, and Vitamin B$_6$ Supplements
    and PMS Symptoms   69

Obesity and Fertility   69
Central Body Fat and Fertility   70
Weight Loss and Fertility   70

Hypothalamic Amenorrhea   70
Nutritional Management of Hypothalamic Amenorrhea   71

The Female Athlete Triad and Infertility   71
Nutritional Management of the Female Athlete Triad   71

Eating Disorders and Fertility   72
Nutritional Management of Women with Anorexia
    Nervosa or Bulimia Nervosa   72

Diabetes Mellitus prior to Pregnancy   72
A Closer Look at Insulin Resistance   72
Nutritional Management of Type 2 Diabetes   73
Other Components of the Nutritional Management of
    Type 2 Diabetes   74
Reducing the Risk of Type 2 Diabetes   75

Polycystic Ovary Syndrome   75
Nutritional Management of Women with PCOS   76

Disoders of Metabolism   76
Phenylketonuria (PKU)   76
Nutritional Management of Women with PKU   76
Celiac Disease   77
Nutritional Management of Celiac Disease   77

Herbal Remedies for Fertility-Related Problems   78

Chapter 4
Nutrition during Pregnancy   83

Introduction   84

The Status of Pregnancy Outcomes   84
Infant Mortality   85
Low Birth Weight, Preterm Delivery, and Infant Mortality   86

Reducing Infant Mortality and Morbidity   87
Health Objectives for the Year 2010   87

Physiology of Pregnancy   87
Maternal Physiology   87
Normal Physiological Changes during Pregnancy   88
The Placenta   91

Embryonic and Fetal Growth and Development   93
Critical Periods of Growth and Development   93
Fetal Body Composition   95
Variation in Fetal Growth   95
Nutrition, Miscarriages, and Preterm Delivery   97

The Fetal-Origins Hypothesis of Later Disease Risk   97
Mechanisms Underlying the Fetal-Origins Hypothesis   98
Limitations of the Fetal-Origins Hypothesis   99

Pregnancy Weight Gain   99
Pregnancy Weight Gain Recommendations   100
Composition of Weight Gain in Pregnancy   101
Postpartum Weight Retention   101

Nutrition Time Line →

1744

First record of ice cream
in America at Maryland
colony

1747

Lind publishes
"Treatise on
Scurvy," citrus
identified as cure

1750
Ojibway and
Sioux war over
control of wild
rice stands

1762
Sandwich invented
by the Earl of
Sandwich

**Nutrition and the Course and Outcome
of Pregnancy 103**
Famine and Pregnancy Outcome 103
Contemporary Prenatal Nutrition Research Results 104
Energy Requirement in Pregnancy 104
Carbohydrate Intake during Pregnancy 105
Alcohol and Pregnancy Outcome 105
Protein Requirement 105
Vegetarian Diets in Pregnancy 106
Maternal Intake of Omega-3 Fatty Acids and
Pregnancy Outcome 107
The Need for Water during Pregnancy 109
Folate and Pregnancy Outcome 109
Folate and Congenital Abnormalities 110
Vitamin A and Pregnancy Outcome 111
Vitamin D Requirement 111
Calcium Requirements in Pregnancy 112
Fluoride 113
Iron Status and the Course and Outcome of Pregnancy 113
Zinc Requirement in Pregnancy 115
Iodine and Pregnancy Outcome 116
Antioxidants from Plants 116
The Need for Sodium during Pregnancy 116
Caffeine Use in Pregnancy 116

Healthy Diets for Pregnancy 117
Effect of Taste and Smell Changes on Dietary Intake
during Pregnancy 117
Pica 117
Assessment of Dietary Intake during Pregnancy 117
Vitamin and Mineral Supplementation during Pregnancy 119
Herbal Remedies and Pregnancy 120

Exercise and Pregnancy Outcome 120
Exercise Recommendations for Pregnant Women 120

Food Safety Issues during Pregnancy 121
Mercury Contamination 121

Common Health Problems during Pregnancy 121
Nausea and Vomiting 121
Heartburn 122
Constipation 122

**Model Nutrition Programs for Risk Reduction
in Pregnancy 123**
The Montreal Diet Dispensary 123
The WIC Program 123

Chapter 5
Nutrition during Pregnancy   129
Conditions and Interventions

Introduction 130

Hypertensive Disorders of Pregnancy 130
Hypertensive Disorders of Pregnancy, Oxidative Stress,
and Nutrition 131
Chronic Hypertension 131
Gestational Hypertension 131
Preeclampsia–Eclampsia 131
Preeclampsia Case Presentation 133
Nutritional Recommendations and Interventions
for Preeclampsia 133

Diabetes in Pregnancy 134
Gestational Diabetes 134
Potential Consequences of Gestational Diabetes 135
Risk Factors for Gestational Diabetes 136
Diagnosis of Gestational Diabetes 136
Treatment of Gestational Diabetes 137
Presentation of a Case Study 137
Exercise Benefits and Recommendations 137
Nutritional Management of Women with Gestational
Diabetes 137
Consumption of Foods with a Low Glycemic Index 139
Postpartum Follow-Up 140
Prevention of Gestational Diabetes 140
Type 1 Diabetes during Pregnancy 140

Multifetal Pregnancies 140
Background Information about Multiple Fetuses 142
Risks Associated with Multifetal Pregnancy 143
Interventions and Services for Risk Reduction 143
Nutrition and the Outcome of Multifetal Pregnancy 144
Dietary Intake in Twin Pregnancy 144
Nutritional Recommendations for Women
with Multifetal Pregnancy 145

HIV/AIDS during Pregnancy 145
Treatment of HIV/AIDS 146
Consequences of HIV/AIDS during Pregnancy 146
Nutritional Factors and HIV/AIDS during Pregnancy 146

Nutrition Time Line ⟶

1771
Potato heralded
as famine food

1774
Americans drink
more coffee in
protest over
Britain's tea tax

Stephanno Bianchetti/CORBIS

1775
Lavoisier ("the father
of the science of
nutrition") discovers
the energy-producing
property of food

1816
Protein and
amino acids
identified
followed by
carbohydrates
and fats in the
mid 1800s

Nutritional Management of Women with HIV/AIDS
during Pregnancy 147

Eating Disorders in Pregnancy 147
Consequences of Eating Disorders in Pregnancy 147
Treatment of Women with Eating Disorders during
Pregnancy 147
Nutritional Interventions for Women with Eating
Disorders 148

Fetal Alcohol Spectrum 148
Effects of Alcohol on Pregnancy Outcome 148
The Fetal Alcohol Syndrome 148

Nutrition and Adolescent Pregnancy 149
Growth during Adolescent Pregnancy 149
Dietary and Other Recommendations for Pregnant
Adolescents 149

Evidence-Based Practice 150

Chapter 6
Nutrition during Lactation   155

Introduction 156

Breastfeeding Goals for the United States 156

Lactation Physiology 157
Functional Units of the Mammary Gland 157
Mammary Gland Development 158
Lactogenesis 158
Hormonal Control of Lactation 159
Secretion of Milk 159
The Letdown Reflex 159

Human Milk Composition 160
Colostrum 160
Water 160
Energy 160
Lipids 162
Protein 162
Milk Carbohydrates 162
Fat-Soluble Vitamins 163
Water-Soluble Vitamins 163
Minerals in Human Milk 163
Taste of Human Milk 164

Benefits of Breastfeeding 164
Breastfeeding Benefits for Mothers 164
Breastfeeding Benefits for Infants 164

Breast Milk Supply and Demand 167
Can Women Make Enough Milk? 167
Does the Size of the Breast Limit a Woman's Ability
to Nurse Her Infant? 167
Is Feeding Frequency Related to the Amount of Milk
a Woman Can Make? 167
Pumping or Expressing Milk 167
Can Women Breastfeed after Breast Reduction or
Augmentation Surgery? 168
Does Silicone from Breast Implants Leach
into the Milk? 168

The Breastfeeding Infant 168
Optimal Duration of Breastfeeding 168
Reflexes 168
Breastfeeding Positioning 168
Identifying Hunger and Satiety 169
Feeding Frequency 169
Vitamin Supplements for Breastfeeding Infants 169
Identifying Breastfeeding Malnutrition 170
Tooth Decay 171

Maternal Diet 172
Energy and Nutrient Needs 172

Maternal Energy Balance and Milk Composition 172
Weight Loss during Breastfeeding 172
Exercise and Breastfeeding 172
Vitamin and Mineral Supplements 173
Vitamin and Mineral Intakes 173
Functional Foods 174
Fluids 174
Alternative Diets 174
Infant Colic 174

**Factors Influencing Breastfeeding Initiation
and Duration 174**
Obesity and Breastfeeding 174
Socioeconomic 174

Breastfeeding Promotion, Facilitation, and Support 175
Role of the Health Care System in Supporting
Breastfeeding 175
Prenatal Breastfeeding Education and Support 175

Nutrition Time Line →

Bettmann/CORBIS

1833
Beaumont's
experiments on
a wounded
man's stomach
greatly expands
knowledge
about digestion

1871
Proteins, carbohydrates,
and fats determined
to be insufficient
to support life; that
there are other
"essential" components

Bettmann/CORBIS

1895
First milk station
providing children
with uncontaminated
milk opens in
New York City

Lactation Support in Hospitals and Birthing Centers   177
Lactation Support after Discharge   177
The Workplace   179
The Community   180

Public Food and Nutrition Programs   180
National Breastfeeding Policy   180
USDA WIC Program   181

Model Breastfeeding Promotion Programs   181
WIC National Breastfeeding Promotion Project—Loving
Support Makes Breastfeeding Work   181
Wellstart International   182

Chapter 7

Nutrition during Lactation   189
Conditions and Interventions

Introduction   190

Common Breastfeeding Conditions   190
Sore Nipples   190
Letdown Failure   190
Hyperactive Letdown   191
Engorgement   191
Plugged Duct   191
Infection   191

Maternal Medications   192

Herbal Remedies   195
Specific Herbs Used in the United States   197

Alcohol and Other Drugs and Exposures   198
Alcohol   198
Nicotine (Smoking Cigarettes)   199
Marijuana   200
Caffeine   200
Other Drugs of Abuse   200
Environmental Exposures   200

Neonatal Jaundice and Kernicterus   201
Bilirubin Metabolism   202
Physiologic versus Pathologic Newborn Jaundice   203
Hyperbilirubinemia and Breastfeeding   204
Prevention and Treatment for Severe Jaundice   206
Information for Parents   206

Breastfeeding Multiples   206

Infant Allergies   207
Food Intolerance   207

Near-Term Infants   208

Human Milk and Preterm Infants   208

Medical Contraindications to Breastfeeding   210
Breastfeeding and HIV Infection   210

Human Milk Collection and Storage   212
Milk Banking   212

Model Programs   213
Breastfeeding Promotion in Physicians' Office Practices
(BPPOP)   213
The Rush Mothers' Milk Club   213

Chapter 8

Infant Nutrition   219

Introduction   220

Assessing Newborn Health   220
Birthweight as an Outcome   220
Infant Mortality   220
Combating Infant Mortality   220
Standard Newborn Growth Assessment   221

Infant Development   221
Motor Development   221
Critical Periods   223
Cognitive Development   223
Digestive System Development   223
Parenting   223

Energy and Nutrient Needs   224
Caloric Needs   224
Protein Needs   224
Fats   225
Metabolic Rate, Calories, Fats, and Protein—How Do
They All Tie Together?   225
Other Nutrients and Non-nutrients   225

Physical Growth Assessment   226
Interpretation of Growth Data   227

Nutrition Time Line ⟶

1896
Atwater publishes
*Proximate Composition
of Food Materials*

Bettmann/CORBIS

1906
Pure Food and Drug Act passed
by President Theodore Roosevelt
to protect consumers against
contaminated foods

1910
Pasteurized milk
introduced

Photo Disc

1912
Funk suggested
scurvy, beriberi,
and pellagra
caused by
deficiency of
"vitamines" in
the diet

Feeding in Early Infancy   228
Breast Milk and Formula   228
Cow's Milk during Infancy   228

Development of Infant Feeding Skill   229
Introduction of Solid Foods   231
The Importance of Infant Feeding Position   232
Preparing for Drinking from a Cup   232
Food Texture and Development   233
First Foods   233
Inappropriate and Unsafe Food Choices   235
Water   235
How Much Food Is Enough for Infants?   235
How Infants Learn Food Preferences   236

Nutrition Guidance   236
Infants and Exercise   236
Supplements for Infants   236

Common Nutritional Problems and Concerns   237
Failure to Thrive   237
Nutrition Intervention for Failure to Thrive   237
Colic   238
Iron-Deficiency Anemia   238
Diarrhea and Constipation   239
Prevention of Baby-Bottle Caries and Ear
    Infections   239
Food Allergies and Intolerances   239
Lactose Intolerance   240

Cross-Cultural Considerations   240

Vegetarian Diets   240

Nutrition Intervention for Risk Reduction   241
Model Program: Newborn Screening and Expanded
    Newborn Screening   241

Energy and Nutrient Needs   247
Energy Needs   247
Protein Requirements   248
Fats   248
Vitamins and Minerals   248

Growth   249
Growth in Preterm Infants   249
Does Intrauterine Growth Predict Growth Outside?   250
Interpretation of Growth   251

**Nutrition for Infants with Special Health
Care Needs**   251
Nutrition Risks to Development   252

Severe Preterm Birth and Nutrition   253
How Sick Babies Are Fed   253
What to Feed Preterm Infants   254
Preterm Infants and Feeding   254

Infants with Congenital Anomalies and Chronic Illness   256
Infants with Genetic Disorders   257

Feeding Problems   258

Nutrition Interventions   260

Nutrition Services   260

Chapter 9
Infant Nutrition   245
Conditions and Interventions

Introduction   246

Infants at Risk   246
Families of Infants with Special Health Care Needs   247

Chapter 10
Toddler and Preschooler   265
Nutrition

Introduction   266
Definitions of the Life-Cycle Stage   266
Importance of Nutrition   266

Tracking Toddler and Preschooler Health   266
Healthy People 2010   266

Normal Growth and Development   266
The 2000 CDC Growth Charts   267
Common Problems with Measuring and Plotting
    Growth Data   268

Physiological and Cognitive Development   268
Toddlers   268
Preschool-Age Children   271

Nutrition Time Line ⟶

Photo Disc

1913
First vitamin
discovered (vitamin A)

1914
Goldberger identifies the
cause of pellagra (niacin
deficiency) in poor
children to be a missing
component of the diet
rather than a germ as
others believed

1916
First dietary guidance
material produced for
the public was released.
It was titled "Food for
Young Children."

1917
First food groups
published, The Five
Food Groups: Milk and
Meat; Vegetables and
Fruits; Cereals; Fats
and Fat Foods; Sugars
and Sugary Foods

Temperament Differences   273
Food Preference Development, Appetite, and
    Satiety   273

Energy and Nutrient Needs   275
Energy Needs   275
Protein   275
Vitamins and Minerals   275

Common Nutrition Problems   276
Iron-Deficiency Anemia   276
Dental Caries   277
Constipation   278
Lead Poisoning   278
Food Security   278
Food Safety   279

Prevention of Nutrition-Related Disorders   279
Overweight and Obesity in Toddlers and
    Preschoolers   279
Prevention and Treatment of Overweight and
    Obesity   280
Nutrition and Prevention of Cardiovascular Disease
    in Toddlers and Preschoolers   280
Vitamin and Mineral Supplements   281
Herbal Supplements   281

Dietary and Physical Activity Recommendations   282
Dietary Guidelines   282
Food Guide Pyramid   282
Recommendations for Intake of Iron, Fiber, Fat, and
    Calcium   283
Fluids   285
Recommended versus Actual Food Intake   285
Cross-Cultural Considerations   286
Vegetarian Diets   286
Child Care Nutrition Standards   286
Physical Activity Recommendations   287

Nutrition Intervention for Risk Reduction   287
Model Program   287

Public Food and Nutrition Programs   288
WIC   288
WIC's Farmers' Market Nutrition Program   288
Head Start and Early Head Start   288
Food Stamps   288

Chapter 11
Toddler and Preschooler Nutrition   293
Conditions and Interventions

Introduction   294

**Who Are Children with Special Health
    Care Needs?   294**

**Nutrition Needs of Toddlers and Preschoolers
    with Chronic Conditions   295**

Growth Assessment   295

Feeding Problems   296
Behavioral Feeding Problems   297
Excessive Fluid Intake   297
Feeding Problems and Food Safety   298
Feeding Problems from Disabilities Involving
    Neuromuscular Control   298

Nutrition-Related Conditions   299
Failure to Thrive   299
Toddler Diarrhea and Celiac Disease   299
Autism   300
Muscle Coordination Problems and Cerebral Palsy   300
Pulmonary Problems   301
Developmental Delay and Evaluations   302

Food Allergies and Intolerance   303

Dietary Supplements and Herbal Remedies   303

Sources of Nutrition Services   303

Chapter 12
Child and Preadolescent Nutrition   307

Introduction   308
Definitions of the Life Cycle Stage   308
Importance of Nutrition   308

Tracking Child and Preadolescent Health   308
Healthy People 2010   308

Nutrition Time Line ⟶

1921

First fortified food produced: Iodized salt. It was needed to prevent widespread iodine deficiency goiter in many parts of the United States

Morton Salt Co.

1928

American Society for Nutritional Sciences and the *Journal of Nutrition* founded

1929

Essential fatty acids identified

Photo Disc

Normal Growth and Development 309
The 2000 CDC Growth Charts 309

**Physiological and Cognitive Development
of School-Age Children 310**
Physiological Development 310
Cognitive Development 311
Development of Feeding Skills 311

Energy and Nutrient Needs of School-Age Children 313
Energy Needs 313
Protein 314
Vitamins and Minerals 314

Common Nutrition Problems 314
Iron Deficiency 314
Dental Caries 314

**Prevention of Nutrition-Related Disorders
in School-Age Children 314**
Overweight and Obesity in School-Age Children 315
Prevention of Overweight and Obesity 316
Nutrition and Prevention of Cardiovascular Disease
in School-Age Children 318
Dietary Supplements 319

Dietary Recommendations 319
Recommendations for Intake of Iron, Fiber, Fat, and Calcium 319
Recommended versus Actual Food Intake 321
Cross-Cultural Considerations 322
Vegetarian Diets 322

Physical Activity Recommendations 322
Recommendations versus Actual Activity 323
Determinants of Physical Activity 323
Organized Sports 324

Nutrition Intervention for Risk Reduction 324
Nutrition Education 324
Nutrition Integrity in Schools 324
Model Programs 325
Model Program: High 5 Alabama 326

Public Food and Nutrition Programs 327
The National School Lunch Program 327
School Breakfast Program 328
Summer Food Service Program 329
Team Nutrition 329

Chapter 13
Child and
Preadolescent Nutrition   335
Conditions and Interventions

Introduction 336

**"Children Are Children First"—What Does
That Mean? 336**

**Nutritional Requirements of Children with
Special Health Care Needs 337**
Energy Needs 337
Protein Needs 337
Other Nutrients 337

Growth Assessment 338
Growth Assessment and Interpretation in Children
with Chronic Conditions 338
Body Composition and Growth 338

Nutrition Recommendations 339
Methods of Meeting Nutritional Requirements 340
Fluids 342

**Eating and Feeding Problems in Children
with Special Health Care Needs 342**
Specific Disorders 342

Dietary Supplements and Herbal Remedies 347

Sources of Nutrition Services 348
USDA Child Nutrition Program 348
Maternal and Child Health Block Program
of the U.S. Department of Health and
Human Services (HHS) 348
Public School Regulations: 504 Accommodation
and IDEA 348
Nutrition Intervention Model Program 349

Chapter 14
Adolescent Nutrition   353

Introduction 354

Nutritional Needs in a Time of Change 354

Nutrition Time Line ⟶

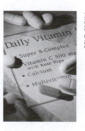

Photo Disc

1937
Pellagra found
to be due to
a deficiency
of niacin

1941
First refined
grain-enrichment
standards developed

1930s
Vitamin C identified
in 1932, followed
by pantothenic acid
and riboflavin in
1933, and vitamin K
in 1934

Normal Physical Growth and Development   355
   Changes in Weight, Body Composition,
     and Skeletal Mass   356

Normal Psychosocial Development   357

**Health and Eating-Related Behaviors during
   Adolescence**   358
   Vegetarian Diets   360
   Dietary Intake and Adequacy Among Adolescents   362

Energy and Nutrient Requirements of Adolescents   362
   Energy   365
   Protein   365
   Carbohydrates   366
   Dietary Fiber   366
   Fat   366
   Calcium   366
   Iron   367
   Zinc   368
   Folate   368
   Vitamin A   369
   Vitamin E   369
   Vitamin C   369

Nutrition Screening, Assessment, and Intervention   369
   Nutrition Education and Counseling   374

Physical Activity and Sports   375
   Factors Affecting Physical Activity   375

**Promoting Healthy Eating and Physical
   Activity Behaviors**   377
   Effective Nutrition Messages for Youth   377
   Parent Involvement   377
   School Programs   377
   Community Involvement in Nutritionally Supportive
     Environments   380

Chapter 15
Adolescent Nutrition 383
Conditions and Interventions

Introduction   384

Overweight and Obesity   384
   Health Implications of Adolescent Overweight   385

Supplement Use   387
   Vitamin–Mineral Supplements   387
   Herbal Remedies   390
   Ergogenic Supplements Used by Teens   390

Nutrition for Adolescent Athletes   391
   Fluids and Hydration   391
   Special Dietary Practices   391

Substance Use   392

Iron-Deficiency Anemia   392

Cardiovascular Disease   393
   Hypertension   393
   Hyperlipidemia   394

Disordered Eating, Dieting, and Eating Disorders   395
   The Continuum of Eating Concerns and Disorders   395
   Prevalence of Eating Disorders   395
   Anorexia Nervosa   395
   Bulimia Nervosa   396
   Binge-Eating Disorder   397
   Disordered Eating Behaviors   398
   Dieting Behaviors   398
   Body Dissatisfaction   399
   Etiology of Eating Disorders   400
   Treating Eating Disorders   400
   Preventing Eating Disorders   401
   Eating Disorders among Adolescents: Summing Things Up   402
   Children and Adolescents with Chronic Health Conditions   402

Chapter 16
Adult Nutrition 407

Introduction   408
   Definition of Adulthood in the Lifecycle   408
   Importance of Nutrition   408

Healthy People 2010 Objectives   408

Physiological Changes of Adulthood   408
   Sensory Changes   409
   Hormonal Changes   409
   Physiological Changes in Males: Climacteric   409
   Physiological Changes in Females: Menopause   410

Nutrition Time Line ⟶

1941
First Recommended
Dietary Allowances (RDAs)
announced by President
Franklin Roosevelt on radio

1946
National School
Lunch Act passed

1947
Vitamin B$_{12}$
identified

Maintaining a Healthy Body 412
Energy for Weight Management 412
Determining Energy Needs 412
Energy for Weight Change 413
Actual Energy Intake 413
Fad Diets 414

Dietary Recommendations 415
Dietary Recommendations to Combat Nutritional
   Concerns 415
Food Advice 416
MyPyramid 417
Food Advice Reflects Cultural Food Patterns 417
Alcohol: Food, Drug, and Nutrient 417
Water and Other Fluids 419
Diuretic Effects of Caffeine and Alcohol 420
International Guidance Related to Fluids 420

Nutrient Recommendations 420
Nutrients of Public Health Concern 420
Actual Intake of Food 422
Supplements: Vitamins and Minerals 424
Supplements and CAM (Complementary and Alternative
   Medicine) 424
Cross-Cultural Considerations 424
Cross-Cultural Dietary Guidance 426
Vegetarian Diets 426

Physical Activity Recommendations 428

Nutrition Intervention for Risk Reduction 430
A Model Health-Promotion Program 430
Public Food and Nutrition Programs 430
Nutrition and Health Promotion 431

Nutrition Intervention 439
Alternative Medicine and Cancer Treatment 439

Cardiovascular Disease: Coronary Heart Disease 439
Definition 440
Prevalence 440
Etiology 440
Effects of CHD 441
Risk Factors 441
Nutrition Interventions 441
Dietary Supplements and Heart Disease 443

Overweight and Obesity 444
Definition 444
Prevalence 445
Etiology 445
Effects of Obesity 445
Nutrition Interventions 446
Bariatric Surgery and Role of the RD 447

Diabetes Mellitus 447
Prevalence 447
Etiology 448
Insulin Resistance 448
Effects of Diabetes 448
Risk Factors 448
Nutrition Interventions for Type 2 Diabetes 448
Glucophage and Statins 449
Metabolic Syndrome 450

HIV/AIDS 450
Definition 450
Prevalence 450
Etiology 450
Effects of HIV/AIDS 452
Nutrition Interventions 452

Chapter 17
Adult Nutrition 437
Conditions and Interventions

Introduction 438

Cancer 438
Definition 438
Prevalence/Etiology 438
Risk Factors 438

Chapter 18
Nutrition and 455
Older Adults

Introduction 456
What Counts As Old? 456
Food Matters: Nutrition Contributes to a Long
   and Healthy Life 456

Nutrition Time Line

1953
Double helix
structure
of DNA
discovered

Photo Disc

1965
Food Stamp Act
passed, Food Stamp
program established

1966
Child Nutrition
Act added school
breakfast to the
National School
Lunch Program

Photo Disc

1968
First national nutrition
survey in United States
launched (the Ten
State Nutrition
Survey)

A Picture of the Aging Population: Vital Statistics 457
Global Population Trends: Life Expectancy
and Life Span 457
Nutrition: A Component of Health Objectives
for the Older Adult Population 458

Theories of Aging 458
Programmed Aging 459
Wear-and-Tear Theories of Aging 459
Calorie Restriction and Longevity 459

Physiological Changes 460
Body Composition Changes 460
Changing Sensual Awareness: Taste and Smell, Chewing
and Swallowing, Appetite and Thirst 461

Nutritional Risk Factors 462

Dietary Recommendations 465
Food-Based Guidance: MyPyramid 465
Actual Food Group Intake 466
Eating Occasions 466

Nutrient Recommendations 467
Estimating Energy Needs 467
Nutrient Recommendations: Macro- and Micronutrients
of Concern 467
Recommendations for Fluid 470
Age-Associated Changes in Metabolism: Nutrients
of Concern 470
Low Dietary Intake: Nutrients of Concern 472
Nutrient Supplements: Why, When, Who, What,
and How Much? 473
Dietary Supplements, Functional Foods, Nutraceuticals:
Special Interest for Older Adults 475
Nutrient Recommendations: Using the Food Label 475
Cross-Cultural Considerations in Making Dietary
Recommendations 477

Food Safety Recommendations 477

Physical Activity Recommendations 478
Exercise Guidelines 478

**Nutrition Policy and Intervention for Risk
Reduction 479**
Nutrition Education 479
Model Programs Exemplify Intervention Goals 480

Community Food and Nutrition Programs 481
Elderly Nutrition Programs 481
Senior Nutrition Program 481
The Promise of Prevention: Health Promotion 482

Chapter 19
Nutrition and Older Adults   487
Conditions and Interventions

Introduction: The Importance of Nutrition 488

Nutrition and Health 488

Heart Disease/Cardiovascular Disease 489
Prevalence 489
Risk Factors 489
Nutritional Remedies for Cardiovascular Diseases 489

Stroke 489
Definition 489
Prevalence 489
Etiology 489
Effects of Stroke 489
Risk Factors 490
Nutritional Remedies 491

Hypertension 491
Definition 491
Prevalence 492
Etiology 492
Effects of Hypertension 492
Risk Factors 492
Nutritional Remedies 492

Cancer 492
Prevalence of Cancer 492
Etiology and Effects of Cancer 492
Risk Factors 493
Nutritional Remedies for Cancer 493

Diabetes 494
Effects of Diabetes 494
Risk Factors 494
Nutritional Interventions 495

Nutrition Time Line ⟶

1970
First Canadian national nutrition survey launched (Nutrition Canada National Survey)

1972
Special Supplemental Food and Nutrition Program for Women, Infants, and Children (WIC) established

1977
Dietary Goals for the United States issued

1978
First Health Objectives for the Nation released

1989
First national scientific consensus report on diet and chronic disease published

Obesity   496
Definition   496
Prevalence   496
Etiology/Effects/Risk Factors of Obesity   496
Nutritional Remedies   496

Osteoporosis   497
Definition   497
Prevalence   497
Etiology   497
Inadequate Bone Mass   497
Increased Bone Loss   498
Effects of Osteoporosis   498
Risk Factors   498
Nutritional Remedies   498
Other Issues Impacting Nutritional Remedes   500

Oral Health   500

Gastrointestinal Diseases   501
Gastroesophageal Reflux Disease (GERD)   501

Changes in Nutrient Availability: Vitamin B$_{12}$ Deficiency   502
Pernicious Anemia   502
Food-bound Vitmain B$_{12}$ Malabsorption   503
Constipation   504

Inflammatory Diseases: Osteoarthritis   504
Definition   504
Etiology   505
Effects of Osteoarthritis   505
Risk Factors   505
Nutritional Remedies   505

Cognitive Disorders: Alzheimer's Disease   506
Prevalence of Dementia   507
Etiology of Cognitive Disorders   507
Effects of Cognitive Disorders   507
Nutrition Interventions for Cognitive Disorders   507

Medications and Polypharmacy   508

Low Body Weight/Underweight   509
Etiology   510
Nutrition Interventions   510

Dehydration   511
Etiology   512
Effects of Dehydration   512
Nutritional Interventions   512
Rehydrate Slowly   513
Dehydration at End-of-Life   513

Bereavement   513

Appendix A
CDC Growth Charts   A-1

Appendix B
Nutrient Intakes of Adults
Aged 70 and Older   A-9

Appendix C
Measurement Abbreviations
and Equivalents   A-11

Appendix D
Body Mass Index (BMI)   A-13

Glossary   G-1

Index   I-1

Nutrition Time Line →

1997
RDAs expanded to
Dietary Reference
Intakes (DRIs)

1998
Folic acid
fortification
of refined grain
products begins

Photo Disc

2003
Sequencing of DNA in the
human genome completed.
Marks beginning of new era
of research innutrient–gene
interactions

2006
Obestity and
diabetes become
global epidemics

Preface

It is with a renewed sense of gratitude and pleasure that we offer you the third edition of *Nutrition Through the Life Cycle*. This text was initially developed, and has been revised, to address the needs of instructors teaching, and students taking, a two-to-four credit course in life-cycle nutrition. It is written at a level that assumes students have had an introductory nutrition course. Overall, the text is intended to give instructors a tool they can productively use to enhance their teaching efforts, and to give students an engaging and rewarding educational experience they will carry with them throughout their lives and carrers.

Authors of *Nutrition Through the Life Cycle* represent a group of experts who are actively engaged in clinical practice, teaching, and research related to nutrition during specific phases of the life cycle. All of us remained totally dedicated to the goals established for the text at its conception: to make the text comprehensive, logically organized, science based, realistic, and relevant to the needs of instructors and students.

Chapter 1 summarizes key elements of introductory nutrition and gives students who need it a chance to update or renew their knowledge. Coverage of the life-cycle phases begins with preconceptional nutrition and continues with each major phase of the life cycle through adulthood and the special needs of the elderly. Each of these 19 chapters was developed based on a common organizational framework that includes key nutrition concepts, prevalence statistics, physiological principles, nutritional needs and recommendations, model programs, case studies, and recommended practices. To meet the knowledge needs of students with the variety of career goals represented in many life-cycle nutrition courses, we include two chapters for each life-cycle phase. The first chapter for each life-cycle phase covers normal nutrition topics, and the second covers nutrition-related conditions and interventions. Every chapter focuses on scientifically based information and employs up-to-date resources and references. Each chapter ends with a list of websites and print resources that lead students to reliable information on scientific and applied aspects of life-cycle nutrition.

New to the Third Edition

Advances in knowledge about nutrition and health through the life cycle are expanding at a remarkably high rate. New research is taking our understanding of the roles played by nutrients, nutrient-gene interactions, body fat, physical activity, and dietary supplements to new levels. The continued escalation of obesity and type 2 diabetes rates are having broad effects on the incidence of disease throughout the life cycle. New knowledge about nutrition and health through the life cycle require that we understand the effects of nutrients and body fat on hormonal activity, nutrient triggers to gene expression, and the roles of nutrients in the development and correction of chronic inflammation, oxidative stress, and endothelial dysfunction. Recommendations for dietary and nutrient supplement intake, and for physical activity in health and disease are changing due to these understandings. You will see these emerging areas of direct relevance to nutrition, and updated information about nutrition incorporated throughout the third edition of *Nutrition Through the Life Cycle*.

The third edition differs from the second in important ways:

- We have added a section at the end of chapters called "Key Points." This section highlights the major points made in each chapter. (These do not appear in Chapter 1.)
- "In Focus" boxes have been added and provide background information on major conditions and disorders address in various chapters throughout the book. In Focus boxes on nutrigenomics, diabetes metabolic syndrome, inflammatory diseases, and celiac disease, for example, now appear.
- Additional case studies, tables, illustrations, margin definitions, and headings have been added to chapters.
- Chapter 1 on Nutrition Basics has been revised in multiple ways to incorporate advances in nutrition knowledge.
- Web and other resources are updated to include interactive and streaming media sites, and newly developed nutrition education and information sites.

Instructor Resources

Updated for the third edition is an *Instructor's Resource CD-ROM* that contains Microsoft PowerPoint™ lecture presentations with artwork; chapter outlines, classroom activities, lecture launchers, Internet exercises, discussion questions, hyperlinks to relevant websites, and case studies. The CD-ROM also includes a Test Bank, expanded and improved, that includes multiple-choice, true/false, matching, and discussion exercises.

Acknowledgments

It takes the combined talents and hard work of authors, editors, assistants, and the publisher to develop a new edition of a textbook and its instructional resources. Thanks to the dedicated efforts of Nedah Rose, Teresa Trego, Kate Franco, Elesha Feldman, and Lan Blomo, this edition is all we hoped it would be. The support of Peter Adams, the Executive Editor, throughout the development and production of the third editon is greatly appreciated. We were fortunate to have many other capable Wadsworth/Thomson professionals contribute to this text, including Jennifer Somerville, Marketing Manager and Lisa Michel, Editorial Assistant. The hard work and dedication of Stephanie Kling in copyediting this edition is very much appreciated.

Reviewers

Many thanks to the following reviewers, whose careful reading and thoughtful comments helped enormously in shaping this second edition.

Betty Alford
Texas Woman's University

Leta Aljadir
University of Delaware

Dea Hanson Baxter
Georgia State University

Janet Colson
Middle Tennessee State University

Shelley R. Hancock
The University of Alabama

Dr. Mary Jacob
California State University,
Long Beach

Pera Jambazian
California State University,
Los Angeles

Tay Seacord Kennedy
Oklahoma State University

Younghee Kim
Bowling Green State University

Barbara Kirks
California State University, Chico

Kaye Stanek Krogstrand
University of Nebraska

Sally Ann Lederman
Columbia University

Richard Lewis
University of Georgia

J. Harriett McCoy
University of Arkansas

Sharon McWhinney
Prairie View A&M University

Robert Reynolds
University of Illinois at Chicago

Sharon Nickols-Richardson
Virginia Polytechnic Institute
and State University

Adria Sherman
Rutgers University

Carmen R. Roman-Shriver
Texas Tech University

Joanne Slavin
University of Minnesota

Joanne Spaide
Professor Emeritus, University
of Northern Iowa

Diana-Marie Spillman
Miami University, Oxford, Ohio

Wendy Stuhldreher
Slippery Rock University
of Pennsylvania

Anne VanBeber
Texas Christian University

Phyllis Moser-Veillon
University of Maryland

Janelle Walter
Baylor University

Doris Wang
University of Minnesota Crookston

Suzy Weems, Ph.D.
Stephen F. Austin State University

Kay Wilder
Point Loma Nazarene College

Chapter 1

Nutrition Basics

Chapter Outline

- Introduction
- Principles of the Science of Nutrition
- Nutritional Labeling
- Meeting Nutritional Needs across the Life Cycle
- Nutritional Assessment
- Monitoring the Nation's Nutritional Health
- Public Food and Nutrition Programs
- Nationwide Priorities for Improvements in Nutritional Health

Prepared by **Judith E. Brown**

Key Nutrition Concepts

1   Nutrition is the study of foods, their nutrients and other chemical constituents, and the effects of food constituents on health.

2   Nutrition is an interdisciplinary science.

3   Nutrition recommendations for the public change as new knowledge about nutrition and health relationships is gained.

4   At the core of the science of nutrition are principles that represent basic truths and serve as the foundation of our understanding about nutrition.

5   Healthy individuals require the same nutrients across the life cycle but in differing amounts. Nutritional needs can be met by a wide variety of cultural and religious food practices.

Introduction

Need to freshen up your knowledge of nutrition? Or, do you need to get up to speed on basic nutrition for the course? This chapter presents information about nutrition that paves the way to greater understanding of specific needs and benefits related to nutrition by life-cycle stage.

Nutrition is an interdisciplinary science focused on the study of foods, *nutrients*, and other food constituents and health. The body of knowledge about nutrition is large and is growing rapidly, changing views on what constitutes the best nutrition advice. You are encouraged to refer to nutrition texts and to use the online resources listed at the end of this chapter to fill in any knowledge gaps. You are also encouraged to stay informed in the future and to keep an open mind about the best nutrition advice for many health-related issues. Scientific evidence that drives decisions about nutrition and health changes with time.

This chapter centers on (1) the principles of the science of nutrition, (2) nutrients and other constituents of food, (3) nutritional assessment, (4) public food and nutrition programs, and (5) nationwide priorities for improvements in the public's nutritional health.

Nutrients   Chemical substances in foods that are used by the body for growth and health.

Food Security   Access at all times to a sufficient supply of safe, nutritious foods.

Food Insecurity   Limited or uncertain availability of safe, nutritious foods.

Calorie   A unit of measure of the amount of energy supplied by food. Also known as the "kilocalorie," or the "large Calorie."

Principles of the Science of Nutrition

Every field of science is governed by a set of principles that provides the foundation for growth in knowledge. These principles change little with time. Knowledge of the principles of nutrition listed in Table 1.1 will serve as a springboard to greater understanding of the nutrition and health relationships explored in the chapters to come.

PRINCIPLE #1   Food is a basic need of humans.

Humans need enough food to live and the right assortment of foods for optimal health (Illustration 1.1 on the next page). People who have enough food to meet their needs at all times experience *food security*. They are able to acquire food in socially acceptable ways—without having to scavenge or steal food. *Food insecurity* exists when the availability of safe, nutritious foods, or the ability to acquire them in socially acceptable ways, is limited or uncertain. About 12% of U.S. households are food insecure.[1]

PRINCIPLE #2   Foods provide energy (calories), nutrients, and other substances needed for growth and health.

People eat foods for many different reasons. The most compelling reason is the requirement for *calories* (energy), nutrients, and other substances supplied by foods for growth and health.

A calorie is a measure of the amount of energy transferred from food to the body. Because calories are a unit of measure and not a substance actually present in food, they are not considered to be nutrients.

Nutrients are chemical substances in food that the body uses for a variety of functions that support growth, tissue maintenance and repair, and ongoing health.

Table 1.1   Principles of human nutrition

Principle #1   Food is a basic need of humans.

Principle #2   Foods provide energy (calories), nutrients, and other substances needed for growth and health.

Principle #3   Health problems related to nutrition originate within cells.

Principle #4   Poor nutrition can result from both inadequate and excessive levels of nutrient intake.

Principle #5   Humans have adaptive mechanisms for managing fluctuations in food intake.

Principle #6   Malnutrition can result from poor diets and from disease states, genetic factors, or combinations of these causes.

Principle #7   Some groups of people are at higher risk of becoming inadequately nourished than others.

Principle #8   Poor nutrition can influence the development of certain chronic diseases.

Principle #9   Adequacy and balance are key characteristics of a healthy diet.

Principle #10   There are no "good" or "bad" foods.

Illustration 1.1 The need for food is part of Maslow's hierarchy of needs.

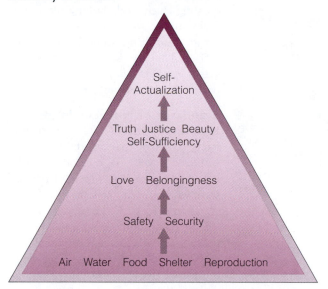

Essentially, every part of our body was once a nutrient consumed in food.

There are six categories of nutrients (Table 1.2). Each category except water consists of a number of different substances.

Essential and Nonessential Nutrients

Of the many nutrients required for growth and health, some must be provided by the diet while others can be made by the body.

Essential Nutrients   Nutrients the body cannot manufacture, or generally produce in sufficient amounts, are referred to as *essential nutrients*. Here *essential* means "required in the diet." All of the following nutrients are considered essential:

- Carbohydrates
- Certain amino acids (the essential amino acids: histidine, isoleucine, leucine, lysine, methionine, phenylalanine, threonine, tryptophan, and valine)
- Linoleic acid and alpha-linolenic acid (essential fatty acids)
- Vitamins
- Minerals
- Water

Nonessential Nutrients   Cholesterol, creatine, and glucose are examples of nonessential nutrients. *Nonessential nutrients* are present in food and used by the body, but they do not have to be part of our diets. Many of the beneficial chemical substances in plants are not considered essential, for example, yet they play important roles in maintaining health.

Table 1.2 The six categories of nutrients

1. **Carbohydrates**   Chemical substances in foods that consist of a single sugar molecule or multiples of sugar molecules in various forms. Sugar and fruit, starchy vegetables, and whole grain products are good dietary sources.
2. **Proteins**   Chemical substances in foods that are made up of chains of amino acids. Animal products and dried beans are examples of protein sources.
3. **Fats (Lipids)**   Components of food that are soluble in fat but not in water. They are more properly referred to as "lipids." Most fats are composed of glycerol attached to three fatty acids. Oil, butter, sausage, and avocado are examples of rich sources of dietary fats.
4. **Vitamins**   Thirteen specific chemical substances that perform specific functions in the body. Vitamins are present in many foods and are essential components of the diet. Vegetables, fruits, and grains are good sources of vitamins.
5. **Minerals**   In the context of nutrition, minerals consist of 15 elements found in foods that perform particular functions in the body. Milk, dark, leafy vegetables, and meat are good sources of minerals.
6. **Water**   An essential component of the diet provided by food and fluid.

Requirments for Essential Nutrients   All humans require the same set of essential nutrients, but the amount of nutrients needed varies based on:

- Age
- Body size
- Gender
- Genetic traits
- Growth

- Illness
- Lifestyle habits (e.g., smoking, alcohol intake)
- Medication use
- Pregnancy and lactation

Amounts of essential nutrients required each day vary a great deal, from cups (for water) to micrograms (for example, for folate and vitamin B_{12}).

Dietary Intake Standards

Dietary intake standards developed for the public cannot take into account all of the factors that influence nutrient needs, but they do account for the major ones of age, gender, growth, and pregnancy and lactation. Intake standards are called *Dietary Reference Intakes (DRIs)*.

Essential Nutrients   Substances required for growth and health that cannot be produced, or produced in sufficient amounts, by the body. They must be obtained from the diet.

Nonessential Nutrients   Nutrients required for growth and health that can be produced by the body from other components of the diet.

Illustration 1.2 Theoretical framework, terms, and abbreviations used in the Dietary Reference Intakes.

- Dietary Reference Intakes (DRIs). This is the general term used for the new nutrient intake standards for healthy people.
- Recommended Dietary Allowances (RDAs). These are levels of essential nutrient intake judged to be adequate to meet the known nutrient needs of practically all healthy persons while decreasing the risk of certain chronic diseases.
- Adequate Intakes (AIs). These are "tentative" RDAs. AIs are based on less conclusive scientific information than are the RDAs.
- Estimated Average Requirements (EARs). These are nutrient intake values that are estimated to meet the requirements of half the healthy individuals in a group. The EARs are used to assess adequacy of intakes of population groups.
- Tolerable Upper Intake Levels (ULs). These are upper limits of nutrient intake compatible with health. The ULs do not reflect desired levels of intake. Rather, they represent total, daily levels of nutrient intake from food, fortified foods, and supplements that should not be exceeded.

DRIs have been developed for most of the essential nutrients and will be updated periodically. (These are listed on the inside front covers of this text.) Current DRIs were developed through a joint U.S.–Canadian effort, and the standards apply to both countries. The DRIs are levels of nutrient intake intended for use as reference values for planning and assessing diets for healthy people. They consist of the Recommended Dietary Allowances (RDAs), which specify intake levels that meet the nutrient needs of over 98% of healthy people, and the other categories of intake standards described in Illustration 1.2. It is recommended that individuals aim for nutrient intakes that approximate the RDAs or Adequate Intake (AI) levels. Estimated Average Requirements (EARs) should be used to examine the possibility of inadequate intakes in individuals and within groups. Additional tests are required to confirm inadequate nutrient intakes and status.[2]

Daily Values (DVs)   Scientifically agreed-upon standards for daily intakes of nutrients from the diet developed for use on nutrition labels.

Tolerable Upper Intake Levels (ULs)   The DRIs include a table indicating levels of daily nutrient intake from foods, fortified products, and supplements that should not be exceeded. They can be used to assess the safety of high intakes of nutrients, particularly from supplements.

Standards of Nutrient Intake for Nutrition Labels   The Nutrition Facts panel on packaged foods uses standard levels of nutrient intakes based on an earlier edition of recommended dietary intake levels.

The levels are known as *Daily Values (DVs)* and are used to identify the amount of a nutrient provided in a serving of food compared to the standard level. The "% DV" listed on nutrition labels represents the percentages of the standards obtained from one serving of the food product. Table 1.3 lists DV standard amounts for nutrients that are mandatory or voluntary components of nutrition labels. Additional information on nutrition labeling is presented later in this chapter.

Carbohydrates

Carbohydrates are used by the body mainly as a source of readily available energy. They consist of the simple sugars (monosaccharides and disaccharides), complex carbohydrates (the polysaccharides), most dietary sources of fiber, and alcohol sugars. Alcohol is a close chemical relative of carbohydrates and is usually considered to be part of this nutrient category. The most basic forms of carbohydrates are single molecules called monosaccharides.

Glucose (also called "blood sugar" and "dextrose"), fructose ("fruit sugar"), and galactose are the most common monosaccharides. Molecules containing two monosaccharides are called disaccharides. The most common disaccharides are:

- Sucrose (glucose + fructose, or common table sugar)
- Maltose (glucose + glucose, or malt sugar)
- Lactose (glucose + galactose, or milk sugar)

Complex carbohydrates (also called polysaccharides) are considered "complex" because they have more elaborate chemical structures than the simple sugars. They include:

- Starches (the plant form of stored carbohydrate)
- Glycogen (the animal form of stored carbohydrate)
- Most types of fiber

Table 1.3 Daily Values (DVs) for nutrition labeling based on intakes of 2000 calories per day in adults and children aged 4 years and above

Mandatory Components of the Nutrition Label	
Food Component	Daily Value (DV)
Total fat	65 g[a]
Saturated fat	20 g
Cholesterol	300 mg[a]
Sodium	2400 mg
Total carbohydrate	300 g
Dietary fiber	25 g
Vitamin A	5000 IU[a]
Vitamin C	60 mg
Calcium	1000 mg
Iron	18 mg

[a]g = grams; mg = milligrams; IU = International Units

Each type of simple and complex carbohydrate, except fiber, provides 4 calories per gram. Fibers do not count as a source of energy, because they cannot be broken down by human digestive enzymes. The main function of fiber is to provide "bulk" for normal elimination. It has other beneficial properties, however. High-fiber diets reduce the rate of glucose absorption (a benefit for people with diabetes) and may help prevent cardiovascular disease and some types of cancer.[3]

Nonalcoholic in the beverage sense, alcohol sugars are like simple sugars, except that they include a chemical component of alcohol. Xylitol, mannitol, and sorbitol are common forms of alcohol sugars. Some are very sweet, and only small amounts are needed to sweeten commercial beverages, gums, yogurt, and other products. Unlike the simple sugars, alcohol sugars do not promote tooth decay.

Alcohol (consumed as ethanol) is considered to be part of the carbohydrate family because its chemical structure is similar to that of glucose. It is a product of the fermentation of sugar with yeast. With 7 calories per gram, alcohol has more calories per gram than do other carbohydrates.

Glycemic Index of Carbohydrates and Carbohydrates in Foods In the not too distant past it was assumed that "a carbohydrate is a carbohydrate is a carbohydrate." It was thought that all types of carbohydrates had the same effect on blood glucose levels and health, so it didn't matter what type was consumed. As is the case with many untested assumptions, this one fell by the wayside. It is now known that some types of simple and complex carbohydrates in foods elevate blood glucose levels more than do others. Such differences are particularly important to people with disorders such as *insulin resistance* and *type 2 diabetes*.[4] (An "In Focus"

box on page 73 expands upon the topics of insulin resistance and diabetes.)

Carbohydrates and carbohydrate-containing foods are now being classified by the extent to which they increase blood glucose levels. This classification system is called the *glycemic index*. Carbohydrates that are digested and absorbed quickly have a high glycemic index and raise blood glucose levels to a higher extent than do those with lower glycemic index values (Table 1.4).

Diets providing low glycemic index carbohydrates have been found to improve blood glucose control in people with diabetes, as well as to reduce elevated levels of blood cholesterol and triglycerides; increase levels of beneficial HDL cholesterol; and decrease the risk of developing type 2 diabetes, some types of cancer, and heart disease.[5]

Recommended Intake Level Recommended intake of carbohydrates is based on their contribution to total energy intake. It is recommended that 45–65% of calories come from carbohydrates. Added sugar should constitute no more than 25% of total caloric intake. It is recommended that adult females consume between 21 and 25 grams, and males 30–38 grams of total dietary fiber daily.[7]

Food Sources of Carbohydrates Carbohydrates are widely distributed in plant foods, while milk is the only important animal source of carbohydrates (lactose). Table 1.5 on page 7 lists selected food sources by type of carbohydrate. Additional information about total carbohydrate and fiber content of foods can be found on nutrition information labels on food packages and at this Web address: www.ars.usda.gov/ba/bhnrc/ndl.

Protein

Protein in foods provides the body with *amino acids* used to build and maintain tissues such as muscle, bone, enzymes, and red blood cells. The body can also use protein as a source of energy—it provides 4 calories per gram. However, this is not a primary function of protein. Of the common types of amino acids, nine must be provided by the diet and are classified as "essential amino acids." (These are listed on page 3.) Many different amino acids obtained from food perform important functions, but since the body can manufacture these from other amino acids, they are classified as "nonessential amino acids."

Insulin Resistance A condition is which cell membranes have a reduced sensitivity to insulin so that more insulin than normal is required to transport a given amount of glucose into cells.

Type 2 Diabetes A disease characterized by high blood glucose levels due to the body's inability to use insulin normally, to produce enough insulin, or both.

Glycemic Index A measure of the extent to which blood glucose is raised by a 50-gram portion of a carbohydrate-containing food compared to 50 grams of glucose or white bread.

Amino Acids The "building blocks" of protein. Unlike carbohydrates and fats, amino acids contain nitrogen.

Table 1.4 Glycemic Index (GI) of selected foods[6]

High GI (70 and higher)		Medium GI (56–69)		Low GI (55 and lower)	
Glucose	100	Breadfruit	69	Honey	55
French bread	95	Orange soda	68	Oatmeal	54
Scone	92	Sucrose	68	Corn	53
Potato, baked	85	Taco shells	68	Cracked wheat bread	53
Potato, instant mashed	85	Angel food cake	67	Orange juice	52
Corn Chex	83	Croissant	67	Banana	52
Pretzel	83	Cream of Wheat	66	Mango	51
Rice Krispies	82	Quaker Quick Oats	65	Potato, boiled	50
Cornflakes	81	Chapati	62	Muesli	48
Corn Pops	80	French bread with		Green peas	48
Gatorade	78	butter and jam	62	Pasta	48
Jelly beans	78	Couscous	61	Carrots, raw	47
Doughnut, cake	76	Raisin bran	61	Cassava	46
Waffle, frozen	76	Sweet potato	61	Lactose	46
French fries	75	Bran muffin	60	Milk chocolate	43
Shredded wheat	75	Just Right cereal	60	All Bran	42
Cheerios	74	Rice, white or brown	60	Orange	42
Popcorn	72	muffin Blueberry	59	Peach	42
Watermelon	72	Coca-Cola	58	Apple juice	40
Grape-Nuts	71	Power bar	56	Plum	39
Wheat bread	70			Apple	38
White bread	70			Pear	38
				Tomato juice	38
				Yam	37
				Dried beans	25
				Grapefruit	25
				Cherries	22
				Fructose	19
				Xylitol	8
				Hummus	6

Kwashiorkor A disease syndrome in children, primarily caused by protein deficiency. It is generally characterized by edema (or swelling), loss of muscle mass, fatty liver, rough skin, discoloration of the hair, growth retardation, and apathy.

Fatty Acids The fat-soluble components of fats in foods.

Food sources of protein differ in quality, based on the types of amino acids they contain. Foods of high protein quality include all of the essential amino acids. Protein from milk, cheese, meat, eggs, and other animal products is considered high quality. Plant sources of protein, with the exception of soybeans, do not provide all nine essential amino acids. Combinations of plant foods, such as grains or seeds with dried beans, however, yield high-quality protein. Amino acids found in these individual foods "complement" each other, thus providing a source of high-quality protein.

Recommended Protein Intake DRIs for protein are shown on the inside front cover of this text. In general, proteins should contribute 10–35% of total energy intake.[7] Protein deficiency, although rare in economically developed countries, leads to loss of muscle tissue, weak-

ness, reduced resistance to disease, kidney and heart problems, and contributes to the deficiency disease *kwashiorkor* in children. Protein deficiency in adults produces a loss of body tissue protein, heart abnormalities, severe diarrhea, and other health problems.

Food Sources of Protein Animal products and dried beans are particularly good sources of protein. These and other food sources of proteins are listed in Table 1.6 on page 8.

Fats (Lipids)

Fats in food share the property of being soluble in fats but not in water. They are actually a subcategory of *lipids*, but this category of macronutrient is referred to as fat in the DRIs.[7] Lipids include fats, oils, and related compounds such as cholesterol. Fats are generally solid at room temperature, whereas oils are usually liquid. Fats and oils are made up of various types of triglycerides (triacylglycerols), which consist of three *fatty acids* attached

Table 1.5 Food sources of carbohydrates

A. Simple Sugars (Mono- and Disaccharides)

	Portion Size	Grams of Carbohydrates		Portion Size	Grams of Carbohydrates
Breakfast Cereals			Apple	1 med	16
Raisin Bran	1 c	18	Orange	1 med	14
Corn Pops	1 c	14	Peach	1 med	8
Frosted Cheerios	1 c	13	**Vegetables**		
Bran Flakes	¾ c	5	Corn	½ c	3
Grape-Nuts	½ c	3	Broccoli	½ c	2
Special K	1 c	3	Potato	1 med	1
Wheat Chex	1 c	2	**Beverages**		
Cornflakes	1 c	2	Soft drinks	12 oz	38
Sweeteners			Fruit drinks	1 c	29
Honey	1 tsp	6	Skim milk	1 c	12
Corn syrup	1 tsp	5	Whole milk	1 c	11
Maple syrup	1 tsp	4	**Candy**		
Table sugar	1 tsp	4	Hard candy	1 oz	28
Fruits			Gumdrops	1 oz	25
Watermelon	1 pc (4" × 8")	25	Caramels	1 oz	21
Banana	1 med	21	Milk chocolate	1 oz	16

B. Complex Carbohydrates (Starches)

	Portion Size	Grams of Carbohydrates		Portion Size	Grams of Carbohydrates
Grain Products			**Dried Beans**		
Rice, white, cooked	½ c	21	White beans, cooked	½ c	13
Pasta, cooked	½ c	15	Kidney beans, cooked	½ c	12
Oatmeal, cooked	½ c	12	Lima beans, cooked	½ c	11
Cheerios	1 c	11	**Vegetables**		
Cornflakes	1 c	11	Potato	1 med	30
Bread, whole wheat	1 slice	7	Corn	½ c	10
			Broccoli	½ c	2

C. Total Fiber

	Portion Size	Grams of Total Fiber		Portion Size	Grams of Total Fiber
Grain Products			Green peas	½ c	4
Bran Buds	½ c	12	Carrots	½ c	3
All Bran	½ c	10	Potato, with skin	1 med	4
Raisin Bran	1 c	7	Collard greens	½ c	3
Bran Flakes	¾ c	5	Corn	½ c	3
Oatmeal, cooked	1 c	4	Cauliflower	½ c	2
Bread, whole wheat	1 slice	2	**Nuts**		
Fruits			Almonds	½ c	5
Avocado	½ med	7	Peanuts	½ c	3
Raspberries	1 c	5	Peanut butter	2 tbs	2
Mango	1 med	4	**Dried Beans**		
Pear, with skin	1 med	4	Pinto beans, cooked	½ c	10
Orange	1 med	3	Black beans, cooked	½ c	8
Banana	1 med	2	Black-eyed peas, cooked	½ c	8
Vegetables			Navy beans, cooked	½ c	6
Lima beans	½ c	5	Lentils, cooked	½ c	5

Table 1.6 Food sources of protein

	Portion Size	Grams of Protein
Meats		
Beef, lean	3 oz	26
Tuna, in water	3 oz	24
Hamburger, lean	3 oz	24
Chicken, no skin	3 oz	24
Lamb	3 oz	22
Pork chop, lean	3 oz	20
Haddock, broiled	3 oz	19
Egg	1 med	6
Dairy Products		
Cottage cheese, low fat	½ c	14
Yogurt, low fat	1 c	13
Milk, skim	1 c	9
Milk, whole	1 c	8
Swiss cheese	1 oz	8
Cheddar cheese	1 oz	7
Grain Products		
Oatmeal, cooked	½ c	4
Pasta, cooked	½ c	4
Bread	1 slice	2
Rice, white or brown	½ c	2

to *glycerol* (Illustration 1.3). The number of carbons contained in the fatty acid component of triglycerides varies from 8 to 22.

Fats and oils are a concentrated source of energy, providing 9 calories per gram. Fats perform a number of important functions in the body. They are precursors for cholesterol and sex hormone synthesis, components of cell membranes, vehicles for carrying certain vitamins that are soluble in fats only, and suppliers of the *essential fatty acids* required for growth and health.

Essential Fatty Acids   There are two essential fatty acids: linoleic acid and alpha-linolenic acid. Because these fatty acids are essential, they must be supplied in the diet. The central nervous system is particularly rich in derivatives of these two fatty acids. They are found in phospholipids, which—along with cholesterol—are the primary lipids in the brain and other nervous system tissue. Biologically active derivatives of essential fatty acids include *prostaglandins, thromboxanes,* and *prostacyclins.*

Linoleic Acid   Linoleic acid is the parent of the omega-6 (or n-6) fatty acid family. One of the major derivatives of linoleic acid is arachidonic acid. Arachidonic acid serves as a primary structural component of the central nervous system. Most vegetable oils, meats, and human milk are good sources of linoleic acid. American diets tend to provide sufficient to excessive levels of linoleic acid, and considerable amounts are stored in body fat.

Alpha-Linolenic Acid   Alpha-linolenic acid is the parent of the omega-3 (n-3) fatty acid family. It is present in many types of dark green vegetables, vegetable oils, and flaxseed. Derivatives of this essential fatty acid include eicosapentaenoic acid (EPA) and docosahexaenoic acid (DHA). Relatively little EPA and DHA are produced in the body from alpha-linolenic acid because the conversion process is slow.[7] EPA and DHA also enter the body through intake of fatty,[9] cold-water fish and shellfish and human milk. The EPA and DHA content of fish provides health benefits. Regular consumption of fish (two or more meals per week of fish low in contaminants such as mercury) not only protects against irregular heartbeat, sudden death, and stroke but also reduces high blood pressure and plaque formation in arteries.[8] DHA is found in large amounts in the central nervous system, the retina of the eye, and the testes. The body stores only small amounts of alpha-linolenic acid, EPA, and DHA.[9]

Omega-6 to Omega-3 Fatty Acid Ratio The ratio of omega-6 to omega-3 fatty acid intake is important because the functions of one are adversely modified by the presence of disproportionately high amounts of the other. Although an exact ratio has not been agreed upon, it is thought that people should consume four times (or less) as much omega-6 fatty acids as omega-3 fatty acids. Many Americans regularly consume vegetable oils but eat fish infrequently. Consequently, the ratio between intake of omega-6 and omega-3 fatty acids is over 9 to 1, indicating a need to increase intake of n-3 fatty acids.[10]

Illustration 1.3 Basic structure of a triglyceride.

Saturated and Unsaturated Fats   Fats (lipids) come in two basic types: *saturated* and *unsaturated*. Whether a fat is saturated or not depends on whether it has one or more double bonds between carbon atoms in one or more of the fatty acid components of the fat. If one double bond is present in one or more of the fatty acids, the fat is considered *monounsaturated*; if two or more are present, the fat is *polyunsaturated.*

Some unsaturated fatty acids are highly unsaturated. Alpha-linolenic acid, for example, contains three double bonds, arachidonic acid four, EPA five, and DHA six. These fatty acids are less stable than fatty acids with fewer double bonds, because double bonds between atoms are weaker than single bonds.

Saturated fats contain no double bonds between carbons and tend to be solid at room temperature. Animal products such as butter, cheese, and meats and two plant oils (coconut and palm) are rich sources of saturated fats. Fat we consume in our diets, whether it contains primarily saturated or unsaturated fatty acids, is generally in the triglyceride (or triacylglycerol) form.

Fats (lipids) also come in these forms:

- Monoglycerides (or monoacylglycerols), consisting of glycerol plus one fatty acid
- Diglycerides (or diacylglycerols), consisting of glycerol and two fatty acids

Although most foods contain both saturated and unsaturated fats, animal foods tend to contain more saturated and less unsaturated fat than plant foods. Saturated fatty acids tend to increase blood levels of LDL cholesterol (the lipoprotein that increases heart-disease risk when present in high levels) whereas unsaturated fatty acids tend to decrease LDL-cholesterol levels.[7]

Hydrogenation and Trans Fats   Oils can be made solid by adding hydrogen to the double bonds of their fatty acids. This process, called hydrogenation, makes some of the fatty acids in oils saturated and enhances storage life and baking qualities. Hydrogenation may alter the molecular structure of the fatty acids, however, changing the naturally occurring *cis* structure to the *trans* form. Trans fatty acids raise blood LDL-cholesterol levels to a greater extent than do saturated fatty acids.[7] Trans fatty acids are naturally present in dairy products and meats, but the primary dietary sources are products made from hydrogenated fats. Due to new nutrition labeling requirements and public uproar, the *trans fat* content of bakery products, chips, fast foods, and other products made with hydrogenated fats is decreasing.

Cholesterol   Dietary *cholesterol* is a fatlike, clear liquid substance found in lean and fat components of animal products. Cholesterol is a component of all animal cell membranes, the brain, and the nerves. It is the precursor of estrogen, testosterone, and vitamin D, which is manufactured in the skin upon exposure to sunlight. The body generally produces only one-third of the cholesterol our bodies use, because more than sufficient amounts of cholesterol are provided in most people's diet. The extent to which dietary cholesterol intake modifies blood cholesterol level appears to vary a good deal based on genetic tendencies.[11] Dietary cholesterol intake affects blood cholesterol level substantially less than does saturated fat intake, however.[7] Leading sources of dietary cholesterol are egg yolks, meat, milk and milk products, and fats such as butter.

Recommended Intake of Fats   Scientific evidence and opinions related to the effects of fat on health have changed substantially in recent years—and so have recommendations for fat intake. In the past, it was recommended that Americans aim for diets providing less than 30% of total calories from fat. Evidence indicating that the *type* of fat consumed is more important to health than is total fat intake has changed this advice. The watchwords for thinking about fat have become "not all fats are created equal: some are better for you than others." Concerns that high-fat diets encourage the development of obesity have been eased by studies demonstrating that excessive caloric intakes—and not just diets high in fat—are related to weight gain. New recommendations regarding fat intake do not encourage increased fat consumption, but rather emphasize that healthy diets include certain types of fat and that total caloric intake and physical activity are the most important components of weight management.[7]

Fats that elevate LDL-cholesterol levels (which increases the risk of heart disease) are regarded as "unhealthful," while those that lower LDL cholesterol and raise blood levels of HDL cholesterol (the one that helps the body get rid of cholesterol in the blood) are considered healthful. The list of unhealthy fats includes trans fats,

Glycerol   A component of fats that is soluble in water. It is converted to glucose in the body.

Essential Fatty Acids   Components of fat that are a required part of the diet (i.e., linoleic and alpha-linolenic acids). Both contain unsaturated fatty acids.

Prostaglandins   A group of physiologically active substances derived from the essential fatty acids. They are present in many tissues and perform such functions as the constriction or dilation of blood vessels, and stimulation of smooth muscles and the uterus.

Thromboxanes   Biologically active substances produced in platelets that increase platelet aggregation (and therefore promote blood clotting), constrict blood vessels, and increase blood pressure.

Prostacyclins   Biologically active substances produced by blood vessel walls that inhibit platelet aggregation (and therefore blood clotting), dilate blood vessels, and reduce blood pressure.

Saturated Fats   Fats in which adjacent carbons in the fatty acid component are linked by single bonds only (e.g., –C–C–C–C–).

Unsaturated Fats   Fats in which adjacent carbons in one or more fatty acids are linked by one or more double bonds (e.g., –C–C=C–C=C–).

Monounsaturated Fats   Fats in which only one pair of adjacent carbons in one or more of its fatty acids is linked by a double bond (e.g., –C–C=C–C–).

Polyunsaturated Fats   Fats in which more than one pair of adjacent carbons in one or more of its fatty acids are linked by two or more double bonds (e.g., –C–C=C–C=C–).

Cholesterol   A fat-soluble, colorless liquid found in animals but not plants.

saturated fats, and cholesterol. Monounsaturated fats, polyunsaturated fats, alpha-linolenic acid, DHA, and EPA are considered healthful fats.

Current recommendations call for consumption of 20–35% of total calories from fat. The AIs for the essential fatty acid linoleic acid are set at 17 grams a day for men and 12 grams for women. AIs for the other essential fatty acid, alpha-linolenic acid, are 1.6 grams per day for men and 1.1 grams for women. It is recommended that people keep their intake of trans fats and saturated fats as low as possible while consuming a nutritionally adequate diet. Americans are being encouraged to increase consumption of EPA and DHA by eating fish more often. They are being urged to reduce saturated fat intake in order to reduce the risk of heart disease.[7]

There is no recommended level of cholesterol intake, because there is no evidence that cholesterol is required in the diet. The body is able to produce enough cholesterol, and people do not develop a cholesterol deficiency disease if it is not consumed. Because blood cholesterol levels tend to increase somewhat as consumption of cholesterol increases, it is recommended that intake should be minimal. Cholesterol intake averages around 250 mg per day in the United States; but a more health-promoting level of intake would be less than 200 mg a day.[12]

Food Sources of Fat   The fat content of many foods can be identified by reading the nutrition information labels on food packages. The amount of total fat, saturated fat, trans fat, and cholesterol in a serving of food is listed on the labels. Table 1.7 lists the total fat, saturated fat, unsaturated fat, trans fat, cholesterol, and omega-3–fatty acid contents (EPA and DHA) of selected foods.

Table 1.7 Food sources of fats

A. Total Fat

	Portion Size	Grams of Fat		Portion Size	Grams of Fat
Fats and Oils			Veggie pita	1	17.0
Mayonnaise	1 tbs	11.0	Subway meatball sandwich	1	16.0
Ranch dressing	1 tbs	6.0	Subway turkey sandwich	1	4.0
Vegetable oils	1 tsp	4.7	**Milk and Milk Products**		
Butter	1 tsp	4.0	Cheddar cheese	1 oz	9.5
Margarine	1 tsp	4.0	Milk, whole	1 c	8.5
Meats, Fish			American cheese	1 oz	6.0
Sausage	4 links	18.0	Cottage cheese, regular	½ c	5.1
Hot dog	2 oz	17.0	Milk, 2%	1 c	5.0
Hamburger, 21% fat	3 oz	15.0	Milk, 1%	1 c	2.9
Hamburger, 16% fat	3 oz	13.5	Milk, skim	1 c	0.4
Steak, rib eye	3 oz	9.9	Yogurt, frozen	1 c	0.3
Bacon	3 strips	9.0	**Other Foods**		
Steak, round	3 oz	5.2	Avocado	½	15.0
Chicken, baked, no skin	3 oz	4.0	Almonds	1 oz	15.0
Flounder, baked	3 oz	1.0	Cashews	1 oz	13.2
Shrimp, boiled	3 oz	1.0	French fries, small serving	1	10.0
Fast Foods			Taco chips	1 oz (10 chips)	10.0
Whopper	8.9 oz	32.0	Potato chips	1 oz (14 chips)	7.0
Big Mac	6.6 oz	31.4	Peanut butter	1 tbs	6.1
Quarter Pounder with Cheese	6.8 oz	28.6	Egg	1	6.0

B. Saturated Fats

	Portion Size	Grams of Fat		Portion Size	Grams of Fat
Fats and Oils			**Milk and Milk Products**		
Margarine	1 tsp	2.9	Cheddar cheese	1 oz	5.9
Butter	1 tsp	2.4	American cheese	1 oz	5.5
Salad dressing, Ranch	1 tbs	1.2	Milk, whole	1 c	5.1
Peanut oil	1 tsp	0.9	Cottage cheese, regular	½ c	3.0
Olive oil	1 tsp	0.7	Milk, 2%	1 c	2.9
Salad dressing, 1000 Island	1 tbs	0.5	Milk, 1%	1 c	1.5
Canola oil	1 tsp	0.3	Milk, skim	1 c	0.3

Table 1.7 Food sources of fats (continued)

B. Saturated Fats (Continued)

	Portion Size	Grams of Fat		Portion Size	Grams of Fat
Meats, Fish			**Fast Foods**		
Hamburger, 21% fat	3 oz	6.7	Croissant w/egg,	1	16.0
Sausage, links	4	5.6	bacon, & cheese		
Hot dog	1	4.9	Sausage croissant	1	16.0
Chicken, fried, with skin	3 oz	3.8	Whopper	1	11.0
Salami	3 oz	3.6	Cheeseburger	1	9.0
Haddock, breaded, fried	3 oz	3.0	Bac'n Cheddar Deluxe	1	8.7
Rabbit	3 oz	3.0	Taco, regular	1	4.0
Pork chop, lean	3 oz	2.7	Chicken breast sandwich	1	3.0
Steak, round, lean	3 oz	2.0	**Nuts and Seeds**		
Turkey, roasted	3 oz	2.0	Macadamia nuts	1 oz	3.2
Chicken, baked, no skin	3 oz	1.7	Peanuts, dry roasted	1 oz	1.9
Prime rib, lean	3 oz	1.3	Sunflower seeds	1 oz	1.6
Venison	3 oz	1.1			
Tuna, in water	3 oz	0.4			

C. Unsaturated Fats

	Portion Size	Grams of Fat		Portion Size	Grams of Fat
Fats and Oils			Chicken, baked, no skin	3 oz	6.0
Canola oil	1 tsp	4.1	Pork chop, lean	3 oz	5.3
Vegetable oils	1 tsp	3.6	Turkey, roasted	3 oz	4.5
Margarine	1 tsp	2.9	Tuna, in water	3 oz	0.7
Butter	1 tsp	1.3	Egg	1	5.0
Milk and Milk Products			**Nuts and Seeds**		
Cottage cheese, regular	½ c	3.0	Sunflower seeds	1 oz	16.6
Cheddar cheese	1 oz	2.9	Almonds	1 oz	12.6
American cheese	1 oz	2.8	Peanuts	1 oz	11.3
Milk, whole	1 c	2.8	Cashews	1 oz	10.2
Meats, Fish					
Hamburger, 21% fat	3 oz	10.9			
Haddock, breaded, fried	3 oz	6.5			

D. Trans Fats

	Portion Size	Grams Trans Fats		Portion Size	Grams Trans Fats
Fats and Oils			**Milk**		
Margarine, stick	1 tsp	1.3	Whole	1 c	0.2
Margarine, tub (soft)	1 tsp	0.1	**Other Foods**		
Shortening	1 tsp	0.3	Doughnut	1	3.2
Butter	1 tsp	0.1	Danish pastry	1	3.0
Margarine, "no trans fat"	1 tsp	0	French fries, small serving	1	2.9
Meats			Cookies	2	1.8
Beef	3 oz	0.5	Corn chips	1 oz	1.4
Chicken	3 oz	0.1	Cake	1 slice	1.0
			Crackers	4 squares	0.5

E. Cholesterol

	Portion Size	Milligrams Cholesterol		Portion Size	Milligrams Cholesterol
Fats and Oils			**Meats, Fish**		
Butter	1 tsp	10.3	Brain	3 oz	1476
Vegetable oils, margarine	1 tsp	0	Liver	3 oz	470

continued

Table 1.7 Food sources of fats (continued)

E. Cholesterol (Continued)

	Portion Size	Milligrams Cholesterol		Portion Size	Milligrams Cholesterol
Egg	1	212	Venison	3 oz	48
Veal	3 oz	128	Wild pig	3 oz	33
Shrimp	3 oz	107	Goat, roasted	3 oz	32
Prime rib	3 oz	80	Tuna, in water	3 oz	25
Chicken, baked, no skin	3 oz	75	**Milk and Milk Products**		
Salmon, broiled	3 oz	74	Ice cream, regular	1 c	56
Turkey, baked, no skin	3 oz	65	Milk, whole	1 c	34
Hamburger, 20% fat	3 oz	64	Milk, 2%	1 c	22
Ostrich, ground	3 oz	63	Yogurt, low fat	1 c	17
Pork chop, lean	3 oz	60	Milk, 1%	1 c	14
Hamburger, 10% fat	3 oz	60	Milk, skim	1 c	7

F. Omega-3 (n-3) Fatty Acids

	Portion Size	Grams EPA + DHA		Portion Size	Grams EPA + DHA
Fish and Seafood			Tuna, white, canned	3.5 oz	1.7
Sardines in oil*	3.5 oz	3.3	Swordfish	3.5 oz	0.8
Mackerel	3.5 oz	2.6	Flounder*	3.5 oz	0.5
Salmon, Atlantic, farmed*	3.5 oz	2.2	Scallops*	3.5 oz	0.5
Lake trout*	3.5 oz	2.0	Carp	3.5 oz	0.3
Herring*	3.5 oz	1.8	Cod*	3.5 oz	0.3
Salmon, sockeye*	3.5 oz	1.5	Crab*	3.5 oz	0.3
Whitefish, lake*	3.5 oz	1.5	Pike, walleye*	3.5 oz	0.3
Anchovies*	3.5 oz	1.4	Catfish, wild*	3.5 oz	0.2
Salmon, chinook*	3.5 oz	1.4	Fish sticks	3.5 oz	0.2
Bluefish	3.5 oz	1.2	Haddock*	3.5 oz	0.2
Halibut*	3.5 oz	1.2	Lobster*	3.5 oz	0.2
Oysters	3.5 oz	1.1	Perch, ocean*	3.5 oz	0.2
Salmon, pink*	3.5 oz	1.0	Salmon, red*	3.5 oz	0.2
Trout, rainbow*	3.5 oz	1.0	Snapper, red*	3.5 oz	0.2
Bass, striped	3.5 oz	0.8	Clams*	3.5 oz	0.1
Oysters*	3.5 oz	0.6	**Others**		
Catfish*	3.5 oz	0.5	Egg yolk*	1	0.4
Pollock*	3.5 oz	0.5	DHA fortified egg*	1	0.2
Shrimp*	3.5 oz	0.5	Human milk*	3.5 oz	0.2

*Mercury content <0.1 ppm as given in: Mercury levels in commercial fish and shellfish, 2006 update, U.S. Environmental Protection Agency, www.epa.gov.

Vitamins

Vitamins are chemical substances in foods that perform specific functions in the body. Thirteen have been discovered so far. They are classified as either fat soluble or water soluble (Table 1.8).

The B-complex vitamins and vitamin C are soluble in water and found dissolved in water in foods. The fat-soluble vitamins consist of vitamins A, D, E, and K and are present in the fat portions of foods. (To remember the fat-soluble vitamins, think of "DEKA" for vitamins D, E, K, and A.) Only these chemical substances are truly vita-mins. Substances such as coenzyme Q_{10}, inositol, provita-min B_5 complex, and pangamic acid (vitamin B_{15}) may be called vitamins, but they are not. Except for vitamin B_{12}, water-soluble vitamin stores in the body are limited and run out within a few weeks to a few months after intake becomes inadequate. Fat-soluble vitamins are stored in the body's fat tissues and the liver. These stores can be sizable and last from months to years when intake is low.

Excessive consumption of the fat-soluble vitamins from supplements, especially of vitamins A and D, produces various symptoms of toxicity. High intake of the water-soluble vitamins from supplements can also produce

Table 1.8 Vitamin solubility

Water-Soluble Vitamins	Fat-Soluble Vitamins
B-complex vitamins	Vitamin A (retinol, beta-carotene)
Thiamin (B_1)	Vitamin D (1,25 dihydroxy-cholecalciferol)
Riboflavin (B_2)	
Niacin (B_3)	
Vitamin B_6	Vitamin E (alpha-tocopherol)
Folate	Vitamin K
Vitamin B_{12}	
Biotin	
Pantothenic acid	
Vitamin C (ascorbic acid)	

adverse health effects. Toxicity symptoms from water-soluble vitamins, however, tend to last a shorter time and are more quickly remedied. Vitamin overdoses are very rarely related to food intake.

Vitamins do not provide energy or serve as structural components of the body. Some play critical roles as *coenzymes* in chemical changes that take place in the body known as *metabolism*. Vitamin A is needed to replace the cells that line the mouth and esophagus, thiamin is needed for maintenance of normal appetite, and riboflavin and folate are needed for the synthesis of body proteins. Other vitamins (vitamins C and E, and beta-carotene—a precursor of vitamin A) act as *antioxidants* and perform other functions. By preventing or repairing damage to cells due to oxidation, these vitamins help maintain body tissues and prevent disease.

Primary functions, consequences of deficiency and overdose, primary food sources, and comments about each vitamin are listed in Table 1.9 starting on the next page.

Recommended Intake of Vitamins Recommendations for levels of intake of vitamins are presented in the tables on the inside front covers of this text. Note that Tolerable Upper Levels of Intake (ULs) for many vitamins are also given; they represent levels of intake that should not be exceeded. Table 1.10 (pp. 19–21) lists food sources of each vitamin.

Other Substances in Food

> "Things don't happen by accident in nature. If you observe it, it has a reason for being there."
>
> Norman Krinsky, Tufts University

There are many substances in foods in addition to nutrients that affect health. Some foods contain naturally occurring toxins, such as poison in puffer fish and solanine in green sections near the skin of some potatoes. Consuming the poison in puffer fish can be lethal; large doses of solanine can interfere with nerve impulses. Some

plant pigments, hormones, and other naturally occurring substances that protect plants from insects, oxidization, and other damaging exposures also appear to benefit human health. These substances in plants are referred to as *phytochemicals*, and knowledge about their effects on human health is advancing rapidly. Consumption of foods rich in specific pigments and other phytochemicals, rather than consumption of isolated phytochemicals, may help prevent certain types of cancer, cataracts, type 2 diabetes, hypertension, infections, and heart disease. High intakes of certain phytochemicals from vegetables, fruits, nuts, seeds, and whole grain products may partially account for lower rates of heart disease and cancer observed in people with high intakes of these foods.[12,13]

Minerals

Humans require the 15 minerals listed in Table 1.11 (p. 21). Minerals are unlike other nutrients in that they consist of single atoms and carry a charge in solution. The property of being charged (or having an unequal number of electrons and protons) is related to many of the functions of minerals. The charge carried by minerals allows them to combine with other minerals to form stable complexes in bone, teeth, cartilage, and other tissues. In body fluids, charged minerals serve as a source of electrical power that stimulates muscles to contract (e.g., the heart to beat) and nerves to react. Minerals also help the body maintain an adequate amount of water in tissues and control how acidic or basic body fluids remain.

The tendency of minerals to form complexes has implications for the absorption of minerals from food. Calcium and zinc, for example, may combine with other minerals in supplements or with dietary fiber and form complexes that cannot be absorbed. Therefore, in general, the proportion of total mineral intake that is absorbed is less than for vitamins.

Functions, consequences of deficiency and overdose, primary food sources, and comments about the 15 minerals needed by humans are summarized in Table 1.12 (pp. 22–25).

Recommended Intake of Minerals Recommendations for intake of minerals are presented in the tables on the inside front covers of this text. Note that Tolerable Upper

Coenzymes Chemical substances that activate enzymes.

Metabolism The chemical changes that take place in the body. The conversion of glucose to energy or body fat is an example of a metabolic process.

Antioxidants Chemical substances that prevent or repair damage to cells caused by exposure to oxidizing agents such as oxygen, ozone, and smoke and to other oxidizing agents normally produced in the body. Many different antioxidants are found in foods; some are made by the body.

Phytochemicals (Phyto = plants) Chemical substances in plants, some of which affect body processes in humans that may benefit health.

Table 1.9 Summary of the vitamins

The Water-Soluble Vitamins

	Primary Functions	Consequences of Deficiency
Thiamin (vitamin B$_1$) AI[a] women: 1.1 mg men: 1.2 mg	• Coenzyme in the metabolism of carbohydrates, alcohol, and some amino acids • Required for the growth and maintenance of nerve and muscle tissues • Required for normal appetite	• Fatigue, weakness • Nerve disorders, mental confusion, apathy • Impaired growth • Swelling • Heart irregularity and failure
Riboflavin (vitamin B$_2$) AI women: 1.1 mg men: 1.3 mg	• Coenzyme involved in energy metabolism of carbohydrates, proteins, and fats • Coenzyme function in cell division • Promotes growth and tissue repair • Promotes normal vision	• Reddened lips, cracks at both corners of the mouth • Fatigue
Niacin (vitamin B$_3$) RDA women: 14 mg men: 16 mg UL: 35 mg (from supplements and fortified foods)	• Coenzyme involved in energy metabolism • Coenzyme required for the synthesis of body fats • Helps maintain normal nervous system functions	• Skin disorders • Nervous and mental disorders • Diarrhea, indigestion • Fatigue
Vitamin B$_6$ (pyridoxine) AI women: 1.3 mg men: 1.3 mg UL: 100 mg	• Coenzyme involved in amnio acid, glucose, and fatty acid metabolism and neurotransmitter synthesis • Coenzyme in the conversion of tryptophan to niacin • Required for normal red blood cell formation • Required for the synthesis of lipids in the nervous and immune systems	• Irritability, depression • Convulsions, twitching • Muscular weakness • Dermatitis near the eyes • Anemia • Kidney stones
Folate (folacin, folic acid) RDA women: 400 mcg men: 400 mcg UL: 1000 mcg (from supplements and fortified foods)	• Required for the conversion of homocysteine to methionine • Methyl (CH$_3$) group donor and coenzyme in DNA synthesis, gene expression and regulation • Required for the normal formation of red blood and other cells	• Megaloblastic cells and anemia • Diarrhea, weakness, irritability, paranoid behavior • Red, sore tongue • Increased blood homocysteine levels • Neural tube defects, low birthweight (in pregnancy); increased risk of heart disease and stroke
Vitamin B$_{12}$ (cyanocobalamin) AI women: 2.4 mcg men: 2.4 mcg	• Coenzyme involved in the synthesis of DNA, RNA, and myelin • Required for the conversion of homocysteine to methionine • Needed for normal red blood cell development	• Neurological disorders (nervousness, tingling sensations, brain degeneration) • Pernicious anemia • Increased blood homocysteine levels • Fatigue • Deficiency reported in 39% of adults in one study
Biotin AI women: 30 mcg men: 30 mcg	• Required by enzymes involved in fat, protein, and glycogen metabolism	• Depression, fatigue, nausea • Hair loss, dry and scaly skin • Muscular pain

[a]AI (Adequate Intakes) and RDAs (Recommended Dietary Allowances) are for 19–30-year-olds; UL (Upper Limits) are for 19–70-year-olds, 1997–2002.

Table 1.9 Summary of the vitamins (continued)

The Water-Soluble Vitamins (Continued)

Consequences of Overdose	Primary Food Sources	Highlights and Comments
• High intakes of thiamin are rapidly excreted by the kidneys. Oral doses of 500 mg/day or less are considered safe.	• Grains and grain products (cereals, rice, pasta, bread) • Ready-to-eat cereals • Pork and ham, liver • Milk, cheese, yogurt • Dried beans and nuts	• Need increases with carbohydrate intake. • There is no "e" on the end of thiamin! • Deficiency rare in the U.S.; may occur in people with alcoholism. • Enriched grains and cereals prevent thiamin deficiency.
• None known. High doses are rapidly excreted by the kidneys.	• Milk, yogurt, cheese • Grains and grain products (cereals, rice, pasta, bread) • Liver, poultry, fish, beef • Eggs	• Destroyed by exposure to light.
• Flushing, headache, cramps, rapid heartbeat, nausea, diarrhea, decreased liver function with doses above 0.5 g per day	• Meats (all types) • Grains and grain products (cereals, rice, pasta, bread) • Dried beans and nuts • Milk, cheese, yogurt • Ready-to-eat cereals • Coffee • Potatoes	• Niacin has a precursor—tryptophan. Tryptophan, an amino acid, is converted to niacin by the body. Much of our niacin intake comes from tryptophan. • High doses raise HDL-cholesterol levels.
• Bone pain, loss of feeling in fingers and toes, muscular weakness, numbness, loss of balance (mimicking multiple sclerosis)	• Oatmeal, bread, breakfast cereals • Bananas, avocados, prunes, tomatoes, potatoes • Chicken, liver • Dried beans • Meats (all types), milk • Green and leafy vegetables	• Vitamins go from B_3 to B_6 because B_4 and B_5 were found to be duplicates of vitamins already identified.
• May cover up signs of vitamin B_{12} deficiency (pernicious anemia)	• Fortified, refined grain products (bread, flour, pasta) • Ready-to-eat cereals • Dark green, leafy vegetables (spinach, collards, romaine) • Broccoli, brussels sprouts • Oranges, bananas, grapefruit • Milk, cheese, yogurt • Dried beans	• Folate means "foliage." It was first discovered in leafy green vegetables. • This vitamin is easily destroyed by heat. • Synthetic form added to fortified grain products is better absorbed than naturally occurring folates.
• None known. Excess vitamin B_{12} is rapidly excreted by the kidneys or is not absorbed into the bloodstream. • Vitamin B_{12} injections may cause a temporary feeling of heightened energy.	• Animal products: beef, lamb, liver, clams, crab, fish, poultry, eggs • Milk and milk products • Ready-to-eat cereals	• Older people and vegans are at risk for vitamin B_{12} deficiency. • Some people become vitamin B_{12} deficient because they are genetically unable to absorb it. • Vitamin B_{12} is found in animal products and microorganisms only.
• None known. Excesses are rapidly excreted.	• Grain and cereal products • Meats, dried beans, cooked eggs • Vegetables	• Deficiency is extremely rare. May be induced by the overconsumption of raw eggs.

continued

Table 1.9 Summary of the vitamins (continued)

The Water-Soluble Vitamins (Continued)

	Primary Functions	Consequences of Deficiency
Pantothenic acid (pantothenate) AI women: 5 mg men: 5 mg	• Coenzyme involved in energy metabolism of carbohydrates and fats • Coenzyme in protein metabolism	• Fatigue, sleep disturbances, numbness, impaired coordination • Vomiting, nausea

Vitamin C (ascorbic acid)
RDA women: 75 mg
       men: 90 mg
      UL: 2000 mg
	• Required for collagen synthesis
	• Acts as an antioxidant; protects LDL cholesterol, eye tissues, sperm proteins, DNA, and lipids against oxidation
	• Required for the conversion of Fe^{++} to Fe^{+++}
	• Required for neurotransmitters and steroid hormone synthesis
		• Bleeding and bruising easily due to weakened blood vessels, cartilage, and other tissues containing collagen
		• Slow recovery from infections and poor wound healing
		• Fatigue, depression
		• Deficiency reported in 9–24% of adults in one study

The Fat-Soluble Vitamins

	Primary Functions	Consequences of Deficiency

Vitamin A
RDA women: 700 mcg
      men: 900 mcg
      UL: 3000 mcg
	• Needed for the formation and maintenance of mucous membranes, skin, bone
	• Needed for vision in dim light
		• Increased susceptibility to infection, increased incidence and severity of infection (including measles)
		• Impaired vision, xerophthalmia, blindness
		• Inability to see in dim light

Vitamin E (alpha-tocopherol)
RDA women: 15 mg
      men: 15 mg
      UL: 1000 mg
	• Acts as an antioxidant, prevents damage to cell membranes in blood cells, lungs, and other tissues by repairing damage caused by free radicals
	• Reduces oxidation of LDL cholesterol
		• Muscle loss, nerve damage
		• Anemia
		• Weakness
		• Many adults may have nonoptimal blood levels

Vitamin D (1,25 dihydroxy-cholecalciferol)
AI   women: 5 mcg
        (200 IU)
        men: 5 mcg
        (200 IU)
        UL: 50 mcg
        (2000 IU)
	• Required for calcium and phosphorus metabolism in the intestines and bone, and for their utilization in bone and teeth formation, nerve and muscle activity	• Weak, deformed bones (children)
		• Loss of calcium from bones (adults), osteoporosis

Table 1.9 Summary of the vitamins (continued)

The Water-Soluble Vitamins (Continued)

Consequences of Overdose	Primary Food Sources	Highlights and Comments
• None known. Excesses are rapidly excreted.	• Many foods, including meats, grains, vegetables, fruits, and milk	• Deficiency is very rare.
• Intakes of 1 g or more per day can cause nausea, cramps, and diarrhea and may increase the risk of kidney stones.	• Fruits: oranges, lemons, limes, strawberries, cantaloupe, honeydew melon, grapefruit, kiwi fruit, mango, papaya • Vegetables: broccoli, green and red peppers, collards, cabbage, tomato, asparagus, potatoes • Ready-to-eat cereals	• Need increases among smokers (to 110–125 mg per day). • Is fragile; easily destroyed by heat and exposure to air. • Supplements may decrease duration and symptoms of colds. • Deficiency may develop within 3 weeks of very low intake.

The Fat-Soluble Vitamins (Continued)

Consequences of Overdose	Primary Food Sources	Highlights and Comments
• Vitamin A toxicity (hypervitamosis A) with acute doses of 500,000 IU, or long-term intake of 50,000 IU per day; limit retinol use in pregnancy to 5000 IU daily • Nausea, irritability, blurred vision, weakness • Increased pressure in the skull, headache • Liver damage • Hair loss, dry skin • Birth defects	• Vitamin A is found in animal products only • Liver, butter, margarine, milk, cheese, eggs • Ready-to-eat cereals	• Beta-carotene is a vitamin A precursor or "provitamin." • Symptoms of vitamin A toxicity may mimic those of brain tumors and liver disease. Vitamin A toxicity is sometimes misdiagnosed because of the similarities in symptoms. • 1 mcg retinol equivalent = 5 IU vitamin A or 6 mcg beta-carotene.
• Intakes of up to 800 IU per day are unrelated to toxic side effects; over 800 IU per day may increase bleeding (blood-clotting time) • Avoid supplement use if aspirin, anticoagulants, or fish oil supplements are taken regularly	• Oils and fats • Salad dressings, mayonnaise, margarine, shortening, butter • Whole grains, wheat germ • Leafy, green vegetables, tomatoes • Nuts and seeds • Eggs	• Vitamin E is destroyed by exposure to oxygen and heat. • Oils naturally contain vitamin E. It's there to protect the fat from breakdown due to free radicals. • Eight forms of vitamin E exist, and each has different antioxidant strengths. • Natural form is better absorbed than synthetic form: 15 IU alpha-tocopherol = 22 IU d-alpha-tocopherol (natural form) and 33 IU synthetic vitamin E.
• Mental retardation in young children • Abnormal bone growth and formation • Nausea, diarrhea, irritability, weight loss • Deposition of calcium in organs such as the kidneys, liver, and heart	• Vitamin D–fortified milk and margarine • Butter • Fish • Eggs • Mushrooms • Milk products such as cheese, yogurt, and ice cream are generally not fortified with vitamin D.	• Vitamin D is manufactured from cholesterol in cells beneath the surface of the skin upon exposure of the skin to sunlight. • Deficiency may be common in ill, homebound, and elderly and hospitalized adults. • Breastfed infants with little sun exposure benefit from vitamin D supplements.

continued

Table 1.9 Summary of the vitamins (continued)

The Fat-Soluble Vitamins (Continued)

	Primary Functions	Consequences of Deficiency
Vitamin K (phylloquinone, menaquinone) AI women: 90 mcg men: 120 mcg	• Regulation of synthesis blood-clotting proteins • Aids in the incorporation of calcium into bones	• Bleeding, bruises • Decreased calcium in bones • Deficiency is rare; may be induced by the long-term use (months or more) of antibiotics

Table 1.10 Food sources of vitamins

Thiamin

Food	Amount	Thiamin (milligrams)	Food	Amount	Thiamin (milligrams)
Meats			Rice	½ c	0.1
Pork roast	3 oz	0.8	Bread	1 slice	0.1
Beef	3 oz	0.4	**Vegetables**		
Ham	3 oz	0.4	Peas	½ c	0.3
Liver	3 oz	0.2	Lima beans	½ c	0.2
Nuts and Seeds			Corn	½ c	0.1
Sunflower seeds	¼ c	0.7	Broccoli	½ c	0.1
Peanuts	¼ c	0.1	Potato	1 med	0.1
Almonds	¼ c	0.1	**Fruits**		
Grains			Orange juice	1 c	0.2
Bran flakes	1 c (1 oz)	0.6	Orange	1	0.1
Macaroni	½ c	0.1	Avocado	½	0.1

Riboflavin

Food	Amount	Riboflavin (milligrams)	Food	Amount	Riboflavin (milligrams)
Milk and Milk Products			Beef	3 oz	0.2
Milk	1 c	0.5	Tuna	3 oz	0.1
2% milk	1 c	0.5	**Vegetables**		
Yogurt, low fat	1 c	0.5	Collard greens	½ c	0.3
Skim milk	1 c	0.4	Broccoli	½ c	0.2
Yogurt	1 c	0.1	Spinach, cooked	½ c	0.1
American cheese	1 oz	0.1	**Eggs**		
Cheddar cheese	1 oz	0.1	Egg	1	0.2
Meats			**Grains**		
Liver	3 oz	3.6	Macaroni	½ c	0.1
Pork chop	3 oz	0.3	Bread	1 slice	0.1

Niacin

Food	Amount	Niacin (milligrams)	Food	Amount	Niacin (milligrams)
Meats			Haddock	3 oz	2.7
Liver	3 oz	14.0	Scallops	3 oz	1.1
Tuna	3 oz	10.3	**Nuts and Seeds**		
Turkey	3 oz	9.5	Peanuts	1 oz	4.9
Chicken	3 oz	7.9	**Vegetables**		
Salmon	3 oz	6.9	Asparagus	½ c	1.5
Veal	3 oz	5.2	**Grains**		
Beef (round steak)	3 oz	5.1	Wheat germ	1 oz	1.5
Pork	3 oz	4.5	Brown rice	½ c	1.2

Table 1.9 Summary of the vitamins (continued)

The Fat-Soluble Vitamins (Continued)

Consequences of Overdose	Primary Food Sources	Highlights and Comments
• Toxicity is a problem only when synthetic forms of vitamin K are taken in excessive amounts. That may cause liver disease.	• Leafy, green vegetables • Grain products	• Vitamin K is produced by bacteria in the gut. Part of our vitamin K supply comes from these bacteria. • Newborns are given a vitamin K injection because they have "sterile" guts and consequently no vitamin K–producing bacteria.

Table 1.10 Food sources of vitamins (continued)

Niacin (Continued)

Food	Amount	Niacin (milligrams)	Food	Amount	Niacin (milligrams)
Noodles, enriched	½ c	1.0	**Milk and Milk Products**		
Rice, white, enriched	½ c	1.0	Cottage cheese	½ c	2.6
Bread, enriched	1 slice	0.7	Milk	1 c	1.9

Vitamin B₆

Food	Amount	Vitamin B₆ (milligrams)	Food	Amount	Vitamin B₆ (milligrams)
Meats			Dried beans, cooked	½ c	0.4
Liver	3 oz	0.8	**Fruits**		
Salmon	3 oz	0.7	Banana	1	0.6
Other fish	3 oz	0.6	Avocado	½	0.4
Chicken	3 oz	0.4	Watermelon	1 c	0.3
Ham	3 oz	0.4	**Vegetables**		
Hamburger	3 oz	0.4	Turnip greens	½ c	0.7
Veal	3 oz	0.4	Brussels sprouts	½ c	0.4
Pork	3 oz	0.3	Potato	1	0.2
Beef	3 oz	0.2	Sweet potato	½ c	0.2
Eggs			Carrots	½ c	0.2
Egg	1	0.3	Peas	½ c	0.1
Legumes					
Split peas	½ c	0.6			

Folate

Food	Amount	Folate (micrograms)	Food	Amount	Folate (micrograms)
Vegetables			Broccoli	½ c	43
Garbanzo beans	½ c	141	**Fruits**		
Navy beans	½ c	128	Cantaloupe	¼ whole	100
Asparagus	½ c	120	Orange juice	1 c	87
Brussels sprouts	½ c	116	Orange	1	59
Black-eyed peas	½ c	102	**Grains**[a]		
Spinach, cooked	½ c	99	Ready-to-eat cereals	1 c (1 oz)	100–400
Romaine lettuce	1 c	86	Oatmeal	½ c	97
Lima beans	½ c	71	Noodles	½ c	45
Peas	½ c	70	Wheat germ	2 tbs	40
Collard greens, cooked	½ c	56	Wild rice	½ c	37
Sweet potato	½ c	43			

[a]Fortified, refined grain products such as bread, rice, pasta, and crackers provide approximately 40 to 60 mcg of folic acid per standard serving.

continued

Table 1.10 Food sources of vitamins (continued)

Vitamin B₁₂

Food	Amount	Vitamin B₁₂ (micrograms)	Food	Amount	Vitamin B₁₂ (micrograms)
Meats			**Milk and Milk Products**		
Liver	3 oz	6.8	Skim milk	1 c	1.0
Trout	3 oz	3.6	Milk	1 c	0.9
Beef	3 oz	2.2	Yogurt	1 c	0.8
Clams	3 oz	2.0	Cottage cheese	½ c	0.7
Crab	3 oz	1.8	American cheese	1 oz	0.2
Lamb	3 oz	1.8	Cheddar cheese	1 oz	0.2
Tuna	3 oz	1.8	**Eggs**		
Veal	3 oz	1.7	Egg	1	0.6
Hamburger, regular	3 oz	1.5			

Vitamin C

Food	Amount	Vitamin C (milligrams)	Food	Amount	Vitamin C (milligrams)
Fruits			Watermelon	1 c	15
Orange juice, vitamin C–fortified	1 c	108	**Vegetables**		
Kiwi fruit	1 or ½ c	108	Green peppers	½ c	95
Grapefruit juice, fresh	1 c	94	Cauliflower, raw	½ c	75
Cranberry juice cocktail	1 c	90	Broccoli	½ c	70
Orange	1	85	Brussels sprouts	½ c	65
Strawberries, fresh	1 c	84	Collard greens	½ c	48
Orange juice, fresh	1 c	82	Vegetable (V-8) juice	¾ c	45
Cantaloupe	¼ whole	63	Tomato juice	¾ c	33
Grapefruit	1 med	51	Cauliflower, cooked	½ c	30
Raspberries, fresh	1 c	31	Potato	1 med	29
			Tomato	1 med	23

Vitamin A (Retinol)

Food Sources of Vitamin A (Retinol)	Amount	Vitamin A (micrograms RE)[b]	Food Sources of Vitamin A (Retinol)	Amount	Vitamin A (micrograms RE)[b]
Meats			2% milk	1 c	139
Liver	3 oz	9124	American cheese	1 oz	82
Salmon	3 oz	53	Whole milk	1 c	76
Tuna	3 oz	14	Swiss cheese	1 oz	65
Eggs			**Fats**		
Egg	1 med	84	Margarine, fortified	1 tsp	46
Milk and Milk Products			Butter	1 tsp	38
Skim milk, fortified	1 c	149			

Vitamin A (Beta-Carotene)

Food Sources of Beta-Carotene	Amount	Vitamin A Value (micrograms RE)[b]	Food Sources of Beta-Carotene	Amount	Vitamin A Value (micrograms RE)[b]
Vegetables			Green peppers	½ c	40
Pumpkin, canned	½ c	2712	**Fruits**		
Sweet potato, canned	½ c	1935	Cantaloupe	¼ whole	430
Carrots, raw	½ c	1913	Apricots, canned	½ c	210
Spinach, cooked	½ c	739	Nectarine	1 med	101
Collard greens, cooked	½ c	175	Watermelon	1 c	59
Broccoli, cooked	½ c	109	Peaches, canned	½ c	47
Winter squash	½ c	53	Papaya	½ c	20

[b]RE (retinol equivalent) = 3.33 IU.

Table 1.10 Food sources of vitamins (continued)

		Vitamin E (IU)[c]			Vitamin E (IU)[c]
Food	Amount		Food	Amount	
Oils			Collard greens	½ c	3.1
Vegetable oil	1 tbs	6.7	Asparagus	½ c	2.1
Mayonnaise	1 tbs	3.4	Spinach, raw	1 c	1.5
Margarine	1 tbs	2.7	**Grains**		
Salad dressing	1 tbs	2.2	Wheat germ	2 tbs	4.2
Nuts and Seeds			Bread, whole wheat	1 slice	2.5
Sunflower seeds	¼ c	27.1	Bread, white	1 slice	1.2
Almonds	¼ c	12.7	**Seafood**		
Peanuts	¼ c	4.9	Crab	3 oz	4.5
Cashews	¼ c	0.7	Shrimp	3 oz	3.7
Vegetables			Fish	3 oz	2.4
Sweet potato	½ c	6.9			

Vitamin D

		Vitamin D (IU)[d]			Vitamin D (IU)[d]
Food	Amount		Food	Amount	
Milk			**Organ meats**		
Milk, whole, low fat, or skim	1 c	100	Beef liver	3 oz	42
Fish and seafood			Chicken liver	3 oz	40
Salmon	3 oz	340	**Eggs**		
Tuna	3 oz	150	Egg yolk	1	27
Shrimp	3 oz	127			

[c]15 mg alpha-tocopherol = 22 IU d-alpha tocopherol (natural form) and 33 IU synthetic vitamin E.
[d]40 IU = 1 mcg.

Table 1.11 Minerals required by humans

Calcium	Fluoride	Chromium
Phosphorus	Iodine	Molybdenum
Magnesium	Selenium	Sodium
Iron	Copper	Potassium
Zinc	Manganese	Chloride

Levels of Intake for many minerals are also given in a separate table. Table 1.13 lists food sources of each mineral.

Water

Water is the last, but not the least, nutrient category. Adults are about 60–70% water by weight. Water provides the medium in which most chemical reactions take place in the body. It plays a role in energy transformation, the excretion of wastes, and temperature regulation.

People need enough water to replace daily losses from perspiration, urination, and exhalation. In normal weather conditions and with normal physical activity levels, adult males need 15–16 cups of water from fluids and foods each day, and females 11 cups. The need for water is generally met by consuming sufficient fluids to satisfy thirst.[14]

The need for water is greater in hot and humid climates, and when physical activity levels are high. People generally consume 75% of their water intake from water and other fluids and 25% from foods.[15] Adequate consumption of water is indicated by the excretion of urine that is pale yellow and normal in volume.[16]

Dietary Sources of Water   The best sources of water are tap and bottled water; nonalcoholic beverages such as fruit juice, milk, and vegetable juice; and brothy soups. Alcohol tends to increase water loss though urine, so beverages such as beer and wine are not as "hydrating" as water is. Caffeinated beverages are hydrating in people who are accustomed to consuming them.[17]

PRINCIPLE #3   Health problems related to nutrition originate within cells.

The functions of each cell are maintained by the nutrients it receives. Problems arise when a cell's need for nutrients differs from the amounts that are available. Cells (Illustration 1.4) are the building blocks of tissues (such as bones and muscles), organs (the heart, kidney, and liver, for example), and systems (such as the circulatory and respiratory systems). Normal cell health and functions are maintained when a nutritional and environmental utopia

Table 1.12 Summary of minerals

	Primary Functions	Consequences of Deficiency
Calcium AI* women: 1000 mg men: 1000 mg UL: 2500 mg	• Component of bones and teeth • Required for muscle and nerve activity, blood clotting	• Poorly mineralized, weak bones (osteoporosis) • Rickets in children • Osteomalacia (rickets in adults) • Stunted growth in children • Convulsions, muscle spasms
Phosphorus RDA women: 700 mg men: 700 mg UL: 4000 mg	• Component of bones and teeth • Component of certain enzymes and other substances involved in energy formation • Required for maintenance of acid-base balance of body fluids	• Loss of appetite • Nausea, vomiting • Weakness • Confusion • Loss of calcium from bones
Magnesium RDA women: 310 mg men: 400 mg UL: 350 mg (from supplements only)	• Component of bones and teeth • Needed for nerve activity • Activates enzymes involved in energy and protein formation	• Stunted growth in children • Weakness • Muscle spasms • Personality changes
Iron RDA women: 18 mg men: 8 mg UL: 45 mg	• Transports oxygen as a component of hemoglobin in red blood cells • Component of myoglobin (a muscle protein) • Required for certain reactions involving energy formation	• Iron deficiency • Iron-deficiency anemia • Weakness, fatigue • Pale appearance • Reduced attention span and resistance to infection • Mental retardation, developmental delay in children
Zinc RDA women: 8 mg men: 11 mg UL: 40 mg	• Required for the activation of many enzymes involved in the reproduction of proteins • Component of insulin, many enzymes	• Growth failure • Delayed sexual maturation • Slow wound healing • Loss of taste and appetite • In pregnancy, low-birth-weight infants and preterm delivery
Fluoride AI women: 3 mg men: 4 mg UL: 10 mg	• Component of bones and teeth (enamel)	• Tooth decay and other dental diseases
Iodine RDA women: 150 mcg men: 150 mcg UL: 1100 mcg	• Component of thyroid hormones that help regulate energy production and growth	• Goiter • Cretinism (mental retardation, hearing loss, growth failure)

*AIs and RDAs are for women and men 19–30 years of age; ULs are for males and females 19–70 years of age, 1997–2004.

Table 1.12 Summary of minerals (continued)

Consequences of Overdose	Primary Food Sources	Highlights and Comments
• Drowsiness • Calcium deposits in kidneys, liver, and other tissues • Suppression of bone remodeling • Decreased zinc absorption	• Milk and milk products (cheese, yogurt) • Broccoli • Dried beans • Calcium-fortified foods (some juices, breakfast cereals, bread, for example)	• The average intake of calcium among U.S. women is approximately 60% of the DRI. • One in four women and one in eight men in the U.S. develop osteoporosis. • Adequate calcium and vitamin D status must be maintained to prevent bone loss.
• Muscle spasms	• Milk and milk products (cheese, yogurt) • Meats • Seeds, nuts • Phosphates added to foods	• Deficiency is generally related to disease processes.
• Diarrhea • Dehydration • Impaired nerve activity due to disrupted utilization of calcium	• Plant foods (dried beans, tofu, peanuts, potatoes, green vegetables) • Milk • Bread • Ready-to-eat cereals • Coffee	• Magnesium is primarily found in plant foods where it is attached to chlorophyll. • Average intake among U.S. adults is below the RDA.
• Hemochromatosis ("iron poisoning") • Vomiting, abdominal pain • Blue coloration of skin • Liver and heart damage, diabetes • Decreased zinc absorption • Atherosclerosis (plaque buildup) in older adults	• Liver, beef, pork • Dried beans • Iron-fortified cereals • Prunes, apricots, raisins • Spinach • Bread • Pasta	• Cooking foods in iron and stainless steel pans increases the iron content of the foods. • Vitamin C, meat, and alcohol increase iron absorption. • Iron deficiency is the most common nutritional deficiency in the world. • Average iron intake of young children and women in the U.S. is low.
• Over 25 mg/day is associated with nausea, vomiting, weakness, fatigue, susceptibility to infection, copper deficiency, and metallic taste in mouth • Increased blood lipids	• Meats (all kinds) • Grains • Nuts • Milk and milk products (cheese, yogurt) • Ready-to-eat cereals • Bread	• Like iron, zinc is better absorbed from meats than from plants. • Marginal zinc deficiency may be common, especially in children. • Zinc supplements may decrease duration and severity of the common cold.
• Fluorosis • Brittle bones • Mottled teeth • Nerve abnormalities	• Fluoridated water and foods and beverages made with it • Tea • Shrimp, crab	• Toothpastes, mouth rinses, and other dental care products may provide fluoride. • Fluoride overdose has been caused by ingestion of fluoridated toothpaste.
• Over 1 mg/day may produce pimples, goiter, and decreased thyroid function	• Iodized salt • Milk and milk products • Seaweed, seafoods • Bread from commercial bakeries	• Iodine deficiency remains a major health problem in some developing countries. • Amount of iodine in plants depends on iodine content of soil. • Most of the iodine in our diet comes from the incidental addition of iodine to foods from cleaning compounds used by food manufacturers.

continued

Table 1.12 Summary of minerals (continued)

	Primary Functions	Consequences of Deficiency
Selenium RDA women: 55 mcg men: 55 mcg UL: 400 mcg	• Acts as an antioxidant in conjunction with vitamin E (protects cells from damage due to exposure to oxygen) • Needed for thyroid hormone production	• Anemia • Muscle pain and tenderness • Keshan disease (heart failure), Kashin-Beck disease (joint disease)
Copper RDA women: 900 mcg men: 900 mcg UL: 10,000 mcg	• Component of enzymes involved in the body's utilization of iron and oxygen • Functions in growth, immunity, cholesterol and glucose utilization, brain development	• Anemia • Seizures • Nerve and bone abnormalities in children • Growth retardation
Manganese AI women: 2.3 mg men: 1.8 mg	• Required for the formation of body fat and bone	• Weight loss • Rash • Nausea and vomiting
Chromium AI women: 35 mcg men: 25 mcg	• Required for the normal utilization of glucose and fat	• Elevated blood glucose and triglyceride levels • Weight loss
Molybdenum RDA women: 45 mcg men: 45 mcg UL: 2000 mcg	• Component of enzymes involved in the transfer of oxygen from one molecule to another	• Rapid heartbeat and breathing • Nausea, vomiting • Coma
Sodium AI adults: 1500 mg UL adults: 2300 mg	• Regulation of acid-base balance in body fluids • Maintenance of water balance in body tissues • Activation of muscles and nerves	• Weakness • Apathy • Poor appetite • Muscle cramps • Headache • Swelling
Potassium AI adults: 4700 mg	• Same as for sodium	• Weakness • Irritability, mental confusion • Irregular heartbeat • Paralysis
Chloride AI adults: 2300 mg	• Component of hydrochloric acid secreted by the stomach (used in digestion) • Maintenance of acid-base balance of body fluids • Maintenance of water balance in the body	• Muscle cramps • Apathy • Poor appetite • Long-term mental retardation in infants

Table 1.12 Summary of minerals (continued)

Consequences of Overdose	Primary Food Sources	Highlights and Comments
• "Selenosis;" symptoms of selenosis are hair and fingernail loss, weakness, liver damage, irritability, and "garlic" or "metallic" breath	• Meats and seafoods • Eggs • Whole grains	• Content of foods depends on amount of selenium in soil, water, and animal feeds. • May play a role in the prevention of some types of cancer.
• Wilson's disease (excessive accumulation of copper in the liver and kidneys) • Vomiting, diarrhea • Tremors • Liver disease	• Bread • Potatoes • Grains • Dried beans • Nuts and seeds • Seafood • Ready-to-eat cereals	• Toxicity can result from copper pipes and cooking pans. • Average intake in the U.S. is below the RDA.
• Infertility in men • Disruptions in the nervous system (psychotic symptoms) • Muscle spasms	• Whole grains • Coffee, tea • Dried beans • Nuts	• Toxicity is related to overexposure to manganese dust in miners.
• Kidney and skin damage	• Whole grains • Wheat germ • Liver, meat • Beer, wine • Oysters	• Toxicity usually results from exposure in chrome-making industries or overuse of supplements. • Supplements do not build muscle mass or increase endurance.
• Loss of copper from the body • Joint pain • Growth failure • Anemia • Gout	• Dried beans • Grains • Dark green vegetables • Liver • Milk and milk products	• Deficiency is extraordinarily rare.
• High blood pressure in susceptible people • Kidney disease • Heart problems	• Foods processed with salt • Cured foods (corned beef, ham, bacon, pickles, sauerkraut) • Table and sea salt • Bread • Milk, cheese • Salad dressing	• Very few foods naturally contain much sodium; processed foods are the leading source. • High-sodium diets are associated with hypertension in "salt-sensitive" people. • Kidney disease, excessive water consumption are related to sodium depletion.
• Irregular heartbeat, heart attack	• Plant foods (potatoes, squash, lima beans, tomatoes, plantains, bananas, oranges, avocados) • Meats • Milk and milk products • Coffee	• Content of vegetables is often reduced in processed foods. • Diuretics (water pills), vomiting, diarrhea may deplete potassium. • Salt substitutes often contain potassium.
• Vomiting	• Same as for sodium (most of the chloride in our diets comes from salt)	• Excessive vomiting and diarrhea may cause chloride deficiency. • Legislation regulating the composition of infant formulas was enacted in response to formula-related chloride deficiency and subsequent mental retardation in infants.

Table 1.13 Food sources of minerals

Magnesium

Food	Amount	Magnesium (mg)	Food	Amount	Magnesium (mg)
Legumes			**Vegetables**		
Lentils, cooked	½ c	134	Bean sprouts	½ c	98
Split peas, cooked	½ c	134	Black-eyed peas	½ c	58
Tofu	½ c	130	Spinach, cooked	½ c	48
Nuts			Lima beans	½ c	32
Peanuts	¼ c	247	**Milk and Milk Products**		
Cashews	¼ c	93	Milk	1 c	30
Almonds	¼ c	80	Cheddar cheese	1 oz	8
Grains			American cheese	1 oz	6
Bran buds	1 c	240	**Meats**		
Wild rice, cooked	½ c	119	Chicken	3 oz	25
Breakfast cereal, fortified	1 c	85	Beef	3 oz	20
Wheat germ	2 tbs	45	Pork	3 oz	20

Calcium*

Food	Amount	Calcium (mg)	Food	Amount	Calcium (mg)
Milk and Milk Products			Ice milk	1 c	180
Yogurt, low fat	1 c	413	American cheese	1 oz	175
Milk shake			Custard	½ c	150
(low-fat frozen yogurt)	1¼ c	352	Cottage cheese	1½ c	70
Yogurt with fruit, low fat	1 c	315	Cottage cheese, low fat	½ c	69
Skim milk	1 c	301	**Vegetables**		
1% milk	1 c	300	Spinach, cooked	½ c	122
2% milk	1 c	298	Kale	1½ c	47
3.25% milk (whole)	1 c	288	Broccoli	½ c	36
Swiss cheese	1 oz	270	**Legumes**		
Milk shake (whole milk)	1¼ c	250	Tofu	½ c	260
Frozen yogurt, low fat	1 c	248	Dried beans, cooked	½ c	60
Frappuccino	1 c	220	**Foods Fortified with Calcium**		
Cheddar cheese	1 oz	204	Orange juice	1 c	350
Frozen yogurt	1 c	200	Frozen waffles	2	300
Cream soup	1 c	186	Soymilk	1 c	200–400
Pudding	½ c	185	Breakfast cereals	1 c	150–1000
Ice cream	1 c	180			

Selenium

Food	Amount	Selenium (mcg)	Food	Amount	Selenium (mcg)
Seafood			Ham	3 oz	29
Lobster	3 oz	66	Beef	3 oz	22
Tuna	3 oz	60	Bacon	3 oz	21
Shrimp	3 oz	54	Chicken	3 oz	18
Oysters	3 oz	48	Lamb	3 oz	14
Fish	3 oz	40	Veal	3 oz	10
Meats			**Eggs**		
Liver	3 oz	56	Egg	1 med	37

Sodium

Food	Amount	Sodium (mg)	Food	Amount	Sodium (mg)
Miscellaneous			Sea salt	1 tsp	1716
Salt	1 tsp	2132	Chicken broth	1 c	1571
Dill pickle	1 (4½ oz)	1930	Ravioli, canned	1 c	1065

*Actually, the richest source of calcium is alligator meat; 3½ ounces contain about 1231 milligrams of calcium. But just try to find it on your grocer's shelf!

Table 1.13 Food sources of minerals (continued)

Sodium (Continued)

Food	Sodium Amount	(mg)	Food	Sodium Amount	(mg)
Spaghetti with sauce, canned	1 c	955	Fish, smoked	3 oz	444
			Bologna	1 oz	370
Baking soda	1 tsp	821	**Milk and Milk Products**		
Beef broth	1 c	782	Cream soup	1 c	1070
Gravy	¼ c	720	Cottage cheese	½ c	455
Italian dressing	2 tbs	720	American cheese	1 oz	405
Pretzels	5 (1 oz)	500	Cheese spread	1 oz	274
Green olives	5	465	Parmesan cheese	1 oz	247
Pizza with cheese	1 wedge	455	Gouda cheese	1 oz	232
Soy sauce	1 tsp	444	Cheddar cheese	1 oz	175
Cheese twists	1 c	329	Skim milk	1 c	125
Bacon	3 slices	303	Whole milk	1 c	120
French dressing	2 tbs	220	**Grains**		
Potato chips	1 oz (10 pieces)	200	Bran flakes	1 c	363
Catsup	1 tbs	155	Cornflakes	1 c	325
Meats, Fish			Croissant	1 med	270
Corned beef	3 oz	808	Bagel	1	260
Ham	3 oz	800	English muffin	1	203
Fish, canned	3 oz	735	White bread	1 slice	130
Meat loaf	3 oz	555	Whole wheat bread	1 slice	130
Sausage	3 oz	483	Saltine crackers	4 squares	125
Hot dog	1	477			

Iron

Food	Amount	Iron (mg)	Food	Amount	Iron (mg)
Meat and Meat Alternates			English muffin	1	1.6
Liver	3 oz	7.5	Rye bread	1 slice	1.0
Round steak	3 oz	3.0	Whole wheat bread	1 slice	0.8
Hamburger, lean	3 oz	3.0	White bread	1 slice	0.6
Baked beans	½ c	3.0	**Fruits**		
Pork	3 oz	2.7	Prune juice	1 c	9.0
White beans	½ c	2.7	Apricots, dried	½ c	2.5
Soybeans	½ c	2.5	Prunes	5 med	2.0
Pork and beans	½ c	2.3	Raisins	¼ c	1.3
Fish	3 oz	1.0	Plums	3 med	1.1
Chicken	3 oz	1.0	**Vegetables**		
Grains			Spinach, cooked	½ c	2.3
Breakfast cereal, iron fortified	1 c	8.0 (4–18)	Lima beans	½ c	2.2
			Black-eyed peas	½ c	1.7
Oatmeal, fortified, cooked	1 c	8.0	Peas	½ c	1.6
Bagel	1	1.7	Asparagus	½ c	1.5

Zinc

Food	Amount	Zinc (mg)	Food	Amount	Zinc (mg)
Meats, Sea Food			Pork	3 oz	2.4
Liver	3 oz	4.6	Chicken	3 oz	2.0
Beef	3 oz	4.0	**Legumes**		
Crab	½ c	3.5	Dried beans, cooked	½ c	1.0
Lamb	3 oz	3.5	Split peas, cooked	½ c	0.9
Turkey ham	3 oz	2.5			

continued

Table 1.13 Food sources of minerals (continued)

Zinc (Continued)

Food	Amount	Zinc (mg)	Food	Amount	Zinc (mg)
Grains			Cashews	¼ c	1.8
Breakfast cereal, fortified	1 c	1.5–4.0	Sunflower seeds	¼ c	1.7
Wheat germ	2 tbs	2.4	Peanut butter	2 tbs	0.9
Oatmeal, cooked	1 c	1.2	**Milk and Milk Products**		
Bran flakes	1 c	1.0	Cheddar cheese	1 oz	1.1
Brown rice, cooked	½ c	0.6	Whole milk	1 c	0.9
White rice	½ c	0.4	American cheese	1 oz	0.8
Nuts and Seeds					
Pecans	¼ c	2.0			

Phosphorus

Food	Amount	Phosphorus (mg)	Food	Amount	Phosphorus (mg)
Milk and Milk Products			**Grains**		
Yogurt	1 c	327	Bran flakes	1 c	180
Skim milk	1 c	250	Shredded wheat	2 large biscuits	81
Whole milk	1 c	250	Whole wheat bread	1 slice	52
Cottage cheese	½ c	150	Noodles, cooked	½ c	47
American cheese	1 oz	130	Rice, cooked	½ c	29
Meats, Fish			White bread	1 slice	24
Pork	3 oz	275	**Vegetables**		
Hamburger	3 oz	165	Potato	1 med	101
Tuna	3 oz	162	Corn	½ c	73
Lobster	3 oz	125	Peas	½ c	70
Chicken	3 oz	120	French fries	½ c	61
Nuts and Seeds			Broccoli	½ c	54
Sunflower seeds	¼ c	319	**Other**		
Peanuts	¼ c	141	Milk chocolate	1 oz	66
Pine nuts	¼ c	106	Cola	12 oz	51
Peanut butter	1 tbs	61	Diet cola	12 oz	45

Potassium

Food	Amount	Potassium (mg)	Food	Amount	Potassium (mg)
Vegetables			Hamburger	3 oz	480
Potato	1 med	780	Lamb	3 oz	382
Winter squash	½ c	327	Pork	3 oz	335
Tomato	1 med	300	Chicken	3 oz	208
Celery	1 stalk	270	**Grains**		
Carrots	1 med	245	Bran buds	1 c	1080
Broccoli	½ c	205	Bran flakes	1 c	248
Fruits			Raisin bran	1 c	242
Avocado	½ med	680	Wheat flakes	1 c	96
Orange juice	1 c	469	**Milk and Milk Products**		
Banana	1 med	440	Yogurt	1 c	531
Raisins	¼ c	370	Skim milk	1 c	400
Prunes	4 large	300	Whole milk	1 c	370
Watermelon	1 c	158	**Other**		
Meats, Fish			Salt substitutes	1 tsp	1300–2378
Fish	3 oz	500			

Illustration 1.4 Schematic representation of the structure and major components of a human cell.

> **Homeostasis** Constancy of the internal environment. The balance of fluids, nutrients, gases, temperature, and other conditions needed to ensure ongoing, proper functioning of cells and, therefore, all parts of the body.

exists within and around cells. This state of optimal, cellular nutrient conditions supports *homeostasis* in the body.

Disruptions in the availability of nutrients, or the presence of harmful substances in the cell's environment, initiate diseases and disorders that eventually affect tissues, organs, and systems. For example, too little folate reduces the conversion of the amino acid methionine to cysteine. This causes a buildup of an intermediary product called homocysteine. High levels of homocysteine disrupt normal cell processes and enhance the deposition of cholesterol and other materials into artery walls.

> **PRINCIPLE #4** Poor nutrition can result from both inadequate and excessive levels of nutrient intake.

Each nutrient has a range of intake levels that corresponds to optimum functioning of that nutrient (Illustration 1.5). Intake levels below and above this range are associated with impaired functions.

Inadequate intake of an essential nutrient, if prolonged, results in obvious deficiency diseases. Marginally deficient diets produce subtle changes in behavior or physical condition. If the optimal intake range is exceeded (usually by overdoses of supplements), mild to severe changes in mental and physical functions occur, depending on the amount of the excess and the nutrient involved. Overt vitamin C deficiency, for example, produces irritability, bleeding gums, pain upon being touched, and failure of bone growth. Marginal deficiency may cause delayed wound healing. The length of time a deficiency or toxicity takes to develop depends on the type and amount of the nutrient consumed and the extent of body nutrient reserves. Intakes of 32 mg/day of vitamin C, or about a third of the RDA for adults (75 mg and 90 mg per day for women and men, respectively), lower blood vitamin C levels to the deficient state within three weeks.[18] On the excessive side, too much supplemental vitamin C causes diarrhea. For nutrients, enough is as good as a feast.

Steps in the Development of Nutrient Deficiencies and Toxicities Poor nutrition due to inadequate diets generally develops in the stages outlined in Illustration 1.6.

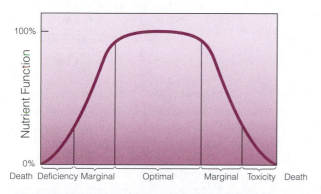

Illustration 1.5 Nutrient function and consequences by level of intake.

Illustration 1.6 Usual steps in the development of nutrient deficiencies and toxicities.

After a period of deficient intake of an essential nutrient, tissue reserves become depleted, and subsequently blood levels of the nutrient decline. When the blood level can no longer supply cells with optimal amounts of nutrients, cell processes change. These changes have a negative effect on the cell's ability to form proteins appropriately, regulate energy formation and use, protect itself from oxidation, or carry out other normal functions. If the deficiency continues, groups of cells malfunction, which leads to problems related to tissue and organ functions. Physical signs of the deficiency may then develop, such as growth failure with protein deficiency or an inability to walk as a result of beriberi (thiamin deficiency). Eventually, some problems produced by the deficiency can no longer be reversed by increased nutrient intake. Blindness that results from serious vitamin A deficiency, for example, is irreversible.

Excessively high intakes of many essential nutrients produce toxicity diseases. Excessive vitamin A, for example, produces hypervitaminosis A, and selenium overdose leads to selenosis. Signs of toxicity stem from an increased level of the nutrient in the blood and the subsequent oversupply of the nutrient to cells. The high nutrient load upsets the balance needed for optimal cell function. These changes in cell function lead to the signs and symptoms of a toxicity disease.

For both deficiency and toxicity diseases, the best way to correct the problem is at the level of intake. Identifying and fixing intake problems prevents related health problems from developing.

Nutrient Deficiencies Are Usually Multiple   Most foods contain many nutrients, so poor diets are generally inadequate in many nutrients. Calcium and vitamin D, for example, are present in milk. Deficiencies of both of these nutrients may develop from a low milk intake and an otherwise poor diet.

The "Ripple Effect"   Dietary changes affect the level of intake of many nutrients. Switching from a high-fat to a low-fat diet, for instance, generally results in a lower intake of calories and higher intake of dietary fiber and vitamins. Consequently, dietary changes introduced for the purpose of improving intake of a particular nutrient produce a "ripple effect" on the intake of other nutrients.

PRINCIPLE #5   Humans have adaptive mechanisms for managing fluctuations in food intake.

Healthy humans have adaptive mechanisms that partially protect the body from poor health due to fluctuations in nutrient intake. These mechanisms act to conserve nutrients when dietary supply is low and to eliminate them when excessively high amounts are present. Dietary surpluses of nutrients such as iron, calcium, vitamin A, and vitamin B_{12} are stored within tissues for later use. In the case of iron and calcium, absorption is also regulated so that the amount absorbed changes in response to the body's need for these nutrients. The body has a low storage capacity for other nutrients, such as vitamin C and water, and excesses are eliminated through urine or stools. Fluctuations in energy intake are primarily regulated by changes in appetite. If too few calories are consumed, however, the body will obtain energy from its glycogen and fat stores. If caloric intakes remain low and a significant amount of body weight is lost, the body downregulates its need for energy by lowering body temperature and the capacity for physical work. When energy intake exceeds need, the extra is converted to fat—and to a lesser extent, to glycogen—and stored for later use.

Although they provide an important buffer, these built-in mechanisms do not protect humans from all the consequences of poor diets. An excessive vitamin A or selenium intake over time, for example, results in toxicity disease; excessive energy intake creates health problems related to obesity; and deficient intakes of other vitamins and minerals compromise health in many ways.

PRINCIPLE #6   Malnutrition can result from poor diets and from disease states, genetic factors, or combinations of these causes.

Malnutrition means "poor nutrition" and results from either inadequate or excessive availability of energy and nutrients. Niacin toxicity, obesity, iron deficiency, and kwashiorkor (protein deficiency in children) are examples of malnutrition.

Malnutrition can result from poor diets as well as from diseases that interfere with the body's ability to use the nutrients consumed. *Primary malnutrition* results when a poor nutritional state is dietary in origin. *Secondary malnutrition*, on the other hand, is precipitated by a disease state, surgical procedure, or medication. Diarrhea, alcoholism, AIDS, and gastrointestinal tract bleeding are examples of conditions that may cause secondary malnutrition.

Malnutrition   Poor nutrition resulting from an excess or lack of calories or nutrients.

Primary Malnutrition   Malnutrition that results directly from inadequate or excessive dietary intake of energy or nutrients.

Secondary Malnutrition   Malnutrition that results from a condition (e.g., disease, surgical procedure, medication use) rather than primarily from dietary intake.

Nutrient–Gene Interactions   Advances in knowledge about nutrient–gene interactions in health and disease are revolutionizing the science and practice of nutrition. This new field of nutrition science is called *nutrigenomics* and is highlighted in the "In Focus" box on page 31. Genes provide the codes for enzyme and other protein synthesis, and consequently affect body functions in a huge number of ways. Although individuals are 99.9% genetically identical, the 0.1% difference in genetic codes makes everyone unique. Variations in gene types (or genotypes) contribute to disease resistance and development, and the way individuals respond to various drugs.[19]

Hundreds of diseases and disorders related to single-gene defects have been identified, and many of these affect nutrient needs. Four examples of such diseases and disorders—including phenylketonuria, celiac disease, lactose intolerance, and hemochromatosis—are described in Table 1.14.[20] Most diseases and disorders related to genetic makeup, however, are due to interactions among environmental factors, genotype, and gene functions.[11]

Food and nutrient intake is a prominent environmental factor that interacts with genotype and gene function. Certain components of foods can turn genes on or off, and others can compensate for the ill effects of certain genotypes on body processes.[11] Newly identified relationships between dietary components and genes are being announced regularly. Here are a few examples of effects of nutrient–gene interactions on health status:

- Consumption of whole oats lowers blood cholesterol level in some people but not others.

- High alcohol intake during pregnancy in some women sharply increases the risk of fetal alcohol syndrome in her fetus, but the fetuses of other women with different genetic traits are not affected by high alcohol intake.

- Regular consumption of green tea reduces the risk of prostate cancer in certain individuals with a particular genetic trait.[11,19,20]

Genetic factors alone cannot explain the rapid rise in obesity and type 2 diabetes in the United States, but they do provide clues about needed preventive and therapeutic measures.

PRINCIPLE #7 Some groups of people are at higher risk of becoming inadequately nourished than others.

Women who are pregnant or breastfeeding, infants, children, people who are ill, and frail elderly persons have a greater need for nutrients than healthy adults and elderly people do. As a result, they are at higher risk of becoming inadequately nourished than others. Within these groups, those at highest risk of nutritional insults are the poor. In cases of widespread food shortages, such as those induced by war or natural disaster, the health of these nutritionally vulnerable groups is compromised the soonest and the most.

PRINCIPLE #8 Poor nutrition can influence the development of certain chronic diseases.

Poor nutrition not only results in deficiency or toxicity diseases; it also plays an important role in the development of heart disease, hypertension, cancer, stroke, osteoporosis, type 2 diabetes, and obesity (Illustration 1.7). Diets high in saturated and trans fats, for example, are related to the development of heart disease; diets low in vegetables and fruits to heart disease and

Table 1.14 Examples of single-gene disorders that affect nutrient need

PKU (phenylketonuria)	A very rare disorder caused by the lack of the enzyme phenylalanine hydroxylase. Lack of this enzyme causes phenylalanine, an essential amino acid, to build up in the blood. High blood levels of phenylalanine during growth lead to mental retardation, poor growth, and other problems. PKU is treated by low-phenylalanine diets.
Celiac disease	An intestinal malabsorption disorder caused by an inherited intolerance to gluten in wheat, rye, and barley. It causes multiple nutrient deficiencies and is treated with gluten-free diets. Celiac disease is also called "nontropical sprue" and "gluten enteropathy."
Lactose intolerance	A common disorder in adults in many countries resulting from lack of the enzyme lactase. Ingestion of lactose in dairy products causes gas, cramps, and nausea due to the presence of undigested lactose in the gut.
Hemochromatosis	A disorder affecting 1 in 200 people that occurs due to a genetic deficiency of a protein that helps regulate iron absorption. Individuals with hemochromatosis absorb more iron than normal and have excessive levels of body iron. High levels of body iron have toxic effects on tissues such as the liver and heart. The disorder can also be produced by excessive levels of iron intake over time and frequent iron injections or blood transfusions.

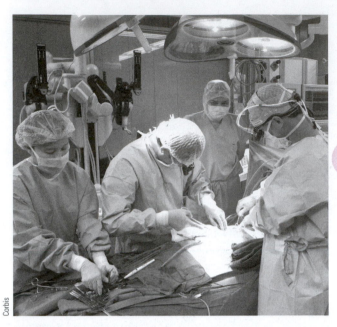

Corbis

Illustration 1.7 Poor nutrition increases the risk of many chronic diseases.

certain types of cancer; diets low in calcium and vitamin D to osteoporosis; and diets high in sugar to tooth decay.[22] Inadequate and excessive nutrient intakes may contribute to the development of more than one disease and produce disease by more than one mechanism. The effects of habitually poor diets on chronic disease development often take years to become apparent.[23]

> **PRINCIPLE #9**   Adequacy and balance are key characteristics of a healthy diet.

Healthy diets contain many different foods that together provide calories, nutrients, and other beneficial substances in amounts that promote the optimal functioning of cells and of health. A variety of food is required to obtain all the nutrients needed because, although no one food contains them all, many different combinations of food can make up a healthy diet.

Adequate diets are most easily obtained by consuming foods that are good sources of a number of nutrients but not packed with calories. Such foods are considered *nutrient-dense foods*. Those that provide calories and low amounts of nutrients are considered *"empty-calorie" foods*. Vegetables, fruits, lean meats, dried beans, breads, and cereals are nutrient dense. Foods such as beer, chips, candy, pastries, sodas, and fruit drinks lead the list of empty-calorie foods.

> **PRINCIPLE #10**   There are no "good" or "bad" foods.

"All things in nutriment are good or bad relatively."

Hippocrates

Unless you are talking about spoiled food or poisonous mushrooms, there is no such thing as a bad food. There are, however, combinations of foods that add up to an unhealthy diet. Occasional consumption of a hot dog, fried chicken, or a chocolate sundae isn't going to shorten a person's life span. But making them dietary staples, along with other empty-calorie foods, will take its toll.

Nutritional Labeling

In 1990 the U.S. Congress passed legislation establishing requirements for nutrition information, nutrient content claims, and health claims presented on food and dietary supplement labels. This legislation, called the Nutrition Labeling and Education Act, requires that almost all multiple-ingredient foods and *dietary supplements* be labeled with a Nutrition Facts panel (Illustration 1.8). The act also requires that nutrient content and health claims appearing on package labels, such as "trans fat–free" and "helps prevent cancer" qualify based on criteria established by the Food and Drug Administration (FDA).

In 2003, the FDA began allowing companies to use "qualified health claims" on dietary supplement and food packaging. These claims are not as strongly supported by scientific findings as are health claims and carry qualifying language that indicates the lack of strong, scientific support.

Nutrition Facts Panel

For foods, the Nutrition Facts panel must list the content of fat, saturated fat, trans fat, cholesterol, sodium, total carbohydrates, fiber, sugars, protein, Vitamins A and C, and calcium and iron in a standard serving. Additional nutrients may be listed on a voluntary basis. If a health claim about a particular nutrient is made for the product, then the product's content of the nutrient addressed in the claim must be shown. Nutrition Facts panels contain a column that lists the % Daily Value (% DV) for each relevant nutrient. This information helps consumers decide, for example, whether the carbohydrate content of a serving of a specific food product is a lot or a little.

Nutrient content claims made on food package labels must meet specific criteria. Products labeled "no trans fat" or "trans fat–free," for example, must contain less than 0.5 grams of trans fat and of saturated fat. Products labeled "low sodium" must contain less than 140 mg sodium per serving.

> **Nutrient-Dense Foods**   Foods that contain relatively high amounts of nutrients compared to their caloric value.
>
> **Empty-Calorie Foods**   Foods that provide an excess of calories relative to their nutrient content.
>
> **Dietary Supplements**   Any product intended to supplement the diet, including vitamin and mineral supplements, proteins, enzymes, amino acids, fish oils, fatty acids, hormones and hormone precursors, and herbs and other plant extracts. In the United States, such products must be labeled "Dietary Supplement."

Nutrition Facts
Serving Size 1 Entree
Serving Per Container 1

Amount Per Serving	
Calories 380 Calories from Fat 170	

	%Daily Value
Total Fat 19g	**29%**
Saturated Fat 10g	**50%**
Trans Fat 2g	
Cholesterol 85g	**28%**
Sodium 810mg	**34%**
Total Carbohydrate 33g	**11%**
Dietary Fiber 3g	**12%**
Sugars 5g	
Protein 20g	

Vitamin A 10%		Vitamin C 0%
Calcium 10%		Iron 15%

Percent Daily Values are based on a 2000 calorie diet. Your daily values may be higher or lower depending on your calorie needs:

		Calories	2000	2500
Total Fat	Less Than		65g	80g
Sat Fat	Less Than		20g	25g
Cholesterol	Less Than		300mg	300mg
Sodium	Less Than		2400mg	2400mg
Total Carbohydrate			300g	375g
Dietary Fiber			25g	30g

Illustration 1.8 Example of a Nutrition Facts panel.

Ingredient Label

Food products must list ingredients in an "ingredient label." The list must begin with the ingredient that contributes the greatest amount of weight to the product, and continue with the other ingredients on a weight basis.

The FDA now requires that ingredient labels include the presence of common food allergens in products. Potential food allergens that must be listed are milk, eggs, fish, shellfish, tree nuts, wheat, peanuts, and soybeans. These eight foods account for 90% of food allergies.

Dietary Supplement Labeling

"You can call anything a dietary supplement, even something you grow in your back yard."

Donna Porter, RD, PhD, Congressional Research Service

Dietary supplements such as herbs, amino acid pills and powders, and vitamin and mineral supplements must show a "Supplement Facts" panel that lists serving size, ingredients, and % DV of essential nutrients contained. Because they do not have to be shown to be safe and effective before they are sold, labels on dietary supplements cannot claim to treat, cure, or prevent disease. They can be labeled with standardized nutrition content claims such as "high in calcium" or "a good source of fiber." They can also be labeled with health claims such as "may reduce the risk of heart disease" if the product qualifies

based on nutrition labeling requirements. Dietary supplements can make other claims on product labels not approved by the FDA, such as "supports the immune system" or "helps maintain mental health," as long as the label doesn't state or imply that the product will prevent, cure, or treat disease. If a health claim is made on a dietary supplement label, the label also must present the FDA disclaimer:

This product has not been evaluated by the FDA. This product is not intended to diagnose, treat, cure, or prevent any disease.

Enrichment and Fortification   Some foods are labeled as "enriched" or "fortified." These two terms have specific definitions developed prior to the Nutrition Labeling and Education Act. *Enrichment* pertains only to refined grain products and covers some of the vitamins and one of the minerals lost when grains are refined. By law, producers of bread, cornmeal, crackers, flour tortillas, white rice, and other products made with refined grains must use flours enriched with thiamin, riboflavin, niacin, and iron.

Any food can be fortified with added vitamins and minerals, and its manufacturers most often do so on a voluntary basis to enhance product sales. However, some foods must be fortified. Refined grain flours must be fortified with folic acid, milk with vitamin D, and low-fat and skim milk with vitamin D and vitamin A. Although fortification is not required for salt, it is often fortified with iodine. *Fortification* of these foods has contributed substantially to reductions in the incidence of diseases related to inadequate intakes.[24]

> **Enrichment**   The replacement of thiamin, riboflavin, niacin, and iron that are lost when grains are refined.
>
> **Fortification**   The addition of one or more vitamins or minerals to a food product.

Herbal Remedies

The FDA considers herbal products to be dietary supplements; they are taken by many people during various stages of the life cycle. Thousands of types of herbal products are available (Illustration 1.9). Some herbal remedies act like drugs and have side effects, but they are not considered to be drugs and are loosely regulated. They do not have to be shown to be safe or effective before they are marketed. Herbs vary substantially in safety and effectiveness—they can have positive, negative, or neutral effects on health. Knowledge of the effects of herbal remedies is far from complete, making it difficult to determine appropriateness of their use in many cases.

The extent to which herbs pose a risk to health depends on the amount taken, the duration of use, and the user's age, stage, and health status.

Illustration 1.9 Herbal products available on the market.

Illustration 1.10 Examples of functional foods.

Functional Foods

Also known as "neutraceuticals," *functional foods* include a variety of products that have theoretically been modified to enhance their contribution to a healthy diet. Foods are made "functional" by:

- Taking out potentially harmful components (e.g., cholesterol in egg yolks and lactose in milk)
- Increasing the amount of nutrients and beneficial non-nutrients (e.g., fiber-fortified liquid meals, calcium- and vitamin C–fortified orange juice)
- Adding new beneficial compounds to foods (e.g., "friendly" bacteria to yogurt and other milk products)

Functional foods, such as those shown in Illustration 1.10, are not regulated, and no specific standards apply to them.[25] Health claims, however, can be made for functional foods given they have been approved by the FDA. Increasingly, the list of functional foods is becoming infiltrated with sports bars, soups, beverages, and cereals spiked with vitamins, minerals, and herbs. Some of the products carry labels with unsubstantiated health claims and may be of little benefit or are potentially unsafe.[26] For these products, the label "functional food" is a marketing term. Others, such as lactose-free milk, xylitol-sweetened gum, and iodized salt, benefit the health of some people.

Prebiotics and Probiotics The terms *prebiotics* and *probiotics* were derived from *antibiotics* due to their probable effects on increasing resistance to various diseases. They are in a class of functional foods by themselves. *Prebiotics* are fiberlike, indigestible carbohydrates that are broken down by bacteria in the colon. The breakdown products foster the growth of beneficial bacteria. The digestive tract generally contains over 500 species of microorganisms and 100 trillion bacteria.[27] Some species of bacteria such as *E. coli* may cause disease; others, such as strains of lactobacillus and bifidobacteria, prevent certain diseases.[28] Because they foster the growth of beneficial bacteria, prebiotics are considered "intestinal fertilizer." *Probiotic* is the term for live, beneficial ("friendly") bacteria that enter food products during fermentation and aging processes. Those that survive digestive enzymes and acids may start colonies of beneficial bacteria in the digestive tract. Table 1.15 lists foods and other sources of pre- and probiotics.

Functional Foods Generally taken to mean food, fortified foods, and enhanced food products that may have health benefits beyond the effects of essential nutrients they contain.

Prebiotics Certain fiberlike forms of indigestible carbohydrates that support the growth of beneficial bacteria in the lower intestine. Nicknamed "intestinal fertilizer."

Probiotics Strains of lactobacillus and bifidobacteria that have beneficial effects on the body. Also called "friendly bacteria."

Prebiotics and probiotics have been credited with important benefits, such as the prevention and treatment of diarrhea and other infections in the gastrointestinal tract; prevention of colon cancer; decreased blood levels of triglycerides, cholesterol, and glucose; and decreased dental caries.[28] Prebiotics and probiotics are generally assumed to be safe because they have been consumed in foods for centuries.[29]

Availability of foods containing prebiotics and probiotics is much more common in Japan and European countries than in Canada or the United States. However, availability of such products is increasing in these countries as research results shed light on their safety and effectiveness.[29]

Table 1.15 Food and other sources of prebiotics and probiotics[29, 36]

Probiotics

Fermented or aged milk and milk products
- Yogurt with live culture
- Buttermilk
- Kefir
- Cottage cheese
- Dairy spreads with added inulin

Other fermented products
- Soy sauce
- Tempeh
- Fresh sauerkraut
- Miso

Breast milk

Probiotic tablets, and powders, and nutritional beverages

Prebiotics

Chicory

Jerusalem artichokes

Wheat

Barley

Rye

Onions

Garlic

Leeks

Prebiotic tablets, powders, and nutritional beverages

Meeting Nutritional Needs across the Life Cycle

Healthy individuals require the same nutrients throughout life, but amounts of nutrients needed vary based on age, growth, and development. Nutrient needs during each stage of the life cycle can be met through a variety of foods and food practices. There is no one, best diet for everyone. Traditional diets defined by diverse cultures and religions provide the foundation for meeting individuals' nutritional needs and the framework for dietary modification when needed.[30] Although it is inaccurate to say that all or most members of a particular cultural group or religion follow the same dietary practices, groups of individuals may share common beliefs about food and food-intake practices.

Dietary Considerations Based on Ethnicity

People immigrating to the United States and other countries both preserve dietary traditions of their cultural group and integrate cross-cultural adaptations into their dietary practices. The extent to which culturally based food habits change depends to some extent on income, food cost, and ethnic food availability. Immigrant families from El Salvador who live in urban areas of the United States, for example, maintain many cultural food practices from their homeland:

- Breakfast generally consists of fried beans, corn tortillas, occasionally eggs, and sweetened coffee with boiled milk.
- Lunch consists of soup, fried meat, rice or rice with vegetables, corn tortillas, and fruit juice.
- Dinner will offer fried beef or chicken, corn tortillas, rice, dried beans, fruit juice, and black coffee.

Cross-cultural adaptations made by a portion of Salvadorans immigrating to the United States include the addition of french fries, hamburgers, American cheeses, salad dressing, tacos, flour tortillas, and peanut butter to their diets.[31]

Sometimes diets of native populations change when their numbers become overwhelmed by other population groups. A primary example of this phenomenon is represented by changes in traditional dietary practices of Native Americans. In general, traditional diets of Indians in the United States consisted of foods such as buffalo, deer, wild berries and other fruits, corn, turnips, squash, wild potato, and wild rice. Loss of land and buffalo, discrimination, poverty, and food programs that offered refined flour, sugar, salt pork, and other high-fat meats drastically changed what Indians ate, how they lived, and their health status. Activities aimed at bringing back traditional foods and dietary practices are under way among many Indian groups.[32]

Food preferences of African Americans vary widely, but may stem from their cultural food heritage. Historically important foods include corn bread, pork, buttermilk, rice, sweet potatoes, greens, cabbage, salt pork, and fried fish. "Soul foods" make up less of the African American diet now than in the past, but remain foods of choice for special occasions and are the foods most likely to be revered.[33]

Dietary Considerations Based on Religion

Many religions have special dietary laws and practices. For example:

- Hindus may not consume foods such as garlic and onions, which are believed to hinder spiritual development.
- Buddhists in certain countries tend to be vegetarian or to eat fish as their only choice of meat. In countries such as Tibet and Japan vegetarianism is rare among Buddhists.
- Alcohol is prohibited as part of Sikhism, and meat prepared by kosher or halal methods is avoided.
- The Church of Latter Day Saints, or the Mormon Church, prohibits alcohol and discourages consumption of caffeine. Mormons may eat meat and prize wheat.
- Seventh-Day Adventists tend to follow a strict lacto-ovo vegetarian diet and exclude alcohol and caffeine.

Whole grains, vegetables, and fruits are considered to be the base of diets; and dried beans, low-fat dairy products, and eggs may be consumed infrequently.

- Jewish dietary laws require that foods consumed must be kosher, or fit to eat according to Judaic law. Organizations are certified as supplying foods that are kosher. The Jewish calendar includes six fasting days that call for total abstinence from food or drink.
- The Muslim religion has dietary laws that require foods to be halal, or permitted for consumption by Muslims. Pork consumption is not allowed, nor is the consumption of animals slaughtered in the name of any god other than Allah. Slaughterhouses must be under the supervision of a halal certifier in order for meat to be considered fit to eat, although some Muslims will eat other meats. Consuming alcohol is prohibited.[34,35]

Additional information about cultural and religious food practices and beliefs can be obtained directly by getting to know people from a variety of cultures and their dietary preferences. This information can be of great benefit in nutrition education and counseling situations.

Nutritional Assessment

Nutritional assessment of groups and individuals is a prerequisite to planning for the prevention or solution of nutrition-related health problems. It represents a broad area within the field of nutrition and is only highlighted here. Resources related to the selection of appropriate nutritional assessment techniques and their implementation are listed at the end of this chapter.

Nutritional status may be assessed for a population group or for an individual. Community-level assessment identifies a population's status using broad nutrition and health indicators, whereas individual assessment provides the baseline for anticipatory guidance and nutrition intervention.

Community-Level Assessment

A target community's "state of nutritional health" can generally be estimated using existing vital statistics data, seeking the opinions of target group members and local health experts, and making observations. Knowledge of average household incomes; the proportion of families participating in the Food Stamp Program, soup kitchens, school breakfast programs, or food banks; and the age distribution of the group can help identify key nutrition concerns and issues. In large communities, rates of infant mortality, heart diseases, and cancer can reveal whether the incidence of these problems is unusually high.

Information gathered from community-level nutritional assessment can be used to develop community-wide programs addressing specific problem areas, such as child-hood obesity or iron-deficiency anemia. Nutrition programs should be integrated into community-based health programs.

Individual-Level Nutritional Assessment

Nutritional assessment of individuals has four major components:

- Clinical/physical assessment
- Dietary assessment
- Anthropometric assessment
- Biochemical assessment

Data from all of these areas are needed to describe a person's nutritional status. Data on height and weight provide information on weight status, for example, and knowledge of blood iron levels tells you something about iron status. It cannot be concluded that people who are normal weight or have good iron status are "well nourished." Single measures do not describe a person's nutritional status.

Clinical/physical Assessment   A clinical/physical assessment involves visual inspection of a person by a trained *registered dietitian* or other qualified professional to note features that may be related to malnutrition. Excessive or inadequate body fat, paleness, bruises, and brittle hair are examples of features that may suggest nutrition-related problems. Physical characteristics are nonspecific indicators, but they can support other findings related to nutritional status. They cannot be used as the sole criterion upon which to base a decision about the presence or absence of a particular nutrition problem.

> **Registered Dietitian**   An individual who has acquired knowledge and skills necessary to pass a national registration examination and who participates in continuing professional education.

Dietary Assessment

Many methods are used for assessing dietary intake. For clinical purposes, 24-hour dietary recalls and food records analyzed by computer programs are most common. Single, 24-hour recalls and food frequency questionnaires are most useful for estimating dietary intakes for groups, whereas multiple recalls and dietary histories are generally used for assessments of individual diets.

24-Hour Dietary Recalls and Records   Becoming proficient at administering 24-hour recalls takes training and practice. Food records, on the other hand, are completed by clients themselves. These are more accurate if the client has also received some training. Generally, the purpose of assessing an individual's diet is to estimate the person's overall diet quality so that strengths and weaknesses can be identified, or to assess intake of specific nutrients that may be involved in disease states.

Information on at least three days of dietary intake (preferably two usual weekdays and one weekend day) is needed to obtain a reliable estimate of intake by food group, calories, and nutrients. A good approach is to have a trained dietary interviewer administer a 24-hour recall and then have the client record her or his own diet on two other days. The experience of thinking about what was eaten, portion sizes, ingredients, and recipes helps train people to complete their food records accurately. Completed records should be reviewed with the client during a telephone call or clinic session to make sure they are accurate.

Dietary History   Dietary histories have been used for decades and represent a quantitative method of dietary assessment. They require an interview that is about 1½ hours long and includes a 24-hour dietary recall modified to represent usual intake, careful deliberations over food types and portions, and a cross-check food frequency questionnaire that confirms 24-hour usual dietary intake information. Results must be coded, checked, and processed. Although expensive, diet histories provide more complete and accurate data than most other dietary assessment methods.[37]

Food Frequency Questionnaires   Food frequency questionnaires are often used in epidemiological studies to estimate food and nutrient intake of groups of people. These tools are considered semiquantitative because they force people into describing food intake based on a limited number of food choices and portion sizes (Illustration 1.11). Validated food frequencies are relatively inexpensive to administer and tabulate, and provide good enough estimates of dietary intake to rank people by their food and nutrient intake levels. They tend to underestimate food intake and provide data that is more likely to fail to identify nutrient and health relationships than are quantitative assessment techniques such as the dietary history.[38]

Web Dietary Assessment Resources   Several high-quality computer programs and Internet resources are available for dietary assessment. MyPyramid Tracker, developed by the U.S. Department of Agriculture (USDA), is an example of a high-quality Internet resource. This interactive program provides an analysis of nutrient and food intake. MyPyramid Tracker can be found at www.mypyramid.gov.

USDA Automated Multiple-Pass Method   A computerized, interactive method for collecting interviewer-administered 24-hour dietary recalls either in person or by telephone has recently been validated.[39] It's called the Automated Multiple-Pass Dietary Recall and is being used in government-sponsored nutritional studies. It utilizes a five-step multiple-pass 24-hour recall. The term *multiple pass* refers to the repeated use of questions that hone-in the accuracy of information provided by interviewees about the food they ate the previous day.

The five-step interview process used by the automated multiple-pass method consists of the following:

1. **The Quick List**   Quickly collect a list of foods and beverages consumed the previous day.
2. **The Forgotten Foods List**   Probe for foods forgotten during development of the Quick List.
3. **Time and Occasion List**   Collect information on the time and eating occasion for each food.
4. **The Detail Cycle**   Collect detailed information on the description and amount of food consumed using USDA's interactive Food Model Booklet and measuring guides.
5. **Final Probe Review**   Review 24-hour recall and ask about anything else consumed.

Food	Frequency of Consumption								
	Never or less than once per month	1–3 per month	1 per week	2–4 per week	5–6 per week	1 per day	2–3 per day	4–5 per day	6+ per day
1. a. Broth type soups, 1 cup									
b. Tap water, 1 cup									
c. Sparkling or mineral water, 1 cup serving									
d. Decaffeinated black tea, iced or hot, 1 cup									
e. Herbal tea, (no caffine) iced or hot, 1 cup									
2. Custard or pudding, 1/2 cup									
3. Onions, 1/4 cup, alone or in combination									

Illustration 1.11 Example component of a food frequency questionnaire.

SOURCE: J. Brown, University of Minnesota; Diana Project form, adapted from W. Willett's Food Frequency Questionnaire.

Additional information on this advanced method for assessing dietary intake can be found at www.ars.usda. gov/Services/docs.htm?docid=7710.

Anthropometric Assessment

Individual measures of body size (height, weight, percent body fat, bone density, and head and waist circumferences, for example) are useful in the assessment of nutritional status—if done correctly. Each measure requires use of standard techniques and calibrated instruments by trained personnel. Unfortunately, *anthropometric* measurements are frequently performed and recorded incorrectly in clinical practice. Training on anthropometric measures is often available through public health agencies and programs such as WIC (Special Supplemental Nutrition Program for Women, Infants, and Children), and courses and training sessions are sometimes presented at universities.

Biochemical Assessment

Nutrient and enzyme levels, DNA characteristics, and other biological markers are components of a biochemical assessment of nutritional status. Which biological markers are measured depends on what problems are suspected, based on other evidence. For example, a young child who tires easily, has a short attention span, and does not appear to be consuming sufficient iron based on dietary assessment results may have blood taken for analyses of hemoglobin and serum ferritin (markers of iron status). Suspected inborn errors of metabolism that may underlie nutrient malabsorption may be identified through DNA or other tests. Such results provide specific information on a component of a person's nutritional status and are very helpful in diagnosing a particular condition.

After the fact-finding phase of nutritional assessment, the nutritionist or other professionals must "apply their brains to their clients' problems." There is no "one size fits all" approach to solving nutrition problems—each has to be figured out individually.

In the future, biochemical assessments will include a nutrigenomics profile to identify health risks due to interactions among an individual's genetic makeup, gene functions, and components of food.[11]

Monitoring the Nation's Nutritional Health

Food availability, dietary intake, weight status, and nutrition-related disease incidence are investigated regularly in the United States by the National Nutrition Monitoring System. This wide-ranging system is primarily responsible for the conduct of *nutrition surveillance* and *nutrition monitoring* studies.

The first U.S. nutrition survey began in 1936 when hunger, poor growth in children, and vitamin and mineral deficiencies were common. Today's surveys monitor rates of obesity, diabetes, and other nutrition-related disorders; the safety of the food supply (for example, the mercury content of fish and pesticide residues in vegetables and fruits); and food and nutrient intake.

The four major, ongoing U.S. studies related to diet, the food supply, and nutritional health are overviewed in Table 1.16. Together with findings from studies conducted by university researchers and others, the results give direction to food and nutrition policies and programs aimed at safeguarding the food supply, improving the population's nutritional health, and maintaining a successful agricultural economy.[40]

Anthropometry   The science of measuring the human body and its parts.

Nutrition Surveillance   Continuous assessment of nutritional status for the purpose of detecting changes in trend or distribution in order to initiate corrective measures.

Nutrition Monitoring   Assessment of dietary or nutrition status at intermittent times with the aim of detecting changes in the dietary or nutritional status of a population.

Table 1.16 U.S. national nutrition monitoring systems

Survey	Purpose
1. National Health and Nutrition Examination Survey (NHANES)*	• Assess dietary intake, health, and nutritional status in a sample of 5000 people of all ages on a yearly basis, it oversamples low-income individuals.
2. Behavioral Risk Factor Surveillance System	• Identifies self-reported dietary behaviors, nutritional knowledge, and health status of thousands of adults by telephone interview.
3. Nationwide Food Consumption Survey (NFCS)	• Estimates food and nutrient intake, and understanding of diet and health relationships among a national sample of individuals.
4. Total Diet Study	• Ongoing assessment of levels of various pesticides. contaminants, and nutrients in foods and diets.

*NHANES was combined with the Continuing Survey of Food Intakes by Individuals (CFSII) in 2001.

Public Food and Nutrition Programs

A variety of federal, state, and local programs are available to provide food and nutrition services to families and individuals. Many communities have nutrition coalitions or partnership groups that collaborate on meeting the food and nutritional needs of community members. Programs representing church-based feeding sites, food shelves, Second Harvest Programs, the Salvation Army, missions, and others are usually a part of local coalitions. These central resources can be identified by contacting the local public health or cooperative extension agency. State-level programs are generally part of large national programs.

Government-sponsored food and nutrition programs are widely available throughout the country. Some programs, such as the School Lunch Program, benefit many children. Other programs are targeted to families and individuals in need. *Need* is generally defined as individual and household incomes below the poverty line. Some programs, such as WIC, have eligibility standards of up to 185% of the poverty line (Table 1.17). Income guidelines change periodically and are several thousand dollars higher for people in Alaska and Hawaii.

WIC

The Special Supplemental Nutrition Program for Women, Infants, and Children (WIC) was established in 1972 and is administered by the USDA. The program provides nutrition education and counseling as well as food vouchers for low-income pregnant, postpartum, and breastfeeding women and for low-income children under the age of 5 years. Food vouchers apply to nutritious foods such as fortified breakfast cereals, iron-fortified infant cereals and formula, milk, cheese, eggs, peanut butter, dried beans, and 100% fruit and vegetable juices. Some WIC programs offer vouchers for the purchase of produce at farmer's markets. WIC staff also provide breastfeeding support and referrals to health care and social service providers.

Eligibility for WIC is based on low-income status and the presence of a nutritional risk, such as iron deficiency or underweight. WIC services are provided through approximately 10,000 clinic sites throughout the United States and in American Samoa, Guam, Puerto Rico, and the Virgin Islands. The program serves more than 7 million women and children each year. Nearly half of all infants and a quarter of all young children in the United States participate in the WIC program.

The WIC program has been shown to have a number of positive effects on the health of participants. Infants born to women participating in WIC while pregnant weigh more, and they are less likely to be small at birth and to be born before term than are the infants of nonparticipating low-income women. Children served by WIC tend to consume more nutritious diets and experience lower rates of iron deficiency than children who are low income but not enrolled in WIC. WIC is cost effective: every dollar invested in WIC prenatal nutrition services saves $3.13 on Medicaid costs for infants during the first 2 years of life.[42]

Table 1.18 presents information on existing federal food and nutrition programs. You can get more information on these programs through the Internet at this address: www.nutrition.gov.

Nationwide Priorities for Improvements in Nutritional Health

Public health initiatives involving population-based improvements in food safety, food availability, and nutritional status have led to major gains in the health status of the country's population. Among the important components of this success story are programs that have expanded the availability of housing; safe food and water; foods fortified with iodine, iron, vitamin D, or folate; fluoridated public water; food assistance; and nutrition education. Since 1900, the average life span of persons in the United States has lengthened by more than 30 years, and 25 years of this gain are estimated to be due to the quiet revolutions that have taken place in public health.[42]

Although the United States spends more on health care than any other country, statistics from the World Health Organization (WHO) indicate that it ranks twenty-sixth among developed countries of the world in life expectancy. Today's priorities for improvements in the public's health and longevity center on reducing obesity, infant mortality, smoking, excessive alcohol consumption, accidents, violence, and physical inactivity. Goals for dietary changes are a central part of the nation's overall plan for health improvements. For the two out of three Americans who do not smoke or drink excessively, dietary intakes represent the major environmental influence on long-term health.[43]

Table 1.17 Income eligibility standards for the WIC program (≤185% of poverty income, 2006–2007)

Household Size	Household Income per Year
1	$18,130
2	24,420
3	30,710
4	37,000
5	43,290
6	49,580
7	55,870
8	62,160
Each additional member	6,290

Table 1.18 Examples of federal food and nutrition programs

Program	Activity
Child and Adult Care Food Program (CACFP)	Reimburses child and adult care organizations in low-income areas for provision of nutritious foods.
Summer Food Service Program	Provides foods to children in low-income areas during the summer.
School Breakfast and Lunch Programs	Provide free breakfasts and reduced-cost or no-cost lunches to children from families who cannot afford to buy them.
Food Stamp Program	Subsidizes food purchases of low-income families and individuals.
WIC	Serves low-income, high-risk, and pregnant and breastfeeding women and children up to 5 years of age. Provides supplemental nutritious foods and nutrition education as an adjunct to health care.
Head Start Program	Includes nutrition education for children and parents and supplies meals for children in the program.
Team Nutrition	A USDA nutrition education program that brings active learning about nutrition to schoolchildren.

Goals for improving the nutritional health of the nation are summarized in the document "Healthy People 2010: Objectives for the Nation" (Table 1.19). Because the seeds of many chronic diseases are planted during pregnancy and childhood, major emphasis is placed on dietary habits early in life.

U.S. Nutrition and Health Guidelines

In the United States, the national guidelines for diet and physical activity are called the "Dietary Guidelines for Americans" and the major educational tool for consumers "MyPyramid" (Illustration 1.12)

Dietary Guidelines for Americans The Dietary Guidelines for Americans provide science-based recommendations to promote health and to reduce the risk for major chronic diseases through diet and physical activity. Due to its credibility and focus on health promotion and disease prevention for the public, the Dietary Guidelines form the basis of federal food and nutrition education programs and policies. Under legislative mandate, the Dietary Guideline for Americans must be updated every five years.

The 2005 edition of the Dietary Guidelines stresses the importance of selecting nutrient-dense foods, balancing caloric intake with output, and increasing physical activity. This document includes the promise that the health of most individuals will be enhanced if the recommendations are followed. Each recommendation is part of an integrated whole—all the recommendations should be implemented for best results.

Focus Areas and Key Recommendations The Dietary Guidelines for 2005 include nine "Focus Areas" and 23 "Key Recommendations." These are highlighted in Table 1.20. "Key Recommendations for Specific Population Groups" are also provided by the Dietary Guidelines. Special population groups addressed are infants and young children, pregnant women, older adults, and people with weakened immune systems. These and other recommendations included in the Dietary Guidelines are available online (see the resources section at the end of this chapter).

Implementation of the Dietary Guidelines The MyPyramid Food Guide is the major how-to tool intended to help the public implement the 2005 Dietary Guidelines. In addition, the Dietary Guidelines report identified the Dietary Approaches to Stop Hypertension (DASH) Eating Plan as being consistent with the recommendations. MyPyramid covers both food selection and physical exercise, and the DASH Eating Plan addresses only dietary intake. Both are valuable tools that provide the framework for planning nutrient-dense, calorically appropriate diets that diminish the risk of chronic disease.

MyPyramid Food Guide Food group guides to the intake of healthful diets have been available in the United States since 1916. Known by names such as "Basic Four Food Groups" and "Food Guide Pyramid," the guides have been periodically updated. New releases of food group guides reflect existing scientific knowledge about nutrition and health and are modified to address emerging health problems.

The U.S. Department of Agriculture (USDA) released its newest version of the food guide in 2005. Called "MyPyramid," it is very different from previous guides. It is Internet-based and resource-filled, and some of the educational resources offered are interactive. It is a popular website.

MyPyramid guidance stresses the importance of lower calorie intakes and increased physical activity levels. Because low intakes increase the risk of certain chronic diseases, the new food guide emphasizes intake of whole grain products, colorful vegetables, low-fat dairy products, and lean meats. MyPyramid uses cups and ounces as the primary measures of how much food to

Table 1.19 2010 Nutrition Objectives for the Nation

√ Increase the proportion of adults who are at a healthy weight from 42 to 60%.

√ Reduce the proportion of adults who are obese from 23 to 15%.

√ Reduce the proportion of children and adolescents who are overweight or obese from 11 to 5%.

√ Reduce growth retardation among low-income children under age 5 years from 8 to 5%.

√ Increase the proportion of persons aged 2 years and older who:

- consume at least 2 daily servings of fruit from 28 to 75%.
- consume at least 3 daily servings of vegetables, with at least one-third being dark green or deep yellow vegetables, from 3 to 50%.
- consume at least 6 daily servings of grain products, with at least three being whole grains, from 7 to 50%.
- consume less than 10% of calories from saturated fat from 36 to 75%.
- consume no more than 30% of total calories from fat from 33 to 75%.
- consume 2400 mg or less of sodium daily from 21 to 65%.
- meet dietary recommendations for calcium from 46 to 75%.

√ Reduce iron deficiency among young children and females of childbearing age from 4–11% to 1–7%.

√ Reduce anemia among low-income pregnant females in their third trimester from 29 to 20%.

√ Increase the proportion of worksites that offer nutrition or weight-management classes or counseling from 55 to 85%.

√ Increase the proportion of physician office visits made by patients with the diagnosis of cardiovascular disease, diabetes, osteoporosis, or hyperlipidemia that include counseling or education related to diet and nutrition from 42 to 75%.

√ Increase food security among U.S. households and in so doing reduce hunger rates.

Other Objectives Related to Nutrition

√ Reduce infant mortality from 7.6 to no more than 5 per 1000 live births.

√ Reduce the incidence of spina bifida and other neural tube defects from 7 to 3 per 10,000 live births.

√ Reduce the incidence of birth defects from 1.7 to 1.2 per 1000 live births.

√ Increase the proportion of women who receive preconceptional counseling.

√ Increase the proportion of pregnant women who begin prenatal care in the first trimester from 81 to 90% or more.

√ Reduce low birthweight (<2500 grams) from 7.3 to 5%.

√ Reduce preterm births (<37 weeks) from 9.1 to 7.6%.

√ Increase abstinence from alcohol use by pregnant women from 79 to 95%.

√ Reduce the incidence of fetal alcohol syndrome.

√ Increase the proportion of women who gain weight appropriately during pregnancy.

√ Increase from 60 to 75% the proportion of women who exclusively breastfeed after delivery.

consume daily, and it gives recommended number of cups or ounces for food within each group. MyPyramid does not provide specific recommendations for infants, pregnant or breastfeeding women, or for strict vegetarians.

MyPyramid.gov: The Website   A large assortment of educational and assessment tools are available at the MyPyramid website (www.mypyramid.gov). Here are a few examples of the types of resources you can find:

- The *For Professionals* link will connect you to detailed information about using MyPyramid educational materials and provides seven days of sample menus that correspond to MyPyramid recommendations for a 2000-calorie food pattern. Four of the seven days included in the menus, and the nutrient analysis for the menus, are shown in Illustration 1.13. Menus offered as examples on MyPyramid may not correspond to individual food preferences and contain relatively few ethnic foods.

- *Inside the Pyramid* explains each food group, discretionary calories, and physical activity recommendations. It provides information on which foods are within the various groups and food measure equivalents so you can convert food amounts into

Dietary Guidelines
for Americans
2005

www.healthierus.gov/dietaryguidelines

www.mypyramid.gov

Illustration 1.12 The major nutritional and physical activity guides in the United States.

cups and ounces, and oils into teaspoons. Table 1.21 lists food measure equivalents for common foods by food group shown in this section of *Inside the Pyramid*.

- *MyPyramid Tracker* is a dietary and physical activity assessment tool that provides information on your diet quality, physical activity status, and links to nutrient and physical activity information. Although the MyPyramid Food Guide is not designed for weight-loss diets, the MyPyramid Tracker tool can be used to assess and monitor caloric intake and physical activity levels.

The DASH Diet Originally published as a diet that helps control mild and moderate high blood pressure in experimental studies, the DASH eating plan also reduces the risk of cancer, osteoporosis, and heart disease. Improvements in blood pressure are generally seen within two weeks of starting this dietary pattern.[41]

The DASH dietary pattern emphasizes fruits, vegetables, low-fat dairy foods, whole grain products, poultry, fish, and nuts. Only low amounts of fat, red meats, sweets, and sugar-containing beverages are included. This dietary pattern provides ample amounts of potassium, magnesium, calcium, fiber, and protein; and limited amounts of saturated and *trans* fats. Recommendations for types and amounts of food included in this eating plan for a 2000-calorie diet are shown in Table 1.22.

Although one calorie level is shown in the table, DASH Eating Plans for 12 calorie levels (1600 to 3200) are available at the website listed in Table 1.22.

Table 1.20 The 2005 Dietary Guidelines for Americans: Focus areas and examples of key recommendations

Adequate Nutrients within Calorie Needs
- Consume the variety of foods and beverages cited in MyPyramid food guide or the DASH eating plan. (Additional information about the DASH eating plan is presented on page 44.)

Weight Management
- Achieve and maintain a healthy body weight.

Physical Activity
- Engage in at least 30 minutes of moderate-intensity physical activity, above usual activity, at work or home on most days of the week.

Food Groups to Encourage
- Choose and consume colorful vegetables, fruits, whole grain products, and low-fat dairy products.

Fats
- Consume less than 10% of calories from saturated fatty acids, less than 300 mg/day of cholesterol, and keep *trans* fatty acids consumption as low as possible.

Carbohydrates
- Choose fiber-rich fruits, vegetables, and whole grains often.

Sodium and Potassium
- Consume less than 2300 mg (approximately 1 tsp of salt) of sodium per day.

Alcoholic Beverages
- Alcoholic beverages, if chosen, should be consumed sensibly and in moderation.

Food Safety
- Avoid food-borne illness by cleaning hands and food preparation surfaces and cooking foods to safe temperatures.

Sample Menus for a 2000 calorie food pattern

Averaged over a week, this seven day menu provides all of the recommended amounts of nutrients and food from each food group. (Italicized foods are part of the dish or food that preceeds it, which is not italicized.)

Food Group	Daily Average Over One Week	
GRAINS	Total Grains (oz eq) 6.0	
	Whole Grains	3.4
	Refined Grains	2.6
VEGETABLES *	Total Veg* (cups) 2.6	
FRUITS	Fruits (cups) 2.1	
MILK	Milk (cups) 3.1	
MEAT & BEANS	Meat/ Beans (oz eq) 5.6	
OILS	Oils (tsp/grams) 7.2 tsp/32.4 g	

*Vegetable subgroups	(weekly totals)
Dk-Green Veg (cups)	3.3
Orange Veg (cups)	2.3
Beans/Peas (cups)	3.0
Starchy Veg (cups)	3.4
Other Veg (cups)	6.6

Nutrient	Daily Average Over One Week
Calories	1994
Protein, g	98
Protein, % kcal	20
Carbohydrate, g	264
Carbohydrate, % kcal	53
Total fat, g	67
Total fat, % kcal	30
Saturated fat, g	16
Saturated fat, % kcal	7.0
Monounsaturated fat, g	23
Polyunsaturated fat, g	23
Linoleic Acid, g	21
Alpha-linolenic Acid, g	1.1
Cholesterol, mg	207
Total dietary fiber, g	31
Potassium, mg	4715
Sodium, mg*	1948
Calcium, mg	1389
Magnesium, mg	432
Copper, mg	1.9
Iron, mg	21
Phosphorus, mg	1830
Zinc, mg	14
Thiamin, mg	1.9
Riboflavin, mg	2.5
Niacin Equivalents, mg	24
Vitamin B6, mg	2.9
Vitamin B12, mcg	18.4
Vitamin C, mg	190
Vitamin E, mg (AT)	18.9
Vitamin A, mcg (RAE)	1430
Dietary Folate Equivalents, mcg	558

* Starred items are foods that are labelled as no-salt-added, low-sodium, or low-salt versions of the foods. They can also be prepared from scratch with little or no added salt. All other foods are regular commercial products that contain variable levels of sodium. Average sodium level of the 7 day menu assumes no-salt-added in cooking or at the table.

Sample Menus for a 2000 Calorie Food Pattern

Averaged over a week, this seven day menu provides all of the recommended amounts of nutrients and food from each food group. (Italicized foods are part of the dish or food that preceeds it.)

Day 1

BREAKFAST

Breakfast burrito
- *1 flour tortilla (7˝ diameter)*
- *1 scrambled egg (in 1 tsp soft margarine)*
- *1/3 cup black beans**
- *2 tbsp salsa*
1 cup orange juice
1 cup fat-free milk

LUNCH

Roast beef sandwich
- *1 whole grain sandwich bun*
- *3 ounces lean roast beef*
- *2 slices tomato*
- *1/4 cup shredded romaine lettuce*
- *1/8 cup sauteed mushrooms (in 1 tsp oil)*
- *1 1/2 ounce part-skim mozzarella cheese*
- *1 tsp yellow mustard*
3/4 cup baked potato wedges*
- *1 tbsp ketchup*
1 unsweetened beverage

DINNER

Stuffed broiled salmon
- *5 ounce salmon filet*
- *1 ounce bread stuffing mix*
- *1 tbsp chopped onions*
- *1 tbsp diced celery*
- *2 tsp canola oil*
1/2 cup saffron (white) rice
- *1 ounce slivered almonds*
1/2 cup steamed broccoli
- *1 tsp soft margarine*
1 cup fat-free milk

SNACKS

1 cup cantaloupe

Day 2

BREAKFAST

Hot cereal
- *1/2 cup cooked oatmeal*
- *2 tbsp raisins*
- *1 tsp soft margarine*
1/2 cup fat-free milk
1 cup orange juice

LUNCH

Taco salad
- *2 ounces tortilla chips*
- *2 ounces ground turkey, sauteed in 2 tsp sunflower oil*
- *1/2 cup black beans**
- *1/2 cup iceberg lettuce*
- *2 slices tomato*
- *1 ounce low-fat cheddar cheese*
- *2 tbsp salsa*
- *1/2 cup avocado*
- *1 tsp lime juice*
1 unsweetened beverage

DINNER

Spinach lasagna
- *1 cup lasagna noodles, cooked (2 oz dry)*
- *2/3 cup cooked spinach*
- *1/2 cup ricotta cheese*
- *1/2 cup tomato sauce tomato bits**
- *1 ounce part-skim mozzarella cheese*
1 ounce whole wheat dinner roll
1 cup fat-free milk

SNACKS

1/2 ounce dry-roasted almonds*
1/4 cup pineapple
2 tbsp raisins

Day 3

BREAKFAST

Cold cereal
- *1 cup bran flakes*
- *1 cup fat-free milk*
- *1 small banana*
1 slice whole wheat toast
- *1 tsp soft margarine*
1 cup prune juice

LUNCH

Tuna fish sandwich
- *2 slices rye bread*
- *3 ounces tuna (packed in water, drained)*
- *2 tsp mayonnaise*
- *1 tbsp diced celery*
- *1/4 cup shredded romaine lettuce*
- *2 slices tomato*
1 medium pear
1 cup fat-free milk

DINNER

Roasted chicked breast
- *3 ounces boneless skinless chicken breast**
1 large baked sweet potato
1/2 cup peas and onions
- *1 tsp soft margarine*
1 ounce whole wheat dinner roll
- *1 tsp soft margarine*
1 cup leafy greens salad
- *3 tsp sunflower oil and vinegar dressing*

SNACKS

1/4 cup dried apricots
1 cup low-fat fruited yogurt

Day 4

BREAKFAST

1 whole wheat English muffin
- *2 tsp soft margarine*
- *1 tbsp jam or preserves*
1 medium grapefruit
1 hard-cooked egg
1 unsweetened beverage

LUNCH

White bean-vegetable soup
- *1 1/4 cup chunky vegetable soup*
- *1/2 cup white beans**
2 ounce breadstick
8 baby carrots
1 cup fat-free milk

DINNER

Rigatoni with meat sauce
- *1 cup rigatoni pasta (2 ounces dry)*
- *1/2 cup tomato sauce tomato bits**
- *2 ounces extra lean cooked ground beef (sauteed in 2 tsp vegetable oil)*
- *3 tbsp grated Parmesan cheese*
Spinach salad
- *1 cup baby spinach leaves*
- *1/2 cup tangerine slices*
- *1/2 ounce chopped walnuts*
- *3 tsp sunflower oil and vinegar dressing*
1 cup fat-free milk

SNACKS

1 cup low-fat fruited yogurt

Illustration 1.13 MyPyramid sample menu (showing 4 of 7 days) and analysis for a 2000 calorie food pattern.

Table 1.21 MyPyramid food measure equivalents

Grains

bagel	1 mini bagel = 1 oz
	1 large bagel = 4 oz
biscuit	1–2" diameter = 1 oz
	1–3" diameter = 2 oz
bread	1 slice = 1 oz
cooked cereal	½ cup = 1 oz
crackers	5 whole wheat = 1 oz
	7 square/round = 1 oz
English muffin	½ muffin = 1 oz
muffin	1–2½" diameter = 1 oz
	1–3½" diameter = 3 oz
pancake	1–4½" diameter = 1 oz
	2–3" diameter = 1 oz
popcorn	3 cups = 1 oz
breakfast cereal	1 cup flakes = 1 oz
	1¼ cups = 1 oz
rice	½ cup = 1 oz
pasta	½ cup = 1 oz
tortilla	1–6" diameter = 1 oz
	1–12" diameter = 4 oz

Vegetables

cooked	1 cup = 1 cup
carrots	2 medium = 1 cup
	12 baby = 1 cup
celery	1 large stalk = 1 cup
corn on the cob	1–6" long = ½ cup
	1–9" long = 1 cup
green/red peppers	1 large = 1 cup
potatoes	1 medium (3" diameter) = 1 cup
raw, leafy	2 cups = 1 cup
tomato	1 large = 1 cup

Milk

milk	1 cup = 1 cup
yoghurt	1 cup = 1 cup
cheese	1½ oz hard = 1 cup
	⅓ cup shredded = 1 cup
	2 oz processed = 1 cup

	½ cup ricotta = 1 cup
	2 cups cottage cheese = 1 cup
pudding	1 cup = 1 cup
frozen yogurt	1 cup = 1 cup
ice cream	1½ cup = 1 cup

Fruits

apple	1 small = 1 cup
	½ large = 1 cup
banana	1 large = 1 cup
cantaloupe	⅛ = 1 cup
grapes	32 = 1 cup
grapefruit	1 = 1 cup
orange	1 = 1 cup
peach	1 = 1 cup
pear	1 = 1 cup
plums	3 = 1 cup
strawberries	8 large = 1 cup
watermelon	1" wedge = 1 cup
dried fruit	½ cup = 1 cup
fruit juice	1 cup = 1 cup

Meat, Fish and Beans

steak	1–3½" × 2½" × ½" = 3 oz
hamburger	1 small = 2 oz
	1 medium = 4 oz
	1 large = 6 oz
chicken	½ breast = 3 oz
	1 thigh = 2 oz
	1 leg = 3½ oz
pork chops	1 medium = 3 oz
fish	1 small can tuna = 3½ oz
	1 small fish = 3 oz
	1 salmon steak = 5 oz
seafood	5 large srimp = 1 oz
	10 medium clams = 3 oz
	½ cup crab = 2 oz
	½ cup lobster = 2½ oz
eggs	1 small = 1 oz
	1 large = 2 oz

Table 1.22 The DASH Eating Plan for a 2000-calorie diet

	Number of servings		Number of servings
Grains*	6–8	Lean meats, poultry, fish	2 or fewer
Vegetables	4–5	Nuts, seeds, legumes	4–5/week
Fruits	4–5	Fats and oils	2–3
Fat-free or low-fat milk and milk poducts	2–3	Sweets	5/week

*whole grain products primarily

SOURCE: Dietary Guidelines for Americans, Appendix A-1: The DASH Eating Plan at 1,600-, 2,000-, 2,600-, and 3,100- Caloric Levels (www.health.gov/dietaryguidelines/dga2005/document/html/appendixA.htm).

Resources

Public Nutrition Programs

Start here if you want to identify government resources related to the Dietary Guidelines and to nutrition information, services, and programs.
Website: www.nutrition.gov

Dietary Assessment

View nutrient profiles for 13,000 foods commonly eaten in the United States in the What's In The Foods You Eat search tool.
Website: www.ars.usda.gov/foodsearch

Find USDA's Automated Multiple-Pass Method for 24-hour dietary recalls at this address.
Website: www.ars.usda.gov/Services/docs.htm?docid=7710

Locate the MyPyramid one-day dietary assessment tool here.
Website: www.mypyramidtracker.gov

National Nutrition Monitoring System

This address links to an 86-slide, informative Power Point presentation of the National Nutrition Monitoring System in the United States.
Website: www.cdc.gov/nchs/nhanes.htm

Nutrition and Health Guides

The U.S. Department of Agriculture's MyPyramid Food Guide main page, with links to dietary and physical activity assessments, menu planning information, interactive learning tools, and much more.
Website: www.mypyramid.gov

Visit the 2006 revised Canadian Food Guide to Healthy Eaing published at Health Canada's website.
Website: www.hc-sc.gc.ca/hpfb-dgpsa/onpp-bppn/revision_food_ guide_e.html

American Dietetic Association

Links to a wide variety of resources on nutrition and health.
Websites: www.eatright.org and www.dietitians.ca

National Academy of Sciences

Want more information on the Dietary Reference Intakes? You can get full reports here.
Website: www.iom.edu/fnb

Medline

Get access to scientific journal articles on a variety of nutrition topics.
Website: www.nlm.nih.gov/medlineplus

Healthy People 2010: Objectives for the Nation and Health, Canada's Health Goals

The home page for "Healthy People 2010."
Website: www.healthypeople.gov

Read about the health status of Canadians and nutrition and other goals for Canada.
Website: www.hc-sc.gc.ca

References

1. Wilde PE, Peterman JN. Individual weight change is associated with household food security status. J Nutr. 2006;1395–1400.

2. Summary report on application of the DRIs for dietary assessment. Washington (DC): National Academy of Sciences; 2000.

3. Seal CJ. Whole grains and CVD risk. Proc Nutr Soc. 2006;65:24–34.

4. Musick T et al. Carbohydrate and calories: it's not what we used to think. Compr Ther. 2006;32:47–50.

5. Livesey G. Low-glycemic diets and health: implications of obesity. Proc Nutr Soc. 2005;64:105–13.

6. Data extracted from: Foster-Powell K, et al. International table of glycemic index and glycemic load values. Am J Clin Nutr. 2002;76:5–56.

7. Dietary reference intakes: energy, carbohydrate, fiber, fat, fatty acids, cholesterol, protein, and amino acids. Washington (DC): National Academies Press; 2002.

8. Gebauer SK et al. n-3 Fatty acid dietary recommendations and food sources to achieve essentiality and cardiovascular benefits. Am J Clin Nutr. 2006;83(suppl):1526S–35S.

9. Arterburn LM et al. Distribution, interconversion, and dose response of n-3 fatty acids in humans. Am J Clin Nutr. 2006;83(suppl):1487S–76S.

10. Hibbeln JR et al. n-3 Healthy intakes of n-3 and n-6 fatty acids: estimations considering worldwide diversity. Am J Clin Nutr. 2006;83(suppl):1483S–93S.

11. Go VLW et al. Nutrient-gene interaction: metabolic genotype-phenotype relationships. J Nutr. 2005;135:3016S–20S.

12. Krauss RM et al. Revision 2000: a statement for healthcare providers from the Nutrition Committee of the American Heart Association. J Nutr. 2001;131:132–46.

13. Usui T. Pharmaceutical prospects of phytoestrogens. Endocr J. 2006;53:7–20.

14. Dietary reference intakes: water and electrolytes. Washington (DC): National Academies Press; 2004.

15. Grandjean AC et al. Hydration: issues for the 21st century. Nutr Rev. 2003;61:261–71.

16. Noakes TD. Too many fluids as bad as too few. BMJ. 2003;327:113–14.

17. Kolasa KM. Water: the recommendations we give—are they scientifically accurate? [conference paper]. Food and Nutrition Conference and Exhibition; 2002 Oct; American Dietetic Association.

18. Johnston CS, Corte C. People with marginal vitamin C status are at high risk of developing vitamin C deficiency. J Am Diet Assoc. 1999;99:854–6.

19. Simopoulos AP. Genetic variation and nutrition. Experimental Biology Annual Meeting; 2003 Apr 12; San Diego, CA.

20. Guttmacher AE et al. Genomic medicine—a primer. N Engl J Med. 2002;347:1512–20.

21. Stover PJ. Influence of genetic variation on nutritional requirements. Am J Clin Nutr. 2006;83(suppl):436S–42S.

22. Simon HB. Diet and exercise. WebMD Scientific American Medicine. 2003 Oct 14:1–36.

23. Heaney RP. Long-latency deficiency disease: insights from calcium and vitamin D. Am J Clin Nutr. 2003;78:912–19.

24. Park YK et al. History of cereal-grain product fortification in the United States. Nutr Today. 2001;36:123–36.

25. Hyman P. Claims for functional foods under the current food regulatory scheme. Nutr Today. 2002;37:217–19.

26. Hasler CM. Functional foods: benefits, concerns and challenges. J Nutr. 2002;132:3772–81.

27. Beale B. Probiotics: their tiny worlds are under scrutiny. The Scientist. 2002 Jul:20–22.

28. Sanders ME. Probiotics: considerations for human health. Nutr Rev. 2003;61:91–97.

29. Brannon CB. Prebiotics: feeding friendly bacteria. Today's Dietitian. 2003 Sep:12–16.

30. Kittler PG, Sucher KP. Diet counseling in a multicultural society. Diabetes Educator. 1989;16:127–32.

31. Romero-Gwynn E et al. Dietary patterns and acculturation among immigrants from El Salvador. Nutr Today. 2003;35:233–40.

32. Conti K. The four winds nutrition model: a new culturally specific food guide and nutrition approach for the Northern Plains Indians seminar presentation, University of Minnesota, Minneapolis; Oct 28, 2002.

33. Dirks RT, Duran N. African American dietary patterns at the beginning of the 20th century. J Nutr. 2001;131:1881–9.

34. McCaffree J. Dietary restrictions of other religions. J Am Diet Assoc. 2003;103:912.

35. Eliasi JR, Dwyer JT. Kosher and halal: religious observances affecting dietary intake. J Am Diet Assoc. 2003;103:911–12.

36. Halsted CH. Dietary supplements and functional foods: 2 sides of a coin? Am J Clin Nutr. 2003;77 (suppl):1001S–7S.

37. Dwyer J. Dietary assessment. In: Shils ME, editor. Modern nutrition in health and disease. Philadelphia: Lippincott, Williams & Wilkins; 1998. pp. 937–62.

38. Horner NK et al. Participant characteristics associated with errors in self-reported energy intake from the Women's Health Initiative food-frequency questionnaire. Am J Clin Nutr. 2002;76:766–73.

39. Conway JM et al. Accuracy of the dietary recall using the USDA five-step multiple-pass method in men: an observational study. J Am Diet Assoc. 2004; 104:595–603.

40. National Nutrition Monitoring System, http://www.cdc.gov/nchs/nhanes.htm, accessed Aug 2006.

41. Nowson CA et al. Blood pressure response to dietary modifications in free-living individuals. J Nutr. 134:2322–2329.

42. Position of the American Dietetics Association: child and adolescent food and nutrition programs. J Am Diet Assoc. 2003;103:887–92.

43. Public health achievements. Nutrition Today. 1999 May/Jun:2.

43. Deckelbaum RJ, Fisher EA, Winston M, et al. Summary of a scientific conference on preventive nutrition: pediatrics to geriatrics. Circulation 1999; 100: 450–6.

Chapter 2

Preconception Nutrition

Chapter Outline

- Introduction
- Preconception Overview
- Reproductive Physiology
- Sources of Disruptions in Fertility
- Nutrition-Related Disruptions in Fertility
- Nutrition and Contraceptives
- Other Preconceptual Nutrition Concerns
- Model Preconceptional Nutrition Programs

Supestock

Photo Disc

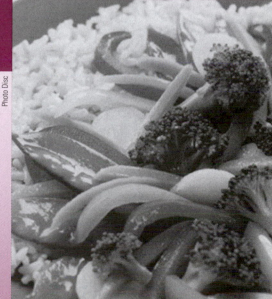

Photo Disc

Prepared by **Judith E. Brown**

Key Nutrition Concepts

1   Fertility is achieved and maintained by carefully orchestrated, complex processes that can be disrupted by a number of factors related to body composition and dietary intake.

2   Oral contraceptives and contraceptive implants can adversely affect some aspects of nutritional status.

3   Optimal nutritional status prior to pregnancy enhances the likelihood of conception and helps ensure a healthy pregnancy and robust newborn.

Introduction

Human reproduction is the result of a superb orchestration of complex and interrelated genetic, biological, environmental, and behavioral processes. Given favorable states of health, these processes occur smoothly in females and males and set the stage for successful reproduction. However, less than optimal states of health, brought about by conditions such as acute undernutrition or high levels of alcohol intake, can disrupt these finely tuned processes and diminish reproductive capacity. Sometimes conception occurs in the presence of poor nutritional or health status. Such events increase the likelihood that fetal growth and development, and the health of the mother during pregnancy, will be compromised.

This chapter first highlights vital statistics related to the preconception period and presents background information on reproductive physiology. Then the focus is on (1) nutrition and the development and maintenance of the biological capacity to reproduce, (2) nutritional effects of contraceptives, (3) preconceptional nutritional status and the course and outcome of pregnancy, and (4) model programs that promote preconceptional nutritional health. The following chapter addresses the role of nutrition in specific conditions—such as premenstrual syndrome, diabetes, eating disorders, obesity, celiac disease, and polycystic ovary syndrome—that affect preconceptional health or very early pregnancy outcomes.

Infertility   Absence of production of children. Commonly used to mean a biological inability to bear children.

Infecundity   Biological inability to bear children after 1 year of unprotected intercourse.

Fertility   Actual production of children. The word best applies to specific vital statistic rates, but is commonly taken to mean the ability to bear children.

Fecundity   Biological ability to bear children.

Miscarriage   Generally defined as the loss of a conceptus in the first 20 weeks of pregnancy. Also called spontaneous abortion.

Endocrine   A system of ductless glands, such as the thyroid, adrenal glands, ovaries, and testes, that produces secretions that affect body functions.

Immunological   Having to do with the immune system and its functions in protecting the body from bacterial, viral, fungal, or other infections and from foreign proteins (i.e., those proteins that differ from proteins normally found in the body).

Subfertility   Reduced level of fertility characterized by unusually long time to conception (over 12 months) or repeated, early pregnancy losses.

Preconception Overview

Approximately 15% of all couples in the Western world are involuntarily childless. They are generally considered to be *infertile,* or more correctly *infecund.*

Fertility refers to the actual production of children, whereas *fecundity* addresses the biological capacity to bear children. *Fertility* best applies to vital statistics on fertility rates, or the number of births per 1000 women of childbearing age (15–44 years in most statistical reports). For example, in 2005 the U.S. fertility rate was 66.7 births per 1000 women aged 15 to 44 years.[1] Even scientists and clinicians rarely use these terms correctly, however, and to do so in this chapter would cause undue confusion. Hence, we will use the familiar meaning of *infertility.*

Infertility is generally defined as the lack of conception after 1 year of unprotected intercourse. This definition leads to a "yes/no" answer about fertility that is misleading. Approximately 40% of couples diagnosed as infertile will conceive a child in 3 years without the help of technology.[2] Given these chances, fertility procedures can be considered successful only if they increase conceptions by over 40%, and many currently do not.[3] Chances of conceiving decrease the longer infertility lasts and as women and men age beyond 35 years.[4]

Healthy couples having regular, unprotected intercourse have a 30–50% chance of a diagnosed pregnancy within a given menstrual cycle. However, many more conceptions probably occur. Studies show that 30–50% of conceptions are lost by resorption into the uterine wall within the first 6 weeks after conception. Another 9% are lost by *miscarriage* in the first 20 weeks of pregnancy.[5]

The most common known cause of miscarriage is the presence of a severe defect in the fetus. Miscarriages can also be caused by maternal infection, structural abnormalities of the uterus, *endocrine* or *immunological* disturbances, and unknown, random events.[5]

Women who experience multiple miscarriages (variously defined as two or three), men who have sperm abnormalities (such as low sperm count or density, malformed sperm or immobile sperm), and women who ovulate infrequently are considered *subfertile.* It is estimated that 18% of married couples in the United States are subfertile due to delayed time to conception (over 12 months) or repeated, early pregnancy losses.[2] A silver lining to subfertility is that the reproductive capacity of one individual can compensate for diminished potential in the other. In addition, subfertility can be diminished by improvements in diet and lifestyle.[6]

2010 Nutrition Objectives for the Nation Related to the Preconceptional Period

National priorities for improvement in health status prior to pregnancy include five related to nutrition (Table 2.1).

Table 2.1 2010 nutrition objectives for the nation related to preconception

- Increase the proportion of adults who are at a healthy weight from 42 to 60%.
- Reduce the proportion of adults who are obese from 23 to 15%.
- Reduce iron deficiency among young children and females of childbearing age from 4–11% to 1–7%.
- Reduce the incidence of spina bifida and other neural tube defects from 7 to 3 per 10,000 live births and birth defects from 1.7 to 1.2 per 1000 live births.
- Increase the proportion of women who receive preconceptional counseling.

Objectives for improving intake of vegetables, calcium, whole grains, and other dietary components apply to men as well as women prior to pregnancy.

Reproductive Physiology

The reproductive systems of females and males (Illustration 2.1 on the next page) begin developing in the first months after conception and continue to grow in size and complexity of function through *puberty*. Females are born with a complement of immature *ova* and males with sperm-producing capabilities. The capacity for reproduction is established during puberty when hormonal changes cause the maturation of the reproductive system over the course of 3 to 5 years.

Approximately 7 million immature ova, or *primordial follicles,* are formed during early fetal development, but only about one-half million per ovary remain by the onset of puberty. During a woman's fertile years, some 400–500 ova will mature and be released for possible fertilization. Very few ova remain by *menopause.*

For men, sperm numbers and viability decrease somewhat after approximately 35 years of age, but sperm are still produced from puberty onward.[4] Because females are born with their lifetime supply of ova, the number with chromosomes damaged by oxidation, radioactive particle exposure, and aging increases with time. Consequently, children born to women older than roughly 35 years of age are more likely to have disorders related to defects in chromosomes than are children born to younger women.[2]

Female Reproductive System

During puberty females develop monthly **menstrual cycles,** the purpose of which is to prepare an ovum for fertilization by sperm and the uterus for implantation of a fertilized egg. Menstrual cycles result from complex interactions among hormones secreted by the hypothalamus, the **pituitary gland,** and the ovary. Knowledge of hormonal changes during the menstrual cycle is expanding, and the process is more complex than indicated in this presentation, which focuses on nutritional effects on hormonal changes in the menstrual cycle and on fertility.

Menstrual cycles are 28 days long on average, but it is not uncommon for cycles to be several days shorter or longer. The first day of the cycle is when menses, or blood flow, begins. The first half of the cycle is called the *follicular phase;* the last 14 days is the *luteal phase.* Hormonal changes during these phases of the menstrual cycle are shown in Illustration 2.2 on page 51.

Hormonal Effects During the Menstrual Cycle At the beginning of the follicular phase, estrogen stimulates the hypothalamus to secrete *gonadotropin-releasing hormone* (GnRH), which causes the pituitary gland to release the *follicle-stimulating hormone* (FSH) and *luteinizing hormone* (LH). (See Table 2.2 on page 52 for definitions of these hormones.) FSH prompts the growth and maturation of 6–20 follicles, or capsules in the surface of the ovary in which ova mature. The presence of FSH stimulates the production of *estrogen* by cells within the follicles. Estrogen and FSH further stimulate the growth and maturation of follicles while rising LH levels cause cells within the follicles to secrete *progesterone*. Estrogen and progesterone also prompt the uterine wall (or endometrium) to store glycogen and other nutrients and to expand the growth of blood vessels and connective tissue. These changes prepare the uterus for nourishing a conceptus after implantation. Just prior to ovulation, which usually occurs on day 14 of a 28-day menstrual cycle, blood levels of FSH and LH peak. The surge in LH level results in the release of an ovum from a follicle, and voilà! Ovulation occurs.

The luteal phase of the menstrual cycle begins after ovulation. Much of the hormonal activity that regulates biological processes during this half of the cycle is initiated by the cells in the follicle left behind when the egg was released. These cells grow in number and size and form the **corpus luteum**

Puberty The period in life during which humans become biologically capable of reproduction.

Ova Eggs of the female produced and stored within the ovaries (singular = *ovum*).

Menopause Cessation of the menstrual cycle and reproductive capacity in females.

Menstrual Cycle An approximately 4-week interval in which hormones direct a buildup of blood and nutrient stores within the wall of the uterus and ovum maturation and release. If the ovum is fertilized by a sperm, the stored blood and nutrients are used to support the growth of the fertilized ovum. If fertilization does not occur, they are released from the uterine wall over a period of 3 to 7 days. The period of blood flow is called the *menses,* or the menstrual period.

Pituitary Gland A pea-sized gland located at the base of the brain. It is connected to the hypothalamus and produces and secretes growth hormone, prolactin, oxytocin, follicle-stimulating hormone, luteinizing hormone, and other hormones in response to signals from the hypothalamus.

Corpus Luteum (*corpus* = body, *luteum* = yellow) A tissue about 12 mm in diameter formed from the follicle that contained the ovum prior to its release. It produces estrogen and progesterone. The "yellow body" derivation comes from the accumulation of lipid precursors of these hormones in the corpus luteum.

OVARY
One of a pair of primary reproductive organs in which oocytes (immature eggs) form and mature; produces hormones (estrogens and progesterone), which stimulate maturation of oocytes, formation of corpus luteum (a glandular structure), and preparation of the uterine lining for pregnancy

OVIDUCT
One of a pair of ciliated channels through which oocytes are conducted from an ovary to the uterus; usual site of fertilization

UTERUS
Chamber in which embryo develops; its narrowed-down portion (the cervix) secretes mucus that helps sperm move into uterus and that bars many bacteria

MYOMETRIUM
Thick muscle layers of uterus that stretch enormously during pregnancy

ENDOMETRIUM
Inner lining of uterus; site of implantation of blastocyst (early embryonic stage); becomes thickened, nutrient-packed, highly vascularized tissue during a pregnancy; gives rise to maternal portion of placenta, an organ that metabolically supports embryonic and fetal development

urinary bladder

urethra

opening of cervix

CLITORIS
Small organ responsive to sexual stimulation

LABIUM MINOR
One of a pair of inner skin folds of external genitals

LABIUM MAJOR
One of a pair of outermost, fat-padded skin folds of external genitals

anus

VAGINA
Organ of sexual intercourse; also serves as birth canal

PROSTATE GLAND
Secretion of substances that become part of semen

EJACULATORY DUCT
One of a pair of sperm-conducting ducts

SEMINAL VESICLE
One of a pair of glands that secrete fructose and prostaglandins, which become part of semen

urinary bladder

URETHRA
Dual-purpose duct; serves as channel for ejaculation of sperm during sexual arousal, also for urine excretion at other times

BULBOURETHRAL GLAND
One of a pair of glands that secrete a lubricating mucus

urethra

anus

VAS DEFERENS
One of a pair of ducts for rapid transport of sperm

EPIDIDYMIS
One of a pair of ducts in which sperm complete maturation; the portion farthest from testis stores mature sperm

erectile tissue

TESTIS
One of a pair of primary reproductive organs; packed with sperm-producing tubules and cells that secrete testosterone and other hormones

PENIS
Organ of sexual intercourse

Illustration 2.1 Mature female and male reproductive systems.

from the original follicle. The corpus luteum secretes large amounts of progesterone and some estrogen. These hormones now inhibit the production of GnRH, and thus the secretion of FSH and LH. Without sufficient FSH and LH, ova within follicles do not mature and are not released. (This is also how estrogen and progesterone in some birth control pills inhibit ova maturation and release.) Estrogen and progesterone secreted by the corpus

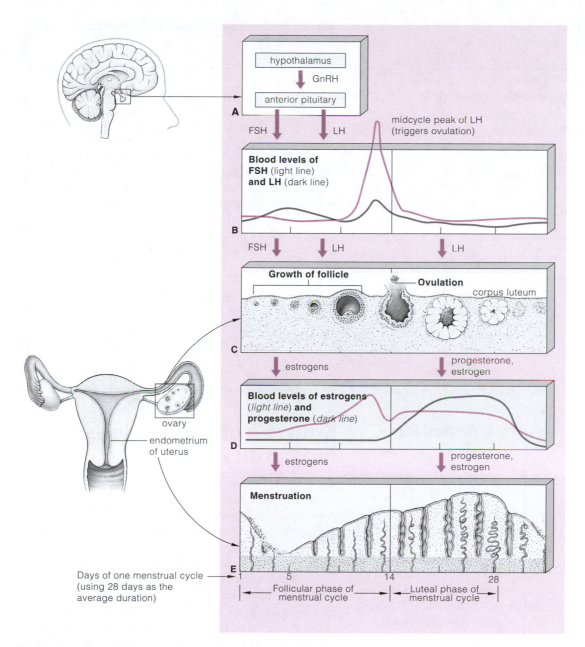

Illustration 2.2 Changes in the ovary and uterus, correlated with changing hormone levels during the follicular and luteal phases of the menstrual cycle.

luteum further stimulate the development of the endometrium. If the ovum is not fertilized, the production of hormones by the corpus luteum declines, and blood levels of progesterone and estrogen fall. This decline removes the inhibitory effect of these hormones on GnRH release, and GnRH is again able to stimulate release of FSH for the next cycle of follicle development, and of LH for the stimulation of progesterone and estrogen production. Decreased levels of progesterone and estrogen also cause blood vessels in the uterine wall to constrict, allowing the uterine wall to release its outer layer in the *menstrual flow*. Cramps and other side effects of menstruation can be traced to the production of *prostaglandins* by the uterus. These substances cause the

uterus to contract and release the blood and nutrients stored in the uterine wall.

If the ovum is fertilized, it will generally implant in the lining of the uterus within 8 to 10 days. Hormones secreted by the dividing, fertilized egg signal the corpus luteum to increase in size and to continue to produce enough estrogen and progesterone to maintain the nutrient and blood vessel supply in the endometrium. The corpus luteum ceases to function within the first few months of pregnancy, when it is no longer needed for hormone production.

Prostaglandins   A group of physiologically active substances derived from the essential fatty acids. They are present in many tissues and perform such functions as the constriction or dilation of blood vessels, and stimulation of smooth muscles and the uterus.

Table 2.2 Hormones that affect reproduction

Hormone	Abbreviation	Source	Action
Gonadotropin-releasing hormone	GnRH	Hypothalamus	Stimulates release of FSH and LH
Follicle-stimulating hormone	FSH	Pituitary	Stimulates the maturation of ova and sperm
Luteinizing hormone	LH	Pituitary	Stimulates secretion of estrogen, progesterone, and testosterone and growth of the corpus luteum
Estrogen (most abundant form is estradiol)		Ovaries, testes, fat cells, corpus luteum, and placenta (during pregnancy)	Stimulates release of GnRH in follicular phase and inhibits in luteal phase; stimulates thickening of uterine wall during menstrual cycle
Progesterone (progestin, progestogen, and gestagon are similar)		Ovaries and placenta	"Progestational": prepares uterus for fertilized ovum and to maintain a pregnancy; stimulates uterine lining buildup during menstrual cycle; helps stimulate cell division of fertilized ova; inhibits action of testosterone
Testosterone		Mostly by testes	Stimulates maturation of male sex organs and sperm, formation of muscle tissue and other functions

Testes Male reproductive glands located in the scrotum. Also called testicles.

Androgens Types of steroid hormones produced in the testes, ovaries, and adrenal cortex from cholesterol. Some androgens (testosterone, dihydrotestosterone) stimulate development and functioning of male sex organs.

Epididymis Tissues on top of the testes that store sperm.

Semen The penile ejaculate containing a mixture of sperm and secretions from the testes, prostate, and other glands. It is rich in zinc, fructose, and other nutrients. Also called seminal fluid.

Pelvic Inflammatory Disease (PID) A general term applied to infections of the cervix, uterus, fallopian tubes, or ovaries. Occurs predominantly in young women and is generally caused by infection with a sexually transmitted disease, such as gonorrhea or chlamydia, or with intrauterine device (IUD) use.

Endometriosis A disease characterized by the presence of endometrial tissue in abnormal locations, such as deep within the uterine wall, in the ovary, or in other sites within the body. The condition is quite painful and is associated with abnormal menstrual cycles and infertility in 30–40% of affected women.

Male Reproductive System

Reproductive capacity in males is established by complex interactions among the hypothalamus, pituitary gland, and *testes*. The process in males is ongoing rather than cyclic. Fluctuating levels of GnRH signal the release of FSH and LH, which trigger the production of testosterone (Table 2.2) by the testes. Testosterone and other *androgens* stimulate the maturation of sperm, which takes 70–80 days. When mature, sperm are transported to the *epididymis* for storage. Upon ejaculation, sperm mix with secretions from the testes, seminal vesicle, prostate, and bulbourethral gland to form *semen*.

Just as some aspects of female reproductive processes remain unclear, scientists have yet to fully elucidate hormonal and other processes involved in male reproduction.

Sources of Disruptions in Fertility

The intricate mechanisms that regulate fertility can be disrupted by many factors, including adverse nutritional exposures, contraceptive use, severe stress, infection, tubal damage and other structural problems, and chromosomal abnormalities (Table 2.3).[5,6] Conditions that modify fertility appear to affect hormones that regulate ovulation, the presence or length of the luteal phase, sperm production, or the tubular passageways that ova and sperm must travel for conception to occur. Sexually transmitted infections, for example, can result in *pelvic inflammatory disease (PID)*, which may lead to scarring and blockage of the fallopian tubes.[7] *Endometriosis* is also a common cause of reduced fertility. It develops when portions of the endometrial wall that build up during menstrual cycles become embedded within other body tissues. Endocrine abnormalities that modify hormonal regulation of fertility are the leading diagnoses related to infertility. "Unknown cause" is the second leading diagnosis, however, and is applied to about one-half of all cases of male and female infertility.[5]

Nutrition-Related Disruptions in Fertility

Disruptions in fertility related to nutritional status include undernutrition, weight loss, obesity, high exercise levels, and intake of specific foods and food components. Nutritional factors generally exert only temporary influence on fertility; normal fertility returns once the problem is corrected.

Table 2.3 Factors related to impaired fertility in women and men

Females and Males	Females	Males
• Weight loss >15% of normal weight	• Recent oral contraceptive use (within 2 months)	• Inadequate zinc status
• Negative energy balance	• Anorexia nervosa, bulimia nervosa	• Inadequate antioxidant status (selenium, vitamins C and E, carotinoids)
• Inadequate body fat	• High coffee/caffeine intake	• Heavy metal exposure (lead, mercury, cadmium, manganese)
• Excessive body fat, especially central fat	• High fiber intake	• Halogen (in some pesticides) and glycol exposure (in antifreeze, de-icers)
• Extreme levels of exercise	• Vegetarian diets (?)[a]	
• High alcohol intake	• Carotenemia (?)[a]	• Estrogen exposure (in DDT, PCBs)
• Endocrine disorders (e.g., hypothyroidism, Cushing's disease)	• Age >35 years	• Chromosomal abnormalities in sperm
• Structural abnormalities of the reproductive tract	• Pelvic inflammatory disease (PID)	• Sperm defects
• Celiac disease	• Endometriosis	• Excessive heat to testes
• Crowding	• Poor iron stores	• Steroid abuse
• Severe stress		• Vitamin D inadequacy (?)[a]
• Infection (sexually transmitted diseases)		• DHA inadequacy (?)[a]
• Diabetes		

[a]The "?" indicates that the relationship to impaired fertility is controversial.

Undernutrition and Fertility

Does undernutrition decrease fertility in populations? The answer depends on whether the undernutrition is long term (chronic) or short term (acute). Chronic undernutrition appears to reduce fertility by only a small amount.[8] Acute undernutrition due to famine or deliberate weight loss in normal-weight women clearly decreases fertility.

Chronic Undernutrition The primary effect of chronic undernutrition on reproduction in women is the birth of small and frail infants who have a high likelihood of death in the first year of life.[9] Infant death rates in developing countries where malnutrition is common often exceed 50 per 1000 live births. By contrast, infant death rates are less than 4 per 1000 live births in countries such as Hong Kong, Japan, and Sweden.[10]

The effect of chronic undernutrition on fertility is difficult to study accurately, and conclusions about relationships will change as more is learned. Investigations of relationships between chronic undernutrition and fertility are complicated by differences in the use of contraception, age of puberty and marriage, breastfeeding duration (longer periods of breastfeeding increase the time to the next pregnancy), access to induced abortions, and social and economic incentives or constraints on family size.[11] In less developed countries with poor access to contraception, births per woman average 6 to 8, whereas in developed countries (where contraception is generally available) they average at or below the replacement level of 2.1 births per woman.[12] Without careful study, it might appear

fertility is *lower*[13] in better-nourished women in developed countries than in poorly nourished women in less developed countries.

Of the environmental factors that influence fertility, education and child survival appear to be the most important. Fertility rates in poor countries decline substantially as women become educated and as child survival increases.[13,14]

Acute Undernutrition Undernutrition among previously well-nourished women is associated with a dramatic decline in fertility that recovers when food intake does.[9] Periods of feast and famine in the nomadic Kung tribe of Botswana and the Turkana people of Kenya, for example, are associated with major shifts in fertility.[15,16] These groups of hunter-gatherers (although relatively few survive now) experience sharp, seasonal fluctuations in body weight depending on the success of their hunting and foraging for plant foods. Birthrates decline substantially during periods of famine and increase with food availability. When these hunter-gatherers become farmers and food supply is more dependable, body weight increases, activity levels decrease, and pregnancy rates go up.

Other evidence also suggests a connection between undernutrition and infertility. Food shortages in Europe in the seventeenth and eighteenth centuries were accompanied by dramatic declines in birthrates. Famine in Holland during World War II led to calorie intakes of about 1000 per day among women. One out of two women in famine-affected areas stopped menstruating, and the birthrate dropped by 53%. Fertility status improved within 4 months after the end of the famine, but for many

Case Study 2.1

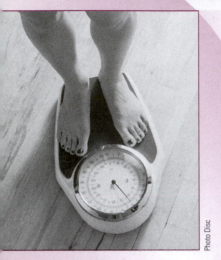

Photo Disc

Cyclic Infertility with Weight Loss and Gain

After four years of experiencing amenorrhea, Tonya seeks medical care to help her become pregnant. She is convinced that her lack of menstrual periods is the cause of her infertility. Tonya's height is 5 feet 5 inches; her weight is 107 pounds, which she has maintained for 4 years (she previously weighed 121 pounds). Her FSH and LH levels are both abnormally low, and she is not ovulating. When the importance and methods of weight gain are explained to her, Tonya agrees to gain some weight. After she regains 7 pounds, her LH level is normal, but her FSH level is still low, and the luteal phase of her cycles is abnormally short. When her weight reaches 119 pounds, Tonya's LH and FSH levels, ovulation, and menstrual cycles are normal.

Questions

1. Was Tonya underweight or normal weight based on BMI when she weighed 107 pounds? (Use the BMI chart on the inside back cover of the text to answer this question.)
2. Can you determine Tonya's body fat content based on her BMI?
3. Why isn't Tonya ovulating?
4. What likely happened to Tonya's average estrogen level when her weight decreased from 121 to 107 pounds?
5. What are two likely reasons Tonya was advised to gain weight to improve her chances of conception rather than being given Clomid or another ovulation-inducing drug?

---

women it took as long as a year for their menstrual cycles to return to normal.[17] Similarly, the 1974–1975 famine in Bangladesh resulted in a 40% decline in births.[8]

Famines are associated with more than disruptions in the food supply. They are usually accompanied by low availability of fuel for heating and cooking, poor living conditions, anxiety, fear, and despair. These factors also probably contribute to the declines in fertility observed with famine.

Acute reduction in food intake appears to reduce reproductive capacity by modifying hormonal signals that regulate menstrual cycles in females. It also appears to impair sperm maturation in males.[8]

Leptin   A protein secreted by fat cells that, by binding to specific receptor sites in the hypothalamus, decreases appetite, increases energy expenditure, and stimulates gonadotropin secretion. Leptin levels are elevated by high, and reduced by low, levels of body fat.

Body Mass Index (BMI)   Weight in kg/height in m². BMIs <18.5 are considered underweight, 18.5–24.9 normal weight, 25–29.9 overweight, and BMIs of 30 and higher obesity.

Body Fat and Fertility

Both low and high amounts of body fat are related to decreased fertility in females and males due to alterations in hormone levels. Fat cells produce estrogen, androgens, and *leptin,* and these hormones affect reproductive processes. In women, estrogen, androgens, and leptin

levels are elevated by high levels of body fat and reduced by low levels of body fat.[18] Obesity lowers testosterone and increases leptin in men.[19] Weight loss through caloric reduction and increased physical activity is the recommended initial treatment for infertility among women and men with high levels of body fat.[20] Infertility treatments, such as drugs that induce ovulation, are less effective in obese than in normal-weight women.[20]

The relationship between low levels of body fat and delayed onset of menstruation and infertility has led to the conclusion that a critical level of body fat is needed to trigger and sustain normal reproductive functions.[21] Specific levels of body fat consistently related to infertility have not been identified. However, fertility is often compromised in women with *body mass indexes* (BMIs) of less than 20 or over 30 kg/m².[22]

Approximately one in three men and women in the United States are obese, making infertility related to excess body fat a common problem and an important health concern.[1] The topic of body fat and fertility is revisited in more detail in the next chapter.

Weight Loss and Fertility in Females   In normal-weight women, weight loss that exceeds approximately 10–15% of usual weight decreases estrogen, LH, and FSH concentrations.[23] Consequences of these hormonal changes include *amenorrhea, anovulatory cycles,* and short or absent luteal phases. It is estimated that about 30% of cases of impaired fertility are related to simple weight loss. In the past, this form of amenorrhea was called "weight-related amenorrhea." It is now called "hypothalamic amenorrhea." Hormone levels tend to return to normal when weight is restored to within 95% of previous weight.[9] Case Study 2.1 provides an example of the effect of weight loss on fertility.

Weight gain is the recommended first-line treatment for amenorrhea related to low body weight. In many cases, however, the advice is more easily given than applied. About 10% of underweight women will not consider weight gain and may change health care providers in search of a different solution to infertility.[24]

Treatment of underweight women with Clomid (clomiphene citrate, a drug that induces ovulation) generally does not improve fertility until weight is regained. Fertility may be improved through the use of GnRH, FSH, and other hormones. However, twice as many infants born to underweight women receiving such therapy are small for gestational age compared to infants born to underweight women who gain weight and experience unassisted conception.[9]

The eating disorder anorexia nervosa is associated with similar, but more severe, changes in endocrine and hypothalamic functions than those seen with weight loss in normal-weight women. (This topic, as well as hypothalamic amenorrhea, is covered in Chapter 3.)

Weight Loss and Fertility in Males   Weight loss decreases fertility in men just as it does in women. In the classic starvation experiments by Keys during World War II,[22] men experiencing a 50% reduction in caloric intake reported substantially reduced sexual drive early in the study. Sperm viability and motility decreased as weight reached 10 to 15% below normal, and sperm production ceased entirely when weight loss exceeded 25% of normal weight. Sperm production and libido returned to normal after weight was regained.

Exercise and Infertility

The adverse effects of intense levels of physical activity on fertility were observed over 40 years ago in female competitive athletes. Since then, a number of studies have shown that young female athletes may experience delayed age at puberty and lack menstrual cycles. Average age of menarche for competitive female athletes and ballet dancers is often delayed by 2 to 4 years. The delay in menarche increases if females begin training for events that require thinness (such as gymnastics) before menarche normally would begin. Very high levels of exercise can also interrupt previously established, normal menstrual cycles. The presence of abnormal cycles reportedly ranges from about 23% in joggers to 86% in female bodybuilders (Table 2.4).[26, 27]

Delays and interruptions in normal menstrual cycles appear to result from hormonal and metabolic changes primarily related to caloric deficits rather than intense exercise.[28] Metabolic and hormonal status generally reverts to normal after high levels of training and caloric deficits end.

Some of the hormones involved in fertility impairments perform other important functions in the body, which may also be disrupted. Reduced levels of estrogen that accompany low levels of body fat and amenorrhea, for example, may decrease bone density and increase the risk of shortness, bone fractures, and osteoporosis.[29]

Diet and Fertility

Certain dietary components appear to influence fertility by modifying estrogen, LH, and other hormone levels in women. Vegetarian diets, low fat intake, lack of iron, and high intakes of dietary fiber, soy, caffeine, and alcohol are related to impaired fertility. The effects of several of these factors on fertility status may be interrelated.

Plant Foods and Fertility   Women who regularly consume plant-based, low-fat, high-fiber ($\geq$ 25 g per day) diets and no meat appear to have lower circulating levels of estrogen and may be more likely to have irregular menstrual cycles than omnivores. These results apply to vegetarians who are thin, normal weight, or overweight.[30, 31] Diets providing less than 20% of calories from fat appear to lengthen menstrual cycles among women in general.[32]

> **Amenorrhea**   Absence of menstrual cycle.
>
> **Anovulatory Cycles**   Menstrual cycles in which ovulation does not occur.

Table 2.4 Incidence of irregular or absent menstrual cycles in female athletes and sedentary women

	Incidence of Irregular or Absent Menstrual Cycles
Joggers (5 to 30 miles per week)	23%
Runners (over 30 miles per week)	34%
Long-distance runners (over 70 miles per week)	43%
Competitive bodybuilders	86%
Noncompetitive bodybuilders	30%
Volleyball players	48%
Ballet dancers	44%
Sedentary women	13%

Consumption of 20–200 mg/day of isoflavones from soy increases menstrual cycle length by about a day. Isoflavones appear to decrease blood levels of gonadotropins, estrogen, and progesterone. Soy intake has not been related to androgen levels nor semen quality in males.[33]

Carotenemia and Fertility   Several studies indicate that women with *carotenemia* may experience amenorrhea and menstrual dysfunction.[35] Women who consume 12 mg or more of beta-carotene daily for over 6 weeks by constant munching on foods such as dried green pepper flakes or carrots, or by taking "tanning pills," tend to develop carotenemia. Skin color and fertility return to normal within 2 to 6 weeks after high levels of intake are discontinued. Why carotenemia may cause infertility is not known, and the relationship must be considered speculative until additional studies are undertaken. It is not clear that the carotene content of the diet, rather than other substances in plentiful supply in plant foods, accounts for the relationship observed.

Carotenemia   A condition, caused by ingestion of high amounts of carotenoids (or carotenes) from plant foods, in which the skin turns yellowish orange.

Preconception Iron Status, Fertility, and Pregnancy Outcomes   Iron status prior to pregnancy is related to fertility and pregnancy outcomes. Results of a large, prospective study of nurses indicate that infertility due to a lack of ovulation is related to iron intake. In this study, women who regularly used iron supplements and consumed plant sources of iron were 60% less likely to develop ovulatory infertility than women who did not. Mechanisms underlying the link between iron status and ovulation are not yet established.[36]

Iron deficiency prior to pregnancy has been shown to increase the risk that iron deficiency anemia will occur during pregnancy and that infants will be born with low stores of iron. Iron deficiency before pregnancy is also related to increased rates of preterm delivery.[37] Iron deficiency occurs in 12% of U.S. women of childbearing age overall, and among 22% of Mexican American and 19% of African American women.[1] About one-half of women in the United State enter pregnancy with inadequate iron stores.

It is easier and more efficient to build up iron stores before pregnancy than during pregnancy.[37] Iron status can generally be improved by taking modest doses of iron supplements (18 mg a day) and by the regular consumption of vitamin C–rich fruits and vegetables along with plant sources of iron, iron-fortified cereals, and lean meats.

Caffeine and Fertility   Should women concerned about infertility consume coffee and other foods with caffeine? It appears that high intakes of caffeine may prolong the time it takes to become pregnant. In a study of European women, researchers found that the chance of conception within a 10-month interval of unprotected intercourse was half as likely among women who consumed over 4 cups of coffee per day (> 500 mg caffeine) versus the conception rate of women who consumed little coffee.[38] Another study reported that intake of over 300 mg of caffeine daily from coffee, sodas, and tea decreased the chance of conceiving by 27% per cycle compared to negligible caffeine intake.[39] In both studies, the effect of caffeine on time to conception was stronger in women who smoked.

It is not known why caffeine appears to prolong time to conception. Nor is it clear if it is caffeine, one or more of the hundreds of other chemical substances in coffee, or an unexamined characteristic of women who consume lots of caffeine that accounts for the apparent effects of caffeine on fertility. The prudent course of action is to advise concerned women who consume more than several cups of regular coffee each day to cut down on coffee and other products high in caffeine. Table 2.5 shows the caffeine content of selected beverages and foods. Women who cut out coffee altogether should be advised to do so gradually, to reduce "caffeine withdrawal headaches" and fatigue.

Alcohol and Fertility   Alcohol may influence fertility by decreasing estrogen and testosterone levels and disrupting normal menstrual cycles. In a study of 430

Table 2.5 Caffeine content of foods and beverages

Foods and Beverages	Caffeine (mg)
Coffee, 1 c	
Drip	137–153
Percolated	97–125
Instant	61–70
Decaffeinated	0.5–4.0
Tea, 1 c	
Brewed 5 minutes	32–176
Instant	40–80
Soft Drinks, 12 oz	
Mountain Dew	54
Coca-Cola	46
Diet Coca-Cola	46
Dr. Pepper	40
Pepsi-Cola	38
Diet Pepsi-Cola	37
Ginger ale	0
7-Up	0
Chocolate Products	
Cocoa, chocolate milk, 1 c	10–17
Milk chocolate, 1 oz	1–15
Chocolate syrup, 2 tb	4

Danish couples attempting pregnancy for 6 months, consumption of from 1 to 5 alcohol-containing drinks per week by women was related to a 39% lower chance of conception. Alcoholic beverage consumption of over 10 drinks per week was related to a 66% reduction in the probability of conception during the 6-month period.[40] Not all studies show alcohol intake affecting fertility, however; the effect, if real, may be weak. Nevertheless, because alcohol should not be consumed during pregnancy due to the risk of fetal malformations and mental retardation, it makes sense that women who are attempting conception should not consume alcohol.

Other Factors Contributing to Infertility in Males

The genesis of infertility in males and females is not well understood, but less is known about factors affecting male reproductive capacity than about those affecting females. Examinations of fertility status in males tend to focus on sperm quality, assessed as sperm number (concentration), motility (movement), and morphology (shape).[2] A number of chromosomal abnormalities (*Klinefelter's syndrome,* for example) and environmental toxins have been related to modification of sperm. Certain drugs used for hypertension and cancer, disease processes such as diabetes and atherosclerosis (hardening of the arteries), and endocrine disorders that disrupt testosterone production affect male fertility primarily in ways that involve sperm production or erectile function.

Sperm production appears to be sensitive to a male's exposure to a number of nutritional and other environmental and lifestyle factors (see Table 2.3). Because sperm development occurs over 70 to 80 days, however, there is a delay of up to 3 months between onset of the exposure (such as use of a medication or nutrient toxicity) and the appearance of sperm abnormalities.

Low Zinc Status   Zinc intakes of 5 mg per day or less are associated with decreased semen volume and testosterone levels in experimental studies. Zinc plays a critical role in male reproduction potential, and the concentration of zinc in semen is high. Because zinc is concentrated in semen, it is lost via ejaculated semen. Zinc in seminal fluid is an essential cofactor for enzymes involved in testosterone production, DNA replication, protein synthesis, and cell division. It appears to protect sperm from bacteria and chromosomal damage.

Zinc also plays important roles in sexual organ development in males. Deficiency of zinc prior to and during early adolescence is associated with *hypogonadism* and lack of sexual development, such as reduced growth of the penis, failure of the voice to deepen, and lack of muscle enlargement. Hypogonadism has been observed in some Middle Eastern males due to diets based on whole grains with infrequent consumption of animal products.

Zinc in whole grains is poorly absorbed, whereas animal products generally contain good amounts of absorbable zinc.

Supplementation with 10 mg of zinc or more per day reverses zinc-related infertility and sexual maturational problems in males over time.[2]

Klinefelter's Syndrome   A congenital abnormality in which testes are small and firm, legs abnormally long, and intelligence generally subnormal.

Hypogonadism   Atrophy or reduced development of testes or ovaries. Results in immature development of secondary sexual characteristics.

Antioxidant Nutrients   Sperm are particularly susceptible to oxidative damage because of their high concentration of polyunsaturated fatty acids.

Antioxidants such as selenium, vitamins C and E, beta-carotene, and other carotenoids protect sperm DNA from oxidative damage and promote normal sperm number, motility, and function.[41]

Vitamin D   Initial evidence indicates that inadequate vitamin D status is related to low sperm count and infertility in men. Adequate levels of vitamin D, whether derived from sunshine or fortified foods, may improve sperm counts in men with habitually poor vitamin D status.[42]

Alcohol Intake   Impaired fertility is common in alcoholics and is related to direct toxic effects of alcohol on the testes. Light to moderate alcohol consumption does not appear to affect fertility in males.[6]

Heavy Metal Exposure   Exposure to high levels of lead is related to decreased sperm production and abnormal sperm motility and shape.[43] Inhaled or ingested lead is transported to the pituitary gland, where it appears to disrupt hormonal communications with the testes. The result is lowered testosterone levels and decreased sperm production and motility. The men most likely to be exposed to excess lead tend to be workers in smelting and battery factories.[44]

Mercury can build up in fish living in contaminated waters. Ingestion of fish from waters contaminated with mercury in Hong Kong has been associated with decreased sperm count and abnormal semen.[45] Consumption of fish from the U.S. Great Lakes does not appear to pose similar problems.[46]

Exposure to excess levels of cadmium, manganese, boron, cobalt, copper, nickel, silver, or tin may also affect male fertility, but evidence from human studies is sparse. These metals may build up in male reproductive systems through the inhalation of fumes or dust containing particles or through contaminated water.[44]

Halogens   Occupational exposure to pesticides made from halogen compounds (such as dibromochloropropane) has been observed to cause sperm count reduction and male infertility.[44]

Case Study 2.2

Negative Lifestyle and Male Subfertility

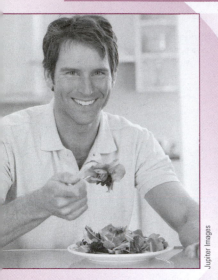

Jupiter Images

Pam and Scott were ready to start a family and thought conception would occur shortly after they stopped using contraception. Two years later they were still childless and had arranged to see a fertility specialist. The specialist could find no physical reason why Pam could not conceive, but found that Scott's sperm count was low and his sperm showed limited motility. Because Scott's BMI put him just into the obese range ($31 kg/m^2$), the doctor sent him to the staff dietitian.

A careful examination of Scott's dietary intake and physical activity level by the dietitian revealed very low intakes of vegetables and fruits (less than one serving a day on average), and a lack of fish, seafood, and milk consumption. His alcohol and coffee intake were low, as was his usual level of physical activity and the time he spent outside in the sun.

Questions

1. Based on the information provided in the case study, list three risk factors for subfertility related to Scott's lifestyle.
2. Assume Scott makes the necessary improvements to his diet and loses 15 pounds. Will those lifestyle improvements correct his subfertility?
3. Assume the primary reason for Scott's subfertility was inadequate availability of antioxidant nutrients for normal sperm development. Approximately how long will it take for improvements in sperm count and motility to occur after his antioxidant intake becomes adequate?

Glycols Glycols are widely used in antifreeze, solvents, and de-icers for airplanes. Occupational exposures to these compounds can decrease fertility in males threefold.[44]

Hormones Exposure to synthetic estrogens in pharmaceutical factories and to the estrogen-like substances of DDT (dichlorodiphenyltrichloroethane—an insecticide) PEs (phthalate exters—used as plasticizers) and PCBs (polychlorinated biphenyls—used in transformers) has been found to reduce libido and increase impotence and breast size in males.[47] It has been speculated that increased exposure to estrogenic compounds and other pollutants may be responsible for the reported decrease in sperm count in men over recent decades.

Is sperm count declining in men? The issue is not settled. Approximately 17% of males have low sperm counts, and several studies have noted declines in the number of sperm produced of about 1 to 2% per year over recent decades.[48] Some researchers have attributed declines in sperm counts to estrogenic environmental pollutants and increasing rates of obesity.[49] Other studies, however, have not identified declining sperm counts in populations studied.[48,50]

Scrotum A muscular sac containing the testes.

Heat Elevating the temperature of the *scrotum* and testes can reduce sperm count. Long-haul truck driving, welding, and foundry work may cause increased temperatures and decreased fertility in some men.[44] Prolonged and frequent exposure to hot-tub water is also thought to reduce sperm count in men.

Steroid Abuse Doses of steroids (e.g., testosterone) used by some bodybuilders are up to 40 times higher than therapeutic doses. Side effects of steroid abuse are multiple and include atrophy (shrinking) of the testicles, absence of sperm, and decreased libido. Fertility generally returns to normal after steroid use ends, but other disease risks may linger.[51]

Nutrition and Contraceptives

The contraceptive revolution emerged in full force in the 1960s when use of pills with heavy doses of estradiol (the most biologically active form of estrogen) and progesterone became widespread. The adverse side effects of these oral contraceptives were plentiful and included increased risk of heart attack and stroke, elevated blood lipids, glucose intolerance, weight gain, and folate and vitamin B_6 deficiencies. New generations of oral contraceptives employed increasingly lower hormone doses, and

Table 2.6 Nutrition-related side effects of contraceptives

Oral Contraceptives
- Increased blood levels of HDL cholesterol (the "good" cholesterol)
- Increased blood levels of triglycerides and LDL cholesterol
- Increased risk of venous thromboembolism (blood clots)
- Decreased blood levels of vitamin B_{12}
- Increased blood levels of copper

Contraceptive Injections (Depo-Provera)
- Weight gain
- Increased blood levels of LDL cholesterol and insulin
- Decreased blood levels of HDL cholesterol
- Decreased bone density

Contraceptive Implants (Norplant)
- Weight gain

side effects diminished substantially. Nevertheless, some remain. Those related to use of the newer contraceptives are listed in Table 2.6.

The current generation of oral contraceptives elevates some blood lipids and carries a slight risk of *venous thromboembolism*. Their continuing effect on the formation of blood clots appears to be related to the progesterone content of the pills.[52]

The newest generation of fertility control products for females includes contraceptive implants, patches, and injections. These, too, have a number of nutritional side effects. Development of hormonal contraceptive methods for control of fertility in males lags far behind advances in female contraception.

Oral Contraceptives and Nutritional Status

Many types of birth control pills containing low doses of different forms of estrogen and progesterone are currently available. Use of these pills by healthy, nonsmoking women is not associated with an increased risk of heart attack, glucose intolerance, or nutrient deficiencies, nor is use related to weight gain. Their use is, however, associated with a twofold increased risk of thromboembolism (blood-clot formation) and an increased risk of cervical cancer.[53]

The new generation of oral contraceptives increases blood levels of triglycerides by about 30% and total cholesterol levels by approximately 6% on average. HDL cholesterol—the "good" blood cholesterol fraction—is increased slightly by these contraceptives.[54]

Oral contraceptives still decrease blood levels of certain nutrients. Blood levels of vitamin B_{12} were found to be an average of 33% lower in oral contraceptive users

versus nonusers in a Canadian study of 14- to 20-year-old females.[55] Serum copper levels were 34 to 55% higher and may be related to the increased risk of blood-clot formation observed among oral contraceptive users.

Venous Thromboembolism A blood clot in a vein.

It is recommended that females who are obese, over the age of 35 years, and smoke, and those who have cardiovascular disease, hypertension, diabetes, or are immobilized, use nonhormonal methods of contraception due to their increased risk of venous thromboembolism.[54] Women who take oral contraceptive pills are cautioned against consuming more than a half-ounce of licorice per day. Genuine black licorice contains glycyrrhizic acid, a substance that gives licorice its distinctive taste. Consumption of several ounces of licorice can increase blood pressure and fluid retention in women using the pill.[57] It is generally also recommended that women stop using oral contraceptive pills about 3 months prior to attempting pregnancy.

Contraceptive Injections

DMPA (depot medroxyprogesterone acetate), or Depo-Provera, is the primary type of injectable contraception used in females. Injections that suppress ovulation are given every 3 months. Although highly effective, Depo-Provera has discontinuation rates that average over 50% within the first year of use. Weight gain is a leading reason for discontinuation (27%); irregular periods (24%), fatigue (23%), headache (25%), and abdominal pain (18%) are also commonly reported reasons for discontinuation.[58,59] Weight gain averages 12 pounds during the first year of Depo-Provera use,[59] but not all studies report weight gain.[60] Long-term use of this contraceptive is related to decreased bone density and blood levels of HDL cholesterol and increased levels of LDL cholesterol and insulin.[61]

Contraceptive Implants

Norplant (levonorgestral), the leading contraceptive implant, prevents contraception for up to 7 years.[62] This contraceptive method is highly effective for normal-weight and underweight women, as evidenced by a 1.9% pregnancy rate.[63] Pregnancy rates are 4.2% in obese women, however.[64] High rates of side effects, especially erratic bleeding (69%), weight gain (41%), and headaches (30%), lead to early removal of the implant in about half of users.[65] Average weight gain 1 year after the implant has been reported to be 9 pounds.[66]

Contraceptive Patches

The contraception patch releases a type of estrogen and progesterone. Tests indicate it is highly effective and easy to use. The patch is placed on the skin for 3 weeks and then taken off for a week.

Contraceptive patches increase blood levels of cholesterol and triglycerides to a greater extent than do oral contraceptives. On the positive side, use of the patch is related to slight increases in HDL-cholesterol levels. The most commonly reported adverse reactions to contraceptive patches are breast soreness, headaches, application-site reactions, and abdominal pain. Contraceptive patches are less effective in women weighing over 198 pounds (90 kg) than in women who weigh less. Women who should not use a contraceptive patch include those with a history of heart disease, stroke, blood clots, and reproductive cancers.[67] Pregnancy should be separated from use of the patch by at least 6 weeks. Contraceptive patches are relatively expensive. Three patches (good for a month) cost about $35, whereas generic oral contraceptive pills cost approximately $5 per month.

Emerging Forms of Contraceptives

Several types of male contraceptives are under development and testing. One of the products being tested consists of periodic injections of progestin and implanted crystallized testosterone. Initial tests show that the hormones effectively suppress sperm production and are 97% effective in preventing conception. Men receiving the progestin/testosterone contraceptive gained an average of 3% of their initial body weight during the 12-month treatment period. Additional studies on the safety of this new contraceptive are under way.[68] Contraceptive pills for males that suppress sperm production are now available.

New types of contraceptives such as the monthly vaginal ring, hormone-releasing IUDs, and contraceptive pills that limit menstrual cycles to four or fewer per year have become available.[69] Little is known about the nutrition side-effects of these forms of contraceptives.

Other Preconceptual Nutrition Concerns

Approximately 8 to 10 days after an ovum is fertilized, it implants into the uterine wall. Within the first month after conception, the developing *embryo* will have grown from a single cell to millions of cells, basic structures of organs will have formed, and the blueprint for future growth and development will have been established. All this often happens before women know they are pregnant or attend a prenatal clinic. The time to establish a state of optimal health and nutritional status is before conception.

Embryo The developing organism from conception through 8 weeks.

Fetus The developing organism from 8 weeks after conception to the moment of birth.

Neural Tube Defects (NTDs) Spina bifida and other malformations of the neural tube. Defects result from incomplete formation of the neural tube during the first month after conception.

Very Early Pregnancy Nutrition Exposures

Table 2.7 summarizes major nutritional exposures that adversely affect the growth and development of the embryo

---

Table 2.7 Nutritional exposures before and very early in pregnancy that disrupt fetal growth and development

Weight Status

- Underweight increases the risk of maternal complications during pregnancy and the delivery of small and early newborns.
- Obesity increases the risk of clinical complications during pregnancy and delivery of newborns with neural tube defects or excessive body fat.

Nutrient Status

- Insufficient folate intake increases the risk of embryonic development of neural tube defects.
- Excessive vitamin A intake (retinol, retinoic acid) increases the risk the fetus will develop facial and heart abnormalities.
- High maternal blood levels of lead increase the risk of mental retardation in the offspring.
- Iodine deficiency early in pregnancy increases the risk that children will experience impaired mental and physical development.
- Iron deficiency increases the risk of early delivery and development of iron deficiency in the child within the first few years of life.

Alcohol

- Regular intake of alcohol increases the risk of *fetal alcohol syndrome* and *fetal alcohol effects,* both of which include impaired mental and physical development.

Diabetes

- Poorly controlled blood glucose levels early in pregnancy increase the risk of fetal malformations, excessive infant size at birth, and the development of diabetes in the offspring later in life.

---

and *fetus.*[70] It is important to be aware that any of these conditions, if present preconceptionally, may impair embryonic and fetal growth and development. This chapter touches on the importance of adequate folate intake prior to conception. Chapter 4, on nutrition during pregnancy, expands the folate discussion and returns to the other topics listed in Table 2.7.

Folate Status Prior to Conception and Neural Tube Defects Folate status prior to conception is an important concern because inadequate folate very early in pregnancy can cause *neural tube defects* (NTDs). These defects develop within 21 days after conception—or before many women even know they are pregnant, and well before prenatal care begins.[71] Knowledge of the folate–neural tube defect relationship, and awareness that folate intake is inadequate in many women of childbearing age, prompted efforts to increase folate intake. In particular, efforts are focused on encouraging women to consume folic acid, a

Table 2.8 MyPyramid food group recommendations for preconception women with various levels of caloric need

	Food Group*					
Calorie need	Grains (ounces)	Vegetable (cups)	Fruits (cups)	Milk (cups)	Meat & Beans (ounces)	Oil (tsp)
1800	6	2.5	1.5	3	5	5
2000	6	2.5	2	3	5.5	5.5
2200	7	3	2	3	6	6
2400	8	3	2	3	6.5	7

*"Other" calories are allotted for desserts, sweets, and fats based on caloric need: 2000 calorie need = 195 other calories; 2,100 calorie need = 250; 2,200 calorie need = 290; and 2,400 calorie need = 360.

highly absorbable, synthetic form of this B vitamin. In 1998, that task was made easier when the Food and Drug Administration mandated that refined grain products such as white bread, grits, crackers, rice, and pasta be fortified with folic acid.

Nearly 40 countries now fortify refined grain products with folic acid, and rates of NTDs have fallen in each of these countries. Rates of NTDs in Nova Scotia, Canada, for example, fell 55% in the 2 years after folic acid fortification became mandatory. Declines in the incidence of NTDs in the United States are reported to be below those achieved in Canada, but reliable data on NTD rates for the United State are not yet available.[72] Countries in the European Union have yet to implement folic acid fortification; rates of NTDs in these countries have not changed.[73]

Intake of folic acid and folate status has increased substantially in most U.S. population groups since fortification. Since 1998, the prevalence of low levels of red blood cell folate (a marker of long-term folate status) has declined from 45 to 7%. Rates of low red cell folate levels continue to be high (20–21%) among non-Hispanic blacks, however.[72] Concern exists that some women are unaware of the importance of consuming folic-acid-fortified refined grain products and other sources of folate prior to pregnancy.[74] Although women of reproductive age have increased their median intake of folic acid by at least 100 mcg a day, only 39% of white women, 26% of black women, and 28% of Mexican American women have reached the 400-mcg-a-day target for folate intake.[74]

Women can get enough folate by consuming a good basic diet and a fully fortified breakfast cereal (Smart Start, Total, or Product 19, for example) or a regular breakfast cereal (Cheerios, Corn Flakes, Raisin Bran, etc.) and 6 to 8 servings of refined grain products each day. Folic acid supplements (400 mcg per day) can also provide folic acid.

Recommended Dietary Intakes for Preconceptional Women

Recommendations for food and nutrient intakes for women who may become pregnant differ from those for adult women in general in several ways. It is recommended that women who may become pregnant (1) consume 400 mcg of folate from fortified grain products, vegetables, fruits, or supplements; (2) take no more than 5000 IU of vitamin A (retinol or retinoic acid) from supplements daily; and (3) limit or omit alcohol-containing beverages.[70] Recommendations for nutrient intake given in the Dietary Reference Intakes (DRIs) should be applied, paying careful attention to the Tolerable Upper Intake Levels (see tables on the inside front covers of this text).

Food selection prior to pregnancy should be based on the MyPyramid recommendations (see Table 2.8). The number of servings recommended from each food group is given in cups, ounces, or teaspoons, and determined primarily by a person's caloric need. An example menu offering several options for food selection for meals and snacks is given in Table 2.9.

Model Preconceptional Nutrition Programs

This section highlights two model programs, one in the United States and one in Indonesia, related to nutrition during the preconceptional period.

Preconceptional Benefits of WIC

Women are eligible to enter the WIC program (Supplemental Nutrition Program for Women, Infants, and Children) when pregnant. In this model program in California, women were experimentally provided WIC food supplements and nutrition education during *and* between consecutive pregnancies. Women who received WIC benefits during one pregnancy and continuing through the first two months of the next pregnancy had better iron status and delivered newborns with higher birth weights and greater lengths than women who received WIC benefits during pregnancy only.[79] The study demonstrates that low-income women at nutritional risk benefit from WIC services before, as well as during, pregnancy.

Decreasing Iron Deficiency in Preconceptional Women in Indonesia

Approximately one in every two women in Indonesia experiences iron-deficiency anemia during pregnancy. In a unique effort to prevent this problem, the Ministry of Health initiated regulations that require a couple applying for a marriage license to receive advice on iron status from those dispensing the license. All women are now advised to take 30–60 mg of iron along with folic acid in a supplement. Of 344 women studied after the program was initiated, 98% reported that they had purchased and taken iron-folate tablets; 56% had taken at least 30 tablets. The incidence of iron deficiency in this group of women dropped by almost half.[76]

Preconception Care: Preparing for Pregnancy

"Each woman, man, and couple should make a reproductive life plan that includes whether and when they want to have children and how they will maintain their reproductive health."

Centers for Disease Control and Prevention, 2006

Increasingly, routine health care visits and educational sessions are being recommended and introduced into health care organizations. Services focus on risk assessment of behaviors such as weight status; dietary and alcohol intake; folate and iron status; and vitamin, mineral, and herbal supplement use, as well as on the presence of diseases such as diabetes, hypertension, infection, and genetic traits that may be transmitted to offspring. Psychosocial needs should also be addressed as part of preconceptional care, and referrals made to appropriate services for issues such as eating disorders, abuse, violence, or lack of food or shelter.[78] The desire of couples planning for pregnancy to have a healthy newborn makes the preconceptional period a prime time for positive behavioral changes. It presents opportunities to make lasting improvements in the health and well-being of individuals and families.

CDC's Preconception Health Initiative Efforts to boost the availability and utilization of preconception health care services now have the backing of the Centers for Disease Control and Prevention (CDC). In 2006, the CDC released a report highlighting recommendations for improving preconception health and health care services. The report was developed in response to the slow progress the United States had made in improving rates of poor pregnancy outcomes and in achieving the 2010 Health Objectives for the Nation related to preconception health. Additionally, the recommendations address the persistent problem of higher rates of poor pregnancy outcomes in African Americans, Hispanic Americans, and other groups compared to Caucasians.

Table 2.9 Example menu for preconceptional women that corresponds to MyPyramid recommendations. The menu works for men, too, if portion sizes are adjusted upward to meet caloric needs.

Breakfast

Fruit juice, 1 c
Fortified, whole grain breakfast cereal (hot or cold), 1 c
Sliced peaches or banana, 1 c
Skim or low-fat milk, 1 c
Coffee or tea

Lunch

Pork almond ding, 1 c, or taco salad with beans, chicken, or beef, 1 c
Rice, 1 c
Skim or low-fat milk, 1 c

Dinner

Veggie burger, lean beef, poultry, or fish, 3–4 oz
Macaroni salad, pasta, or potato, ½ c
Kale, turnip greens, collards, or spinach, 1 c
Tortilla, pita bread, corn-bread muffin, or whole grain roll, 1
Pumpkin or sweet potato pie, 1 slice, or ice cream, 1 c
Skim or low-fat milk, 1 c

Snacks

Yogurt with fruit
Graham crackers
Dried fruit
Nuts
Seeds
Apple, pear, mango, pineapple, or other fruit

The report concludes that preconception health care should be delivered at regularly scheduled primary care visits and include education about preconception health and pregnancy outcome; screening for vaccination, weight, iron and folate status; assessment of alcohol use; and management of disorders such as diabetes and celiac disease. The CDC recommends that preconception services include counseling to modify individual health behaviors that, if left unchanged, would negatively impact fertility and pregnancy outcomes. Initiatives such as this one are increasing public awareness of the importance of preconception health and of planning pregnancy.[77]

Starting pregnancy in the best health status possible can make an important difference to reproductive outcomes. It should be recognized, however, that even in ideal conditions, continued infertility, early pregnancy loss, fetal malformations, and maternal complications will sometimes occur.

Key Points

1. Preconceptional nutritional status influences maternal health and the course and outcome of pregnancy.

2. Dietary intake and body-fat composition affect the development and maintenance of a person's biological capacity to reproduce.

3. The primary affect of undernutrition on reproduction in women is the birth of small and frail infants who have a high risk of mortality early in life.

4. Acute undernutrition in previously well-nourished women is associated with a dramatic decline in their biological capacity to conceive.

5. High and low levels of body fat; high alcohol, soy, coffee, and fiber intake; and iron deficiency may impair fertility in females.

6. High and low levels of body fat; inadequate zinc, antioxidant, vitamin D, and DHA status; and heavy-metal exposure appear to reduce fertility in males.

7. Oral contraceptives increase levels of HDL-cholesterol, triglycerides, and LDL-cholesterol; they decrease blood levels of vitamin B_{12} and copper somewhat. Contraceptive injections and implants are associated with weight gain.

8. Adequate folate status prior to pregnancy substantially reduces the risk of neural tube defects in newborns. Over 60% of preconception women in the United States fail to consume the recommended amount of folate.

9. Low iron stores prior to pregnancy increase the risk of iron deficiency during pregnancy, preterm delivery, and low iron stores in the infant. Iron stores can be more effectively accumulated prior to rather than during pregnancy.

10. Adequate and healthy preconceptional diets for women and men are described by the MyPyramid Food Guide.

11. Preconception health services should be a part of primary health care and would likely improve fertility and pregnancy outcomes. Preconception health care components should include screening for folate, iron, and weight status, as well as identification and management of nutrition-related disorders such as diabetes and celiac disease. Interventions that improve poor lifestyle habits should be provided as a part of the care.

Resources

Vital Statistics

The National Center for Health Statistics provides information on fertility and birthrates, and other vital statistics data.
Website: www.cdc.gov/nchs

Merck Manual of Diagnosis and Therapy

This ever-popular and useful guide is available online and includes information on infertility and related conditions.
Website: www.merck.com/mrkshared/mmanual/home.jsp

Medscape Women's Health Journal

Receive automatic updates on women's health, fertility, and contraception topics by subscribing to this free online journal.
Website: www.medscape.com/medscape/WomensHealth/journal

The Babycenter Company

A popular source of consumer advice on preconception planning, nutrition, and related topics that is, however, heavily infiltrated with advertisements.
Website: www.babycenter.com

Journal Articles

Look up research articles related to any of the topics covered in this chapter through these cites.
Websites: www.ncbi.nlm.nih.gov/pubmed
www.nlm.nih.gov/medlineplus/femalereproductivesystem.html
www.cdc.gov/reproductivehealth/index.htm
http://familydoctor.org/x5042.xml
www.nlm.nih.gov/medlineplus/infertility.html

References

1. Faststats A to Z. www.cdc.gov/nchs/fastats, accessed 11 /06.

2. Wong WY et al. Male factor subfertility: possible causes and the impact of nutritional factors. Fertil Steril. 2000;73:435–42.

3. Panagopoulou E et al. Emotionally expressive coping reduces pregnancy rates in patients undergoing in vitro fertilization. Fertil Steril. 2006;86:672–7.

4. Ford WC et al. Increasing paternal age is associated with delayed conception in a large population of fertile couples: evidence for declining fecundity in older men. The ALSPAC Study Team (Avon Longitudinal Study of Pregnancy and Childhood). Hum. Reprod. 2000;15:1703–8.

5. Kovac P. Recurrent pregnancy loss. www.medscape.com/viewprogram/5293, accessed 4/06.

6. Hassan MA et al. Negative lifestyle is associated with a significant reduction in fecundity. Fertil Steril. 2004;81:384–92.

7. Powers HJ. Interaction among folate, riboflavin, genotype, and cancer, with reference to colorectal and cervical cancer. J Nutr. 2005;135:2960S–6S.

8. Bongaarts J. Does malnutrition affect fecundity? A summary of evidence. Science. 1980;208:564–9.

9. van der Spuy ZM. Nutrition and reproduction. Clin Obstet Gynecol. 1985;12:579–604.

10. MacDorman MF et al. Annual summary of vital statistics—2001. Pediatrics 2002;110:1037–50.

11. Rao VG, Sugunan AP, Sehgal SC. Nutritional deficiency disorders and high mortality among children of the Great Andamanese tribe. Natl Med J India. 1998;11:65–8.

12. Wood JW. Maternal nutrition and reproduction: why demographers and physiologists disagree about a fundamental relationship. Ann NY Acad Sci. 1994;709:101–16.

13. Hirschman C. Why fertility changes. Annu Rev Sociol. 1994;20:203–33.

14. Ramakrishnan U et al. Early childhood nutrition, education and fertility milestones in Guatemala. J Nutr. 1999;129:2196–202.

15. Pena R et al. Fertility and infant mortality trends in Nicaragua 1964–1993. The role of women's education. J Epidemiol Community Health. 1999;53:132–7.

16. Kolata GB. Kung hunter-gatherers: feminism, diet, and birth control. Science. 1974;185:932–4.

17. Leslie PW et al. Evaluation of reproductive function in Turkana women with enzyme immunoassays of urinary hormones in the field. Hum Biol. 1996;68:95–117.

18. Stein Z et al. Famine and human development: the Dutch hunger winter of 1944–1945. New York: Oxford University Press; 1975: pp. 119–48.

19. Pasquali R. Obesity and androgens: facts and perspectives. Fertil Steril. 2006;85:1319–40.

20. Maturitas. Obesity, fat distribution and infertility, 2006;54:363–71.

21. Reichlin S. Female fertility and the body fat connection (review). N Engl J Med. 2203;348:869–70.

22. Moran LJ, Norman RJ. The obese patient with infertility: a practical approach to diagnosis and treatment. Nutr Clin Care. 2002;5:290–7.

23. Williams NL et al. Evidence for a causal role of low energy availability in the induction of menstrual cycle disturbances during strenuous exercise training. J Clin Endocrinol Metab. 2001;86:5184–93.

24. Zaadstra BM et al. Fat and female fecundity: prospective study of effect of body fat distribution on conception rates. BMJ. 1993;306:484–7.

25. Bates GW. Body weight control practice as a cause of infertility. Clin Obstet Gynecol. 1985;28:632–44.

26. Keys A et al. The biology of human starvation. Minneapolis: University of Minnesota Press; 1950.

27. Cumming DC et al. Physical activity, nutrition, and reproduction. Ann NY Acad Sci. 1994;709:55–76.

28. Lloyd T et al. Interrelationships of diet, athletic activity, menstrual status, and bone density in collegiate women. Am J Clin Nutr. 1987;46:681–4.

29. Loucks AB et al. The female athlete triad: do female athletes need to tale special care to avoid low energy availability? Med Sci Sports Exerc. 2006;38:1694–1700.

30. Ireland ML, Nattiv A. The female athlete. Philadelphia: W.B. Saunders; 2003.

31. Hill PB et al. Gonadotropin release and meat consumption in vegetarian women. Am J Clin Nutr. 1986; 43:37–41.

32. Wyshak G, Snow RC. Fiber consumption and menstrual regularity in young women. J Women's Health, 1993;2:295–9.

33. Reichman ME et al. Effect of dietary fat on length of the follicular phase of the menstrual cycle in a controlled diet setting. J Clin Endocrinol Metab. 1992;74:1171–5.

34. Kurzer M. Hormonal effects of soy in premenopausal women. J Nutr. 2002; 132:570S–3S.

35. Kemmann E et al. Amenorrhea associated with carotenemia. JAMA. 1983;249:926–9.

36. Wentz AC. Body weight and amenorrhea. Obstet Gynecol. 1980;56:482–7.

37. Chavarro et al. Iron consumption and the risk of infertilty. Obstet Gynecol. 2006; 108:1145–1152.

38. Allen LH. Multiple micronutrients in pregnancy and lactation: an overview. Am J Clin Nutr. 2005;81:1 206S–1212S.

39. Bolumar F et al. Caffeine intake and delayed conception: a European multicenter study on infertility and subfecundity. European study group on infertility subfecundity. Am J Epidemiol. 1997;145:324–34.

40. Hatch EE, Bracken MB. Association of delayed conception with caffeine consumption. Am J Epidemiol. 1993;138:1082–92.

41. Jensen J et al. The prevalence and etiology of impotence in 101 male hypertensive outpatients. Am J Hypertens. 1999;12:271–5.

42. Eskenazi B et al. Antioxidant intake and sperm count and motility. Hum Repro 2005;20:1006–12.

43. Corbett S. Investigating vitamin D receptor activity in sperm from fertile and infertile men. Presentation at the 60th Annual Meeting of the American Society for Reproductive Medicine, Philadelphia, Oct. 17, 2004.

44. Benoff S et al. Blood lead levels and infertility in males. Hum Reprod. 2003; 18:374–83.

45. Lamb EJ, Bennett S. Epidemiologic studies of male factors in infertility. Ann NY Acad Sci. 1994;709:165–78.

46. Choy CM et al. Infertility, blood mercury concentrations and dietary seafood consumption: a case-control study. Brit J Obstet Gynecol. 2002;109:1121–5.

47. Buck GM et al. Consumption of contaminated sport fish from Lake Ontario and time-to-pregnancy. New York State Angler Cohort. Am J Epidemiol. 1997;146:949–54.

48. Rozata R et al. Role of environmental estrogens in the deterioration of male factor fertility. Fertil Steril. 2002;78:1187–94.

49. Sokol RZ et al. Comparison of two methods for the measurement of sperm concentration. Fertil Steril. 2000;73:591–4.

50. Sallmen M et al. Reduced fertility among overweight and obese men. Epidemiology. 2006;17:520–3.

51. Joffe M. Time trends in biological fertility in Britain. Lancet. 2000;355:1961–5.

52. Lloyd FH et al. Anabolic steroid abuse by body builders and male subfertility. BMJ. 1996;313:100–1.

53. Vandenbroucke JP et al. Oral contraceptives and the risk of venous thrombosis. N Engl J Med. 2001;344:1527–35.

54. Beral V et al. Oral contraceptives and the risk of cervical cancer. Lancet. 2003; 361:1159–67.

55. Crook D. The metabolic consequences of treating postmenopausal women with non-oral hormone replacement therapy. Br J Obstet Gynaecol. 1997;104Suppl16:4–13.

56. Green TJ et al. Oral contraceptives did not affect biochemical folate indexes and homocysteine concentrations in adolescent females. J Am Diet Assoc. 1998;98:49–55.

57. Berg G et al. Use of oral contraceptives and serum betacarotene. Eur J Clin Nutr. 1997;51:181–7.

58. Sellerberg U et al. Women who take the pill should limit consumption of licorice. www.medscape.com, accessed 2/02.

59. Polaneczky M et al. Early experience with the contraceptive use of depot medroxyprogesterone acetate in an inner-city clinic population. Fam Plann Perspect. 1996;28:174–8.

60. Matson SC et al. Physical findings and symptoms of depot medroxyprogesterone acetate use in adolescent females. J Pediatr Adolesc Gynecol. 1997;10:18–23.

61. Pelkman C. Hormones and weight change. J Repro Med. 2002;47 (suppl):791–4.

62. Shaarawy M et al. Effects of the long-term use of depot medroxyprogesterone acetate as hormonal contraceptive on bone mineral density and biochemical markers of bone remodeling. Contraception. 2006;74:297–302.

63. Sivin I et al. Prolonged effectiveness of Norplant® capsule implants: a 7-year study. Contraception. 2000;61:187–94.

64. Meirik O, Farley TM, Sivin I. Safety and efficacy of levonorgestrel implant, intrauterine device, and sterilization. Obstet Gynecol. 2001;97:539–47.

65. Kaunitz AM. Long-acting hormonal contraception: assessing impact on bone density, weight, and mood. Int J Fertil Womens Med. 1999;44:110–7.

66. Haugen MM et al. Patient satisfaction with a levonorgestrel-releasing contraceptive implant. Reasons for and patterns of removal. J Reprod Med. 1996;41:849–54.

67. Berenson AB et al. Contraceptive outcomes among adolescents prescribed Norplant implants versus oral contraceptives after one year of use. Am J Obstet Gynecol. 1997;176:586–92.

68. Sicat BL. Ortha Evra, a new contraceptive patch. Pharmacotherapy. 2003;23:472–80.

69. Turner L et al. Safety and efficacy of a male contraceptive. J Clin Endocrinol Meta. 2003;88:4659–67.

70. Shulman LP. Contraception today: risks and benefits in scientific, clinical, and socio-cultural context, www.medscape.com/viewprogram/6049, accessed 9/06.

71. Institute of Medicine. Nutrition during pregnancy. Washington, DC:National Academies Press; 1990.

72. Eskes TK. Open or closed? a world of difference: a history of homocysteine research. Nutr Rev. 1998;56:236–44.

73. Ganji V et al. Trends in serum folate, RBC folate, and circulating total homocysteine concentrations in the US. J Nutr. 2006;138:153–8

74. Czernichow S et al. Case for folic acid and vitamin B12 fortification in Europe. Semin Vasc Med. 2005;5:156–62.

75. Kuntz KM et al. Folate intake in U.S. falling short of established targets, Am J Public Health. 2006;96. www.medscape.com/viewartide/545509, 10/06.

76. Caan B et al. Benefits associated with WIC supplemental feeding during the inter-pregnancy interval. Am J Clin Nutr. 1987; 45:29–41.

77. Jus'at I et al. Reaching young Indonesian women through marriage registries: an innovative approach for anemia control. J Nutr. 2000;130:456S–8S.

78. Recommendations to improve preconception health and health care–United States, MMWR April 21, 2006.

79. Dixon D et al. The first prenatal visit. Clin Rv. 2000;10:53–74.

> "Women's nutritional status before conception influences physiologic events during pregnancy, and nutrition during pregnancy sets the stage for meeting nutritional needs during lactation."
>
> J. C. King[1]

Chapter 3

Preconception Nutrition:
Conditions and Interventions

Chapter Outline

- Introduction
- Premenstrual Syndrome
- Obesity and Fertility
- Hypothalamic Amenorrhea
- Female Athlete Triad and Infertility
- Eating Disorders and Fertility
- Diabetes Mellitus prior to Pregnancy
- Polycystic Ovary Syndrome
- Disorders of Metabolism
- Herbal Remedies for Fertility-Related Problems

Prepared by **Judith E. Brown**

Key Nutrition Concepts

1 Nutritional and health status before and during the first 2 months after conception influences embryonic development and the risk of complications during pregnancy.

2 Nutrition therapy plays an important role in the management of a number of conditions that affect preconceptional health and reproductive outcomes.

Introduction

This chapter addresses specific nutrition-related conditions of women before conception and during the *periconceptional period,* and the interventions that address them. Conditions presented here have important implications for health and well-being, and for reproductive outcomes. We begin with a discussion of premenstrual syndrome and progress to obesity, hypothalamic amenorrhea, the female athlete triad, eating disorders, diabetes, polycystic ovary syndrome, inborn errors of metabolism, and celiac disease. Then we address the role of herbal remedies in the treatment of disorders of menstruation and fertility.

Premenstrual Syndrome

It wasn't until 1987 that PMS, or *premenstrual syndrome,* moved from the psychogenic disorder section of medical textbooks to chapters on physiologically based problems. It is diagnosed according to criteria stipulated in the fourth version of the *Diagnostic and Statistical Manual of Mental Disorders (DSM IV).* A standard questionnaire, rather than physical examination or laboratory tests, is used to diagnose PMS. Common physical signs and psychological symptoms of PMS are listed in Table 3.1. For a diagnosis, at least five signs or symptoms of PMS intense enough to disrupt work or social life must be present in three consecutive luteal phases.

PMS is characterized by life-disrupting physiological and psychological changes that begin in the luteal phase of the menstrual cycle and end with menses (menstrual bleeding). Some symptoms of PMS occur in about 40% of women of childbearing age, and they are severe enough to be considered PMS in about 5 to 10% of these women.[2,3] Although up to 70% of women suffer through monthly bouts of lower abdominal cramps, bloating, back pain,

Periconceptional Period   Around the time of conception, generally defined as the month before and the month after conception.

Premenstrual Syndrome   (*premenstrual* = the period of time preceding menstrual bleeding; *syndrome* = a constellation of symptoms) A condition occurring among women of reproductive age that includes a group of physical, psychological, and behavioral symptoms with onset in the luteal phase and subsiding with menstrual bleeding. Also called premenstrual dysphoric disorder (PMDD).

Dysmenorrhea   Painful menstruation due to abdominal cramps, back pain, headache, and/or other symptoms.

headache, food cravings, or irritability, these conditions do not qualify as PMS. They are considered to represent *dysmenorrhea,* a condition related to prostaglandin release near to and during menses.[4]

Premenstrual dysphoric disorder (PDD) is a severe form of PMS. It is characterized by marked mood swings, depressed mood, irritability, anxiety, and physical symptoms (breast tenderness, headache, joint or muscle pain). PDD is diagnosed when three or more of these symptoms occur during at least two consecutive menstrual cycles. The treatment strategy for PDD is the same as that for PMS.[5]

The cause of PMS is unknown, but it is thought to be related to abnormal serotonin activity following ovulation.[2] Almost all remedies tested for PMS show about a 30% decline in symptoms with placebo, so an effective treatment must bring relief to an even higher proportion of women receiving it in experimental studies. Serotonin reuptake inhibitors, which are the active ingredient in some types of antidepressants, effectively reduce PMS symptoms. Decreased caffeine intake; exercise and stress reduction; magnesium, calcium, or vitamin B_6 supplements; and a number of herbal remedies are also used to treat PMS.[5]

Caffeine Intake and PMS

It is commonly recommended that women reduce their intake of coffee and other beverages high in caffeine to decrease PMS symptoms. This recommendation is holding up with time. A study at the University of Oregon demonstrated that PMS symptoms in college women increased in severity as coffee intake increased from 1 cup to 8 to 10 cups a day.[6] Risk of severe symptoms was eight times higher in women consuming the highest average daily amount of coffee (8 to 10 cups) compared to non-coffee drinkers.

Exercise and Stress Reduction

Increasing daily physical activity and reducing daily stressors appear to diminish PMS symptoms in many women.[7] Regular physical activity tends to improve energy level, mood, and feelings of well-being in women with PMS.

Table 3.1 Common signs and symptoms of PMS

Physical Signs	Psychological Symptoms
• Fatigue	• Craving for sweet or salty foods
• Abdominal bloating	• Depression
• Swelling of the hands or feet	• Irritability
• Headache	• Mood swings
• Tender breasts	• Anxiety
• Nausea	• Social withdrawal

Relieving stress—by such techniques as sitting comfortably and quietly with eyes closed while relaxing deep muscles, breathing through the nose, and exhaling while silently saying a word such as "one"—appears to decrease symptoms. When done for 15 to 20 minutes twice daily over 5 months, this exercise was associated with a 58% improvement in PMS symptoms.[7]

Magnesium, Calcium, Vitamin D, and Vitamin B₆ Supplements and PMS Symptoms

Magnesium, calcium, vitamin D, and vitamin B₆ supplements appear to decrease symptoms of PMS in many women. Mechanisms underlying improvements are incompletely understood.

Magnesium   Magnesium supplements of 200 mg per day given during two cycles have been shown to decrease swelling, breast tenderness, and abdominal bloating symptoms of PMS. The beneficial response to magnesium was seen during the second month of treatment.[8] The 200-mg daily dose of magnesium is below the Tolerable Upper Intake Level (UL) for magnesium of 350 mg daily and is therefore considered safe.

Calcium and Vitamin D   Calcium supplements of 1200 mg per day for three cycles were found to reduce the PMS symptoms of irritability, depression, anxiety, headaches, and cramps by 48%, versus a reduction of 30% in the placebo group. The effect of calcium increased with duration of supplement use.[9]

A recent study of women with PMS and control women demonstrated that vitamin D in addition to calcium may affect risk.[10] Women with PMS were found to have lower blood levels of vitamin D and calcium, as well as lower intakes, than women without the syndrome. The risk of developing PMS was 40% lower in women with average vitamn D intakes of 706 IU (17.7 mcg) per day compared to women consuming 112 IU (2.8 mcg). Daily calcium intakes of 1,283 mg were related to a 30% risk reduction compared to women consuming 529 mg. The UL for vitamn D for women is 50 mcg (2,000 IU), and 2,500 mg for calcium.

Vitamin B₆   The effectiveness of vitamin B₆ in alleviating symptoms of PMS is a matter of controversy. Initial studies were promising; but later studies showed no improvements in symptoms compared to placebo.[11] To help straighten out the controversy, a meta-analysis of studies on the use of B₆ for PMS was undertaken. This study concluded that daily doses of 50–100 mg appear to reduce the severity of premenstrual depressive symptoms.[12] Because of the potential benefit, and the low risk of harm, vitamin B₆ supplements are sometimes used in clinical practice to diminish PMS symptoms. The UL for vitamin B₆ is 100 mg/day; doses recommended should not exceed this level.

Obesity and Fertility

Rates of reproductive health problems related to excess body fat are increasing in the United States and other countries along with rising rates of obesity. In the United States, 24.2% of men and 23.5% of women are obese. The proportion of the population who are massively obese (BMI $\geq$40 kg/m^2) is increasing and now stands at 3%.[13]

Men who are obese are at risk of infertility due to low levels of testosterone and *sex hormone binding globulin (SHBG)*, and low sperm count (Table 3.2).[14] Obesity in women increases the risk of infertility due to highly irregular or anovulatory menstrual cycles. These conditions are likely related to high levels of estrogen and free testosterone that can accompany obesity. *Insulin resistance* frequently occurs along with obesity and contributes to adverse hormonal changes that affect fertility. Insulin can bind to specific receptors on the ovary and stimulate testosterone production. Androgens such as testosterone suppress follicular growth, leading to ovulatory dysfunction.[15]

In addition to infertility, obese women have higher rates of metabolic syndrome, polycystic ovary syndrome, gestational diabetes and hypertension, fetal overgrowth, cesarean delivery, and stillbirth than women who are not obese.[16] Metabolic syndrome is a common condition in obese men and women. It is referred to again in this and other chapters and is overviewed in the "In Focus" box on p. 70.

Irregular menstrual cycles contribute to a lack of early prenatal care in some obese women. Women may not be aware they have become pregnant because delayed onset of menstrual cycles may not be unusual.

> **Sex Hormone Binding Globulin (SHBG)**   A protein that binds with the sex hormones testosterone and estrogen. Also called steroid hormone binding globulin, because testosterone and estrogen are produced from cholesterol and are thus considered to be steroid hormones. These hormones are inactive when bound to SHBG, but are available for use when needed. Low levels of SHBG are related to increased availability of testosterone and estrogen in the body.
>
> **Insulin Resistance**   A condition in which cells "resist" the action of insulin in facilitating the passage of glucose into cells.

Table 3.2 Biological bases of infertility in obese men and women

Men	Women
• Low testosterone levels	• High estrogen and free-testosterone levels
• Reduced levels of sex hormone binding globulin	• Reduced levels of sex hormone binding globulin
• Elevated leptin levels	• Elevated leptin levels
• Low sperm count	• Irregular or anovulatory menstrual cycles

Central Body Fat and Fertility

The presence of central body obesity, indicated by a waist circumference of 35 inches or greater, is a risk factor for impaired fertility.[19] In a study of women attending an artificial insemination clinic due to infertility in their partners, conception occurred in just half as many women with high central body fat as in other women.[20] In general, it takes women with high central body fat stores longer to become pregnant than it does women with low levels of central fat.[21]

Weight Loss and Fertility

Weight loss should be the first therapy option for men and women who are obese and infertile.[22] (Read about one woman's experience with weight loss and fertility in Case Study 3.1 on the next page.)

Studies of both women and men have demonstrated that weight loss of 7 to 22 pounds in women with BMIs over 25 kg/m², and of 100 pounds in massively obese men, are related to a return of fertility in most study participants.[16,23] Weight loss in the studies of obese women was accomplished by diet and exercise; the study of massively obese men used gastric bypass surgery. Weight loss produced an increase in SHBG in males; it increased SHBG and decreased estrogen and insulin resistance in women.

Weight loss is considered the first therapeutic option for infertility in obese people, in part because it is less costly than medications and has many health benefits. This option is also recommended because hormone therapy often does not work in the presence of obesity.[24]

Pregnancy After Gastric Bypass Surgery   Gastric bypass surgery results in high levels of weight loss and restores fertility in many formerly very obese women. Women and men undergoing this surgery often do not become normal weight, but rather reach weights that place them within the overweight to slightly obese ranges. On average, women who attempt to conceive deliver a liveborn infant 27 months after the surgery. Rates of gestational diabetes and pregnancy induced hypertension are higher in post-surgery women than other women, and infants tend to weigh somewhat less than average at birth.[25]

Chronic Inflammation   Low-grade inflammation that lasts weeks, months, or years. Inflammation is the first response of the body's immune system to infection or irritation. Inflammation triggers the release of biologically active substances that promote oxidation and other potentially harmful reactions in the body.

Hypothalamic Amenorrhea   Cessation of menstruation related to changes in hypothalamic signals that maintain the secretion of hormones required for ovulation. Changes in hypothalamic signals appear to be triggered by an energy deficit. Also called functional amenorrhea and weight-related amenorrhea.

Hypothalamic Amenorrhea

One of the most common causes of anovulation and loss of menstrual cycles is *hypothalamic amenorrhea*.[26]

Previously called "weight-related amenorrhea," women affected by hypothalamic amenorrhea have no abnormalities of the hypothalamus, but its functions appear to be disrupted by an energy deficiency. Deficits in energy and possibly nutrients appear to disrupt hypothalamic signal that lead to normal secretion of gonadotropin-releasing hormone and luteinizing hormone. (These hormones are described in Chapter 2.) Women with hypothalamic amenorrhea may also be leptin-deficient due to low levels of body fat. These hormonal changes prevent ovulation.

The onset of hypothalamic amenorrhea is related to underweight, weight loss, or weight loss accompanied by intense exercise. It usually occurs in women engaged in intellectual professions or those exposed to social stress. A minority of cases of hypothalamic amenorrhea appear to be related to high levels of social stress accompanied by subtle deficits in calorie intake. It is often preceded by menstrual irregularities lasting months to years.[27]

Case Study 3.1

Anna Marie's Tale

Exercise can be bad for you—or at least it is for Anna Marie. She and her husband Mark already have two delightful children, full-time jobs, and hectic schedules. Mark wants more children, but Anna Marie is dead set against it. Mark refuses to use contraception and has made Anna Marie promise not to use any, either. Anna Marie makes the promise because she thinks she can avoid pregnancy by staying at her weight of 210 pounds. At this weight, Anna Marie seldom has a menstrual period and figures the odds of conception are slim. For 2 years, Anna Maria's plan for avoiding conception has worked.

Now that the children are older, Anna Maria finds she has a bit of free time, which she uses to indulge her love of swimming. Within months of beginning her program of swimming regularly, however, Anna Marie abandons it. Her menstrual periods have become regular, and her contraception method is lost.

Anna Marie's weight at 210 pounds has remained stable during the months she has been swimming. It appears that her improved level of physical fitness and body fat has improved her fertility status.

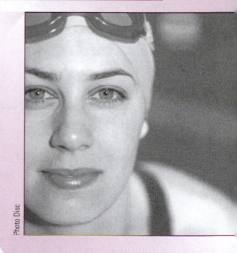

Photo Disc

Questions

1. What was likely the reason for Anna Marie's infertility when she was inactive?
2. Give an example of a hormonal change that may have occurred after Anna Marie began exercising regularly.
3. Name a possible health consequence related to Anna Marie's high weight and lack of menstrual cycles.

Nutritional Management of Hypothalamic Amenorrhea

"...let's not tread into new expensive treatments without correction of nutrient deficiencies or without first attempting to modify the dietary intake to meet all the energy and nutrient needs of the patient."

Fima Lifshitz, MD[28]

Fertility can be restored in underweight and energy-deficient women by hormonal therapy. Risks of pregnancy and newborn complications, as well as health care costs are higher, however, if this approach is taken. The first treatment approach should be weight gain accomplished by the consumption of a healthful diet. Weight gains of 6 to 11 pounds (3 to 5 kg) are usually sufficient to restore fertility and improve the outcome of pregnancy.[16]

The Female Athlete Triad and Infertility

Improved opportunities for female participation in sports in recent decades have been followed by an upsurge in the number of females who are competitive athletes. Although a healthy trend, some who are involved in sports that emphasize a lean body type are compromising their health. Very high levels of physical activity combined with negative caloric balance can place women at risk of developing the "female athlete triad." It is called a triad because it consists of three major conditions: amenorrhea, disordered eating, and osteoporosis. Amenorrhea associated with the female athlete triad appears to be triggered when energy intake is about 30% less than energy requirement.[29] This level of energy deficit leads to a loss of normal secretion of luteinizing hormone, a subsequent lack of estrogen production, and other hormonal changes also seen in hypothalamic amenorrhea. Metabolic changes triggered by hormonal shifts result in decreased bone density and an increased susceptibility to stress fractures in affected athletes.[30]

Nutritional Management of the Female Athlete Triad

Treatment of the female athlete triad focuses on correction of the negative energy balance and associated eating disorders, and on restoration of bone mass accretion. Peak bone mass is established before age 30, so it is particularly important that interruptions in bone development be short in duration. Vitamin D, calcium, and other supplements may be needed in addition to a balanced and adequate diet to facilitate bone development.[31]

Eating Disorders and Fertility

Both *anorexia nervosa* and *bulimia nervosa* are related to menstrual irregularities and infertility.[32] These disorders affect about 3 to 5% of young women, and likely twice that many have clinically important symptoms related to these eating disorders.[33] Amenorrhea is a cardinal manifestation of anorexia nervosa, and little bleeding during menses (oligomenorrhea) or amenorrhea may occur in women with bulimia nervosa. Amenorrhea in anorexia nervosa is related to irregular release of GnRH and very low levels of estrogen. Menses generally returns upon weight gain, but some cases of infertility persist even after normal weight is attained. This effect may be related to continued low levels of body fat, low dietary fat intake, excessive exercise, or other factors.[33]

Food binges and crash diets associated with bulimia nervosa are related to low FSH and LH levels, menstrual disturbances, and infertility.[32]

Anorexia Nervosa (*anorexia* = poor appetite; *nervosa* = mental disorder) A disorder characterized by extreme underweight, malnutrition, amenorrhea, low bone density, irrational fear of weight gain, restricted food intake, hyperactivity, and disturbances in body image.

Bulimia Nervosa (*bulimia* = ox hunger) A disorder characterized by repeated bouts of uncontrolled, rapid ingestion of large quantities of food (binge eating) followed by self-induced vomiting, laxatives or diuretic use, fasting, or vigorous exercise in order to prevent weight gain. Binge eating is often followed by feelings of disgust and guilt. Menstrual cycle abnormalities may accompany this disorder.

Teratogenic Exposures that produce malformations in embryos or fetuses.

Congenital Abnormality A structural, functional, or metabolic abnormality present at birth. Also called congenital anomalies. These may be caused by environmental or genetic factors, or by a combination of the two. Structural abnormalities are generally referred to as congenital malformations, and metabolic abnormalities as inborn errors of metabolism.

Insulin A hormone produced by the pancreas that regulates the movement of glucose from the bloodstream into cells within organs and muscles.

Nutritional Management of Women with Anorexia Nervosa or Bulimia Nervosa

The primary therapeutic goal for anorexia nervosa is normalization of body weight, and for bulimia nervosa, normalization of eating behaviors.[34] Recommended treatment for anorexia nervosa involves long-term, multidisciplinary services. Hospitalization may be required in severe cases. Certain psychotherapeutic medications are moderately effective for treating bulimia nervosa, but cognitive-behavioral therapy is the best, established approach.

Diabetes Mellitus prior to Pregnancy

Many women with diabetes mellitus are unaware that the disorder increases the risk of maternal and fetal complications, and they fail to get blood glucose under excellent control prior to conception.[35] High blood glucose levels during the first 2 months of pregnancy are *teratogenic;* they are associated with a two- to threefold increase in *congenital abnormalities* in newborns. Exposure to high blood glucose during the first 2 months in utero is related to malformations of the pelvis, central nervous system, and heart in newborns, as well as to higher rates of miscarriage.[36]

Management approaches to blood glucose control in diabetes depends, in part, on the type of diabetes present. Women may have *type 1 diabetes* (characterized by onset before the adult years) or *type 2 diabetes,* which most often occurs in adults. Once thought of as a disease of older adults, type 2 diabetes is becoming increasingly common in young adults due to the obesity epidemic. Approximately 4 in 5 U.S. adults with type 2 diabetes are overweight or obese.[37] Background information on diabetes is presented in "In Focus" box on page 73.

A Closer Look at Insulin Resistance

Insulin is a hormone produced by the beta cells of the pancreas. It performs a variety of functions, one of which is to increase the passage of glucose from blood into cells by attaching to and stimulating receptors on cell membranes. When insulin is bound to these receptors, enzymes are activated that open cell membrane "doors" to the passage of glucose. Each cell membrane normally contains 20,000 active insulin receptors. With insulin resistance, only about 5000 receptors are functional.[38] With insulin resistance, cells "resist" this process, and that lowers the amount of glucose that is transported into cells. The limited transport of blood glucose into cells triggers the pancreas to produce and release additional insulin. If the capacity of the pancreas to produce increasing quantities of insulin needed for glucose transport into cells is exceeded, blood glucose levels become and remain abnormally high.

The incidence of insulin resistance in adults and children is increasing in the United States and many other countries and is now considered to be a major public health problem.[39] Most cases of insulin resistance result from an interaction of genetic and environmental factors. Obesity (especially large stores of central body fat), physical inactivity, and small size at birth are related to insulin resistance. The condition is associated with the

In Focus

Diabetes

Definition: There are three major types of diabetes mellitus: type 1, type 2, and gestational. All types of diabetes are characterized by abnormally high blood glucose levels, or fasting levels of 126 mg/dL (7 mmol/L) or higher. People with diabetes are considered to be "carbohydrate intolerant" because carbohydrate consumption tends to raise blood glucose levels.

Type 1 diabetes is carbohydrate intolerance resulting from destruction of insulin-producing cells of the pancreas. Individuals with type 1 diabetes require an external supply of insulin. It is considered an *autoimmune disease*. About 10% of cases of diabetes are of this type.

Type 2 diabetes is carbohydrate intolerance due to the body's inability to use insulin normally, to produce enough insulin, or both. About 90% of cases of diabetes are of this type.

Gestational diabetes is carbohydrate intolerance that begins or is first recognized during pregnancy. It is closely related to type 2 diabetes. (Additional information about gestational diabetes is covered in Chapter 5.)

Prevalence: 9.4% of U.S. adults have type 2 diabetes and less than 1% of youth and adults have type 1. Gestational diabetes is diagnosed in 3–7% of pregnancies.

Rates of all types of diabetes are increasing worldwide. Increased rates of type 2 and gestational diabetes are related to esclating, global rates of obesity.

Major physiological aspects and consequences: Major signs and symptoms of diabetes are frequent urination, increased thirst and fluid intake, increased appetite, and elevated blood glucose levels.

People with type 2 and gestational diabetes are generally obese and have insulin resistance. Insulin is the principal hormone that regulates the transfer of glucose from blood into muscle and fat cells. Its presence causes the liver to produce more glucose than normal and it decreases body fat breakdown.

Elevated blood glucose levels in people with diabetes have many adverse consequences in the short- and long-terms. Such consequences can be limited or postponed by tightly managed blood glucose levels. Diabetes increases the risk of coronary heart disease, kidney disease, vision problems and blindness, nerve problems, and loss of limbs.

Type 1 and 2 diabetes are chronic diseases that require lifelong management. Patient education, nutritional support, and self-glucose monitoring are important for long-term blood glucose control. Some people with type 2 diabetes can control the disease through dietary and physical activity changes and weight loss.

Risk factors: Exposure to certain infectious diseases, drugs, other environmental agents, and vitamin D inadequacy early in life can trigger the onset of Type 1 diabetes in genetically susceptible individuals. Obesity, especially central obesity, is the main risk factor for type 2 and gestational diabetes.

---

development of *polycystic ovary syndrome (PCOS)*, *metabolic syndrome*, type 2 diabetes, *gestational diabetes*, and heart disease.[40]

Nutritional Management of Type 2 Diabetes

"There is a substantial gap between recommended diabetes care and the care patients actually receive in the health care setting."[41]

Some people with type 2 diabetes can manage their glucose levels with diet and exercise, whereas others will need an oral medication that increases insulin production or sensitivity, or insulin to further boost glucose absorption into cells. Some types of oral medications cannot be used during pregnancy because they also increase insulin in the fetus and cause excessive fetal growth and fat gain.[36]

Individualized diet and exercise recommendations and an educational and follow-up program developed and implemented by registered dietitians, certified diabetes educators (CDE), physicians, and nurses are preferred for diabetes management. Carefully planned and monitored dietary recommendations are a major component of the management of type 2 diabetes.

Individual blood glucose levels vary a good deal in response to diet composition, so dietary prescriptions must be tailored for every person.

For patients with type 2 diabetes, the American Diabetes Association's guidelines recommend:

- Weight loss of 7% of body weight or more
- 150 or more minutes per week of physical activity. Resistance training should be included.
- Percent of total calories from the energy nutrients: 15–20% protein, <30% fat, and carbohydrates, approximately 50%

Autoimmune Diseases Diseases that result from a failure of an organism to recognize its own constituent parts as "self." The organism attempts to defend itself from the foreign substance through actions of its immune system. These actions can damage molecules, tissues, and organs. Type 1 diabetes, lupus, and rheumatoid arthritis are examples of autoimmune diseases.

Polycystic Ovary Syndrome (PCOS) (*polycysts* = many cysts; i.e., abnormal sacs with membranous linings) A condition in females characterized by insulin resistance, high blood insulin and testosterone levels, obesity, polycystic ovaries, menstrual dysfunction, amenorrhea, infertility, hirsutism (excess body hair), and acne.

- Percent of total calories from saturated fat: <7%
- Percent of total calories from *trans* fat: as low as possible
- Restriction of cholesterol intake to 200 mg per day or less
- 14 g fiber per 1000 calories of food intake
- Whole grains should comprise half of all grain intake
- Low glycemic-index foods that are rich in fiber and other important nutrients should be encouraged.

There is no evidence to support prescribing diets such as "no concentrated sweets" or "no added sugar."[42]

Glycemic Index and Glycemic Load *Glycemic index (GI)* is a measure of the extent to which 50 grams (about 1¾ ounces) of carbohydrate-containing foods raise 2-hour postprandial blood glucose levels compared to 50 grams of glucose or white bread. Not all expert committees on diabetes recommend the use of low-GI foods as a primary strategy in the dietary management of diabetes.[37,43] However, mounting evidence indicates that low-GI diets are beneficial for the control of blood glucose and insulin levels. Diets that provide low-GI carbohydrates and approximately 30 grams of fiber daily are associated with reduced blood levels of glucose, insulin, and triglycerides versus lower-fiber (15 g/day), high-GI diets. (Table 5.7 in Chapter 5 lists the GI of a variety of foods.) Foods with high GI raise blood glucose and insulin levels more than do foods with low GI, and high-GI foods lead to more episodes of hyperglycemia (high blood glucose level) than do diets providing mainly low-GI carbohydrates. Long-term consumption of diets providing mainly high-GI carbohydrate foods appear to increase the risk of obesity, type 2 diabetes, and coronary heart disease.[44] Although some people have trouble adhering to low-GI diets, they appear to be practical and acceptable in motivated adults with type 2 diabetes.[37]

Glycemic Load The blood-glucose-raising effect of carbohydrate-containing foods also depends on the amount of food consumed. So, consuming a high-GI food such as instant mashed potatoes, in small amounts, will raise blood glucose level less than consuming a larger amount of this type of potato. *Glycemic load (GL)* was established to adjust GI values according to the amount of food consumed.

> **Glycemic Index (GI)** A measure of the extent to which blood glucose levels are raised by a specific amount of carbohydrate-containing food conpared to the same amount of glucose and white bread.
>
> **Glycemic Load (GL)** A measure of the extent to which blood glucose levels are raised by a specific amount of carbohydrate-containing food. It is calculated by multiplying the carbohydrate content of an amount of food consumed by the glycemic index of the food, and dividing the result by 100.

The GL of a food is obtained by multiplying the GI of the food by the grams of carbohydrate contained in the amount of food consumed, and then dividing the result by 100. A blueberry muffin, for example, has a GI of 59. If a person ate a 1-ounce muffin containing 11 grams of carbohydrate, the GL would equal (59 × 11)/100, or 6.5. If she ate a 3-ounce muffin with 33 grams of carbohydrate, then the GL would be (59 × 33)/100, or 19.5. The higher the GL of a food, the greater the expected elevation in blood glucose and insulin due to consumption of the food.

Other Components of the Nutritional Management of Type 2 Diabetes

Individuals with diabetes are at risk for heart disease due to abnormal blood lipid levels. Consequently, diets recommended include foods that improve blood lipid concentrations. Diets that help lower high-LDL cholesterol without lowering concentrations of beneficial HDLcholesterol, and that lower elevated triglyceride levels, follow these guidelines:

- Provide fat mainly in the form of monounsaturated fatty acids (vegetables, olive oil, peanuts and peanut oils, nuts, and seeds).
- Keep intake of saturated fats from animal products below 7% of total calories.
- Limit cholesterol intake to less than 200 mg per day.
- Minimize intake of trans fats from bakery products, fried foods, and snack foods.

Fish intake is often encouraged among people with diabetes. Consuming 2–3 servings of fish per week, or taking fish oil supplements, lowers blood triglycerides in people with elevated levels.[37] A sample menu that incorporates nutritional criteria for type 2 diabetes is given in Chapter 5.

Chromium plays critical roles in glucose and lipid metabolism through its potentiation of insulin action. Trial runs of supplementation with 500–1000 micrograms of chromium daily may be used to test its effects on glucose tolerance. Magnesium supplementation is indicated for people with diabetes who have low serum magnesium levels.[37]

Weight Loss and Type 2 Diabetes Weight loss in overweight and obese individuals with diabetes lowers blood glucose levels and blood pressure, improves blood lipid concentrations, and increases insulin sensitivity. The most effective approach to weight loss in people with diabetes tested so far combines reduced caloric intake with exercise and behavioral therapy. Both aerobic and strength-building exercises are recommended. Behavioral therapy generally includes self-monitoring of body weight, diet, and physical activities; stress management; and problem-solving skill building. Selection of acceptable foods and dietary patterns, realistic goals for weight loss, enjoyable physical activities, and helpful behavioral therapies are critical to the long-term success of weight-control efforts. Diet drugs and surgery may be indicated for individuals who are unable to lose weight through behavioral changes.[45]

A concerted effort will be needed to translate new knowledge about diet and nutrition into improved treatment for and prevention of diabetes and its complications.[46]

Reducing the Risk of Type 2 Diabetes

Women who are overweight and have from borderline to high blood glucose levels (or *impaired glucose tolerance*) may be able to reduce their risk of developing type 2 diabetes or gestational diabetes during pregnancy. Modest levels of weight loss are consistently associated with reduced risk of type 2 diabetes.[47] One large study taking place over 3 years found that losses in body weight of about 7% and 150 minutes per week of exercise reduced the risk of developing type 2 diabetes by 50%.[48]

Whether weight loss and dietary changes before pregnancy in women at risk of type 2 diabetes reduce the chances that the disorder will develop during pregnancy is not known. However, improvements in body weight, dietary intake, and physical activity level would confer health benefits and are low-risk interventions. They should be considered for women who are overweight and have impaired glucose tolerance prior to pregnancy.

Polycystic Ovary Syndrome

Case scenario: Lupe is a 28-year-old woman who is 5 feet 3 inches tall and weighs 208 pounds. She and her husband want to start a family, but her highly irregular periods and a failure to become pregnant as soon as desired have brought her to see her doctor. At the clinic it is determined that Lupe's waist circumference is 38 inches, and her body mass index (BMI) is 37 kg/m². Laboratory tests show that her blood level of insulin and triglycerides are high, and that she is not ovulating. Lupe is diagnosed as having *polycystic ovary syndrome,* or PCOS.

About 10% of women of reproductive age have PCOS, and it is the leading cause of female infertility.[49] PCOS is not a disease but rather a syndrome that consists of a variety of clinical signs. It is considered by some experts to be a sex-specific form of metabolic syndrome.[50] Most women with PCOS are infertile due to the absence of ovulation, and they have menstrual irregularities. Characteristically, the outer layer of the ovaries of these women is thick and hard, and it may look yellowish.

Due to the hard covering on the ovaries, follicles don't break open to release the egg, so ovulation does not occur.[49]

Many women with PCOS are obese, and even in women with PCOS who are not obese, levels of intra-abdominal fat are usually high.[38] Some women with PCOS have excess body hair (hirsutism), acne, and high blood levels of insulin, triglycerides, and androgens; and they have low levels of HDL cholesterol (Table 3.3). PCOS is sometimes difficult to diagnose (and may therefore not be treated) because signs and symptoms of the disorder vary considerably among individual women.[51]

The cause of PCOS is still debated, but it looks more and more as though insulin resistance is a pivotal factor in most cases regardless of body weight.[51] Less commonly, PCOS is caused by androgen-secreting tumors in the ovaries or adrenal gland, other disorders, and certain medications.[52] High blood levels of insulin stimulate the ovaries to produce androgens (such as testosterone), and excess androgens disrupt development of follicles.[15] High blood levels of androgens also lead to excess hair growth on the face and other parts of the body, while high insulin levels raise triglyceride and lower HDL-cholesterol levels.[54]

PCOS appears to have a genetic component, its development is influenced by environment-gene interactions. It tends to run in families where females have a history of infertility, menstrual problems, type 2 diabetes, central obesity, and hirsutism. Inutero exposures that affect fetal gene programming may be a factor in its development. Although obesity does not cause PCOS, it exacerbates reproductive and metabolic problems associated with it. Rates of PCOS in teens and women increase with increasing rates of overweight and obesity in females.[50] Women with PCOS are at increased risk of gestational and type 2 diabetes and cardiovascular disease.[51]

Table 3.3 Variation in clinical signs associated with PCOS[38,51,53]

Clinical Sign	Percent of Women with PCOS Affected
Menstrual irregularities	90%
Polycystic ovaries	67–86%
Excess abdominal fat	80%
Insulin resistance	80%
Overweight, Obesity	80%
Abnormal facial and body hair	70%
High testosterone levels	70%
Infertility	70%
Low HDL-cholesterol levels	64%
High triglycerides	47%

Nutritional Management of Women with PCOS

"Many women with PCOS have their symptoms ignored for years and have never received the proper diagnosis and treatment. She may have been told 'just lose weight!' and admonished when she found that very hard to do."

Martha McKittrick, RD, CDN, CDE[54]

The primary goal in the treatment of PCOS is to increase insulin sensitivity. A number of insulin-sensitizing drugs, such as metformin and rosiglitazone, can be used to lower blood insulin levels and reduce the excess production of androgens by the ovaries.[55] Other drugs may be used to stimulate ovulation. For women with PCOS who are obese, weight loss and exercise are primary components of therapy. Weight loss and exercise improve insulin sensitivity, benefit blood lipids and insulin levels, and lower fasting glucose and testosterone levels in women with PCOS. However, women with PCOS have a more difficult time losing weight than the general population. Care must be taken to individualize eating and exercise plans if weight loss is to succeed. PCOS is a long-term health problem that requires a sustainable approach to weight loss and exercise. In addition, women may benefit from knowing more about PCOS, long-term health risks, and why weight loss and exercise are needed.[54]

Dietary recommendations for teens and women with PCOS emphasize marine sources of the omega-3 fatty acids eicosapentaenoic acid (EPA) and docosahexaenoic acid (DHA) or fish oils, whole grains, fruits, and vegetables high in antioxidants and fiber, regular meals, nonfat dairy products, and low-GI-carbohydrates.[38,56] EPA and DHA increase insulin sensitivity. Basic foods such as whole grains and high-fiber, low-GI-carbohydrates are encouraged to limit blood glucose surges and insulin production. Weight loss is recommended (if needed) as is aerobic and strength-building exercise (30 minutes or more per day). If drugs are used to treat the symptoms of PCOS they should be used in conjunction with diet and exercise.[16]

Clinicians have observed that women with PCOS crave sweets and, because they are highly desired, recommend that they not be totally excluded.[51] Females with PCOS may be scared to eat and develop eating disorders out of fear of getting diabetes.[38]

Symptoms of PCOS tend to improve substantially with weight loss of 5–10% of initial body weight.[54] Symptoms generally worsen if diet, weight loss, and exercise

Phenylketonuria (PKU) An inherited error in phenylalanine metabolism most commonly caused by a deficiency of phenylalanine hydroxylase, which converts the essential amino acid phenylalanine to the nonessential amino acid tyrosine.

Celiac Disease An autoimmune disease that causes malabsorption due to an inherited sensitivity to the gliadin portion of gluten in wheat, rye, and barley. It is often responsible for iron, folate, zinc, and other deficiencies. Also called celiac sprue and nontropical sprue.

recommendations are not followed.[51] Many women with PCOS are able to modify eating and exercise behaviors and have children.[38]

Counseling for women with PCOS should be supportive and convey a real understanding of the difficulties the woman has probably been through in getting appropriate advice and help in the past.[54]

Disorders of Metabolism

Two metabolic disorders that affect embryonic development or fertility are covered here: *phenylketonuria (PKU)* and *celiac disease.*

Phenylketonuria (PKU)

Phenylketonuria derives its name from the characteristic presence of phenylalanine in the urine of people with this condition. PKU is an inherited problem that causes elevation in blood phenylalanine levels due to low levels or lack of the enzyme phenylalanine hydroxylase. Lack of this enzyme diminishes the conversion of the essential amino acid phenylalanine to tyrosine, a nonessential amino acid, and causes phenylalanine to accumulate in blood. If present during very early pregnancy, high blood levels of phenylalanine impair normal central nervous system development of the embryo. Elevated phenylalanine levels in the first 8 weeks of pregnancy increase the risk of heart defects. The risk increases if high blood levels of phenylalanine are combined with low-protein diets early in pregnancy.[57] Untreated women with PKU have a 92% chance of delivering a newborn with mental retardation, and a 73% chance that the infant will be born with an abnormally small head (microcephaly).[58,59]

People with undiagnosed PKU may self-select low-protein foods because meat and other rich sources of protein make them light-headed and easily confused. They may find it difficult to comprehend all the information they receive after the inborn error is diagnosed.[60]

Nutritional Management of Women with PKU

PKU can be successfully managed by a low-phenylalanine diet, instituted and monitored with the help of an experienced registered dietitian. This diet should be followed throughout life, but it is critical that it be adhered to prior to conception and maintained throughout pregnancy. It usually takes about 4 to 6 months to learn and implement the PKU diet and to lower blood phenylalanine levels.[60] Women who establish normal blood phenylalanine levels before pregnancy and maintain them throughout pregnancy tend to deliver infants with normal intelligence. The later in pregnancy that control is achieved, the more severe mental retardation and other problems become.[58,59]

Case Study 3.2

Celiac Disease

Chloe, age 30, has not had a period for over 2 years. Her gynecological exam turns up no abnormalities, but the hormones she is given to stimulate her menstrual periods do not work. Since the age of 10, Chloe has had painful stomach cramps, frequent diarrhea or constipation, and periodic iron-deficiency anemia. Multiple visits to doctors have failed to find the cause of Chloe's health problems. At around the age of 20, Chloe had begun to wonder if she was a medical anomaly or a hypochondriac.

Alamy

Still bothered by her health problems and about to be married, Chloe seeks care again. This time she is seen by a nurse practitioner who has just read an article on celiac disease. The nurse sends Chloe to a registered dietitian, who advises Chloe on a gluten-free diet. After faithfully following the diet for a week, Chloe feels better. The cramps, diarrhea, and constipation are much improved, and later on, her menstrual cycles return. She returns to her doctor for a checkup and requests a test for celiac disease. By that time, however, her intestinal biopsy comes back normal because she has been on the diet for months.

Questions

1. What should have been the first clue that Chloe might have celiac disease?
2. What facts provide other clues to the possibility of celiac disease?
3. How long will Chloe have to stay on a gluten-free diet?

Celiac Disease

Celiac disease, highlighted in the "In Focus" box on page 78, is becoming increasingly recognized as a cause of infertility and reproductive health problems in men and women. In males, celiac disease is related to alterations in the actions of androgens, delayed sexual maturation, and hypogonadism. Hypogonadism is marked by a deficiency of sex hormones and poor development and functioning of the reproductive system. In females, celiac disease is associated with amenorrhea, increased rates of miscarriage, fetal growth restriction, low birth-weight deliveries, and a short duration of lactation. It is hypothesized that the effects of celiac disease on reproductive functions in males and females is related to malabsorption-induced deficiencies of nutrients such as vitamins A, E, and D; folate; and iron.[61]

Specific causes of the abnormal reproductive functions observed in people with celiac disease are yet to be clarified. Normal reproductive functions return after celiac disease has been stabilized with a gluten-free diet.[61]

Not all individuals with celiac disease have overt symptoms, so it may be missed as a underlying cause of infertility. It should be considered in unexplained cases of infertility, early pregnacy loss, and poor pregnancy outcomes. Case Study 3.2 describes the experience of a woman ultimately diagnosed with celiac disease.

Nutritional Management of Celiac Disease

Treatment of celiac disease centers on the goals of elimination of gluten from the diet, correction of vitamin and mineral deficiences, and the long-term maintenance of health. The cornertone of treatment—elimination of gluten—can be challenging. Gluten is found in a variety of nongrain foods including hot dogs, deli meats, some vitamin and mineral supplements, flavored potato chips, bouillon, and salad dressing. An example of a gluten-free, one-day diet for an adult with celiac disease is shown in Table 3.4.

People with celiac disease are generally avid food label readers and eventually become skilled at selecting gluten-free foods. This skill will become easier to master in the future. Beginning in 2008, the Food and Drug Administration will allow gluten-free foods to be labelled "gluten-free" if they qualify according to a standard definition.[26]

Foods labeled "gluten-free" have been available in large grocery and speciality stores for years, and product options are increasing. Institution of labeling standards for gluten-free products will help guarantee that these products are actually free of gluten. Reliable Internet resources for people with celiac disease are listed at the end of this chapter.

Consultation with a Registered Dietitian Nutrition counselling by an experienced registered dietitian has

Table 3.4 Example of one day's diet and snack option for an adult with celiac disease

Breakfast	Dinner
Gluten-free bagel with nut butter	Lamb stew (thickened with potato starch) with carrots and lentils
Sliced bananas in yogurt	Rice
Tea	Gluten-free cake
	Low-fat milk
Lunch	**Snack Options**
Gluten-free pasta salad with chicken, broccoli, and tomatoes	Popcorn
Oil and vinegar dressing	Spring rolls with rice paper
Gluten-free roll with margarine	Ice cream
Fresh fruit	Fruit
Low-fat milk	Dark chocolate
	Gluten-free cookies
	String cheese
	Rice cakes

In Focus

Celiac Disease

Definition: Celiac disease is an autoimmune disease that occurs in people with a genetic susceptibility to the protein gliadin found in the gluten component of wheat, rye, and barley. Oats do not contain gluten, but commercial oats and oat products are often cross-contaminated by grains that do. Celiac disease is also called *celiac sprue* and *nontropical sprue*.

Prevalence: Celiac disease occurs in about 2 million people in the United States and in 0.5–1% of the population worldwide.[26,64] Although awareness of, and screening for, celiac disease is increasing, far more individuals have the disease than have been diagnosed with it.

Major physiological aspects and consequences: The presence of gliadin in the small intestine triggers an *autoimmune response* that causes an inflammatory reaction to occur in the inside lining of the small intestine. Over time, the inflammation causes the lining of the small intestine to become flattened and to absorb nutrients poorly. The damage produced by chronic inflammation in the small intestine leads to a variety of vitamin and mineral deficiencies and other health consequences.

Signs and symptoms of celiac disease range from very mild to severe and vary by age and sex. Diarrhea, nutrient malabsorption, bloating, weight loss, iron deficiency anemia, infertility, fatigue, and growth failure in children often characterize the disease. Some cases of celiac disease are clinically silent—presenting no clear, related symptoms. Long-standing, untreated celiac disease predisposes individuals to other autoimmune diseases. The only effective treatment is a lifelong gluten-free diet.

Risk factors: The primary risk factor for celiac disease is a genetic predisposition toward reacting to gladin as a foreign protein. Repeated exposure to certain types of infectious agents, such as *rotavirus*, may also trigger the onset of the disease in genetically susceptible people.[65]

been demonstrated to be a primary, effective means for helping individuals manage celiac disease. In general, the word *celiac* is being used by dietitians to guide the management of this disorder:

Autoimmune response Chemicals released by the immune system that attack its own molecules, cells, and tissues.

Rotavirus A virus that is the most common cause of severe diarrhea among children. Diarrhea caused by rotavirus generally lasts 2 days, and recovery is full in otherwise healthy children. The rotavirus is generally spread from an infected peron's stools to food.

C Consultation with a skilled dietitian
E Education about the disease
L Lifelong adherence to a gluten-free diet
I Identification and treatment of nutritional deficiencies
A Access to an advocacy group
C Continuous long-term follow-up by a multldisciplinary team.[62]

Successful management of celiac disease is almost always accompanied by improvements in the quality of life.[63]

Herbal Remedies for Fertility-Related Problems

Women may use herbs and supplements for chronic gynecologic conditions such as premenstrual syndrome, menstrual problems, and infertility. Therapies that carry some level of support from randomized controlled trials indicate that thiamin and vitamin E may be helpful for painful menstrual periods; and that calcium, magnesium, Vitamin D, and vitamin B_6 may help relieve premenstrual syndrome symptoms. Herbal extracts from the berries of the chaste tree (named after the plant monks were said to have chewed to inspire chasity), also show promise for safely relieving PMS symptoms. Chaste tree berry extracts are not yet considered safe for women who may become or who are pregnant, or who are taking oral contraceptive pills.[66,67] Evening primrose oil, which contains high amounts of essential fatty acids, does not appear to beat a placebo in relation to PMS relief.[68]

There are too few trials involving herbs and supplement use for the treatment of infertility to warrant solid recommendations, but chaste berry, antioxidants, and Fertility Blend (a proprietary supplement containing chaste berry, green tea extracts, L-arginine, vitamins and minerals) have some preliminary support of safety and effectiveness.[66] Coenzyme Q_{10} given at a dose of 400 mg per day for 6 months in males with poor sperm motility for unknown reasons increased sperm motility and pregnancy rates. Coenzyme Q_{10} acts as an antioxidant and is a coenzyme in energy formation. It may help protect sperm membranes from oxidation and facilitate energy formation by sperm. Coenzyme Q_{10} appears to be well tolerated.[69] Bee propolis, which consists of plant resins collected by honeybees, appears to have anti-inflammatory activity and has been found to increase pregnancy rates in women with mild endometriosis. Use of the compound was not related to adverse side effects in one study.[70]

Women and men using herbs for fertility problems should inform their health care providers. They should avoid herbs and other dietary supplements that have not been demonstrated to be safe for use during early pregnancy.

Key Points

1. Dietary intake, supplement use, weight status, and exercise levels before conception affect fertility in women and men, and the course and outcome of pregnancy in women.

2. Symptoms of PMS can be modified in some females by specific changes in dietary intake and supplement use.

3. Obesity is related to a number of hormonal and metabolic changes that compromise fertility and health status in men and women.

4. Modest levels of weight loss in obese women and men, and weight gain in underweight individuals, improve fertility.

5. Rates of diseases and disorders associated with obesity such as metabolic syndrome, polycystic ovary syndrome, and type 2 and gestational diabetes are increasing the United States and other countries. Obesity-related health problems affecting fertility and the course and outcome of pregnancy are being seen increasingly in clinical practice.

6. Chronic inflammation is an important component of disorders such as metabolic syndrome and polycystic ovary syndrome. Antioxidant nutrients and the omega-3 fatty acids EPA and DHA play key roles in reducing adverse effects of chronic inflammation.

7. Insulin resistance is a key feature of obesity, type 2 and gestational diabetes, metabolic syndrome, and polycystic ovary syndrome. Exercise (particularly resistance exercise), weight loss, whole grain products and high-fiber foods, EPA and DHA, and low-GI-carbohydrates tend to reduce insulin resistance.

8. Energy deficits in individuals with hypothalamic amenorrhea, eating disorders, and the female athlete triad are related to hormonal changes that reduce fertility.

9. Energy and estrogen deficits in women with the female athlete triad lead to reduced fertility and lower bone formation in young women.

10. Some herbal remedies and other dietary supplements appear to reduce symptoms of premenstrual syndrome and may aid fertility. Due to limited knowledge of the safety of herbs and other dietary supplements taken before conception and during early pregnancy, they are best used under medical supervision.

Resources

Health Topics

High-quality information on eating disorders, PMS, and other conditions as presented in this chapter can be found here.
Website: **www.healthfinder.gov**

Nutrition and Women's Health

This FDA site represents the pooled resources of several government agencies and provides updated reports on a wide assortment of nutrition and women's health topics.
Website: **www.fda.gov/womens/informat.html**

PCOS Support

Chat rooms for women with PCOS include the following:
Website: **www.obgyn.net/PCOS/PCOS.asp**
Website: **www.PCOSupport.org**

Celiac Disease Resources

Support and other resources for people with celiac disease are available from the following *Websites:*
The Celiac Disease Foundation: **www.celiac.org**
Celiac Sprue Association, USA: **www.csaceliacs.org**

Gluten Intolerance Group: **www.gluten.net**
Raising Our Celiac Kids: **www.celiackids.com**
Resource guide for gluten-free diets: **www.glutenfreediet.ca**

Diabetes Resources

Support and other resources related to diabetes are available from the following *Websites:*

American Diabetes Association: **www.diabetes.org**

The National Institutes of Health: **www.niddk.nih.gov**
Canadian Diabetes Association: **www.diabetes.ca**

Herbal and Other Remedies

Check out the safety and effectiveness of complementary and alternative medical treatments for fertility-related problems from this site.

Website: **www.cfscan.fda.gov/dms/supplmnt.html**

References

1. King, JC. Preface. Am J Clin Nutr. 2000;71(suppl):1217S.

2. Freeman EW et al. Differential response to antidepressants in women with premenstrual syndrome/premenstrual dysphoric disorder: a randomized controlled trial. Arch Gen Psychiatry. 1999;56:932–9.

3. Ugarriza DN et al. Premenstrual syndrome: diagnosis and intervention. Nurse Pract. 1998;23:40, 45, 49–52.

4. Vaitukaitis JL. Premenstrual syndrome (editorial). N Engl J Med. 1984;311:1371–3.

5. Grady-Weliky T. Premenstrual dysphoric disorder. N Engl J Med. 2003;348:433–8.

6. Daugherty JE. Treatment strategies for premenstrual syndrome. Am Fam Physician. 1998;58:183–92, 197–8.

7. Goodale et al. Alleviation of premenstrual syndrome symptoms with the relaxation response. Obstet Gynecol. 1990;75: 649–55.

8. Walker AF et al. Magnesium supplementation alleviates premenstrual symptoms of fluid retention. J Women's Health. 1998;7: 1157–65.

9. Thys-Jacobs S et al. Calcium carbonate and the premenstrual syndrome: effects on premenstrual and menstrual symptoms. Premenstrual Syndrome Study Group. Am J Obstet Gynecol. 1998;179:444–52.

10. Bertone-Johnson et al. Calcium and vitamin D intake and risk of incident premenstrual syndrome. Arch Intern Med. 2005; 165:1246–52.

11. Berman MK et al. Vitamin B-6 in premenstrual syndrome. J Am Diet Assoc. 1990;90:859–61.

12. Wyatt KM et al. Efficacy of vitamin B_6 in the treatment of premenstrual syndrome: a systematic review. BMJ. 1999;318: 1275–81.

13. State-specific prevalence of obesity among adults—United States, 2005. MMWR. 2006;55:985–8.

14. Pasquali R. Obesity and androgens: facts and perspectives, Fertil Steril. 2006;85: 1319–40.

15. Kovac P. Recurrent pregnancy loss. www.medscape.com/viewprogram/5293, accessed 4/06.

16. Kulkarni S. Lifestyle management before infertility treatment. www.medscape.com/viewarticle/52224, accessed 1/06.

17. Grundy SM et al. Definition of metabolic syndrome: report of the National Institute of Heart, Lung, and Blood Institute/American Heart Association Conference on Scientific Issues Related to Definition. Circ. 2004;109: 433–8.

18. Azadbakht L et al. Beneficial effects of a Dietary Approaches to Stop Hypertension eating plan on features of the metabolic syndrome. Diabete Care. 2005;28:2823–31.

19. Hu FB. Overweight and obesity in women: health risks and consequences. J Women's Health. 2003;12:163–72.

20. Zaadstra BM et al. Fat and female fecundity: prospective study of effect of body fat distribution on conception rates. BMJ. 1993;306:484–7.

21. Kaye SA et al. Associations of body mass and fat distribution with sex hormone concentrations in postmenopausal women. International J Epidemiol. 1990;131: 794–803.

22. Moran LJ, Norman RJ. The obese patient with infertility: a practical approach to diagnosis and treatment. Nutr Clin Care. 2003;5:290–7.

23. Pasquali R et al. Achievement of near-normal body weight as the prerequisite to normalize sex hormone-binding globulin concentrations in massively obese men. Int J Obes Relat Metab Disord. 1997;21:1–5.

24. Steinkampf MP et al. Subcutaneous FSH administration in obese women. Fertil Steril. 2003;80:99–102.

25. Bar-Zohar D et al. Pregnancy after laparoscopic adjustable gastric banding: perinatal outcome is favorable also for women with relatively high gestational weight gain. Surg Endosc. 2006;20:1580–3.

26. Bren L. Food labels identify allergens more clearly. FDA Consumer Magazine. Mar–Ap 2006.

27. Ahima RS. Body fat, leptin, and hypothalamic amenorrhea. N Engl J Med. 2004;352;95962.

28. Lifshitz F, quoted in: Welt CK et al. Recombinant human leptin in women with hypothalarnic amenorrhea. Growth, Genetics & Hormones. 2005;21:lssue 1, www.GGHjoumal.com, accessed 3/05.

29. Loucks AB et al. The female athlete triad: do female athletes need to take special care to avoid low energy availability? Med Sci Sports Exerc. 2006;38:1694–1700.

30. Stafford DE. Altered hypothalamic-pituitary-ovarian axis function in young female athletes: implications and recommendations for management. Treat Endocrinol. 2005;4:l47–54.

31. Waldrop J. Early identification and interventions for female athlete triad. J Pediatr Health Care. 2005;19:213–20.

32. Resch M et al. Bulimia from a gynecological view: hormonal changes.

33. Becker AE et al. Eating disorders. N Engl J Med. 1999;340:1092–8.

34. Agras WS et al. A multicenter comparison of cognitive behavioral therapy and interpersonal psychotherapy for bulimia nervosa. Arch Gen Psychiatry. 2000;57: 459–66.

35. Lorber DL. Preconception counseling of women with diabetes. Pract Diabetol. 1995;14;12–16.

36. Carr SR. Effect of maternal hyperglycemia on fetal development. American Dietetic Association Annual Meeting and Exhibition, Kansas City, MO, October 10, 1998.

37. Neff LM. Evidence-based dietary recommendations for patients with type 2 diabetes mellitus. Nutr Clin Care. 2003;6: 51–61.

38. Pape J. Polycyctic ovary syndrome: a personal and professional experience. Food and Nutrition Exhibition and Conference, St. Louis, Oct. 23, 2005.

39. Health, United States, 2003. www.cdc.gov/nchs.hus.htm, accessed 10/03.

40. Brunzell JD, Hokanson JE. Dyslipidemia of central obesity and insulin resistance. Diabetes Care. 1999;22(suppl)3:C10–3.

41. Saaddine JB et al. A diabetes report card for the United States: quality of care in the 1990s. Ann Intern Med. 2002;136: 565–74.

42. Albright A et al. Revised nutrition guidelines for diabetes prevention. Diabetes Care. 2006;29:2140–2157.

43. American Diabetes Association. Evidence-based nutrition principles and recommendations for the treatment and prevention of diabetes and related complications: position statement. Diabetes Care. 2003;26:S51–S61.

44. Foster-Powel K et al. International table of glycemic index and glycemic load values: 2002. Am J Clin Nutr. 2002;76:5–56.

45. Bedno SA. Weight loss in diabetes management. Nutr Clin Care. 2003;6:62–72.

46. Bowman BA, Vinicor F. The knowledge gap in diabetes. Nutr Clin Care. 2003;6: 49–50.

47. Danaei G et al. Raised blood glucose levels responsible for more than three million deaths each year. Lancet. http://www.medscape.com/viewarticle/547594, accesed 11/06.

48. Diabetes Prevention Program Research Group. Reduction in the incidence of type 2 diabetes with lifestyle intervention or metaformin. N Engl J Med. 2002;346: 393–403.

49. Obesity and polycystic ovary syndrome. Clin Endocrinol. 2006;65:137–145.

50. Ehrmann DA. Polycystic ovary syndrome. N Engl J Med. 2005;352:1223–36.

51. Azziz R. Diagnostic criteria for polycystic ovary syndrome: a reapprasial. Fert Steril. 2004;83:1343–6.

52. Tolstoi LG, Josimovich JB. Weight loss and medication in polycystic ovary syndrome therapy. Nutr Today. 2002;37:57–62.

53. Stein K. Polycystic ovary syndrome: what it is and why registered dietitians need to know. J Am Diet Assoc. 2006;106: 1738–41.

54. McKittrick M. Diet and polycystic ovary syndrome. Nutr Today. 2002;37: 63–69.

55. Ghazeeri G et al. Effect of rosiglitazone on spontaneous and clomiphene citrate-induced ovulation in women with PCOS. Fertil Sterlit. 2003;79:562–6.

56. Arpino L. Polycystic ovary syndrome. Food and Nutrition Exhibition and Conference, St. Louis, Oct. 23, 2005.

57. Michals-Matalon K et al. Maternal phenylketonuria, low protein intake, and congenital heart defects. Am J Obstet Gynecol. 2002;187:221–4.

58. Waisbren SE et al. Outcome at age 4 years in offspring of women with maternal phenylketonuria: the Maternal PKU Collaborative Study. JAMA. 2000; 283:756–62.

59. Friedman EG et al. The International Collaborative Study on maternal phenylketonuria: organization, study design and description of the sample. Eur J Pediatr. 1996;155(suppl)1:S158–61.

60. Isaacs, J. Metabolic dietitian. Personal communication, 12/10/03.

61. Stazi AV et al. Reproductive aspects of celiac disease. Ann Ital Med Int. 2005;20: 143–57.

62. Palmer S. Help clients hurdle gluten-free diet dilemmas. Today's Dietitian, 2005; Aug:28–32.

63. Green PHR et al. Characteristics of adult celiac disease in the USA: results of a national survey. Am J Gastroent. 2001;96:126–31.

64. Fasano A. Systemic autoimmune disorders in celiac disease. Curr Opin Gastroenterol. 2006;22;674–9.

65. Rewers M et al. Rotavirus infections and the development of celiac disease, Am J Gastroenterol. 2006;101:2333–2340.

66. Dennehy CE. The use of herbs and dietary supplements in gynecology: an evidence-based review. J Midwifery Womens Health. 2006;51:402–9.

67. Vitex or chaste tree; it helps balance women's cycles. Environ Nutr. 1999;22:8.

68. Collins A et al. Essential fatty acids in the treatment of premenstrual syndrome. Obstet Gynecol. 1993;81:93–8.

69. Balercia G et al. Coenzyme Q and the treatment of male infertility. Fertil Steril. 2004;81:93–8.

70. Hitt E. Bee propolis may improve infertility associated with mild endometriosis. American Society of Reproductive Medicine Annual Meeting, abstract 0-84, presented Oct. 13, 2003.

Chapter 4

Nutrition during Pregnancy

Chapter Outline

- Introduction
- The Status of Pregnancy Outcomes
- Reducing Infant Mortality and Morbidity
- Physiology of Pregnancy
- Embryonic and Fetal Growth and Development
- The Fetal-Origins Hypothesis of Later Disease Risk
- Pregnancy Weight Gain
- Nutrition and the Course and Outcome of Pregnancy
- Healthy Diets for Pregnancy
- Exercise and Pregnancy Outcome
- Food Safety Issues during Pregnancy
- Common Health Problems during Pregnancy
- Model Nutrition Programs for Risk Reduction in Pregnancy

Prepared by **Judith E. Brown**

Key Nutrition Concepts

1 Many aspects of nutritional status, such as dietary intake, supplement use, and weight change, influence the course and outcome of pregnancy.

2 The fetus is not a parasite; it depends on the mother's nutrient intake to meet its nutritional needs.

3 Periods of rapid growth and development of fetal organs and tissues occur during specific times throughout pregnancy. Essential nutrients must be available in required amounts during these times for fetal growth and development to proceed optimally.

4 The risk of heart disease, diabetes, hypertension, and other health problems during adulthood may be influenced by maternal nutrition during pregnancy.

Introduction

The 9 months of pregnancy represent the most intense period of growth and development humans ever experience. How well these processes go depends on many factors, most of which are modifiable. Of the factors affecting fetal growth and development that are within our control to change, nutritional status stands out. At no other time in life are the benefits of optimal nutritional status more obvious than during pregnancy.

This chapter addresses the status of pregnancy outcomes in the United States and other countries. It covers physiological changes that take place to accommodate pregnancy, and the impact of these changes on maternal nutritional needs. The chapter presents the roles of nutrition in fostering fetal growth, development, and long-term health; it also covers dietary supplement use and weight-gain recommendations. The discussion goes on to consider common problems during pregnancy that can be addressed with nutritional remedies. Chapter 5 addresses clinical conditions and nutrition interventions during pregnancy. We begin this chapter by highlighting vital statistic reports that clearly show a need for improving pregnancy outcomes in the United States.

The Status of Pregnancy Outcomes

The status of reproductive outcomes in the United States and other economically developed countries is routinely assessed through examination of a particular set of vital statistics data called *natality statistics* (*natality* means "related to birth"). Natality statistics summarize important information about the occurrence of pregnancy complications and harmful behaviors, in addition to infant mortality (death) and morbidity (illness) rates within a specific population. These data are used to identify problems in need of resolution and to identify progress in meeting national goals for improvement in the course and outcome of pregnancy.

Illustration 4.1 presents a time line of key intervals and events before, during, and after pregnancy. Specific time points and periods are labeled. Table 4.1 shows terms characterizing and changing rates of different natality statistics. They are frequently referred to in this chapter. Also referred to often in this chapter are weights in grams (g) and kilograms (kg), as well as in pounds and ounces. There are 448 grams in a pound and 2.2 pounds in a kilogram. (FYI: A table of measurement abbreviations and equivalents is located in Appendix C.)

^{a}LMP = last menstrual period

^{b}Perinatal definition varies from 20 to 24 weeks gestation to 7 to 28 days after birth.

Illustration 4.1 Time-related terms before, during, and after pregnancy.

Table 4.1 Natality statistics: Rates, definitions, and trends in the rates in the United States[1]

	Rates		Definition
	1995	2004/2005	
Maternal mortality	7.1	7.1	Deaths/100,000 live births
Fetal deaths (stillbirths)	7.0	6.6	Deaths/1000 pregnancies over 20 weeks gestation
Perinatal mortality	7.6	6.9	Deaths/1000 deliveries over 20 weeks gestation to 7 days after birth
Neonatal mortality	4.9	4.5	Deaths from delivery to 28 days/1000 live births
Postneonatal mortality	2.7	2.3	Deaths from 28 days after birth to 1 year/1000 live births
Infant mortality	7.6	6.8	Deaths from birth to age 1 year/1000 live births
Preterm	11.0	12.7	Births <37 weeks gestation/100 live births
Very preterm	1.9	2.0	Births <34 weeks gestation/100 live births
Low birth weight	7.3	8.2	Newborn weights <2500 g (5 lb 8 oz)/100 live births
Very low birth weight	1.4	1.5	Newborn weights <1500 g (3 lb 4 oz)/100 live births
Multifetal pregnancies			
Twins	1 in 40	1 in 31	Number of twin births/total live births
Triplets+	1 in 784	1 in 588	Number of triplets plus higher-order multiple births/total live births
Adolescent pregnancies	56.8	40.1	Births/1000 females aged 15 to 19 years

Infant Mortality

> "Infant mortality is a mirror of a population's physical health and socioeconomic status."

Infant mortality reflects the general health status of a population to a considerable degree because so many of the environmental factors that affect the health of pregnant women and newborns also affect the health of the rest of the population. The graph presented in Illustration 4.2 demonstrates this point. This historical view of infant mortality rates indicates that population-wide improvements in social circumstances, infectious disease control, and availability of safe and nutritious foods have corresponded to greater reductions in infant mortality than have technological advances in medical care.[2] Small improvements in infant mortality in the past few decades in the United States are largely due to technological advances in medical care that save ill newborns. High-level medical care has not favorably affected the need for such care, however. The United States spends more money on health care than any other nation, yet it ranks twenty-seventh in the international comparison of infant mortality rates (Table 4.2 on the next page).

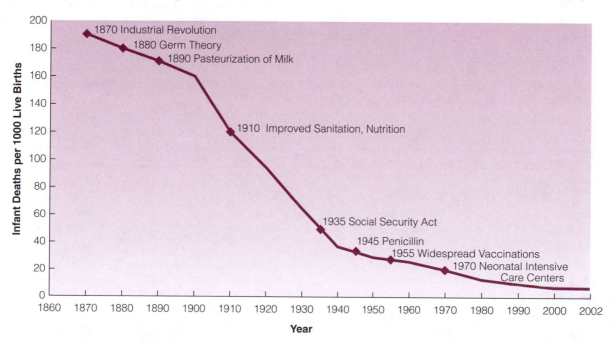

Illustration 4.2 Chronology of events related to declines in infant mortality in the United States.

SOURCE: Judith E. Brown, 2001.

Table 4.2 Infant mortality rates (deaths per 1000 live births) in 2004 for a sampling of countries[3]

Country	Infant Mortality Rate
Singapore	2.3
Sweden	2.8
Hong Kong	3.0
Japan	3.2
Finland	3.5
Norway	3.7
Czech Republic	3.9
Germany	4.1
France	4.2
Switzerland	4.3
Spain	4.4
Denmark	4.5
Austria	4.6
Australia	4.6
Belgium	4.6
Canada	4.7
Netherlands	5.0
Portugal	5.0
United Kingdom	5.1
Ireland	5.3
Greece	5.4
Italy	5.8
New Zealand	5.8
Cuba	6.2
United States	6.8
Israel	6.9
Slovakia	7.3

Infant mortality reflects deaths during the first year of life. However, two-thirds of deaths to *liveborn infants* occur within the first month after birth, or during the neonatal period. Deaths in the first month after birth and fetal deaths (or stillbirths) result largely from health problems that develop in the mother or fetus during pregnancy. Together they constitute the eleventh leading cause of death in the United States.[4]

Liveborn Infant   The World Health Organization developed a standard definition of *liveborn* to be used by all countries when assessing an infant's status at birth. By this definition, a liveborn infant is the outcome of delivery when a completely expelled or extracted fetus breathes, or shows any sign of life such as beating of the heart, pulsation of the umbilical cord, or definite movement of voluntary muscles, whether or not the cord has been cut or the placenta is still attached.

Low Birth Weight, Preterm Delivery, and Infant Mortality

Infants born low birth weight or preterm are at substantially higher risk of dying in the first year of life than are larger and older newborns. Low-birth-weight infants, for example, make up 8.2% of all births, yet comprise 66% of all infant deaths. The 12.7% of newborns delivered prior to 37 weeks of pregnancy similarly account for a disproportionately large number of infant deaths.[1] Low-birth-weight and preterm infant outcomes are intertwined in that the shorter the pregnancy, the less newborns tend to weigh. Table 4.3 shows increases in birth weight with the duration of pregnancy and birth-weight-specific infant mortality rates.

Rates of low birth weight in the United States have trended slowly upward since 1983 and have remained approximately two times higher in African American infants than in other infants. The high rate of low birth weight in African American infants clearly represents a problem in need of resolution.

Table 4.3 Range of birth weights by gestational age, U.S.[4]

Birthweight		Weeks Gestation	Infant Mortality Rate
Pounds (lb) and Ounces (oz)	Grams		
<1 lb 2 oz	<500	<22	846
1 lb 2 oz–2 lb 3 oz	500–999	22–27	316
2 lb 3 oz–3 lb 5 oz	1000–1499	27–29	62
3 lb 5 oz–4 lb 6 oz	1500–1999	29–31	28
4 lb 6 oz–5 lb 8 oz	2000–2499	31–33	12
5 lb 8 oz–6 lb 10 oz	2500–2999	33–36	4.6
6 lb 10 oz–7 lb 11 oz	3000–3499	36–40	2.4
7 lb 11 oz–8 lb 13 oz	3500–3999	40+	1.7
8 lb 13 oz–9 lb 14 oz	4000–4499	40+	1.5
9 lb 14 oz–11 lb	4500–4999	40+	2.5
>11 lb	5000+	40+	—

Reducing Infant Mortality and Morbidity

Deaths and illnesses associated with low-birth weight and preterm infants can be reduced through improvements in the birth weight of newborns. Infants weighing 3500 to 4500 grams at birth (or 7 lb 12 oz to 10 lb) are least likely to die within the first year of life, as well as in the perinatal, neonatal, and postneonatal periods.[1] Newborns weighing 3500 to 4500 grams are also at an advantage as a group in relation to overall health status and subsequent mental development.[5] They are less likely to develop heart disease, diabetes, lung disease, hypertension, and other disorders later in life.[6] Reducing the proportion of infants born small or early would clearly decrease infant mortality.

Health Objectives for the Year 2010

National health objectives for pregnant women and newborns focus on the reduction of low birth weight, preterm delivery, and infant mortality. A number of the objectives are related to improvements in nutritional status (Table 4.4).

Physiology of Pregnancy

Conception triggers thousands of complex and sequenced biological changes that transform two united cells into a member of the next generation of human beings. The rapidity with which structures and functions develop in mother and fetus and the time-critical nature of energy and nutrient needs make maternal nutritional status a key element of successful reproduction.

Pregnancy begins at conception; that occurs approximately 14 days before a woman's next menstrual period is scheduled to begin. Assessed from conception, pregnancy averages 38 weeks, or 266 days, in length. Most commonly, however, pregnancy duration is given as 40 weeks (280 days) because it is measured from the date of the first day of the last menstrual period (LMP). Consequently, the common way of measuring pregnancy duration includes two nonpregnant weeks at the beginning. The anticipated date of delivery is denoted by the ancient terminology of "estimated date of confinement," or EDC. Assessment of duration of pregnancy as weeks from conception is correctly termed *gestational age,* whereas time in pregnancy estimated from LMP reflects *menstrual age.* It is particularly important to get these terms straight during early fetal development, when a 2-week error in duration of pregnancy may mean miscalculating the timing of nutrient-related events in pregnancy.

Maternal Physiology

Changes in maternal physiology during pregnancy are so profound that they were previously considered abnormal and in need of correction. Doctors routinely advised pregnant women to follow low-sodium diets to reduce fluid retention, restricted their patients' weight gain and dietary intake to prevent complications at delivery, and prescribed excessive levels of iron and other supplements to bring blood nutrient levels back up to "normal." We now know that what is considered normal physiological status of nonpregnant women cannot be considered normal for women who are pregnant. Fortunately, it is now understood that attempts to bring maternal physiological changes back to nonpregnant levels may cause more harm than good to the pregnancy.

Changes in maternal body composition and functions occur in a specific sequence during pregnancy. The order of the sequence is absolute because the successful completion of each change depends on the one before it. Because maternal physiological changes set the stage for fetal growth and development, they begin in earnest within a week after conception.[7]

The sequence of physiological changes taking place during pregnancy is listed in Table 4.5. The table indicates the timing of maximal rates of change in maternal tissues, the *placenta,* and fetal weight across pregnancy. To provide the fetus with sufficient energy, nutrients, and oxygen for growth, the mother must first expand the volume of plasma that can be circulated. Maternal nutrient stores are accumulated next. These stores are established in advance of the time they will be needed to support large gains in fetal weight. Similarly, the maximal rate of placental growth is timed to precede that of fetal weight gain. This sequence of events ensures that the placenta is fully prepared for the high level of functioning that will be needed as fetal weight increases most rapidly. Fetuses depend on the functioning

Table 4.4 Health objectives for the nation related to pregnant women and infants

- Reduce anemia among low-income pregnant females in their third trimester from 29 to 20%.
- Reduce infant mortality from 7.6 to no more than 5 per 1000 live births.
- Reduce the incidence of spina bifida and other neural tube defects from 7 to 3 per 10,000 live births.
- Reduce low birth weight (<2500 g) from 7.3 to 5%.
- Reduce preterm births (<37 weeks) from 9.1 to 7.6%.
- Increase abstinence from alcohol use by pregnant women from 79 to 95%.
- Reduce the incidence of fetal alcohol syndrome.
- Increase the proportion of women who gain weight appropriately during pregnancy.

Placenta   A disk-shaped organ of nutrient and gas interchange between mother and fetus. At term, the placenta weighs about 15% of the weight of the fetus.

Table 4.5 Sequence of tissue development and approximate gestational week of maximal rates of change in maternal systems, the placenta, and fetus during pregnancy[7]

Tissue	Sequence of Development	Gestational Week of Maximal Rate of Growth
Maternal plasma volume	1	20
Maternal nutrient stores	2	20
Placental weight	3	31
Uterine blood flow	4	37
Fetal weight	5	37

of multiple systems, established well in advance of their maximal rates of growth and development. Abnormalities in the development of any of these physiological systems can modify fetal growth and development.

Normal Physiological Changes during Pregnancy

Physiological changes in pregnancy can be divided into two basic groups: those occurring in the first half of pregnancy and those in the second half. In general, physiological changes in the first half are considered "maternal anabolic" changes because they build the capacity of the mother's body to deliver relatively large quantities of blood, oxygen, and nutrients to the fetus in the second half of pregnancy. The second half is a time of "maternal catabolic" changes in which energy and nutrient stores, and the heightened capacity to deliver stored energy and nutrients to the fetus, predominate (Table 4.6). Approximately 10% of fetal growth is accomplished in the first half of pregnancy, and the remaining 90% occurs in the second half.[9]

The list of physiological changes that normally occur during pregnancy is extensive (Table 4.7), and such changes

Edema Swelling (usually of the legs and feet, but can also extend throughout the body) due to an accumulation of extracellular fluid.

affect every maternal organ and system. Changes that are most directly related to maternal energy and nutrient needs are discussed further.

Body Water Changes A woman's body gains a good deal of water during pregnancy, primarily due to increased volumes of plasma and extracellular fluid, as well as amniotic fluid.[11] Total body water increases in pregnancy range from 7 to 10 liters (approximately 7 to 10 quarts, or about 2 to 2½ gallons). About two-thirds of the expansion is intracellular (blood and body tissues) and one-third is extracellular (fluid in spaces between cells).[7] Plasma volume begins to increase within a few weeks after conception and reaches a maximum at approximately 34 weeks. Early pregnancy surges in plasma volume appear to be the primary reason that pregnant women feel tired and become exhausted easily when undertaking exercise performed routinely prior to pregnancy. Fatigue associated with plasma-volume increases in the second and third months of pregnancy declines as other compensatory physiological adjustments are made.

Gains in body water vary a good deal among women during normal pregnancy. High gains are associated with increasing degrees of *edema* and weight gain. If not accompanied by hypertension, edema generally reflects a healthy expansion of plasma volume. Birth weight is strongly related to plasma volume: generally, the greater the expansion, the greater the newborn size.[7] The increased volume of water in the blood is responsible for the "dilution effect" of pregnancy on blood concentrations of some vitamins and minerals. Blood levels of fat-soluble vitamins tend to increase in pregnancy, whereas levels of the water-soluble vitamins tend to decrease. Vitamin supplement use can modify these relationships.[10]

Hormonal Changes Many physiological changes in pregnancy are modulated by hormones produced by the

Table 4.6 Summary of maternal anabolic and catabolic phases of pregnancy[8–10]

Maternal Anabolic Phase 0–20 Weeks	Maternal Catabolic Phase 20+ Weeks
Blood volume expansion, increased cardiac output	Mobilization of fat and nutrient stores
Buildup of fat, nutrient, and liver glycogen stores	Increased production and blood levels of glucose, triglycerides, and fatty acids; decreased liver glycogen stores
Growth of some maternal organs	Accelerated fasting metabolism
Increased appetite, food intake (positive caloric balance)	Increased appetite and food intake decline somewhat near term.
Decreased exercise tolerance	Increased exercise tolerance
Increased levels of anabolic hormones	Increased levels of catabolic hormones

Table 4.7 Normal changes in maternal physiology during pregnancy[7,8]

Blood Volume Expansion
- Blood volume increases 20%
- Plasma volume increases 50%
- Edema (occurs in 60–75% of women)

Hemodilution
- Concentrations of most vitamins and minerals in blood decrease

Blood Lipid Levels
- Increased concentrations of cholesterol, LDL cholesterol, triglycerides, HDL cholesterol

Blood Glucose Levels
- Increased insulin resistance (increased plasma levels of glucose and insulin)

Maternal Organ and Tissue Enlargement
- Heart, thyroid, liver, kidneys, uterus, breasts, adipose tissue

Circulatory System
- Increased cardiac output through increased heart rate and stroke volume (30–50%)
- Increased heart rate (16% or 6 beats/min)
- Decreased blood pressure in the first half of pregnancy (−9%), followed by a return to nonpregnancy levels in the second half

Respiratory System
- Increased tidal volume, or the amount of air inhaled and exhaled (30–40%)
- Increased oxygen consumption (10%)

Food Intake
- Increased appetite and food intake; weight gain
- Taste and odor changes, modification in preference for some foods
- Increased thirst

Gastrointestinal Changes
- Relaxed gastrointestinal tract muscle tone
- Increased gastric and intestinal transit time
- Nausea (70%), vomiting (40%)
- Heartburn
- Constipation

Kidney Changes
- Increased glomerular filtration rate (50–60%)
- Increased sodium conservation
- Increased nutrient spillage into urine; protein is conserved
- Increased risk of urinary tract infection

Immune System
- Suppressed immunity
- Increased risk of urinary and reproductive tract infection

Basal metabolism
- Increased basal metabolic rate in second half of pregnancy
- Increased body temperature

Hormones
- Placental secretions of large amounts of hormones needed to support physiological changes of pregnancy

placenta (Table 4.8 and Illustration 4.3 on page 90). The placenta serves many roles, but a key one is the production of *steroid hormones,* such as progesterone and estrogen. The placenta is also the main supplier of many other hormones needed to support the physiological changes of pregnancy.

Maternal Nutrient Metabolism   Adjustments in maternal nutrient metabolism are apparent within the first few weeks after conception and progress throughout pregnancy.[9] The changes are sequenced and interrelated in complex ways that are appreciated but not yet fully understood.

Pregnancy: A Pro-oxidative State   An important change in maternal metabolism is taking center stage in recent research reports. It is now clear that normal pregnancy is a pro-oxidative state, and some of the abnormalities that arise during pregnancy are related, in part, to excessively high levels of oxidation.

Pregnancy is a time of high oxidative stress for the body largely because it's a busy time for placental and maternal *mitochondria.* Energy formation in mitochondria leads to

the production of free radicals that can initiate oxidation reactions and therefore oxidative stress. Oxidative damage to maternal and fetal cellular structures and functions occur when the body's protective mechanisms are overwhelmed, and the external supply of antioxidants is too low. Inflammatory disorders such as insulin resistance, diabetes, pre-eclampsia, obesity, and infection increase free radical formation and contribute to oxidative stress during pregnancy. Excess intake of supplemental iron and iron deficiency also increase oxidative stress.

Protective mechanisms that guard against free radical damage normally increase during pregnancy, but additional protection is required from antioxidant nutrients in foods. Adequate levels of intake of antioxidants such as vitamins C and E from foods, and of a wide variety of naturally occurring antioxidants in vegetables and fruits, provide protection against damage due to oxidative stress.[12]

Steroid Hormones   Hormones such as progesterone, estrogen, and testosterone produced primarily from cholesterol.

Mitochondria   Specialized compartments ("organelles") located within cells that produce energy for use by cells. Mitochondria are rich in unsaturated fats, proteins, and enzymes.

Table 4.8 Key placental hormones and examples of their roles in pregnancy[10,14]

Human chorionic gonadotropin (hCG)

Maintains early pregnancy by stimulating the corpus luteum to produce estrogen and progesterone. It stimulates growth of the endometrium. The placenta produces estrogen and progesterone after the first 2 months of pregnancy.

Progesterone

Maintains the implant; stimulates growth of the endometrium and its secretion of nutrients; relaxes smooth muscles of the uterine blood vessels and gastrointestinal tract; stimulates breast development, promotes lipid deposition.

Estrogen

Increases lipid formation and storage, protein synthesis, and uterine blood flow; prompts uterine and breast duct development; promotes ligament flexibility.

Human chorionic somatotropin (hCS)

Increases maternal insulin resistance to maintain glucose availability for fetal use; promotes protein synthesis and the breakdown of fat for energy for maternal use.

Leptin

May participate in the regulation of appetite and lipid metabolism, weight gain, and utilization of fat stores.

Carbohydrate Metabolism   Many adjustments in carbohydrate metabolism are made during pregnancy that promote the availability of glucose to the fetus. Glucose is

the fetus's preferred fuel, even though fats can be utilized for energy. Continued availability of a fetal supply of glucose is accomplished primarily through metabolic changes that promote maternal insulin resistance. These changes, sometimes referred to as the *diabetogenic effect of pregnancy,* make normal pregnant women slightly carbohydrate intolerant in the third trimester of pregnancy.[13] Illustration 4.4 provides an example of the normal levels of plasma glucose and insulin during late pregnancy compared to prepregnancy levels.

Carbohydrate metabolism in the first half of pregnancy is characterized by estrogen- and progesterone-stimulated increases in insulin production and conversion of glucose to glycogen and fat. In the second half, rising levels of hCS and prolactin from the mother's pituitary gland inhibit the conversion of glucose to glycogen and fat.[10] At the same time, insulin resistance builds in the mother, increasing her reliance on fats for energy. Decreased conversion of glucose to glycogen and fat, lowered maternal utilization of glucose, and increased liver production of glucose help to ensure that a constant supply of glucose for fetal growth and development is available in the second half of pregnancy.

Fasting maternal blood glucose levels decline in the third trimester due to increased utilization of glucose by the rapidly growing fetus. However, postmeal blood glucose concentrations are elevated and remain higher longer than before pregnancy.[13]

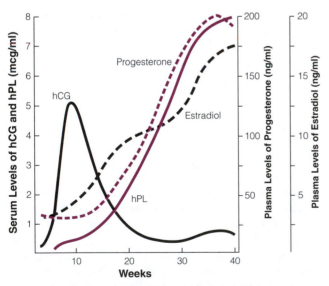

Illustration 4.3 Changes in maternal plasma concentration of hormones during pregnancy.

SOURCE: From Pedro Rosso. *Nutrition and Metabolism in Pregnancy: Mother and Fetus.* Copyright © 1990 by Oxford University Press, Inc. Used by permission of Oxford University Press, Inc.

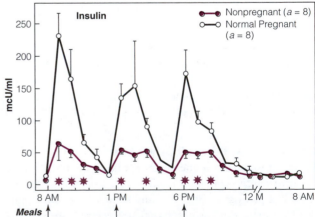

Illustration 4.4 Plasma glucose and insulin levels in nonpregnant women and in women near term.

SOURCE: Reprinted from *American Journal of Obstetrics and Gynecology* 140(6): 730–736. R. L. Phelps et al., © 1981, with permission from Elsevier.

Accelerated Fasting Metabolism   Maternal metabolism is rapidly converted toward *glucogenic amino acid* utilization, fat oxidation, and increased production of *ketones* with fasts that last longer than 12 hours. Decreased levels of plasma glucose and insulin, and increased levels of triglycerides, free fatty acids, and ketones are seen hours before they occur in nonpregnant fasting women. The rapid conversion to fasting metabolism allows pregnant women to use primarily stored fat for energy while sparing glucose and amino acids for fetal use.[13]

Although these metabolic adaptations help ensure a constant fetal supply of glucose, fasting eventually increases the dependence of the fetus on ketone bodies for energy. Prolonged fetal utilization of ketones, such as occurs in women with poorly controlled diabetes or in those who lose weight during part or all of pregnancy, is associated with reduced growth and impaired intellectual development of the offspring.[15]

Protein Metabolism   Nitrogen and protein are needed in increased amounts during pregnancy for synthesis of new maternal and fetal tissues. It is estimated that 925 grams (2 pounds) of protein is accumulated during pregnancy.[16] To some extent the increased need for protein is met through reduced levels of nitrogen excretion and the conservation of amino acids for protein tissue synthesis. There is no evidence, however, that the mother's body stores protein early in pregnancy in order to meet fetal needs for protein later in pregnancy. Maternal and fetal needs for protein are primarily fulfilled by the mother's intake of protein during pregnancy.[9]

Fat Metabolism   Multiple changes occur in the body's utilization of fats during pregnancy. Overall, changes in lipid metabolism promote the accumulation of maternal fat stores in the first half of pregnancy and enhance fat mobilization in the second half.[13] In addition to seeing increasing maternal reliance on fat stores for energy as pregnancy progresses, we see blood levels of many lipoproteins increase dramatically (Table 4.9). Plasma triglyceride levels increase first and most dramatically, reaching three times nonpregnant levels by term.[9,13] Cholesterol-containing lipoproteins, phospholipids, and fatty acids also increase, but to a lesser extent than do triglycerides. The increased cholesterol supply is used by the placenta for steroid hormone synthesis, and by the fetus for nerve and cell membrane formation.[13] Small increases in HDL cholesterol in pregnancy appear to decline within a year postpartum and remain lower than prepregnancy levels. It is speculated that declines in HDL cholesterol after pregnancy may contribute to an increased risk of heart disease in women. Other changes in serum lipids appear to revert to prepregnancy levels postpartum.[18]

By the third trimester of pregnancy most women have a lipid profile that would be considered atherogenic, if

Table 4.9 Changes in cholesterol and triglyceride levels during pregnancy[17,18]

Trimester	Cholesterol mmol/L	Triglycerides mmol/L
1	5.78	1.19
2	6.88	1.32
3	8.14	2.58
Nonpregnant	5.11	0.80

*To convert mmol/L to mg/dL, see the Measurement Abbreviations and Equivalents table located in Appendix C.

not for pregnancy. These blood lipid changes are normal, however, which is why blood lipid screening is not recommended during pregnancy.[18] Normal changes in blood lipid levels during pregnancy appear to be unrelated to maternal dietary intake.[19]

> **Glucogenic Amino Acids**   Amino acids such as alanine and glutamate that can be converted to glucose.
>
> **Ketones**   Metabolic by-products of the breakdown of fatty acids in energy formation. β-hydroxybutyric acid, acetoacetic acid, and acetone are the major ketones, or "ketone bodies."

Mineral Metabolism   Impressive changes in mineral metabolism occur during pregnancy. Calcium metabolism is characterized by an increased rate of bone turnover and reformation.[11] Elevated levels of body water and tissue synthesis during pregnancy are accompanied by increased requirements for sodium and other minerals. Sodium metabolism is delicately balanced during pregnancy to promote an accumulation of sodium by the mother, placenta, and fetus. This is accomplished by changes in the kidneys that increase aldosterone secretion and the retention of sodium. This normal change in pregnancy renders ineffective and potentially harmful any attempts to prevent and treat high blood pressure in pregnancy by reducing sodium intake. Sodium restriction may overstress mechanisms that act to conserve sodium and lead to functional and growth impairments due to sodium depletion.[20]

The Placenta

The word *placenta* is derived from the Latin word for *cake*. The placenta, with its round, disklike shape, looks somewhat like a cake—the more so the more active the imagination. The placenta develops from embryonic tissue and is larger than the fetus for most of pregnancy. Development of the placenta precedes fetal development.

Functions of the placenta include hormone and enzyme production, nutrient and gas exchange between the mother and fetus, and removal of waste products from the fetus. Its structure, including a double lining of cells separating maternal and fetal blood, acts as a barrier to some harmful compounds, and it governs the rate of passage of nutrients and other substances into and out of the fetal

Fetal side   **Maternal side**

Umbilical vein

Umbilical artery

Umbilical cord

Maternal artery

Pool of maternal blood

Maternal vein

Illustration 4.5 Structure of the placenta. Maternal arteries and veins are part of the maternal circulation, whereas umbilical arteries and veins are part of the fetal circulation. Blood enters the fetus through umbilical veins and exits through umbilical arteries.

circulation (Illustration 4.5). The barrier role of the placenta is better described as a fence than as a filter that guards the fetus against all things harmful. Many potentially harmful substances (alcohol, excessive levels of some vitamins, drugs, and certain viruses, for example) do pass through the placenta to the fetus. The placenta is a barrier to the passage of maternal red blood cells, bacteria, and many large proteins. The placenta also prevents the mixing of fetal and maternal blood until delivery, when ruptures in blood vessels may occur.

Nutrient Transfer   The placenta uses 30–40% of the glucose delivered by the maternal circulation. If nutrient supply is low, the placenta fulfills its needs before nutrients are made available to the fetus. If nutrient supplies fall short of meeting placental needs, functioning of the placenta is compromised to sustain the nutrient supply and health of the mother.[21]

Nutrient transfer across the placenta depends on a number of factors, including:

- The size and the charge of molecules available for transport
- Lipid solubility of the particles being transported
- The concentration of nutrients in maternal and fetal blood

Small molecules with little or no charge (water, for example) and lipids (cholesterol and ketones, for instance) pass through the placenta most easily, while large molecules (e.g., insulin and enzymes) aren't transferred at all. Nutrient exchange between the mother and fetus is unregulated for some nutrients, oxygen, and carbon dioxide; it is highly regulated for other nutrients. Nutrient transfer based on concentration gradients determined by the levels of the nutrient in the maternal and the fetal blood is unregulated. In these cases, nutrients cross placenta membranes by simple diffusion from blood with high concentration of the nutrient to blood with lower concentration.

Three primary mechanisms regulate nutrient transfer: facilitated diffusion, active transport, and endocytosis (or pinocytosis). Table 4.10 summarizes mechanisms of nutrient transfer across the placenta and provides examples of

Table 4.10 Mechanisms of nutrient transport across the placenta[10,22]	
Mechanism	**Examples of Nutrients**
Passive diffusion (also called *simple diffusion*) Nutrients transferred from blood with higher concentration levels to blood with lower concentration levels	Water, some amino acids and glucose, free fatty acids, ketones, vitamins E and K,[a] some minerals (sodium, chloride), gases
Facilitated diffusion Receptors ("carriers") on cell membranes increase the rate of nutrient transfer	Some glucose, iron, vitamins A and D
Active transport Energy (from ATP) and cell membrane receptors	Water-soluble vitamins, some minerals (calcium, zinc, iron, potassium) and amino acids required for transfer
Endocytosis (also called *pinocytosis*) Nutrients and other molecules are engulfed by placenta membrane and released into fetal blood supply	Immunoglobulins, albumin

[a]Vitamin K crosses the placenta slowly and to a limited degree.

nutrients transported by each specific mechanism as they are known.

The fetus receives small amounts of water and other nutrients from ingestion of *amniotic fluid.* By the second half of pregnancy the fetus is able to swallow and absorb water, minerals, nitrogenous waste products, and other substances in amniotic fluid.

The Fetus Is Not a Parasite The fetus is not a "parasite"—it cannot take whatever nutrients it needs from the mother's body at the mother's expense. When maternal nutrient intakes fall below optimum levels or adjustment thresholds, fetal growth and development are compromised more than maternal health.[23] In general, nutrients will first be used to support maternal nutrient needs for her health and physiological changes, and next for placental development, before they become available at optimal levels to the fetus. For example:

- Underweight women gaining the same amount of weight as normal-weight women tend to deliver smaller infants and to retain more of the weight gained during pregnancy at the expense of fetal growth.[10]
- Fetal growth tends to be reduced in pregnant teenagers who gain height during pregnancy compared to fetal growth in teens who do not grow during pregnancy.[24]
- Vitamin and mineral deficiencies and toxicities in newborns have been observed in women who showed no signs of deficiency or toxicity diseases during pregnancy.[10]

If the fetus did act as a parasite, it would harm the mother for its own benefit. Rather, the fetus is generally harmed more by poor maternal nutritional status than is the mother.[22]

Embryonic and Fetal Growth and Development

The rate of human *growth* and *development* is higher during gestation than at any time thereafter. If the rate of weight gain achieved in the 9 months of gestation continued after delivery, infants would weigh about 160 pounds at their first birthdays and be 20 feet tall by age 20! Table 4.11 provides an overview of embryonic and fetal growth and development during pregnancy.

Critical Periods of Growth and Development

Fetal growth and development proceed along genetically determined pathways in which cells are programmed to multiply, *differentiate,* and establish long-term functional levels during set time intervals. Such time intervals are known as *critical periods* and are most intense during the first 2 months after conception, when a majority of organs and tissues form. On the whole, critical periods represent a "one-way street," because it is not possible to reverse directions and correct errors in growth or development that occurred during a previous critical period. Consequently, adverse effects of nutritional and other insults occurring during critical periods of growth and development persist throughout life.[26]

Hyperplasia Critical periods of growth and development are characterized by hyperplasia, or an increase in cell multiplication. Because every human cell has a specific amount of DNA, periods of hyperplasia can be determined by noting times during gestation when the DNA content of specific organs and tissues increases sharply. The critical period of rapid cell multiplication of the forebrain, for example, is between 10 and 20 weeks of gestation (Illustration 4.6 on p. 95).

The brain is the first organ that develops in humans, and along with the rest of the central nervous system, it is given priority access to energy, nutrient, and oxygen supplies. Thus, in conditions of low energy, nutrient, and oxygen availability, the needs of the central nervous system will be met before those of other fetal tissues such as the liver or muscles. The heart and adrenal glands come next after the central nervous system in the hierarchy of targets for preferential nutrient delivery.[21]

Deficits or excesses in nutrients supplied to the embryo and fetus during critical periods of cell multiplication can produce lifelong defects in organ and tissue structure and function. The organ or tissue undergoing critical periods of growth at the time of the adverse exposure will be affected most.[27] For example, the neural tube develops into the brain and spinal cord during weeks 3 and 4 after conception. If folate supplies are inadequate during this critical period of growth, permanent defects in brain or spinal cord formation occur, regardless of folate availability at other times. Other tissues—such as the pancreas, which does not undergo rapid cell multiplication until the third trimester of pregnancy—do not appear to be affected by the early shortage of folate.

Some degree of hyperplasia takes place in a number of organs and tissues in

Amniotic Fluid The fluid contained in the amniotic sac that surrounds the fetus in the uterus.

Growth Increase in an organism's size through cell multiplication (hyperplasia) and enlargement of cell size (hypertrophy).

Development Progression of the physical and mental capabilities of an organism through growth and differentiation of organs and tissues, and integration of functions.

Differentiation Cellular acquisition of one or more characteristics or functions different from that of the original cells.

Critical Periods Preprogrammed time periods during embryonic and fetal development when specific cells, organs, and tissues are formed and integrated, or functional levels established. Also called *sensitive periods.*

Table 4.11 Notes on normal embryonic and fetal growth and development[10,25]

Day 1	Conception; one cell called the zygote exists.	**Week 9**	Embryo now considered a fetus.
Day 2–3	Eight cells have formed (called the morula) and enter the uterine cavity.	**Month 3**	Weighs 1 oz; primitive egg and sperm cells developed, hard palate fuses, breathes in amniotic fluid.
Day 6–8	The morula becomes fluid filled and is renamed the blastocyst. The blastocyst is comprised of 250 cells, and cell differentaion begins.	**Month 4**	Weighs about 6 oz; placenta diameter is 3 inches.
Day 10	Embryo implants into the uterine wall, where glycogen is accumulating.	**Month 5**	Weighs about 1 lb, 11 inches long; skeleton begins to calcify, hair grows.
Day 12	Embryo is composed of thousands of cells; differentiation well under way. Utero placental ciculation being formed.	**Month 6**	14 inches long; fat accumulation begins, permanent teeth buds form; lungs, gastrointestinal tract, and kidneys formed but are not fully functional.
Week 4 (21–28 days)	¼ inch long; rudimentary head, trunk, arms; heart "practices" beating; spinal cord and two major brain lobes present.		
Week 5 (28–35 days)	Rudimentary kidney, liver, circulatory system, eyes, ears, mouth, hands, arms, and gastrointestinal tract; heart beats 65 times per minute circulating its own, newly formed blood.	**Month 7**	Gains ½–1 oz per day.
Week 7 (49–56 days)	½ inch long, weighs 2–3 g; brain sends impulses, gastrointestinal tract produces enzymes, kidney eliminates some waste products, liver produces red blood cells, muscles work. (Approximately 25% of blastocysts and embryos will be lost before 7 weeks.)	**Months 8 and 9**	Gains about 1 oz per day; stores fat, glycogen, iron, folate, B_6 and B_{12}, riboflavin, calcium, magnesium, vitamins A, E, D; functions of organs continue to develop. Growth rate declines near term. Placenta weighs 500–650 g (1–1½ lb) at term.

the first year or two after birth and during the adolescent growth spurt. Cells of the central nervous system, for instance, continue to multiply for about two years after birth, but at a much slower pace than early in pregnancy. Skeletal and muscle cells increase in number during the adolescent growth spurt.[28]

In utero and early life changes in DNA content of the brain have been investigated in fetuses, infants, and young children dying from non-nutritional causes and from undernutrition. Illustration 4.7 presents results from one such study that show deficits in DNA content (or cell number) in the brains of children dying of protein-energy malnutrition versus those dying from accidents. Deficits in DNA were apparent a few months after birth, indicating that severe malnutrition early in pregnancy reduced brain cell number in utero.[29]

Hyperplasia and Hypertrophy   Cell multiplication continues at a lower rate after critical periods of cell multiplication and is accompanied by increases in the size of cells. This phase of growth can be seen in Illus-

tration 4.6, where it begins around 20 weeks in the forebrain when the rate of increase in DNA content slows. Cell size increases mainly due to an accumulation of protein and lipids inside of cells. Consequently, increases in cell size can be determined by measuring the protein or lipid content of cells. Specialized functions of cells, such as production of digestive enzymes by cells within the small intestine or neurotransmitters by nerve cells, occur along with increases in cell number and size.[10]

Hypertrophy   Periods of hyperplasia-hypertrophy are followed by hypertrophy only. During this phase cells continue to accumulate protein and lipids, and functional levels continue to grow in sophistication, but cells no longer multiply. Reductions in cell size caused by unfavorable nutrient environments or other conditions are associated with deficits in organ and tissue functions, such as reduced mental capabilities or declines in muscular coordination. Such functional changes can often be reduced or reversed later if deficits are corrected.[26]

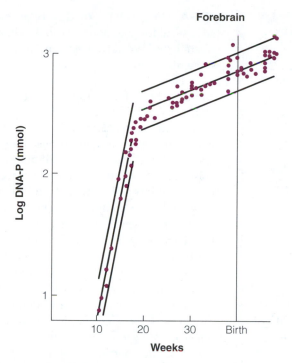

Illustration 4.6 The critical period of cell multiplication of the forebrain. Growth in cell numbers is indicated by increases in DNA content of a given amount of tissue.

SOURCE: From "Quantitative Growth and Development of Human Brain" by J. Dobbing and J. Sands, in *Archives of Disease of Children*, 48(10):757–767. © 1973 BMJ Publishing Group. Reprinted with permission.

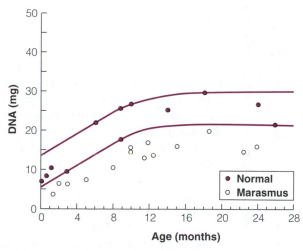

Illustration 4.7 DNA content of the cerebellum of the human brain in young children dying from non-nutritional causes and from undernutrition.

SOURCE: Reprinted from M. Winick, "Malnutrition and Brain Development," *Journal of Pediatrics* 74(6):667–679, © 1969, with permission from Elsevier.

Maturation   The last phase of growth and development is maturation—the stabilization of cell number and size. This phase occurs after tissues and organs are fully developed later in life.

Fetal Body Composition

The fetus undergoes marked changes in body composition during pregnancy (Table 4.12). The general trend is toward progressive increases in fat, protein, and mineral content. Some of the most drastic changes take place in the last 5 weeks of pregnancy, when fat and mineral content increase substantially.

Variation in Fetal Growth

Given a healthy mother and fetal access to needed amounts of energy, nutrients, and oxygen and freedom from toxins, fetal genetic growth potential is achieved.[31] However, as evidenced by the relatively high rate of low birth weight in the United States, optimal conditions required for achievement of genetic growth potential often do not exist during pregnancy. Variations in fetal growth and development are not generally due to genetic causes but rather to environmental factors such as energy, nutrient, and oxygen availability, and to conditions that interfere with genetically programmed growth and development. Insulin-like growth factor-1 (IGF-1) is the primary growth stimulator of the fetus. It promotes uptake of nutrients by the fetus and inhibits fetal tissue breakdown. Levels of IGF-1 are sensitive to maternal nutrition; it is decreased by undernutrition. Low levels of IGF-1 decrease muscle and skeletal mass and produce asymmetrical growth.[21] Factors such as prepregnancy underweight and shortness, low weight gain during pregnancy, poor dietary intakes, smoking, drug abuse, and certain clinical complications of pregnancy are associated with reduced fetal growth.[32]

Risk of illness and death varies substantially with size at birth and is particularly high for newborns experiencing intrauterine growth retardation (IUGR).[27] For a portion of newborns, smallness at birth is normal and may reflect familial genetic traits. Because IUGR is complicated to determine, it is usually approximated by assessment of size for gestational age using a reference standard (Table 4.13 on the next page). Infants are generally considered likely to have experienced intrauterine growth retardation if their weight for gestational age or length is low. Newborns whose weight is less than the 10th percentile for gestational age are considered *small for gestational age,* or *SGA*. This determination is further categorized into *disproportionately small for gestational age (dSGA)* and *proportionately small for gestational age (pSGA)*. Newborns who weigh less than the 10th percentile of weight for gestational age but have normal length and head circumference for age are

Small for Gestational Age (SGA)
Newborn weight is ≤10th percentile for gestational age. Also called *small for date* (SFD).

Disproportionately Small for Gestational Age (dSGA)   Newborn weight is ≤10th percentile of weight for gestational age; length and head circumference are normal. Also called *asymmetrical SGA*.

Proportionately Small for Gestational Age (pSGA)   Newborn weight, length, and head circumference are ≤10th percentile for gestational age. Also called *symmetrical SGA*.

Table 4.12 Estimated changes in body composition of the fetus by time in pregnancy[10,30]

Component	10 Weeks	20 Weeks	30 Weeks	40 Weeks
Body weight, g	10	300	1667	3450
Water, g	<9	263	1364	700
Protein, g	<1	22	134	446
Fat, g	<1	26	66	525
Sodium, meq	<1	32	136	243
Potassium, meq	<1	12	75	170
Calcium, g	<1	1	10	28
Magnesium, mg	<1	5	31	76
Iron, mg	<1	17	104	278
Zinc, mg	<1	6	26	53

Table 4.13 Percentiles of weight in grams for newborn gestational age

Gestational Age (wk)	5th Pctl	10th Pctl	50th Pctl	90th Pctl	95th Pctl
20	249	275	412	772	912
21	280	314	433	790	957
22	330	376	496	826	1023
23	385	440	582	882	1107
24	435	498	674	977	1223
25	480	558	779	1138	1397
26	529	625	899	1362	1640
27	591	702	1035	1635	1927
28	670	798	1196	1977	2237
29	772	925	1394	2361	2553
30	910	1085	1637	2710	2847
31	1088	1278	1918	2986	3108
32	1294	1495	2203	3200	3338
33	1513	1725	2458	3370	3536
34	1735	1950	2667	3502	3697
35	1950	2159	2831	3596	3812
36	2156	2354	2974	3668	3888
37	2357	2541	3117	3755	3956
38	2543	2714	3263	3867	4027
39	2685	2852	3400	3980	4107
40	2761	2929	3495	4060	4185
41	2777	2948	3527	4094	4217
42	2764	2935	3522	4098	4213
43	2741	2907	3505	4096	4178
44	2724	2885	3491	4096	4122

NOTE: Pctl = percentile
SOURCE: From *Obstetrics and Gynecology*, Vol. 87, No. 2, 1996, pp. 163–168, table 2. Copyright © 1996. Reprinted by permission of Lippincott, Williams & Wilkins.

dSGA Infants who are disproportionately small for gestational age look skinny, wasted, and wrinkly. They tend to have small abdominal circumferences, reflecting a lack of glycogen stores in the liver, and little body fat. It appears that these infants have experienced in utero malnutrition in the third trimester of pregnancy and that it compromised liver glycogen and fat storage. Short-term episodes of malnutrition, such as maternal weight loss or low weight gain late in pregnancy that compromise energy, nutrient, or oxygen availability appear to be related to dSGA.[31] These infants generally have smaller organ sizes but the normal number of cells in organs and tissues. Infants who are dSGA are at risk of developing the "hypos" after birth (hypoglycemia, hypocalcemia, hypomagnesiumenia, and hypothermia). If the period of maternal undernutrition was short, dSGA infants tend experience good catch-up growth with nutritional rehabilitation.[21] Unfortunately, disproportionately small infants tend to perform less well academically and are at greater risk than other infants for heart disease, hypertension, and type 2 diabetes in the adult years.[5] (We return to this issue later in this chapter.)

pSGA Proportionately SGA newborns look small but well proportioned. It is believed that these infants have experienced long-term malnutrition in utero, due to factors such as prepregnancy underweight, consistently low rates of maternal weight gain in pregnancy and the corresponding inadequate dietary intake, or chronic exposure to alcohol.[10]

Because nutritional insults existed during critical periods of growth early in pregnancy, pSGA infants generally have a reduced number of cells in organs and tissues. These babies tend to exhibit fewer health problems at birth than do dSGA infants, but catch-up growth is poorer, even with nutritional rehabilitation. On average, pSGA infants remain shorter and lighter, and have smaller head circumferences throughout life than do infants born *appropriate for gestational age (AGA)* or *large for gestational age (LGA)*.[33]

The goal of nutritional rehabilitation for pSGA infants should be catch-up in weight and length, and not just weight. This goal appears to be easier to reach if

considered dSGA. If weight, length, and head circumference are less than the 10th percentile for gestational age, then the newborn is considered pSGA. Approximately two-thirds of SGA newborns in the United States are disproportionately small, and one-third are proportionately small.[31]

pSGA infants are breastfed. Excessive weight gain by pSGA infants appears to increase the risk of obesity and insulin-resistance-related disorders, such as hypertension and type 2 diabetes, later in life.[34]

LGA   Newborns with weights greater than the 90th percentile for gestational age are considered to be large for gestational age. About 1–2% of U.S. newborns are LGA. Although it is difficult to predict LGA, it appears to be related to prepregnancy obesity, poorly controlled diabetes in pregnancy, excessive weight gain in pregnancy (over 44 pounds), and possibly other factors.

Except for infants born to women with poorly controlled diabetes during pregnancy or other health problems, LGA newborns experience far lower illness and death rates than do SGA infants, and tend to be taller later in life.[35] Delivery and postpartum complications in mothers, however, tend to be higher with LGA newborns, and include increased rates of operative delivery, *shoulder dystocia,* and postpartum hemorrhage.

Ponderal Index   Increasingly, ponderal index (PI) is being used to assess the appropriateness of newborn size. Like body mass index, it is a measure of weight-for-length, but it is calculated by dividing weight by the cube of length (PI = weight in grams/cm^3 × 1000). Values between approximately 23 and 25 reflect normal weight-for-length; lower values represent thinness and higher values heaviness at birth. Wider adoption of PI to assess newborn size in practice awaits development of standard values indicative of increased health risk related to thinness and heaviness at birth.

Nutrition, Miscarriages, and Preterm Delivery

Several other pregnancy outcomes are related in part to maternal nutrition. Highlighted here are the roles played by nutrition in miscarriage and preterm delivery.

Miscarriages   Over 30% of implanted embryos are lost by reabsorption into the uterus or expulsion before 20 weeks of conception. Roughly a third of these losses are recognized as a miscarriage. Such early losses of embryos and fetuses are thought to be primarily caused by genetic, uterine, or hormonal abnormalities, reproductive tract infections, or to tissue rejection due to immune system disorders.[36] Intakes of caffeine over 500 mg per day (or the equivalent of about 4 cups of brewed coffee), especially in women with nausea and vomiting of pregnancy, sharply increase the risk of miscarriage.[37] The presence of nausea and vomiting is otherwise related to a very low risk of miscarriage.[37] Women who enter pregnancy underweight are at substantially higher risk of miscarriage than are normal and overweight women.[38]

Preterm Delivery   Infants born preterm are at greater risk than other infants of death, neurological problems reflected later in low IQ scores, congenital malformations, and chronic health problems such as *cerebral palsy.* The risk for these outcomes increases rapidly as gestational age at birth decreases. Infants born very preterm (<34 weeks) commonly have problems related to growth, digestion, respiration, and other conditions due to immaturity.[42] Low stores of fat, essential fatty acids, glycogen, and other nutrients in very preterm infants may also interfere with growth and health after delivery. Additionally, breast-milk content of riboflavin and vitamins A, C, and B$_{12}$ may be low in women who have inadequate intake of these vitamins during the third trimester of pregnancy.[43–45]

Although preterm delivery is a major health problem in the United States, its etiology remains unclear, and the search for effective prevention programs continues. A portion of preterm deliveries appears to be related to genital tract infections, insufficient uterine-placental blood flow, placental abruption (bleeding into the uterus), prepregnancy underweight, low weight gain in pregnancy, short interpregnancy interval (<6 months), and high levels of psychological or social stress. It is also fairly common in women who have previously delivered preterm.[46,47] Improvements in prenatal care for women at risk of preterm delivery—such as close supervision of the pregnancy, inclusion of nutritional counseling, encouragement of adequate weight gain in underweight and normal-weight women, and home visits—appear to decrease the risk of preterm delivery somewhat. Use of multivitamin supplements before pregnancy has been shown in one study to reduce the risk of preterm delivery (from 12% to 5%).[48] Obesity and inadequacy of folate during pregnancy also increases the risk of preterm delivery.[49,50]

> **Appropriate for Gestational Age (AGA)**   Weight, length, and head circumference are between the 10th and 90th percentiles for gestational age.
>
> **Large for Gestational Age (LGA)**   Weight for gestational age exceeds the 90th percentile for gestational age. Also defined as birth weight greater than 4500 g (≥10 lb) and referred to as *excessively sized for gestational age, or macrosomic.*
>
> **Shoulder Dystocia**   Blockage or difficulty of delivery due to obstruction of the birth canal by the infant's shoulders.
>
> **Cerebral Palsy**   A group of disorders characterized by impaired muscle activity and coordination present at birth or developed during early childhood.

The Fetal-Origins Hypothesis of Later Disease Risk

"The implications of the associations between fetal nutrition and adult disease are immense, and if substantiated, demand intense scrutiny of current prenatal nutrition policies."

J. King, 2000[51]

In the last decade, thinking about chronic disease risk has changed substantially. In contrast to the earlier idea that disease risk begins during childhood or in the adult years,

Fetal-Origins Hypothesis The theory that exposures to adverse nutritional and other conditions during critical or sensitive periods of growth and development can permanently affect body structures and functions. Such changes may predispose individuals to cardiovascular diseases, type 2 diabetes, hypertension, and other disorders later in life. Also called *metabolic programming* and the *Barker Hypothesis*.

Nutrition Programming The process by which exposure of the fetus to certain levels of energy and nutrients modify the function of genes in ways that affect metabolism and the development of diseases later in life.

studies testing the *fetal-origins hypothesis* indicate that risks may begin in utero. The concept that chronic disease risk may be established in utero is strongly supported by animal studies and by epidemiological investigations in humans.[6] Much of the evidence that relates in utero exposures to later disease in humans comes from studies showing increased risk for diseases such as heart disease, hypertension, type 2 diabetes, gestational diabetes, and chronic bronchitis in small, short, and thin newborns (Table 4.14). Maternal nutrition is hypothesized to play a key role in mechanisms that lead to later disease risk because it is a major factor affecting fetal growth and development.[40] Although smallness and thinness at birth are recognized as risk factors for later disease development, specific aspects of maternal nutrition unrelated to size at birth have also been related to disease development later in life.

Relatively small reductions in weight or disproportions in newborn size have been related to increased later disease risk. Risk of cardiovascular disease (heart disease and stroke), for instance, is associated with birthweights below 7.5 pounds (3360 g)—weights that are often considered "normal." Results of the U.S. Nurses Study, which compared newborn birth weight to risk of cardiovascular disease in adults are provided in Table 4.15 and illustrate this point. Infants at risk for later disease include those born at weights below that genetically programmed, even if birthweights are considered "normal."[53] Additionally, disease risk in people born small, short, or thin may be exacerbated by later, excessive weight gain.[54]

Table 4.15 Association of birthweight with the risk of cardiovascular disease in the U.S. Nurses Study[52]

Birthweight	Relative Risk of:	
	Heart Disease	Stroke
<5 lb (2240 g)	1.5	2.3
5–5½ lb (2240–2500 g)	1.3	1.4
5½–7 lb (2500–3136 g)	1.1	1.3
7–8½ lb (3136–3808 g)	1.0	1.0
8½–10 lb (3808–4480 g)	1.0	1.0
>10 lb (>4480 g)	0.7	0.7

Mechanisms Underlying the Fetal-Origins Hypothesis

Given less than optimal growing conditions during gestation, fetal tissues make functional adaptations to cope with energy and nutrient shortages and excesses. The ability to modify functions in response to available energy or nutrient supply is advantageous to the fetus in several ways. Adaptations could help the fetus survive by changing its requirements for energy or nutrients, and they may also biologically prepare the fetus for similar nutritional circumstances after birth. Fetal adaptations that lower energy or nutrient needs in response to a low supply, for example, may promote survival in an environment of limited food after birth.

Nutrition Programming Contemporary research is shedding light on how adaptive, functional changes may be accomplished by the fetus. Changes in function appear to be due to *nutrition programming* of gene functions.

Energy and nutrient availability during fetal development, and very early in life, can program the function of genes. These programmed changes in function last a lifetime. Although the structure of genes is not changed, energy and nutrient availability can modify which genes are expressed (or can be turned on), and which remain inactive. Genes that are active produce proteins that influence metabolism and health in certain ways.[55] Nutrition programming of gene function represents the major mechanism that underlies the relationship between maternal nutrition and later disease development.[40]

Modifications in gene expression can be prompted by other in utero exposures including infection, inflammation, and hypertension.[56]

Examples of Proposed Nutrition Programming Effects Results of observational studies during human pregnancy, and laboratory studies involving animals,

Table 4.14 Examples of diseases and other conditions in adults related to smallness or thinness at birth[39–41]

Allergies	Mood disorders
Autoimmune diseases	Obesity
Bronchitis	Ovarian cancer
Cardiovascular disease	Polycystic ovary syndrome
Decreased bone mineral content	Schizophrenia
Gestational diabetes	Short stature
Hypertension	Stroke
Irritable bowel syndrome	Subfertility in males
Kidney disease	Suicide
Metabolic syndrome	Type 2 diabetes

point to ways in which in utero energy and nutrient exposures may prompt fetal adaptations that affect gene programming and later disease risk. Energy availability has been a major focus of many of these studies because of its importance to fetal growth and development. In human studies, energy available for fetal growth and development is often assessed by pregnancy weight change and newborn birth size.

An inadequate availability of glucose during fetal growth and development would hinder central nervous system (CNS) development and threaten fetal survival. Mechanisms are set in place, however, that triage available glucose to the central nervous system (CNS). This change represents an adaptation by the body to how glucose utilization is programmed to operate.

What adaptations are made to ensure the CNS gets priority access to glucose? Animal studies indicate that the expression of genes that produce insulin receptors on muscle cell membranes may be suppressed in response to a low availability of glucose. This increases insulin resistance and decreases uptake of glucose by muscle cells, and reduces their growth. It also increases the availability of glucose for CNS development.

Adaptations that decrease muscle utilization of glucose and reduce muscle size may serve the offspring well later in life if food availability and intake are limited. If food is abundant and food intake is high, however, such adaptations may lead to elevated blood levels of glucose and insulin. These changes may increase the risk of obesity, type 2 diabetes, gestational diabetes, and other disorders associated with insulin resistance.[57]

Increased susceptibility to insulin resistance and weight gain in infants experiencing nutritional insults in utero has been attributed to a "thrifty phenotype," or genetic functional types programmed in utero that act to conserve energy.[54]

The function of genes involved in cholesterol metabolism appear to be modified in males with birth weights less than 3.2 kg (7 lb). In these individuals, production of "good" cholesterol, HDL, tends to decrease in response to high-fat, high-saturated-fat diets. HDL cholesterol production generally increases in males with higher birth weights in response to this type of diet. High blood levels of HDL-cholesterol are protective against heart disease.[58]

Some studies have shown a link between maternal nutritional exposures during pregnancy and later disease risk in infants with a wide range of birth weights. Low weight gain around mid-pregnancy, for example, has been associated with higher blood pressure in children, and low levels of maternal body fat during pregnancy with increased risk of heart disease in offspring.[59] Children born to women receiving a calcium supplement during pregnancy have been found to have lower blood pressure than children born to women given a placebo.[60]

Limitations of the Fetal-Origins Hypothesis

The hypothesis that maternal and fetal nutrition exposures influence later disease risk is gaining support and recognition. Many questions are unanswered, however. Which specific nutritional exposures are responsible for changes in gene function and increased disease risk? When do the vulnerable periods of fetal sensitivity to poor nutrition occur? What levels of energy and nutrient availability are related to the optimal functioning of genes? The implications of the associations between maternal and fetal nutrition and adult disease risk are immense, and if substantiated, demand intense scrutiny of current prenatal nutrition policies and recommendations.

Pregnancy Weight Gain

"Any obstetrician who allows a woman to lose her attractiveness (i.e., gain too much weight) is depriving her of many things that make for her mental well-being, her husband's contentment, and her own personal satisfaction."

Loughran, *American Journal of Obstetrics and Gynecology*, 1946

Weight gain during pregnancy is an important consideration because newborn weight and health status tend to increase as weight gain increases. Birth weights of infants born to women with weight gains of 15 pounds (7 kg) for example, average 3100 grams (6 lb 14 oz). This weight is about 500 grams less than the average birth weight of 3600 grams (8 lb) in women gaining 30 pounds (13.6 kg). Rates of low birth weight are higher in women gaining too little weight during pregnancy.[22] Weight gain during pregnancy is an indicator of plasma volume expansion and positive calorie balance, and provides a rough index of dietary adequacy.[61]

Multiple studies show broad agreement on amounts of weight gain that are related to the birth of infants with weights that place them within the lowest category of risk for death or health problems.[32] Yet how much weight should be gained during pregnancy remains a hotly debated topic. Earlier in the last century, when gains were routinely restricted to 15 or 20 pounds, weight gain in pregnancy was seen as the cause of pregnancy hypertension, difficult deliveries, and obesity in women. Pregnant women would be placed on low-calorie diets and given diuretics and amphetamines and urged to use saccharin to limit weight gain.[62]

Although none of these notions have been shown to be true, weight gain during pregnancy still represents a prickly issue. Weight gain and body weight are not only a matter of health, but are also closely linked to some people's view of what is socially acceptable.

Psychological and sociological biases related to body weight and shape in women are an important reason to apply recommendations for weight gain in pregnancy based on scientific studies and consensus.

Pregnancy Weight Gain Recommendations

Current recommendations for weight gain in pregnancy are based primarily on gains associated with the birth of healthy-sized newborns (approximately 3500–4500 g or 7 lb 13 oz to 10 lb).[31] As shown in Illustration 4.8, however, prepregnancy weight status influences the relationship between weight gain and birth weight. The higher the weight before pregnancy, the lower the weight gain needed to produce healthy-sized infants. Recommended weight gains for women entering pregnancy underweight, normal weight, overweight, and obese are displayed in Table 4.16.[32]

Because underweight women tend to retain some of the weight gained in pregnancy for their own needs, they need to gain more weight in pregnancy than do other women. Overweight and obese women, on the other hand, are able to use a portion of their energy stores to support fetal growth, so they need to gain less. A separate and higher pregnancy weight-gain recommendation is given for women expecting twins.

Young adolescents and women who are at the low end of the prepregnancy weight ranges are encouraged to gain near the upper end of their weight-gain range, and short women and those at the higher end of the prepregnancy weight ranges ought to gain near the lower end of the range. Because factors such as duration of gestation, smoking, maternal health status, *gravida,* and *parity* also influence birth weight, gaining a certain amount of weight during pregnancy does not guarantee that newborns will be a healthy size. It does improve the chances that this will happen, however.

> **Gravida**   Number of pregnancies a woman has experienced.
>
> **Parity**   The number of previous deliveries experienced by a woman; *nulliparous* = no previous deliveries, *primiparous* = one previous delivery, *multiparous* = two or more previous deliveries. Women who have delivered infants are considered to be "parous."

Table 4.16 Pregnancy weight gain recommendations[32]	
Prepregnancy Weight Status Body Mass Index[a]	Recommended Weight Gain
Underweight, <18.5 kg/m²	28–40 lb (12.7–18.2 kg)
Normal weight, 18.5–24.9 kg/m²	25–35 lb (11.4–15.9 kg)
Overweight, 25–29.9 kg/m²	15–25 lb (6.8–11.4 kg)
Obese, 30 kg/m² or higher	15 lb (6.8 kg) at least
Twin pregnancy	35–45 lb (15.9–20.5 kg)

[a]Body Mass Index categories modified based on 1997 changes from the National Institutes of Health. Young adolescents should achieve gains at the upper end of ranges, and short women at the lower end.

Approximately 40% of U.S. women gain within the recommended weight ranges during pregnancy.[64] For all except the obese, women who gain within the recommended ranges are approximately half as likely to deliver low-birth-weight or SGA newborns as are women who gain less. Rates of LGA newborns, Caesarean-section deliveries, and postpartum weight retention tend to be higher when pregnancy weight gain exceeds that recommended.[64] It is suggested that insulin resistance may be related to excessive weight gains during pregnancy and some of the adverse neonatal outcomes.[63]

Restriction of pregnancy weight gain to levels below the recommended ranges is not recommended. It does not decrease the risk of pregnancy-related hypertension and is associated with increased infant death and low birth weight, and poorer offspring growth and development.[20] In addition, low weight gain in pregnancy may increase the risk that infants will develop heart disease, type 2 diabetes, hypertension, and other types of chronic disease later in life.[27]

Rate of Pregnancy Weight Gain   Rates at which weight is gained during pregnancy appear to be as important to newborn outcomes as is total weight gain. Low rates of gain in the first trimester of pregnancy may downregulate fetal growth and result in reduced birth weight and thinness.[65] For underweight and normal-weight women, rates of gain of less than 0.5 pound (0.25 kg) per week in the second half of pregnancy, and of less than 0.75 pound (0.37 kg) per week in the third trimester of pregnancy, double the risk of preterm delivery and SGA newborns. For overweight and obese women, rates of gain of less than 0.5 pound (0.25 kg) per week in the third trimester also double the risk of preterm birth.[66] Third-trimester weight gains exceeding approximately 1.5 pounds a week (0.7 kg), however, add little to birth weight in normal-weight and heavier women, and may increase postpartum weight retention.[67]

Rate of weight gain is generally highest around midpregnancy—which is prior to the time the fetus gains

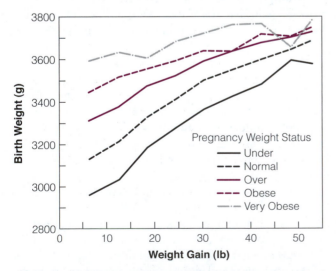

Illustration 4.8 Pregnancy weight gain by prepregnancy weight status and birth weight.

SOURCE: From *Clinical Nutrition*, Vol. 7, Fig. 1, p. 186, J. E. Brown, © 1988. Reprinted by permission from Elsevier.

Illustration 4.9 Rates of maternal and fetal weight gain during pregnancy.

SOURCE: Curves drawn by Judith E. Brown, 2002.[65]

most of its weight (Illustration 4.9). In general, the pattern of gain should be within a few pounds of that represented by the weight-gain curves shown in Illustration 4.10.[32] Some weight (3 to 5 pounds) should be gained in the first trimester, followed by gradual and consistent gains thereafter. The rate of weight gain often slows a bit a few weeks prior to delivery, but as is the case for the rest of pregnancy, weight should not be lost until after delivery.[65]

Composition of Weight Gain in Pregnancy

A question often asked by pregnant women is, "Where does the weight gain go?" Where the weight gain generally goes by time in pregnancy is shown in Table 4.17. The fetus actually comprises only about a third of the total weight gained during pregnancy in women who enter pregnancy at normal weight or underweight. Most of the rest of the weight is accounted for by the increased weight of maternal tissues.

Body Fat Changes Pregnant women store a significant amount of body fat in normal pregnancy in order to meet their own and the fetus's energy needs, and quite likely to prepare for the energy demands of lactation. Body fat stores increase the most between 10 and 20 weeks of pregnancy, or before fetal energy requirements are highest. Levels of stored fat tend to decrease before the end of pregnancy. Only 0.5 kg of the approximately 3.5 kg of fat stored during pregnancy is deposited in the fetus.[9,26]

Postpartum Weight Retention

Concern about the role of pregnancy weight gain in fostering long-term maternal obesity has increased in the United States, along with the rising incidence of obesity in adults. Increased weight after pregnancy appears to be related to a variety of factors, including excessively high weight gain in pregnancy (over 45 lb, or 20.5 kg), weight gain after delivery, and low activity levels.[68] High blood levels of insulin early in pregnancy, and levels of leptin, have been related to increased weight gain during pregnancy. Levels of both hormones are related to diet.[69,70]

Women tend to lose about 15 pounds the day of delivery, but subsequent weight loss is highly variable.[68] On average, however, women who gain within the recommended ranges of weight gain are 2.0 pounds (0.9 kg) heavier 1 year after delivery than they were before pregnancy.[71] This gain is slightly above the amount of weight women tend to gain with age.[72] Postpartum weight retention tends to be slightly less in women who breastfeed for at least 6 months after pregnancy.[73] Women who gain less than the recommended amount of weight gain in pregnancy do not retain less weight on average after pregnancy than do women who gain within the ranges.[71] Postpartum weight can be reduced by identifying high weight gainers during pregnancy and getting the women identified involved in an exercise and healthy eating program.[74]

Table 4.17 Components of weight gain during pregnancy for healthy, normal-weight women delivering a 3500-g (about 8 lb) infant at term[7,9,10,30]

Component	Weight Gain, grams			
	10 Weeks	20 Weeks	30 Weeks	40 Weeks
Fetus	5	300	1500	3550
Placenta	20	170	430	670
Uterus	140	320	600	1120
Amniotic fluid	30	350	750	896
Breasts	45	180	360	448
Blood supply	100	600	1300	1344
Extracellular fluid	0	265	803	3200
Maternal fat stores	315	2135	3640	3500
Total weight gain at term = 14.7 kg or 32 lb				

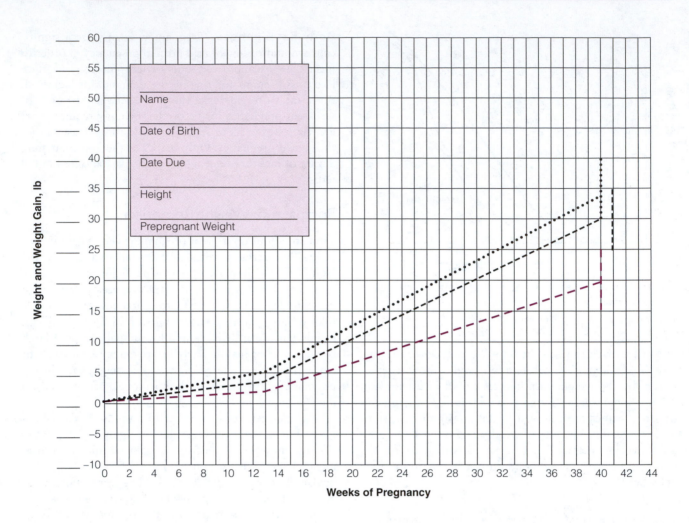

Name

Date of Birth

Date Due

Height

Prepregnant Weight

Weekly of Pregnancy

Weight and Weight Gain, lb

Weight Record			
Date	Weeks of Gestation	Weight	Notes

Please bring this chart with you to each prenatal visit.

Illustration 4.10 The Institute of Medicine's prenatal weight-gain graph and weight-recording form for women.

SOURCE: From *Nutrition during Pregnancy and Lactation*. National Academy of Sciences, 1992. Reprinted with permission of National Academy Press.

Nutrition and the Course and Outcome of Pregnancy

> "A mother who wishes her child to have black eyes should frequently eat mice."
>
> Prenatal diet folklore from Ancient Rome[75]

The history of beliefs about the effects of maternal diet on the course and outcome of pregnancy is rife with superstition, ill-founded and hazardous conclusions, and unhelpful suggestions. Societies have shared a belief in the importance of "eating right" during pregnancy for the child's sake, but actual knowledge about maternal nutrition and the course and outcome of pregnancy has been acquired only relatively recently.

Famine and Pregnancy Outcome

Much of the scientific interest in the effects of maternal nutrition on the course and outcome of pregnancy comes from studies done in the first half of the twentieth century. Ecological studies on effects of famines in Europe and Japan during World War II on the course and outcome of pregnancy demonstrated potential negative, as well as positive, effects of food intake on fertility and newborn outcomes.

The Dutch Hunger Winter, 1943–1944

As mentioned briefly in Chapter 2, people in many parts of Holland experienced severe food shortages for an 8-month period during World War II due to enemy occupation of major cities. Although people in Holland were generally well nourished and had a reasonable standard of living before the disaster, conditions rapidly deteriorated during the famine. In addition to intakes that averaged only about 1100 Kcal and 34 grams of protein per day, fuel was in low supply and the winter harsh.

Carefully kept records by health officials showed a sharp decline in pregnancy rates of over 50% during the famine, an effect attributed to absent and irregular menstrual periods. Average birth weight declined by 372 grams (13 oz), delivery of low-birth-weight infants increased by 50%, and rates of infant deaths increased. Birth weight did not fully "catch up" in infants born to women exposed to famine early in pregnancy, even if they received enough food later in pregnancy. This result supports the notion that the fetal growth trajectory may be established early in pregnancy and that early nutritional deprivations limit fetal growth regardless of food intake later in pregnancy.[76]

Although the Dutch famine was associated with major declines in fertility and newborn health and survival, the rather good nutritional status of women prior to the famine likely protected pregnant women and their infants from more severe disruptions in health. Normal fertility status and newborn outcomes returned within a year after the famine ended.[77]

Studies undertaken in the last 30 years on adults who were born to women during the hunger winter (the Dutch famine cohort) show relationships between the timing of famine during pregnancy and adult offspring health outcomes. Examples of relationships identified are shown in Table 4.18.

The Seige of Leningrad, 1942

Unlike people in Holland, the population in Leningrad (now called St. Petersburg) had experienced moderate deprivations in nutritional status and quality of life prior to the famine. As was the case for pregnant women in Holland, the famine in Leningrad resulted in average intakes of approximately 1100 Kcal per day. Infertility and low-birth-weight rates increased over 50%, infant death rates rose, and birth weights dropped by an average of 535 grams (1.2 lb) during the famine.[80] Rates of pSGA newborns also increased, suggesting that the poor nutritional status of women coming into pregnancy and persistent undernutrition during pregnancy interfered with critical periods of fetal growth.

Food Shortages in Japan

Effects of World War II–associated food shortages on reproductive outcomes in Japan were similar to those observed in Holland. Japanese women tended to be well nourished prior to the shortages. Lack of food before and during pregnancy was reflected in decreased fertility status among women and in reductions in birth weight that averaged 200 grams.

Social and economic improvements occurring in Japan after the war led to increased availability of many foods, including animal products. This higher plane of nutrition achieved during the postwar years in Japan was accompanied by major increases in newborn size and the "growing up" of Japanese children. In a trend that continues today, subsequent generations of Japanese adults averaged 2 inches

Table 4.18 Exposure to the Dutch World War II famine by time in pregnancy and adult offspring health risks[78,79]

	Period of Famine	
First Trimester	First and/or Second Trimester	Second Half of Pregnancy
Schizophrenia	Antisocial personality disorder	Decreased glucose tolerance
High LDL and low HDL cholesterol		
High body weight and central body fat		
Infertility		
Neural tube defects		

taller than the previous generation.[81] Infant mortality in Japan, which ranked among the highest for industrialized nations prior to World War II, declined incredibly after the war and remains well below rates in the United States and in a number of other developed countries.[82]

Food shortages continue to occur in various parts of the world and to adversely affect fertility and the course and outcome of pregnancy. Effects have become predictable, such that declines in fertility and newborn size and vitality are viewed as part of the consequences of such disasters. For example, the siege of Sarajevo, which decreased food availability during 1993–1994, led to reduced caloric and nutrient intakes during pregnancy, reduced maternal weight gain and newborn weights, and increased rates of perinatal mortality and congenital anomalies.[83] Birth weight did not fully catch up in infants born to women exposed to famine early in pregnancy, even if they received enough food later in pregnancy. This result supports the notion that the fetal growth trajectory may be established early in pregnancy and that early nutritional deprivations limit fetal growth regardless of food intake later in pregnancy.[76]

Contemporary Prenatal Nutrition Research Results

> "Faulty nutrition, not just malnutrition, can influence fetal health."
>
> Bertha Burke, 1948[84]

Carefully conducted studies of diet and pregnancy outcome in the first half of the twentieth century began the era of scientifically based recommendations on nutrition and pregnancy. The now-classic studies conducted by Bertha Burke at Harvard in the 1940s were particularly influential.[84] These studies showed that diet quality during pregnancy, assessed using diet histories, was strongly related to newborn health status. Newborns assessed as having optimal physical condition by pediatricians were found to be much more common among women consuming high-quality diets, whereas those with the poorest physical condition were born to women with the poorest-quality diets. Average birth weight of newborns assessed as being in optimal physical condition was 7 pounds, 15 ounces in females, and 8 pounds, 8 ounces in males.[85] Although Burke's studies did not show that high-quality pregnancy diets by themselves were responsible for robust newborn health, they provided some of the first evidence that prenatal diet quality may strongly influence pregnancy outcome.

Thousands of other studies on the effects of nutrition on the course and outcome of pregnancy are now available. The following sections highlight research results and recommendations related to calories, key nutrients, and other substances in food that influence the course and outcome of pregnancy.

Energy Requirement in Pregnancy

Energy requirements increase during pregnancy, mainly due to increased maternal body mass and fetal growth. The additional requirements can be allocated to different maternal and fetal tissues by estimating the amount of oxygen used (or "consumed") by the various tissues. Illustration 4.11 shows the results of work on oxygen consumption during pregnancy undertaken by Hytten and Chamberlain. Approximately one-third of the increased calorie need in pregnancy is related to increased work of the heart, and another third to increased energy needs for respiration and accretion of breast tissue, uterine muscles, and the placenta. The fetus accounts for about a third of the increased energy needs of pregnancy.[7]

The increased need for energy in pregnancy averages 300 Kcal a day, or a total of 80,000 Kcal.[7] The DRIs for energy intake for pregnancy are +340 Kcal per day for the second trimester and +452 Kcal per day for the third trimester of pregnancy. Caloric intake recommendations represent a rough estimate that by no means applies to every woman.

Additional energy requirements of women have been found by different studies to range from 210–570 Kcal a day.[9] The need for additional calories during pregnancy may be a good deal lower in women who perform little exercise, and higher in women who are very active. Low levels of energy expenditure from physical activity are common in the first trimester of pregnancy, and the energy savings may produce a positive caloric balance even though a woman's caloric intake hasn't changed much. Contrary to a previous belief, energy needs of pregnant women do not appear to be affected by "metabolic efficiencies" of pregnancy that decrease caloric need.[86]

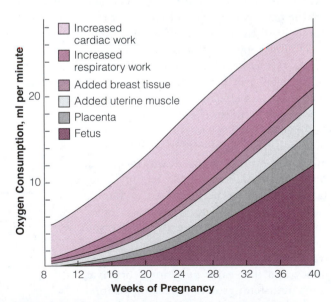

Illustration 4.11 Components of increased oxygen consumption in normal pregnancy.

SOURCE: F. Hytten and G. Chamberlain, eds., *Clinical Physiology in Obstetrics*. Reprinted by permission of Blackwell Science Ltd.

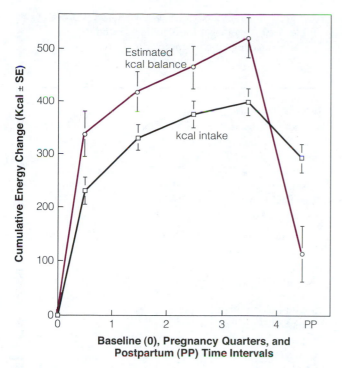

Illustration 4.12 Estimated caloric balance in pregnancy through 6–8 weeks postpartum.

SOURCE: From *Clinical Perinatology*, 24(2):433–449, by J. E. Brown and E. S. B. Khan, © 1997. Reprinted by permission of W. B. Saunders Co.

Illustration 4.12 shows the difference between caloric (Kcal) intake and estimated caloric balance throughout pregnancy in a group of women served by a health maintenance organization.[86] The graph indicates that estimated caloric balance is higher than caloric intake throughout pregnancy and becomes negative postpartum. The positive caloric balance observed during pregnancy is due to the fact that women consumed more calories than they expended in physical activity and basal metabolism.

Assessment of Caloric Intake   Adequacy of calorie intake is most easily assessed in practice by pregnancy weight gain. Rates of gain in women who do not have noticeable edema are a good indicator of caloric balance. Women who are losing weight are in negative balance, and those gaining weight are in positive caloric balance. Adjustments in rate of weight gain can be approached by modifying physical activity level, calorie intake, or both. Women should maintain a positive caloric balance and rate of weight gain throughout pregnancy. Meal skipping or fasting during pregnancy are not advised. Whether women should restrict their food and fluid intake during labor, however, is hotly debated.

Carbohydrate Intake during Pregnancy

Approximately 50–65% of total caloric intake during pregnancy should come from carbohydrate. Women should consume a minimum of 175 grams carbohydrate to meet the fetal brain's need for glucose. On average, women in the United States consume 53% of calories (269 g) from carbohydrates during pregnancy.[87] High-fiber foods generally provide a variety of beneficial phytochemicals and a hefty measure of protection against constipation.[88]

Glycemic Index of Carbohydrates   The glycemic index of carbohydrate-containing foods consumed may be of particular relevance during pregnancy because maternal blood glucose levels affect fetal growth. Habitual consumption during pregnancy of meals that provide high-glycemic carbohydrates do appear to increase fat accumulation in the fetus, and that may increase the risk of obesity and chronic disease.[89]

Artificial Sweeteners   There is no evidence that consumption of aspartame (Nutrasweet) or acesulfame K (Sunette) is harmful in pregnancy.[90] Diet soft drinks and other artificially sweetened beverages and foods are often poor sources of nutrients, however, and may displace other, more nutrient-dense foods in the diet.

Alcohol and Pregnancy Outcome

Alcohol ingested by a pregnancy woman readily passes through the placenta to the fetus where it can interrupt normal growth and development. Adverse effects of high amounts of alcohol intake (such as several drinks per day or more) are strongly related to abnormal mental development and growth in the offspring, and the deficits are lifelong. Adverse effects of alcohol intake during pregnancy are mild or undetectable when intakes are low or when alcohol intake exists but is infrequent. However, because no clearly safe level of alcohol intake has been identified, it is strongly advised that women who are pregnant do not drink. It is further recommended that women who *may* become pregnant not drink alcohol. In nutero alcohol exposure during the first, critical months of pregnancy may impair organ development.[91]

Frequent consumption of high amounts of alcohol from early pregnancy onward is related to the development of fetal alcohol syndrome. This topic is addressed in the next chapter.

Protein Requirement

The recommended protein intake for pregnancy is +25 grams per day, or 71 grams daily for females aged 14 and older. On average, pregnant women in the United States consume 78 grams of protein daily.[87]

Protein content of nonvegetarian diets can be simply estimated by evaluating women's usual daily intake of major sources of protein. A tool for estimating protein intake is shown in Table 4.19.

Table 4.19 Tool for estimating protein intake

Food	Protein, grams	How much protein is there in this usual day's diet?	
Milk, 1 c	8	2 slice toast	6
Cheese, 1 oz	7	1 c milk	8
Egg, 1	7	3 oz tuna	21
Meat, 1 oz	7	2 sl bread	6
Dried beans, 1 c	13	2 oz chicken	14
Bread, 1 slice or oz	3	1 oz cheese	7
		2 tortillas	6
		½ c refried beans	7
		Total g protein =	75

Vegetarian Diets in Pregnancy

"The topic of vegetarian dietary practices often brings with it a variety of images and attitudes regarding those who follow such practices. Those attitudes may have limited, if any, basis in actual fact."

Patricia Johnston, 1988[92]

Nutrient needs in pregnancy may be met by many different types of diets, including those that omit animal products.[90] It is the type and amount of food consumed, not the label placed on it, that determines the appropriateness of dietary intake during pregnancy.

A food guide for pregnant women who exclude animal products from their diet can be found in Table 4.20. Diets of pregnant vegetarians are sometimes low in vitamins B_{12} and D, calcium, zinc, omega-3 fatty acids (or n-3 fatty acids), and riboflavin due to the lack of consumption of rich food sources of these nutrients. Table 4.21 on the next page provides a list of plant-based food sources of four of these key nutrients. Vitamin B_{12} deficiency during pregnancy may not become apparent until after delivery. Two cases of neurological impairment and growth failure due to maternal B_{12} deficiency were identified in 4- to 8-month-old infants in Georgia in 2001. Both infants were born to women who followed a vegetarian diet during pregnancy.[96] Overall, the incidence of iron deficiency is the same in vegetarian and nonvegetarian pregnant women.[93]

Protein intake is generally adequate in the vegetarian diet, but may be low in vegans. Protein needs are met by vegetarians who regularly consume a variety of plant sources of protein and meet energy needs. In pregnant women who consume no animal products, the variety of plant protein sources needs to include complementary sources of protein daily. Protein sources that complement each other, or provide a complete source of protein, include legumes (such as lentils, chickpeas, black-eyed peas, black beans, and lima beans) and grains (corn, rice, bulgur, and barley, for example). Protein need is about 30% higher in vegetarians

Table 4.20 Vegetarian food guide adapted for pregnant women[93–95]

Food Group	Servings per Day
A. Grains	6–11
Whole grain bread, 1 slice	
Cooked grains, ½ c	
Fortified cold cereals, 1 oz	
Fortified cooked cereals, ½ c	
Corn, ½ c	
Pasta, ½ c	
Tortillas, 1 small	
Crackers, 4 small	
B. Legumes, Nuts, Seeds, Dairy	5–7
Dried beans, cooked, ½ c	
Peas, ½ c	
Soy products, ½ c or 2–3 oz	
Soynuts, ¼ c	
nut and seed butter, 2 Tbsp	
Nuts and seeds, ¼ c	
Eggs, 1	
Cow's milk, 1 c	
Cheese, 1 oz	
Yogurt, ½ c	
Fortified soymilk, 1 c	
C. Vegetables	4
Cooked vegetables, ½ c	
Raw vegetables, 1 c	
Vegetable juice, ½ c	
D. Fruits	2
Medium-sized fruit, 1	
Cut-up raw or cooked, ½ c	
Fruit juice, ½ c	
Dried fruit, ¼ c	
E. Fats, Oils, and Sweets	2+ depending on caloric need
Mayonaise, oil margarine, 1 Tbsp	
Honey, syrup, jams, jellies, sugar, 1 Tbsp	

than nonvegetarians due to the lower essential amino acid content and digestibility of plant protein (except for soy).[93]

Availability of vegetarian food products in large grocery and organic-food stores has expanded substantially in the past few years. Vegetarians can now select veggie burgers, meat analog entrees, meals-in-a-cup, and frozen desserts from food-store shelves. Fortified juice, soymilks, breakfast cereal, and meat substitutes are available and can contribute substantially to vegetarians' intake of vitamins B_{12} and D and calcium. DHA derived from algae can be used to provide a source of this omega-3 fatty acid in diets of vegetarian pregnant women who do not consume fish or seafood.[93] (Additional information on the omega-3 fatty acids follows.)

Table 4.21 Examples of sources of iron, calcium, vitamin D, and vitamin B$_{12}$ in vegetarian diets

Iron	
Food	Iron, mg
A. Legumes, Cooked	
Tofu, firm, ½ c	6.6
Soybeans, ½ c	4.4
Lentils, ½ c	3.3
Kidney beans, ½ c	2.6
Chickpeas, ½ c	2.4
Tempeh, ½ c	2.2
Black beans, ½ c	1.7
B. Nuts and Seeds	
Dried pumpkin and squash seeds, ¼ c	5.2
Sesame tahini, 2 Tbsp	2.7
Toasted sunflower seeds, ¼ c	2.3
Cashews, ¼ c	2.1
C. Fortified Breakfast Cereals	
Ready-to-eat cereals, 1 oz	2.1–18
Cream of wheat, cooked, ½ c	5.1
Instant oatmeal, cooked, ½ c	4.2
Quinoa, cooked, ½ c	2.1

Calcium	
Food	Calcium, mg
A. Legumes	
Fortified soy yogurt, ½ c	367
Tofu, firm, calcium set, ½ c	120–430
Fortified soymilk, 1 c	200–400
Soybean greens, ½ c	130
Tempeh, ½ c	92
Soybeans, cooked, ½ c	88
Navy beans, cooked, ½ c	64
Black beans, cooked, ½ c	46
B. Nuts and Seeds	
Sesame tahini, 2 Tbsp	128
Almonds, ¼ c	88
Almond butter, 2 Tbsp	86
Filberts (hazelnuts), chopped, ¼ c	64
Brazil nuts, ¼ c (2)	62

Calcium (continued)	
Food	Calcium, mg
C. Vegetables, Cooked	
Spinach, ½ c	122
Kale, ½ c	47
Broccoli, ½ c	36
Bok choy (Chinese cabbage), ½ c	36
D. Dairy Products	
Yogurt, ½ c	207
Cow's milk, 1 c	288
Swiss cheese, 1 oz	270
Frappuccino, 1 c	220
Cheddar cheese, 1 oz	204
Ice Cream, 1 c	180

Vitamin D	
Food	Vitamin D, mcg
Cow's milk, 1 c	2.5
Fortified soymilk, 1 c	1–3
Fortified breakfast cereals, 1 oz	0.5–1.0
Egg yolk, 1 large	0.6

Vitamin B$_{12}$	
Food	Vitamin B$_{12}$, mcg
Fortified breakfast cereals, 1 oz	0.6–6.0
Fortified soymilk, 1 c	0.8–3.2
Nutritional yeast, 1 Tbsp	1.5
Fortified veggie soy meats, 3 oz	1.5–3.6
Cow's milk, 1 c	0.8–1.0
Egg, 1 large	0.5

Computerized nutrient analysis of several days of usual food intake may be especially helpful in vegetarian diets due to the variability of dietary practices.[93] Evaluation of rate of weight gain in pregnancy is generally a good way to assess the adequacy of energy intake. Case Study 4.1 on page 108 is related to the dietary assessment results of a pregnant, vegan woman.

Maternal Intake of Omega-3 Fatty Acids and Pregnancy Outcome

EPA (eicosapentaenoic acid) and DHA (doceosahexaenoic acid) are two derivatives of the essential fatty acid alpha-linolenic acid. These fatty acids belong to the omega-3 fatty acid family and are also referred to as n-3 long-chained, polyunsaturated fatty acids. The term *omega-3* comes from the fact that the double bonds in the fatty acids begin at the third carbon from the end of the carbon chain (Illustration 4.13). Alpha-linolenic acid is found in good quantities in flaxseed, walnut oil, soybean oil, canola oil, and leafy green vegetables. However, the conversion of alpha-linolenic acid to EPA and DHA by the body is limited (9% in women).[97] Consequently, adequate intakes of EPA and DHA depend on the consumption of foods that contain these fatty acids. EPA is converted to DHA by the body.

EPA and DHA perform specific functions in the body that are particularly important during pregnancy and

Case Study 4.1

Photo Disc

Vegan Diet during Pregnancy

Ms. Lederman, a healthy 32-year-old woman entering her thirteenth week of pregnancy, asks her doctor for a referral to a dietitian to discuss her vegan diet. She receives the referral, and while making an appointment with the nutrition consulting service, is asked to record her food intake for 3 days prior to the appointment. Ms. L follows the instructions she was given and carefully completes a 3-day food record. Prior to the appointment, she sends her food record to the dietitian she will be seeing.

During the appointment, the dietitian learns that Ms. L started pregnancy at normal weight, has gained 3 pounds so far in pregnancy, has no history of iron or another nutrient deficiency, and is experiencing a normal course of pregnancy. Ms. L has been a vegan since the age of 16, and although she believes it is good for her health, she worries that her baby may not be getting the nutrients she or he needs. Ms. L wears sunscreen whenever she goes outside, so she makes little or no vitamin D in her skin. She makes sure to combine plant sources of protein (usually dried beans and grains), so she'll consume complete sources of protein every day.

Results of the dietary analysis performed by the dietitian showed the following average calorie and nutrient intake levels:

Kcal: 2237
Protein, g: 71
Linoleic acid (n-6 fatty acids), g: 15.2
Alpha-linolenic acid (n-3 fatty acids), g: 0.54
Vitamin B_{12}, mcg: 2.1
Vitamin D, mcg: 3 (120 IU)
Zinc, mg: 15

Questions

1. Is Ms. L consuming enough protein?
2. Based on the information presented, which nutrients are consumed in amounts that are below the DRI standard for pregnancy?
3. Suggest three types of food Ms. L could consume to bring up her intake of the nutrients identified in question 2.

---

lactation. Derivatives of EPA reduce inflammation, dilate blood vessels, and reduce blood clotting. DHA is a major structural component of phospholipids in cell membranes in the central nervous system, including retinal photoreceptors. High amounts of DHA are also found in sperm.

Optimal functioning of the central nervous system appears to depend on the availability of sufficient amounts of DHA during critical phases of growth and development when central nervous system tissues are being formed.[98]

Illustration 4.13 The structure of the omega-3 fatty acid DHA, showing the "alpha" end on the left and the "omega" end on the right.

Women who consume adequate amounts of EPA and DHA during pregnancy and lactation tend to deliver infants with somewhat higher levels of intelligence, better vision, and otherwise more mature central nervous system functioning than do women who consume low amounts of these fatty acids. Sufficient intake of EPA and DHA during pregnancy prolongs gestation by an average of 4 days, and decreases the risk of preterm delivery.[99,100]

Dietary Intake Recommendations for EPA and DHA   An adequate intake of EPA and DHA during pregnancy and lactation is estimated to be 300 mg per day.[101] Most women in the United States consume about a third of this amount, and vegan women are at particularly high risk of poor EPA and DHA status.[87] The Food and Drug Administration recommends that intake of EPA and DHA does not exceed 2 grams per day.

EPA and DHA are found together in fish, fish oils, and seafood (it turns out that fish really is "brain food"). DHA is available from egg yolk and DHA-fortified eggs and other products, and in certain types of algae that produce it. Human milk from women with adequate intakes is an excellent source of DHA. Prenatal supplements are becoming a source of DHA. Many types of prenatal supplements prescribed to pregnant women provide DHA in addition to vitamins and minerals. Table 1.7 in Chapter 1 includes a list of food sources of EPA and DHA.

Due to the presence of mercury and other contaminants in some types of fish, it is recommended that women who are pregnant or breastfeeding consume no more than 12 ounces of fish per week. Fish consumed should be good sources of EPA and DHA and contain low amounts of mercury and other contaminants. Fish known to generally contain high levels of mercury (swordfish, king mackerel, tilefish, and shark) should not be consumed. No more than 6 ounces per week of albacore tuna (labeled as "white tuna" on cans) should be consumed each week.[102] The Environmental Protection Agency provides information on the safety of locally caught fish at the website: www.epa.gov/waterscience/fish/states.htm.

Many pregnant women avoid eating fish during pregnancy due to a concern that its content of mercury and other pollutants may harm the baby. Avoidance of modest fish consumption due to confusion regarding risks and benefits could result in suboptimal neurodevelopment in children.[103] For women who do not like fish, fish oil (not fish liver oil which may contain high levels of vitamins D and A) supplementation appears safe and beneficial.[104]

The Need for Water during Pregnancy

The large increase in water need during pregnancy is generally met by increased levels of thirst. On average, women consume about 9 cups of fluid daily during pregnancy.[105] Women who engage in physical activity in hot and humid climates should drink enough to keep urine light colored and normal in volume. Water, diluted fruit juice, iced tea, and other unsweetened beverages are good choices for staying hydrated.

Folate and Pregnancy Outcome

Inadequate folate during pregnancy has long been associated with anemia in pregnancy and reduced fetal growth.[106] Only during the last two decades, however, has the broad spectrum of effects of folate been recognized. Discoveries of the multiple effects of inadequate folate intake on the development of congenital abnormalities and clinical complications of pregnancy represent some of the most important advances in our knowledge about nutrition and pregnancy.

Folate Backgrounder   The term *folate* encompasses all compounds that have the properties of folic acid and includes monoglutamate and polyglutamate forms of the vitamin. The monoglutamate form of folate is represented primarily by folic acid, a synthetic form of folate used in fortified foods and supplements. A similar monoglutamate form of folate naturally occurs in a few foods. Food sources of folate contain primarily the polyglutamate form of folate. The two major types of folates are often distinguished by referring to the monoglutamates as folic acid and the polyglutamates as dietary folate.

Bioavailability of folic acid and dietary folate differs substantially. Folic acid is nearly 100% bioavailable if taken in a supplement on an empty stomach, and 85% bioavailable if consumed with food or in fortified foods. Naturally occurring folates are 50% bioavailable on average.[107]

Folate requirements increase dramatically during pregnancy due to the extensive organ and tissue growth that takes place.

Functions of Folate   Folate is a methyl group (CH_3) donor and enzyme cofactor in metabolic reactions involved in the synthesis of DNA, gene expression, and gene regulation. Deficiency of folate impairs these processes, leading to abnormal cell division and tissue formation.[108] Folate serves as a methyl donor in the conversion of homocysteine to the amino acid methionine. The conversion of homocysteine to methionine depends primarily on three enzymes and folate, vitamin B_{12}, and vitamin B_6 cofactors. Lack of folate in particular, and less commonly a lack of vitamin B_{12}, as well as genetic abnormalities in the enzymes can lead to an accumulation of homocysteine. This may result in methionine shortage at a crucial stage of fetal development. High cellular and plasma levels of homocysteine may increase the risk of rupture of the placenta, stillbirth, preterm delivery, *preeclampsia,* structural abnormalities (congenital defects) in the newborn, and reduced birth weight. Folic

Preeclampsia   A pregnancy-specific condition that usually occurs after 20 weeks of pregnancy (but may occur earlier). It is characterized by increased blood pressure and protein in the urine and is associated with decreased blood flow to maternal organs and through the placenta.

acid supplements (500–600 mcg per day) in the second and third trimesters of pregnancy decrease homocysteine levels and improve pregnancy outcomes.

A common genetic defect has been identified in the enzyme 5,10-methylene tetrahydrofolate reductase (MTHFR). This variant of the normal enzyme reduces the level of activity of MTHFR by about half. Variant forms of MTHFR and defects in methionine synthase are thought to be present in approximately 30% of the population.[109]

Folate and Congenital Abnormalities

"There is a widespread belief that congenital malformations are always the result of defective genes. Perhaps a nutritional deficiency resulting in a defective gene leads to the same congenital abnormality."

R. D. Mussey, 1949[75]

Researchers have known since the 1950s that low and high intakes of certain vitamins and minerals cause congenital abnormalities in laboratory animals. They have also known that neural tube defects, brain and heart defects, and cleft palate can be caused by feeding pregnant rats folate-deficient diets.[110] Firmly held beliefs that only severe malnutrition affects fetal growth and that genetic errors are the sole cause of congenital abnormalities delayed recognition of the importance of folate to human pregnancy.[111]

Neural tube defects (NTDs) are malformations of the spinal cord and brain. There are three major types of NTDs:

- Spina bifida is marked by the spinal cord failing to close, leaving a gap where spinal fluid collects during pregnancy (see Illustration 4.14). Paralysis below the gap in the spinal cord occurs in severe cases.

- Anencephaly is the absence of the brain or spinal cord.

- Encephalocele is characterized by the protrusion of the brain through the skull.

Illustration 4.14 A newborn child with spina bifida.

SOURCE: Photo Researchs, Inc.

It is now well accepted that inadequate availability of folate between 21 and 27 days after conception (when the embryo is only 2–3 mm in length) can interrupt normal cell differentiation and cause NTDs.[112] Neural tube defects are among the most common types of congenital abnormalities identified in infants, with approximately 4000 pregnancies affected each year in the United States.[113] NTDs are among the most preventable types of congenital abnormalities that exist.[114] Approximately 70% of cases of NTDs can be prevented by consumption of adequate folate before and during very early pregnancy.[106]

Folate Status of Women in the United States

Folate status is assessed by serum and red cell folate levels. Of the two measures, red cell folate levels are the preferred indicator because they represent long-term folate intake, whereas serum folate levels reflect only recent intake. Levels of red cell folate of over 300 ng/mL (or 680 nmol/L) are associated with very low risk of NTDs.[115] These levels of red cell folate can generally be achieved by folic acid intakes that average 400 mcg daily.[116] As shown in Illustration 4.15, red cell folate levels are higher among women who consume folic-acid-fortified cereals or supplements compared to women consuming folate from food only.[116]

Folate status in women of childbearing age in the United States has improved since the advent of folic-acid fortification of refined grain products in 1998. Average levels of red cell folate in U.S. women have increased from 181 to 235 ng/mL since fortification began.[114] Low levels of intake of folic-acid-fortified grain products and breakfast cereals still leave some women with too little folate, however.[117]

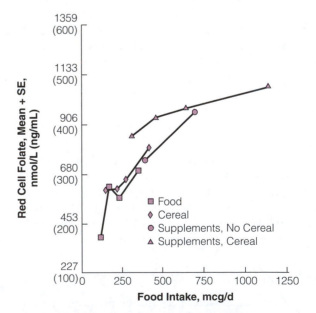

Illustration 4.15 Mean red cell folate level in preconceptional women by level of intake of various sources of folate.

SOURCE: *JAMA*, Vol. 277, No. 6, p. 551 (Feb 19, 1997).

Dietary Sources of Folate Many vegetables and fruits are good sources of folate (see Table 1.9 in Chapter 1), but only a few foods contain the highly bioavailable form of folate. Table 4.22 lists some foods that naturally contain the highly bioavailable, monoglutamate form of folate and foods that provide folic acid through fortification.

Adequacy of folic acid intake before and during pregnancy can be estimated by adding up the amount of folic acid in foods typically consumed in the daily diet using the data in the table. Whole grain products including breads and pastas, brown rice, oatmeal, shredded wheat, and organic grain products may or may not be fortified with folic acid. You have to check food labels to find that out.

Recommended Intake of Folate Due to variation in folate bioavailability, the DRI for folate takes into consideration a measure called *dietary folate equivalents,* or DFE. One DFE equals any of the following:

- 1 mcg food folate
- 0.6 mcg folic acid consumed in fortified foods or a supplement taken with food
- 0.5 mcg of folic acid taken as a supplement on an empty stomach

Folic acid taken in a supplement without food provides twice the dietary folate equivalents as does an equivalent amount of folate from food.

It is recommended that women consume 600 mcg DFE of folate per day during pregnancy and include 400 mcg folic acid from fortified foods or supplements.[107] The remaining 200 mcg DFE should be obtained from vegetables and fruits. These nutrient-dense foods provide an average of 40 mcg of folate per serving.[116] Because NTDs develop before women may realize they are pregnant, adequate folate should be consumed several months prior to, as well as throughout, pregnancy.

Table 4.22 Food sources of folic acid	Amount	Folic Acid mcg
A. Foods		
Orange	1	40
Orange juice	6 oz	82
Pineapple juice	6 oz	44
Papaya juice	6 oz	40
Dried beans	½ c	50
B. Fortified Foods		
Highly fortified breakfast cereals[a]	1 c or 1 oz	400
Breakfast cereals	1 c or 1 oz	100
Bread, roll	1 slice or 1 oz	40
Pasta	½ c	30
Rice	½ c	30

[a]Includes Product 19, Smart Start, Special K, and Total.

Women who have previously delivered an infant affected by an NTD are being urged to take 4000 mcg (4.0 mg) of folic acid in a supplement to reduce the risk of recurrence.[113] This dose, however, may be much higher than needed based on results of clinical trials.[118] The upper limit for intake of folic acid from fortified foods and supplements is set at 1000 mcg per day. There is no upper limit for folate consumed in its naturally occurring form in foods. The 1000 mcg level represents an amount of folic acid that may mask the neurological signs of vitamin B_{12} deficiency. If left untreated, vitamin B_{12} deficiency leads to irreversible neurological damage.[113]

Vitamin A and Pregnancy Outcome

Vitamin A is a key nutrient in pregnancy because it plays important roles in reactions involved in cell differentiation. Deficiency of this vitamin is rare in pregnant women in industrialized countries, but it is a major problem in many developing nations. Vitamin A deficiency that occurs early in pregnancy can produce malformations of fetal lungs, urinary tract, and heart.[119]

Of more concern than vitamin A deficiency in the United States are problems associated with excessive intakes of vitamin A in the form of retinol or retinoic acid (but not beta-carotene). Intakes of these forms of vitamin A of over 10,000 IU per day, and the use of medications such as Accutane and Retin-A for acne and wrinkle treatment, increase the risk of fetal abnormalities. Effects are particularly striking in infants born to women using Accutane or Retin-A early in pregnancy (see Illustration 4.16). Fetal exposure to the high doses of retinoic acid in these drugs tends to develop "retinoic acid syndrome." Features of this syndrome include small ears or no ears, abnormal or missing ear canals, brain malformation, and heart defects.[10]

Due to the potential toxicity of retinol, it is recommended that women take no more than 5,000 IU of vitamin A as retinol from supplements during pregnancy.[120] Most supplements made today contain beta-carotene rather than retinol. High intakes of beta-carotene have not been related to birth defects.[50] Although women are issued strong warnings not to take Retin-A or Accutane if pregnancy is possible, ill-timed use continues to occur to some extent, as does the retinoic acid syndrome.

Vitamin D Requirement

Vitamin D supports fetal growth, the addition of calcium to bone, and tooth and enamel formation. Lack of it compromises fetal growth and bone development, and this may be happening during many pregnancies. About 42% of African American and 4% of Caucasian women have low blood levels of vitamin D.[50] Vegan women are at risk for poor vitamin D status because vitamin D is naturally present only in animal products. These and other risk

Illustration 4.16 An 8-month-old infant exposed to high levels of retinoic acid in utero. Note the high forehead, flat nasal bridge, and malformed ear.

SOURCE: Used by permission of Harcourt Health Sciences, Inc. Lott IT et al., Fetal hydrocephalus and ear abnormalities associated with maternal use of isotretinoin, *J Pediatr* 1984;105:597–600.

factors for inadequate vitamin D status during pregnancy are listed in Table 4.23

The primary effect of insufficient vitamin D during pregnancy is poor fetal bone formation due to poor utilization of calcium.[19] Infants born to women with inadequate vitamin D status in pregnancy tend to be small, have poorly calcified bones and abnormal enamel, and show low blood levels of calcium after birth.[12]

An intake of 5 mcg (200 IU) vitamin D daily is officially recommended for pregnancy. This amount of vitamin D can be obtained by consuming 3 cups of vitamin D–fortified milk a day, or by exposing the skin to sunshine. Two 15-minute sunbathing sessions per week lead to the production of about 1250 mcg (50,000 IU) vitamin D and a low risk of sunburn in most people. Individuals with dark skin need two to five times this length of sun exposure to produce that much vitamin D. Winter sunlight in northern climates is too weak to produce vitamin D formation in the skin. There is no evidence that vitamin D overdose occurs due to sun exposure.[121]

Some credible experts assert that more vitamin D than 5 mcg daily during pregnancy is needed. It would not be surprising to see recommended intake levels for vitamin D increase.[123] Intakes of vitamin D from foods and supplements should not exceed 50 mcg (2000 IU) daily.

Table 4.23 Risk factors for vitamin D inadequacy during pregnancy

Vegan diet

Consumption of small amounts of vitamin D–fortified milk or of raw milk

Limited exposure of the skin to the direct rays of the sun

Consistent use of sun block

Dark skin

Calcium Requirements in Pregnancy

Calcium metabolism changes meaningfully during pregnancy. Absorption of calcium from food increases, excretion of calcium in urine likewise increases, and bone mineral turnover takes place at a higher rate.[123] The additional requirement for calcium in the last quarter of pregnancy is approximately 300 mg per day and may be obtained by increased absorption and by release of calcium from bone.[9] (Calcium is not taken from the teeth, however.) Calcium lost from bones appears to be replaced after pregnancy in women with adequate intakes of calcium and vitamin D.[51] Inadequate calcium intake has been related to increased blood pressure during pregnancy, decreased subsequent bone remineralization, increased blood pressure of infants, and decreased breast-milk concentration of calcium.[123]

Calcium and the Release of Lead from Bones Lead is transferred from the mother to the fetus and can disrupt normal development of the central nervous system.[124] Pregnant women who do not consume enough calcium show greater increases in blood lead levels than women who consume 1000 mg (the DRI for calcium) or more per day. Bone tissues contain about 95% of the body's lead content, and the lead is released into the bloodstream when bones demineralize. Bone tissues demineralize to a greater extent in pregnant women who fail to consume adequate calcium.[124]

Calcium needs during pregnancy can be met by drinking 3 cups of milk or calcium-fortified soymilk, or 2 cups of calcium-fortified orange juice plus a cup of milk, or by choosing a sufficient number of other good sources of calcium daily. (Table 1.12 in Chapter 1 lists food sources of calcium.)

Fluoride

Teeth develop in utero, so why isn't it recommended that pregnant women consume sufficient fluoride so that the fetus builds cavity-resistant teeth? This is a logical question, but it has the answer "because only trace amounts of fluoride pass through the placenta to the fetus." Children of pregnant women given fluoride supplements during pregnancy have the same rates of dental caries as do children of women who did not receive supplements.[125]

Iron Status and the Course and Outcome of Pregnancy

Iron status is a leading topic of discussion in prenatal nutrition because the need for iron increases substantially; women require about 1000 mg (1 g) of additional iron during pregnancy:

- 300 mg is used by the fetus and placenta.
- 250 mg is lost at delivery.
- 450 mg is used to increase red blood cell mass.

Maternal iron stores get a boost after delivery when iron liberated during the breakdown of surplus red blood cells is recycled.[126]

Approximately 12% of women enter pregnancy with *iron deficiency* and little stored iron, and consequently are at risk of developing *iron-deficiency anemia* in pregnancy.[127]

Iron-Deficiency Anemia in Pregnancy Over recent decades, rates of iron-deficiency anemia in pregnancy have remained high in women in developing as well as developed countries (Table 4.24). Iron-deficiency anemia at the beginning of pregnancy increases the risk of preterm delivery and low-birth-weight infants by two to three

Table 4.24 Estimates of the incidence of iron-deficiency anemia in women in developing and developed countries[126,127]

	% with Iron-Deficiency Anemia	
	Developing Countries	Developed Countries
Nonpregnant	43	12
Pregnant	56	18

times.[128] Iron deficiency during pregnancy is related to lower scores on intelligence, language, gross motor, and attention tests in affected children at the age of 5 years. The mechanisms underlying these effects are unknown, but they may be related to decreased oxygen delivery to the placenta and fetus, increased rates of infection, and altered neurotransmitter function or nerve formation in the fetal brain.[129] Iron deficiency often occurs toward the end of pregnancy even among women who enter pregnancy with some iron stores. It is far more common than iron-deficiency anemia.

Iron Deficiency A condition marked by depleted iron stores. It is characterized by weakness, fatigue, short attention span, poor appetite, increased susceptibility to infection, and irritability.

Iron-Deficiency Anemia A condition often marked by low hemoglobin level. It is characterized by the signs of iron deficiency plus paleness, exhaustion, and a rapid heart rate.

Assessment of Iron Status Red cell mass increases substantially (30%) in pregnancy. However, plasma volume expands more (by about 50%). The higher increase in plasma volume compared to red cell mass makes it appear that amounts of hemoglobin, ferritin, and packed red blood cells have decreased.[8] They have not decreased but rather have become diluted by the large increase in plasma volume. Hemoglobin concentration normally decreases until the middle of the second trimester and then rises somewhat in the third. It is not necessary to prevent normal declines in hemoglobin level during pregnancy.[130]

Due to the dilution effects of increased plasma volume, changes in hemoglobin levels tend to be more indicative of plasma volume expansion than of iron status.[128] Low levels of hemoglobin or serum ferritin may be associated with high plasma volume expansion (hypervolemia), and high hemoglobin levels are related to low plasma volume expansion (hypovolemia). Low levels of plasma volume expansion are associated with reduced fetal growth, whereas newborns tend to be larger in women with higher levels of plasma volume expansion.[131]

The Centers for Disease Control have developed standard hemoglobin levels to be used in the identification of iron-deficiency anemia in pregnant women. These standards (shown in Table 4.25) represent levels below the 5th percentile of hemoglobin values in pregnancy.[132]

Table 4.25 CDC's gestational age-specific cutoffs for anemia in pregnancy[132]

Gestational Weeks	Hemoglobin (g/dL) Indicating Anemia[a]
12	<11.0
16	<10.6
20	<10.5
24	<10.5
28	<10.7
32	<11.0
36	<11.4
40	<11.9

[a]For women living in high altitudes, hemoglobin values should be increased by 0.2 g/dL for every 1000 feet above 3000 and by 0.3 g/dL for every 1000 feet above 7000. For cigarette smokers, hemoglobin values should be adjusted upward by 0.3 g/dL.

By trimester, hemoglobin levels indicative of iron-deficiency anemia are:

- <11.0 g/dL in the first and third trimesters
- <10.5 g/dL in the second trimester

Serum ferritin cut-points indicative of iron-deficiency anemia in pregnancy have also been developed:[132]

	Serum Ferritin, ng/mL
Normal	>35
Depleted Stores	<20
Iron Deficiency	≤15

Hemoglobin and serum ferritin are the most commonly employed measures of iron status in pregnant women.[32]

The diagnosis of iron-deficiency anemia is more complicated than often thought. No single test of iron status is totally accurate, because (1) many factors, including infection and inflammatory disease, affect iron status; and (2) each test measures a different aspect of iron status. It is best to base the diagnosis of iron-deficiency anemia on results of several tests.[132]

Women entering pregnancy with adequate iron stores tend to absorb about 10% of total iron ingested; those with low stores absorb more—about 20% of the iron consumed. The largest percentage of iron absorption, 40%, occurs in women who enter pregnancy with iron-deficiency anemia.

Iron absorption from foods and supplements is enhanced in women with low iron stores during pregnancy, and absorption increases as pregnancy progresses.[32] Absorption is highest after the thirtieth week of pregnancy, when the greatest amount of iron transfer to the fetus occurs. The amount of iron the placenta can transfer to the fetus, however, is limited in women who are iron deficient. Maternal iron depletion in pregnancy decreases fetal iron stores, increases the risk that infants will develop iron-deficiency anemia, and is associated with development of maternal postpartum depression.[50]

Pros and Cons of Iron Supplementation Absorption of iron from multimineral supplements is substantially lower than is iron absorption from supplements containing iron only. For example, women given a multimineral supplement containing iron, calcium, and magnesium absorb less than 5% of the iron, whereas women given a similar dose of iron in a supplement containing iron only absorb over twice that much.[133]

The amount of iron absorbed from supplements depends primarily on women's need for iron and the amount of iron in the supplement. As can be seen in Illustration 4.17, the amount of iron absorbed from

Illustration 4.17 Effect of dose of supplemental iron on iron absorption in women during pregnancy.

SOURCE: From Pedro Rosso, *Nutrition and Metabolism in Pregnancy: Mother and Fetus.* Copyright © 1990 by Oxford University Press, Inc. Used by permission of Oxford University Press, Inc.

supplements decreases substantially as the dose of iron increases. Although controversial, one theory states that the small level of improvement in iron absorption that occurs with doses of iron over 30 or 60 mg may be too little to justify their use.[130] In addition, the acceptance of high levels of iron supplementation by women is often poor, due to side effects related to the ingestion of high amounts of iron. Nausea, cramps, gas, and constipation are associated with the presence of free iron in the intestines, and these side effects increase as doses of supplemental iron increase (Table 4.26). Side effects experienced in using iron supplements are a major reason that women fail to take them.[128] Difficulties related to building iron stores during pregnancy provide a strong rationale for screening women for iron status prior to pregnancy and establishing good levels of stored iron before pregnancy if needed.[50]

A relatively new concern about high-dose iron supplements is emerging. Iron supplements providing 60 mg or more iron per day regularly expose the intestinal mucosa to free iron radicals. The oxidizing effects of iron radicals cause inflammation and mitochondrial damage in cells.[12] In addition, iron doses over 30 mg per day decrease zinc absorption and lower zinc status.[32]

Amounts of elemental iron in supplements vary depending on the form of the iron compound in the supplement (Table 4.27). The proportion of iron absorbed from a constant amount of iron from each type of supplement listed in Table 4.27 is approximately equal.[136]

Since 1997 the FDA has required that the iron content of multivitamin and mineral supplements not exceed 30 mg. This action limits the excess iron that will remain in the gut after ingestion of multivitamin and mineral supplements with iron. Higher doses of iron have to be provided as single-ingredient iron supplements.

Table 4.26 Increased occurrence of side effects in women by supplemental iron dose[134,135]

Dose of Iron, mg/day	Side Effects
60	32%
120	40%
240	72%

Table 4.27 Percent of elemental iron by weight in various types of iron supplements

Supplement Type	Iron Content
Ferrous sulfate	20%
Ferrous gluconate	12%
Ferrous fumarate	32%

Recommendations Related to Iron Supplementation in Pregnancy   It is generally recommended that pregnant women in the United States take a 30-mg iron supplement daily after the twelveth week of pregnancy.[137] Women with iron-deficiency anemia are often given 60–180 mg of iron per day.[138] Iron supplementation during pregnancy, and usual clinical practices related to it, are subjects of heated debate.

Although iron supplements increase iron stores and help prevent anemia in many women who take them, some pregnant women do not benefit from iron supplements because they do not need them. Unused iron supplements, when stored and later found by young children, pose a risk of iron poisoning. Because iron status early in pregnancy is associated with adverse newborn outcomes, iron supplements may be given too late in pregnancy to achieve optimal improvements in outcomes.[130]

Proposed Alternatives to Routine Iron Supplementation   It has been suggested that women's iron status be assessed at the first prenatal visit to determine if there is a need for iron supplements. A 30-mg iron supplement would be indicated when hemoglobin levels are <11 g/dL, or if serum ferritin levels are <30 mcg/L. Women with higher values would be monitored for iron status but not given a supplement.[137]

Recommended Intake of Iron during Pregnancy
The increased need for iron can be met by intakes that lead to an additional 3.7 mg absorbed iron per day on average throughout pregnancy. This is a large increase, especially considering that nonpregnant women consuming the DRI for iron (18 mg) absorb only around 1.8 mg of iron daily. Given an ongoing need for 1.8 mg of absorbed iron a day, and the additional need of 3.7 mg of iron daily for pregnancy, the total need for absorbed iron during pregnancy is 5.5 mg daily. Assuming 20% of iron consumed is absorbed, average iron consumption of 27 mg per day (the DRI for iron for pregnancy) will meet the iron needs of pregnancy. The Upper Limit for iron intake during pregnancy is set at 45 mg per day.

Zinc Requirement in Pregnancy

Zinc functions as a cofactor for many enzymes, including those involved in protein synthesis. Like iron, the bioavailability of zinc is higher in meats and low in plants, especially whole grains. Low levels of zinc intake, reliance on whole grain cereals for dietary zinc, and doses of iron supplements over 30 mg daily all decrease zinc status.[139]

A firm consensus on effects of zinc deficiency in pregnancy has not been reached, although animal studies clearly show that it is related to growth retardation and malformations.[140] Levels of serum zinc representative of marginal deficiency have been associated with preterm delivery, intrapartum hemorrhage, infections, and prolonged labor in human studies. It is difficult to know if these results reflect true relationships, however, because serum zinc does not appear to be a very good marker of zinc status.[139]

Iodine and Pregnancy Outcome

Iodine is needed for the synthesis of thyroid hormones and protein tissues, and deficiency of it early in pregnancy can lead to *hypothyroidism* in the offspring. Hypothyroidism in infants is endemic in parts of southern and eastern Europe, Asia, Africa, and Latin America.[141] The incidence of infant hypothyroidism has been found to decrease by over 70% when at-risk women in developing countries are given iodine supplements before or in the first half of pregnancy.

> **Hypothyroidism**   A condition characterized by growth impairment and mental retardation and deafness when caused by inadequate maternal intake of iodine during pregnancy. Used to be called *cretinism*.

Rates of infant deaths are also substantially improved, as is the psychomotor development of the offspring.[146] Iodine supplementation in the second half of pregnancy does not improve infant outcomes.[141]

About half of pregnant women in the United States consume less than the recommended 220 mcg of iodine daily, and 7% have low urinary iodine levels.[142] The most reliable source of iodine is iodized salt. One teaspoon contains 400 mcg iodine. Fish, shellfish, seaweed, and some types of tea provide iodine. Iodine also ends up in foods manufactured in plants that use iodine-containing solutions to clean equipment. Women who consume iodized salt are not likely to need supplemental iodine. Usual iodine intake should not exceed 1100 mcg daily during pregnancy.

Antioxidants from Plants

Normal pregnancy is a pro-oxidative state and is accompanied by an increased requirement for antioxidants. Many plant pigments act as antioxidants and help protect fetal DNA from damage due to exposure to oxygen and other oxidizing chemicals produced in the body. Antioxidants also reduce maternal tissue damage associated with inflammation and oxidation. Vitamins C and E (also found in plant foods) likewise perform important antioxidant roles during pregnancy.[143] High intakes during pregnancy of foods rich in Vitamin E, for example, appear to reduce the risk of asthma in children. Children born to women consuming diets providing 26 mg of vitamin E daily on average are less likely to develop wheezing and asthma during early childhood than are children born to women who consume 14 mg of vitamin E daily.[144] It is suggested that vitamin E intake during pregnancy may reduce asthma by decreasing lung inflammation in the offspring.[145]

Foods rich in antioxidants advertise that fact by their color. Red, orange, dark green, deep yellow, and blue-purple fruits and vegetables generally provide good amounts of antioxidants. Many of these same foods are rich in vitamin C. Women who are pregnant should consume at least 5 cups of vegetables and fruits daily.

The Need for Sodium during Pregnancy

Sodium plays a critical role in maintaining the body's water balance. Requirements for it increase markedly during pregnancy due to plasma volume expansion. But the need for increased amounts of sodium in pregnancy hasn't always been appreciated. Thirty years ago in the United States, it was accepted practice to put all pregnant women on low-sodium diets. (Routine sodium restriction is still practiced in some European countries.) It was then thought that sodium increased water retention and blood pressure, and that sodium restriction would prevent edema and high blood pressure. We now know this isn't accurate and that inadequate sodium intake can complicate the course and outcome of pregnancy.[147] Sodium restriction during pregnancy may exhaust sodium conservation mechanisms and lead to excessive sodium loss.[148]

Sodium restriction is not indicated in normal pregnancy or for the control of edema or high blood pressure that develops in pregnancy. Women should be given no advice on modifying sodium intake; if they ask about it, they should be advised to consume salt "to taste."[149]

Caffeine Use in Pregnancy

Caffeine has long been suspected of causing adverse effects in pregnant women because it increases heart rate, acts as a diuretic, and stimulates the central nervous system. It easily passes from maternal to fetal blood and lingers in the fetus longer than in maternal blood because the fetus excretes it more slowly.[150]

Due to its high caffeine level, coffee has generally been the target of investigations on the effects of caffeine intake on pregnancy outcome. Coffee, however, contains hundreds of substances, and some of these have effects similar to those of caffeine. So, although conclusions about caffeine's effects on pregnancy are largely based on coffee intake, it is possible that other components of coffee are responsible for effects observed. Coffee is by far the largest contributor to caffeine intake in most people,[151] and pregnant women consume on average 170 mg caffeine from coffee per day in pregnancy.[152] (Table 2.4 in Chapter 2 provides a list of the caffeine content of beverages and foods.)

Despite the possibilities, caffeine and coffee intake during pregnancy have not been related to an increased risk of fetal malformations, reduced fetal growth, labor or delivery complications, preterm delivery, or developmental problems in offspring.[150,151]

No long-term consequences of coffee intake during pregnancy have been observed in children 7 years later. Children of women who drank coffee during pregnancy have been found to have similar levels of intellectual and neuromotor development when compared to non-coffee drinkers.[152] High levels of coffee and caffeine intake (over approximately 500 mg per day) have, however, been related to miscarriage.[150] It is generally concluded that intake of up to 4 cups of coffee per day during pregnancy is safe.

Healthy Diets for Pregnancy

Healthy diets for women during pregnancy are described in terms of calories and nutrient intake, and by food choices. Such diets have a number of characteristics in common (Table 4.28).

Adequacy of caloric intake during pregnancy is generally based on rate of weight gain, but for nutrients it is based on the DRIs (Table 4.29 on the next page). Nutrient intakes during pregnancy should approximate those given in the DRI table, and food intake should correspond to the recommended types and quantity of food recommended in MyPyramid. (MyPyramid recommendations are presented in an upcoming section on dietary assessment in pregnancy.)

Effect of Taste and Smell Changes on Dietary Intake during Pregnancy

No inner voice directs women to consume foods that provide needed nutrients during pregnancy. Pregnant women may, however, develop food preferences and aversions due to changes in the sense of taste and smell; and they may experience *pica*.

Table 4.28 Basics of a good diet for normal pregnancy

Good pregnancy diets:

1. Provide sufficient calories to support appropriate rates of weight gain.
2. Follow the MyPyramid food group recommendations.
3. Provide all essential nutrients at recommended levels of intake from the diet (with the possible exception of iron).
4. Include 600 mcg folate, of which 400 mcg is folic acid, daily.
5. Include 300 mg EPA and DHA daily.
6. Provide sufficient dietary fiber (28 g/day)
7. Include 9 c fluid daily.
8. Include salt "to taste."
9. Exclude alcohol and limit coffee intake to ≤4 cups per day.
10. Are satisfying and enjoyable.

Changes in the way certain foods taste, and the odor of foods and other substances, affect two out of three women during pregnancy. If asked to recall, many previously pregnant women could tell you which foods tasted really good to them, and which odors made them feel queasy to even think about. Increased preference for foods such as sweets, fruits, salty foods, and dairy products are common.[17] The odors of meat being cooked, coffee, perfume, cigarette smoke, and gasoline are common nasal offenders and may stimulate episodes of nausea.[136] The biological bases for such changes are not known, but they are suspected of being related to hormonal changes of pregnancy.

Pica

"Pica permits the mind no rest until it is satisfied."

F. W. Craig, 1935

Classified as an eating disorder, pica affects over half of pregnant women in some locations of the southern part of the United States. It is more common in African Americans than in other ethnic groups, and it is common enough to be considered a normal behavior in some countries. Historically, one type of pica—*geophagia*—was thought to provide women with additional minerals and to ease gastrointestinal upsets. The cause of pica remains a mystery.[153]

Nonfood items most commonly craved and consumed by pregnant women with pica include ice or freezer frost (*pagophagia*), laundry starch or cornstarch (*amylophagia*), baking soda and powder, and clay or dirt (geophagia). Women experiencing pica are more likely to be iron deficient than those who don't, and iron-deficiency anemia is especially common among pregnant women who compulsively consume ice or freezer frost.[154] It is not clear, however, whether iron deficiency leads to pica or if pica leads to iron deficiency.

> **Pica**   An eating disorder characterized by the compulsion to eat substances that are not food.
>
> **Geophagia**   Compulsive consumption of clay or dirt.
>
> **Pagophagia**   Compulsive consumption of ice or freezer frost.
>
> **Amylophagia**   Compulsive consumption of laundry starch or cornstarch.

Pica does not appear to be related to newborn weight or preterm delivery. It can, however, complicate control of gestational diabetes if starch is eaten, and it has caused lead poisoning, intestinal obstruction, and parasitic infestation of the gastrointestinal tract.[154] Women with amylophagia sometimes accept powdered milk as an alternative to laundry starch or cornstarch, and treating anemia often stops the craving for ice or freezer frost.

Assessment of Dietary Intake during Pregnancy

Routine assessment of dietary practices is recommended for all pregnant women to determine the need for an

Table 4.29 Dietary Reference Intakes (DRIs) for pregnant and nonpregnant women aged 19–30 years*

	Pregnant	Nonpregnant	Upper Limit (UL)
Energy, kcal			
2nd trimester	+350	2403	—
3rd trimester	+452		
Protein, gm	71	46	—
Linoleic acid, g	13	12	—
Alpha-linolenic acid, g	1.4	1.1	—
Vitamin A, mcg	770	700	3000
Vitamin C, mg	85	75	2000
Vitamin D, mcg[a]	5	5	50
Vitamin E, mg	15	15	1000[c]
Vitamin K, mcg	90	90	—
Thiamin, mg	1.4	1.1	—
Riboflavin, mg	1.4	1.1	—
Niacin, mg	18	14	35[c]
Vitamin B$_6$	1.9	1.3	100
Folate, mcg[b]	600	400	1000[c,d]
Vitamin B$_{12}$, mcg	2.6	2.4	—
Pantothenic acid, mcg	6	5	—
Biotin, mcg	30	30	—
Choline, g	450	425	3.5
Calcium, mg	1000	1000	2500
Chromium, mcg	30	25	—
Copper, mcg	1000	900	10,000
Fluoride, mg	3	3	10
Iodine, mcg	220	150	1100
Iron, mg	27	18	45
Magnesium, mg	350	310	350[c]
Manganese, mg	2	1.8	11
Molybdenum, mcg	50	45	2000
Phosphorus, mg	700	700	3500
Selenium, mcg	60	55	400
Zinc, mg	11	8	40

*DRIs for females <19 and >30 years are listed inside the front covers of this book.

[a] 1 mcg = 40 IU vitamin D; DRI applies in the absence of adequate sunlight.

[b] As Dietary Folate Equivalent (DFE). 1 DFE = 1 mcg food folate = 0.6 mcg folic acid from fortified food or supplement consumed with food = 0.5 mcg of a supplement taken on an empty stomach.

[c] UL applies to intake from supplements or synthetic form only.

[d] Applies to intake of folic acid.

improved diet or vitamin and mineral supplements.[32] Dietary assessment in pregnancy should cover usual dietary intake, dietary supplement use, and weight-gain progress. For best results, several days of accurately recorded, usual intake should be used.

Several levels of dietary assessment can be undertaken. (Internet resources for dietary assessment are listed at the end of the chapter.) Which assessment level is best primarily depends on the skill level of the health professional responsible for interpreting the results. Results of food-based assessments are rather straightforward to interpret, whereas computerized assessments of levels of nutrient intake are more complex.

The MyPyramid food guide (addressed in Chapter 1) does not include specific recommendations for pregnant women. Nonetheless, this food guide provides a good way to assess the overall quality of a typical day of food intake. Table 4.30 presents this guide based on a caloric need of 2400, which is not unusual for pregnant women in the United States. Daily amounts recommended for each food group can be compared to that recorded by a pregnant woman and the results used to identify the general quality of the diet. Table 4.31 shows how closely the example of a day's diet based on MyPyramid recommendations matches recommended levels of nutrient intake for pregnancy. With the exception of vitamin E and iron (which are supplied in rather low amounts by the example diet), nutrient levels correspond to recommended intakes for pregnancy.

Computerized analysis, given accurate records and entry of dietary intake and a high-quality nutrient database, provides results useful for estimating the quantity of calories and nutrients consumed. Detailed knowledge of dietary intake is particularly useful for women at risk of nutrient inadequacies or excesses, and for women with conditions such as gestational diabetes, food intolerances, multiple fetuses, and other special dietary needs.

Evaluation of dietary supplement use includes an examination of types and amounts of supplements consumed. Levels of intake exceeding the Upper Limits (ULs) for pregnancy, as well as the use of herbs and other dietary supplements not known to be safe for pregnancy, should be identified.

Cultural Considerations People tend to be attached to existing food preferences, many of which may have deep cultural roots. Dietary recommendations will differ for Native Alaskans accustomed to a diet based on wild game; for Cambodians, Vietnamese, and Somalis who may think no meal is complete without rice; and for lactose-intolerant individuals.

The belief that consumption of certain foods "marks" the baby is common in many cultures. People may think, for example, that a woman who loves mangos and eats lots of them during pregnancy may have a baby born with a "mango-shaped" birthmark. Some cultures would hold that the baby will also have learned to love mangos because mom ate them often while pregnant.

Dietary recommendations that are not consistent with a person's usual dietary practices and beliefs, or that are not viewed as acceptable or even preferred by the woman, are least likely to be effective. For best results,

Table 4.30 MyPyramid Food Guide for pregnant women[a]

Food Group	Ounces/cups recommended per day	Examples of Equivalent Measures
Grains	8 oz	• 1 slice bread = 1 oz • 1 c cold cereal = 1 oz • 1 c cooked rice, pasta, or cereal = 2 oz
Vegetables	3 c	• 2 c tossed salad = 1 c
Fruits	2 c	• 1 c fruit juice = 1 c
Milk	3 c	• ⅓ c shredded cheese = 1 c • 2 slices American cheese = 1 c • 1½ oz hard cheese = 1 c • 1½ c ice cream = 1 c
Meat and beans	6½ oz	• 1 small egg = 1 oz • 1 Tbsp peanut butter = 1 oz • ¼ c dried beans = 1 oz • ½ oz nuts = 1 oz
Oils	7 tsp	• 1 Tbsp mayonnaise = 2½ tsp oil • 1 Tbsp salad dressing = 1 tsp oil

[a]Based on MyPyramid food guide for women consuming 2400 calories a day. Serving numbers will vary based on actual caloric need.

Table 4.31 An example of 1 day's typical diet for a pregnant women based on MyPyramid food intake recommendations for a 2400 calorie diet, and results of an energy and nutrient assessment[a] of the day's diet

1 Day Typical Diet

Grains, 8 oz	Cheerios, 1 c
	Cracked wheat bread, 2 slices
	Brown rice, 1 c
	Rolls, 2 oz
	Corn tortilla, 1 oz
Vegetables, 3 c	Carrots, cooked, 1 c
	Tomato slices, 1 c
	Potatoes, boiled, 1 c
Fruit, 2 c	Orange, 1
	Banana, 1
Milk, 3 c	1% milk, 2 c
	American cheese, 1½ oz
Meat and beans	Light tuna in water, 2½ oz
	Chicken, baked, no skin, 4 oz
Fats and oils, 7 tsp	soybean oil, 4 tsp
	Italian dressing, 2 Tbsp
360 Kcal	Coke, 12 oz
	sugar, 2 tsp
	margarine, 4 tsp
Energy and nutrient analysis of the 1 day's diet[a]	

Energy (Kcal) in the Diet   2376

	Nutrient Analysis	Pregnancy RDA/AI
Protein, g	108	71
Fiber, g	26	28
Total fat, g	24	
Vitamin A, mcg	2120	770
Vitamin C, mg	125	85
Vitamin E, mg	8.8	15
Thiamin (B_1), mg	1.8	1.4
Riboflavin (B_2), mg	2.4	1.4
Niacin(B_3), mg	43	18
Folate, mcg	488	600
Vitamin B_6, mg	3.4	1.9
Vitamin B_{12}, mcg	3.4	2.6
Calcium, mg	1350	1000
Magnesium, mg	421	350
Iron, mg	18	27
Zinc, mg	13	11
Selenium, mcg	177	60
Potassium, mg	3720	4700

[a]Analyzed using the mypyramid.gov program.

dietary adjustments recommended for each individual pregnant woman should take into account her usual practices and preferences.

Vitamin and Mineral Supplementation during Pregnancy

The only supplement routinely recommended for pregnant women is iron.[32] Vitamin and mineral supplements may be indicated for specific individuals, however. An appropriate vitamin or mineral supplement is recommended when dietary and clinical assessments determine that a woman is deficient in a particular nutrient. It is also recommended that certain groups of women at high risk for poor dietary intake or increased nutrient need be given a multivitamin and mineral supplement in place of the iron supplement. Individuals in these groups include women who:

• Have a poor-quality, unchangeable diet
• Are pregnant with multiple fetuses

- Are vegan
- Smoke cigarettes
- Have iron-deficiency anemia
- Use illicit drugs or abuse alcohol[32]

The composition of prenatal vitamin and mineral supplements varies considerably. Prescription prenatal supplements contain 1 mg folic acid—an amount that can only be obtained by prescription. Supplements provided to pregnant women should contain only essential nutrients most likely to be lacking in their diets. These nutrients include vitamin B_6, folic acid, vitamin D, iron, zinc, and calcium. Nutrient amounts should approximate recommended intake levels and not exceed Tolerable Upper Intake Levels for pregnant women. Supplement use should be accompanied by nutritional counseling that helps women select and consume foods that add up to a healthful diet.

Herbal Remedies and Pregnancy

Herbal remedies are becoming commonly used during pregnancy, although very little is known about their safety and effectiveness. Little is known because herbs are rarely tested in pregnant women due to concerns about potential damage to the fetus. About one-third of commonly used herbal remedies have been deemed unsafe for use by pregnant women.[155] Table 4.32 provides a list of some of these herbs.

Advice to use herbal remedies during pregnancy appears to be based primarily on their traditional use in different societies. This strategy for assessing the safety of herbs doesn't always work. Some herbs considered safe based on traditional use have been found to produce malformations in animal studies.[155] Others, such as blue cohosh, which was previously thought to safely induce uterine contractions, may increase the risk of heart failure in the baby.[156] Ginseng, the most commonly used herb in the world, has been found to cause malformations in rat embryos[157] and ginkgo may promote excessive bleeding.[158]

Table 4.32 Herbs to avoid in pregnancy[156,157]	
Aloe vera	Ergot
Anise	Feverfew
Black cohosh	Ginkgo
Black haw	Ginseng
Blue cohosh	Juniper
Borage	Kava
Buckthorn	Licorice
Comfrey	Pennyroyal
Cotton root	Raspberry leaf
Dandelion leaf	Saw palmetto
Ephedra, ma huang	Senna

Peppermint tea and ginger root, taken for nausea, appear to be safe.[155]

Ginger, given in oral doses of 1 gram daily for 4 days, has been found to decrease the severity of nausea and vomiting during pregnancy in a majority of women. Ginger use in this study involving 70 women was not related to complications of pregnancy or poor pregnancy outcomes.[159]

Manufacturers of herbal remedies do not have to prove they are safe for use by pregnant women. However, the FDA does advise that claims related to pregnancy not be made for herbal supplements.

Exercise and Pregnancy Outcome

"Exercise is no longer simply being allowed during pregnancy, it is actively being encouraged."

K. Johnson[160]

There is no evidence that moderate or vigorous exercise undertaken by healthy women consuming high-quality diets and gaining appropriate amounts of weight is harmful to mother or fetus.[87] The bulk of evidence indicates that exercise during pregnancy benefits both the mother and her fetus. Women who exercise regularly during pregnancy feel healthier and have an enhanced sense of well-being, and their labors appear to be somewhat shorter than is the case for women who do not exercise.[161]

Researchers and practitioners are beginning to focus more on the advantages of exercise than on possible disadvantages. Women who exercise regularly during pregnancy reduce their risk of developing gestational diabetes, pregnancy-induced hypertension, low back pain, excessive weight gain, and blood clots.[162]

Exercise during pregnancy can reduce fetal growth in women who are poorly nourished and gain little weight in pregnancy. It is also important for women to avoid dehydration by drinking plenty of fluids while exercising and not to become overheated during physical activity.[87]

Is it safe to begin an exercise program during pregnancy? Not only is it generally safe, it is being encouraged.[160] Beginning an exercise program during pregnancy may improve fetal growth. This effect was shown in a study involving nonexercising pregnant women who began to exercise at 8 weeks of pregnancy. Women participated in three to five weight-bearing exercise sessions a week until delivery. Placenta function was better, and newborn weight and length greater, in exercising women compared to women who did not exercise.[163]

Exercise Recommendations for Pregnant Women

Exercise recommendations for pregnant women are similar to those for other healthy women. Pregnant women should exercise three to five times a week for 30 minutes at a heart

Table 4.33 Target heart rates for healthy pregnant women[87]	
Age, years	Heart Rate
<20	140–155
20–29	135–150
30–39	130–145
40+	125–140

rate that achieves 60–70% VO_2 max (Table 4.33). Exercise should begin with about 5 minutes of warm-up stretches and movements and end with the same length of cooldown activities. Recommended types of exercise include walking, cycling, swimming, jogging, and dancing. Better left until after pregnancy are activities such as water and snow skiing, surfing, mountain climbing, scuba diving, and horseback riding. Switching to non-weight-bearing exercises is advised toward the end of pregnancy.[87]

Food Safety Issues during Pregnancy

Certain foodborne illness can be devastating during pregnancy. Increased progesterone levels that normally occur decrease pregnant women's ability to resist infectious diseases, so they are more susceptible to the effects of foodborne infections.[154] One particularly important foodborne illness is caused by *Listeria monocytogenes*. The placenta does not protect the fetus from listeria infection in the mother. Listerosis during pregnancy is associated with spontaneous abortion and stillbirth.[164] To prevent this foodborne infection, pregnant women should not eat raw fish, oysters, unpasteurized cheese, raw or undercooked meat, or unpasteurized milk. Luncheon meats, hot dogs, and other processed meats should be stored correctly.

The protozoan *Toxoplasma gondii* also causes serious effects in pregnant women and their fetuses. This protozoan can be transferred from mother to fetus and cause mental retardation, blindness, seizures, and death.[155] Sources of *T. gondii* include raw and undercooked meats, the surface of fruits and vegetables, and cat litter. Cats that eat wild animals and undercooked meats can become infected and transfer the infection through the air and via stools left in their litter boxes.[165]

Mercury Contamination

Fish have come under fire as a potential source of mercury overload due to contamination of waters and fish by fungicides, fossil fuel exhaust, and products used in smelting plants, pulp and paper mills, leather-tanning facilities, and chemical manufacturing plants. Mercury, which passes from the mother's blood to the fetus, is a fetal neurotoxin that can produce mild to severe

effects on fetal brain development. Fetuses exposed to high amounts of mercury can develop mental retardation, hearing loss, numbness, and seizures. Pregnant women are generally only slightly affected by the mercury overload. However, it accumulates in the mother's tissues and may increase fetal exposure to mercury during pregnancy and lactation.[166]

L. Monocytogenes, or Listeria A foodborne bacterial infection that can lead to preterm delivery and stillbirth in pregnant women. Listeria infection is commonly associated with the ingestion of soft cheeses, unpasteurized milk, ready-to-eat deli meats, and hot dogs.

T. Gondii, or Toxoplasmosis A parasitic infection that can impair fetal brain development. The source of the infection is often hands contaminated with soil or the contents of a cat litter box; or raw or partially cooked pork, lamb, or venison.

High levels of mercury are most likely to be present in the muscles of large, long-lived predatory fish such as shark, swordfish, tilefish, albacore tuna, walleye, pickerel, and bass. Mercury content of bottom feeders, such as carp, channel catfish, and white sucker is generally less than half the amount found in predatory fish. Other fish that tend to have low mercury content include "light" (not white) tuna, haddock, tilapia, salmon, cod, pollack, and sole. Shrimp, lobster, and crab generally have low mercury content, too.[167] Recommendations for fish intake during pregnancy are given on page 109.

Common Health Problems during Pregnancy

Some of the physiological changes that occur in pregnancy are accompanied by side effects that can dull the bliss of expecting a child by making women feel physically miserable. Common ailments of pregnancy, such as nausea and vomiting, heartburn, and constipation, are generally more amenable to prevention than to treatment, but often can be relieved through dietary measures.

Nausea and Vomiting

Nausea occurs in about seven in ten pregnancies, and vomiting in four of ten. Symptoms of nausea generally begin around week 5 of gestation and generally disappear by week 12. Up to 15% of pregnant women will experience nausea and vomiting throughout pregnancy. The conditions are so common that they are considered a normal part of pregnancy. Unless severe or prolonged, nausea and vomiting during pregnancy are associated with a reduction in risk of miscarriage of greater than 60% and with healthy newborn outcomes. The cause of nausea and vomiting is not yet clear, but they are thought to be related to increased levels of human chorionic gonadotropin, progesterone, estrogen, or other hormones early in pregnancy.[168]

In the past, the nausea and vomiting of pregnancy was called "morning sickness," because it was thought to occur mostly after waking up. It actually occurs at all times of day—a mere 17% of women experience nausea and vomiting only in the morning.[168] Iron supplements may aggravate nausea and vomiting when taken in the first trimester of pregnancy.[127]

Hyperemesis Gravidarum   Between 1 and 2% of pregnant women with nausea and vomiting develop *hyperemesis gravidarum* (more commonly called hyperemesis).[168] Hyperemesis is characterized by severe nausea and vomiting that last throughout much of pregnancy. It can be debilitating. In addition to the mother feeling very sick, frequent vomiting can lead to weight loss, electrolyte imbalances, and dehydration. Women with hyperemesis who gain weight normally during pregnancy (about 30 pounds total) are not at increased risk of delivering small infants, but women who gain less (21–22 pounds) are.[169]

Management of Nausea and Vomiting   Many approaches to the treatment of nausea and vomiting are used in clinical practice, but only a few are considered safe and effective. Dietary interventions represent the safest method, primarily because the short- and long-term safety of many drugs and herbal remedies early in pregnancy is unclear.[170] Here are some general recommendations for women experiencing nausea and vomiting:

- Continue to gain weight.
- Separate liquid and solid food intake.
- Avoid odors and foods that trigger nausea.
- Select foods that are well tolerated.

Many women find that hard-boiled eggs, potato chips, popcorn, yogurt, crackers, and other high-carbohydrate foods are well tolerated. Personal support and understanding are important components of counseling women with nausea and vomiting. Care should be taken to individualize dietary advice based on each woman's food preferences and tolerances. Women with hyperemesis may require rehydration therapy to restore fluids and electrolyte balance.

Periodically, articles will appear in the popular press claiming that nausea and vomiting are caused by certain foods, and that women should avoid them to protect their fetus from harmful substances in the foods. Not too long ago it was claimed that bitter-tasting vegetables, for example, should be avoided. When put to the test, this notion was found to be groundless.[171] Theoretical claims that certain foods elicit nausea and vomiting in order to protect the fetus from harmful effects of the food should be considered unreliable until proven in scientific studies.

Dietary Supplements for the Treatment of Nausea and Vomiting   Three types of dietary supplements have been found to decrease the symptoms of nausea and vomiting of pregnancy:

- Vitamin B_6 (pyridoxine) supplements given in a 25 mg dose every 8 hours or 75 mg per day reduce

the severity of nausea in many women. Vitamin B_6 doses of 10 mg three times a day have also been found to lessen nausea and vomiting episodes.[168,172] The upper limit for vitamin B_6 intake during pregnancy is 100 mg per day.

- Multivitamin supplements taken prior to and early in pregnancy decreased the occurrence of nausea and vomiting.[173]
- Ginger in doses of 1 gram a day for 4 days decreases nausea and vomiting.[159]

Use of moderate doses of vitamins in a multivitamin supplement, and vitamin B_6 in doses under the Tolerable Upper Intake Level, appear safe. Further research is needed before a definitive statement can be made regarding the safety of ginger use during pregnancy.

Heartburn

Pregnancy is accompanied by relaxation of gastrointestinal tract muscles. This effect is attributed primarily to progesterone. Relaxation of the muscular valve known as the cardiac or lower esophageal sphincter at the top of the stomach is thought to be the principal reason for the 30–50% incidence of heartburn in women during pregnancy. The loose upper valve may allow stomach contents to be pushed back into the esophagus.[174]

Management of Heartburn   Dietary advice for the prevention and management of heartburn includes:

- Ingest small meals frequently.
- Do not go to bed with a full stomach.
- Avoid foods that seem to make heartburn worse.

Elevating the upper body during sleep, and not bending down so your head is below the waist, also reduce gastric reflux. Antacid tablets, which act locally in the stomach, are often recommended, but heartburn pills are not.[174]

Constipation

Relaxed gastrointestinal muscle tone is thought to be primarily responsible for the increased incidence of constipation and hemorrhoids in pregnancy. The best way to prevent these maladies is to consume approximately 30 grams of dietary fiber daily.[88] (Food sources of fiber are listed in Table 1.5C in Chapter 1, page 7.) Laxative pills are not recommended for use by pregnant women, but bulk-forming fiber in products such as Metamucil, Citrucel, and Perdiem are considered safe and effective for the prevention and treatment of constipation.[174] Women should drink a cup or more of water along with the fiber supplement.

Model Nutrition Programs for Risk Reduction in Pregnancy

"Pregnancy may be the most sensitive period of the lifecycle in which intervention may reap the greatest benefits."

A. Prentice[123]

Two programs that have been shown to substantially improve pregnancy outcomes are highlighted in this section. First is the intervention program offered by the Montreal Diet Dispensary (MDD); second is the Supplemental Nutrition Program for Women, Infants, and Children (WIC).

The Montreal Diet Dispensary

The Montreal Diet Dispensary (MDD) has served low-income, high-risk pregnant females with nutritional assessment and intervention services since the early 1900s. Part of the rationale for the WIC program in the United States was based on the successes of the MDD program. The program is located in a large, comfortable house (see Illustration 4.18) in urban Montreal. Clients are warmly welcomed into a nonthreatening, relaxed setting.

Developed as an adjunct to routine prenatal care, the MDD intervention strategy has four major components:

1. Assess the usual dietary intake and risk profile of each pregnant woman, including calories, protein, and selected vitamin and mineral adequacy; also assess stress level.

2. Determine individual nutritional rehabilitation needs based on results of the assessment.

3. Teach clients the importance of optimal nutrition and about changes that should be made through practical examples.

4. Provide regular follow-up and supervision.

Illustration 4.18 The Montreal Diet Dispensary.

The MDD dietitians are carefully trained and hold the interests of their clients first in their hearts. They treat clients with respect, openness, and affection; they also address client needs, such as transportation or emergency food or housing. Staff interactions with clients are nonjudgmental in nature and include positive feedback and praise for dietary changes and other successes of clients.

The initial client visit to the MDD takes about 75 minutes, and follow-up visits are scheduled at 2-week intervals for 40 minutes each. Women are identified as undernourished if their protein intake falls below that recommended for pregnancy, and an additional protein allowance is added to the diet. Women who are underweight are given an additional daily allowance of 20 grams protein and 200 calories for each additional pound of weight gain needed to achieve a maximum of 2 pounds per week. Women identified as being under excessive stress (such as having a partner in jail, being homeless, or being abused) receive an additional allowance of protein and calories and lots of positive attention. Food supplements, including milk and eggs, and vitamin supplements are provided to women who need them.

Impact of MDD Services   Multiple studies have shown that women receiving MDD services have higher-birth-weight infants (+107 grams), fewer low-birth-weight infants (−50%), and infants with lower rates of perinatal mortality than is the case for similar women not receiving MDD services.[175,176]

The program is cost effective in relation to savings on newborn critical care, and programs based on MDD services have spread across Canada. Expenditures per client average $450. The program is primarily supported by Centraide of Greater Montreal, provincial and federal programs, and other contributions.[177]

The WIC Program

The WIC program represents an outstanding example of a successful public program intended to serve the nutritional needs of low-income women and families. It is cited as a model program in several other chapters and is described in Chapter 1.

In operation since 1974, WIC provides nutritional assessment, education and counseling, food supplements, and access to health services to over 6 million participants. WIC serves low-income pregnant, postpartum, and breastfeeding women; and children up to 5 years of age who are at nutritional risk. Supplemental food provided to women includes milk, ready-to-eat cereals, dried beans, fruit juice, and cheese; some programs offer vouchers for farmer's markets.

Participation in WIC is related to reduced rates of iron-deficiency anemia in pregnancy, higher-birth-weight infants, decreased low-birth-weight infants, and lower rates of iron-deficiency anemia in women after delivery. For each dollar invested in WIC, approximately $3 in health care costs are saved. Internet addresses leading to additional information about WIC are listed in the resources section at the end of this chapter.

Key Points

1. Nutritional status before and during pregnancy can modify the health of women during pregnancy, as well as the current and future health of infants.

2. The United States spends more money on health care than any other country, yet its birth outcomes are far from the best internationally. Improved maternal nutrition could help improve the health status of U.S. newborns.

3. A woman's body prepares in advance for upcoming physiological events related to placental growth and fetal growth and development (such as proliferation of cells in organs and tissues of the placenta and fetus, and rapid increases in fetal weight). Consequently, nutritional needs must be met *prior* to the physiological changes.

4. Functions of the placenta include hormone and enzyme production, nutrient and gas exchange between mother and fetus, and removal of waste products from the fetus.

5. The placenta does *not* block all harmful substances from entering the fetus.

6. The fetus is *not* a parasite. It cannot take whatever nutrients it needs from the mother's body.

7. Variations in fetal growth and development are generally *not* due to genetic causes but rather to environmental factors such as energy, nutrient, and oxygen availability, and to conditions that interfere with genetically programmed growth and development.

8. Energy and nutrient availability is considered the major intrauterine environmental factor that alters expression of fetal genes. This phenomenon represents the major mechanism that underlies the relationship between maternal nutrition and later disease risk.

9. Pregnancy weight gain affects birth weight and long-term health outcomes. Weight gain recommendations are based on prepregnancy weight status.

10. Excessive weight gain is related to postpartum weight retention.

11. Caloric adequacy during pregnancy can be estimated by weight gain.

12. High-quality vegetarian diets promote a healthy course and outcome of pregnancy.

13. Consumption of the omega-3 fatty acids, EPA and DHA, promote visual and intellectual development in offspring, and increase gestational duration somewhat. Most U.S. women consume too little EPA and DHA during pregnancy.

14. Key nutrients of particular importance during pregnancy are folate, vitamins A and D, calcium, iron, zinc, iodine, and EPA and DHA. Antioxidants from plant food also play key roles in maintaining maternal and fetal health.

15. Not all pregnant women need a multivitamin and mineral supplement during pregnancy. Women at risk of deficiencies do.

16. In general, exercise is beneficial to the course and outcome of pregnancy.

17. Certain foodborne illnesses in pregnant women can threaten fetal survival.

18. Some of the common discomforts of pregnancy, such as nausea and vomiting and constipation, can be ameliorated by nutritional measures.

Resources

Pregnancy Resources and Information
Visit the Women's Health Resource Center for access to journal articles and summaries; information on pregnancy, growth, and development; and health care and diversity information.
Website: **www.medscape.com**
The Bureau of Maternal and Child Health website provides information on programs for pregnant women, hot topics, and announcements of new publications.
Website: **http://mchlibrary.info**
The National Library of Medicine website offers extensive coverage of scientific journal articles, summaries, and educational resources from a variety of reputable organizations on pregnancy, nutrition, diet, and disorders of pregnancy.
Website: **www.nlm.nih.gov/medlineplus**

Fish Advisories
This site links to local freshwater fish advisories.
Website: **www.epa.gov/waterscience/fish/states.htm**

Health Canada
Nutrition website that provides access to Health Canada's policy statement on nutrient needs of pregnant women. The revised edition of Canada's Food Guide will be released in 2007.
Website: **www.hc-sc.gc.ca/fn-an/food-guide-aliment/index_e.html**

U.S. Government

This USDA site provides links to vast resources related to education materials for WIC, vegetarian diets, health fraud, and other topics.
Website: www.nal.usda.gov/fnic
The U.S. government's site, which provides information on food and nutrition programs and eligibility; links to scientific references; and information about the nutrition needs of infants, children, adults, and seniors.
Website: www.nutrition.gov

Vegetarian Diets

Useful information on vegetarian diets, resources, and organizations can be obtained through these websites:
Website: www.vegetariannutrition.net
Website: www.vrg.org

Web Conference

Access a Medscape Web Conference on "Milestones in Prenatal Nutrition: The Emerging Role of Omega-3 Fatty Acids" presented Dec. 27, 2006. The conference consists of a multimedia presentation featuring synchronized audio and slides. (You may have to register with medscape.com to obtain access to this site. Registration is free.)
Website: www.medscape.com/viewprogram/6390?src=mp

WIC

USDA provides access to information about the WIC program, the WIC Works Food Safety Resource List, and other resources.
Website: www.fns.usda.gov/wic/aboutwic

Dietary Analysis

Select "MyPyramid Tracker" and run dietary intake records one day at a time. Analyzes diet by food groups and selected nutrients.
Website: www.mypyramid.gov
This USDA site is the best one for food composition data.
Website: www.ars.usda.gov/ba/bhnrc.ndl

References

1. Births: Preliminary Data for 2005, www.cdc.gov/nchs/products/pubs/pubd/hestats/prelimbirths05/prelimbirths05.htm, accessed 12/06.

2. Wegman ME. Infant mortality in the 20th century, dramatic but uneven progress. J Nutr. 2001;131:401S–8S.

3. The C.I.A. World Factbook. International infant mortality rates. www.cia.gov/cia/publications/factbook/geos/xx.html, accessed 10/06.

4. Health United States, 2000. Vol. 2001; and Deaths: Preliminary Data for 2002; www.cdc.gov/nchs, accessed 2/14/04.

5. Kirkegaard I et al. Size at birth and later school performance. Pediatrics. 2006;118: 1600–1606.

6. Godfrey KM. Maternal regulation of fetal development and health in adult life. Eur J Obstet Gynecol Repro Biol. 1998; 78:141–50.

7. Hytten FE, Leitch I. The physiology of human pregnancy. Oxford: Blackwell Scientific Publications; 1971.

8. Cruikshank DP et al. Maternal physiology in pregnancy. In: Gabbe SG et al., eds. Obstetrics: normal and problem pregnancies. New York: Churchill Livingstone; 1996: pp. 91–109.

9. King JC. Physiology of pregnancy and nutrient metabolism. Am J Clin Nutr. 2000; 71:1218S–25S.

10. Rosso P. Nutrition and metabolism in pregnancy. New York: Oxford University Press; 1990: pp. 117–118, 125, 150–151. Winick M. Nutrition, pregnancy, and early infancy. Baltimore: Williams & Wilkins; 1989: p. 182.

11. King JC. Effect of reproduction on the bioavailability of calcium, zinc and selenium. J Nutr. 2001;131:1355S–8S.

12. Casanueva E et al. Iron and oxidative stress in pregnancy. J Nutr. 2003;133: 1700S–1708S.

13. Butte NF. Carbohydrate and lipid metabolism in pregnancy: normal compared with gestational diabetes mellitus. Am J Clin Nutr. 2000;71:1256S–61S.

14. Fazleabas AT. Update and advances in reproductive biology. Experimental Biology Annual Meeting, San Diego, April 11, 2003.

15. Naeye RL, Chez RA. Effects of maternal acetonuria and low pregnancy weight gain on children's psychomotor development. Am J Obstet Gynecol. 1981;139:189–93.

16. Duggleby SL, Jackson AA. Higher weight at birth is related to decreased amino acid oxidation during pregnancy. Am J Clin Nutr. 2002;76:852–7.

17. van Stiphout WAHJ, Hofman A, de Bruijn AM. Serum lipids in young women before, during, and after pregnancy. Am J Epidemiol. 1987;126:922–8.

18. Martin U et al. Is normal pregnancy atherogenic? Clin Sci (Colch). 1999;96: 421–5.

19. Ortega RM, Gaspar MJ, Cantero M. Influence of maternal serum lipids and maternal diet during the third trimester of pregnancy on umbilical cord blood lipids in two populations of Spanish newborns. Int J Vitam Nutr Res. 1996;66:250–7.

20. American College of Obstetricians and Gynecologists. Nutrition during pregnancy. ACOG Technical Bulletin No. 179. 1993:1–7.

21. Gluckman PD. Fetal origins of adult disease: insulin resistance. Medscape Ob/Gyn & Women's Health. 2003;8:4–5.

22. Rosso P, Cramoy C. Nutrition and pregnancy. In: Winick M, ed. Human nutrition: pre- and post-natal development. New York: Plenum Press; 1979: pp. 133–228.

23. King JC. The risk of maternal nutritional depletion and poor outcomes increases in early or closely spaced pregnancies. J Nutr. 2003;133:1732S–6S.

24. Scholl TO, Stein TP, Smith WK. Leptin and maternal growth during adolescent pregnancy. Am J Clin Nutr. 2000;72:1542–7.

25. Norwitz ER et al. Implantation and the survival of early pregnancy. N Engl J Med. 2001;345:1400–8.

26. Rozovski SJ, Winick M. Nutrition and cellular growth. In: Winick M, ed. Human nutrition: pre- and postnatal development. New York: Plenum Press; 1979: pp. 61–102.

27. Godfrey KM, Barker DJ. Fetal nutrition and adult disease. Am J Clin Nutr 2000; 71:1344S–52S.

28. Smith DW. Growth and its disorders: basics and standards, approach and classifications, growth deficiency disorders, growth excess disorders, obesity. In: Schaeffer AJ, Markowtiz M, eds. Major problems in clinical pediatrics. Vol. 15. Philadelphia: W. B. Saunders Company; 1979: pp. 134–169.

29. Winick M. Malnutrition and brain development. J Pediatr. 1969;74:667–79.

30. Ziegler EE et al. Body composition of the reference fetus. Growth. 1976;40:329–41.

31. Rosso P, Winick M. Intrauterine growth retardation. A new systematic approach based on the clinical and biochemical characteristics of this condition. J Perinat Med. 1974;2:147–60.

32. National Academy of Sciences. Nutrition during pregnancy. I. Weight gain. II. Nutrient supplements. Washington, DC: National Academy Press; 1990.

33. Hediger ML et al. Growth of infants and young children born small or large for gestational age: findings from the Third National Health and Nutrition Examination

Survey. Arch Pediatr Adolesc Med. 1998; 152:1225–31.

34. Lucas A et al. Breastfeeding and catch-up growth in infants born small for gestational age. Acta Paediatr. 1997;86:564–9.

35. Luo ZC, Albertsson-Wikland K, Karlberg J. Length and body mass index at birth and target height influences on patterns of postnatal growth in children born small for gestational age. Pediatrics. 1998;102:E72.

36. Kovac P. Recurrent pregnancy loss. www.medscape.com/viewprogram/5293, accessed 4/06.

37. Wen W et al. The associations of maternal caffeine consumption and nausea with spontaneous abortion. Epidemiology. 2001;12:38–42.

38. Maconochile N et al. Underweight and the risk of miscarriage. Online edition of BJOG: An International Journal of Obstetrics and Gynecology, www.medicinenet.com/script/main/art.asp?articlekey=77914, accessed 12/06.

39. Prentice AM. Early influences on human energy regulation: thrifty genotypes and thrifty phenotypes. Physiol Behav. 2005; 86:640–5.

40. Wu G et al. Maternal nutrition and fetal development. J Nutr. 2004;134:2l69–72.

41. Bengston et al. Restricted fetal growth linked to irritable bowel syndrome, Gut 2006, www.medscape.com/viewarticle/545241, accessed.

42. Kramer MS et al. The contribution of mild and moderate preterm birth to infant mortality. Fetal and Infant Health Study Group of the Canadian Perinatal Surveillance System. JAMA. 2000;284:843–9.

43. Chery C et al. Hyperhomocysteinemia is related to a decreased level of vitamin B_{12} in the second and third trimesters of pregnancy. Clin Chem Lab Med. 2002;40:1105–8.

44. Ortega RM et al. Riboflavin levels in maternal milk: the influence of vitamin B_2 status during the third trimester of pregnancy. J Am Coll Nutr. 1999;18:324–9.

45. Sommer A et al. Assessment and control of vitamin A deficiency: the Annecy Accords. J Nutr. 2002;132:2845S–50S.

46. Goldenberg RL, Rouse DJ. Prevention of premature birth. N Eng J Med. 1998; 339:313–20.

47. Hobel C, Culhane J. Role of psychosocial and nutritional stress on poor pregnancy outcome. J Nutr. 2003;133:1709S–17S.

48. Vahratian A et al. Multivitamin use and the risk of preterm birth. Am J Epidemiol. 2004;1 60:886–92.

49. Kulkarni S. Lifestyle management before infertility treatment. www.medscape.com/viewarticle/52224, accessed 1/06.

50. Allen LH. Multiple micronutrients in pregnancy and lactation: an overview. Am J Clin Nutr. 2005;81:1206S–1212S.

51. King JC. Preface. Am J Clin Nutr. 2000; 71:1217S.

52. Rich-Edwards JW et al. Birth weight and risk of cardiovascular disease in a cohort of women followed up since 1976. BMJ. 1997;315:396–400.

53. Gluckman PD, Pinal CS. Regulation of fetal growth by the somatotrophic axis. J Nutr. 2003;133:1741S–6S.

54. Vivian EM. Type 2 diabetes in children and adolescents—the next epidemic? Curr Med Res Opin. 2006;22:297–306.

55. Stover PJ. Influence of human genetic variation on nutritional requirements. Am J Clin Nutr. 2006:83(suppl):438S–42S.

56. Fernandez-Twinn DS et al. Mechanisms by which poor early growth programs type-2 diabetes, obesity and the metabolic syndrome. Physiol Behav. 2006;88:234–43.

57. Barker DJP et al. Trajectories of growth among children who have coronary events as adults. N Engl J Med. 2005;353:1802–9.

58. Robinson SM et al. Combined effects of dietary fat and birth weight on serum cholesterol concentrations: the Hertfordshire Cohort Study. Am J Clin Nutr. 2006;84:237–44.

59. Kuzawa CW, Adair LS. Lipid profiles in adolescent Filipinos: relation to birth weight and maternal energy status during pregnancy. Am J Clin Nutr. 2003;77:960–6.

60. Belizan JM et al. Long-term effect of calcium supplementation during pregnancy on the blood pressure of offspring; follow up of a randomized controlled trial. BMJ. 1997;315:281–5.

61. Villar J, Cossio TG. Nutritional factors associated with low birth weight and short gestational age. Clin Nutr. 1986;5:78–85.

62. Eastman NJ. Expectant motherhood. Boston: Little, Brown and Company; 1947: p. 198.

63. Stotland NE et al. Excess gestational weight gain associated with adverse neonatal outcomes. Obstet Gynecol. 2006;108:635–43.

64. Parker JD, Abrams B. Prenatal weight gain advice: an examination of the recent prenatal weight gain recommendations of the Institute of Medicine. Obstet Gynecol. 1992;79:664–9.

65. Brown JE et al. Variation in newborn size by trimester weight change in pregnancy. Am J Clin Nutr. 2002;76:205–9.

66. Carmichael SL, Abrams B. A critical review of the relationship between gestational weight gain and preterm delivery. Obstet Gynecol. 1997;89:865–73.

67. To WW, Cheung W. The relationship between weight gain in pregnancy, birthweight and postpartum weight retention. Aust N Z J Obstet Gynaecol. 1998;38:176–9.

68. Abrams B. Prenatal weight gain and postpartum weight retention: a delicate balance (editorial). Am J Public Health. 1993; 83:1082–4.

69. King JC, Mukherjea R. Leptin targeted in research on obesity, pregnancy. ARS News Service, USDA, www.ars.usda.gov, accessed 10/02.

70. Scholl TO et al. Insulin and the "thrifty" woman: the influence of insulin during pregnancy on gestational weight gain and postpartum weight retention. Maternal Child Health Journal. 2002;6:255–61.

71. Keppel KG, Taffel SM. Pregnancy related weight gain and retention: implications of the 1990 Institute of Medicine guidelines. Am J Public Health. 1993;83:1100–3.

72. Brown JE, Kaye SA, Folsom AR. Parity-related weight change in women. Int J Obes. 1992;16:627–31.

73. Gigante DP, Victora CG, Barros FC. Breast-feeding has a limited long-term effect on anthropometry and body composition of Brazilian mothers. J Nutr. 2001;131:78–84.

74. Polley A et al. Randomized controlled trial to prevent weight gain in pregnant women. Int J Obes. 2002;26:1494–1502.

75. Mussey RD. Nutrition and human reproduction: an historical review. Am J Obstet Gynecol. 1949;58:1037–48.

76. Stein AD, Ravelli AC, Lumey LH. Famine, third-trimester pregnancy weight gain, and intrauterine growth: the Dutch Famine Birth Cohort Study. Hum Biol. 1995;67:135–50.

77. Smith CA. Effects of maternal undernutrition upon the newborn infant in Holland (1944–1945). J Pediatr. 1947;30:229–43.

78. Susser M, Stein Z. Timing in prenatal nutrition: a reprise of the Dutch Famine Study. Nutr Rev. 1994;52:84–94.

79. Roseboom TJ et al. Plasma lipid profiles in adults after prenatal exposure to the Dutch famine. Am J Clin Nutr. 2000; 72:1101–6.

80. Antonov AN. Children born during the siege of Leningrad in 1942. J Pediatr. 1947; 30:250–9.

81. Gruenwald P et al. Influence of environmental factors on fetal growth in man. Lancet. 1967;I:1026–8.

82. Guyer B et al. Annual summary of vital statistics—1997. Pediatrics. 1998;102: 1333–49.

83. Simic S et al. Nutritional effects of the siege on newborn babies in Sarajevo. Eur J Clin Nutr. 1995;49(suppl)2:S33–6.

84. Burke BS. Nutritional needs in pregnancy in relation to nutritional intakes as shown by dietary histories. Obstetr Gynecol Survey. 1948;3:716–30.

85. Burke BS, Harding VV, Stuart HC. Nutrition studies during pregnancy. IV. Relation of protein content of mother's diet during pregnancy to birth length, birth weight, and condition of infant at birth. J Pediatr. 1943; 23:506–15.

86. Brown JE, Kahn ESB. Maternal nutrition and the outcome of pregnancy: a renaissance in research. Clin Perinatol. 1997;24:433–49.

87. Dietary reference intakes: energy, carbohydrate, fiber, fat, fatty acids, cholesterol, protein, and amino acids. Washington, DC: National Academies Press; 2002.

88. Anderson JW. Health implications of wheat fiber. Am J Clin Nutr. 1985;41: 1103–12.

89. Moses RG et al. Effect of a low-glycemic-index diet during pregnancy on obstetrical outcomes. Am J Clin Nutr. 2006;84;807–12.

90. Duffy VB, Anderson GH. Position of the American Dietetic Association: use of nutritive and nonnutritive sweeteners. J Amer Diet Assoc. 1998;98:580–7.

91. American Academy of Pediatrics. Committee on Substance Abuse and Committee on Children with Disabilities. Fetal alcohol syndrome and alcohol-related neurodevelopmental disorders. Pediatrics. 2000;106:358–61.

92. Johnston PK. Counseling the pregnant vegetarian. Am J Clin Nutr. 1988;48:901–5.

93. Position of the American Dietetic Association and Dietitians of Canada: vegetarian diets. J Am Diet Assoc. 2003;103:748–65.

94. Messina V et al. A new food guide for North American vegetarians. J Am Diet Assoc. 2003;103:771–5.

95. Haddad EH et al. Vegetarian food guide pyramid: a conceptual framework. Am J Clin Nutr. 1999;70(suppl):615S–19S.

96. Neurologic impairment in children associated with maternal dietary deficiency of cobalamin—Georgia, 2001. MMWR. 2003;52(04):61–4.

97. Williams CM et al. Long-chain n-3 PUFA: plant v. marine sources. Proc Nutr Soc. 2006;65:42–50.

98. Alessandri JM et al. Polyunsaturated fatty acids in the central nervous system: evolution of concepts and nutritional implications throughout life. Reprod Nutr Dev. 2004;44:509–38.

99. Jensen CL Effects of n-3 fatty acids during pregnancy and lactation. Am J Ciin Nutr. 2006;83(6 Suppl):1452S–1457S.

100. Szajewska H et al. Effect of n-3 long-chain polyunsaturated fatty acid supplementation of women with low-risk pregnancies on pregnancy outcomes and growth measures at birth: a meta-analysis of randomized controlled trials. Am J Clin Nutr. 2006;83: 1337–44.

101. Simopoulos AP. Human requirement for n-3 polyunsaturated fatty acids. Poult Sci. 2000;79:961–70.

102. What you need to know about the safety of fish and shellfish: 2004 EPA and FDA advice for women who might become pregnant, women who are pregnant, nursing mothers, young children. www.cfsan.fda.gov/~dms/admehg3.html, accessed 1/07.

103. Mozaffarian D et al. Health benefits of eating fish outweigh risks. JAMA. 2006;296: 1885–18.

104. Dunstan JA et al. Fish-oil-supplements during pregnancy and eye and hand coordination in children. Arch Dis Child Fetal Neonatal Ed, published online December 21, 2006.

105. Ershow AG, Brown LM, Cantor KP. Intake of tapwater and total water by pregnant and lactating women. Am J Public Health. 1991;81:328–34.

106. Lumley J et al. Periconceptional supplementation with folate and/or multivitamins for preventing neural tube defects. Cochrane Database Syst Rev. 2000;2.

107. Suitor CW, Bailey LB. Dietary folate equivalents: interpretation and application. J Amer Diet Assoc. 2000;100:88–94.

108. Bailey LB. Evaluation of a new Recommended Dietary Allowance for folate. J Amer Diet Assoc. 1992;92:463–71.

109. Fodinger M et al. Recent insights into the molecular genetics of the homocysteine metabolism. Kidney Int. 2001;59(suppl)78: S238–42.

110. Warkany J. Production of congenital malformations by dietary measures. JAMA. 1958;168:2020–3.

111. Smithells RW. Availability of folic acid. Lancet. 1984;1:508.

112. Eskes TKAB. From birth to conception. open or closed. Eur J Obstet Gynecol Reprod Biol. 1998;78:169–77.

113. Folic acid. MMWR Morb Mortal Wkly Rep. 2001;50:185–9.

114. Folate status in women of childbearing age—United States, 1999. MMWR Morb Mortal Wkly Rep. 2000;49:962–5.

115. Daly LE et al. Folate levels and neural tube defects. JAMA. 1995;274:1698–702.

116. Brown JE et al. Predictors of red cell folate level in women attempting pregnancy. JAMA. 1997;277:548–52.

117. Folate Intake in U.S. Falling Short of Established Targets. MMWR. 2007;55: 1377–80.

118. Shaw GM et al. Periconceptional vitamin use, dietary folate, and the occurrence of neural tube defects. Epidemiology. 1995;6:219–26.

119. Zile MH. Function of vitamin A in vertebrate embryonic development. J Nutr. 2001;131:705–8.

120. Smithells D. Vitamins in early pregnancy. BMJ. 1996;313:128–9.

121. Specker BL. Do North American women need supplemental vitamin D during pregnancy or lactation? Am J Clin Nutr. 1994;59:484S–91S.

122. Hollis BW et al. Vitamin D deficiency during pregnancy: an ongoing epidemic. Am J Clin Nutr. 2006;84:273.

123. Prentice A. Maternal calcium metabolism and bone mineral status. Am J Clin Nutr. 2000;71:1312S–6S.

124. Hertz-Picciotto I et al. Patterns and determinants of blood lead during pregnancy. Am J Epidemiol. 2000;152:829–37.

125. Leverett DH et al. Randomized clinical trial of the effect of prenatal fluoride supplements in preventing dental caries. Caries Res. 1997;31:174–9.

126. Fairbanks VF. Iron in medicine and nutrition. In: Modern Nutrition in Health and Disease, Shils ME et al., eds. 9th ed. Philadelphia: Lippincott, Williams & Wilkins; 1999. pp. 193–221.

127. Cogswell ME et al. Iron supplement use among women in the United States: science, policy, and practice. J Nutr. 2003;133: 1974S–7S.

128. Allen LH. Biological mechanisms that might underlie iron's effects on fetal growth and preterm birth. J Nutr. 2001; 131:581S–9S.

129. Tamura T et al. Cord serum ferritin concentrations and mental and psychomotor development of children at five years of age. J Pediatrics. 2002;140:165–70.

130. Beard JL. Iron requirements in adolescent females. J Nutr. 2000;130:440S–2S.

131. Allen LH. Anemia and iron deficiency: effects on pregnancy outcome. Am J Clin Nutr. 2000;71:1280S–4S.

132. Recommendations to prevent and control iron deficiency in the United States. Centers for Disease Control and Prevention. MMWR Morb Mortal Wkly Rep. 1998; 47:1–29.

133. Seligman PA et al. Measurements of iron absorption from prenatal multivitamin–mineral supplements. Obstet Gynecol. 1983;61:356–62.

134. Dawson EB et al. Iron in prenatal multivitamin/multimineral supplements. Bioavailability. J Reprod Med. 1998; 43:133–40.

135. Singh K, Fong YF, Kuperan P. A comparison between intravenous iron polymaltose complex (ferrum hausmann) and oral ferrous fumarate in the treatment of iron deficiency anaemia in pregnancy. Eur J Haematol. 1998;60:119–24.

136. Casanueva E, Viteri FE. Iron and oxidative stress in pregnancy. J Nutr. 2003; 133:1700S–8S.

137. Iron deficiency anemia: recommended guidelines for prevention, detection, and management among US children and women of childbearing age. Washington, DC: National Academy Press; 1993.

138. Cogswell ME et al. Iron supplementation during pregnancy, anemia, and birth weight: a randomized controlled trial. Am J Clin Nutr. 2003;78:773–81.

139. King JC. Determinants of maternal zinc status during pregnancy. Am J Clin Nutr. 2000;71:1334S–43S.

140. Tamura T et al. Maternal plasma zinc concentrations and pregnancy outcome. Am J Clin Nutr. 2000;71:109–13.

141. Xue-Yi C et al. Timing of vulnerability of the brain to iodine deficiency in endemic cretinism. N Eng J Med. 1994;331: 1739–44.

142. Caldwell KL et al. Urinary iodine concentration: United States National Health And Nutrition Examination Survey 2001–2002. Thyroid, 2005;15:692–9.

143. Agarwal A et al. Role of oxidative stress in female reproduction. Reprod Biol Endocrinol. 2005;14;3:28.

144. Litonjua AA et al. Maternal antioxidant intake in pregnancy and wheezing illnesses in children at 2 y of age. Am J Clin Nutr. 2006;84:903–11.

145. Devereux G et al. Low intake of vitamin E during pregnancy may increase risk for childhood asthma. Am J Respir Grit Care Med. 2006;174:499–507.

146. Mahomed K, Gulmezoglu AM. Maternal iodine supplements in areas of deficiency (Cochrane Review). The Cochrane Library. Issue 3. Oxford: Update Software, 1999.

147. Delemarre FM et al. Eclampsia despite strict dietary sodium restriction. Gynecol Obstet Invest. 2001;51:64–5.

148. Pike RL, Gursky DS. Further evidence of deleterious effects produced by sodium restriction during pregnancy. Am J Clin Nutr. 1970;23:883–9.

149. Duley L, Henderson-Smart D. Reduced salt intake compared to normal dietary salt, or high intake, in pregnancy. Cochrane Database Syst Rev. 2000;2.

150. Nehlig A, Debry G. Consequences on the newborn of chronic maternal consumption of coffee during gestation and lactation: a review. J Am Coll Nutr. 1994;13:6–21.

151. Hinds TS et al. The effect of caffeine on pregnancy outcome variables. Nutr Rev. 1996;54:203–7.

152. Barr HM, Streissguth AP. Caffeine use during pregnancy and child outcome: a 7-year prospective study. Neurotoxicol Teratol. 1991;13:441–8.

153. Johns T, Duquette M. Detoxification and mineral supplementation as functions of geophagy. Am J Clin Nutr. 1991;53:448–56.

154. Rainville AJ. Pica practices of pregnant women are associated with lower maternal hemoglobin level at delivery. J Amer Diet Assoc. 1998;98:293–6.

155. FDA statement concerning structure/function rule and pregnancy claims. HHS Statement, U.S. Department of Health and Human Services 2000; 65 FR 1000; January 6, 2000; Docket No. 98N-0044.

156. Jones FA. Herbs: useful plants. J R Soc Med. 1996;89:717–9.

157. Meisler JG. Toward optimal health: the experts discuss the use of botanicals by women. J Women's Health. 2003;12:847–52.

158. Dugoua JJ et al. Safety and efficacy of ginkgo (Ginkgo biloba) during pregnancy and lactation. Can J Clin Pharmacol. 2006;13:e277–84. Epub 2006 Nov 3.

159. Vutyavanich T et al. Ginger for nausea and vomiting in pregnancy: randomized, double-masked, placebo-controlled trial. Obstet Gynecol. 2001;97:577–82.

160. Johnson K. Pregnancy exercise recommendations growing more liberal. Medscape Ob/Gyn & Women's Health. 2003;8:1–2.

161. Artal R, Sherman C. Exercise during pregnancy—safe and beneficial for most. Physician and Sports Medicine. 1999;27:51–56.

162. Dempsey JC et al. No need for a pregnant pause: physical activity may reduce the occurrence of gestational diabetes mellitus and preeclampsia. Exerc Sport Sci Rev. 2005;33:141–9.

163. Clapp JF III et al. Beginning regular exercise in early pregnancy: effect on fetoplacental growth. Am J Obstet Gynecol. 2000;183:1484–8.

164. Position of the American Dietetics Association: Food Safety. J Am Diet Assoc. 2003;103:1203–18.

165. Soto C. Toxoplasmosis in pregnancy. Clin Rev. 2002;12:51–6.

166. EPA fish advisory update. Vol. 2001, 2001.

167. Schober SE et al. Fish consumption of blood mercury levels in women of childbearing age in the United States. JAMA. 2003; 289:1667–74.

168. Badell ML et al. Treatment options for nausea and vomiting during pregnancy. Pharmacotherapy. 2006;26:1273–87.

169. Gross S et al. Maternal weight loss associated with hyperemesis gravidarum: a predictor of fetal outcome. Am J Obstet Gynecol. 1989;160:906–9.

170. Scialli A. Revived focus on treatment for morning sickness. Medscape Wire. 2000.

171. Brown JE, Kahn ES, Hartman TJ. Profet, profits and proof: do nausea and vomiting of early pregnancy protect women from "harmful" vegetables? Am J Obstet Gynecol. 1997;176:179–81.

172. Jewel D, Young G. Interventions for nausea and vomiting in early pregnancy. Cochrane Database Syst Rev. 2002; (1):CD000145.

173. Emelianova S et al. Prevalence and severity of nausea and vomiting of pregnancy and effect of vitamin supplementation. Clin Invest Med. 1999;22:106–10.

174. Baron TH, Ramirez B, Richter JE. Gastrointestinal motility disorders during pregnancy. Ann Intern Med. 1993;118:366–75.

175. Higgins AC et al. Impact of the Higgins Nutrition Intervention Program on birth weight: a within-mother analysis. J Amer Diet Assoc. 1989; 89:1097–103.

176. Dubois S et al. Ability of the Higgins Nutrition Intervention Program to improve adolescent pregnancy outcome. J Amer Diet Assoc. 1997; 97:871–8.

177. Duquette MP, Director of the Montreal Diet Dispensary. Personal communication, 2000.

Chapter 5

Nutrition during Pregnancy:

Conditions and Interventions

Chapter Outline

- Introduction
- Hypertensive Disorders of Pregnancy
- Diabetes in Pregnancy
- Multifetal Pregnancies
- HIV/AIDS during Pregnancy
- Eating Disorders in Pregnancy
- Fetal Alcohol Spectrum
- Nutrition and Adolescent Pregnancy
- Evidence-Based Practice

Photo Disc

Ayleen and Jeremy Perez-Marty

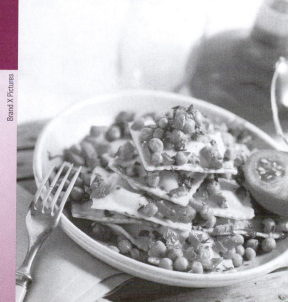

Brand X Pictures

Prepared by **Judith E. Brown**

Key Nutrition Concepts

1   Some complications of pregnancy are related to women's nutritional status.

2   Nutritional interventions for a number of complications of pregnancy can benefit maternal and infant health outcomes.

3   Nutritional interventions during pregnancy should be based on scientific evidence that supports their safety, effectiveness, and affordability.

Introduction

"Practice is science touched with emotion."

Stephen Paget, 1909, *Confessio Medici*

Almost all healthy women expect that their pregnancies will proceed normally and that they will be rewarded at delivery with a healthy newborn. For the vast majority of pregnancies, this expectation is fulfilled. For other women, however, the path to a healthy newborn is strewn with obstacles in the form of health problems that women bring into or develop during pregnancy. This chapter addresses a number of these health conditions and the role of nutrition in their etiology and management. The specific health conditions presented are hypertensive disorders of pregnancy, preexisting and gestational diabetes, obesity, multifetal pregnancy, HIV/AIDS, eating disorders, fetal alcohol spectrum, and adolescent pregnancy.

Hypertensive Disorders of Pregnancy

Hypertensive disorders of pregnancy are the second leading cause of maternal mortality in the United States. They affect 6 to 8% of pregnancies and contribute significantly to stillbirths, fetal and newborn deaths, and other adverse outcomes of pregnancy. The causes of most cases of hypertension during pregnancy remain unknown, and cures for these disorders remain elusive.[1]

Several types of hypertensive disorders in pregnancy have been identified (Table 5.1). In the past, the major types of hypertensive disorders in pregnancy were grouped under the heading "pregnancy-induced hypertension," or PIH. This terminology is being phased out in favor of the classification scheme for hypertensive disorders of pregnancy presented in Table 5.1.

Table 5.1   Definitions and features of hypertensive disorders of pregnancy[*2]

Chronic Hypertension

Hypertension that is present before pregnancy or diagnosed before 20 weeks of pregnancy. Hypertension is defined as blood pressure $\geq$140 mm Hg systolic or $\geq$90 mm Hg diastolic blood pressure.

Hypertension first diagnosed during pregnancy that does not resolve after pregnancy is also classified as chronic hypertension.

Gestational Hypertension

This condition exists when elevated blood pressure levels are detected for the first time after mid-pregnancy. It is not accompanied by proteinuria. If blood pressure returns to normal by 12 weeks postpartum, the condition is considered to be transient hypertension of pregnancy. If it remains elevated, then the woman is considered to have chronic hypertension.

Women with gestational hypertension are at lower risk for poor pregnancy outcomes than are women with preeclampsia.

Preeclampsia–Eclampsia

A pregnancy-specific syndrome that usually occurs after 20 weeks gestation (but that may occur earlier) in previously normotensive women. It is determined by increased blood pressure during pregnancy to $\geq$140 mm Hg systolic or $\geq$90 mm Hg diastolic and is accompanied by proteinuria. In the absence of proteinuria, the disease is highly suspected when increased blood pressure is accompanied by headache, blurred vision, abdominal pain, low platelet count, and abnormal liver enzyme values.

- Proteinuria is defined as the urinary excretion of $\geq$0.3 grams of protein in a 24-hour urine specimen. This usually correlates well with readings of $\geq$30 mg/dL protein, or $\geq$2 on dipstick readings taken in samples from women free of urinary tract infection. In the absence of urinary tract infection, proteinuria is a manifestation of kidney damage.
- Eclampsia is defined as the occurrence of seizures that cannot be attributed to other causes in women with preeclampsia.

Preeclampsia Superimposed on Chronic Hypertension

This disorder is characterized by the development of proteinuria during pregnancy in women with chronic hypertension. In women with hypertension and proteinuria before 20 weeks of pregnancy, it is indicated by a sudden increase in proteinuria, blood pressure, or abnormal platelet or liver enzyme levels.

*Blood pressure values used to determine status should be based on two or more measurements of blood pressure in relaxed settings.

Hypertensive Disorders of Pregnancy, Oxidative Stress, and Nutrition

All forms of hypertension in pregnancy (as well as many other disorders such as diabetes) are related to inflammation, *oxidative stress,* and damage to the *endothelium.* Over time, oxidative stress within the endothelium leads to endothelial dysfunction. Consequences of endothelial dysfunction include impaired blood flow, an increased tendency of blood to clot, and plaque formation.[3]

A number of nutritional and other environmental factors are related to oxidative stress (Table 5.2)[4,5] Except for weight loss, pregnant women with hypertensive disorders may benefit from lifestyle and dietary changes that lower oxidative stress.

Chronic Hypertension

The incidence of chronic hypertension—or that diagnosed prior to pregnancy or before 20 weeks after conception—ranges from 1 to 5% depending on the population studied. The condition is more likely to occur in African Americans, obese women, women over 35 years of age, and women who experienced high blood pressure in a previous pregnancy.[7]

Women with mild hypertension may be taken off antihypertension medications preconceptionally or early in pregnancy, because the drugs do not appear to improve the course or outcome of pregnancy.[8] Mild hypertension in healthy women that does not become worse during pregnancy appears to pose few risks to maternal and newborn health. Pregnancies among women with blood pressures ≥160/110 mm Hg—either or both values—are associated with an increased risk of fetal death, preterm delivery, and fetal growth retardation. Selection of the proper antihypertension medicines for women during pregnancy reduces these risks somewhat. Some antihypertension medicines reduce maternal blood sodium levels and limit plasma volume expansion. This, in turn, decreases fetal growth.

Nutritional Interventions for Women with Chronic Hypertension in Pregnancy Preconceptional and pregnancy diets of women with hypertension

Table 5.2 Environmental factors that increase oxidative stress[4–6]

- *Trans* fats
- Inadequate intake of antioxidants nutrients (vitamins C, E, carotenoids, selenium) and antioxidant phytochemicals from plants
- Habitual consumption of high-glycemic-index carbohydrates, elevated blood glucose levels
- Excess body fat
- Physical inactivity
- Smoking

should be carefully monitored with the aim of achieving adequate and balanced diets for pregnancy. Weight-gain recommendations are the same as for other pregnant women.

Women with salt-sensitive hypertension, or hypertension that responds to dietary sodium intake, must be managed along a fine line between consuming too much sodium for good blood pressure control, and consuming too little at the potential cost of impaired fetal growth.[8] For women with hypertension that was managed successfully in part by a low-sodium diet prior to pregnancy, continuing that dietary approach is generally recommended.[2]

> **Oxidative Stress** A condition that occurs in the endothelium when it is exposed to a surplus of oxidizing agents relative to its supply of antioxidants.
>
> **Endothelium** The layer of cells lining the inside of blood vessels.

Gestational Hypertension

Gestational hypertension is usually diagnosed after 20 weeks of pregnancy. Unlike women with preeclampsia, women with gestational diabetes have neither elevated serum insulin levels nor proteinuria.[9] Women with this disorder are at greater risk for hypertension and stroke later in life.[1]

Women with gestational hypertension tend to be overweight or obese and have excess central body fat. There is some evidence that antioxidants such as vitamin C help reduce the severity of this form of hypertension.[1]

Preeclampsia–Eclampsia

Preeclampsia–eclampsia represents a syndrome characterized by:

- Oxidative stress, inflammation, and endothelial dysfunction
- Blood vessel spasms and constriction
- Increased blood pressure
- Adverse maternal immune system responses to the placenta
- Platelet aggregation and blood coagulation due to deficits in *prostacyclin* relative to *thromboxane*
- Alterations of hormonal and other systems related to blood volume and pressure control
- Alterations in calcium regulatory hormones
- Reduced calcium excretion[2,10]

Virtually all maternal organs can be affected in preeclampsia. Organs most affected by small blood clots, vasoconstriction, and reduced blood flow are the placenta and the mother's kidney, liver, and brain.[10]

> **Prostacyclin** A potent inhibitor of platelet aggregation and a powerful vasodilator and blood pressure reducer derived from n-3 fatty acids.
>
> **Thromboxane** The parent of a group of thromboxanes derived from the n-6 fatty acid arachidonic acid. Thromboxane increases platelet aggregation and constricts blood vessels, causing blood pressure to increase.

Eclampsia can be a life-threatening condition and one that is difficult to predict. Eclamptic seizures appear to be related to hypertension, the tendency of blood to clot, and spasms of and damage to blood vessels in the brain. It complicates about 1 in 2000 pregnancies.[11]

Signs and symptoms of preeclampsia range from mild to severe (Table 5.3), as do the health consequences (Table 5.4). The cause of preeclampsia is unknown but appears to originate from abnormal implantation and vascularization of the placenta, and poor blood flow through the placenta.[10] Abnormal blood flow through the placenta is an important characteristic of preeclampsia because it decreases the delivery of nutrients and gases to the fetus. It appears be related to oxidative stress, reduced antioxidant defenses, and endothelial dysfunction.[17] Insulin resistance is also a common characteristic of preeclampsia and contributes to some of the negative consequences observed.[1] The only cure is delivery.[2] Signs and symptoms of preeclampsia generally disappear rapidly after delivery, but eclampsia may occur within 12 days following delivery.[18]

Women with preeclampsia are at increased risk for developing gestational diabetes during pregnancy, and type 2 diabetes, hypertension, heart disease, and strokes later in life.[19] About 15% of women with gestational diabetes and 30% of those with type 2 diabetes prior to pregnancy will develop preeclampsia.[15] A history of preeclampsia increases the risk that it will occur in subsequent pregnancies.[2]

Risk Factors for the Development of Preeclampsia   The roots of preeclampsia lie very early in pregnancy, but as yet there is no reliable means of identifying women who will develop it before the condition is established.[18] However, women with insulin resistance, obesity, or other characteristics listed in Table 5.5 are at increased risk for developing the disease. Characteristics listed confer at least a twofold increase in risk.

Increased rates of preterm delivery and low birthweight in infants born to women with preeclampsia are partly related to clinical decisions to deliver fetuses early in order to treat the disease. Most infants born to women with this disorder are normal weight, however, and some newborns are large for gestational age. Variations in birth weight associated with preeclampsia appear be related to the severity of the disease in individual women.[20]

The risk of developing preeclampsia is higher in women who were born small for gestational age (SGA). It appears that growth restriction in utero may impair mechanisms involved in the regulation of blood pressure and increase the probability that high blood pressure will develop with the physiological stresses of pregnancy.[23]

Table 5.3 Signs and symptoms of preeclampsia[12–14]

- Hypertension
- Increased urinary protein (albumin)
- Decreased plasma volume expansion (hemoglobin levels >13 g/dL)
- Low urine output
- Persistent and severe headaches
- Sensitivity of the eyes to bright light
- Blurred vision
- Abdominal pain
- Nausea
- Increased platelet aggregation, vasoconstriction related to increased thromboxane levels and decreased levels of prostacyclin

---

Table 5.4 Outcomes related to the existence of preeclampsia during pregnancy[7,15,16]

Mother
- Early delivery by cesarean section
- Acute renal (kidney) dysfunction
- Increased risk of gestational diabetes, hypertension, and type 2 diabetes later in life
- Abruptio placenta (rupture of the placenta)

Newborn
- Growth restriction
- Respiratory distress syndrome

Table 5.5 Risk factors for preeclampsia[2,7,21,22]

- First pregnancy (nulliparous)
- Obesity, especially high levels of central body fat
- Underweight
- Mother's smallness at birth
- African Americans, American Indians
- History of preeclampsia
- Preexisting diabetes mellitus
- Age over 35 years
- Multifetal pregnancy
- Insulin resistance
- Chronic hypertension
- Renal disease
- History of preeclampsia in the mother's or father's mother
- High blood levels of homocysteine
- Inadequate diet (possibly related to inadequate vitamins C and E, other antioxidants, and calcium)

Interestingly, women or men whose mothers had preeclampsia while pregnant with them are more likely to have children born from pregnancies complicated by preeclampsia.[24]

Nutrient Intake and Preeclampsia   Inadequate nutrient status has been investigated as a potential cause of preeclampsia, and various nutrients have been given to women with the aim of reducing its occurrence. Some of the results are promising. Table 5.6 provides an overview of nutritional and other remedies that have been tested for their usefulness in preventing preeclampsia.

Several studies have shown highly significant reductions in the incidence of preeclampsia in women who took supplemental vitamin C (200 mg daily) and E (400 IU daily) from midpregnancy onward.[25,26] Dietary calcium, calcium supplements, and omega-3 fatty acids have also been studied for the effects on the development and severity of preeclampsia. Calcium, whether obtained from the diet or supplements (1000–2000 g daily) has been related to a 50% or higher reduction in the risk of preeclampsia and a decreased severity of the syndrome.[27,28] Other studies, however, have not been able to identify benefits related to the use of vitamin or calcium supplements.[29] Omega-3 fatty acids given after the first trimester do not appear to influence the onset or course of preeclampsia.[30]

Preeclampsia Case Presentation

Signs, symptoms, severity, and causes of preeclampsia vary from woman to woman. Therefore, appropriate interventions for women presenting differing aspects

of the syndrome are best designed on a case-by-case basis.[20] By way of example, Case Study 5.1 describes the course of preeclampsia in one woman experiencing the condition.

Nutritional Recommendations and Interventions for Preeclampsia

In the best of circumstances, dietary interventions for preeclampsia would begin prior to pregnancy. This approach might give women the opportunities to decrease body weight and stores of central body fat, become physically fit, and consume a diet that reduces the need for insulin. Short of those circumstances, dietary recommendations and interventions should begin as early in pregnancy as possible and target at-risk women.

Nutritional and physical activity recommendations that may benefit women at risk of preeclampsia are within the normal scope of practice:

- 1000–2000 mg per day of dietary or supplemental calcium
- 400 IU vitamin E and 200 mg vitamin C daily
- Five or more servings of colorful vegetables and fruits daily
- No restriction of sodium intake (with the possible exception of some women with chronic hypertension)
- Consumption of the assortment of other basic foods recommended in MyPyramid
- Three regular meals and snacks daily
- Consumption of low-glycemic-index rather than high-glycemic-index carbohydrate foods[25,32,33]
- The glycemic index (GI) of various food sources of carbohydrates is shown in Table 1.4 in chapter 1
- Moderate exercise (for example, walking, swimming, noncompetitive tennis, or dancing for 30 minutes three times a week) unless medically contraindicated
- Weight gain that follows recommendations based on prepregnancy weight status

Iron supplements, especially if taken in high doses, may aggravate inflammation by increasing the body's free radical load.[34] Women with preeclampsia should not be given high-dose iron supplements.

Whether supplemental calcium or vitamins C and E should be recommended to women at risk of preeclampsia isn't clear, but doses below the Tolerable Upper Level of Intake (UL) for pregnancy are likely safe. Additional studies of their benefits to outcomes such as intrauterine growth retardation and perinatal death associated with preeclampsia are needed.

Table 5.6 Status of effectiveness of preventive measures for preeclampsia[2,8,31]

- Antihypertensive medications: Controversial, may not improve pregnancy outcomes
- Low-dose aspirin: May reduce risk of preeclampsia and improve outcomes to some extent
- Calcium supplementation: May reduce the incidence of preeclampsia in high-risk women
- Magnesium supplementation: Does not appear to prevent preeclampsia
- Fish oils (n-3 fatty acids): No reduction in preeclampsia
- Vitamins C and E: May help prevent preeclampsia
- High- or low-protein diets: Ineffective for prevention of preeclampsia
- Low-sodium diets: Ineffective for the prevention of preeclampsia
- Weight loss during pregnancy: Ineffective for the prevention of preeclampsia

Case Study 5.1

A Case of Preeclampsia

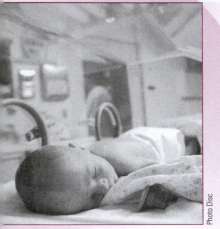

Photo Disc

Susan is a 19-year-old "meat-and-potato" type eater who rarely consumes vegetables, fruits, or dairy products. She likes monosyllabic vegetables (beans, corn, and peas), bananas and oranges, and chocolate milk. She generally consumes one of these vegetables and fruits each day, and always has a glass of chocolate milk. Susan consumes sweetened iced tea throughout the day and twice a week she eats rice with meat rather than potatoes. She finds this type of diet satisfying and rarely consumes foods other than those mentioned.

Her first 17 weeks of pregnancy were uneventful. At week 18 she was found to have proteinuria. By week 22, her blood pressure had increased to 150/100 mm Hg, and she was diagnosed with preeclampsia. Laboratory studies indicated that her blood glucose level was on the high side of normal and that she was insulin resistant. She was lost to follow-up after her week 22 visit.

A bit overweight prior to pregnancy, Susan did not gain weight and restricted her salt intake after midpregnancy. She believed these actions would help lower her weight and blood pressure. Although she had a supply of a prenatal vitamin and mineral supplement, she rarely remembered to take them. Her baby, weighing 5 pounds 5 ounces (2380 grams), was delivered by cesarean section at week 36.

Questions

1. List four ways in which Susan's dietary intake likely contributes to oxidative stress.
2. Identify two other characteristics of her diet that are contraindicated for women with preeclampsia. Answers should be different than those for question 1.
3. List three health problems Susan is at increased risk of developing due to her history of preeclampsia.

Diabetes in Pregnancy

Diabetes is the second leading complication in pregnancy. It has several forms:

> **Gestational Diabetes**   Carbohydrate intolerance with onset or first recognition in pregnancy.

- Gestational diabetes
- Type 2 diabetes
- Type 1 diabetes
- Other specific types[35]

Gestational diabetes and type 2 diabetes are part of the same disease process. This chapter focuses on gestational diabetes, which develops during pregnancy. Information related to type 1 diabetes, or insulin-dependent diabetes, is presented as well.

Gestational Diabetes

> "The difference between gestational and type 2 diabetes may be the moment of detection."
>
> Branchtein[36]

Approximately 3–7% of pregnant women develop *gestational diabetes* during pregnancy, and the incidence is increasing along with obesity.[37] It is not clear whether it develops primarily due to increasing insulin resistance and exaggerated liver production of glucose late in pregnancy, or to limits on the ability of beta cells to produce enough insulin to overcome the effects of increased insulin resistance in pregnancy.[38] Gestational diabetes in underweight and normal-weight women appears to be related to insulin resistance combined with a deficit in insulin production, whereas insulin resistance—not inadequate insulin production—may underlie it in obese women.[39] Women who develop gestational diabetes appear to enter pregnancy with a predisposition to insulin resistance and type 2 diabetes that is expressed due to physiological changes that occur during pregnancy.

The insulin resistance brought into pregnancy, or the tendency to develop it, may be clinically silent in that glucose levels may not be elevated and blood pressure may be normal. However, high blood levels of glucose and other signs related to increased insulin resistance develop as pregnancy progresses. Women with gestational diabetes develop elevated levels not only of blood glucose but also of triglycerides, fatty acids, and sometimes blood pressure. In some cases, gestational diabetes

Illustration 5.1 Concentrations of fetal blood glucose following an intravenous dose of glucose to the mother.

SOURCE: F. Hytten and G. Chamberlain, eds., *Clinical Physiology in Obstetrics,* Blackwell, 1980. Reprinted by permission; and based on data from Coultart, et al. Reprinted by permission.

appears to be related to exaggerated metabolic changes favoring elevated blood glucose levels[38] or reduced insulin output.[40] High maternal blood glucose levels reach the fetus (Illustration 5.1) and cause the fetus to increase insulin production to lower it. The higher the level of blood glucose received, the larger the fetal output of insulin.

Potential Consequences of Gestational Diabetes

Potential consequences associated with gestational diabetes are summarized in Table 5.7. Elevated *hemoglobin A1c* levels, a long-term marker of blood glucose concentrations, indicate poor glucose control and higher risk of adverse outcomes. Specifically, hemoglobin A1c levels over 8% are associated with higher rates of spontaneous abortion, stillbirth, neonatal death, and *congenital anomalies* than are values below 7%.[41] Exposure to high insulin levels in utero leads to increased glucose uptake into cells and the conversion of glucose to triglycerides. These changes increase fetal formation of fat and muscle tissue and may program metabolic adaptations, increasing the likelihood that

Hemoglobin A1c A form of hemoglobin used to identify blood glucose levels over the lifetime of a red blood cell (120 days). Glucose molecules in blood will attach to hemoglobin (and stay attached). The amount of glucose that attaches to hemoglobin is proportional to levels of glucose in the blood. The normal range of hemoglobin A1c is 4 to 5.9%. Also called glycosylated hemoglobin and glycated hemoglobin.

Congenital Anomalies Structural, functional, or metabolic abnormalities present at birth. Also called congenital abnormalities.

insulin resistance, type 2 diabetes, high blood pressure, and obesity will develop later in life.[38] The chances that these disorders will occur increase with higher maternal levels of glucose and triglycerides during pregnancy.[42]

Effects of high maternal levels of glucose and triglycerides are particularly striking in the Pima Indians of Arizona. Fetal exposure to poorly controlled maternal diabetes incurs a tenfold increase in the risk that children will develop type 2 diabetes. Offspring of diabetic Pima mothers are heavier at birth, have higher body mass index (BMI) throughout childhood, and have 7–20 times greater incidence of type 2 diabetes in early adulthood. Although risks of these conditions increase in offspring of women with poorly controlled diabetes in general, the pronounced effect in Pima Indians is likely due to a strong genetic tendency toward insulin resistance and obesity.[43]

The end of pregnancy initially restores insulin sensitivity in most women with gestational diabetes. However, a degree of insulin resistance often remains.[15] Close to half of women with gestational diabetes in a previous pregnancy will develop it in a subsequent pregnancy.[44] The cumulative incidence of type 2 diabetes in women who experienced gestational diabetes is approximately 50% after 5 years. Women with weight gain after pregnancy and repeated pregnancies continue to experience insulin insufficiency and resistance; this group is even at higher risk of developing type 2 diabetes later in life. Among women who have experienced gestational diabetes, those requiring insulin therapy have higher blood pressure than women whose gestational diabetes was controlled with diet and exercise.[15]

Table 5.7 Adverse outcomes associated with gestational diabetes[15,38,40]

Mother
- Cesarean delivery to prevent shoulder dystocia
- Increased risk for preeclampsia during pregnancy
- Increased risk of type 2 diabetes, hypertension, and obesity later in life
- Increased risk for gestational diabetes in a subsequent pregnancy

Offspring
- Stillbirth
- Spontaneous abortion
- Congenital anomalies
- Macrosomia (>10 lb or >4500 g)
- Neonatal hypoglycemia, death
- Increased risk of insulin resistance, type 2 diabetes, high blood pressure, and obesity later in life

Risk Factors for Gestational Diabetes

Both type 2 and gestational diabetes are linked to multiple inherited predispositions and their environmental triggers, such as excess body fat and low physical activity levels.[15] Results of a large, prospective study indicate that the risk for gestational diabetes decreases 26% for each 10 grams of fiber consumed daily from plant sources. Diets both low in cereal fiber and high in glycemic load are associated with a 2.15-fold increased risk compared to diets high in cereal fiber and low in glycemic load.[45] About half of women who develop gestational diabetes have no identified risk for the disease.[46] Risk factors for gestational diabetes are outlined in Table 5.8.

Diagnosis of Gestational Diabetes

The diagnosis of gestational diabetes is based on abnormal blood glucose levels. These are detected by blood glucose tests. Glucose screening is recommended for women at high risk of gestational diabetes at the initial visit or as soon as possible thereafter. High risk is identified in women who have one or more of the following:

- Marked obesity
- Diabetes in a mother, father, sister, or brother
- History of glucose intolerance
- Previous macrosomic infant
- Current glucosuria

A 50-gram oral glucose challenge test is generally used for blood glucose screening. This test can be done without fasting. Blood is collected 1 hour after the glucose load is consumed and tested for glucose content. This test should be followed by an oral glucose tolerance test if glucose level is high, or ≥130 mg/dL (7.2 mmol/L). (You can convert mg/dL to millimoles per liter, or mmol/L, by multiplying mg/dL by 0.05551.)

The oral glucose tolerance test (OGTT) is the basis for the diagnosis of most cases of gestational diabetes. It can be bypassed among women with very high glucose screening results and treatment started. A 100-gram glucose, 3-hour test is used for the OGTT. The practice of "loading women up" with a high-carbohydrate diet 3 days in advance of the test is no longer recommended.[47] New glucose-load products, such as flavored soft drinks, jellybeans, and gelatin, are also being used for blood glucose tests.[48,49] A diagnosis of gestational diabetes is made when two or more values for venous serum or plasma glucose concentrations exceed these levels:

Overnight fast	95 mg/dL
One hour after glucose load	180 mg/dL
Two hours after glucose load	155 mg/dL
Three hours after glucose load	140 mg/dL

Because of their increased risk for preeclampsia, women with gestational diabetes should be closely monitored for preeclampsia.[48]

A plasma glucose screening between 24 and 28 weeks of pregnancy is recommended for women at "average risk" and for high-risk women not determined by glucose screen to have elevated glucose levels earlier. Average risk is defined as women who fit neither the low- nor the high-risk profile.

Glucose screens are not recommended for women at low risk, defined as:

- Age < 25 years
- Member of a low-risk ethnic group (those other than Hispanic, African American, South or East Asian, Pacific Islander, Native American, or indigenous Australian)
- No diabetes in first-degree relatives
- Normal prepregnancy weight and weight gain during pregnancy
- No history of glucose intolerance
- No prior poor obstetrical outcomes.[48,49]

Women with gestational diabetes may notice an increased level of thirst (especially in the morning), an increased volume of urine, and other signs related to high blood glucose levels.[50] Urinary glucose cannot be used to diagnose nor monitor gestational diabetes, because the results do not correspond to blood glucose levels.[51]

Table 5.8 Risk factors for gestational diabetes[44–46]

- Obesity, especially high levels of central body fat
- Weight gain between pregnancies
- Underweight
- Age over 35 years
- Native American, Hispanic, African American, South or East Asian, Pacific Islander, indigenous Australian ancestry
- Family history of gestational diabetes
- History of delivery of a macrosomic newborn (>4500 g or >10 lb)
- Chronic hypertension
- Mother was SGA at birth
- History of gestational diabetes in a previous pregnancy
- Diabetes in pregnant women's mothers during the pregnancy with them and LGA at birth
- Low fiber intake, high-glycemic-load diets

Treatment of Gestational Diabetes

A team approach to caring for women with diabetes in pregnancy is advised. Such teams often consist of an obstetrician, a registered dietitian who is also a certified diabetes educator, a nurse educator, and an endocrinologist.[52] The primary approach to treatment is medical nutrition therapy that begins with attempts to normalize blood glucose levels with diet and exercise.

If postprandial glucose levels remain high 2 weeks after institution of the diet and exercise plan, then insulin injections will be added. Postprandial blood glucose rather than fasting glucose levels are related to fetal overgrowth and are the main indicators of adequacy of blood glucose control.[38]

Medical nutrition therapy has been shown to effectively normalize blood glucose levels and to decrease the risk of adverse perinatal outcomes.[52] Results shown in Table 5.9 demonstrate the effect and the usefulness of identifying and intervening upon women with gestational diabetes. It can also be noted from the results that a higher proportion of large newborns occurs even with medical nutrition therapy, but that the incidence is substantially less than in women with untreated gestational diabetes.

Blood glucose levels can be brought down by low caloric intakes. However, accelerated rates of starvation metabolism during pregnancy, as well as potentially deleterious effects of resulting ketonemia on fetal development, exclude this approach to blood glucose control.[54] Correspondingly, restriction of pregnancy weight gain to below recommended amounts is not advised.[55] Aggressive treatment of gestational diabetes that excessively limits caloric intake and weight gain increases the risk of SGA newborns.[56] On the other hand, excessively high caloric balances and weight gains are of concern because they increase the risk of macrosomia.[57]

Type 2 diabetes in nonpregnant individuals is often treated with sulfonylurea oral medications. These drugs cannot be used in pregnancy, because they cross the placenta and stimulate fetal insulin production. Other types of oral medications such as metformin are being tested for use among women with gestational diabetes.[58]

Presentation of a Case Study

No two women with gestational diabetes share the same history, risks, needs, and response to treatment. Case Study 5.2, represents an individual's experience with the disorder.

Exercise Benefits and Recommendations

Insulin resistance is decreased and blood glucose control enhanced by regular aerobic exercise such as walking, jogging, biking, golfing, hiking, swimming, and moderate weight lifting. This appears to be the case as well in women with gestational diabetes. Weight lifting with the arms 3 days a week for 20 minutes per session for 6 weeks, and exercising on a recumbent bicycle at 50% VO$_2$ max for 45 minutes three times a week, have been found to normalize blood glucose levels in some women.[59]

Levels of exercise that approximate 50–60% of VO$_2$ max, or maximal oxygen uptake, are most often recommended for women with gestational diabetes. These levels are estimated in practice using a formula for heart rates associated with various levels of VO$_2$ max. The formula is 220 − age × 0.50 (for 50% of VO$_2$ max) = heartbeats per minute. In the case of a 29-year-old, the estimated heart rate at 50% of VO$_2$ max would be 220 − 29 × 0.50, or 96 beats per minute. This would be the maximum heart rate she should experience while exercising. Levels of exercise should make women become slightly sweaty but not overheated, dehydrated, or exhausted.[55]

Nutritional Management of Women with Gestational Diabetes

Primary outcome goals for women with gestational diabetes are well-controlled blood glucose levels or hemoglobin A1c < 7%, and a healthy newborn.[60] Other goals include the normalization of carbohydrate metabolism and a reduction in the mother's and offspring's subsequent risk of diabetes, hypertension, heart disease, and obesity.[55] For most women, diet and exercise changes will be the primary way to achieve these goals. In other women, supplemental insulin will also help achieve these goals.

The following are components of the nutritional management of women with gestational diabetes:

- Assessing dietary and exercise habits
- Developing an individualized diet and exercise plan for blood glucose control

Table 5.9 Comparison of outcomes of unrecognized and diet-treated gestational diabetes

| Outcome | Gestational Diabetes | | |
	Unrecognized	Diet-Treated	Controls
LGA (>90th percentile)	44%	9%	5%
Macrosomia (>4500 g)	44%	15%	8%
Shoulder dystocia	25%	3%	3%
Birth trauma	25%	0%	0%

SOURCE: Data from Adams, 1998.[51]

Case Study 5.2

Photo Disc

Elizabeth's Story: Gestational Diabetes

Elizabeth is a 36-year-old who entered pregnancy with a BMI of 23.5 kg/m². She began receiving prenatal care at 32 weeks gestation and was screened for gestational diabetes the next day. Results of her oral glucose tolerance test revealed the following blood glucose levels:

Fasting: 90 mg/dL
1-hour: 195 mg/dL
2-hour: 163 mg/dL
3-hour: 135 mg/dL

Elizabeth's health care provider advised her to consume a no-sugar, low-carbohydrate diet and to keep her weight gain low throughout the rest of pregnancy. She delivered a large infant (4750 grams) at 39 weeks gestation.

Questions

1. Did Elizabeth have gestational diabetes?
2. Was she insulin resistant?
3. What's the most likely reason Elizabeth delivered an abnormally large newborn?
4. What was wrong with the dietary advice Elizabeth was given?
5. List three components of appropriate dietary advice for women with gestational diabetes.

- Monitoring weight gain
- Interpreting blood glucose and urinary ketone results
- Ensuring follow-up during pregnancy and postpartum[53,59]

Women with type 2 diabetes coming into pregnancy are managed in much the same fashion as are women with gestational diabetes, only nutritional care begins earlier. Ideally, normal blood glucose levels should be established prior to conception and then maintained in good control through pregnancy. Diet and exercise plans for women with type 2 diabetes can often be based on what has worked in the past, thus simplifying planning for needs associated with pregnancy.

The Diet Plan   In general, diets developed for women with gestational diabetes emphasize:

- Whole grain breads and cereals, vegetables, fruits, and high-fiber foods

- Limited intake of simple sugars and foods and beverages that contain them
- Low-GI foods, or carbohydrate foods that do not greatly raise glucose levels
- Monounsaturated fats
- Three regular meals and snacks daily

Dietary planning is based around a calculated level of caloric need. These initial estimates of caloric need are intended to meet both maternal and fetal demand for energy while limiting increases in blood glucose levels. They are based on the pregnant woman's current weight and her need to gain weight during pregnancy. Estimated levels of caloric need according to women's current weight status are shown in Table 5.10 on the next page.

Calories are generally distributed among the meals and three snacks. Lunch serves as the largest meal, and breakfast and snacks are limited to 10 to 15% of total calories.[55]

Caloric levels and meal and snack plans are considered to be starting points and often require modifications after results of blood glucose home monitoring tests are known. Caloric levels for overweight and obese women are particularly likely to initially fall short of need,[61] and morning blood glucose levels are most likely to be high.[62] Reduction of calories from carbohydrate is indicated at breakfast for women with high morning glucose levels.

Specific recommendations for the distribution of calories from carbohydrate and fat primarily depend on individual eating habits and the effect of carbohydrate and fat on blood glucose levels. The distribution of calories utilized in practice, however, commonly falls into these ranges:

- 40–50% of calories from carbohydrate
- 30–40% from fat
- 20% from protein

The relatively low carbohydrate, high-fat diet decreases the need for insulin by lowering the amount of glucose absorbed from food, and blunts postprandial increases in blood glucose and insulin levels (Table 5.11). The addition of high-fiber foods to diet plans may also enhance blood glucose control. These changes in turn reduce fetal overgrowth and other adverse effects of insulin resistance and high levels of glucose and insulin.[59]

Consumption of Foods with a Low Glycemic Index

Whether low-GI foods benefit women with diabetes in pregnancy has been much debated and is somewhat controversial. Low-GI foods help women sustain modest improvements in blood glucose levels and decrease insulin requirements.[32,63] Illustration 5.2 on the following page demonstrates this point by showing blood glucose levels after a meal containing white bread (GI = 70) or spaghetti (GI = 48) is consumed.

Example Meal Plans   Individualized diet plans for women with gestational diabetes include a variety of foods that correspond to the preferences and needs of women. Two examples of such diet plans are shown in Table 5.12 on page 141. One menu provides approximately 2200 calories, the other 2400. Both menus include low-GI food sources of carbohydrate and the nutrients needed by women during pregnancy.

Urinary Ketone Testing   Women with gestational diabetes are generally instructed to monitor urinary ketone levels using dipsticks. In the past all dipsticks used were insensitive to β-hydroxybutyrate, the primary ketone spilled into urine. Sticks that detect β-hydroxybutyrate

---

Table 5.10 Estimating levels of caloric need in women with gestational diabetes[59]

Current Weight Status	Definition	Body Weight, kcal/kg
Under weight	<80% of average weight	35–40
Normal weight	80–120% of average weight	30
Overweight	120–150% of average weight	24
Obese	>150% of average weight	12–15

Table 5.11 Effects of 6 weeks of low- and high-carbohydrate diets on maternal and newborn outcomes in women with gestational diabetes[65]

Outcome	Diet	
	Low Carbohydrate (<42% of total calories)	High Carbohydrate (>45% of total calories)
Postprandial glucose values	110 mg/dL	132 mg/dL
Fasting glucose	92 mg/dl	94 mg/dL
Insulin requirement	5%	33%
Urinary ketones	10%	0%
Birthweight, g	3694	3890
LGA	9%	42%
Cesarean delivery due to LGA fetus	3%	48%

are becoming increasingly available and should be used. When interpreting results of urinary ketone tests, keep in mind that 10–20% of pregnant women spill ketones after an overnight fast.[55] This means the severity and consistency of positive findings for urinary ketones should be considered.

Postpartum Follow-Up

The relatively low carbohydrate, high-fat diet instituted during pregnancy is generally changed to a higher-carbohydrate, lower-fat diet after delivery. Emphasis on low-GI and high-fiber foods, along with gradual weight loss, may also be components of postpartum nutrition counseling.

Prevention of Gestational Diabetes

Reducing overweight and obesity, increasing physical activity, and decreasing insulin resistance prior to pregnancy are important components of reducing the risk of gestational diabetes.[61] Screening programs that identify women with insulin resistance or glucose intolerance prior to pregnancy have also been advocated for risk reduction.[64] The risk of type 2 diabetes after pregnancy can be reduced substantially by healthful eating, aerobic and resistance exercise, and maintenance of normal weight.[66]

Type 1 Diabetes during Pregnancy

Women with type 1 diabetes have deficient insulin output and must rely on insulin injections or an insulin pump to meet their need for insulin. Type 1 diabetes represents a potentially more hazardous condition to mother and fetus than do most cases of gestational diabetes.

Type 1 diabetes places women at risk of kidney disease, hypertension, and other complications of pregnancy.[67] Newborns of women with this type of diabetes are at increased risk of mortality, of being SGA or LGA, and of experiencing hypoglycemia and other problems within 12 hours after birth. Hypoglycemia occurs in about half of macrosomic infants.[68] Coming into pregnancy with this type of diabetes also increases by threefold (from 2–3% to 6–9%) the risk of congenital malformations of the pelvis, central nervous system, and heart in offspring. Good control of blood glucose and diets rich in antioxidants reduce the risk of malformations. Maintenance of normal glucose levels from the start of pregnancy decreases the risk of fetal malformations and macrosomia.[69]

Blood glucose control from the beginning of pregnancy is also important because the fetal growth trajectory may be largely determined in the first half of pregnancy. Exposure to high amounts of glucose and insulin when the fetal growth trajectory is being established may set the "metabolic stage" for fetal accumulation of fat and lean tissue later in pregnancy.[70] Even relatively low elevations

Illustration 5.2 Blood glucose response in people with type 2 diabetes to meals containing white bread or spaghetti.

SOURCE: G. Riccardi and A. A. Rivellese, Diabetes: Nutrition in Prevention and Management. *Nutr Metab Cardiovasc Dis* 1999;9:33–6. Reproduced with permission of Medikal Press S.r.l.

in blood glucose levels can meaningfully increase birth weight.[52] Unfortunately, only 10% of women with type 1 diabetes receive preconceptional care.[69]

Availability of a variety of new insulins, the insulin pump, and self-monitoring technology has revolutionized the care of type 1 diabetes during pregnancy.

Nutritional Management of Type 1 Diabetes in Pregnancy   Primary goals for the nutritional management of type 1 diabetes in pregnancy are continual control of blood glucose levels, nutritional adequacy of dietary intake, achievement of recommended amounts of weight gain, and a healthy mother and newborn. Careful home monitoring of glucose levels and adjustments in dietary intake, exercise, and insulin dose based on the results are key events that increase the likelihood of reaching these goals. Monitoring urinary ketones is particularly important in women with type 1 diabetes because they are more prone to developing ketosis than are women with gestational diabetes.[35] Inclusion of ample amounts of dietary fiber (25–35 g per day) reduces insulin requirements in many women with type 1 diabetes in pregnancy.[62]

Multifetal Pregnancies

Rates of multifetal pregnancy in the United States have increased markedly since 1980. Twin births, which accounted for 1 in 56 births in 1980, constituted 1 in 31 births in 2005. Rates of triplet and higher-order multiple births (referred to as *triplet+* births) have increased from 1 in 2941 to 1 in 558 births.[71] The leading reason for the increased prevalence of multifetal pregnancies in the United States and other developed countries is the use of

Table 5.12 Examples of three-meal, three-snack 1-day menus at two caloric and carbohydrate levels for women with gestational diabetes

2220-Calorie Diet	Carbohydrates, g	Calories	2400-Calorie Diet	Carbohydrates, g	Calories
Breakfast			**Breakfast**		
All Bran, ½ c	22	80	Complete Wheat		
2% milk, ½ c	6	61	Bran Flakes, ¾ c	23	90
Mozzarella cheese			2% milk, ½ c	6	61
stick, 1 oz	1	78	Egg, 1	1	74
Black coffee, tea			Black coffee, tea		
Morning Snack			**Morning Snack**		
Oat bran bagel, ½	19	98	Peanuts, 2 oz	10	326
Sugar-free, low-fat			Carrot, 1	7	31
yogurt, 1 c	17	155	Graham crackers,		
Lunch			4 small or 1 sheet	11	59
Tuna salad, ½ c	19	192	**Lunch**		
Whole grain			Beef or chicken		
bread, 2 slices	24	130	burrito, 1	33	255
Carrot and celery			Salsa, ½ c	7	33
sticks, 1 c	7	31	Black beans, 1 c	40	228
Potato salad, ½ c	14	179	Apple, 1	21	81
Orange, 1	15	62	Black coffee, tea,		
Black coffee, tea,			water, or diet soda		
water, or diet soda			**Midday Snack**		
Midday Snack			Banana, ½	28	55
Peaches canned in			2% milk, 1 c	12	121
juice, ½ c	29	109	**Dinner**		
2% milk, 1 c	12	121	Lean pork chop, 4 oz	0	263
Dinner			Pinto beans, 1 c	22	116
Lean roast beef, 3 oz	0	188	Corn bread, 1 oz	12	92
Broccoli, 1 c with:	10	50	Margarine, 1 tsp	1	33
Cheese (melted), 1 oz	1	105	Garden salad, 2 cup	0	10
Roll, 2 oz	30	167	Feta cheese, 1 oz	1	74
Margarine, 2 tsp	0	67	Salad dressing, 2 Tbsp	3	104
Grapes, 15	14	53	Black coffee, tea,		
Black coffee, tea, water,			water, or diet soda		
or diet soda			**Bedtime Snack**		
Bedtime Snack			Peanut butter, 2 Tbsp	12	190
Hard-boiled egg, 1	1	74	Rice cake, 1	8	35
Saltine crackers, 4	8	50	2% milk, 1 c	12	121
2% milk, 1 c	12	121	**Total:**	**270 g**	**2442**
Total:	**261 g**	**2171**	or 44% of total calories		
or 48% of total calories					

SOURCE: Developed by Judith Brown.

assisted reproductive technology.[71] Rates of twin and higher-order births are highest by far in women 45–54 years old (1 in 5 births), the age group most likely to receive assisted reproductive technological interventions to achieve pregnancy.[72]

The progressively older ages at which U.S. women are bearing children also contribute to rising rates of multifetal pregnancies. The chances of a spontaneous multifetal pregnancy increase with age after about 35 years. Rates of spontaneous multifetal pregnancy also increase with increasing weight status. For example, the rate of twin pregnancy is about two times higher in obese than in underweight women.[73] Rates of triplet+ pregnancies are headed downward due to improved assisted

Assisted Reproductive Technology (ART) An umbrella term for fertility treatments such as *in vitro* fertilization (IVF, a technique in which egg cells are fertilized by sperm outside the woman's body), artificial insemination, and hormone treatments.

reproductive technologies that reduce higher-order, multifetal pregnancies.[74]

Upward trends in low birth weight and preterm delivery in the United States over recent years have been strongly influenced by the upsurge in multiple births. Only 3% of newborns are from multifetal pregnancies, yet they account for 21% of all low-birth-weight newborns, 14% of preterm births, and 13% of infant deaths.[75]

Background Information about Multiple Fetuses

The most common type of multifetal pregnancies, those with twin fetuses, come in several types and levels of risk. Twins are dizygotic (DZ) if two eggs were fertilized, and monozygotic (MZ) if one egg was. Monozygotic twins result when the fertilized and rapidly dividing egg splits in two within days after conception. The term *identical* is often used to describe MZ twins, and *fraternal* denotes DZ twins. These terms are misleading, so the preferred terms are *monozygotic* and *dizygotic*.[76] About 70% of twins are DZ, and 30% are MZ.

Monozygotic twins are always the same sex, whereas DZ twins are the same sex half the time and different sexes half the time. Monozygotic twins are genetically identical in almost all ways, but they are seldom absolutely identical. Genetic differences in pairs of MZ twins can result from chromosome abnormalities in one twin, unequal genetic expression of maternally and paternally derived genes, and environmental effects on gene expression. Rates of MZ twins are remarkably stable across population groups and do not appear to be influenced by heredity.[77]

Dizygotic twins represent individuals with differing genetic "fingerprints." The incidence of DZ twin pregnancies is influenced both by inherited and environmental factors. Rates of DZ twins vary among racial groups and by country. Rates tend to decrease in populations during famine and to increase when food availability and nutritional status improve.[78] Periconceptional vitamin and mineral supplement use has also been related to an increased incidence of DZ twin pregnancy.[79]

Twins also vary in the number of placentas; some are born having used the same placenta, but more commonly each fetus has its own. Twins may share a common amniotic sac and one of the membranes around the sac (the chorion), or have separate amniotic sacs and membranes (Illustration 5.3). Twins at highest risk of death, malformations, growth retardation, short gestation, and other serious problems are those that share the same amniotic sac and chorion, and to a lesser degree, MZ twins in general.[77,80]

Determining twin type is not always an easy task during or after pregnancy. Definitive diagnoses of tough cases can be made through DNA fingerprinting.[80]

In Utero Growth of Twins and Triplets   Fetal growth patterns of twins and triplets compared to singleton fetuses are shown in Illustration 5.4. Rates of weight gain for each group of fetuses are the same until about 28 weeks of gestation. Rates of weight gain begin to decline in twin and triplet fetuses after that point, however, and remain lower until delivery. Variations in birth weight of twin and triplet newborns appear to be related to factors that affect fetal growth after 28 weeks of pregnancy.[80]

The Vanishing Twin Phenomenon   The disappearance of embryos within 13 weeks of conception is not unusual.

(a)   **(b)**   **(c)**

Illustration 5.3 Twins (a) with two amniotic sacs, two chorions, and two placentas; (b) with one amniotic sac, chorion, and placenta; and (c) with two amniotic sacs, one chorion, and fused placentas.

SOURCE: From G. Martins in Hebammen-Lelubuch, 4th Ed. Copyright © 1983 by Thieme. Reprinted with permission.

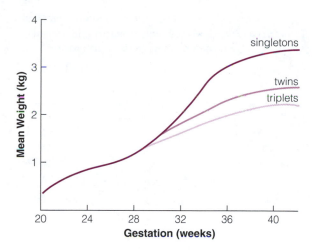

Illustration 5.4 Rates of fetal weight gain in singleton, twin, and triplet fetuses.

SOURCE: From "Multifetal Pregnancies: Epidemiology, Clinical Characteristics, and Management," by M. Smith-Levitin, et al., in *Maternal-Fetal Medicine: Principles and Practice*, 3rd Ed., R. K. Creasy and R. Resnick (Eds.), p. 589–601. Copyright © 1994. Reprinted by permission of W. B. Saunders Co.

It has been estimated that 6 to 12% of pregnancies begin as twins, but that only about 3% result in the birth of twins. Most fetal losses silently occur by absorption into the uterus within the first 8 weeks after conception. The prognosis for continued viability of a pregnancy associated with a vanishing twin tends to be good.[81]

Risks Associated with Multifetal Pregnancy

Singleton pregnancy is the biological norm for humans, so it may be expected that multifetal pregnancy would be accompanied by increasing health risks (Table 5.13).[82] Multifetal pregnancies present substantial risks to both mother and fetuses, and the risks increase as the number of fetuses increase (Table 5.14). Newborns from twin pregnancies at lowest risk of death in the perinatal period weigh between 3000 and 3500 grams (6.7 to 7.8 lb) at birth and are born at 37–39 weeks gestation. Triplets tend to do best when they weigh over 2000 grams (4.5 lb) and are born at 34–35 weeks gestation.[82]

Table 5.13 Risks to mother and fetuses associated with multifetal pregnancy[82]

Pregnant Women
- Preeclampsia
- Iron-deficiency anemia
- Gestational diabetes
- Hyperemesis gravidarum
- Placenta previa
- Kidney disease
- Fetal loss
- Preterm delivery
- Cesarean delivery

Newborns
- Neonatal death
- Congenital abnormalities
- Respiratory distress syndrome
- Intraventricular hemorrhage
- Cerebral palsy

Unfortunately, these outcomes do not represent the usual. Data presented in Table 5.15 on the next page show that median weights of twins born at 37, 38, and 39 weeks gestation fall below the 3000- to 3500-gram range. However, the 3000- to 3500-gram birth weight range for twins, and the >2000-gram mark for triplet newborns, can serve as goals for the provision of nutrition services.

Interventions and Services for Risk Reduction

Special multidisciplinary programs that offer women with multifetal pregnancy a consistent, main provider of care; preterm prevention education; increased attention to nutritional needs; and intensive follow-up achieve better pregnancy outcomes than does routine prenatal care.[83] Rates of very low birth weight (≤1500 g or ≤3.3 lb) have been reported to be substantially lower (6% versus 26%),

Table 5.14 Average birth weight and gestational age at delivery, and low-birth-weight rates, of singleton, twin, and triplet newborns[82,85]

	Mean Birth Weight	Mean Gestational Age	Low-Birth-Weight Rate
Singletons	3440 g (7.7 lb)	39–40 weeks	6%
Twins	2400 g (5.4 lb)	37 weeks	54%
Triplets	1800 g (4.0 lb)	33–34 weeks	90%

Table 5.15 Median birth weight for gestational age at delivery of twins

Gestational Age, weeks	Birth Weight
28	995 g (2.2 lb)
29	1145 g (2.6 lb)
30	1300 g (2.9 lb)
31	1445 g (3.2 lb)
32	1580 g (3.5 lb)
33	1750 g (3.9 lb)
34	1905 g (4.3 lb)
35	2165 g (4.8 lb)
36	2275 g (5.1 lb)
37	2430 g (5.4 lb)
38	2565 g (5.7 lb)
39	2680 g (6.0 lb)
40	2810 g (6.3 lb)
41	2685 g (6.0 lb)

SOURCE: Data from Cohen SB, 1997.[91]

Table 5.16 Prepregnancy weight status and weight-gain relationships in twin pregnancy[74]

Prepregnancy Weight Status	Weight Gain Related to Birth Weights of >2500 g (5.5 lb)
Underweight	44.2 lb (20.1 kg)
Normal weight	40.9 lb (18.6 kg)
Overweight	37.8 lb (17.2 kg)
Obese	37.2 lb (16.9 kg)
Very obese	29.2 lb (13.3 kg)

neonatal intensive care admissions three times lower (13% versus 38%), and perinatal mortality strikingly lower (1% versus 8%) among women who receive such services.[84] Interventions offered by the Montreal Diet Dispensary, which focuses on improving the nutritional status and well-being of the pregnant women served, have been shown to substantially reduce poor outcomes compared to those for similar women not receiving the services. Improvements include a 27% reduction in the rate of low birth weight, 47% decline in very low birth weight, 32% lower rate of preterm delivery, and a 79% drop in mortality during the first 7 days after birth.[86]

Nutrition and the Outcome of Multifetal Pregnancy

Nutritional factors are suspected of playing a major role in the course and outcome of multifetal pregnancy, but much remains to be learned. Of the nutritional factors that influence multifetal pregnancy, weight gain during twin pregnancy has been studied most.

Weight Gain in Multifetal Pregnancy   As with singleton pregnancy, weight gain in multifetal pregnancy is linearly related to birth weight, and weight gains associated with newborn weight vary based on prepregnancy weight status (Table 5.16).[83] The Institute of Medicine recommends that women with twins gain 35 to 45 pounds (15.9 to 20.5 kg). It is advised that underweight women gain toward the upper end of this range, and overweight and obese women near the lower end.[87]

Rate of Weight Gain in Twin Pregnancy   A positive rate of weight gain in the first half of twin pregnancy is

strongly associated with increased birth weight.[88,89] On the other hand, weight loss after 28 weeks of pregnancy increases the risk of preterm delivery by threefold.[90]

Recommended rates of weight gain for women with twin pregnancy are:

- 0.5 pounds (0.2 kg) per week in the first trimester
- 1.5 pounds (0.7 kg) per week in the second and third trimesters[82]

Weight Gain in Triplet Pregnancy   Several studies have examined the relationship between weight gain and birth weight in women with triplets. The general result is that weight gains of about 50 pounds (22.7 kg) correspond to healthy-sized triplets. Rates of gain related to a total weight gain of 50 pounds in women who will average 33 to 34 weeks of gestation are 1.5 pounds (0.7 kg) per week or more, starting as early in pregnancy as possible.[82]

Dietary Intake in Twin Pregnancy

Ensuring "adequate nutrition" is widely acknowledged to be a key component of prenatal care for women with multifetal pregnancy. However, it is not clear what constitutes adequate nutrition. Energy and nutrient needs clearly increase during multifetal pregnancy due to increased levels of maternal blood volume, extracellular fluid, and uterine, placental, and fetal growth. The normally high expansion in extracellular volume and its side effect of leg and ankle edema can be seen in the healthy woman with a twin pregnancy shown in Illustration 5.5. Increases in energy and nutrient needs place demands on the mother in terms of the nutritional costs of building and maintaining these tissues. Although their newborns are smaller, women with twins still produce around 5000 g (11.2 lb) of fetal weight, and women with triplets 5400 g (13.4 lb) or more.

Evidence of higher caloric need for tissue maintenance and growth in multifetal than singleton pregnancy comes from studies that show increased weight gain and a quicker

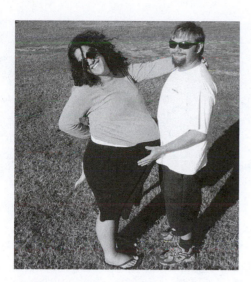

Illustration 5.5 The woman pregnant with twins shown in this photo labeled it "My poor feet!" The leg and ankle swelling are an expected and normal part of a healthy twin pregnancy.

SOURCE: Photo courtesy of Alyeen and Jeremy Perez-Marty.

onset of starvation metabolism in women expecting more than one newborn. Reduced rates of twin deliveries, as well as the higher incidence of twins in overweight and obese women, imply that energy status is an important factor in multifetal pregnancy.[82] Whereas it is obvious that energy and nutrient needs are higher in multifetal than singleton pregnancy, levels of energy balance and nutrient intake associated with optimal outcomes of multifetal pregnancy have not been quantitated.

Results of a large prospective study indicate that women with twins enter pregnancy with higher average caloric intakes (2030 versus 1789 cal per day) and consume an average of 265 cal more per day during pregnancy than women with singleton pregnancy. Nutrient intakes during pregnancy are also higher in women bearing twins than with singleton pregnancy.[73]

Several studies have concluded that the need for specific nutrients is increased during multifetal pregnancy. The need for essential fatty acids (linoleic and alpha-linolenic acid) appears to be increased in multifetal pregnancy. Poor essential fatty acid status is related to neurologic abnormalities and vision impairments in twin offspring.[92] Requirements for iron and calcium have also been found to be increased based on the magnitude of physiological changes that take place in multifetal pregnancy. Levels of essential fatty acids, iron, or calcium required by women to meet these increased needs are unknown, however.[82]

Vitamin and Mineral Supplements and Multifetal Pregnancy Benefits and hazards of multivitamin and mineral supplement use in multifetal pregnancy have not been reported. Consequently, the extent to which they may be required is unknown. Levels of nutrient intake

exceeding the DRI Tolerable Upper Intake Levels should be avoided.

Nutritional Recommendations for Women with Multifetal Pregnancy

Due to the lack of study results, nutritional recommendations for women with multifetal pregnancy are largely based on logical assumptions and theories (Table 5.17). It is reasoned, for example, that caloric needs for twin pregnancy can be extrapolated from weight gain. Theoretically, to achieve a 40-pound (18.2 kg) weight gain, or 10 pounds (4.5 kg) more than in singleton pregnancy, women with twins would need to consume approximately 35,000 cal more during pregnancy than do women with singleton pregnancies. This increase would amount to about 150 cal per day above the level for singleton pregnancy, or an average of 450 cal more per day than prepregnancy. To achieve higher rates of gain, underweight women may need a higher level of intake, and overweight and obese women lower levels. Energy needs will also vary by energy expenditure levels. As for singleton pregnancy, adequacy of caloric intake can be estimated by weight-gain progress.[82]

Food-intake recommendations for women with multifetal pregnancy are primarily estimated based on assumptions related to caloric and nutrient needs. Women with multifetal pregnancy likely benefit from diets selected from the MyPyramid groups and nutrient intakes that somewhat exceed the RDAs/AIs.

Although higher risk than singleton pregnancies, outcomes of twin pregnancy can be excellent. Illustration 5.6 shows healthy, term newborn twins. Their mother remained in good health during pregnancy while consuming the type of diet and supplements, and gaining weight as recommended by her health care providers.

Recommendations from the Popular Press Websites, books, and pamphlets are available that provide ample amounts of scientifically unsupported "guesses" about food and nutrient requirements of women with multifetal pregnancy. Even if presented with steely resolution, any advice that strays from current scientifically based wisdom about nutritional needs of women during pregnancy should be sidestepped.

HIV/AIDS during Pregnancy

The world first became aware of acquired immunodeficiency syndrome (AIDS) in the summer of 1981. It was caused by a newly recognized microbe, the human immunodeficiency virus (HIV). Since then, over 100,000 women of childbearing age in the United States have been diagnosed with AIDS,[91] and 15 million worldwide. Transmission of the virus during pregnancy and delivery is a major route to the spread of the infection. Approximately

Table 5.17 "Best practice" recommendations for nutrition during multifetal pregnancy[82]

Weight Gain

Twin pregnancy: Overall gain of 35–45 lb (15.9–20.5 kg). Underweight women should gain at the upper end of this range, and overweight and obese women at the lower end.

- First trimester: 4–6 lb (1.8–2.7 kg)
- Second and third trimesters: 1.5 lb (0.7 kg) per week

Triplet pregnancy: Overall gain of approximately 50 lb (22.7 kg)

- Gain of 1.5 lb (0.7 kg) per week through pregnancy

Daily Food Intake

Twin pregnancy (2400–2800 calories a day)

- Grains: 8–10 oz
- Vegetables: 3–3.5c
- Fruits: 2–2.5c
- Meat and beans: 6.5–7 oz
- Milk: 3c
- Oil: 7–8 tsp
- Discretionary calorie allowance: 362–426

Triplet pregnancy

- Foods intake from the MyPyramid groups should be consumed at a level that promotes targeted weight gain.

Caloric Intake

Twin pregnancy

- 450 calories above prepregnancy intake, or the amount consistent with targeted weight gain progress

Triplet pregnancy

- Caloric intake levels should promote targeted weight gain progress.

Nutrient Intake

Twin and triplet pregnancy

- RDA or AI levels or somewhat more than these levels
- Intakes should be lower than ULs.

Vitamin and Mineral Supplements

Twin pregnancy

- Use a prenatal vitamin and mineral supplement.

Triplet pregnancy

- Provide a prenatal vitamin and mineral supplements; avoid excessively high amounts of nutrients.

Illustration 5.6 The outcome of a healthy twin pregnancy: Isa weighed 6 pounds, 8 ounces (2912 grams) and Manu, 6 pounds, 7 ounces (2884 grams). The twins were delivered by a scheduled cesarean section at 38 weeks, 4 days, and were above average weight for their gestational age.

SOURCE: Photo courtesy of Alyeen and Jeremy Perez-Marty.

20% of children with HIV/AIDS are infected during pregnancy or delivery, and 14–21% during breastfeeding.[93]

Treatment of HIV/AIDS

The primary focus of care for pregnant women with HIV/AIDS is the prevention of transmission of the virus to the fetus and infant. The transmission of the AIDS virus from mother to child is negligible if treatment is provided before, during, and after pregnancy.[93] In developing countries where there is not enough money to purchase the drugs, transmission rates can be substantially reduced by giving the mother a short course of a specific anti-HIV drug before delivery.[94]

Consequences of HIV/AIDS during Pregnancy

Disease processes such as compromised immune system functions related to HIV/AIDS progress during pregnancy, but it does not appear that the infection itself is related to adverse pregnancy outcomes. Although adverse pregnancy outcomes such as preterm delivery, fetal growth retardation, and low birth weight tend to be higher in women with HIV/AIDS, differences are most closely related to poverty, poor food availability, compromised health status, and the coexistence of other infections.[95]

Nutritional Factors and HIV/AIDS during Pregnancy

HIV/AIDS is related to poor nutritional status that further compromises the body's ability to fight infections. The disease can lead to nutrient losses and fat malabsorption due to diarrhea, and inflammatory responses to the infection cause the loss of lean muscle mass. Loss of calcium from bones and decreased bone density is a common finding in individuals with HIV/AIDS.[96] Risks of inadequate nutrient status of a wide variety of vitamins and minerals increase as the disease progresses. Nutritional needs increase the most during the later stages of HIV/AIDS as diarrhea, wasting, and reductions in CD4 counts (a measure of white blood cells that help the body fight infection) increase. New drugs used to treat HIV/AIDS are associated with increased insulin resistance and the accumulation of central body fat.[97]

The compromised immune status of women with HIV/AIDS, and further decreases in immune response during pregnancy, mean that women with the disease are at high risk of developing foodborne infections during pregnancy. Risk of infection originating from foods can be decreased if raw or uncooked meats and seafood and unpasteurized milk products and honey are not consumed. Safe food-handling practices at home can also reduce the risk of foodborne infection.[98]

Nutritional Management of Women with HIV/AIDS during Pregnancy

Goals for the nutritional management of women with HIV/AIDS include:

- Maintenance of a positive nitrogen balance and preservation of lean muscle and bone mass
- Adequate intake of energy and nutrients to support maternal physiological changes and fetal growth and development
- Correction of elements of poor nutritional status identified by nutritional assessment
- Avoidance of foodborne infections
- Delivery of a healthy newborn[98]

Nutrient requirements of HIV-infected pregnant women are not known, but it is suspected that energy and nutrient needs will be somewhat higher due to the effects of the virus on the body.

Insufficient information exists to provide specific standards for nutritional care for women with HIV/AIDS during pregnancy. Consequently, nutritional recommendations for women with HIV/AIDS are consistent with recommendations for pregnant women in general. As is the case for other pregnant women, foods should be the primary source of nutrients in women with HIV/AIDS.[98]

Eating Disorders in Pregnancy

Eating disorders represent relatively rare conditions in pregnancy because many women with such disorders are subfertile or infertile. Such disorders can have far-reaching effects on both mother and fetus, however, when they do occur. The eating disorder most commonly observed among pregnant women in the United States is bulimia nervosa, or a condition marked by both severe food restriction and bingeing and purging.[99] It is estimated that 1–3% of adolescents and young women in the United States have this condition.[100] Women with bulimia nervosa exhibit poorly controlled eating patterns marked by recurrent episodes of binge eating. To prevent weight gain, women will induce vomiting, use laxatives, exercise intensely, or fast after binges. Self-worth in women with bulimia nervosa is usually closely tied to their weight and shape. A history of sexual abuse is common among women with this eating disorder, as well as in women with anorexia nervosa.[101]

Pregnancy is rarely suspected in women with anorexia nervosa because amenorrhea is a diagnostic criteria for the disorder. Nonetheless, women with anorexia nervosa may occasionally ovulate and become at risk for conception. To women with anorexia nervosa, body weight is of utmost importance, and they are generally fully dedicated to achieving extreme thinness. Adolescents and women with this condition will refuse to eat, even when ravenously hungry; limit their food choices to low-calorie foods only; and exercise excessively.[101]

Eating disorder symptoms often subside during the second and third trimesters of pregnancy, but they rarely vanish altogether. Symptoms tend to return after delivery, sometimes to a more severe extent than was the case prior to pregnancy.[102] Information on eating disorders in nonpregnant individuals is included in Chapter 15.

Consequences of Eating Disorders in Pregnancy

Women with eating disorders during pregnancy are at higher risk for spontaneous abortion, hypertension, and difficult deliveries than are women without an eating disorder. Pregnancy weight gain is generally below the recommended amounts, and newborns tend to be smaller and to experience higher rates of neonatal complications.[103]

Treatment of Women with Eating Disorders during Pregnancy

It is recommended that pregnant women with eating disorders be referred to an eating disorders clinic or specialist. Most large communities have special clinics and programs for women with eating disorders, and they commonly use a team approach to problem solving around the eating disorder. Nutritionists or dietitians often participate in these services because they are knowledgeable about the woman's individual nutritional needs and those of pregnancy.

Health professionals serving women with eating disorders in pregnancy can facilitate open communication and behavioral change by gently encouraging women to talk about their eating disorder, fears, and concerns.[102]

Nutritional Interventions for Women with Eating Disorders

Behavioral changes required for improvements in nutritional status and weight gain in women with eating disorders are most likely to work when the changes are considered acceptable to the women with the disorder. Frequently, the health professional presents the types of changes that need to be made and explains why, and then works with the woman to develop specific plans accomplishing these changes.

Fetal Alcohol Spectrum

New terminology related to the harmful effects of alcohol on fetal growth and development has recently been introduced by the Centers for Disease Control and Prevention (CDC).[104] The term *fetal alcohol spectrum* is now being used to describe the range of effects of fetal alcohol exposure on mental development and physical growth. This range includes behavioral problems, mental retardation, aggressiveness, nervousness, short attention span, and growth-stunting and birth defects.[105] A diagnosis of *fetal alcohol syndrome* is made when infants or young children exhibit a specific set of characteristics included in the fetal alcohol spectrum.

Fetal exposure to alcohol during pregnancy is a leading, preventable cause of birth defects, mental retardation, and developmental disorders in children and adults. Approximately 1 in 12 women in the United States consume alcohol during pregnancy, and 1 in 30 consume 5 or more drinks on one occasion at least monthly.[106] It is estimated that 1 in every 1000 newborns in the United States are affected by the fetal alcohol syndrome.[104]

Effects of Alcohol on Pregnancy Outcome

Alcohol consumed by a woman easily crosses the placenta to the fetus. Because the fetus has yet to fully develop enzymes that break it down, alcohol lingers in the fetal circulation. This situation, combined with the fact that the fetus is smaller and has far less blood than the mother does, increases the harmful effects of alcohol on the fetus as compared to the mother. Alcohol exposure during critical periods of growth and development can permanently impair organ and tissue formation, growth, health, and mental development.

Poor dietary intakes of some women who consume alcohol regularly in pregnancy, as well as the negative effect of alcohol on the availability of certain nutrients, may also contribute to the harmful effects of alcohol exposure during pregnancy.[107]

Consumption of 4 or more drinks a day, or occasional episodes of consumption of 5 or more drinks in a row, is considered to represent heavy alcohol intake during pregnancy. Heavy drinking during pregnancy increases the risk of miscarriage, stillbirth, and infant death within the first month after delivery.[109] Approximately 40% of the fetuses

Table 5.18 Approximate incidence of structural abnormalities of the brain in 5- to 14-week-old fetuses exposed to alcohol	
Maternal Alcohol Intake, drinks	Abnormal Brain Structure
13 to 31/day	100%
6.3 to 13/day	83%
2 to 6.3 occasionally	29%
≤ 2/day	0%

SOURCE: Table is based on information presented by Konovalov et al., 1997.[108]

born to women who drink heavily early in pregnancy will develop fetal alcohol syndrome (FAS). The likelihood that the fetus will be affected by FAS increases as the number of drinks consumed early in pregnancy increases (Table 5.18). Because a "safe" dose of alcohol consumption during pregnancy has not been identified, it is recommended that women do not drink alcohol while pregnant.

The Fetal Alcohol Syndrome

FAS was first described in 1973 and consists of pSGA, mental retardation, and a set of common malformations (Illustration 5.7). Diagnosis of FAS is difficult, however, because many of the facial features are not unique to the

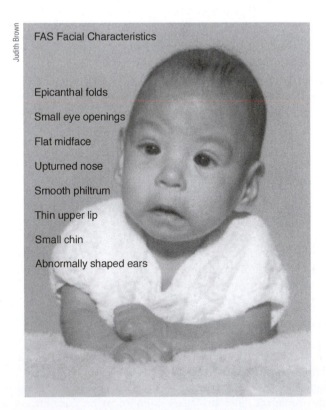

FAS Facial Characteristics

Epicanthal folds

Small eye openings

Flat midface

Upturned nose

Smooth philtrum

Thin upper lip

Small chin

Abnormally shaped ears

Judith Brown

Illustration 5.7 Features of FAS in children.

syndrome. Short noses, flat nasal bridges, and thin upper lips, for example, are also sometimes normal facial features. New criteria developed by the CDC are aimed at limiting this confusion. To be diagnosed with FAS, children must have the characteristic smooth *philtrum*, thin *vermillion border,* small *palpebral fissures,* and a documented neurological disorder (for example, a small head circumference, problems with physical movements, seizures, or cognitive delay), and reduced growth.[104]

Children with FAS tend to have poor coordination, short attention span, and behavioral problems, and they remain small for their age. Adults with FAS often find it difficult to hold a job or live independently.[109]

> **Philtrum** The vertical groove between the bottom of the nose and the upper lip. The philtrum is smooth, or flat, when there is no groove.
>
> **Vermillion Border** The exposed pink or reddish margin of a lip. A thin vermillion border in FAS denotes a thin upper lip.
>
> **Palpebral Fissure** The space between the top and bottom eyelid when the eye is open. This opening is small in children with FAS.

Approximately 10 times the number of infants born with FAS have lesser degrees of alcohol-related damage. These effects are characterized by mental and behavioral abnormalities but not malformations.[105]

Nutrition and Adolescent Pregnancy

Between 1991 and 2004, rates of pregnancy among teens 15–19 years old fell 28%. Adolescent pregnancies hit a record low of 41.2 births per 1000 females aged 15–19 years in 2004. Although the rates are declining, the United States continues to have one of the highest rates of adolescent pregnancy of all developed countries.[110] Adolescents are at higher risk for a number of clinical complications and other unfavorable outcomes compared to adult women (Table 5.19). The downward shift in birth weight observed in newborns of adolescent mothers is related to increased rates of perinatal mortality, and suggests the existence of unfavorable nutrition and health conditions during pregnancy.[111]

The extent to which increased rates of poor outcomes in pregnant teens are associated with biological immaturity or with lifestyle factors such as drug use, smoking, and poor dietary intakes (that influence health status) is unclear. Age-related differences in outcome diminish substantially when potentially harmful lifestyle factors are taken into account, diminishing the theory that biological immaturity accounts for the differences. Very young adolescents becoming pregnant within a few years after the onset of menstruation may be at risk due to biological immaturity. They tend to have shorter gestations and a higher likelihood of cephalopelvic disproportion.[111] Poorly nourished, growing adolescent mothers may compete with the fetus for calories and nutrients—and win.[112]

Table 5.19 Risks associated with adolescent pregnancy[111,116]

- Low birth weight
- Perinatal death
- Cesarean delivery
- Cephalopelvic disproportion (head too large for birth canal)
- Preeclampsia
- Iron-deficiency anemia
- Delayed, reduced educational achievement
- Low income

Growth during Adolescent Pregnancy

Young teens who are growing when pregnancy occurs continue to gain height and weight during pregnancy—but at the expense of fetal growth. Teens who continue to grow during pregnancy give birth, on average, to infants that weigh 155 grams less than infants of adult women, even if they gain more weight than adults do.[112] Rates of spontaneous abortion, preterm birth, and low birth weight are also higher in growing than nongrowing adolescents.[113] Young adolescents gain more maternal fat tissue during the last trimester of pregnancy and retain more weight postpartum than do nongrowing teens. Growing teens experience a surge in blood leptin levels during the last trimester, which may decrease maternal use of fat stores and increase utilization of glucose by the mother. Increased use of glucose by the mother appears to decrease energy ability to the fetus.[115]

Obesity, Excess Weight Gain, and Adolescent Pregnancy Increasing rates of overweight and obesity among adolescents appear to be placing additional teens at risk of poor pregnancy outcomes. Adolescents entering pregnancy overweight or obese are at increased risk for cesarean delivery, hypertensive disorders of pregnancy, gestational diabetes, and the delivery of excessively large newborns.[114] In obese adolescents, weight gains during pregnancy that exceed those recommended are related to decreased placental growth and the birth of infants disproportionately small for gestational age.[113]

Dietary and Other Recommendations for Pregnant Adolescents

Recommendations for pregnant adolescents are basically the same as for older pregnant women, with a few exceptions. Recommendations for weight gain and protein intake are the same, but young adolescents may need more calories to support their own growth as well as that of the fetus. Caloric need should be met by a nutrient-dense diet and lead to rates of weight gain that follow those recommended. Pregnant adolescents have a higher requirement for calcium. The AI for pregnant teens for calcium is 1300 mg per day, or 300 mg higher than for adult pregnant women. This increased need can be met by the consumption of 4 daily servings of milk and milk products, combined with a varied, basic diet.

The importance of lifestyle and other environmental factors to pregnancy outcome in teens emphasizes the need for special, comprehensive teen pregnancy health care programs. Nutrition counseling is an important component of the multidisciplinary services that should be offered to pregnant adolescents. Nutrition services that include individualized nutrition assessment, intervention, education, guidance on weight gain, and follow-up enhance birth-weight outcomes. Additional, specialized services that focus on the psychosocial needs of pregnant adolescents, support/discussion groups, and home visits also contribute to improved maternal and infant outcomes for adolescents.[41]

Because most pregnant adolescents have low income, referral to appropriate food and nutrition programs and other assistance related to health care, housing, and education should be core components of services.

Evidence-Based Practice

"Enormous amounts of new knowledge are barreling down the information highway, but they are not arriving at the doorsteps of our patients."

Claude Lenfant, National Institutes of Health[117]

The clinical, nutritional management of the conditions covered in this chapter, as well as other complications during pregnancy, is not entirely evidence based. Such practices are a problem when they burden women and families with costs, call for dietary changes not known to work, or potentially cause harm. Practices not based on evidence, which likely pose little risk or burden and may potentially be of help, should nonetheless be carefully evaluated. Outdated practices often linger far too long, at the expense of missed opportunities for real improvements.

Use of practices not supported by scientific evidence should always be questioned and confirmed to represent "best practice" insofar as that can be determined. To know what best practice is requires vigilant attention to scientific developments related to the nutritional management of clinical conditions during pregnancy.

Key Points

1. Maternal diet, weight status, and physical activity levels influence the development of a number of conditions that adversely effect the course and outcome of pregnancy.

2. Oxidative stress and endothelial dysfunction appear to be related to the development and progression of hypertensive disorders of pregnancy.

3. Oxidative stress occurs when the body's exposure to oxidizing agents is greater than its supply of antioxidants. Oxidative stress damages the endothelium (the inside lining of blood vessels), and that contributes to endothelial dysfunction. Endothelial dysfunction impairs blood flow, causes an increased tendency for blood to clot, increases plaque formation, and causes other problems.

4. Excess body fat, low levels of physical activity, *trans* fats, lack of antioxidants, high-glycemic-index carbohydrates, and elevated blood glucose increase oxidative stress. It is lowered by weight loss in overweight individuals, increased levels of exercise, exclusion of *trans* fats from the diet, normal blood glucose levels, and sufficient intake of antioxidants.

5. Nutritional management of gestational diabetes focuses on individually based dietary and exercise plans that help maintain blood glucose levels within the normal range and that foster maternal health and appropriate weight gain.

6. Multifetal pregnancies are classified as "high risk" because of above-average rates of complications and less-than-optimal pregnancy outcomes. Energy, nutrient, and weight gain needs are somewhat higher in multifetal than singleton pregnancies.

7. Pregnant women with HIV/AIDS may benefit from nutrition interventions that conserve lean muscle and bone mass, correct nutrient deficiencies, and lead to healthful dietary intake and weight gain in pregnancy. Foodborne illnesses can severely affect people with HIV/AIDS, making food safety a priority concern.

8. Women with eating disorders should be closely monitored during pregnancy to facilitate appropriate dietary intake and weight gain.

9. High levels of alcohol intake during pregnancy are associated with a broad range of mental and physical disorders. The range of disorders is called the *fetal alcohol spectrum*. It is recommended that women do not drink alcohol-containing beverages during pregnancy.

10. Growing adolescents are at risk of poor pregnancy outcomes. Pregnancy outcomes of adolescents who are not growing primary depend on the health status of the teen. Dietary quality and weight status are important components of the health status of adolescents and influence the course and outcome of their pregnancies.

Resources

Clinical Conditions of Pregnancy

The U.S. Preventive Health Services Task Force periodically reviews and updates recommendations for gestational and other forms of diabetes. Updates can be downloaded from this site:
Website: www.ahrq.gov/clinic/serfiles.htm
Visit the Women's Health Resource Center for access to journal articles and summaries and health care information.
Website: www.medscape.com

The National Library of Medicine's website offers extensive coverage of scientific journal articles, summaries, and educational resources from a variety of reputable organizations on disorders of pregnancy.
Website: www.nlm.nih.gov/medlineplus
The Canadian Diabetes Association's website section "About Diabetes" clearly presents facts on gestational diabetes.
Website: www.diabetes.ca

References

1. Banerjee M et al. Pregnancy as the prodrome to vascular dysfunction and cardiovascular risk. Nat Clin Pract Cardiovasc Med. 2006;3:596–603.

2. Report of the National High Blood Pressure Education Program Working Group on High Blood Pressure in Pregnancy. Am J Obstet Gynecol. 2000;183:S1–S22.

3. Caballero E. Dyslipidemia and vascular function: the rationale for early and aggressive intervention, www.medscape.com. accessed 4/18/06

4. Sies H et al. Nutritional, dietary and postprandial oxidative stress. J Nutr. 2005;135:969–72.

5. Hu Y et al. Relations of glycemic index and glycemic load with plasma oxidative stress markers. Am J Clin Nutr. 2006;84:70–6.

6. Brown AA et al. Dietary modulation of endothelial function: implications for cardiovascular disease. Am J Clin Nutr. 2001;73:673–86.

7. Sibai BM et al. Risk factors for preeclampsia, abruptio placentae, and adverse neonatal outcomes among women with chronic hypertension. National Institute of Child Health and Human Development Network of Maternal-Fetal Medicine Units. N Engl J Med. 1998;339:667–71.

8. Haddad B, Sibai BM. Chronic hypertension in pregnancy. Ann Med. 1999;31:246–52.

9. Solomon CG et al. Higher cholesterol and insulin levels in pregnancy are associated with increased risk for pregnancy-induced hypertension. Am J Hypertens. 1999;12:276–82.

10. Roberts JM et al. Nutrient involvement in preeclampsia. J Nutr. 2003;133:1684S–92S.

11. Prevent recurrent eclamptic seizures with magnesium sulfate, an unconventional anticonvulsant. Drug Ther Perspect. 2000;16:6–8.

12. Galtier-Dereure F et al. Obesity and pregnancy: complications and cost. Am J Clin Nutr. 2000;71:1242S–8S.

13. Ritchie LD, King JC. Dietary calcium and pregnancy-induced hypertension: is there a relation? Am J Clin Nutr. 2000;71:1371S–4S.

14. Velzing-Aarts FV et al. Umbilical vessels of preeclamptic women have low contents of both n-3 and n-6 long-chain polyunsaturated fatty acids. Am J Clin Nutr. 1999;69:293–8.

15. Berkowitz KM. Insulin resistance and preeclampsia. Clin Perinatol. 1998;25:873–85.

16. Hauth JC et al. Pregnancy outcomes in healthy nulliparas who developed hypertension. Calcium for Preeclampsia Prevention Study Group. Obstet Gynecol. 2000;95:24–8.

17. Holmes VA et al. Could antioxidant supplementation prevent preeclampsia? Proc Nutr Soc. 2005;64:491–501.

18. Pipkin FB. Risk factors for preeclampsia (editorial). N Engl J Med. 2001;344:925–6.

19. Germain AM et al. Endothelial dysfunction: a link among preeclampsia, recurrent pregnancy loss, and future cardiovascular events? Hypertension. 2007;43:90–5. Epub 2006 Nov 20.

20. Xiong X et al. Association of preeclampsia with high birth weight for age. Am J Obstet Gynecol. 2000;183:148–55.

21. Villar J et al. Nutritional interventions during pregnancy for the prevention or treatment of maternal morbidity and preterm delivery: an overview or randomized controlled trials. J Nutr. 2003;133:1606S–25S.

22. Coomarasamy A et al. Effect of low-dose aspirin on the incidence of preeclampsia. Obstet Gynecol. 2003;101:1319–32.

23. Klebanoff MA et al. Maternal size at birth and the development of hypertension during pregnancy: a test of the Barker hypothesis. Arch Intern Med. 1999;159:1607–12.

24. Esplin MS et al. Paternal and maternal components of the predisposition to preeclampsia. N Engl J Med. 2001;344:867–72.

25. Rumiris D et al. Lower rate of preeclampsia after antioxidant supplementation in pregnant women with low antioxidant status. Hypertens Pregnancy. 2006;25:241–53.

26. Rumbold AR et al. Vitamins C and E and the risks of preeclampsia and perinatal complications. N Engl J Med. 2006;354:1796–806.

27. Hofmeyr GJ et al. Calcium supplementation during pregnancy for preventing hypertensive disorders and related problems. Cochrane Database Syst Rev. 2006;3:CD001059.

28. Villar J et al. Calcium supplementation and the severity of preeclampsia. Am J Obstet Gynceol. 2006;194:639–49.

29. Spinnato JA 2nd. New therapies in the prevention of preeclampsia. Curr Opin Obstet Gynecol. 2006;18:601–4.

30. Jensen CL. Effects of n-3 fatty acids during pregnancy and lactation. Am J Clin Nutr. 2006;83:1452S–1457S.

31. Duley L et al. Antiplatelet drugs for prevention of pre-eclampsia and its consequences: systematic review. BMJ. 2001;322:329–33.

32. Brand-Miller J et al. Low-glycemic-index diets in the management of diabetes: a meta-analysis of randomized controlled trials. Diabetes Care. 2003;26:2261–67.

33. Roberts JM et al. Nutrient involvement in preeclampsia. J Nutr. 2003;133:1684S–92S.

34. Rayman MP et al. Iron supplementation in preeclampsia. Am J Obstet Gynecol. 2002;187:412–18.

35. Carr SR. Effect of maternal hyperglycemia on fetal development, American Dietetic Association Annual Meeting and Exhibition, Kansas City, MO, Oct. 10, 1998.

36. Branchtein L et al. Waist circumference and waist-to-hip ratio are related to gestational glucose tolerance. Diabetes Care. 1997;20:509–11.

37. Ferrara A et al. An Increase in the incidence of gestational diabetes mellitus: Northern California, 1991–2000. Obstet Gynecol. 2004;103:526–33.

38. Butte NF. Carbohydrate and lipid metabolism in pregnancy: normal compared with gestational diabetes mellitus. Am J Clin Nutr. 2000;71:1256S–61S.

39. Carducci AA et al. Glucose tolerance and insulin secretion in pregnancy. Diabetes, Nutrition & Metabolism—Clinical & Experimental. 1999;12:264–70.

40. Homko CJ et al. Fuel metabolism during pregnancy. Semin Reprod Endocrinol. 1999;17:119–25.

41. Nielsen JN et al. Interventions to improve diet and weight gain among pregnancy adolescents and reccommendations for future research. J Am Diet Assoc. 2006; 106:1825–40.

42. Silverman BL et al. Impaired glucose tolerance in adolescent offspring of diabetic mothers. Relationship to fetal hyperinsulinism. Diabetes Care. 1995;18:611–7.

43. Lindsay RS et al. Secular trends in birth weight, BMI, and diabetes in the offspring of diabetic mothers. Diabetes Care. 2000;23: 1249–54.

44. MacNeill S et al. Rates and risk factors for recurrence of gestational diabetes. Diabetes Care. 2001;24:659–62.

45. Zhang C et al. Dietary fiber intake, dietary glycemic load, and the risk for gestational diabetes mellitus. Diabetes Care. 2006; 29:2223–30.

46. McMahon MJ et al. Gestational diabetes mellitus. Risk factors, obstetric complications and infant outcomes. J Reprod Med. 1998;43:372–8.

47. Crowe SM et al. Oral glucose tolerance test and the preparatory diet. Am J Obstet Gynecol. 2000;182:1052–4.

48. Kjos SL, Buchanan TA. Gestational diabetes mellitus. N Engl J Med. 1999;341: 1749–56.

49. Gestational diabetes. http://my.webmd.com, accessed 10/02.

50. Coleman T. Patient centered care of diabetes in general practice. Study failed to measure patient centeredness of GPs' consulting behavior. BMJ. 1999;318:1621–2.

51. Adams KM et al. Sequelae of unrecognized gestational diabetes. Am J Obstet Gynecol. 1998;178:1321–32.

52. Mello G et al. What degree of maternal metabolic control in women with type 1 diabetes is associated with normal body size and proportions in full-term infants? Diabetes Care. 2000;23:1494–8.

53. Reader D et al. Impact of gestational diabetes mellitus nutrition practice guidelines implemented by registered dietitians on pregnancy outcomes. J Am Diet Assoc. 2006; 106:1426–33.

54. Jovanovic L. Time to reassess the optimal dietary prescription for women with gestational diabetes [editorial]. Am J Clin Nutr. 1999;70:3–4.

55. Jovanovic L. Management of diet and exercise in gestational diabetes mellitus. Prenat Neonat Med. 1998;3:534–541.

56. Ecker JL et al. Gestational diabetes. N Engl J Med. 2000;342:896–7.

57. Pezzarossa A et al. Effects of maternal weight variations and gestational diabetes mellitus on neonatal birth weight. J Diabetes Complications. 1996;10:78–83.

58. Koren G et al. Safety of exposure to metformin during first trimester of pregnancy. Fertil Steril. 2006;88:658–63.

59. Fagen C et al. Nutrition management in women with gestational diabetes mellitus: a review by ADA's Diabetes Care and Education Dietetic Practice Group. J Am Diet Assoc. 1995;95:460–7.

60. Nielsen G et al. Hemogloblin A1c levels and outcomes of pregnancy in women with type 1 diabetes, Diabetes Care. 2006;29: 2612–16.

61. Butte NF. Dieting and exercise in overweight, lactating women [editorial]. N Engl J Med. 2000;342:502–3.

62. Kalkwarf HJ et al. Dietary fiber intakes and insulin requirements in pregnant women with type 1 diabetes. J Am Diet Assoc. 2001; 101:305–10.

63. Riccardi G, Rivellese AA. Diabetes: nutrition in prevention and management. Nutr Metab Cardiovasc Dis. 1999;9:33–6.

64. Jovanovic L. Aberrations in diabetes metabolism throughout pregnancy, 60th Scientific Sessions of the American Diabetes Association, San Antonio, TX, June 11, 2000.

65. Major CA et al. The effects of carbohydrate restriction in patients with diet-controlled gestational diabetes. Obstet Gynecol. 1998;91:600–4.

66. Haynes RB. What's new in medicine? Evidence-based medicine hits a home run for diabetes care. WebMD Scientific American Medicine, posted 12/08/03, www.medscape.com.

67. Temple RC et al. Glycemic control during the second trimester of pregnancy and the risk of preeclampsia in women with type 1 diabetes. BJOG. 2006;113:1329–32.

68. Landon MB. Diabetes mellitus and other endocrine disorders. In: Gabbe SG et al., eds. Obstetrics, normal and problem pregnancies. New York: Churchill Livingstone; 1996. pp. 1037–81.

69. Reece EA et al. The role of free radicals and membrane lipids in diabetes-induced congenital malformations. J Soc Gynecol Investig. 1998;5:178–87.

70. Raychaudhuri K, Maresh MJ. Glycemic control throughout pregnancy and fetal growth in insulin-dependent diabetes. Obstet Gynecol. 2000;95:190–4.

71. Births: Preliminary Data for 2005, www.cdc.gov/nchs/products/pubs/pubd/hestats/prelimbirths05/prelimbirths05.htm, accessed 12/06.

72. Health, United States, 2003. www.cdc.gov/nchs, accessed 10/03.

73. Brown JE, Schloesser PT. Prepregnancy weight status, prenatal weight gain, and the outcome of term twin gestations. Am J Obst Gynecol. 1990;162:182–6.

74. Rebar RW et al. Assisted reproductive technology in the United States. N Engl J Med. 2004;350:1603–4.

75. Guyer B et al. Annual summary of vital statistics—1998. Pediatrics. 1999;104: 1229–46.

76. Hoskins RE. Zygosity as a risk factor for complications and outcomes of twin pregnancy. Acta Genet Med Gemellol. 1995;44:11–23.

77. Redline RW. Non-identical twins with a single placenta—disproving dogma in perinatal pathology. N Engl J Med. 2003;349: 111–14.

78. Imaizumi Y. A comparative study of twinning and triplet rates in 17 countries, 1971–1996. Acta Genet Med Gemellol. 1998;47:101–14.

79. Czeizel AE et al. Higher rate of multiple births after periconceptional vitamin supplementation. New Engl J Med. 1994;330: 1687–8.

80. Maclennan AH. Multiple gestation. Clinical characteristics and management. In: Creasy RK, Resnick R eds. Maternal-fetal medicine: principles and practice. Philadelphia: W. B. Saunders; 1994. pp. 589–601.

81. Chasen S et al. Pregnancy outcomes with a vanishing twin. Am J Obstet Gynecol. 2006;195:814–7.

82. Brown JE, Carlson M. Nutrition and multifetal pregnancy. J Am Diet Assoc. 2000;100:343–8. Updated based on MyPyramid, 2/07.

83. Kogan MD et al. Trends in twin birth outcomes and prenatal care utilization in the United States, 1981–1997. JAMA. 2000; 284:335–41.

84. Ellings JM et al. Reduction in very low birth weight deliveries and perinatal mortality in a specialized, multidisciplinary twin clinic. Obstet Gynecol. 1993;81:387–91.

85. Ventura SJ et al. Births: final data for 1999. Vol. 2001: www.cdc.gov/nchs, 2001.

86. Dubois S et al. Twin pregnancy: the impact of the Higgins Nutrition Intervention Program on maternal and neonatal outcomes. Am J Clin Nutr. 1991;53:1397–403.

87. National Academy of Sciences. Nutrition during pregnancy. I. Weight gain. II. Nutrient supplements. Washington, DC: National Academy Press, 1990.

88. Luke B et al. Critical periods of maternal weight gain: effect on twin birth weight. Am J Obstet Gynecol. 1997;177:1055–62.

89. Brown JE et al. Early pregnancy weight change and newborn size (Abstract), FASEB, Orlando, FL, March 31–April 4, 2000.

90. Konwinski T et al. Maternal pregestational weight and multiple pregnancy duration. Acta Genet Med Gemellol. 1973;22:44–7.

91. Cohen SB et al. New birth weight nomograms for twin gestation on the basis of accurate gestational age. Am J Obstet Gynecol. 1997;177:1101–4.

92. Zeijdner EE et al. Essential fatty acid status in plasma phospholipids of mother and neonate after multiple pregnancy. Prostaglandins Leukotrienes and Essential Fatty Acids. 1997;56:395–401.

93. Fawzi WW et al. Randomized trial of effects of vitamin supplements on pregnancy outcomes and T cell counts in HIV-1-infected women in Tanzania. Lancet. 1998;351: 1477–82.

94. French M et al. Highly active antiretroviral therapy. Lancet. 1998;351:1056–7.

95. Rich KC et al. Maternal and infant factors predicting disease progression in human immunodeficiency virus type 1-infected infants. Women and Infants Transmission Study Group. Pediatrics. 2000;105:E8.

96. HIV infection on the rise worldwide. Press release from the World Health Organization, www.medscape.com/viewarticle/548140, accessed 11/06.

97. Fields-Gardner C, Ayoob KT. Position of the American Dietetic Association and Dietitians of Canada: nutrition intervention in the care of persons with human immunodeficiency virus infection. J Am Diet Assoc. 2000;100:708–17.

98. Zullig L et al. Nutrition for HIV positive women: practical guidelines. Women's Health and Reproductive Nutrition Report. 2005;Winter:1,4,14–16.

99. Bonne OB et al. Delayed detection of pregnancy in patients with anorexia nervosa: two case reports. Int J Eat Disord. 1996;20: 423–25.

100. Herzog DB et al. Mortality in eating disorders: a descriptive study. Int J Eat Disord. 2000;28:20–6.

101. Becker AE et al. Eating disorders. N Engl J Med. 1999;340:1092–8.

102. Little L, Lowkes E.J. Midwifery. Womens Health. 2000;45:301–7.

103. Morrill ES et al. Bulimia nervosa during pregnancy: a review. J Am Diet Assoc. 2001; 102:448–54.

104. CDC issues guidelines for the Identification of fetal alcohol syndrome. MMWR. 2005;54(RR-11):1–15.

105. American Academy of Pediatrics. Committees on Substance Abuse and Children with Disabilities, Fetal alcohol syndrome and alcohol-related neurodevelopmental disorders. Pediatrics. 2000;106:358–61.

106. Ventura SJ et al. Births: final data for 1998. Natl Vitat Stat Rep. 2000;48:1–100.

107. Shankar K et al. Physiologic and genomic analyses of nutrition ethanol interactions during gestation: implications for fetal ethanol toxicity. Exp Biol Med. (Maywood). 2006;231:1379–97.

108. Konovaloov HV et al. Disorders in brain development in the progeny of mothers who used alcohol during pregnancy. Early Human Dev. 1997;48:153–66.

109. March of Dimes Birth Defect Foundation. Public health education information sheet: Drinking alcohol during pregnancy, 1997.

110. Teen births, www.cdc.gov/nchs/fastats/teenbrth.htm, accessed 1/07.

111. Rosso P. Nutrition and metabolism in pregnancy. New York: Oxford University Press; 1990. pp. 117–118, 125, 150–151.

112. Scholl TO et al. Maternal growth during pregnancy and the competition for nutrients. Am J Clin Nutr. 1994;60:183–8.

113. Wallace JM et al. Nutritionally mediated placental growth restriction in the growing adolescent: consequences for the fetus. Biol Reprod. 2004;71:1055–62.

114. Sukalich S et al. Obstetric outcomes in overweight and obese adolescents. Am J Obstet Gynecol. 2006;195:851–5.

115. Scholl TO et al. Leptin and maternal growth during adolescent pregnancy. Am J Clin Nutr. 2000;72:1542–7.

116. Lenders CM et al. Curr Opin Pediatr. 2000;12:291–6.

117. Lenfant C. Clinical research to clinical practice—lost in translation. N Engl J Med. 2003;868–74.

> "One of the best things that only you can do is breastfeed your baby for as long as possible. The longer a mom and baby breastfeed, the greater the benefits are for both mom and baby."
>
> The National Women's Health Information Center, 2002

Chapter 6

Nutrition during Lactation

Chapter Outline

- Introduction
- Breastfeeding Goals for the United States
- Lactation Physiology
- Human Milk Composition
- Benefits of Breastfeeding
- Breast Milk Supply and Demand
- The Breastfeeding Infant
- Maternal Diet
- Maternal Energy Balance and Milk Composition
- Factors Influencing Breastfeeding Initiation and Duration
- Breastfeeding Promotion, Facilitation, and Support
- Public Food and Nutrition Programs
- Model Breastfeeding Promotion Programs

brand X pictures

Getty

Getty

Prepared by **Maureen A. Murtaugh** and **Carolyn Sharbaugh** with **Denise Sofka**

Key Nutrition Concepts

1   Human milk is the best food for newborn infants for the first year of life or longer.

2   Maternal diet does not significantly alter the protein, carbohydrate, fat, and major mineral composition of breast milk, but it does alter the fatty acid profile and the amounts of some vitamins and trace minerals.

3   When maternal diet is inadequate, the quality of milk is preserved over the quantity for the majority of nutrients.

4   Health care policies and procedures and the knowledge and attitudes of health care providers affect community breastfeeding rates.

Introduction

The benefits of breastfeeding to mothers and infants are well established. Federal breastfeeding promotion efforts and the greater understanding of the advantages of breastfeeding have contributed to the resurgence of breastfeeding in the United States since the 1970s. Nevertheless, racial and ethnic disparities in breastfeeding initiation rates remain, and despite the knowledge that the benefits increase with longer duration, there has been little increase in the duration of lactation among all women.

The health care system, the workplace, and the community can either hinder or facilitate the initiation and continuation of breastfeeding. Health programs can play a significant role in increasing breastfeeding rates to optimize maternal and infant nutrition. Health care professionals who wish to manage and promote breastfeeding should understand the physiology of lactation, the composition of human milk, and the benefits to mothers and infants. Helping women achieve appropriate nutritional status to optimize breastfeeding requires consideration of energy and nutrient needs, weight goals, effects of exercise during breastfeeding, and vitamin and mineral supplement needs.

Multilevel (health care system, community, workplace, and family) support is critical for women who suffer from common breastfeeding challenges and medical conditions. Human milk is the preferred food for premature and sick newborns. It is rarely necessary to discontinue breastfeeding to manage medical problems or medication use. However, adequately experienced and informed health care professionals are needed to provide support for successful breastfeeding. Common breastfeeding conditions and interventions are discussed in Chapter 7.

Breastfeeding Goals for the United States

"During the twentieth century, infant feeding practices have undergone dramatic changes that reflect shifts in values and attitudes in the U.S. society as a whole. They have tended to occur first among those women at the forefront of changes in dominant social values and among those with resources (whether it is time, energy or money) to permit adoption of new feeding practices."

Institute of Medicine, Subcommittee on
Nutrition During Lactation, 1991[1]

In the early 1900s, almost all infants in the United States were breastfed. As safe human milk substitutes (HMSs) became widely available, breastfeeding rates steadily declined, reaching levels below 30% in the 1950s and 1960s, and then rose dramatically in the 1970s.[2] In the early 1980s levels peaked above 60. In recognition of the health and economic benefits of breastfeeding, national goals for breastfeeding rates have been established and revised since 1980.[3] Nonetheless, breastfeeding rates declined until the early 1990s.[2]

Healthy People 2010 contains broad-reaching national health goals for the new decade, focusing on two major themes: (1) increasing the quality and years of healthy life and (2) eliminating racial and ethnic disparities in health status.[3] In addition to adding specific breastfeeding goals for black or African Americans and Hispanic or Latina Americans, Healthy People 2010 places increased emphasis on the duration of breastfeeding. By the year 2010, the goal is for at least 75% of American women to breastfeed their infants in the early postpartum period, at least 50% at 6 months, and 25% at 1 year (Table 6.1). Increasing the rates of breastfeeding is a compelling public health goal, particularly among the racial and ethnic groups who are less likely to initiate and sustain breastfeeding throughout the infant's first year.

Recent evidence suggests steady progress toward meeting the Healthy People 2010 goals for breastfeeding initiation, breastfeeding duration, and exclusive breastfeeding.[3] In 2005, the national breastfeeding initiation rate fell just short of the 75% goal, but 21 states achieved the goal of 75% or more. At 6 months, overall ever-breastfeeding rates fell 11% short of the 50% mark, and only 5 states achieved the goal rate. At 1 year overall rates fell only 5% below the 2010 goal, and 11 states had rates meeting or exceeding 25%. Despite the American Academy of Pediatrics recommendation that an infant be exclusively breastfed until 6 months of age,[4,5] rates of exclusive breastfeeding continue to be lower than 50% (Table 6.1). In the early postpartum period (7 days), exclusive breastfeeding is 59.4%. Rates fall to

44.3% at 2 months, to 29% by 4 months, and 13.9% at 6 months. Oregon is the only state with exclusive breastfeeding meeting or exceeding 25%.

Despite these increases in breastfeeding rates, gaps in breastfeeding remain—geographically, and among racial groups, educational groups, and income groups.[6,7] Breastfeeding rates are almost 25% lower among mothers participating in the Special Supplemental Nutrition Program for Women, Infants, and Children (WIC) compared with women who do not participate in WIC.[8] Breastfeeding rates are highest in the western states with California, Washington, Oregon, and Hawaii meeting the 75% breastfeeding initiation rate. Some of the lowest rates persist among southern states. Initiation rates remain lowest among African American and American Indian women. Rates of breastfeeding among Hispanic or Latino women now equal or exceed rates among white women. Rates remain lower among younger and unmarried women and among women with lower income.

If the United States is to reach these breastfeeding goals, health professionals must take an active role in promoting and supporting lactation. Becoming a breastfeeding advocate requires a thorough understanding of lactation physiology and thorough knowledge of clinical and community resources for support.

Lactation Physiology

Functional Units of the Mammary Gland

The functional units of the *mammary gland* are the *alveoli* (Illustration 6.1). Each alveolus is composed of a cluster of cells (*secretory cells*) with a duct in the center, whose job it is to secrete milk. The ducts are arranged like branches of a tree, each smaller duct leading to a larger duct. These branchlike ducts lead to the nipple. *Myoepithelial cells* surround the secretory cells. Myoepithelial cells can contract under the influence of *oxytocin* and cause milk to be ejected into the ducts. Dilations in the *lactiferous sinuses* behind the nipple allow for storage of milk.

> **Mammary Gland**   The source of milk for offspring, also commonly called the breast. The presence of mammary glands is a characteristic of mammals.
>
> **Alveoli**   A rounded or oblong-shaped cavity present in the breast.
>
> **Secretory Cells**   Cells in the acinus (milk gland) that are responsible for secreting milk components into the ducts.
>
> **Myoepithelial Cells**   Specialized cells that line the alveoli and that can contract to cause milk to be secreted into the duct.
>
> **Oxytocin**   A hormone produced during letdown that causes milk to be ejected into the ducts.
>
> **Lactiferous Sinuses**   Larger ducts for storage of milk behind the nipple.

Table 6.1	Healthy People 2010 breastfeeding objectives for the nation	

Objective: Increase the proportion of mothers who breastfeed their babies.

	1998 Baseline Percent (%)	2010 Goal Percent (%)
In early postpartum period		
All women	64	75
Black or African American	45	75
Hispanic or Latino	66	75
White	68	75
At 6 months		
All women	29	50
Black or African American	19	50
Hispanic or Latino	28	50
White	31	50
At 1 year		
All women	16	25
Black or African American	9	25
Hispanic or Latino	17	25
White	19	25

SOURCE: U.S. Department of Health and Human Services. Healthy People 2010: Conference Edition—volumes I and II. Washington, DC: U.S. Department of Health and Human Services, Public Health Service, Office of the Assistant Secretary for Health, January 2000.[3]

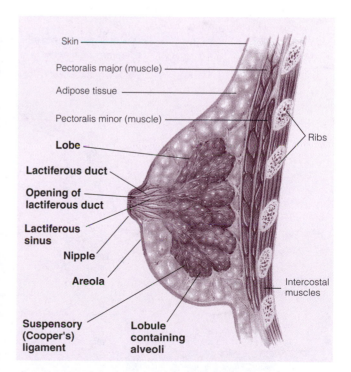

Skin

Pectoralis major (muscle)

Adipose tissue

Pectoralis minor (muscle)

Lobe

Lactiferous duct

Opening of lactiferous duct

Lactiferous sinus

Nipple

Areola

Suspensory (Cooper's) ligament

Lobule containing alveoli

Ribs

Intercostal muscles

Illustration 6.1   Breast of a lactating female.

This cut away view shows the mammary glands and ducts.

Mammary Gland Development

During puberty, the ovaries mature and the release of estrogen and progesterone increases (Table 6.2). The cyclic release of these two hormones governs pubertal breast development (Illustration 6.2). The mammary gland develops its lobular structure (*lobes*) under the cyclic production of proges-terone and is usually com-plete within 12 to 18 months after menarche. As the ductal system matures, cells that can secrete milk develop, the nipple grows, and its pigmentation changes. Fibrous and fatty tissues increase around the ducts.

Lobes Rounded structures of the mammary gland.

Lactogenesis Another term for human milk production.

In pregnancy, the luteal and placental hormones (pla-cental lactogen and chorionic gonadotropin) allow fur-ther preparation for breastfeeding (Illustration 6.2). Estrogen stimulates development of the glands that will make milk. Progesterone allows the tubules to elongate and the cells that line the tubules (epithelial cells) to duplicate.

Lactogenesis

Breast milk production, or *lactogenesis*, is classically described as occurring in three stages.[9] The first stage, or lactogenesis I, begins during the last trimester of preg-nancy; the second and third stages (lactogenesis II and III) occur after birth.

- *Lactogenesis I.* During the first stage of milk pro-duction, milk begins to form, and the lactose and protein content of milk increases. This stage extends through the first few postpartum days when suckling is not necessary for initiating milk production.

- *Lactogenesis II.* This stage begins 2–5 days postpar-tum and is marked by increased blood flow to the mammary gland. Clinically, it is considered the on-set of copious milk secretion, or "when milk comes

Table 6.2 Hormones contributing to breast development and lactation

Hormone	Role in Lactation	Stage of Lactation
Estrogen	Ductal growth	Mammary gland differentiation with menstruation
Progesterone	Alveolar development	After onset of menses and during pregnancy
Human growth hormone	Development of terminal end buds	Mammary gland development
Human placental lactogen	Alveolar development	Pregnancy
Prolactin	Alveolar development and milk secretion	Pregnancy and breastfeeding (from the third trimester of pregnancy to weaning)
Oxytocin	Letdown: ejection of milk from myoepithelial cells	From the onset of milk secretion to weaning

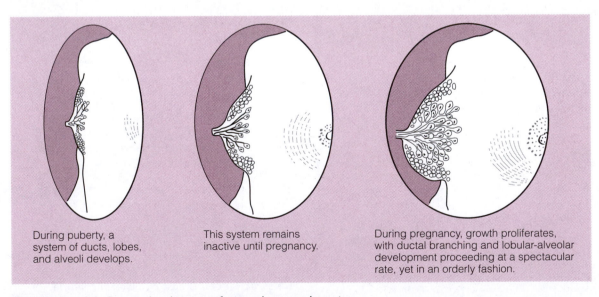

During puberty, a system of ducts, lobes, and alveoli develops.

This system remains inactive until pregnancy.

During pregnancy, growth proliferates, with ductal branching and lobular-alveolar development proceeding at a spectacular rate, yet in an orderly fashion.

Illustration 6.2 Breast development from puberty to lactation.

in." Significant changes in both the milk composition and the quantity of milk that can be produced occur over the first 10 days of the baby's life.

- *Lactogenesis III.* This stage of breast milk production begins about 10 days after birth and is the stage in which the milk composition becomes stable.[10]

Hormonal Control of Lactation

Prolactin and oxytocin are necessary for establishing and maintaining a milk supply. *Prolactin* is a hormone that stimulates milk production. Suckling is a major stimulator of prolactin secretion: prolactin levels double with suckling.[11] Stress, sleep, and sexual intercourse also stimulate prolactin levels. To prevent milk production in the last 3 months of pregnancy, prolactin activity is suppressed by a prolactin-inhibiting factor that is released by the hypothalamus. This inhibition of prolactin allows the mother's body to prepare for milk production during pregnancy. The actual level of prolactin in the blood is not related to the amount of milk made, but prolactin is necessary for milk synthesis to occur.[12]

Oxytocin release is also stimulated by suckling or nipple stimulation. Its main role is in letdown, or the ejection of milk from the milk gland (acinus) into the milk ducts and lactiferous sinuses. Women may experience tingling or sometimes sharp shooting pain that lasts about a minute and corresponds with contractions in the milk ducts. Oxytocin also acts on the uterus, causing it to contract, seal blood vessels, and shrink its size.

Secretion of Milk

Although the process of milk production is complex, understanding the basic mechanisms of milk secretion is important to understanding how factors such as nutritional status, supplementation, medications, and disease may affect breastfeeding or milk composition. As described by Neville et al.,[13] the secretory cell in the breast uses five pathways for milk secretion (Illustration 6.3). Briefly, some components like lactose are made in the secretory cells and secreted into ducts. Water, sodium, potassium, and chloride are able to pass through alveolar cell membranes in either direction (pas-

sive diffusion). Other components are processed within the cells of the breast. Milk fat comes from triglycerides from the mother's blood and from new fatty acids produced in the breast. Fats are made soluble in milk by addition of a protein carrier to form milk-fat globules.[14] These milk-fat globules are then secreted into the ducts. Immunoglobulin A and other plasma proteins are captured from the mother's blood and taken into the alveolar cells. These proteins are then secreted into the milk ducts.

Prolactin A hormone necessary for milk production.

The Letdown Reflex

The letdown reflex stimulates milk release from the breast. An infant at the nipple usually causes letdown. The stimuli from the infant suckling are passed through nerves to the hypothalamus, which responds by promoting oxytocin release from the posterior pituitary gland (Illustration 6.4). The oxytocin causes contraction of the myoepithelial cells surrounding the secretory cells. As a result, milk is released through the ducts to the lactiferous sinus, making it available to the infant. Other stimuli, such as hearing a baby cry, sexual arousal, and thinking about nursing, can also cause letdown, and milk will leak from the breasts.

Illustration 6.3 The pathways for secretion of milk components.

SOURCE: From *Lactation: Physiology, Nutrition, & Breastfeeding* by J.C. Allen and C. Watters, pp. 49–102. Copyright © 1983 Klumer Academics. Reprinted with permission.

Both hormones act on the mammary glands:
- Prolactin stimulates milk production.
- Oxytocin stimulates milk ejection.

Both hormones act on the reproductive organs:
- Prolactin inhibits ovulation.
- Oxytocin promotes uterine contractions.

Illustration 6.4 The letdown reflex.
An infant suckling at the breast stimulates the pituitary to release the hormones prolactin and oxytocin.

Table 6.3 Compositions of 100 ml colostrum (days 1–5 postpartum) and mature milk (day 15 postpartum)

Contents	Colostrum	Mature Milk
Calories (kcal)	55	67
Fat (g)	2.9	4.2
Lactose (g)	5.3	7.0
Total Protein (g)	2.0	1.1
secretory IgA	0.5[a]	0.1
lactoferrin	0.5	0.2
casein	0.5	0.4
Calcium (mg)	28	30
Sodium (mg)	48	15
Vitamin A (µg retinol equivalents)	151	75
Vitamin B_1 (µg)	2	14
Vitamin B_2 (µg)	30	40
Vitamin C (µg)	6	5

[a]Concentration is considerably higher at 1–3 days postpartum than days 4 and 5.
SOURCE: Adapted from Prentice A, Constituents of Human milk. *Food and Nutrition Bulletin.* The United Nations University Press, Vol. 17, No. 4, December 1996.[16]

Human Milk Composition

"Thus, the complexity of milk as a system designed to deliver nutrients and nonnutritive messages to the neonate has increased."

R.G. Jensen, *Handbook of Milk Composition*[15]

Human milk is an elegantly designed natural resource. It is the only food needed by the majority of healthy infants for approximately 6 months. The composition of milk is designed not only to nurture, but also to protect infants from infectious and certain chronic diseases. Human milk composition is changeable over a single feeding, over a day, according to the age of the infant or gestation at delivery, presence of infection in the breast, with menses, and maternal nutritional status.

As our ability to measure and identify novel components increases, we recognize that the composition of human milk is complex. Hundreds of components of human milk have been identified, and their nutritive and non-nutritive roles are under investigation. The basic nutrient composition of colostrum and mature milk is provided in Table 6.3, and a comparison of mature human milk with cow's milk follows in Table 6.4. *The Handbook of Milk Composition*[15] and *Breastfeeding: A Guide for the Medical Professional*[11] provide more detailed descriptions of the composition of human and other milks.

Colostrum The milk produced in the first 2–3 days after the baby is born. Colostrum is higher in protein and lower in lactose than milk produced after a milk supply is established.

Colostrum

The first milk, *colostrum*, is a thick, often yellow fluid produced during lactogenesis II (days 1–3 after infant birth). Infants may drink only 2 to 10 ml (1.5–2 tsp) of colostrum per feeding in the first 2–3 days. Colostrum provides about 580–700 Kcal/L and is higher in protein and lower in carbohydrate and fat than mature milk (produced 2 weeks after infant birth). Secretory immunoglobulin A and lactoferrin are the primary proteins present in colostrum, but other proteins present in mature milk are not present. The concentration of mononuclear cells (a specific type of white blood cell from the mother that provides immune protection) is highest in colostrum. Colostrum has higher concentrations of sodium, potassium, and chloride than more mature milk.

Water

Breast milk is isotonic with maternal plasma. This biological design of milk means that babies do not need water or other fluids to maintain hydration, even in hot climates.[18] As a major component of human milk, water allows suspension of the milk sugars, proteins, Immunoglobulin A, sodium, potassium, citrate, magnesium, calcium, chloride, and water-soluble vitamins.

Energy

Human milk provides approximately 0.65 cal/ml, although the energy content varies with its fat (and to a lesser degree,

Table 6.4 Human and cow's milk composition

Nutrient	Units	Human Milk (1 fl oz)	Whole Cow's Milk (1 fl oz)	Nutrient	Units	Human Milk (1 fl oz)	Whole Cow's Milk (1 fl oz)
Water	g	26.95	26.94	Lipids (*Continued*)			
Energy	kcal	22	18	12:0	g	0.079	0.023
Energy	kj	90	76	14:0	g	0.099	0.091
Protein	g	0.32	0.98	16:0	g	0.283	0.000
Total lipid (fat)	g	1.35	0.99	18:0	g	0.090	0.253
Carbohydrate	g	2.12	1.38	Fatty acids, total monounsaturated	g	0.511	0.000
Fiber, total dietary	g	0.0	0.0				
Sugars, total	g	2.12	1.60	Fatty acids, total polyunsaturated	g	0.153	0.000
Minerals				Cholesterol	mg	4	0.000
Calcium, Ca	mg	10	34				
Iron, Fe	mg	0.01	0.01	**Amino acids**			0.000
Magnesium, Mg	mg	1	3	Tryptophan	g	0.005	0.000
Phosphorus, P	mg	4	28	Threonine	g	0.014	0.000
Potassium, K	mg	16	44	Isoleucine	g	0.017	3
Sodium, Na	mg	5	12	Leucine	g	0.029	
Zinc, Zn	mg	0.05	0.12	Lysine	g	0.021	0.023
Copper, Cu	mg	0.016	0.003	Methionine	g	0.006	0.044
Manganese, Mn	mg	0.008	0.001	Cystine	g	0.006	0.050
Selenium, Se	mcg	0.6	1.1	Phenylalanine	g	0.014	0.081
				Tyrosine	g	0.016	0.043
Vitamins				Valine	g	0.019	0.023
Vitamin C, total ascorbic acid	mg	1.5	0.0	Arginine	g	0.013	0.005
				Histidine	g	0.007	0.045
Thiamin	mg	0.004	0.013	Alanine	g	0.011	0.046
Riboflavin	mg	0.011	0.056	Aspartic acid	g	0.025	0.059
Niacin	mg	0.055	0.033	Glutamic acid	g	0.052	0.023
Pantothenic acid	mg	0.069	0.110	Glycine	g	0.008	0.023
Vitamin B_6	mg	0.003	0.011	Proline	g	0.025	0.031
Folate, DFE	mcg_DFE	2	2	Serine	g	0.013	0.072
Vitamin B_{12}	mcg	0.02	0.13				
Vitamin A, RAE	mcg_RAE	19	9	**Other**			0.198
Vitamin E (alpha-tocopherol)	mg	0.02	0.02	Alcohol, ethyl	g	0.0	0.023
				Caffeine	mg	0	0.104
Vitamin D	IU	1	12	Theobromine	mg	0	0.033
Vitamin K (phylloquinone)	mcg	0.1	0.1	Carotene, beta	mcg	2	
				Carotene, alpha	mcg	0	0.0
Lipids				Cryptoxanthin, beta	mcg	0	0
Fatty acids, total saturated	g	0.619	0.569	Lycopene	mcg	0	0
4:0	g	0.000	0.023	Lutein + zeaxanthin	mcg	0	2
6:0	g	0.000	0.023				
8:0	g	0.000	0.023				
10:0	g	0.019	0.023				

SOURCE: Adapted from USDA National Nutrient Database for Standard Reference, Release 18, 2005.[17]

protein and carbohydrate) composition. Breastfed infants consume fewer calories than infants fed HMS.[19,20] It is not known whether this difference in energy intake of breast-fed infants has to do with the composition of human milk, the inability to see the volume of feedings when providing human milk, the differences in the suckling at the breast compared to an artificial nipple, or other factors. Infants who are breastfed are thinner for their weight at 8–11 months than infants fed HMS, but these differences disappear by 12–23 months of age and few differences are notable by 5 years of age.[21]

Lipids

Lipids are the second largest component of breast milk by concentration (3–5% in mature milk). Lipids provide half of the energy of human milk.[15] Human milk fat is low at the beginning of a feeding in foremilk, and higher at the end in the hindmilk that follows.[22]

Effect of Maternal Diet on Fat Composition   The fatty acid profile, but not the fat content, of human milk varies with the diet of the mother.[22] When diets rich in polyunsaturated fats are consumed, more polyunsaturated fatty acids are present in the milk.[15] When a mother is losing weight, the fatty acid profile of her fat stores is reflected in the milk.[23] When very low-fat diets with adequate calories from carbohydrate and protein are fed, more medium chain fatty acids are synthesized in the breast.

DHA   Recent interest in lipids in human milk stems from studies showing developmental advantages provided by docosahexaenioic acid (DHA).[24] DHA has been added to HMS in recent years. This addition stems from identification of developmental advantages provided to infants by DHA.[24] DHA is essential for retinal development and accumulates during the last months of pregnancy. The advantages of human milk seem particularly important to premature infants born before 37 weeks, perhaps because the concentrations of DHA are higher in the milk of mothers delivering preterm infants as compared to full-term infants.[25] Advantages for term infants have been demonstrated as well. For example, a Norwegian study suggests that cod liver oil supplementation during pregnancy was associated with higher IQ scores at 4 years of age in breast versus HMS-fed infants.[26] Cod liver oil contains high levels of DHA as well as high levels of vitamin A and vitamin D, so it should be used with caution.

***Trans* Fatty Acids**   *Trans* fatty acids stemming from the mother's diet are present in human milk.[27] *Trans* fat concentrations are similar in American and Canadian women, but lower in the milk of women from European and African countries. Removal of *trans* fatty acids from many food products in the United States should translate into lower levels of *trans* fat in human milk.

Cholesterol   Cholesterol, an essential component of all cell membranes, is needed for growth and replication of cells. Cholesterol concentration ranges from 10–20 mg/d and varies depending on the time of day.[15] Breastfed infants have higher intakes of cholesterol and higher levels of serum cholesterol than infants fed HMS.[28] Early consumption of cholesterol through breast milk appears to be related to lower blood cholesterol levels later in life.[29]

Protein

The protein content of mature human milk is relatively low (0.8–1.0%) compared to other mammalian milks such as cow's milk (Table 6.4). The concentration of proteins synthesized in the breast are more affected by the age of the infant (time since delivery) than maternal intake and maternal serum proteins. Proteins synthesized by the breast are more variable because hormones that regulate gene expression and guide protein synthesis change with time.[30] Despite the relatively low concentration, human milk proteins have important nutritive and non-nutritive value. Proteins and their digestive products, such as peptides, exhibit a variety of antiviral and antimicrobial effects.[31] Enzymes present in human milk might also provide protection by facilitating actions that prevent inflammation.

Casein   Casein is the major class of protein in mature milk from women who deliver either at term or preterm.[32] Casein, calcium phosphate, and other ions such as magnesium and citrate appear as an aggregate and are the source of milk's white appearance.[33] Casein's digestive products, casein phosphopeptides, keep calcium in soluble form and facilitate its absorption.

Whey Proteins   Whey proteins are the proteins that remain soluble in water after casein is precipitated from milk by acid or enzymes. Whey proteins include milk and serum proteins, enzymes, and immunoglobulins, among others. Several mineral-, hormone-, or vitamin-binding proteins are also identified as components of whey proteins. These include lactoferrin that carries iron in a form that is easy to absorb and has bacteriostatic activity. The enzymes present in whey proteins aid in digestion and protection against bacteria.

Nonprotein Nitrogen   Nonprotein nitrogen provides 20–25% of the nitrogen in milk.[33] Urea accounts for 30–50% of nonprotein nitrogen and nucleotides for 20%, depending on the stage of lactation and the diet of the mother. Some of this nitrogen is available for the infant to use for producing nonessential amino acids. Some of the nonprotein nitrogen is used to produce other proteins with biological roles such as hormones, growth factors, free amino acids, nucleic acids, nucleotides, and carnitine. The role of individual nucleotides in human milk is under investigation; however, nucleotides appear to play important roles in growth and disease resistance.

Milk Carbohydrates

Lactose is the dominant carbohydrate in human milk. Other carbohydrates—including monosaccharides, such as glucose, polysaccharides, oligosaccharides, and protein-bound carbohydrates—are also present.[34] Lactose enhances calcium absorption. As the second largest carbohydrate

Osmolality A measure of the concentration of particles in solution.

component, oligosaccharides contribute calories at low *osmolality,* stimulate the growth of bifidus bacteria in the gut, and inhibit the growth of *E. coli* and other potentially harmful bacteria.

Oligosaccharides Oligosaccharides are medium-length carbohydrates containing lactose on one end. Oligosaccharides can be free, or bound to proteins as glycoproteins, or bound to lipids as glycolipids, or they can bind to other structures. The conjugated and unconjugated oligosaccharides are classified as glycans. Over 130 different oligosaccharides are present as functional ingredients of human milk.[35,36] Oligosaccharides in human milk prevent the binding of pathogenic microorganisms to the gut, thereby preventing infection and diarrhea.

Fat-Soluble Vitamins

Vitamin A Colostrum has approximately twice the concentration of vitamin A as mature milk does. Some of the vitamin A in human milk is in the form of beta-carotene. Its presence is responsible for the characteristic yellow color of colostrum. In mature milk, vitamin A is present at 75 µg/dl or 280 IU/dl.[37] These levels are adequate to meet infant needs.[38]

Vitamin D Vitamin D is present in both lipid and aqueous (water) compartments of human milk. Most vitamin D is in the form of 25-OH$_2$ vitamin D and vitamin D$_3$. Vitamin D levels of human milk vary with maternal diet and exposure to sunshine.[39] Maternal exposure to sunlight has been reported to increase the vitamin D$_3$ level in milk ten fold.[40]

Vitamin E The level of total tocopherols in human milk is related to the milk's fat content. Human milk contains 40 µg of vitamin E per gram of lipid in the milk.[41] Levels of alpha-tocopherol decrease from colostrum to transitional milk and to mature milk, whereas beta and gamma tocopherols remain stable throughout each stage of lactation. The level of vitamin E present in human milk is adequate to meet the needs of full-term infants for muscle integrity and resistance of red blood cells to hemolysis (breaking of red blood cells). The levels of vitamin E in preterm milk have been reported to be the same[42] and higher[43] than in term milk. However, in both reports, the levels present were not considered adequate to meet the needs of preterm infants.

Vitamin K Vitamin K is present in human milk at levels of 2.3 µg/dL.[41] Approximately 5% of breastfed infants are at risk for vitamin K deficiency based on vitamin K–dependent clotting factors. There are cases of vitamin K deficiency among exclusively breastfed infants who did not receive vitamin K at birth.

Water-Soluble Vitamins

Water-soluble vitamins in human milk are generally responsive to the content of the maternal diet or supplements (vitamin C, riboflavin, niacin, B$_6$, and biotin). Problems with vitamin levels of these nutrients in human milk are related to their deficiency in the mother's diet. Clinical problems relating to water-soluble vitamins are rare in infants nursed by mothers with inadequate diets.[11] Vitamin B$_6$ is considered most likely to be deficient in human milk; levels of B$_6$ in human milk directly reflect maternal intake.[44]

Vitamin B$_{12}$ and Folic Acid Vitamin B$_{12}$ and folic acid are bound to whey proteins in human milk; therefore, their content in milk is less influenced by maternal intake of these vitamins than are the other water-soluble vitamins. Factors that influence protein secretion (hormones and the age of the infant, or time since delivery) are more likely to alter the human milk levels of B$_{12}$ and folate than is dietary intake.[1,45] Infant illness associated with low folate levels in milk has not been reported. Folate levels increase with the duration of lactation despite a decrease in maternal serum and red blood cell folate levels.[1] B$_{12}$ deficiency or low levels of B$_{12}$ in milk has been reported for women who have had gastric bypass, hypothyroidism, consume vegan diets, have latent pernicious anemia, or are generally malnourished.[46]

Minerals in Human Milk

The minerals in human milk contribute substantially to the osmolality of human milk. *Monovalent ion* secretion is managed closely by the alveolar cells, in balance with lactose, to maintain the isosmotic composition of human milk.

Mineral content in milk is related to the growth rate of the offspring. The mineral content of human milk is much lower than the concentration

Monovalent Ion An atom with an electrical charge of +1 or –1.

in cow's milk and the milk of other animals whose offspring grow faster. With the exception of magnesium, the concentration of minerals decreases over the first 4 months postpartum. This decline in the mineral content of milk during the period of rapid growth is not what one would expect, but infant growth is well supported.[47] The lower mineral concentration of human milk is easier for the kidneys to handle. This reduced load on the kidneys is considered a significant benefit of human milk.

Bioavailability An important feature of several of the minerals (magnesium, calcium, iron, zinc) in human milk is the packaging that makes them highly available (bioavailable) to the infant.[48] Packaging minerals so that the infant can use them efficiently also reduces the burden to the mother because less of the mineral is needed in the milk. For example, zinc is 49% available from human milk, but only 10% available from cow's milk and cow's milk–based HMS.[49] Exclusively breastfed infants have little risk of anemia,[50] despite the seemingly low concentration of iron in human milk. One study suggests that infants who are exclusively breastfed for 6.5 months are less likely to be anemic than those nursed exclusively for 5.5 months.[51]

Zinc   The importance of zinc to human growth is well established. Human milk zinc is bound to protein and is highly available, in comparison to cow's milk and cow's milk–based HMS. Both the zinc intake (per kg) and the zinc requirements of infants decline after the first few months.[52] Normally, zinc homeostasis and human milk zinc levels are maintained even in the face of low maternal zinc intake.[53] Rare cases of zinc deficiency, which appears as intractable diaper rash, have been noted in exclusively breastfed infants, however.[49] A defect in the mammary gland uptake of zinc has been described as the cause of low milk concentration when maternal serum zinc concentrations are normal.[49] In these cases, infants seem to respond to zinc supplementation.

Trace Minerals   Trace minerals (copper, selenium, chromium, manganese, molybdenum, nickel, and fluorine) are present in the human body in small concentrations and are essential for growth and development. Less is known about trace minerals and infant health than about other nutrients. In general, however, the levels of trace minerals in human milk are not altered by the mother's diet or supplement use, excepting fluoride. The DRI for fluoride is 0.1 mg daily for infants less than 6 months.[54] Fluoride provided in community water is safe for breastfeeding women and their infants. Most infants who live in areas with fluoridated water do not need an additional supplement.[55,56] If bottled water is used, water with fluoride added should be purchased.[55]

Taste of Human Milk

> "... too full o' th' milk of human kindness to
> catch the nearest way."
>
> Shakespeare's *Macbeth*, Act I, Scene V

This line from Shakespeare reflects the centuries-old belief that a breastfeeding woman's diet influences the composition of her milk and has a long-lasting influence on the child. The flavor of human milk is an important taste experience for newborn infants, but flavor of human milk is often ignored when the benefits of human milk or its composition is considered. Human milk is slightly sweet[57] and it carries the flavors of compounds ingested such as mint, garlic, vanilla, and alcohol.[58]

Infant responses to flavors in milk seem to depend on the length of time since the mother consumed the food, the amount of the flavor that the mother consumed, and the frequency with which the flavor is consumed (new versus repeated exposure). Infants seem more interested in their mother's milk when flavors are new to them. Researchers found that infants nursed at the breast longer if a flavor (garlic) was new to them than if the mother had taken garlic tablets for several days.[59] Infants who were exposed to carrot juice flavor in their mother's milk ate less of a carrot-flavored cereal and spent less time feeding at the breast than infants who had not been exposed to the carrot flavor. Thus, exposing infants to a variety of flavors in human milk may contribute to their interest in and consumption of human milk as well as their acceptance of new flavors in solid foods.[60]

Benefits of Breastfeeding

Breastfeeding Benefits for Mothers

Breastfeeding women experience hormonal, physical, and psychosocial benefits.[61] Breastfeeding immediately increases levels of oxytocin, a hormone that stimulates uterine contractions, minimizes maternal postpartum blood loss, and helps the uterus to return to nonpregnant size.[62]

After the birth, the return of fertility (through monthly ovulation) is delayed in most women during breastfeeding, particularly with exclusive breastfeeding.[62] This delay in ovulation results in longer intervals between pregnancies. Breastfeeding alone, however, is not as effective as other available birth control methods. Consequently, many health care professionals in the United States do not offer breastfeeding as an option for birth control.

Many women experience psychological benefits including increased self-confidence and facilitated bonding with their infants.[63] Many still consider faster return to prepregnancy weight a benefit of breastfeeding; however, women may lose or gain weight while nursing. The impact of breastfeeding on maternal weight is discussed in more detail later in this chapter. In addition to these short-term benefits, women who nurse at a younger age and for longer duration have lower risk of breast and ovarian cancer[64–67] and rheumatoid arthritis.[68]

Breastfeeding Benefits for Infants

> "Breastfeeding—the main source of active and passive immunity in the vulnerable early months and years of life— is considered to be the most effective preventive means of reducing the death rate of children under five."[69]
>
> Lubbock MH, Clark D, Goldman, AS. Breastfeeding: maintaining an irreplaceable immunological resource. Nature Reviews Immunology. 2004; 4(Jul) 565–572.

Nutritional Benefits   The value of the composition of human milk is widely recognized. Companies that make HMS often use human milk as the standard, recognizing the many unique properties of human milk:

- With its dynamic composition and the appropriate balance of nutrients, human milk provides optimal nutrition to the infant.[11,15]

- The balance of nutrients in human milk matches human infant requirements for growth and development closely; no other animal milk or HMS meets infant needs as well.

- Human milk is isosmotic (of similar ion concentration; in this case human milk and plasma are of similar ion concentration) and therefore meets the requirements for infants without other forms of food or water.

- The relatively low protein content of breast milk compared to cow's milk meets the infants needs without overloading the immature kidneys with nitrogen.

- Whey protein in human milk forms a soft, easily digestible curd.

- Human milk provides generous amounts of lipids in the form of essential fatty acids, saturated fatty acids, medium-chain-triglycerides, and cholesterol.

- Long-chain polyunsaturated fatty acids, especially docosahexaenoic acid (DHA), which promotes optimal development of the central nervous system, are present in human milk and are present in only some of the HMS marketed in the United States.

- Minerals in breast milk are largely protein bound and balanced to enhance their availability and meet infant needs with minimal demand on maternal reserves.

Immunological Benefits

One of the most important realizations about breastfeeding in the last few decades is the ability of human milk to protect against infections. Cells (T- and B-lymphocytes), *secretory immunoglobins* (sIgA, sIgG, sIgM, sIgE, sIgD), histocompatibility antigens, T-cell products, many nonspecific factors (e.g., complement, bifidus factor), carrier proteins (lactoferrin, transferring, vitamin B_{12}-binding protein, and corticoid-binding protein), and enzymes (lysozyme, lipoprotein lipase, leukocyte enzymes) are components of milk that confer immunological benefits.

Cellular components in human milk (*macrophages, neutrophils, T- and B-lymphocytes,* and *epithelial cells*) are especially high in colostrum but are also present for months in mature human milk in lower concentrations. The function of macrophages in human milk includes phagocytosis of fungi and bacteria, killing of bacteria and production of the complement proteins, lysozome, and lactoferrin and immunoglobin A and G.[11]

Leukocyte function appears to offer more protection to the breast than to immunocompetence of the infant. Neutrophils, however, appear to be activated and contribute to phagocytosis at the mucosa of the infant's gastrointestinal tract.[70] Both T- and B- lymphocytes provide the infant with protection against organisms in the gastrointestinal tract. This protection may extend beyond acute infection to allergy, necrotizing enterocolitis, tuberculosis, and neonatal meningitis.[11]

Immunoglobins are thought to be transported from maternal plasma across secretory epithelium to create secretory immunoglobins.[71] The predominant (90%) immunoglobulin in human milk, *secretory immunoglobin A(sIgA),* also appears to be most important in terms of the protection conferred to the infant. sIgA and sIgM protect the infant by blocking colonization with pathogens and limiting the number of antigens that cross the mucosal barrier. sIgA protects against enteroviruses, cytomegalovirus, herpes simplex virus, respiratory syncytial virus, rubella, retrovirus, and rotavirus.[11,72] sIgM protects against cytomegalovirus, respiratory syncytial virus, and rubella.

Bifidus factor supports growth of lactobacillus bifidus. Lactobacillus is a probiotic bacteria that stimulates antibody production and enhances phagocytosis of antigens.[11]

Lysozyme protects against enterobacteria and other gram-positive bacteria. Lysozyme is secreted by neutrophils and macrophages.

Binding proteins in human milk bind iron and vitamin B_{12}, making the nutrients unavailable for pathogens to grow in the infant's gastrointestinal tract. Such factors are also responsible for the some of the differences in intestinal flora (natural bacteria of the gastrointestinal tract) of breastfed infants versus HMS-fed infants.

Individual fatty acids and other milk components (oligosaccharides, gangliosides, and glycoconjugates) resulting from digestion of human milk are antimicrobial.[73,74] The digestive products of triacylglycerides and the lipid globule appear to protect against *Escherichia coli* 0157:H7, *Campylobacter jejuni, Listeria monocytogenes,* and *Clostridium perfringens.*[75,76] Monoacylglycerides are able to lyse enveloped viruses, bacteria, and protozoa. Glycoconjugates (glycoproteins, glycolipids, glycoaminoglycans, and oligosaccharides) may bind pathogens directly, thus preventing infection. Nucleotides are reported to increase resistance to *Staphylococcus aureus* and *Candida albicans* and may increase response to vaccine antigens.[77]

Growth factors and hormones in human milk, such as insulin, enhance the maturation of the infant's gastrointestinal tract. These substances also help to protect the infant, especially neonates, against viral and bacterial pathogens.

Immunoglobin A specific protein that is produced by blood cells to fight infection.

Macrophages A white blood cell that acts mainly through phagocytosis.

Neutrophils Class of white blood cells that are involved in the protection against infection.

T-lymphocyte A white blood cell that is active in fighting infection. (May also be called T-cell; the t in T-cell stands for thymus.) These cells coordinate the immune system by secreting hormones that act on other cells.

B-lymphocytes White blood cells that are responsible for producing immunoglobulins.

Epithelial Cells Cells that line the surface of the body.

Secretory Immunoglobin A A protein found in secretions that protect the body's mucosal surfaces from infections. The mode of action may be by reducing the binding of a microorganism with cells lining the digestive tract. It is present in human colostrum but not transferred across the placenta.

Lower Infant Mortality in Developing Countries

In the developing world, 10 million children die each year and 60% are believed to be preventable deaths.[78] Improving breastfeeding practices could save approximately 1.3 million lives annually and continuing breastfeeding with complementary foods could save an additional 600,000.[78] This protection of lives is at the center of the World Health Organization (WHO) and UNICEF's joint efforts called the Global Strategy for Infant and Young Child Feeding to remind the international community of the impact of feeding practices (including breastfeeding) on children's health outcomes.[78]

Breastfeeding may also play a role in reducing the risk of sudden infant death syndrome (SIDS), but this is still under investigation. Researchers disagree about whether breastfeeding has a primary effect in reducing risk of SIDS. An analysis of available studies found that bottle-feeding increases the risk of SIDS, but other factors related to feeding choice may be responsible for this finding.[79]

Fewer Acute Illnesses

Reduced infant illness is evident in countries with high infant illness (*morbidity*) and death (*mortality*) rates, poor sanitation, and questionable water supplies. Even in the United States and other developed nations, where modern health care systems, safe water, and proper sanitation are commonplace, there is a clear relationship between breastfeeding and reduced rates of illness in infants. In U.S. samples, the *incidence* of diarrhea is estimated to be 50% lower in exclusively breastfed infants.[80] Internationally, gastrointestinal infection was lower among infants exclusively breastfed for 6 months when compared to those exclusively breastfed for only 3 months.[80,81] Ear infections are 19% lower, and the number of prolonged episodes of ear infection was 80% lower among breastfed infants than among infants fed HMS. In a U.S. population study, breastfed infants experienced 17% less coughing and wheezing and 29% less vomiting than infants fed HMS.[82]

Morbidity The rate of illnesses in a population.

Mortality Rate of death.

Cognitive Function The process of thinking.

Reductions in Chronic Illness

In addition to the lower rate of acute illnesses in breastfed children, breastfeeding also seems to protect against chronic childhood diseases. Breastfeeding may reduce the risk of celiac disease,[83] inflammatory bowel disease,[84] and neuroblastoma.[85] HMS feeding results in an increase in the risk of allergy (30%) and asthmatic disease (25%).[86] These reductions in acute and chronic infant illness increase with greater use of human milk.[87] For example, infants who receive some human milk and some HMS are at 60% greater risk of ear infection than those fed exclusively human milk. The risks, particularly for allergy and asthmatic disease, are reduced for the duration of breastfeeding and for months to years after weaning.[86]

Breastfeeding and Childhood Overweight

Considerable attention has been paid to the role of breastfeeding in preventing obesity, but this relationship is still a topic of controversy. Breastfed infants typically are leaner than HMS-fed infants at 1 year of age without any difference in activity level or development.[88] A large body of literature suggests that there is a small reduction in risk of overweight in children greater than 3 years of age who were breastfed.[89] The effect of breastfeeding on the incidence of overweight was greater with longer duration of breastfeeding in some,[90] but not all studies.[91] Several potential mechanisms have been identified for the modest reduction in obesity in children who breastfed including metabolic programming, possibly related to biological messages in human milk, learned self-regulation of energy intake, and other characteristic of the families or parents such as healthy lifestyle. Given the epidemic of obesity in the United States, it is likely that the debate over this issue will continue.

Cognitive Benefits

Several reports have linked breastfeeding, and especially duration of breastfeeding, with cognitive benefits, assessed by IQ.[26,92] The increases in cognitive ability associated with breastfeeding is significant even after adjusting for family environment.[93] Cognitive development gains increase with the duration of breastfeeding.[92] In addition, higher intelligence quotients (IQ) of infants breastfed for 6 months appear to be greater among infants born small for gestational age (11 points) than among infants born appropriate weight for age (3 points).[94] The differences in *cognitive function* are also greater in premature infants fed human milk than in those fed HMS.[95] Recognition that the fatty acid composition of milk plays an important role in neuropsychological development bolsters the credibility of psychological or cognitive benefits from breastfeeding.

Analgesic Effects

Breastfeeding seems to work as analgesic to infants. Breastfeeding during venipucture seems to reduce infant pain as well as 30% glucose solution followed by pacifier use.[96] However, breastfeeding before a heel prick[97] did not seem to reduce infant pain response. Breastfeeding may be used to reduce infant discomfort during minor invasive procedures.[98]

Socioeconomic Benefits

A decrease in medical care for breastfed infants is the primary socioeconomic benefit of breastfeeding. Medicaid costs for breastfed WIC infants in Colorado were $175 lower than for infants who were fed HMS.[99] Never-breastfed infants have an excess of care for lower respiratory tract illness, otitis media, and gastrointestinal disease compared to infants breastfed for at least 3 months.[100] Each 1000 never-breastfed infants had 2033 more sick care visits, 212 days of hospitalization, and 609 more prescriptions. In addition, in one study of two companies with established lactation programs, the 1-day maternal absenteeism from work due to infant illness was approximately two-thirds lower in breastfeeding women than nonbreastfeeding women.[101] Companies benefit through lower medical costs and greater employee productivity.

Breast Milk Supply and Demand

Can Women Make Enough Milk?

Typical milk production averages approximately 600 ml (240 ml = 1 cup, or 8 ounces) in the month postpartum and continues to increase to approximately 750–800 ml per day by 4–5 months postpartum.[1] Milk production can range from 450–1200 ml per day in women who are nursing one infant.[20] Infant weight, the caloric density of milk, and the infant's age contribute significantly to the infant demand for milk. Milk increases to meet the demand of twins, triplets, or infants and toddlers suckling simultaneously; it can also be increased by pumping the milk.

Traditionally factors such as how vigorously an infant nurses, how much time the infant is at the breast, and how many times he or she nurses in a day were thought to control milk production. We now know that milk synthesis (rate of accumulation of milk in the breast) is related to infant demand.[102] That is, the removal of milk from the breast seems to be the signal to make more milk, and most women are able to increase their milk production to meet infant demand.[103]

An average of 24% of milk is left in the breast after feeding.[104] Thus, the short-term milk storage of the breast does not seem to be a limiting factor to infant milk intake. The average rate of synthesis in a day was only 64% of the highest rate of milk synthesis, suggesting that milk synthesis could be increased considerably. Comparisons of milk production between mothers of singletons and twins shows that the breasts have the capacity to synthesize much more than a singleton infant usually drinks.[105]

Does the Size of the Breast Limit a Woman's Ability to Nurse Her Infant?

The size of a woman's breast does not determine the amount of milk production tissue (clusters of alveoli containing secretory cells that produce the milk).[106] Much of the variation in breast size is due to the amount of fat in the breast. The size of the breast does limit storage because of limitations in the expansion of the ducts and lactiferous sinuses. Daily milk production is not related to the total milk storage capacity within the breast, however.[104]

This means that women with small breasts can produce the same amount of milk as women with large breasts, although the latter woman may be able to feed her infant less frequently to deliver the same volume of milk compared with a woman with smaller breasts.

Is Feeding Frequency Related to the Amount of Milk a Woman Can Make?

Feeding frequency is not consistently related to milk production. The rate of milk synthesis is highly variable between breasts and between feedings.[107] However, the amount of milk produced in 24 hours and the total milk withdrawn in that 24 hour period is highly related.[104] Milk synthesis is able to quickly respond to infant demand.

The breast responds to the degree of emptying during a feeding, and this response is a link between maternal milk supply and infant demand. Daly proposed that the breast responds to the infant's need by measuring how completely the infant empties the breast.[100,104,107] For example, if a lot of milk is left in the breast, then milk synthesis will be low to prevent engorgement; if the breast is fully emptied, synthesis will be high to replenish the milk supply.

Exact mechanisms of milk supply and demand are not well understood, but they seem to be related to a protein called feedback inhibitor of lactation (FIL).[107] FIL is an active whey protein that inhibits milk secretion. This protein inhibits all milk components equally according to their concentration in milk. Therefore, this protein seems to affect milk quantity only, not milk composition.

Pumping or Expressing Milk

Pumping or expressing milk may be needed for many reasons, including maternal or infant illness or separation. Women can express milk using several different methods: manually, hand pumps, commercial electric pumps, or hospital-grade electric pumps. A pump that allows mothers to pump both breasts at the same time (10 minutes per session) can save time over single pumping (20 minutes per session).[108] Electric pumps are efficient and may increase prolactin more than hand expression or hand pumping.[108,109] Insufficient milk production is a common problem among women who express milk. Researchers working with women who pumped their breasts report that 8 to 12 or more milk expressions per day were necessary to stimulate an adequate production of milk.[108] The optimal number of expressions in a 24-hour period is likely to differ for women according to how well they empty the breast and the storage capacity of the breast. Women who are able to establish an adequate volume of milk (>500 ml per day) in the first 2 weeks postpartum are more likely to still have enough milk for their infant at 4–5 weeks postpartum.[108] This recommendation is consistent with the advice to nurse the infant (or pump) early and often to build a good milk supply. See Chapter 7 (pages 212–213) for information on storage of milk.

Can Women Breastfeed after Breast Reduction or Augmentation Surgery?

Information regarding breastfeeding rates after breast surgery is scarce, but women generally have more problems after breast reduction surgery than after augmentation surgery. Accumulating evidence does suggest that women who undergo breast reduction surgery may be at risk for unsuccessful lactation.[110] The type of surgery appears to be an important determinant of ability to breastfeed. Some now recommend incision around the lower part of the breast to avoid damage to the ductal system caused by incision in the midst of the breast. Women with periareolar (around the nipple) incisions experience greater difficulties with breastfeeding.[111] Lactation consultants recommend that the surgery date, type, and incision used, as well as prior breastfeeding experience, be ascertained. Infants should be closely followed to prompt intervention when needed.

Does Silicone from Breast Implants Leach into the Milk?

The American Academy of Pediatrics does not consider silicone implants a reason not to breastfeed.[112] Nearly a million women in the United States have breast implants containing silicone. Early reports introduced concern of esophageal dysfunction in children of women who had silicone implants, but more recent research found no evidence to support such claims, and silicone concentrations in milk from women with implants are not elevated[113] and, in fact, are lower than those in formula and cow's milk.[114] It is possible that immunological responses could cause unfavorable effects; however, there is no evidence suggesting direct toxicity to the infant.

The Breastfeeding Infant

Optimal Duration of Breastfeeding

The gap between "best breastfeeding practice" and the norms for breastfeeding in the United States makes study of the optimal duration particularly important for health care professionals. The health benefits to the mother–child pair should be the primary criteria to determine the optimal duration of breastfeeding—and not simply whether the cultural environment makes such duration practical. The American Academy of Pediatrics (AAP) has taken a clear stand on this issue, saying that breastfeeding should continue for a year or longer.[4,5] The U.S. Surgeon General recommends human milk feeding exclusively for 6 months, noting further it is better to breastfeed for 6 months and

best to breastfeed for 12 months, with solid foods being introduced at 4–6 months.[61]

Infants who are breastfed for 6 months experience fewer illnesses from gastrointestinal infection than infants who are given HMS and breast milk at 3 or 4 months of age. Deficits in growth have not been demonstrated among infants in developing or developed countries who are exclusively breastfeed for 6 months or longer.[115] Breastfeeding can prevent intestinal blood loss in infants—a factor that should be considered when determining its optimal duration. Infants fed cow's milk before the age of 6 months suffer nutritionally significant losses of iron via intestinal blood[116]—an observation that supports the AAP's recommendation.[4,5] Through 1 year, breastfed infants suffer fewer acute infections than formula-fed infants do, a finding that supports breastfeeding beyond the introduction of solids.

Reflexes

Healthy term infants are born with several reflexes that enable newborns to nourish themselves. Observations show that 18-week-old fetuses start sucking. By 34 weeks gestation, the suck has adequate pace and rhythm to be nutritive. The gag reflex is the reflex that prevents taking food and fluids into the lungs. This reflex is developed by 28 weeks gestation. These reflexes allow term infants to suck and swallow in a coordinated pattern that protects the airways.

Two other reflexes describe the infant's ability to position herself to breastfeed. The *oral search reflex* is described as the infant opening his or her mouth wide in proximity to the breast while thrusting the tongue forward. The *rooting reflex* results in the infant turning to the side when stimulated on the side of the upper or lower lip. Infants come forward, open their mouth, and extend the tongue when the center of either upper or lower lip is stimulated.

The presence of these reflexes is important to the success of breastfeeding. However, successful nursing also requires appropriate positioning of the infant at the breast and adequate maternal letdown and milk production. Appropriate positioning and maternal assessment of infant nursing behaviors must be learned. Support from lactation consultants and/or other health care professionals who are trained in lactation may be necessary.

Breastfeeding Positioning

Proper positioning of the infant at the breast is important to breastfeeding success.[117,118] Mothers need to learn from health professionals experienced in optimal positioning because improper positioning causes pain and possible damage to nipple and breast tissue. The mother may need to use cushions, pillows, or a footstool to be comfortable and positioned well to nurse the infant (see Illustration 6.5). Once the mother is comfortable,

a. Cradle hold

b. Football or clutch hold

c. Cross-cuddle hold

Illustration 6.5 Positions for breastfeeding.

she should hold the baby so the mouth is directly in front of the nipple and stimulate the oral search reflex by touching the baby's bottom lip with her nipple. Then infants will open their mouth wide and should be brought to the breast with the nipple centered in their mouth (Illustration 6.6). This process is called *latching on*. Infants who are properly attached at the breast have all or most of the areola (the dark pigmented skin around the nipple) in their mouth. If the mother pulls down the infant's lower lip she should see the tongue lying around the lower gum line. The baby's nose should be close to the breast with breathing unrestricted. The mother should hear swallowing, but not smacking, clicking, or slurping. Women who consistently have pain when the infant is suckling should consult a professional trained in lactation to correct the attachment position.[117,118] Infants who are positioned correctly at the breast start to suckle almost immediately, change from quick short sucks to slow deep sucks, and remain relaxed.

Identifying Hunger and Satiety

When infants are hungry they begin to bring their hands to their mouth, suck on them, and start moving their head from side to side with their mouth open (rooting reflex). Infants should be fed when these signs of hunger are displayed rather than waiting for crying, a late sign of hunger. Recognizing early hunger behaviors and initiating feeding before the infant becomes very upset helps mothers and infants who have difficulty nursing.

Nutritive and non-nutritive sucks are different. Feedings begin with non-nutritive sucking. The infant sucks quickly and not particularly rhythmically. Nutritive sucking is slower and more rhythmic as the infant begins to suck and swallow. A mother can hear the infant suck in a quiet room.

Infants should be allowed to nurse as long as they want at one breast. Infants who are fed for shorter periods from both breasts can get larger amounts of foremilk. The high lactose content of foremilk can cause diarrhea.[119] Allowing the infant to nurse at one breast until satisfied creates a pattern that assures that the infant gets both foremilk and hindmilk. The higher fat content of hindmilk may help in signaling satiety. Infants will stop nursing when full. If they are still hungry, after burping, they can be offered the other breast.

Feeding Frequency

Stomach emptying occurs in about 1½ hours for breastfed infants. Ten to 12 feedings per day are normal for newborn infants. As infants develop, feeding frequency will depend, in part, on maternal storage. Different feeding patterns can meet infant needs. In one study, infants who did not feed from midnight through early morning consumed more in the other feedings, particularly in the morning. Milk intake and weight gain of these infants in the first 4 months of life were similar to those of infants whose feedings were distributed over 24 hours.[120]

Vitamin Supplements for Breastfeeding Infants

All infants, whether they are fed human milk or HMS, are vulnerable to vitamin K–deficiency bleeding (VKDB). All infants in the United States receive a vitamin K supplement (1.0 mg by injection) at birth because it is known to decrease the risk of VKDB.[5] Questions regarding potential health risks including cancer related to vitamin K injections in the newborn persist. Therefore, in 2003, the American Academy of Pediatrics called for further research regarding the "efficacy, safety, and bioavailability of oral

a. Touching the baby's bottom lip with the nipple stimulates the oral search reflex.

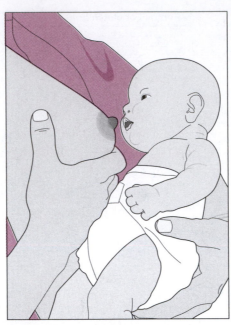

b. The infant opens his mouth wide.

c. The infant is brought to the breast with the nipple centered in his mouth.

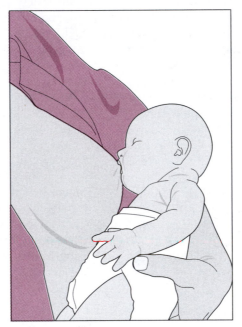

d. The infant is properly latched at the breast and has all of the areola in his mouth.

Illustration 6.6   Attachment.

formulations and optimal dosing regimens of vitamin K to prevent late VKDB."

Exclusively breastfed infants should be given a supplement of 200 IU of vitamin D per a day beginning in the first 2 months of life.[5] Vitamin D supplementation should continue until the infant (or child) consumes 500 ml of vitamin D–fortified HMS or cow's milk. This recommendation by the AAP was made in response to an increasing incidence of vitamin D–deficiency rickets among infants who are exclusively breastfed.

However, the AAP no longer recommends routine fluoride supplements during the first 6 months of life. After 6 months, the decision to supplement with fluoride should be made based on individual situations.[5] If the water supply contains 0.7 to 1.0 ppm of fluoride, no supplement is needed. If the water, food, and toothpaste contains less than 0.3 ppm fluoride, then 0.25 mg of fluoride is recommended. When supplementation is indicated for breastfed infants, maternal supplementation may be the best route.

Breastfed infants do not need iron-fortified HMS or supplements[11] because they rarely experience iron deficiency. The excess iron in HMS might bind with lactoferrin in human milk, resulting in a loss of the protective activity of lactoferrin.

Identifying Breastfeeding Malnutrition

A normal newborn weight loss of up to 7% can occur in the first week postpartum. A loss of 10% should trigger an evaluation of milk transfer to the infant by a lactation consultant or other trained professional and provide support needed to maintain breastfeeding. Malnourished infants become sleepy and nonresponsive, and have a weak cry and few wet diapers. The clinician can use the diagnostic flowchart in Illustration 6.7[11] to help diagnose failure to thrive in breastfed infants. By the fifth to seventh day postpartum, infants who are getting adequate nourishment have wet diapers approximately 6 times a day and have 3–4 soft, yellowish stools per day.[11] Infants who are slow gainers and not malnourished are alert, bright, responsive, and develop normally.[11] Their urine is pale yellow and dilute while stools are loose and seedy (some small particles are present in the stool). In contrast, infants who are failing to thrive are apathetic, hard to arouse, and have a weak cry. They have few wet diapers, and their urine is concentrated. Their stools are infrequent.

Case Study 6.1

Breastfeeding and Adequate Nourishment

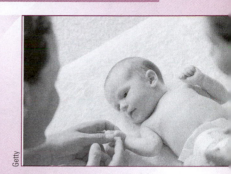

Molly G. is a 24-year-old office manager and part-time aerobics instructor who has delivered vaginally, without complications, a healthy, full-term son, Daniel. With a birth weight of 3200 grams (7 lb), Daniel is the first child for Molly and her husband. Molly is 162 cm (5 ft 4 in) tall with a prepregnancy weight of 56.8 kg (125 lb). She gained 25 kg (55 lb) during her uncomplicated pregnancy and has been a lacto-ovo vegetarian for 5 years. After a 12-hour stay in a birthing center, Molly and her husband bring Daniel home.

At 4 days postpartum Molly, her husband, and her mother-in-law bring the baby to the Health Care Center for his first follow-up visit. At this visit, Daniel weighs 3000 grams (6 lb 6 oz). Molly and her husband are very concerned about whether their son is getting adequate nourishment. They report that Daniel nurses vigorously about every 1½ to 2 hours and never sleeps for more than a couple of hours. Molly says that her milk "came in" on the second postpartum day and that she feels like all she does is nurse. Her nipples are tender, but not uncomfortably sore. She reports Daniel has at least six to eight wet diapers and two to three very loose stools each day. She wonders if she has enough milk and worries about how she will ever return to work in 2 months. She also wants to lose the excess weight she gained during the pregnancy and is eager to return to her aerobics classes. Her husband and mother-in-law are supportive, but they worry about the baby. The pediatrician's examination of the infant concludes Daniel is a healthy infant with no medical problems.

Questions

1. What factors put Molly at high risk for early termination of breastfeeding?
2. What factors indicate that Daniel is getting adequate nourishment?
3. What concerns do you have about Molly's diet? What advice would you give her about her weight-loss plans and eagerness to return to exercise? Do Molly or Daniel need any vitamin–mineral supplements?
4. If Molly lived in your community, what resources would be available for help and support for breastfeeding mothers?
5. What steps can Molly take to continue successful breastfeeding when she returns to work in 2 months?

Mothers of slow-gaining infants in particular should be advised to let the infant nurse at one breast until it is empty or the infant stops nursing, rather than switching breasts after a specific amount of time. This regimen assures that the infant gets hindmilk with a higher fat and calorie content. Lawrence[11] recommends evaluation of slow-gaining infants when the slow-gaining pattern is first recognized (Illustration 6.7).

Tooth Decay

Human milk has infection-fighting components that inhibit the formation of dental caries. Nevertheless, caries can occur in children who are breastfed.[121] Frequent nursing at night after 1 year of age is a risk factor for dental caries. Nevertheless, the prevention of dental caries is not justification for advising early weaning. Rather, the mother should be instructed on the prevention and treatment of early childhood dental caries. All children should be seen by a qualified dentist 6 months after the first tooth erupts or by 12 months of age. Mothers are the primary source of bacteria that cause early childhood caries. Therefore, mothers or primary care givers should be given information on oral hygiene, diet, fluoride, caries removal, and prevention of caries.

Breastfed babies have straighter teeth due to the development of a well-rounded dental arch.[122] A well-rounded arch may also help prevent sleep apnea later in life.

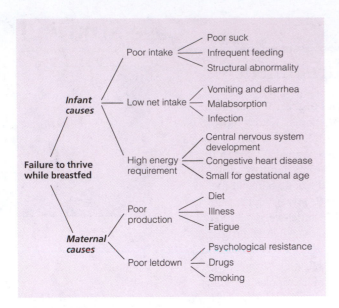

Illustration 6.7 Diagnostic flowchart for failure to thrive.

SOURCE: Lawrence, R. A. and Lawrence, R. M. Breastfeeding: A Guide for the Medical Profession. 5th ed. St. Louis: Mosby, 2006.[11]

Maternal Diet

The U.S. Department of Agriculture's MyPyramid Food Guide and the 2005 Dietary Guidelines (presented in Chapter 1) do not provide specific advice on how breastfeeding women can best meet their nutritional needs.[123] The Dietary Guidelines indicate, however, that moderate weight reduction can be achieved by the breastfeeding mom without compromising the weight gain of the nursing infant.

Diets formed around a MyPyramid food plan provide a healthy assortment of nutrients at specified caloric levels. For a breastfeeding woman requiring 2400 calories daily, her MyPyramid food plan would include:

- Grains, 8 ounces
- Vegetables, 3 cups
- Fruits, 2 cups
- Milk, 3 cups
- Meat and Beans, 6.5 ounces
- Oil, 7 teaspoons

Additional calories for desserts, fats, and sweets for the 2400-cal food plan is 360.

Energy and Nutrient Needs

The Food Guide Pyramid released by the U.S. Department of Agriculture in 1992 estimates that energy needs for breastfeeding women are 1900 if she is inactive, 2500 kcal per day if she is active, and 3100 kcal if she is very active.[123] Until recently, when safe methods of studying energy metabolism became available, determining the energy needs of lactating women was very difficult. As a

result, the energy needs of lactation are estimated using the factorial method, which adds the estimated requirements of lactation to the requirements of nonlactating women.

The DRIs for normal-weight lactating women assume that the energy spent for milk production is 500 cal per day in the first 6 months and 400 cal afterward.[124] The 2002 DRI is 330 additional cal to support 0.8 kg per month weight loss (170 cal per day) for the first 6 months and 400 cal per day afterward. However, a recent review of a state-of-the-art study of energetics in exclusively breastfeeding women calls for a review of recommendations, citing a total energy cost of approximately 623 cal per day assuming 750 grams of milk produced at 0.67 kcal/g and 80% efficiency.[125] With mobilization of approximately 170 kcal per day, net energy needs were estimated at approximately 450 kcal per day.

We now understand that women use several mechanisms to meet the energy needs of lactation. Adjustments in energy intake and energy expenditure must be balanced to meet those needs. Goldberg et al.[126] found that women increased food intake (56% of the need for milk production) and decreased physical activity (44% of the energy need for milk production) to meet the increase in energy needs for lactation. Doubly labeled water studies suggest that the components of energy expenditure vary greatly, and measurements of dietary intake can be unreliable.[127] Therefore, a single recommendation for energy for lactating women could never address all of the individual ways that women meet their energy needs. Assessment of adequacy of energy intake of breastfeeding women should always be made within the context of the mother's overall nutritional status and weight changes and the adequacy of the infant's growth.

Maternal Energy Balance and Milk Composition

The composition of breast milk depends on maternal nutritional status. Protein–calorie malnutrition results in an energy deficit that reduces the volume of milk produced but does not usually compromise the composition of the milk. Several studies show that milk production is maintained when there is a modest level of negative calorie balance. Animal models first identified a potential threshold effect of energy restriction. Baboons fed 60% of their voluntary intake significantly reduced milk production.[133] Yet, baboons fed 80% of their usual intake maintained milk production. Randomized studies such as the one performed on baboons cannot ethically be done with humans. However, a series of human studies on weight loss during breastfeeding (discussed below) support a threshold effect of energy limitations on lactation.

Weight Loss during Breastfeeding

Current DRIs are written assuming a weight loss of 0.8 kg/month.[124] In addition, mechanisms that favor use of maternal fat stores and delivery of nutrients to the breast seem to occur during lactation. Despite these mechanisms that should favor weight loss in breastfeeding women, loss by 12 months postpartum is on average less than the amount needed to return to prepregnancy weight.[134] Even more surprising, postpartum weight changes (−0.1 kg/mo) are smaller in developing countries than in industrialized nations (−0.8 kg/mo). The failure to return to prepregnancy weight may be due to changes in energy intake, energy expenditure, and fat mobilization that easily meet energy needs.

A number of small studies in the United States have addressed the issue of whether weight loss influences milk production. Strode et al.[135] first observed women who voluntarily reduced their energy intake to 68% of their estimated needs for 7 days. No differences in infant intake or milk composition were observed. In the week following the diet, women who consumed fewer than 1500 cal tended to experience a decrease in milk volume.

Women who consumed over 1500 cal per day experienced no decrease. Despite its voluntary nature and the lack of milk samples throughout the day, this study provides important support for the idea that modest and/or short-term reductions in energy are not associated with decreases in milk production. In addition, 22 healthy postpartum women who participated in a 10-week weight-loss program, which reduced energy intake by 23%, maintained milk production.[136] The women lost an average of about a pound a week during the 10 weeks.

Studies of weight loss during lactation also suggest that energy restriction can be accomplished without large decreases in the quality of the maternal diet. Maternal reduction of energy intake by 500 cal per day through lowered intake of sugary and high-fat foods resulted in similar micronutrient intakes to women who did not decrease intake, with the exceptions of calcium and vitamin D.[137] None of the women consumed adequate vitamins C or E. Therefore, careful attention to consumption of calcium and vitamin D–rich foods such as low-fat and fat-free dairy products as well as fruit and vegetable consumption is needed.

Exercise and Breastfeeding

Studies[138,139] examining the effect of increasing energy expenditure on weight and lactation suggests that it is safe. In a *cross-sectional study,* vigorous exercise increased energy expenditure, but these women also increased their energy intake so that the calorie deficit was similar in the two groups.[138] There were no significant differences in milk volume between groups, although the group of exercising women tended to have higher milk volumes. A later 12-week exercise intervention trial studied women who were 6 weeks postpartum and exclusively breastfeeding, to examine the effects of exercise on body composition and energy expenditure during breastfeeding.[139] The women were randomly assigned to two groups. One group followed an aerobic exercise regimen

for 45 minutes, 5 days a week. The other group did no exercise. The exercising women increased energy expenditure by 400 kcal per day and increased energy intake to compensate for this use. Both groups experienced similar weight changes, milk volume, composition, and infant weight gain and serum prolactin levels. These studies suggest that lactating women efficiently balance their energy intake to support energy expenditure.

The available evidence suggests that modest energy restriction combined with increases in activity may be effective at helping women to lose weight, while improving their metabolic profile and increasing fat losses. Despite the still small numbers of women who have been studied, a consistent lack of effect on milk production (infant intake), milk energy output, and infant growth is encouraging.

> **Cross-Sectional Study**   A study that measures the current disease and exposure status of all individuals in a sample at a single timepoint (i.e., a study that measures the prevalence of breastfeeding at 6 months of age).

The mechanisms responsible for the maintenance of milk production may be different for exercise and energy restriction. It appears that aerobic activity enhances fatty acid mobilization to meet the needs of milk production. Restricted energy intake requires increases in prolactin levels to promote use of dietary fatty acids or to promote mobilization from fat. These prolactin increases explain how milk production can be maintained despite negative energy balance. An additional study suggests that immunological factors in milk are not reduced by exercise in breastfeeding women.[140]

Vitamin and Mineral Supplements

A 1991 Institute of Medicine report, *Nutrition during Lactation,* stated that well-nourished breastfeeding women do not need routine vitamin or mineral supplementation.[1] Instead, supplementation should target specific nutritional needs of individual women. Supplementation strategies should take into account how nutrients are secreted into human milk and the potential for nutrient–nutrient interactions in mothers and their infants. For example, women who avoid dairy products completely should use calcium (1200 mg) and vitamin D (10 µg) supplements.

Vitamin and Mineral Intakes

Vitamin and mineral intakes that do not meet recommended levels (of folate, thiamin, vitamin A, calcium, iron, and zinc) have been reported for lactating women.[37,54,131,132] Ten percent of lactating women have thiamin intakes below the recommended levels,[131] whereas fewer than 5% of nonlactating women have intakes below the 1998 DRI. However, these reports of inadequate intake have not been followed by reports of deficits in nutritional status of the mother–infant pair. Nor is there a recommendation for vitamin and mineral supplements for all breastfeeding women. A careful balance is needed between concern over inadequate maternal dietary intake and causing women not to breastfeed because they do not have an optimal diet.

Functional Foods

Concern has been expressed about possible ill effects of high intakes of fortified foods in addition to supplements. Although this is an important issue, studies to date have not identified adverse reactions related to fortified food consumption and RDA levels of nutrients in supplements.

Fluids

There is no evidence that increasing fluid intake will increase milk production or that a short-term fluid deficit results in a decrease in milk production. Fluid demands rise during breastfeeding, however, so women should drink fluids to thirst. Once a mother and her infant have the nursing routine down, she may find it convenient to have something to drink while she nurses. Although many women want to know how many glasses to drink per day, the amount needed varies depending on climate, milk production, body size, and other factors. Therefore, women are advised to drink enough fluids to keep their urine pale yellow. The current RDI for water for lactating women sums the recommendation for nonpregnant and nonlactating women (2.7 L for women 19–30) and the water content of the average milk output during the first 6 months (0.78 L milk × 87% water = 0.68 L) for a total need of 3.4 L for women 19–30.[141]

Alternative Diets

Breastfeeding women can follow alternative eating patterns and be well nourished. The goal is to adequately nourish the mother and child, not to force women to use supplements and/or products that are not part of their normal eating patterns. Incorporation of soy products, vegetarian diets of various sorts, and other alternative diet choices can be followed as long as they meet maternal nutritional needs. Vegans who do not consume dairy products and eggs, however, may need to plan carefully to consume adequate amounts of calories, protein, calcium, vitamin D, vitamin B_{12}, iron, and zinc. Vegetarians' intakes of protein are generally adequate as long as energy intake is adequate. Breastfeeding women who consume no animal products should use plant foods with bioavailable B_{12} from sources such as yeasts, seaweed, and fortified soy products. Women who are unable to get adequate B_{12} from foods should take a vitamin B_{12} supplement. The 1991 Institute of Medicine[1] report recommends a multiple vitamin–mineral supplement for vegetarians because human milk may be low in

vitamin B_{12} even when mothers do not exhibit deficiency symptoms. See Chapter 7 for information on the use of herbals during breastfeeding.

Infant Colic

It is widely believed that components of maternal diet are related to infant colic. Information is growing to support this idea.[128–130] A randomized trial assessing maternal avoidance of cow's milk, eggs, peanuts, tree nuts, wheat, soy, and fish resulted in a reduction in colic symptoms of their infants in the first 6 weeks of life.[128]

Factors Influencing Breastfeeding Initiation and Duration

Obesity and Breastfeeding

Overweight and obesity prior to pregnancy and excess weight gain during pregnancy are associated with shorter duration of breastfeeding.[142] This association is independent of socioeconomic status and other factors also known to affect breastfeeding duration. This influence of obesity may be related to lower prolactin responses early postpartum and resulting difficulty in establishing adequate milk supply.[143,144] Therefore, maintenance of ideal body weight is important for lactation success.

Socioeconomic

All new mothers, both low-income and more affluent, need support for breastfeeding. However, low-income women often lack the education, support, and confidence to interpret the abundant and pervasive mixed messages on infant feeding practices. Consider the strikingly different context for pregnancy, birth, and parenting for low-income women and their more affluent counterparts:

Profile of a low-income pregnant woman: She says she wants to do what is best for her baby and, in fact, knows the breast is best. However, she is afraid breastfeeding will cause her baby to be too "clingy." She feels extremely uncomfortable about nursing around family, much less in public. She is certainly not up for the pain she has heard breastfeeding causes. To make things more difficult, in the hospital, she is separated from her baby soon after delivery and is given little assistance for getting started. She is sent home from the hospital with samples of free formula.

Profile of an affluent pregnant woman: The affluent expectant mother has friends who have breastfed and have helped build her confidence that she can breastfeed successfully. She may have been able to choose her birth setting and select a hospital with knowledgeable staff allowing mother and baby to stay together around the clock. Because she knows there may be bumps in the road getting started, she seeks out support from friends or the doctor after discharge. At home, she has a supportive husband who is proud of her for offering the best for their baby. If she returns to work, she knows she can still breastfeed to keep that special closeness with her baby even after returning to work.

Common barriers to breastfeeding initiation expressed by expectant mothers are:

- Embarrassment
- Time and social constraints, and concerns about loss of freedom (particularly issues of working moms)
- Lack of support from family and friends
- Lack of confidence
- Concerns about diet and health practices
- In adolescents, fear of pain[145–147]

Additional obstacles to the initiation and continuation of breastfeeding include:

- Insufficient prenatal breastfeeding education
- Health cares provider apathy and misinformation
- Inadequate health care provider lactation management training
- Disruptive hospital policies
- Early hospital discharge
- Lack of routine follow-up care and postpartum home health visits
- Maternal employment, especially in the absence of workplace facilities and support for breastfeeding
- Lack of broad societal support
- Media portrayal of bottle-feeding as the norm
- Commercial promotion of infant formula through distribution of hospital discharge packs, coupons for free or discounted formula, and television and general magazine advertising[5,148]

Breastfeeding Promotion, Facilitation, and Support

"Significant steps must be taken to increase breastfeeding rates in the United States and to close the wide racial and ethnic gaps in breastfeeding. This goal can only be achieved by supporting breastfeeding in the family, community, workplace, health care sector, and society."

Health and Human Services Blueprint for Action on Breastfeeding, 2000[61]

The support a woman receives from those around her directly impacts on her capability to breastfeed optimally.[61] A number of factors in the health care system, workplace, and community facilitate the initiation and continuation of breastfeeding.

Role of the Health Care System in Supporting Breastfeeding

Health care providers and facilities exert tremendous influence over the mother–infant dyad with the power to promote and model optimal breastfeeding practices during prenatal care, at delivery, and after discharge. There is evidence for the effectiveness of lay[149] and professional support[150,151] on the duration of any breastfeeding.

The AAP is working in partnership with the Maternal and Child Health Bureau (MCHB), maternal and pediatric health professionals, residency programs, public health representatives, and other breastfeeding personnel to strengthen the AAP/MCHB Breastfeeding Promotion in Physicians Office Practices (BPOPIII). This program aims to: (1) provide training in breastfeeding promotion to pediatrics practices and individuals though webcasts or teleconferencing technology; (2) develop model residency program curricula for obstetrics, pediatrics, and family medicine; (3) provide technical assistance and resources to physicians, residents, public health representatives, and families; (4) strengthen and expand national collaborative networks and action groups at local, state, and regional levels to implement effective breastfeeding strategies and initiatives in underserved populations; and (5) assess changes in breastfeeding rates in physicians' practices after the practice implements breastfeeding education, counseling interventions, and ongoing support for mothers.

Prenatal Breastfeeding Education and Support

Culturally competent prenatal breastfeeding education that is given frequently in person can have a significant positive influence on breastfeeding rates.[152,153] Best Start Social Marketing developed an effective three-step counseling strategy (Table 6.5) that quickly identifies a woman's particular barriers to choosing breastfeeding and provides targeted education while affirming the woman's ability to breastfeed.[154]

This Best Start approach replaces forced choice questions, such as "Are you going to breast or bottle-feed?" with open-ended questions, such as "What have you heard about breastfeeding?" or "What questions do you have about breastfeeding?" This provides the women with an opportunity to begin a dialogue with her provider about the infant feeding decision. The use of follow-up probes can help the counselor understand the woman's specific concern. Counselors should avoid the common pitfall of overwhelming women with too much information, which can give the impression that breastfeeding is difficult.[147]

Another effective strategy utilizes peer counselors and peer group discussions with at least one or two woman who have successfully breastfed.[155,156] Exposure to mothers nursing their babies increases a woman's level of comfort

Table 6.5 The Best Start three-step breastfeeding counseling strategy[147]

1. Ask open-ended questions to identify the woman's concerns.
   - Dietitian: "What have you heard about breastfeeding?"
   - Client: "I hear it's best for my baby, but all my friends say it really hurts!"
2. Affirm her feelings by reassuring her that these feelings are normal.
   - Dietitian: "You know, most women worry about whether it will hurt."
3. Educate by clarifying how other women like her have dealt with her concerns. Avoid overeducating or giving the impression that breastfeeding is hard to master.
   - Dietitian: "Did you know that it is not supposed to be painful, and if you are having discomfort, there are people who can help make it better?"

SOURCE: Adapted from McCamman, S and Page-Goertz, S Breast-feeding Success: You Can Make the Difference. *The Perinatal Nutrition Report* Vol 4: Winter 1998, 2.[154]

Table 6.6 Key teaching points prior to birth[147]

In the hospital or birthing center, mothers should:

Request early first feeding

Practice frequent, exclusive breastfeeding

Ask to be taught swallowing indicators

Learn indicators of sufficient intake

Ask for help if it hurts

Know sources for help

Understand postpartum rest and recovery needs

Avoid supplements unless medically indicated

SOURCE: Adapted from S. Page-Goertz and S. McCamman, Breast-feeding Success: You Can Make the Difference. *The Perinatal Nutrition Report* Vol. 4: Winter 1998, 2.[154]

Table 6.7 World Health Organization's International/UNICEF Code on the marketing of breast milk substitutes

- No advertising of breast milk substitutes
- No free samples or supplies
- No promotion of products through health care facilities
- No company sales representative to advise mothers
- No gifts or personal samples to health workers
- No gifts or pictures idealizing formula feeding, including pictures of infants, on the labels of the infant milk containers
- Information to health workers should be scientific and factual
- All information on artificial feeding, including labels, should explain the benefits of breastfeeding and the costs and hazards associated with formula feeding
- Unsuitable products should not be promoted for babies
- Manufacturers and distributors should comply with the Code's provisions even if countries have not adopted laws or other measures

SOURCE: Adapted from World Health Organization: Contemporary patterns of breast-feeding. Report on the WHO Collaborative Study on Breastfeeding, Geneva; World Health Organization, 1981.[165]

with breastfeeding and provides a forum for informal discussion with family and friends. Discussion of an individual's personal experience can be an effective way to help women see and believe that others like her share her concerns.[145]

Toward the end of pregnancy, women need information on what to expect in the hospital or birthing center and practical tips for initiating breastfeeding. Because fathers,[157–159] grandmothers,[160] *doulas*, friends, and social networks[161] all have a powerful role on infant feeding decisions, it is important to include these influential people as often as possible in breastfeeding promotion efforts.[145,158,161] An Italian study demonstrates the power of educating fathers.[157] Higher breastfeeding rates (25%) were achieved at 6 months in the group where fathers received education on their role in supporting breastfeeding than in the control group (15%). Differences were even greater when women reported difficulties with breastfeeding (24.5% versus 4.5%). Several key points are shown in Table 6.6.

The environment for the delivery of prenatal care is an important barrier or facilitator of breastfeeding. It should provide positive messages about breastfeeding, such as posters that promote breastfeeding and magazines and literature in the waiting room; there should be no advertisements or promotions of formula.

Doula An individual who surrounds, interacts with, and aids the mother at any time within the period that includes pregnancy, birth, and lactation; may be a relative, friend, or neighbor and is usually but not necessarily female. One who gives psychological encouragement and physical assistance to a new mother.[11]

Patient education materials that include formula advertising, samples, and business reply cards for free formula[161–163] *and* in direct violation of the World Health Organization's International Code on the Marketing of Breast Milk Substitutes (Table 6.7). Women who have been exposed to materials and products from formula companies prenatally are more likely to stop breastfeeding in the first 2 weeks.[163] Use of these materials provides a subtle message that infant formula is equivalent to breast milk.

Although not all women will choose to breastfeed, the goal of prenatal breastfeeding education is to empower every woman with sufficient knowledge to make an informed decision about how to feed her baby. Some professionals view breastfeeding as a personal choice rather than a public health issue and voice concern that breastfeeding promotional efforts may cause women who choose HMS feeding to feel guilty. On the other hand, some women who formula-feed report feeling angry about not getting enough breastfeeding information during their pregnancy.[146] In recognition of the benefits of breastfeeding and the important role of health professionals in promoting and supporting breastfeeding, the leading health and professional organizations in the United States that provide perinatal care have established policies in support of breastfeeding as the preferred infant feeding method.[5,148]

Successful maternal education program examples are highlighted in The *CDC Guide to Breastfeeding Interventions*.[152] These include health insurance plans that provide breastfeeding education for their members, Baby Friendly Hospital Initiatives offering patient infant-feeding classes, and health departments offering training programs for persons who provide breastfeeding education.

Lactation Support in Hospitals and Birthing Centers

Hospital policies and routines significantly impact on critical early experience with breastfeeding, with effects extending far beyond the short stay.[166] Illustration 6.8 provides examples of practices that influence this pivotal initiation experience. As in prenatal care settings, the distribution of free samples of infant formula coupons and hospital discharge packs is discouraged because of the detrimental effects of this practice on breastfeeding success, particularly among vulnerable groups such as new mothers and low-income women.[166] Model policies for hospitals and physicians are available from the Breastfeeding Coalition (www.inlandempirebreastfeedingcoalition.org/index.htm) and the Academy of Breastfeeding Medicine (http://bfmed.org).

In an effort to promote, protect, and support breastfeeding in hospitals and birthing centers worldwide, the World Health Organization (WHO) and UNICEF established the Baby Friendly Hospital Initiative in 1992.[167] This initiative focuses on 10 evidence-based components of hospital care that impact on breastfeeding success (Table 6.8). The Baby Friendly USA program designates facilities within the United States who meet the guidelines. As of September 2006, there were 55 hospitals and birthing centers in the United States that had met all of the criteria in Table 6.9 and were designated as Baby Friendly (www.babyfriendlyusa.org/eng/03.html), whereas in 2003 there were only 38. Evidence suggests that the Baby Friendly Hospital Initiative is responsible for an increase in breastfeeding rates in Switzerland.[169] Further monitoring of the Baby Friendly Hospital Initiative is needed to document its successes.

In 2002 WHO and UNICEF came together to try to revitalize the international community in breastfeeding promotion with the Global Strategy for Infant and Young Child Feeding.[78] This report builds on the *Innocenti Declaration*[167,170] and the Baby Friendly Hospital Initiative by recognizing the importance of feeding in all children including those in difficult circumstances such as emergency situations, low-birth-weight infants, and infants of mothers with HIV. The initiative includes (1) a call on governments to develop and implement policy on infant and child feeding within the context of the national policy for nutrition, child and reproductive health as well as poverty reduction; (2) access to skilled support for initiation and maintenance of breastfeeding exclusively for 6 months and with safe complementary foods for up to 2 years or beyond; (3) empowerment of health care professionals to provide breastfeeding support and extend their services into the community; (4) review of progress in implementation of the International Code of Marketing of Breast Milk Substitutes and consideration of new measures to protect families from commercial interests and influence; and (5) enactment of legislation to protect the breastfeeding rights of working women in accordance with international labor standards. The strategy also specifies responsibilities of international organizations, nongovernmental organizations, and other interested parties. This report provides an important framework for acceleration of support of appropriate feeding for infants and children worldwide by linking resources and intervention areas available in many sectors.

Innocenti Declaration   On the Protection, Promotion, and Support of Breastfeeding: Policy statement adopted by participants at the World Health Organization UNICEF policymakers' meeting on Breastfeeding in the 1990s; A Global Initiative held in Italy in 1990. The policy established exclusive breastfeeding from birth to 4-6 months of age as a global goal for optimal maternal and child health.

Lactation Support after Discharge

Breastfeeding support is essential in the first few weeks after delivery, as lactation is being established.[61] Younger women and women with lower socioeconomic status are more likely to stop breastfeeding by 4 weeks postpartum and cite sore nipples, inadequate milk supply, feeling that the infant is not satisfied, and infant problems as reasons for stopping.[171] A study of inner-city Baltimore WIC program participants[145] provides strong evidence that 7 to 10 days postpartum is the critical window for providing breastfeeding support; 35% of mothers who initiated breastfeeding in the hospital had stopped by 7 to 10 days.

A pediatrician, nurse, or other knowledgeable health care practitioner (home visit or in the office) should see all breastfeeding mothers and their newborns when the newborn is 2 to 4 days of age. Breastfeeding should be observed and evaluated for evidence of successful breastfeeding behavior. This is also an important time to revisit

Hospital Practices That Influence Breastfeeding Initiation

	← Strongly Encouraging →	← Encouraging →	← Discouraging →	← Strongly Discouraging →
Physical Contact	• Baby put to breast immediately in delivery room • Baby not taken from mother after delivery • Woman helped by staff to suckle baby in recovery room • Rooming-in; staff help with baby care in room, not only in nursery	• Staff sensitivity to cultural norms and expectations of woman	• Scheduled feedings regardless of mother's breastfeeding wishes	• Mother-infant separation at birth • Mother-infant housed on separate floors in postpartum period • Mother separated from baby due to bilirubin problem • No rooming-in policy
Verbal Communication	• Staff initiates discussion re: woman's intention to breastfeed pre- and intrapartum • Staff encourages and reinforces breastfeeding immediately on labor and delivery • Staff discusses use of breast pump and realities of separation from baby, re: breastfeeding	• Appropriate language skills of staff, teaching how to handle breast engorgement and nipple problems • Staff's own skills and comfort re: art of breastfeeding and time to teach woman on one-to-one basis	• Staff instructs woman "to get good night's rest and miss the feed" • Strict times allotted for breastfeeding regardless of mother/baby's feeding "cycle"	• Woman told to "take it easy," "get your rest" . . . impression that breastfeeding is effortful/tiring • Woman told she doesn't "do it right," staff interrupts her efforts, corrects her re: positions, etc.
Nonverbal Communication	• Staff (doctors as well as nurses) give reinforcement for breastfeeding (respect, smiles, affirmation) • Nurse (or any attendant) making mother comfortable and helping to arrange baby at breast for nursing • Woman sees others breastfeeding in hospital	• Pictures of woman breastfeeding • Literature on breastfeeding in understandable terms • Closed-circuit TV show in hospital on breastfeeding	• Pictures of woman bottle-feeding • Staff interrupts her breastfeeding session for lab tests, etc. • Woman doesn't see others breastfeeding	• Woman given infant formula kit and infant food literature • Woman sees official-looking nurses authoritatively caring for babies by bottle-feeding (leads to woman's insecurities re: own capability of care)
Experiential	• If breastfeeding not immediately successful, staff continues to be supportive • Previous success with breastfeeding experience in hospital			• Previous failure with breastfeeding experience in hospital

Illustration 6.8 Hospital practices that influence breastfeeding initiation.[61,147]

Table 6.8 The Baby-Friendly Hospital Initiative 10 steps to successful breastfeeding

1. Have a written breastfeeding policy that is regularly communicated to all health care staff.
2. Train all staff in skills necessary to implement this policy.
3. Inform all pregnant women about the benefits and management of breastfeeding.
4. Help mothers initiate breastfeeding within half an hour of birth.
5. Show mothers how to breastfeed and how to sustain lactation, even if they should be separated from their infants.
6. Feed newborn infants nothing but breast milk, unless medically indicated, and under no circumstances provide breast milk substitutes, feeding bottles, or pacifiers free of charge or at low cost.
7. Practice rooming-in, which allows mothers and infants to remain together 24 hours a day.
8. Encourage breastfeeding on demand.
9. Give no artificial pacifiers to breastfeeding infants.
10. Help start breastfeeding support groups and refer mothers to them.

SOURCE: World Health Organization, Protecting, Promoting and Supporting Breast-feeding: The Special Role of Maternity Services. A Joint WHO/UNICIF statement. Geneva, Switzerland, 1989.[167]

Table 6.9 Important elements of worksite lactation support programs

- Prenatal lactation education tailored for working women
- Corporate policies providing information for all employees on the benefits of breastfeeding and on why their breastfeeding co-workers need support
- Education for personnel about the services available to support breastfeeding women
- Adequate breaks, flexible work hours, job sharing, and part-time work
- Private "Mother's Rooms" for expressing milk in a secure and relaxing environment
- Access to hospital-grade, autocycling breast pumps at the workplace
- Small refrigerators for the safe storage of breast milk
- Subsidization or purchase of individually owned portable breast pumps for employees
- Access to lactation professional on-site or by phone to give breastfeeding education, counseling, and support during pregnancy, after delivery, and when the mother returns to work
- Coordination with on-site or near-site child care programs so the infant can be breastfed during the day
- Support groups for working mothers with children

SOURCE: U.S. Department of Health and Human Services. HHS Blueprint for Action on Breastfeeding, Washington, D.C. U.S. Department of Health and Human Services, Office on Women's Health, 2000.[61]

the major concerns the mother identified during pregnancy, as well as discussing any new concerns. Mothers should be armed with information on sources of trained, skilled, and available help in the community such as *lactation consultants,* peer counselors, the WIC program, or *La Leche League,* should questions or complications arise.[11,145] Follow-up telephone calls, as necessary, provide additional support to mothers who are not fully confident in their ability to breastfeed successfully.[11,145]

The Workplace

The increase in the proportion of women working that began after World War II has been one of the most significant social and economic trends in modern U.S. history.[172] In 1940, one in four U.S. workers was a woman; by 1998 almost one in two workers was a woman. In 2005, nearly 60% of mothers with children younger than 3 years of age worked.[173]

Barriers to breastfeeding and employment have been recognized by the Surgeon General for over 20 years.[173] These barriers include lack of on-site day care, insufficiently paid maternity leave, rigid work schedules, and employers who lack knowledge about breastfeeding. Current law ensures a woman's right to breastfeed her infant anywhere on federal property that she and her child are authorized to be. Still in process is further legislation to require that women cannot be fired or discriminated against if they breastfeed or express milk during their own lunchtime or break time, to provide a tax credit for employers who provide lactation services, and to develop minimum standards for breast pump safety.

Planning to return to work full time does not appear to impact on breastfeeding initiation rates substantially.[2] Breastfeeding duration, however, is adversely affected by employment. At 6 months 35% of nonemployed women are still breastfeeding, compared with 33.4% of mothers working part-time, and only 23% of those employed full-time. Occupation influences duration of breastfeeding. Women in professional occupations breastfeed longer than do women in sales, clerical, or technical occupations.[175] Part-time work is more conducive to breastfeeding; the

Lactation Consultant Health care professional whose scope of practice is focused on providing education and management to prevent and solve breastfeeding problems and to encourage a social environment that effectively supports the breastfeeding mother–infant dyad. Those who successfully complete the International Board of Lactation Consultant Examiners (IBLCE) certification process are entitled to use the IBCLC (International Board Certified Lactation Consultant) after their names (www.iblee.org).

La Leche League International, nonprofit, nonsectarian organization dedicated to providing education, information, support, and encouragement to women who want to breastfeed. Founded in 1956 by seven women who had learned about successful breastfeeding while nursing their own babies, it currently has approximately 7100 accredited lay leaders to facilitate more than 3000 monthly mother-to-mother breastfeeding support group meetings around the world (www.lalecheleague.org).

number of hours mothers work per day is inversely associated with the likelihood that the mother will continue to breastfeed.[176] However, workplace lactation programs may be able to help meet the Healthy People 2010 goals for breastfeeding. A lactation program (including an employee's choice of a class on the benefits of breastfeeding), services of a lactation consultant, and a private room in the workplace with equipment for pumping resulted in an average weaning age of 9 months.[177] Importantly, more than 50% continued to breastfeed to 6 months, and the majority were working full time (84.2%).

Studies indicate that women who continue to breastfeed once returning to work miss less time from work because of baby-related illnesses, and have shorter absences when they do miss work, compared with women who do not breastfeed.[99] Worksite programs that support breastfeeding facilitate the continuation of breastfeeding after mothers return to their jobs and offer additional advantages to employers: employee morale and loyalty, image as family-friendly, recruiting for personnel, and retention of employees after childbirth all improved.[176,178] Companies that have adopted breastfeeding support programs have noted cost savings of $3 per $1 invested in breastfeeding support in addition to lower health care costs over the first year.[179] Key elements of worksite lactation support programs are presented in Table 6.9.

Women planning to return to work have several choices. Breast milk can be expressed during the day into sterile containers, refrigerated or frozen, and then used for subsequent bottle feedings when they are at work. (See Chapter 7 for storage guidelines.) With on-site child care, it is possible to breastfeed during breaks and lunch hours. Another possibility is to train the body to produce milk only when the mother is home during the evenings and during the night. To accomplish this, a woman should omit one feeding at a time during the periods of the day when she will not be feeding or expressing milk. This will help her to reduce her milk supply without experiencing engorgement. She gradually weans to the feedings at the appropriate time of the day. This method works because removal of milk is the stimuli for milk production. Generally, at least two feedings per day are needed for women to continue making milk. No evidence indicates that it is necessary to introduce a bottle sooner than 10 days before returning to work.[180] Unless a mother is returning to work immediately after delivery, a bottle should not be introduced before lactation is well established, which is usually at least 4 weeks. Information about hospital-grade breast pumps is readily available on the Internet.

Women should seek child care providers or facilities that are supportive of breastfeeding. Supportive facilities provide accommodations for mothers who wish to breastfeed their children at the facility or have their children fed expressed milk. Facilities should also follow established child care standards, including standards for the storage and handling of expressed breast milk.

The Community

To increase breastfeeding rates in a community, it is important to identify community attitudes and obstacles to breastfeeding, and to solicit the support for breastfeeding from community leadership. Establishment of a multidisciplinary breastfeeding task force with representatives from physicians, hospitals and birthing centers, public health, home visitors, La Leche League, government, industry, school boards, and journalists can be an effective vehicle for assessing community breastfeeding support needs and sponsoring collaborative efforts to overcoming community obstacles to breastfeeding.[145] Barriers to breastfeeding may include lack of access to reliable and culturally appropriate sources of information and social support, cultural perception of bottle feeding as the norm, aggressive marketing of breast milk substitutes, and laws that prohibit breastfeeding in public. For example, in 2006 there was public outrage when the image of an infant feeding at a woman's breast was portrayed on the front of a magazine about babies. This reaction clearly demonstrates that breastfeeding is not perceived as a cultural norm.

In the past decade legislative efforts have been made to protect a woman's right to breastfeed. States vary widely in not only whether there is legislation relating to breastfeeding, but also the depth and breadth of the legislation. Legislation is used to protect a woman's right to breastfeed, consider breastfeeding in family law situations, to regulate breast pumps, and to provide incentives to employers who provide breastfeeding support. Legislation addresses issues such as a woman's right to breastfeed in public and on federal property, express milk at work, and to be exempt from jury duty. A state-by-state listing of legislation is maintained on the La Leche League International website (www.lalecheleague.org/LawBills.html).

To facilitate breastfeeding support advocacy among health professionals in health care facilities and in the community the National Alliance of Breastfeeding Advocacy (NABA) maintains a searchable database on the Web. The database includes a state-by-state listing of all known breastfeeding coalitions and task forces, AAP chapter presidents, International Lactation Consultant Association affiliates, La Leche League International contacts, WIC breastfeeding directors and Lamaze state contacts, and Title V Maternal and Child Health Directors.

Public Food and Nutrition Programs

National Breastfeeding Policy

The U.S. Department of Health and Human Services (DHHS) is the lead federal agency for policy development to advance the promotion, protection, and support of breastfeeding for families in the United States. Over the past several decades, the Office of the Surgeon General and the

Maternal and Child Health Bureau have highlighted the public health importance of breastfeeding though numerous workshops and publications (Table 6.10). In 2000, the HHS Blueprint for Action on Breastfeeding outlined a comprehensive framework to increase breastfeeding rates in the United States and to promote optimal breastfeeding practices. The action plan is based on education, training, awareness, support, and research.[61]

In the 1990s, DHHS through its Maternal and Child Health Bureau and the Centers for Disease Control supported the establishment of the United States Breastfeeding Committee (USBC), to fulfill one of the goals identified in the Innocenti Declaration:[170] Each nation should establish "a multisectoral national breastfeeding committee composed of representatives from relevant government departments, nongovernmental organizations, and health professional associations." The USBC is a collaborative partnership of organizations working to develop a strategic plan for protecting, promoting, and supporting breastfeeding in the United States.[186]

Many HHS agencies have breastfeeding initiatives. The Title V Maternal and Child Health programs of the Health Resources and Services Administration provide substantial support services, training, and research for breastfeeding (www.mchb.hrsa.gov). The Centers for Disease Control and Prevention plays a major role in supporting breastfeeding nationally through applied research, program evaluation, and surveillance (www.cdc.gov).

USDA WIC Program

WIC, the Special Supplemental Nutrition Program for Women, Infants, and Children, is a federal program operated by the U.S. Department of Agriculture Food and Nutrition Service in partnership with state and local health departments. Created in 1972, WIC is designed to provide nutrition education, supplementary foods, and referrals for health and social services to economically disadvantaged women who are pregnant, postpartum, or are caring for infants and children under 5. WIC operates through a network of state health departments, Indian tribal organizations, U.S. territories, and local agencies providing services to more than 8 million program participants per year.

In 1989, Congress mandated (Public Law 101-147) a specific portion of each state's WIC budget allocation to be used exclusively for the promotion and support of breastfeeding among its participants and authorized the use of WIC administrative funds to purchase breastfeeding aids such as breast pumps. Reauthorization legislation, the Healthy Meals for Healthy Americans Act of 1994 (P.L. 103-448), increased the budget allocation to $21 for each pregnant and breastfeeding woman in support of breastfeeding promotion. Through this legislation, each state has a breastfeeding coordinator and a plan to coordinate operations with local agency programs for breastfeeding promotion. WIC's 1999 budget appropriation increased this amount further to $23.53. In 2005, WIC expenditures were more than 3 billion dollars (www.fns.usda.gov/pd/wisummary). The average monthly cost per participant has risen from $15.68 in 1972 to $37.42 in 2005.

The USDA Food and Nutrition Information Center supports a WIC Works website to serve health and nutrition professionals working in the WIC Program. The WIC Works site, www.nal.usda.gov/wicworks includes an e-mail discussion group, links to training materials on breastfeeding promotion, and information on how to share resources and recommendations.

Table 6.10 Landmark U.S. breastfeeding policy statements and conferences

- Report of the Surgeon General's Workshop on Breastfeeding and Human Lactation[174]
- Follow-up Report: Surgeon General's Workshop on Breastfeeding and Human Lactation[181]
- Healthy People 2000 Breastfeeding Goals for the Nation[182]
- DHHS Maternal and Child Health Bureau National Workshop: Call to Action: Better Nutrition for Mothers, Children, and Families. Washington, DC: National Center for Education in Maternal and Child Health[183]
- Second Follow-up Report: Surgeon General's Workshop on Breastfeeding and Human Lactation[184]
- National Breastfeeding Policy Conference. Presented by the UCLA Center for Healthier Children, Families and Communities, Breastfeeding Resource Program in cooperation with the U.S. Department of Health and Human Services, Health Resources and Services Administration, Maternal and Child Health Bureau, and the Centers for Disease Control and Prevention[185]
- Healthy People 2010 Breastfeeding Goals for the Nation[3]
- DHHS Blueprint for Action on Breastfeeding[61]
- U.S. Breastfeeding Committee Strategic Plan[186]

Model Breastfeeding Promotion Programs

WIC National Breastfeeding Promotion Project—Loving Support Makes Breastfeeding Work

In 1995, the USDA Food and Nutrition Service entered into a cooperative agreement with Best Start *Social Marketing* (a not-for-profit

Social Marketing Combines the principles of commercial marketing with health education to promote a socially beneficial idea, practice, or product.[146]

Focus Group Interviews Small group discussions guided by a trained moderator, which provide insights into the participant's perceptions, attitudes, and opinions on a designated topic.[144]

organization assisting public health, education, and social service organizations with social marketing services) to develop a national WIC breastfeeding promotion project to implement at the state level. The project has four goals: (1) to encourage WIC participants to begin and to continue breastfeeding, (2) to increase referrals to WIC clinics for breastfeeding support, (3) to increase general public acceptance and support of breastfeeding, and (4) to provide support and technical assistance to WIC professionals in promoting breastfeeding.

Qualitative research data were collected in 10 pilot states through a series of observations, personal and telephone interviews, and *focus group interviews* with WIC participants and individuals who might influence their infant feeding decisions such as mothers, boyfriends and husbands, health care providers, and WIC staff.[187] Motivations and perceived barriers related to breastfeeding as well as social network influences on feeding choice were identified.

Results from the formative research were used to develop a marketing plan and program material. In contrast to the traditional public health approach of addressing breastfeeding as a medical health decision, the marketing plan repositioned breastfeeding as a way for a family to establish a special relationship with their child from the very onset of its life.[147,186] The campaign slogan, Loving Support Makes Breastfeeding Work, capitalizes on the concept that everyone is important to women's breastfeeding success—family, friends, doctors, and the community. Campaign materials and a counseling program were developed to help mothers work through individual barriers and constraints to breastfeeding

In 1997, the campaign was implemented at the state level. Participating states had the opportunity to purchase campaign materials including pamphlets, posters, and radio and television public service announcements that address barriers and encourage breastfeeding. Training was provided on coalition building, utilizing local media, promoting effective breastfeeding counseling strategies and techniques (Table 6.5), and managing peer counseling programs. Technical assistance on implementing the campaign was provided. Participating states implemented all or parts of the campaign in their states. Since 1997, the Loving Support Makes Breastfeeding Work campaign has expanded to other states; currently 72 state agencies, Indian tribal organizations, and territories are participating at various levels.[188,189]

Preliminary results from an evaluation of the program's impact in Mississippi show evidence of the program's success in both breastfeeding rates and changed attitude and awareness toward breastfeeding. Prior to initiation of the Loving Support campaign in 1997, Mississippi ranked fiftieth in the nation in breastfeeding initiation and duration rates. Mississippi's campaign included a public awareness component with television and radio spots, newspaper ads, and billboards in high traffic areas; a patient and family education program in health departments and WIC clinics; extensive outreach with health providers in the community and 25 hospitals; and community outreach including training to child care centers and worksites. In 1999, the state moved into forty-eighth place nationally in breastfeeding initiation rates at hospital discharge. Duration rates at 6 months, which climbed from 7.0% to 15.4% from 1998 to 1999, moved Mississippi WIC from fiftieth to thirty-third in the country. Preliminary data from focus groups conducted across the state in 1999, with support form USDA, revealed that the campaign had a significant impact on WIC program participants' breastfeeding knowledge and attitudes.[189]

Other states are currently assessing their individual campaigns.[187] Building on the idea that fathers are important, a father-to-father breastfeeding promotion program in Texas resulted in increased breastfeeding rates at WIC clinics using peer dads.[190] Based on data from the Ross Breastfeeding Survey,[2] breastfeeding initiation rates in Iowa increased from 57.8% to 65.1% after a year of the campaign. The rates of women still nursing at 6 months after birth also increased; before the start of the campaign, 20.4% of women were still breastfeeding at 6 months compared with 32.2% a year after the program's start. Increased breastfeeding support from relatives and friends was also documented from data collected in a mail survey. Black women in New York say that WIC program staff influence their decision to breastfeed.[191] Women reported that free formula influenced formula feeding, but that personalized breastfeeding promotion and trusting relationships with WIC staff encouraged breastfeeding. Unfortunately, rates of breastfeeding among WIC recipients remain considerably less than segments of the population in higher socioeconomic levels.[8]

Wellstart International

Wellstart International is an independent, nonprofit organization headquartered in San Diego, California, that is dedicated to supporting the health and nutrition of mothers and infants worldwide through the promotion of breastfeeding. Wellstart focuses on the development of local leadership and teamwork by providing education and technical assistance to perinatal health care providers and educators around the world, enabling them to promote maternal and child health in their own settings through the support of breastfeeding. Wellstart faculty and staff offer in-depth clinical and programmatic expertise, and both domestic and international experience to hospitals, clinics, and university schools of medicine, nursing, and nutrition, as well as to a wide variety of governmental and nongovernmental health and population agencies. Wellstart is a designated World Health Organization Collaborating Center on Breastfeeding Promotion, with international and domestic efforts actively taking place.

Since 1977 Wellstart has assisted health care professionals establish self-sustaining breastfeeding promotion programs worldwide. One example is seen in the establishment of the National Training and Technical Support Center for Breastfeeding in Cairo, Egypt, in collaboration with the Ministry of Health and Population of Egypt. The Support Center trains health workers and provides technical support for other related program areas. It is part of the countrywide Technical Support Collaboration involving applied research, evaluation of university curricula, countrywide training of health workers, work to achieve Mother–Baby Friendly Hospital status, community outreach, behavior change activities, and monitoring and evaluation. Lessons learned were an important outcome, as well as recommendations for remaining self-sustaining. Documents such as Counseling Guidelines on Infant Feeding for Use in Egypt and the Final Report on the Technical Support Collaboration on Infant Feeding were created to be disseminated to all service providers and community workers.

The combined approach of community outreach activities, coupled with the use of Information, Education, and Communication (IEC) materials targeting key behaviors, and coordination and referral to trained service providers, produce significant behavior changes at the local, regional, and national levels:[188]

- *Initiation of Breastfeeding.* Behaviors related to initiation of the first breastfeed improved (p < 0.001), with the optimal behavior of initiating breastfeeding within the first hour increasing from 46% to 72% within the demonstration areas.
- *Exclusive Breastfeeding.* Exclusive breastfeeding for all age groups increased (p < 0.001), with levels increasing from 24% to 74% for the 0–3 month age group, from 8% to 55% for the 4–6 month age group, and from 15.5% to 65% for the combined 0–6 month age group.
- *Infants receiving water and herbal teas.* The use of water or herbal teas within the 0–3 and 4–6 month age groups declined (p < 0.001), as would be expected if exclusive breastfeeding was increasing.
- *Bottle and Pacifier Use.* For infants 0–12 months of age bottle use decreased from 49% to 12%; pacifier use decreased from 42% to 13.6% (p < 0.001).
- *Complementary Feeding.* Data related to timely introduction of complementary foods, as well as types of complementary foods received by infants also showed substantial improvements (p < 0.001). The percentage of infants 7–9 months of age receiving no complementary food decreased from 16% to 3%.

On a global level, we must do everything we can

"to increase women's confidence in their ability to breastfeed. Such empowerment involves the removal of constraints and influence that manipulate perceptions and behavior towards breastfeeding, often by subtle and indirect means. Furthermore, obstacles to breastfeeding within the health system, the workplace, and the community must be eliminated."

These words are from the WHO/UNICEF Innocenti Declaration on the Protection Promotion and Support of Breastfeeding, 1990.[170] As we have seen, breastfeeding is best for the vast majority of infants and is physiologically possible for the vast majority of women. The challenge is to overcome barriers and provide support systems at the local, national, and international level so that the initiation and duration of breastfeeding will continue to increase.

Key Points

1. With rare exceptions, human milk is the optimal food for infants, exclusively for 6 months and with supplemental foods for a year or longer. Benefits to infants include protection from iron deficiency, better gains in cognitive ability, fewer acute respiratory and gastrointestinal illnesses, and lower risk of sudden infant death syndrome, celiac disease, inflammatory bowel disease, neuroblastoma, allergies, and asthma.

2. A thorough understanding of the anatomy and physiology of lactation is key to enabling health care providers in providing effective lactation support.

3. Maternal benefits of breastfeeding include minimization of postpartum blood loss, delayed fertility, greater self-confidence and bonding with their baby, and reduced risk of ovarian and breast cancers.

4. Newborns who are getting adequate human milk have about six wet diapers and three to four soft, yellowish stools per day by 5–7 days postpartum. Even slow gainers are alert, bright, and responsive, whereas infants who are failing to thrive are apathetic, hard to arouse, and have a weak cry.

5. Removal of milk from the breast is the stimulus for making more milk. Most women can make enough milk for their infant, although storage capacity (size of the breast) differences may determine how often the infant feeds in 24 hours.

6. Breastfed infants should be supplemented with vitamin K at birth and with 200 IU of vitamin D per day beginning at 2 months. In areas where water is not fluoridated, infant supplementation after the age of 6 months may be the best choice.

7. Breastfeeding women can lose modest amounts of weight while breastfeeding by choosing a diet following the MyPyramid food guide. Breastfeeding women need an additional 330 calories per day in the first 6 months, and 400 cal thereafter. Energy intake should be adjusted for activity level and achievement and maintenance of healthy weight.

8. Maternal diet does not significantly alter the protein, carbohydrate, fat, and major mineral composition of breast milk, but does affect the fatty acid profile, amounts of some vitamins, and some, but not all, trace minerals. For a majority of nutrients, the quality of the milk is preserved over the quantity of milk when maternal diet is inadequate.

9. Maternal diet may be associated with infant colic. Avoidance of cow's milk and cow's milk products, eggs, peanuts, tree nuts, wheat, soy, and fish have been associated with a reduction in infant colic symptoms in the first 6 weeks of life.

10. Successful breastfeeding is possible for women who follow vegetarian diets. Careful attention to vitamin B_{12} supplementation is important for women who are vegan.

11. Support for breastfeeding women from husbands, mothers, sisters, health care providers, communities, employers, and policy makers is critical to breastfeeding success and impacts breastfeeding rates in the community.

Resources

International Lactation Consultants Association
Telephone: 919-787-5181
Website: www.ilca.org

La Leche League
Telephone: 847-519-7730
Website: www.lalecheleague.org

Baby-Friendly USA Hospital Initiative
Telephone: 508-888-8044
Website: www.babyfriendlyusa.org

Best Start Social Marketing
Telephone: 813-971-2119
Website: www.beststartinc.org

DHHS HRSA Maternal and Child Health Bureau
Telephone: 301-443-0205
Website: http://mchb.hrsa.gov/whusa_05/pages/0428breastfeed.htm

Food and Nutrition Information Center, Lifecycle, Nutrition
Website: http://fnic.nal.usda.gov/nal_display/index.php?info_center=4&tax_level=2&tax_subject=257&topic_id=1357&placement_default=0

USDA Women, Infants, and Children (WIC) Program
Telephone: 703-305-2736
Website: www.fns.usda.gov/wic

Wellstart International
Telephone: 619-295-5192
Website: www.usbreastfeeding.org/breastfeeding/compend-Wellstart.htm

Academy of Breastfeeding Medicine
Telephone: 800-990-4ABM (toll free)
Website: www.bfmed.org

References

1. Institute of Medicine. Nutrition During Lactation. Washington, DC: National Academy Press; 1991.

2. Ross Products Division, Abbott laboratories Breastfeeding Trends through 2000; 2003.

3. U.S. Department of Health and Human Services. Healthy People 2010: Conference Edition—volumes I and II. Washington, DC: U.S. Department of Health and Human Services, Public Health Service, Office of the Assistant Secretary for Health; 2000.

4. Breastfeeding and the use of human milk. Pediatrics. 1997;100(6):1035–39.

5. Gartner LM et al. Breastfeeding and the use of human milk. Pediatrics. Feb 2005; 115(2):496–506.

6. Centers for Disease Control and Prevention Department of Heatlh and Human Services. Breastfeeding: Data and Statistics:

Breastfeeding Practices—Results from the 2005 National Immunization Survey. Atlanta, GA: Centers for Disease Control and Prevention, U.S. Department of Health and Human Services; 2005.

7. Phares TM et al. Surveillance for disparities in maternal health-related behaviors—selected states, Pregnancy Risk Assessment Monitoring System (PRAMS), 2000–2001. MMWR Surveill Summ. Jul 2 2004;53(4): 1–13.

8. Ryan AS, Zhou W. Lower breastfeeding rates persist among the Special Supplemental Nutrition Program for Women, Infants, and Children participants, 1978–2003. Pediatrics. Apr 2006;117(4):1136–46.

9. Hartmann PE. Changes in the composition and yield of the mammary secretion of cows during the initiation of lactation. J Endocrinol. 1973;59(2):231–47.

10. Vorherr H. Human lactation and breastfeeding. In: Larson BL, ed. The mammary

gland/human lactation/milk synthesis. Vol 4. New York, NY: Academic Press; 1978.

11. Lawrence RA, Lawrence RM. Breastfeeding: a guide for the medical professional. 6th ed. Minneapolis: Mosby; 2005.

12. Cox DB et al. Blood and milk prolactin and the rate of milk synthesis in women. Exp Physiol. 1996;81(6):1007–20.

13. Neville MC. The physiological basis of milk secretion. Part I. Basic physiology 1990;5861–8. Ann NY Acad Sci. 1990;5: 861–68.

14. Jensen RG. The lipids of human milk. Boca Raton, FL: CRC Press; 1989.

15. Jensen RG, ed. Handbook of Milk Composition. New York: Academic Press; 1995.

16. Prentice A. Constituents of human milk. Food and Nutrition Bulletin. The United Nations University Press, Volume 17, 1996.

17. Agriculture USDA. National Nutrient Database for Standard Reference: United

States Department of Health and Human Development; 2005.

18. Williams HG. 'And not a drop to drink'—why water is harmful for newborns. Breastfeeding Review. 2006;14(2):5–9.

19. Axelson I et al. Protein and energy intake during weaning: I. Effects on growth. Acta Paediatrica Scandinavica. 1987; 76(2):321–27.

20. Butte NF et al. Longitudinal changes in milk composition of mothers delivering preterm and term infants. Early Hum Dev. 1984;9(2):153–62.

21. Hediger ML et al. Early infant feeding and growth status of US-born infants and children aged 4–71 mo: analyses from the third National Health and Nutrition Examination Survey, 1988–1994. Am J Clin Nutr. 2000;72(1):159–67.

22. Connor WE, Lowensohn R, Hatcher L. Increased docosahexaenoic acid levels in human newborn infants by administration of sardines and fish oil during pregnancy. Lipids. 1996;31 Suppl:S183–87.

23. Insull W, Ahrens EH. The fatty acids of human milk from mothers on diets taken ad libitum. Biochem J. 1959;72:27.

24. Agostoni C et al. Docosahexaenoic acid status and developmental quotient of healthy term infants. Lancet. 1995;346(8975):638.

25. Kovacs A et al. Fatty acids in early human milk after preterm and full-term delivery. J Pediatr Gastroenterol Nutr. 2005;41:454–59.

26. Helland I, Smith L, Saarem K. Maternal supplementation with very-long-chain n-3 fatty acids during pregnancy and lactation augments children's IQ at 4 years of age. Pediatr. 2003;111(1):e39–44.

27. Mosley EE et al. trans Fatty acids in milk produced by women in the United States. Am J Clin Nutr. Dec 2005;82(6): 1292–97.

28. Wong WW et al. Effect of dietary cholesterol on cholesterol synthesis in breast-fed and formula-fed infants. J Lipid Res. 1993; 34(8):1403–11.

29. Owen C et al. Infant feeding and blood cholesterol: a study in adolescents and a systematic review. Pediatr. 2003;110(3): 597–608.

30. Rosen JM et al. Regulatory sequences involved in the hormonal control of casein gene expression. Ann N Y Acad Sci. 1986; 464:87–99.

31. Phadke S et al. Antimicrobial peptides in mucosal secretions: the importance of local secretions in mitigating infection. J Nutr. 2005;135:1289–93.

32. Velona T et al. Protein profiles in breast milk from mothers delivering term and preterm babies. Pediatr Res. 1999;45 (5 Pt 1):658–63.

33. Lonnerdal B, Atkinson S. Nitrogenous components of milk. In: Jensen RG, ed. Handbook of Milk Composition. San Diego, CA: Academic Press; 1995:351–68.

34. Newburg DS, Neubauer SH. Carboyhdrates in milks: analysis, quantities, and signficance. In: Jensen RG, ed. Handbook of Milk Composition. San Diego, CA: Academic Press; 1995:273–349.

35. Bode L. Recent advances on structure, metabolism, and funciton of human milk oligosaccharides. J Nutr. 2006;136:2127–30.

36. Morrow A et al. Human-milk glycans that inhibit pathogen binding protect breast-feeding infants agains diarrhea. J Nutr. 2005;135(1304–1207).

37. Food and Nutrition Board, Institute of Medicine. Dietary Reference Intakes for Vitamin A, Vitamin K, Arsenic, Boron, Chromium, Copper, Iodine, Iron, Manganese, Molybdenum, Nickel, Silicon, Vanadium, and Zinc. Washington, DC: National Academy Press; 2001.

38. Canfield LM et al. beta-Carotene in breast milk and serum is increased after a single beta-carotene dose. Am J Clin Nutr. 1997;66(1):52–61.

39. Rothberg AD et al. Maternal-infant vitamin D relationships during breast-feeding. J Pediatr. 1982;101(4):500–503.

40. Ala-Houhala M et al. 25-Hydroxyvitamin D and vitamin D in human milk: effects of supplementation and season. Am J Clin Nutr. 1988;48(4):1057–60.

41. Lammi-Keefe CJ, Jensen RG. Fat-soluble vitamins in human milk. Nutr Rev. 1984;42 (11):365–71.

42. Haug M et al. Vitamin E in human milk from mothers of preterm and term infants. J Pediatr Gastroenterol Nutr. 1987; 6(4):605–9.

43. Chappell JE, Francis T, Clandinin MT. Vitamin A and E content of human milk at early stages of lactation. Early Hum Dev. 1985;11(2):157–67.

44. Andon MB et al. Dietary intake of total and glycosylated vitamin B-6 and the vitamin B-6 nutritional status of unsupplemented lactating women and their infants. Am J Clin Nutr. 1989;50(5):1050–58.

45. Salmenpera L, Perheentupa J, Simes MA. Folate nutrition is optimal in exclusively breast-fed infants but inadequate in some of their mothers and in formula fed infants. J Pediatr Gastroenterol Nutr. 1986; 5(2):283–89.

46. Specker BL et al. Vitamin B-12: low milk concentrations are related to low serum concentrations in vegetarian women and to methylmalonic aciduria in their infants. Am J Clin Nutr. 1990;52(6):1073–76.

47. Butte NF et al. Macro- and trace-mineral intakes of exclusively breast-fed infants. Am J Clin Nutr. 1987;45(1):42–48.

48. Fransson GB, Lonnerdal B. Zinc, copper, calcium, and magnesium in human milk. J Pediatr. 1982;101(4):504–8.

49. Atkinson SA et al. Abnormal zinc content in human milk. Risk for development of nutritional zinc deficiency in infants. Am J Dis Child. 1989;143(5):608–11.

50. Duncan B et al. Iron and the exclusively breast-fed infant from birth to six months. J Pediatr Gastroenterol Nutr. 1985;4(3): 421–25.

51. Pisacane A et al. Iron status in breast-fed infants. J Pediatr. 1995;127(3):429–421.

52. Krebs NF, Hambidge KM. Zinc requirements and zinc intakes of breast-fed infants. Am J Clin Nutr. 1986;43(2):288–92.

53. Sian L et al. Zinc homeostasis during lactation in a population with a low zinc intake. Am J Clin Nutr. Jan 2002;75(1): 99–103.

54. Food and Nutrition Board Institute of Medicine. Dietary Reference Intakes for Calcium, Phosphorus, Magnesium, Vitamin D, and Fluoride. Washington, DC: National Academy Press; 1999.

55. Story MHK, Sofka D, eds. Bright Futures in Practice: Nutrition. Arlington, VA: National Center for Education and Child Health; 2000.

56. Palmer C, Wolfe SH. Position of the American Dietetic Association: the impact of fluoride on health. J Am Diet Assoc. Oct 2005;105(10):1620–28.

57. McDaniel MR, Barker E, Lederer CL. Sensory characterization of human milk. J Dairy Sci. 1989;72(5):1149–58.

58. Menella JA. Mother's milk: a medium for early flavor experiences. J Hum Lactation. 1995;11(39–45).

59. Menella JA, Beauchamp GK. Experience with a flavor in mother's milk modifies the infants acceptance of flavored cereal. Developmental Psychobiol. 1999;34:197–203.

60. Gerrish CJ, Menella JA. Flavor variety enhances food acceptance in formula-fed infants. Am J Clin Nutr. 2001;73:1080–85.

61. U.S. Department of Health and Human Services. Health and Human Services, Blueprint for Action on Breast-feeding. Washington, DC: U.S. Department of Health and Human Services, Office on Women's Health; 2000.

62. Heinig MJ, Dewey KG. Health effects of breastfeeding for mothers: a critical review. Nutrition Reviews. 1997;10:59–73.

63. Kuzela AL, Stifter CA, Worobey J. Breastfeeding and mother-infant interactions. J Reprod Infant Psychol. 1990;8:185–94.

64. Newcomb PA et al. Lactation and a reduced risk of premenopausal breast cancer. N Engl J Med. 1994;330(2):81–87.

65. Rosenblatt KA, Thomas DB. Lactation and the risk of epithelial ovarian cancer. The WHO Collaborative Study of Neoplasia and Steroid Contraceptives. Int J Epidemiol. 1993;22(2):192–97.

66. Narod SA. Modifiers of risk of heredity of breast cancer: Onqugene 2006;25:5832-6.

67. Kim y Choi JY, Lee KM, Park SM, Ahn SH, Noh DY, Hong YC, Kang D, Yoo KY. Dose-dependent protective effect of breast-feeding against breast cancer among ever lactated Women. Eur J Cancer Prev 2007; 16(2):124–9.

68. Karlson EW. Do breast-feeding and other reproductive factors influence future risk of rheumatoid arthritis? Results from the Nurses' Health Study. Arthritis Rheum. Nov 2004;50(11):3458–67.

70. Keeney SE et al. Activated neutrophils and neutrophil activators in human milk: increased expression of CD11b and decreased expression of L-selectin. J Leukoc Biol. Aug 1993;54(2):97–104.

71. Brandtzaeg P. Mucosal immunity: integration between mother and the breast-fed infant. Vaccine. Jul 28 2003;21(24): 3382–88.

72. Goldman AS. The immune system of human milk: antimicrobial, antiinflammatory and immunomodulating properties. Pediatr Infect Dis J. Aug 1993;12(8):664–71.

73. German JB, Dillard CJ. Composition, structure and absorption of milk lipids: a source of energy, fat-soluble nutrients and bioactive molecules. Crit Rev Food Sci Nutr. 2006;46(1):57–92.

74. Isaacs CE. Human milk inactivates pathogens individually, additively, and synergistically. J Nutr. May 2005;135(5):1286–88.

75. Sprong RC, Hulstein MF, Van der Meer R. Bactericidal activities of milk lipids. Antimicrob Agents Chemother. Apr 2001; 45(4):1298–1301.

76. Sprong RC, Hulstein MF, Van der Meer R. High intake of milk fat inhibits intestinal colonization of Listeria but not of Salmonella in rats. J Nutr. Jul 1999;129(7):1382–89.

77. Pickering LK et al. Modulation of the immune system by human milk and infant formula containing nucleotides. Pediatrics. Feb 1998;101(2):242–49.

78. World Health Organization. Global Strategy for Infant and Young Child Feeding. Geneva, Switzerland: World Health Organization; 2003.

79. McVea KL, Turner PD, Peppler DK. The role of breastfeeding in sudden infant death syndrome. J Hum Lact. 2000;16(1): 13–20.

80. Dewey KG, Heinig MJ, Nommsen-Rivers LA. Differences in morbidity between breast-fed and formula-fed infants. J Pediatr. 1995;126(5 Pt 1):696–702.

81. Kramer MS et al. Infant growth and health outcomes associated with 3 compared with 6 mo of exclusive breastfeeding. Am J Clin Nutr. Aug 2003;78(2):291–95.

82. Raisler J, Alexander C, O'Campo P. Breastfeeding: A dose-response relationship? Am J Pub Health. 1999;89:25–30.

83. Ivarsson A et al. Epidemic of coeliac disease in Swedish children. Acta Paediatr. 2000;89(2):165–71.

84. Koletzko S et al. Role of infant feeding practices in development of Crohn's disease in childhood. BMJ. 1989;298(6688): 1617–18.

85. Daniels JL et al. Breast-feeding and neuroblastoma, USA and Canada. Cancer Causes Control. Jun 2002;13(5):401–5.

86. Oddy WH et al. Association between breastfeeding and asthma in 6 year old children: findings of a prospective birth cohort study. BMJ. 1999;319(7213):815–19.

87. Scariati PD, Grummer-Strawn LM, Fein SB. A longitudinal analysis of infant morbidity and the extent of breastfeeding in the United States. Pediatrics. 1997;99(6):E5.

88. Dewey KG et al. Growth of breast-fed and formula-fed infants from 0 to 18 months: the DARLING Study. Pediatrics. 1992;89 (6 Pt 1):1035–41.

89. Dewey KG. Is breastfeeding protective against child obesity? J Hum Lact. Feb 2003;19(1):9–18.

90. Grummer-Strawn LM, Mei Z. Does breastfeeding protect against pediatric overweight? Analysis of longitudinal data from the Centers for Disease Control and Prevention Pediatric Nutrition Surveillance System. Pediatrics. Feb 2004;113(2):e81–86.

91. Owen CG et al. The effect of breastfeeding on mean body mass index throughout life: a quantitative review of published and unpublished observational evidence. Am J Clin Nutr. Dec 2005;82(6):1298–1307.

92. Rogan WJ, Gladen BC. Breast-feeding and cognitive development. Early Hum Dev. 1993;31(3):181–93.

93. Evenhouse E, Reilly S. Improved estimates of the benefits of breastfeeding using sibling comparisons to reduce selection bias. Health Serv Res. Dec 2005;40(6 Pt 1): 1781–1802.

94. Rao MR et al. Effect of breastfeeding on cognitive development of infants born small for gestational age. Acta Paediatr. 2002;91(3):267–74.

95. Smith MM et al. Initiation of breastfeeding among mothers of very low birth weight infants. Pediatrics. Jun 2003;111 (6 Pt 1):1337–42.

96. Ors R et al. Comparison of sucrose and human milk on pain response in newborns. Eur J Pediatr. Jan 1999;158(1):63–66.

97. Shah PS, Aliwalas LL, V. S. Breastfeeding or breast milk for procedural pain in neonates. Cochrane Database of Systematic Reviews. 2006(3).

98. Carbajal R, Veerapen S, Couderc S, Jugie M, Villey. Analgesic effect of Breastfeeding in term neonates: randomised controlled trial. BMJ 2003;326(7379):13.

99. Splett PL, Montgomery DL. The economic benefits of breastfeeding an infant in the WIC program twelve month follow-up study. Washington, DC: Food and Consumer Service, US Department of Agriculture; 1998.

100. Ball TM, Wright AL. Health care costs of formula-feeding in the first year of life. Pediatrics. 1999;103(4 Pt 2):870–76.

101. Cohen R, Mrtek MB, Mrtek RG. Comparison of maternal absenteeism and infant illness rates among breast-feeding and formula-feeding women in two corporations. Am J Health Promot. Nov–Dec 1995;10(2): 148–53.

102. Daly SE, Hartmann PE. Infant demand and milk supply. Part 1: Infant demand and milk production in lactating women. J Hum Lact. 1995;11(1):21–26.

103. Dewey KG, Lonnerdal B. Infant self-regulation of breast milk intake. Acta Paediatr Scand. 1986;75(6):893–98.

104. Daly SE, Owens RA, Hartmann PE. The short-term synthesis and infant-regulated removal of milk in lactating women. Exp Physiol. 1993;78(2):209–20.

105. Saint L, Maggiore P, Hartmann PE. Yield and nutrient content of milk in eight women breastfeeding twins and one woman breastfeeding triplets. Br J Nutr. 1986; 56:49–58.

106. Newton M. Human lactation. In: Kon SK, ed. Milk: The Mammary Gland and It's Secretion. New York, NY: Academic Press; 1961.

107. Daly SEJ, Hartmann PE. Infant demand and milk supply. Part 2: The short-term control of milk synthesis in lactating women. J Hum Lactation. 1995;11:27–37.

108. Hill PD, Aldag JC, Chatterton RTJ. Breastfeeding experience and milk weight in lactating mothers pumping for preterm infants. Birth. 1999;26(4).

109. Neifert M, Seacat JM. Practical aspects of breastfeeding the premature infant. Perinatal Neonatol. 1988;12:24.

110. Souto GC et al. The impact of breast reduction surgery on breastfeeding performance. J Hum Lact. Feb 2003; 19(1):43–49; quiz 66–49, 120.

111. Hurst N. Breastfeeding after breast augmentation. J Hum Lact. Feb 2003; 19(1):70–71.

112. Transfer of drugs and other chemicals into human milk. Pediatrics. Sep 2001; 108(3):776–89.

113. Berlin CM, Jr. Silicone breast implants and breast-feeding. Pediatrics. Oct 1994;94 (4 Pt 1):547–49.

114. Semple JL et al. Breast milk contamination and silicone implants: preliminary results using silicon as a proxy measurement for silicone. Plast Reconstr Surg. Aug 1998; 102(2):528–33.

115. Kramer MS, Kauma R. Optimal duration of exclusive breastfeeding. Chochrane Review. 2003;2:1–63.

116. Ziegler EE et al. Cow milk feeding in infancy: further observations on blood loss from the gastrointestinal tract. J Pediatr. 1990;116:11–18.

117. Inch S. Breastfeeding problems. Prevention and Management. Community Practitioner. 2006;79(5):165–67.

118. Morland-Schultz K, Hill PD. Prevention of and therapies for nipple pain: a systematic review. J Obstet Gynecol Neonatal Nurs. Jul–Aug 2005;34(4):428–37.

119. Wooldrige MS, Fischer C. Colic, "overfeeding" and symptoms of lactose malabsorption in the breast-fed baby. Lancet. 1988;2:382–84.

120. Butte NF et al. Feeding patterns of exclusively breast-fed infants during the first four months of life. Early Hum Dev. Dec 1985;12(3):291–300.

121. Brams M, Maloney J. "Nursing bottle caries" in breast-fed children. J Pediatr. 1983;103:415–16.

122. Palmer B. The influence of breastfeeding on the development of the oral cavity: a commentary. J Hum Lact. 1998;14(2):93–98.

123. U.S. Department of Health and Human Services, United States Department of Agriculture. Dietary Guidelines for Americans. Washington, DC; 2005. Home and Garden Bulletin No. 232.

124. Food and Nutrition Board, Institute of Medicine. Dietary Reference Intakes for energy, carbohydrate, fiber, fat, fatty acids, cholesterol, protein and amino acids (macronutrients). Washington, DC: National Academy Press; 2002.

125. Butte NF, King JC. Energy requirements during pregnancy and lactation. Public Health Nutr. Oct 2005;8(7A):1010–27.

126. Goldberg GR et al. Longitudinal assessment of energy expenditure in pregnancy by the doubly labeled water method. Am J Clin Nutr. 1993;57:494–505.

127. Prentice AM et al. Energy balance in pregnancy and lactation. In: Allen LH, King J, Lonnerda B, eds. Nutrient Regulation During Pregnancy, Lactation and Infant Growth. New York, NY: Plenum Press; 1994.

128. Hill DJ et al. Effect of a low-allergen maternal diet on colic among breastfed infants: a randomized, controlled trial. Pediatrics. Nov 2005;116(5):e709–15.

129. Lust KD, Brown JE, Thomas W. Maternal intake of cruciferous vegetables and other foods and colic symptoms in exclusively breast-fed infants. J Am Diet Assoc. 1996; 96(1):46–48.

130. Hill DJ et al. A low allergen diet is a significant intervention in infantile colic: results of a community-based study. J Allergy Clin Immunol. Dec 1995;96(6 Pt 1):886–92.

131. Food and Nutrition Board Institute of Medicine. Dietary reference intakes for thiamin riboflavin, niacin, vitamin B_6, folate vitamin B_{12}, pantothenic aci, biotin, and choline. Washington, DC: National Academy Press; 1998.

132. Food and Nutrition Board Institute of Medicine. Dietary Reference Intakes for Vitamin C, Vitamin E, Selenium, and Carotenoids. Washington, DC: National Academy Press; 2000.

133. Roberts S, Cole T, Coward W. Lactational performance in relation to energy intake in the baboon. Am J Clin Nutr. 1985;41:1270–76.

134. Butte NF, Hopkinson JM. Body composition changes during lactation are highly variable among women. J Nutr. 1998; 128:381S–385.

135. Strode MA, Dewey KG, Lonnerdal B. Effects of short-term caloric restriction on lactational performance of well-nourished women. Acta Paediatr Scand. 1986; 75(2):222–29.

136. Dusdieker LB, Hemingway DL, Stumbo PJ. Is milk production impaired by dieting during lactation? Am J Clin Nutr. 1994; 59(4):833–40.

137. Lovelady CA et al. The effects of dieting on food and nutrient intake of lactating women. J Am Diet Assoc. Jun 2006;106(6): 908–12.

138. Lovelady CA, Lonnerdal B, Dewey KG. Lactation performance of exercising women. Am J Clin Nutr. 1990;52(1):103–9.

139. Dewey KG, McCrory MA. Effects of dieting and physical activity on pregnancy and lactation. Am J Clin Nutr. 1994;59(2 Suppl): 446S–452S; discussion 452S–453S.

140. Lovelady CA, Hunter CP, Geigerman C. Effect of exercise on immunologic factors in breast milk. Pediatrics. Feb 2003;111(2): E148–52.

141. Food and Nutrition Board, Institute of Medicine. Dietary Reference Intakes of water, potassium, sodium, chloride, and sulfate. Washington, DC: National Academy Press; 2004.

142. Oddy WH et al. The association of maternal overweight and obesity with breastfeeding duration. J Pediatr. Aug 2006;149(2): 185–91.

143. Rasmussen KM, Kjolhede CL. Prepregnant overweight and obesity diminish the prolactin response to suckling in the first week postpartum. Pediatrics. May 2004; 113(5):e465–71.

144. Rasmussen KM, Hilson JA, Kjolhede CL. Obesity as a risk factor for failure to initiate and sustain lactation. Adv Exp Med Biol. 2002;503:217–22.

145. Lazarov M, Evans A. Breastfeeding—encouraging the best for low-income women. Zero to Three. 2000(August/September): 15–23.

146. Bryant C et al. A strategy for promoting breastfeeding among economically disadvantaged women and adolescents. NAACOG's Clin Issu Perinat Women's Health Nurs. 1992;3(4):723–730.

147. Bryant C, Roy M. Best Start's Three Step Counseling Strategy. Tampa, FL: Best Start, Inc.; 1997.

148. James DC, Dobson B. Position of the American Dietetic Association: promoting and supporting breastfeeding. J Am Diet Assoc. May 2005;105(5):810–18.

149. Lewin SA et al. Lay health workers in primary and community health care. Cochrane Database of Systematic Reviews. 2005(1).

150. Bonuck KA et al. Randomized, controlled trial of a prenatal and postnatal lactation consultant intervention on duration and intensity of breastfeeding up to 12 months. Pediatrics. Dec 2005; 116(6):1413–26.

151. Bonuck K et al. Breast-feeding promotion interventions: good public health and economic sense. J Perinatol. Jan 2002; 22(1):78–81.

152. Shealy KR et al. The CDC Guide to Breastfeeding Intervention. Atlanta: U.S. Department of Health and Human Services, Centers for Disease Control and Prevention; 2005.

153. Dyson L MF, Renfrew MJ. Interventions for promoting the initiation of breastfeeding. Cochrane Database of Systematic Reviews. 2005(2).

154. McCamman S, Page-Goetz S. Breastfeeding success: you can make the difference. Perinatal Nutr Report. 1998; 4(Winter):2–4.

155. Cadwell K. Reaching the goals of "Healthy People 2000" regarding breastfeeding. Clin Perinatol. 1999;26(2): 527–37.

156. Moreland JC et al. A new teaching model to prolong breastfeeding among Latinos. J Hum Lact. 2000;16(4):337–41.

157. Pisacane A et al. A controlled trial of the father's role in breastfeeding promotion. Pediatrics. Oct 2005;116(4):e494–98.

158. Bentley ME et al. Sources of influence on intention to breastfeed among African-American women at entry to WIC. J Hum Lact. 1999;15(1):27–34.

159. United States Department of Agriculture, Service FaN. Fathers Supporting Breastfeeding. Washington, DC: United States Department of Agriculture; 2002.

160. Barron SP et al. Factors influencing duration of breastfeeding among low-income women. J Am Diet Assoc. 1988; 88(12):1557–61.

161. Freed GL, Fraley JK, Schanler RJ. Attitudes of expectant fathers regarding breastfeeding. Pediatrics. 1992;90(2 Pt 1):224–27.

162. Valaitis RK, Sheeshka JD, O'Brien MF. Do consumer infant feeding publications and products available in physicians' offices protect, promote, and support breastfeeding? J Hum Lact. 1997;13(3):203–8.

163. Howard C et al. Office prenatal formula advertising and its effect on breast-feeding patterns. Obstet Gynecol. 2000;95(2):296–303.

164. Heiser B, Walker M. Selling out Mothers and Babies: Marketing of Breast Milk Substitutes in the USA. Ellicott City, MD: National Alliance for Breastfeeding Promotion; 2003.

165. World Health Organisation International Code of marketing of Breast milk Substitutes. Geneva, Switzerland; 1981.

166. Perez-Escamilla R et al. Infant feeding policies in maternity wards and their effect on breastfeeding success: an analytical overview. Am J Public Health. 1994; 84(1):89–97.

167. World Health Organization. Protecting, promoting, and supporting breastfeeding. The special role of maternity services (a joint WHO/UNICEF statement). Geneva, Switzerland: WHO/UNICEF; 1989.

168. UNICEF/WHO. The State of the World's Children. New York, NY: Oxford University Press; 1998.

169. Merten S, Dratva J, Ackermann-Liebrich U. Do baby-friendly hospitals influence breastfeeding duration on a national level? Pediatrics. Nov 2005;116(5):e702–8.

170. UNICEF/WHO. Innocenti Declaration on the Protection, Promotion, and Support of Breast-Feeding. Forence, Italy: UNICEF and WHO; 1990.

171. Ahluwalia IB, Morrow B, Hsia J. Why do women stop breastfeeding? Findings from the Pregnancy Risk Assessment and Monitoring System. Pediatrics. Dec 2005;116(6):1408–12.

172. Naylor AJ. Baby-Friendly Hospital Initiative. Protecting, promoting, and supporting breastfeeding in the twenty-first century. Pediatr Clin North Am. Apr 2001;48(2):475–83.

173. Division of Labor Statistics UDoL. Women in the Labor Force: A Databook. Washington, DC: US Department of Labor; 2004.

174. U.S. Department of Health and Human Services. Report of the Surgeon General's Workshop on Breastfeeding and Human Lactation. Washington, DC: U.S. Department of Health and Human Services; 1984.

175. Kurinu N et al. Does maternal employment affect breast-feeding? Am J Public Health. 1989;79:1247–50.

176. Meek JY. Breastfeeding in the workplace. Pediatr Clin North Am. Apr 2001;48(2):461–74, xvi.

177. Ortiz J, McGilligan K, Kelly P. Duration of breast milk expression among working mothers enrolled in an employer-sponsored lactation program. Pediatr Nurs. Mar–Apr 2004;30(2):111–19.

178. Healthy Mothers Healthy Babies Coalition. Workplace Models of Excellence 2000: Outstanding programs supporting working women that breastfeed. Alexandria, VA: National Healthy Mothers, Healthy Babies Coalition; 2000.

179. United States Breastfeeding Committee. Workplace Breastfeeding Support. Raleigh, NC 2002.

180. Ryan AS. The resurgence of breastfeeding in the United States. Pediatrics. 1997;99(4):e12.

181. U.S. Department of Health and Human Services. Followup Report: The Surgeon General's Workshop on Breastfeeding and Human Lactation. Washington, DC: US Department of Health and Human Services, Public Health Service, Health Resources and Services Administration, 1985.

182. U.S. Department of Health and Human Services. U.S. Department of Health and Human Services Healthy People 2000: National Health Promotion and Disease Prevention Objectives. Washington, DC: U.S. Department of Health and Human Services, Public Health Service, Office of the Assistant Secretary for Health.

183. Sharbaugh CS. Call to Action: Better Nutrition for Mothers, Children, and Families. Washington, DC: National Center for Education in Maternal and Child Health; 1990.

184. Spisak SG, S.S. Second Followup Report: The Surgeon General's Workshop on Breastfeeding and Human Lactation. Washington, DC: National Center for Education in Maternal and Child Health; 1991.

185. Skisser W, Thomas S. Report of the national breastfeeding policy conference: UCLA Center for Healthier Communities, Families and Children; 1999.

186. United States Breastfeeding Committee. Breastfeeding in the United States: A National Agenda. Rockville, MD: Department of Health and Human Services, Maternal and Child Health Bureau; 2001.

187. Andreason A. Marketing Social Change: Changing Behavior to Promote Health, Social Development, and the Environment. San Francisco: Jossey-Bass; 1995.

188. Carothers C. Social Marketing Institute Success Stories: National WIC breastfeeding promotion project.: Best Start Social Marketing, Inc.; 2001.

189. Khoury A. Social Marketing Institute Success Stories: National WIC breastfeeding promotion project. www.social-marketing.org/success/cs-nationalwic.html: Mississippi WIC Program; 2001.

190. Stremler J, Lovera D. Insight from a breastfeeding peer support pilot program for husbands and fathers of Texas WIC participants. J Hum Lact. Nov 2004;20(4):417–22.

191. Cricco-Lizza R. The milk of human kindness: environmental and human interactions in a WIC clinic that influence infant-feeding decisions of Black women. Qual Health Res. Apr 2005;15(4):525–38.

> "The establishment of breastfeeding for at least 6 months, but optimally for at least a year, as a cultural norm supported by medical, social, and economic practices, is a fundamental cornerstone of true wellness promotion."
> American Dietetic Association, 2001

Chapter 7

Nutrition during Lactation:
Conditions and Interventions

Chapter Outline

- Introduction
- Common Breastfeeding Conditions
- Maternal Medications
- Herbal Remedies
- Alcohol and Other Drugs and Exposures
- Neonatal Jaundice and Kernicterus
- Breastfeeding Multiples
- Infant Allergies
- Near-Term Infants
- Human Milk and Preterm Infants
- Medical Contraindications to Breastfeeding
- Human Milk Collection and Storage
- Model Programs

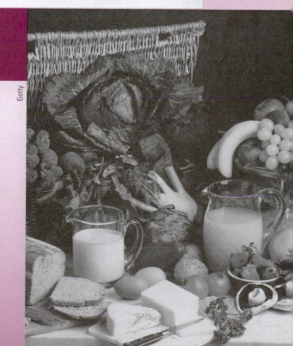

Prepared by **Carolyn Sharbaugh** and **Maureen A. Murtaugh** with **Denise Sofka**

Key Nutrition Concepts

1 Human milk is the preferred feeding for all premature and sick newborns with rare exceptions.

2 Breastfeeding women need consistent, informed, and individualized care in the hospital and at home after discharge.

3 It is usually not necessary to discontinue breastfeeding to manage medical problems of the mother or infant; any medical decision to limit a mother's breastfeeding must be justified by the fact that the risk to her baby clearly outweighs the benefits of breastfeeding.

4 Feeding infants early in the postdelivery period whenever possible is important to successful breastfeeding. Early intervention to address questions or problems is equally important for maintaining breastfeeding.

5 Most medications (including over-the-counter as well as prescription drugs), drugs of abuse, alcohol, nicotine, and herbal remedies taken by nursing mothers are excreted in breast milk.

6 Twins and other multiples can be successfully breastfed without formula supplementation.

Introduction

The key to successful breastfeeding management is for the mother–infant breastfeeding dyad to receive support and informed, consistent, and individualized care from health care professionals both in the hospital and after discharge. The vast majority of women do not experience significant problems with breastfeeding, and many of the more common problems that do arise can be prevented through prenatal breastfeeding education and a positive, supportive breastfeeding initiation period.

This chapter discusses prevention and treatment of common breastfeeding conditions. Issues related to maternal use of medications, herbal remedies, drugs of abuse, and environmental contaminants are addressed. The chapter presents important considerations for breastfeeding multiples, preterm infants, and infants with medical problems. It provides information on the safe collection and storage of human milk and milk banks. The chapter concludes with case studies providing examples of management of challenging breastfeeding problems and with examples of model programs promoting support for breastfeeding in the health care system.

Common Breastfeeding Conditions

Sore Nipples

Early, mild nipple discomfort is common among women initiating breastfeeding. In most women, the discomfort is transient and usually subsides by the end of the first week.[1] Severe nipple pain, the presence of nipple cracks or fissures, pain that persists throughout a feeding, or pain that is not improved by the end of the first week should not be considered normal and requires evaluation.[2] Painful nipples can lead women to become discouraged and lead to the early cessation of breastfeeding.

The best prevention of nipple pain and soreness is proper positioning of the baby on the breast. In order to feed effectively, the baby needs to draw the breast deeply into the mouth, so that the mother's nipple approximates to the junction of the hard and soft palate.[3] This enables the baby to use its tongue smoothly and rhythmically against the undersurface of the breast and remove milk from the ducts. With a good mouthful of breast, the mother's nipple is so far back in the baby's mouth that it is beyond the reach of the compression wave of the tongue, so no pain is caused, and no damage is done. If a woman is experiencing pain, a lactation consultant or a health care professional well trained in lactation should observe the mother nursing her baby. The lactation consultant can determine whether the pain is simply related to early breastfeeding, or if a problem exists.

Common causes of persistent nipple pain include trauma from poor positioning of the infant at the breast, improper release of suction after a feed, infection (thrush or staphyloccus aureus), pumping with too much suction, a problem with the infant's suck, and dermatologic abnormalities.[2] Breast care and cleaning rituals can also contribute to nipple soreness. Proper cleaning of the lactating breast involves only daily washing with warm water. Soaps and other cleansing products can irritate the nipple, and some creams and lotions can cause an allergic reaction and skin irritation. Plastic-backed breast pads used to prevent milk leakage can trap moisture and inhibit air flow to the nipple.[2]

Women can take simple steps to manage nipple pain. Recommended strategies include letting breasts air dry after nursing, rubbing expressed milk or an all-purpose ointment (not petroleum based) on nipples, and using warm compresses on sore nipples.[1,2] The common belief that limiting the frequency or duration of feedings will prevent or heal sore nipples is not substantiated by the literature.[4] Use of a pump to express milk can help to maintain supply if the nipples are so sore that the mother cannot nurse.[3] However, the suction on the breast pump should be adjusted carefully. High suction can make nipples sore and red.

Letdown Failure

"After she latches on, take deep, long breaths—think yoga not Lamaze when you nurse. As you exhale, visualize the milk letting down through your breasts into the baby's mouth."

C. Martin and N. F. Krebs[5]

Letdown failure is not common, but because letdown is necessary to successful breastfeeding, it is important to address the matter. Oxytocin nasal spray can be prescribed by a physician for letdown failure. The synthetic oxytocin is sprayed into the nose and stimulates letdown, but it can only be used for a few days to help women get through a tough time. Other methods should be used at the same time to stimulate letdown. Martin and Krebs recommend a number of techniques to help women relax and enhance letdown:[5]

- Play soothing music that the mother can focus on while nursing.
- Have the partner rub his knuckles down her spine.
- Try different nursing positions.
- Get out of the house. Most babies enjoy a walk.
- Arrange for some time alone (a few hours).
- Switch from caffeinated to decaffeinated beverages and water for a few weeks.

Hyperactive Letdown

Hyperactive letdown can also be a problem, especially among first-time mothers. When letdown is overactive, milk streams from the breast as feeding begins. Milk may also leak from the breast that the infant is not being nursed from. The milk streams quickly, and the infant may be overwhelmed by the volume. The infant may choke or gulp to keep up with the flow. When the infant gulps, the infant may take in air, develop gas pain, and then become fussy.

Management includes removing the infant from the breast when letdown occurs and waiting for the milk flow to slow down before putting the infant back to the breast.[6] The mother can also express the milk until the flow slows. (Expressed milk can be frozen for later use.) Expressing milk also allows infants to get hind milk and prevents gas and colic that may result from a large volume of relatively low-fat milk with high lactose content.

Engorgement

Engorgement occurs when breasts are overfilled with milk, and it is common in first-time mothers. Engorgement occurs when the supply and demand process is not yet established, and the milk is abundant. It also occurs with infrequent or ineffective removal of milk from the breast because of mother–infant separation, a sleepy baby, sore nipples, or improper breastfeeding technique.[2] The best way to prevent engorgement is to nurse the infant frequently. (Newborn infants will often nurse every hour and a half.) If the infant is not available to nurse, expressing milk every few hours will prevent engorgement while helping to build and maintain a milk supply.

The peak time for engorgement varies among women and can occur any time from day 2 through day 14. Engorgement can result in discomfort, difficulty in establishing milk flow, and difficulty in latch on.[6] Severe engorgement inhibits milk flow because the swollen tissue is compressing the milk ducts—not because the mother is

failing to experience the letdown or milk ejection reflex. Once engorgement occurs, there are several simple treatments to help ease the discomfort. It is important for the mother to express milk until her breasts are no longer hard before putting the infant to breast. This will make it more comfortable for her and easier for her baby to latch on. When an infant is unable to extract milk effectively, the use of hand expression or an electric breast pump can help establish milk flow and soften the breast to make it easier for the infant to attach properly and further extract milk. Women can use analgesics to reduce pain from engorgement. A warm shower, warm compresses with massage before feedings, and expressing milk will help to relieve pressure and help trigger milk flow. Application of cold compresses between feedings helps to reduce pain and swelling.[2,7]

Plugged Duct

An obstructed, or "plugged," duct is a localized blockage of milk resulting from milk stasis (milk remaining in the ducts).[2] The mother may feel a painful knot in one breast and usually does not have a fever or other signs of illness. Treatment for plugged ducts is gentle massage, warm compresses, and complete emptying of the breast.[2,6] Women should consider changing nursing positions to facilitate emptying the breast. For example, if the woman is nursing while lying down, she may try a sitting position, or she may switch the position of holding the infant (see Illustration 6.5 in Chapter 6). When plugging occurs repeatedly, a gentle manual massage before nursing often results in the plug being expelled.

Infection

Mastitis is a bacterial infection of the breast most commonly found in breastfeeding women. It occurs in about 1–5% of breastfeeding women,[7] with the highest incidence 2 to 3 weeks after birth, although it may occur at anytime.[2] Some women get mastitis after having cracked or sore nipples, and some get it without any noticeable problem on the surface of the breast, probably from a blood-borne source of infection. Missing a feeding or the infant sleeping through the night may precipitate engorgement, plugged ducts, and then mastitis. Symptoms of mastitis are similar to those of a plugged duct (Table 7.1). In both conditions, there is a painful, enlarged, hard area in the breast, and often an area of redness on the surface of the breast. Cases of mastitis are usually accompanied by a fever and flu-like symptoms.

It is important for the mother to seek early treatment and to continue nursing through mastitis, unless it is too painful. The techniques used to minimize pain from engorgement may also be used for mastitis. Acetaminophen is commonly recommended to help with the pain. The pairing of antibiotics with emptying of the breast to treat bacterial mastitis is important.[2] Adequate rest and fluids are also

Table 7.1 Comparison of symptoms of engorgement, plugged duct, and mastitis

Characteristics	Engorgement	Plugged Duct	Mastitis
Onset	Gradual, immediately postpartum	Gradual, after feedings	Sudden, after 10 days
Site	Both breasts	One breast	Usually one breast
Swelling and Heat	Generalized	May shift, little or no heat	Localized, red, hot, swollen area on breast
Pain	Generalized	Mild, but localized	Intense, but localized
Body Temperature	No fever	No fever	Fever (>101°F)
Other Symptoms	Feels well	Feels well	Flu-like symptoms

SOURCE: Adapted from *Breastfeeding: A Guide for the Medical Profession*, 6th ed., by R. A. Lawrence, copyright 2005, with permission from Elsevier.

important. In a randomized trial, half of 55 women treated only with antibiotics had breast abscess, recurrent mastitis, or symptoms lasting greater than 2 weeks compared to only 2 of 55 who also emptied their breasts (breasts can be emptied by feeding the baby or pumping).[8] Significant delays in seeking treatment for mastitis are associated with the development of abscess and recurrent mastitis[2]. See also Case Study 7.1.

Maternal Medications

"It is equally inappropriate to discontinue breastfeeding when it is not medically necessary to do so as it is to continue breastfeeding while taking contraindicated drugs."

R. A. Lawrence[9]

The single most common medical issue health care providers face in managing breastfeeding patients is maternal medication use.[9] Ninety to 99 percent of breastfeeding women receive some type of medication during their first week postpartum.[10] Most medications—including over-the-counter as well as prescription drugs—taken by nursing mothers are excreted in breast milk, yet data on drug safety may not be readily available to women and health care providers.[11] There is data to suggest that for mothers who discontinue breastfeeding prematurely, concern about the use of medications is a major reason.[10] Recommending that a mother discontinue breastfeeding to take a medication is almost never required and should only be done as a last resort. For most maternal conditions, required drug therapy choices are available that will not cause harm to the infant.[10,12]

Two key questions to address in the analysis of the risk of an infant's exposure to a drug excreted in breast milk are: How much of the drug is excreted in milk, and at the level of excretion, what is the risk of adverse effects?[13,14] Among the numerous variables to examine to answer these questions are:

- The pharmacokinetic properties of the drug
- Time-averaged breast *milk/plasma ratio* of the drug
- The drug *exposure index*
- The infant's ability to absorb, detoxify, and excrete the agent
- The dose, strength, and duration of dosing
- The infant's age, feeding pattern, total diet, and health[9,13]

Additional considerations are the well-established interethnic and racial differences in drug responsiveness, exposure of the infant to the drug during pregnancy, whether the drug can be safely given to the infant directly,[9] and the relative infant dose.[10] The ultimate test of drug safety is the measurement of the infant plasma drug concentration and any pharmacodynamic effects on the infant.[13] With so many active variables, carefully controlled studies on large enough samples to validate the results are rare but have increased during the last decade.

Fortunately, numerous resources (Table 7.2) based on a thorough evaluation of available evidence can assist the health care provider and mother in identifying which drugs are safe and which are not. The American Academy of Pediatrics (AAP) Committee on Drugs publishes guidelines for practitioners.[15,16] The guidelines provide a list of drugs divided into the following seven categories according to risk factors in relationship to breastfeeding:

1. Cytotoxic drugs that may interfere with the cellular metabolism of the nursing infant
2. Drugs of abuse for which adverse effects on the infant during breastfeeding have been reported

Milk to Plasma Drug Concentration Ratio (M/P Ratio)   The ratio of the concentration of drug in milk to the concentration of drug in maternal plasma.[10] Since the ratio varies over time, a time-averaged ratio provides more meaningful information than data obtained at a single time point. It is helpful in understanding the mechanisms of drug transfer and should not be viewed as a predictor of risk to the infant as it is the concentration of the drug in milk, and not the M/P ratio, that is critical to the calculation of infant dose and assessment of risk.[10,14,15]

Exposure Index   The average infant milk intake per kilogram body weight per day × (the milk to plasma ratio divided by the rate of drug clearance) × 100 . It is indicative of the amount of the drug in the breast milk that the infant ingests and is expressed as a percentage of the therapeutic (or equivalent) dose for the infant.[14]

Case Study 7.1

Chronic Mastitis

Photo Disc

This was the first and unremarkable pregnancy for 29-year-old Barbara Ann. Barbara Ann has reported experiencing "a little" breast enlargement during her pregnancy.

Her infant is first put to the breast at 2 hours postpartum, and the infant latches well and suckles vigorously. The infant nurses every 2 hours over the first 3 to 4 days postpartum. Barbara Ann's breasts became noticeably fuller during the third postpartum day, and by the fourth postpartum day they are painfully engorged. In addition, Barbara Ann reports painful, burning, cracked nipples. The engorgement makes it difficult for her baby to latch at the breast. The baby becomes irritable, and Barbara Ann experiences a significant amount of pain. A lactation consultant gives Barbara Ann guidelines for engorgement management.

On day 5, the engorgement is still causing discomfort. Barbara Ann's nipples have become more cracked and painful. The lactation consultant notes that the infant's latch has become shallow and tight, probably in an attempt to control the flow of milk. However, the infant shows all the signs of adequate intake, including 10 very wet and 3 soiled diapers during the 24 hours prior to the consultation.

By day 7 postpartum, Barbara Ann has mastitis. She is treated with a 7-day course of dicloxacillin. A lactation consultant assists her in achieving a proper infant latch.

By day 14, Barbara Ann is feeling much better. The mastitis has resolved, and her nipples are healing. She still has tenderness during infant feedings and a healing crack on the right side. Her breasts are still uncomfortably full and are occasionally swollen and tender.

At 3 weeks postpartum, Barbara Ann develops an inflamed area on the right breast that remains red and tender despite applying warmth and massage to the area. The lactation consultant helps Barbara Ann to position the infant in a way that allows drainage of the inflamed area. Barbara Ann is treated with dicloxacillin. The crack on the right nipple has improved, but is still not completely healed. Barbara Ann continues to show signs of oversupply, such as breasts feeling uncomfortably full, even after feeding, and excessive milk leakage between feedings.

After 10 days of persistent burning pain in the nipple area, Barbara Ann is treated with fluconazole for a yeast infection. Seven days after starting the fluconazole, a topical nystatin ointment is prescribed for her nipples and an oral suspension for her infant.

At 7 weeks postpartum, Barbara Ann calls the lactation consultant to report another flare-up of bacterial mastitis. Her health care provider prescribes a 10-day course of dicloxacillan. Barbara Ann is still treating her nipples with nystatin ointment. At 8 weeks postpartum her mastitis resolved; her nipple pain is still present, but improving. Barbara Ann is nursing the infant on one side only per feeding and reports that the infant latches better when she is in a more reclined position.

SOURCE: Adapted from: Anonymous. Case management of a breastfeeding mother with persistent oversupply and recurrent breast infections. J Hum Lact, 2000, 16:221–5.

Questions

1. Name the causes of engorgement.
2. Name at least two recommendations the lactation consultant might make to decrease engorgement.
3. What measures (other than using the medications prescribed by her physician) should Barbara Ann take to manage the mastitis?
4. How might Barbara Ann decrease the flow rate of her milk?

Table 7.2 Resources on drugs, medications, and contaminants in human milk

- American Academy of Pediatrics, Committee on Drugs. The Transfer of Drugs and Other Chemicals Into Human Milk (RE9403). Pediatrics 108:776, 2001.[15] Website: www.aap.org. http://aappolicy.aappublications.org/cgi/content/full/pediatrics;108/3/776
- Briggs GG, Freeman RK, Yaffe SJ. 2001 *Drugs in Pregnancy and Lactation* (6th ed.). Baltimore, MD: Williams and Wilkins.[21]
- Hale TW. 2004. *Medications and Mothers' Milk* (11th ed.). Amarillo, TX: Pharmasoft Medical Publishing.[22] Dr. Hale will answer questions from health professionals posed to the Pharmasoft website: www.iBreastfeeding.com.
- *Breastfeeding and Maternal Medication: Recommendations for Drugs in the Eleventh WHO Model List of Essential Drugs*, 2003.[23] UNICEF World Health Organization. Website: www.who.int/child-adolescent-health/New_Publications/ NUTRITION/BF_Maternal_Medication.pdf
- Blumenthal M, Busse W, Goldberg A, et al. (Eds). *The Complete German Commission E Monographs: Therapeutic Guide to Herbal Medicines.* Boston: Integrative Medicine Communications, 1998.[24]
- The Breastfeeding and Human Lactation Center, University of Rochester. This service is available for complex medication questions (9:30 am to 4 pm EST, Monday to Friday at 585–275–0088).
- HerbMed—www.herbmed.org an interactive, electronic herbal database with links to scientific publications.
- TOXNET—National Library of Medicine Drug and Lactation database containing summaries of published literature on the effects of over 400 drugs on lactation.
- REPROTOX—An online proprietary reproductive toxicology database, http://reprotox.org.

3. Radioactive compounds that require temporary cessation of breastfeeding

4. Drugs for which the effect on nursing infants is unknown, but may be of concern

5. Drugs that have been associated with significant effect on some nursing infants and should be given to nursing mothers with caution

6. Maternal medications usually compatible with breastfeeding

7. Food and environmental agents having no effect on breastfeeding

The list, which is updated periodically, includes only those drugs about which there is published information. Other useful monographs and review articles provide additional information on a wide array of medications.[10–13,17]

The Breastfeeding and Human Lactation Study Center at the University of Rochester (see Table 7.2) continually updates its database of more than 3000 references on drugs, medications, and contaminants in human milk and is a resource for complex questions on the risks to the breastfed infant. The *Physician's Desk Reference (PDR)* is not a good source for information about drugs and breastfeeding because the information is derived directly from pharmaceutical companies whose first concern is avoiding liability. When there are no studies that prove beyond a doubt that a drug is safe for nursing mothers, the drug companies must advise against use while breastfeeding—even if what is known about the drug suggests that there is little cause for concern.

Only a few medications are thought to be contraindications to breastfeeding: antineoplastic agents, radioactive isotopes, drugs of abuse, and drugs that suppress lactation.[9,10,15] Fortunately, safer alternative medications can

be recommended as a substitute for most other drugs with known adverse effects on infants. Specific knowledge about a medication's safety during breastfeeding will allow proper treatment and avoid unnecessary maternal anxiety and undue risk.

Many women have questions on the safety of oral contraceptive use during lactation. There is currently no evidence of harm but few women have been studied. There is evidence that suggests combined oral contraceptives may reduce the volume of breast milk.[10,18] The American College of Obstetricians and Gynecologists (ACOG) and the World Health Organization recommend against using combined oral contraceptives in the first 6 weeks postpartum. If lactation is well established at 6 weeks, ACOG recommends monitoring the infant's nutritional status if combined oral contraceptives are initiated.[18] WHO does not recommend using combined oral contraceptives from 6 weeks to 6 months unless other forms of contraceptives are not available. The La Leche League International recommends avoiding combined oral contraceptives in breastfeeding because other forms on contraception are available.[18] Progestin-only oral contraceptives and implants are safe and effective during lactation.[10,19] Implants that deliver orally active steroids should only be used after 6 weeks postpartum to avoid transferring of steroids to the newborn.[19]

If a drug or surgery is elective, a mother may be able to delay it until the baby is weaned. If a breastfeeding mother needs a specific medication, and the hazards to the infant are minimal, she should be instructed to take the medication after breastfeeding, at the lowest effective dose, and for the shortest duration.[9,20] Other important steps can be taken to further minimize the effects (Table 7.3). It is also sometimes possible to choose alternative routes for administration of a medication to reduce exposure. For example,

Table 7.3 Minimizing the effect of maternal medication[9]

1. Avoid long-acting forms: Accumulation in the infant is a genuine concern because the infant may have more difficulty excreting a long-acting form of a drug, which usually requires detoxification in the liver.

2. Schedule doses carefully: Check usual absorption rates and peak blood levels of the drug, and schedule the doses so that the least amount possible gets into the milk. In order to minimize milk levels of most drugs, the safest time for a mother to take the drug is usually immediately after her infant nurses.

3. Evaluate the infant: Watch for any unusual signs or symptoms, such as changes in feeding pattern or sleeping habits, fussiness, or rash.

4. Choose the drug that produces the least amount in the milk.

Table 7.4 Herbs traditionally used to affect milk production[* 25,26,41]

Herbs to Promote Milk Flow

Anise
Astralagus
Milk thistle
Caraway
Celery root and seed
Chaste tree berry (Chasteberry) or monk's pepper
Fennel
Fenugreek
Goat's rue
Hollyhoc
Hibiscus flower
Lemongrass
Marshmallow
Stinging nettle
Raspberry
Rauwolfia
Verbena

Herbs that Reduce Milk Flow

Castor bean
Jasmine flower
Sage

*Many traditional galactologues are not currently considered appropriate during lactation (see Table 7.5).

prescribing an inhalant instead of a drug taken by mouth, or a topical application rather than oral dosing, reduces infant exposure. If a drug is to be taken for diagnostic testing (such as a radioactive agent) a mother may need to withhold breastfeeding for a short period of time, pumping and discarding her milk. Discontinuing breastfeeding due to maternal medications is a last resort but may be necessary for the health and well-being of the mother, for example if she needs chemotherapy or radioactive treatment. Any decision to limit a mother's breastfeeding must be justified by the fact that the risk to her baby clearly outweighs the benefits of breastfeeding.

Herbal Remedies

Numerous herbs have been used in folk and traditional systems of healing to affect the flow of milk (Table 7.4), or to treat mastitis, infant colic, and thrush.[24–27] However, scientific information about herb use during lactation, particularly recent studies, is sparse. The limited pertinent safety data are based on traditional use, animal studies, and knowledge of the pharmacologic activities of the products' constituents. Medicinal herbs should be viewed as drugs, with evaluation of both their pharmacological and toxicological potential.[21,28]

A mother may perceive herbs as natural and, therefore, safe and even preferable to conventional beverages, over-the-counter medicines, or prescription drugs. However, the risks of using some herbal remedies may outweigh the potential benefits. Many herbs are far from benign, and many are contraindicated during lactation (Table 7.5). Because little is known about the amount secreted in human milk or the effects on preterm or term infants, herbs that are central nervous system stimulants, cathartic laxatives, hepatotoxic, carcinogenic, cytotoxic,

or mutogenic, or that contain potentially toxic essential oils, are not recommended during lactation.[25,29]

The toxic effects of herbs are often not the fault of the herb itself, but are caused by products containing misidentified plants or contaminants such as heavy metals, synthetic drugs, microbial toxins, and toxic botanicals.[30] *Medicinal herbs* in the United States are regulated as dietary supplements and are not tested for safety or efficacy.

Medicinal Herbs Plants used to prevent or remedy illness.[27]

Herbal teas that are safe for the infant and mother during lactation are presented in Table 7.6. Lawrence[9] recommends using only herbal teas "that are prepared carefully, using only herbs for essence (e.g., Celestial Seasonings brand tea) and avoiding heavy doses of herbs with active principles." Careful attention should also be given to preparation, avoiding long steeping times.

Some culinary herbs may lead to problems when used extensively. In lactation and herbal texts, sage has a folk reputation for lowering milk supply,[25] as do parsley and peppermint, especially if the oil is taken by mouth in large doses.[31] Consumed on occasion, however, in small amounts as part of a reasonably varied diet, peppermint, parsley, sage, and other culinary herbs currently have no documented negative effect on lactation.

Although various herbal gels, ointments, or creams are often suggested for use on the nipples, any substance applied to the breast or nipples could easily be ingested by the nursing child. The use of herbal oils is

Table 7.5 Medicinal herbs considered not appropriate for use during pregnancy or lactation[29,35,41,42]

Agnus castus	Ephedra (ma huang)	Motherwort
Alkanet	Eucalyptus	Myrrh
Aloes	Eupatorium	Nettle
Angelica	Euphorbia	Osha
Apricot kernel	Fennel	Passionflower
Aristolchia	Feverfew	Pennyroyal
Asafoetida	Foxglove	Petasites
Avens	Frangula	Plantain
Basil	Fucus	Pleurisy root
Bladderwrack	Gentian	Pokeroot
Blue flag	German chamomile	Poplar
Bogbean	Germander	Prickly ash
Boldo	Ginkgo biloba	Pulsatilla
Bonese	Ginseng, eleuthero	Queen's delight
Borage	Ginseng, panax	Ragwort
Broom	Golden seal	Red clover
Buchu	Ground ivy	Roman chaparral
Buckthorn	Groundsel	Sassafras
Bugleweed	Guarana	Senna
Burdock	Hawthorne	Shepherd's purse
Calamus	Heliotropium	Skullcap
Calendula	Hops	Skunk cabbage
Cascara	Horehound, black	Squill
Cayenne pepper	Horehound, white	St. John's wort
Chamomile	Horsetail	Stephania
Chasteberry	Hydrocotlye	Stillingia
Chinese rhubarb	Jamaica dogwood	Tansy
Cohosh, black	Joe-pye weed	Tonka bean
Cohosh, blue	Juniper	Uva-ursi
Coltsfoot	Kava kava	Valerian root
Comfrey	Licorice	Vervain
Cornsilk	Liferoot	Wild carrot
Cottonroot	Lobelia	Willow
Crotalaria	Male fern	Wormwood
Darniana	Mandrake	Yarrow
Devil's claw	Mate	Yellow dock
Dogbane	Meadowsweet	Yohimbine
Dong quai	Melilot	
Echinacea	Mistletoe	

*Exclusion from this list should not be a recommendation for safety.

not recommended.[31] In one infant, severe breathing difficulties were documented after the mother used menthol, a significant component of peppermint oil, on her nipples.[32]

Although health care practitioners may wish otherwise, some mothers may refuse prescription drugs and insist on using herbal alternatives. If a mother is consuming a large amount of any herbal product, its contents should be checked. Important information to obtain on the product includes its name, a list of all ingredients, the names of the plants or other components (include the plants' Latin names if possible), details of the preparation, and the amount consumed.[31] Reliable sources of herbal information[23,24,33,34] (Table 7.2) or the regional poison control center may be able to identify potentially harmful pharmacological and toxicological ingredients.

Table 7.6 Herbal teas considered safe during lactation[9]

Tea	Origin/Use
Chicory	Root/caffeine-free coffee substitute
Orange spice	Mixture/flavoring
Peppermint	Leaves/flavoring
Raspberry	Fruit/flavoring
Red bush tea	Leaves, fine twigs/beverage
Rose hips	Fruits/vitamin C

SOURCE: Reprinted from *Breastfeeding: A Guide for the Medical Profession*, 6th ed., by R. A. Lawrence and R. M. Lawrence, copyright 2005, with permission from Elsevier.

In balancing the risks and benefits in a given situation, consideration should be given to the benefits of continued breastfeeding to the baby and the mother. It is also important to consider the varied nature of lactation: newborns face different risks than older babies or toddlers because of immaturity; infants consume varying amounts of human milk; mothers may be looking forward to many months of lactation yet need or desire the benefits of medicinal herbs. A few of the widely used herbs in the United States are discussed below.

Specific Herbs Used in the United States

Echinacea   Echinacea is used for the common cold and to enhance the immune system. Insufficient reliable data is available on its entry into breast milk and effects on the infant. It is available in many forms; the tincture form contains 15–90% alcohol.[29] Gastrointestinal distress in some women has been reported. Consumption of echinacea during lactation is not recommended.[35]

Ginseng Root   Ginseng root, widely believed to increase capacity for mental work and physical activity and to reduce stress, contains dozens of steroid-like glycosides, sterols, coumarins, flavonoids, and polysaccharides.[9] It is reported to have estrogen-like effects in some women, with mastalgia common with extended use and mammary nodularity also reported. While there has been considerable animal experimentation with ginseng root, human data is not extensive. There is no information on transfer to human milk.[29] The lack of standardized preparations, information on dosage, and accurate recording of side effects is a problem. Because of the reported breast effects and occasional reports of vaginal bleeding, the use of ginseng during lactation may not be advisable.[4,9]

St. John's Wort   St. John's wort, widely used in the United States and Europe as a mood stabilizer and antidepressant, has the potential to suppress lactation. Although the plant contains at least 10 classes of biologically active compounds, only hypericin and hyperforin, and their metabolites pseudohypericin and adhyperforin, seem to

be the most important for their neuropharmocologic properties.[36] There is growing evidence that hyperforin may be the key constituent responsible for the antidepressant property of this herb. Composition of St. John's wort preparations may vary based on climate and other vairables related to growth, harvesting, and processing. Large doses decrease prolactin levels, which could potentially reduce breast milk supply.[9] There are also concerns about the numerous potential herb–drug interactions because of the ability of St. John's wort to induce the metabolic activity of cytochrome P450 (CYP), particularly drugs used to treat human immunodeficiency virus (HIV).[36]

There is limited information on the excretion of active components in breast milk and effect on the infant. In a study of 5 mothers who were taking 300 mg of St. John's wort 3 times daily, hyperforin was excreted into breast milk at very low levels and the relative infant doses of 0.9–2.5%. This level of infant exposure to hyperforin through milk is comparable to levels reported in most studies assessing antidepressants or neuroleptics. No side effects were seen in the mothers or infants.[36] A recent clinical trial compared 33 breastfeeding women taking St. John's wort with 101 disease-matched controls and 33 age- and parity-matched controls. No differences between groups were found in maternal adverse side effects, maternal report of decreased milk production, or in infant weight in the first year. In the group taking the St. John's wort, 5 of the infants reported either colic, drowsiness, or lethargy, compared with only 1 infant in each of the control groups.[37] The symptoms were not severe, and specific medical treatment of the affected infants was not required. Until more long-term studies of outcome in infants are available, antidepressants with an established safety profile during breastfeeding may be preferable alternatives. Mothers who are breastfeeding while taking St. John's wort should be alert to changes in infant behavior and should be closely supervised by a pediatrician. If the mother is taking other medications or herbal supplements, she should be informed that drug–herb interactions are possible.

Ephedra (Ma Huang)   This nervous system stimulant is typically found in herbal diet aids and is contraindicated during lactation as it may accelerate heart rate, raise blood pressure, and cause infant colic.[35]

Fenugreek   This spice is used as an artificial flavor for maple syrup, in teas, poultices and ointments, and as an ingredient in East Indian cooking. It is also the most commonly used herbal galactogogue (milk production stimulate). While there is limited scientific evidence to back this claim, there are anecdotal reports of its successful use to increase milk supply. In one account of 1200 women taking fenugreek, almost all reported an increase in milk production within 24 to 72 hours.[38] Rare maternal adverse effects include diarrhea; a maple-like aroma in urine, breast milk, or sweat; and exacerbation of asthmatic symptoms.[39] This herb is derived from a plant in the same family as peanuts and chickpeas and has potential for allergy in sensitive infants. There are reports of colic, abdominal upset, and diarrhea among babies whose mothers took fenugreek.[29] Transfer into milk is assumed because the infant's urine smells of maple syrup. The usual dosage (2 to 3 capsules three times daily or one cup of tea three times a day) is considered compatible with breastfeeding.[39]

Cabbage Leaves Many believe that cabbage leaves (either cool or at room temperature) have been reported to reduce discomfort and swelling, although it is not known how the effects are mediated. In recent randomized trials cabbage leaves and gel packs were equally effective in the treatment of engorement, as were cabbage extract and a placebo cream.[40]

Alcohol and Other Drugs and Exposures

> "Avoid prescribing or proscribing it [alcohol] and . . . assist the mother in appropriately adjusting her alcohol consumption in both timing and volume."
>
> R. A. Lawrence[9]

Alcohol

The harmful effects of alcohol consumption during pregnancy are well documented, and drinking during pregnancy is clearly not recommended. Recommendations on alcohol consumption during lactation, however, are less clearcut and are controversial. Alcohol consumed by the mother passes quickly into her breast milk, and the effects on the breastfeeding baby are directly related to the amount the mother consumes.[43,44]

The level of alcohol in breast milk matches the maternal plasma levels at the time of the infant feeding. Peak maternal plasma and breast milk levels are reached 30 to 60 minutes after alcohol consumption and at 60 to 90 minutes when taken with food.[9] As the alcohol clears from a mother's blood, it clears from her milk. It takes a 120-pound woman about 2 to 3 hours to eliminate from her body the alcohol in one serving of beer or wine (Table 7.7).[45] The common practice of pumping the breasts and then discarding the milk immediately after drinking alcohol does not hasten the disappearance of alcohol from the milk as the newly produced milk still will contain alcohol as long as the mother has measurable blood alcohol levels.[43]

In many cultures folklore passed down for generations encourages the use of alcohol as a galactologue that facilitates milk letdown and rectifies milk insufficiency as well as sedating and calming the "fussy" infant.[46] In contrast to

Table 7.7 Alcohol and breastfeeding: Time (h:min) until zero level in milk is reached for women at different body weights[45]

Maternal Body Weight		Time Alcohol Remains		
lb	(kg)	1 Drink	2 Drinks	3 Drinks
100	(45.4)	2:42	5:25	8:08
120	(54.4)	2:30	5:00	7:30
140	(63.5)	2:19	4:38	6:58
160	(72.6)	2:10	4:20	6:30
180	(81.6)	2:01	4:03	6:05

NOTE: Time is calculated from beginning of drinking. Assumptions made: alcohol metabolism is constant at 15 mg/dl; height of the women is 162.56 cm (5 feet, 4 inches). 1 drink = 12 oz of 5% beer or 5 oz of 11% wine or 1.5 oz of 40% liquor. Example: For a 100-lb woman who consumed 2 drinks in 1 h, it would take 5 h 25 min for there to be no alcohol in her breast milk, but for a 180-lb woman drinking the same amount, it would take 4 h 3 min.
SOURCE: Adapted from E. Ho and A. Collantes et al., Alcohol and breastfeeding: calculation of time to zero level in milk. Biol Neonate 2001;80:219–22.

this folkore, there is now strong evidence of a negative dose-related impact of alcohol on milk supply and the milk letdown reflex. In 2005 Mennella[46] found that during the immediate hours after alcohol consumption by lactating women, the hormonal milieu underlying lactation performance is disrupted. Oxytocin levels significantly decrease, whereas prolactin levels increase significantly. The diminished oxytocin response was significantly related to decreases in milk yield and milk ejection (letdown). In contrast, changes in prolactin were related to self-reported feelings of drunkenness. Recommending alcohol to women as an aid to lactation may be counterproductive; mothers may feel more relaxed, but the hormonal disruption diminished the infant's milk supply.

Maternal alcohol consumption affects the odor of breast milk and the volume consumed by the infant. Breastfed infants consumed, on average, 20% less breast milk during the 3 to 4 hours following their mothers' consumption of an alcoholic beverage (0.3 g ethanol/kg).[43,47] Compensatory increases in intake were then observed during the 8 to 16 hours after exposure when mothers refrained from drinking.[47]

Recent studies on the impact of maternal alcohol ingestion during lactation on infant sleep patterns and psychomotor development have raised concerns about regular consumption of alcohol while lactating. In one study, 11 of 13 breastfed infants had a reduction of more than 40% in active sleep after consuming their mothers' expressed breast milk flavored with alcohol (32 mg) on one testing day and expressed breast milk alone on the other.[48] All infants spent significantly less total time sleeping after consumption of the breast milk with alcohol (56.8 minutes with alcohol compared to 78.2 minutes without). A follow-up study replicated this finding and showed that infants can compensate for this deficit by increasing the amount of time spent in active (rapid eye movement) sleep during the 20.5 hours after the sleep deficit period.[49] The investigators concluded that short-term exposure to small amounts of alcohol in breast milk produces distinctive changes in the infant's sleep–wake patterning. Since both the observed reduced milk consumption and the infant sleep deficits occur only when breastfeeding follows shortly after the mother's alcohol consumption, a nursing woman who drinks occasionally can limit her infant's exposure to alcohol by timing breastfeeding in relation to her drinking.[50]

Epidemiologic data on the effects of moderate drinking throughout lactation period on the human infant are limited. In a study of 400 infants born to members of a health insurance plan, no differences in the infant's cognitive development scores were found at 1 year of age between infants whose mothers consumed alcohol while nursing and those that did not. However, Bayley Psychomotor Development Index scores at 1 year of age were slightly lower among infants who were exposed to alcohol through breast milk than among those who were not exposed.[51] For example, if a mother consumed two drinks daily while nursing, the score decreased by nearly 0.5 SD. The association between maternal drinking and delayed motor development persisted even after the investigators controlled for more than 100 potentially confounding variables, such as maternal tobacco, marijuana, and heavy caffeine use during pregnancy and the first 3 months after delivery.[52] However, the investigators were not able to replicate the motor development findings in a later study that assessed a larger cohort of infants at 18 months with the Griffiths Scales of Mental Development.[53]

Since current research does not show that occasional use (one to two drinks) of alcohol is harmful to the baby, La Leche League continues to support the opinion that the occasional use of alcohol in limited amounts is compatible with breastfeeding.[54] The American Academy of Pediatrics places alcohol in the category "Maternal Medication Usually Compatible with Breastfeeding."[15,16] It lists possible side effects if consumed in large amounts, including drowsiness, deep sleep, weakness, and abnormal weight gain in the infant. The Institute of Medicine Subcommittee on Nutrition During Lactation recommends that lactating women should be advised that if alcohol is consumed, intake should be limited to "no more than 0.5 grams of alcohol per kilogram of maternal body weight per day."[55] For a 60-kilogram (132 pound) woman, 0.5 grams of alcohol per kilogram of body weight corresponds to approximately 2 to 2.5 ounces of liquor, 8 ounces of table wine, or 2 cans of beer.[56] There is concern that prohibiting alcohol may be too restrictive, especially when the research does not support any serious impact on the baby when a mother has an occasional drink. Many feel that nursing mothers are already placed under too many restrictions and may be discouraged from initiating or continuing

to breastfeed because they feel they will face too many limitations.

If a mother does choose to have a drink or two, she can wait for the alcohol to clear her system before nursing according to the times given in Table 7.7. She can plan ahead and have alcohol-free expressed milk stored for the occasion. If she becomes engorged, she can pump her breasts as a means of comfort, and discard her alcohol-containing milk. Drinking water, resting, or pumping and discarding breast milk will not hasten the removal of alcohol from the milk, as the alcohol content of milk matches the maternal plasma alcohol levels. Mothers who are intoxicated should not breastfeed until they are completely sober.

Nicotine (Smoking Cigarettes)

Regardless of feeding choice (breast or bottle), the health risks for infants posed by having a mother who smokes are many, including otitis media, exacerbations of asthma, respiratory infections, and gastrointestinal dysregulation such as colic and acid reflux.[57] It is not ideal to smoke and breastfeed, but it is worse to smoke and not breastfeed. Well-documented data provide clear evidence that children of smoking mothers do better if breastfed in regard to general health, respiratory illness, and risk of sudden infant death syndrome (SIDS)[9] than if bottle-fed. Unfortunately, women who smoke cigarettes are less likely to breastfeed than nonsmokers, are less likely to seek help with breastfeeding difficulties than nonsmokers, and are at increased risk for stopping breastfeeding by 3 months.[58] While lower milk output has been reported among smoking mothers,[58,59] it is unknown which components of cigarette smoke are responsible for the reduced milk production, and several studies provide evidence that smoking does not necessarily hinder breastfeeding.[58] There are also changes in the odor and flavor of the breast milk that might affect the breastfed infant.[59] Substantial epidemiological evidence suggests that social and behavioral factors and not physiological factors are largely responsible for the lower rates of breastfeeding found among smokers.[58] Some women believe that smoking is a barrier to breastfeeding; they do not believe they could, or should adhere to the kinds of healthy practices they think are required of mothers to breastfeed.

Nicotine levels in breast milk of women who smoke are between 1.5 and 3.0 times higher than the level in the mother's blood,[15,58] and the mean 24-hour nicotine concentrations in breast milk rise as cigarette consumption increases. There is no evidence to document whether this amount of nicotine presents a health risk to the nursing infant, and because breastfeeding and smoking may be less detrimental to the child than bottle-feeding and smoking, the AAP Committee on Drugs removed nicotine (and thus smoking) from its 2001 list of drugs of abuse with adverse effects on the infant during breastfeeding.[15]

Dahlstrom et al.[60] estimated that the dose of nicotine in breastfeeding infants was 1 μg per kilogram per feeding, based on data on nicotine concentrations in breast milk within 30 minutes after smoking. Women who smoke 10 to 20 cigarettes per day have 0.4 to 0.5 mg of nicotine/L in their milk. The total daily systematic infant dose from breast milk is estimated at less than 6 or 10 μg/kg/day or 50 times less than the exposure of a 70 kg adult smoking 20 cigarettes per day or using a 21 mg nicotine patch.[61] With gradual intake over a day's time, the infant can metabolize nicotine in the liver and excrete the chemical in the kidney. Numerous studies of nicotine and cotinine concentrations in the nursing mother and her infant confirm that although bottle-fed infants born to smoking mothers and raised in a smoking environment have significant levels of nicotine and metabolites in their urine, breastfed infants have higher levels.[59]

Tobacco smoking also increases the exposure of infants to organochloride pesticides, PCBs, and hexachlorobenzene through breast milk and secondhand smoke.[62] Women should be counseled not to smoke while nursing or in the infant's presence. Mothers who are not willing to stop smoking should cut down, consider low-nicotine cigarettes, and delay feedings as long as possible after smoking. The half-life of nicotine is 95 minutes.

When used as directed, smoking cessation aids that replace nicotine do not appear to pose any more problems for the breastfeeding infant than maternal smoking does.[58,63] Since transdermal nicotine (nicotine patch) provides a steady level of nicotine in plasma and thus in breast milk, the mother cannot control the level of nicotine in breast milk except by changing the strength of the patch. As the mother progresses through to lower patch strengths during smoking cessation therapy, the transfer of nicotine to the infant via milk decreases as much as 70 percent.[58] Mothers who use nicotine replacement therapy intermittently (gum, nasal spray, or inhalation) might minimize the nicotine in their milk by prolonging the duration between nicotine administration and breastfeeding.[61]

Marijuana

Delta-9-tetrhydrocannabinol (THC), an active ingredient in marijuana, transfers and concentrates in breast milk and is absorbed and metabolized by the nursing infant. There is evidence from animal studies of structural changes in the brain cells of newborn animals nursed by mothers whose milk contained THC. Impairment of deoxyribonucleic acid (DNA) and ribonucleic acid (RNA) formation and neurotransmitter systems essential for proper growth and development has been described.[64] In one study following breastfeeding mothers and their infants for 12 months, marijuana exposure via the mother's milk during the first month postpartum appeared to be associated with a decrease in infant motor development at 1 year of age.[65] Concerns about marijuana use during lactation include the amount of the drug the infant ingests while nursing and the amount inhaled from the environment. The possible effect on DNA and RNA metabolism should discourage any maternal use, especially since brain cell development is still taking place in the first months of life. The American Academy of Pediatrics classifies THC as a drug of abuse that is contraindicated during lactation.[15]

Caffeine

Although caffeine ingestion is a frequent concern of breastfeeding mothers, moderate intake causes no problems for most breastfeeding mothers and babies. A dose of caffeine equivalent to a cup of coffee results in breast milk levels of 1% of the level in maternal plasma and, consequently, low levels in the infant.[9] However, because the infant's ability to metabolize caffeine does not fully develop until 3 to 4 months of age, caffeine does accumulate in the infant. Cases of caffeine excess in breastfed infants have been documented.[66] Symptoms, which include infants being wakeful, hyperactive, and fussy, did not require hospitalization and disappeared over a week's time after caffeine was removed from the maternal diet. No long-term effects of caffeine exposure during lactation have been documented.[67]

While most breastfed infants can tolerate a maternal caffeine intake equivalent to five or fewer 5-ounce cups of coffee per day, or less than 750 ml per day, some babies may be more sensitive than others. If a mother suspects her baby is reacting to caffeine, she may try avoiding caffeine from all sources (coffee, tea, soft drinks, over-the-counter medications, chocolate) for 2 to 3 weeks.[54]

Other Drugs of Abuse

Amphetamines, cocaine, heroin, and phencyclidine hydrochloride (angel dust, PCP) are classified by the AAP Committee on Drugs[15] as drugs of abuse that are contraindicated during lactation. The AAP guidelines strongly state that these compounds and all other drugs of abuse are hazardous not only to the nursing infant, but also to the mother's physical and emotional health. In addition to their adverse pharmacological effects on the mother and infant, street drugs lack standardization and may be contaminated with other active ingredients, bacteria, heavy metals, or pesticides.[56]

Environmental Exposures

"The advantages of breastfeeding far outweigh the potential risks from environmental pollutants. Taking into account breastfeeding's short- and long-term health benefits for infants and mothers, the World Health Organization (WHO) recommends breastfeeding in all but extreme circumstances."

World Health Organization[68]

There is now unambiguous data that breast milk accumulates and harbors potentially toxic environmental pollutants.[62,68–71] Persistant organohalogens, including

Table 7.8 Environmental pollutants that may be found in human milk

Pollutant	Potential Health Effect
DDT, DDE	Estrogenic, antiandrogenic activity
PCB/PCDF	Ectodermal defects, developmental delay
TCDD (Dioxin)	Choloracne
Chlordane	Neurotoxicity
Heptachlor	Neurotoxicity
Hexachlorobenzine	Hypotonia, seizures, rash
Volatile organic compounds	
Tetracholorethylene	Hepatotoxicity
Trichlorethylene	Hepatotoxicity
Halothane	Hepatotocicity
Carbon disulfide	Neurotoxicity
Lead	Renal, central nervous system
Mercury	Central nervous system
Brominated flame retardants	Thyroid disorders, brain development

SOURCE: Adapted from U.S. Department of Health and Human Services, Blueprint for action on breastfeeding.[74]

persistent organic pollutants (POPs), heavy metals, and volatile solvents, are among the toxic chemicals most often found in breast milk. A woman comes in contact with environmental chemicals as a matter of course in daily life,[70] through air pollution, drinking water, and diet. A woman also comes in contact with a wide range of environmental chemicals in the home that have the potential to appear in her breast milk, such as household cleaning and personal care products, paints, furniture strippers, and pesticides. Exposure to environmental chemicals can also be occupational.

While the presence of low levels of environmental chemicals, such as those in Table 7.8, in most human milk samples is well documented, the significance of the presence of these contaminants on the well-being of the mother and the infant is unknown. Several recent reviews[62,68–72] address concerns about potential impact on duration of lactation, neurodevelopmental and immunologic outcomes, and carcinogenic effects. While the body of research is growing, huge gaps in the current knowledge of any ill effects remain. The body of evidence supports the benefits of breastfeeding over any potential risks from environmental chemicals. In fact, other factors in breast milk may have a protective effect on normal neurologic development and immunologic outcomes.[62] At this time, there are no established "normal" or "abnormal" levels in breast milk for clinical interpretation, and breast milk is not routinely tested for environmental exposures.[62,74]

Unless the mother has a high-level of occupational exposures, extreme dietary exposures (e.g., from fish in contaminated waters), or unusual residential exposures to hazardous or toxic chemicals, breastfeeding remains overwhelmingly the preferred choice compared with breast milk substitutes.[74–76] The World Health Organization,[75] the American Academy of Pediatrics,[76] the U.S. Department of Health and Human Services,[74] and other major health organizations overwhelmingly support the importance of breastfeeding even in a contaminated world. The benefits of breastfeeding, which include high levels of antioxidants, may prove to be essential to compensate for and outweigh the risk of toxic effects from the environment.

Currently the focus of scientific concerns is being directed toward removing potentially toxic substances from the environment and on establishment of a U.S. Breast Milk Monitoring Program to track trends in exposure levels over time.[71,77] Encouragingly, data from several countries show a decline in the level of DDT and dioxin metabolites in human milk.[77] Unfortunately, the presence of other persistent chemicals, such as flame-retardant chemicals is increasing.[70]

Women should be advised about how to reduce exposures that may affect breast milk quality rather than abandoning breastfeeding for artificial methods (Table 7.9). Women should avoid fish such as swordfish and shark or freshwater fish from waters reported as contaminated by local health agencies and limit exposure to chemicals such as pesticides and solvents found in paints, non-water-based glues, furniture strippers, nail polish, and gas fumes.

Persistant Organic Pollutants (POPs) A family of chemicals manufactured either for a specific purpose (e.g., pesticides or flame retardants in electrical equipment or furniture) or produced as byproducts of incinerated waste. The POP family includes dioxins, polychlorinated biphenyuls (PCBs), poybrominated didpheyl ether (PBDE), and organochlorine pesticides.

Hyperbilirubinemia Elevated blood levels of bilirubin, a yellow pigment that is a byproduct from the breakdown of fetal hemoglobin.

Neonatal Jaundice and Kernicterus

"The AAP discourages the interruption of breastfeeding in healthy term newborns and encourages continued and frequent breastfeeding (at least eight to twelve times every 24 hours)."

American Academy of Pediatrics[78]

Neonatal jaundice is a yellow discoloring of the skin caused by too much bilirubin in the blood (*hyperbilirubinemia*). It is a common and usually benign condition that resolves on its own or with minimal intervention. At least 60% of full-term and 80% of preterm infants will become visibly jaundiced[79] with their serum bilirubin levels exceeding 5 to 7 mg/dl (85 to 199 μmol/L).[80] If hyperbilirubinemia does not resolve and becomes sufficiently severe, the elevated bilirubin levels can cause permanent neurological damage.[79–81]

Table 7.9 Steps a breastfeeding mother can take to reduce exposure to environmental chemicals

1. Avoid smoking cigarettes and drinking alcohol. Some studies have found levels of contaminants are higher in those who smoke and drink alcoholic beverages.

2. Be aware in purchasing homes that some houses or apartments, especially those built before 1978, might have lead-based paints.

3. In general, eat a variety of foods low in animal fats; remove skin and excess fat from meats and poultry. Avoiding high-fat dairy products may reduce the potential burden of fat-soluble contaminants. Avoid processed foods made from ground meat and animal parts such as sausage and hot dogs.

4. Increase consumption of grains, fruits, and vegetables. Thoroughly wash and peel fruits and vegetables to help eliminate the hazard of pesticide residues on the skin. When available, eat food grown without fertilizer or pesticide application. Eat organically grown food, if available.

5. Avoid fish that may have high mercury levels, such as swordfish, shark, tuna, king mackerel, tilefish, and locally caught fish from areas with fish advisories.

6. Limit exposure to chemicals such as solvents found in paints, non-water-based glues, furniture strippers, nail polish, and gasoline fumes.

7. If you work with solvents (in the work place or at home), postpone breastfeeding for several hours after exposure.

8. Run tap water through a home filter before drinking. Filters can reduce levels of common tap water pollutants.

9. Remove the plastic cover of dry cleaned clothing. Air out the garments in a room with open windows, or hang dry cleaning outside for 12–24 hours.

10. Attempt to avoid occupational exposure to chemical contaminants in the workplace, and seek improved chemical safety standards for all employees, especially pregnant and lactating women. Workers should be diligent in following their workplaces' safety recommendations.

11. Alert other family members to be sensitive to contaminant residue they may inadvertently bring into the home. It is possible, for example, to carry PCBs home on clothes, body, or tools. If this is the case, the individual should shower and change clothing before leaving work and keep and launder work clothes separate from other clothing.

12. Review additional suggestions for avoiding chemical exposures in and around the home from the report from the Enviornmental Working Group (see www.ewg.org/reports/mothersmilk/part5.php).

SOURCE: Condon, M. Breast is best, but it could be better: what is in breast milk that should not be? Pediatr. Nurs. 2005;31:333–338.

In recent years, the overall incidence of infant jaundice has risen[81] and hyperbilirubinemia is the most frequent cause for hospital readmission during the first 2 weeks of life in the United States.[82] More infants are becoming jaundiced, and their jaundice is more severe. The Joint Commission on Accreditation of Healthcare Organizations (JCAHO),[83] the Centers for Disease Control, and the American Academy of Pediatrics[78] have all noted the increasing rates and the need for prevention, early detection, and prompt treatment. Risk factors for the development of severe hyperbiliruminia have been identified (see Table 7.10). Higher rates of breastfeeding in conjunction with shorter postpartum hospital stays is the leading explanation for the higher prevalence of neonatal jaundice.[81]

It is important for all health professionals to understand the causes of, risk factors for, and early signs of hyperbilirubinemia. Preventing toxicity from excessive jaundice and protecting and ensuring successful breastfeeding require an understanding of the normal and abnormal patterns and mechanisms of jaundice in the newborn period, particularly the mechanisms related to human milk intake.[84]

Bilirubin Metabolism

Bilirubin is a byproduct of the normal physiologic degradation of hemoglobin. Most hemoglobin in the neonate is derived from fetal erythrocytes. Since higher levels of hemoglobin are necessary in utero to carry the oxygen delivered to the fetus by the placenta, the normal full-term infant has a hematocrit of 50–60%. As soon as the infant is born and begins to breathe room air, the need for high levels of hemoglobin is gone, and excess erythrocytes are destroyed. The released hemoglobin is broken down by the reticuloendothelial system; bilirubin, an insoluble byproduct of the breakdown of hemoglobin, is released into the circulation bound to albumin or another transport protein. The insoluble form of bilirubin is removed from circulation by the liver, which conjugates bilirubin to a water-soluble form and excretes it via the bile to the stool. The balance between liver cell uptake of bilirubin, the rate of bilirubin production, and the rate of bilirubin resorption through the intestines determines the total serum bilirubin (TSB) level.

Table 7.10 Risk factors for severe hyperbilirubinemia[78]

Major Risk Factors

Predischarge total serum bilirubin(TSB) or transcutaneous bilirubin (TcB) in the high-risk zone

Jaundice observed in the first 24 hours of life

Blood group incompatibility with positive direct antiglobulin test, other known hemolytic disease, elevated ETCO

Gestational age 35–36 weeks

Previous siblings received phototherapy

Cephalohematoma or significant bruising

Exclusive breastfeeding, particularly if nursing is not going well and weight loss is excessive

East Asian race

Minor Risk Factors

Predischarge total serum bilirubin(TSB) or TcB in the high intermediate risk zone

Gestational age 37–38 weeks

Jaundice observed before discharge

Previous sibling with jaundice

Macrosomic infant of a diabetic mother

Maternal age ≥ 25 years

Male gender

Decreased Risk Factors

TSB or TcB level in the low-risk zone

Gestational age ≥ 41 weeks

Exclusive bottle feeding

Black race

Discharge from hospital after 72 hours

Oxytocin used in labor

ABO incompatibility

SOURCE: American Academy of Pediatrics Subcommittee on Hyperbilirubinemia. Management of hyperbilirubinemia in the newborn infant 35 or more weeks of gestation. *Pediatrics*, 2004;114:297–316.

Before birth, the maternal liver is responsible for metabolism and clearance of fetal bilirubin. After birth, uniquie developmental factors that control the production, conjugation, and excretion of bilirubin predispose the neonate to hyperbilirubinemia:[81,85]

- Bilirubin production in the neonate is double that of an adult because of breakdown of fetal erythrocytes.
- Uptake of insoluble bilirubin by the liver is limited because of a reduction in the concentration of ligandin, a bilirubin-binding protein in the liver cell.
- Conjugation to a water-soluble form is limited in the liver because of deficient activity of uridine diphosphoglucuronosyl transferase (UDPGT), a liver enzyme responsible for bilirubin conjugation.

- Excretion of bilirubin is delayed because of an enzyme present in the intestine of the newborn, beta glucuronidase, which converts conjugated bilirubin back into its unconjugated state, which is reabsorbed.

Physiologic versus Pathologic Newborn Jaundice

After the first 24 hours, rising bilirubin levels in healthy term infants are reflective of the physiological breakdown of fetal hemoglobin, the increased resorption of the bilirubin from the intestines, and the limited ability of the newborn's immature liver to process large amounts of bilirubin as effectively as a mature liver. Neonates tend to produce more bilirubin than they can eliminate. Prematurity magnifies this imbalance. Retention of unconjugated bilirubin by the newborn is known as normal newborn jaundice or physiologic jaundice of the newborn.[84] Excessive bilirubin is deposited in various tissues, including the skin, muscles, and mucous membranes of the body, causing the skin to take on a yellowish color. In healthy newborns, this condition is temporary and usually resolves within a few days without treatment. In the typical newborn population, the bilirubin level rises steadily in the first 3 to 4 days of life, peaking around the fifth day, and then declining. Bilirubin levels of healthy preterm infants peak later (day 6 to day 7) and take longer to resolve. Bilirubin levels in physiologic jaundice are usually less than 12 mg of bilirubin per dL of blood of infants of white or black mothers, and average 10 to 14 mg in infants of Asian ancestry, including Chinese, Japanese, and Korean, and Native Americans.[86] Levels in white and black mothers also peak earlier than in Asian or Native American mothers.[85]

In contrast to physiologic jaundice of the newborn, pathologic jaundice begins earlier (sometimes before 24 hours of age), rises faster, and lasts longer. A TSB greater than 8 mg/dL in the first 24 hours should be investigated for pathologic origin. Causes of pathologic jaundice include the following:[80]

- Hemolytic disease (immune disorders (Rh isoimunization, ABO or minor blood type incompatibility
- Erythrocyte disorders (glucose-6-phosphate-dehudrogenase deficiency, hereditary spherocytosis)
- Extravasation of blood (cephalohematoma, subgaleal hemorrhage, bruising)
- Inborn errors of metabolism/conjugation defects (galactosemia, Crigler-Najjar syndrome types I and II, Gilbert's syndrome, Lucey-Driscoll syndrome)
- Hypothyroidism
- Polycythemia
- Macrosomic infant of diabetic mother
- Intestinal obstruction; delayed passage of mechonium
- Sepsis

Kernicterus or Bilirubin Encephalopathy   The chronic and permanent clinical sequelae that are the end result of very high untreated bilirubin levels. Excessive bilirubin in the system is deposited in the brain, causing toxicity to the basal ganglia and various brainstem nuclei.[78]

In most cases of pathological jaundice, frequent breast-feeding (10 to 12 times every 24 hours) can continue during diagnosis and treatment of pathological jaundice.[78] An advantage of colostrum and mature human milk is stimulation of bowel movements, speeding elimination of bilirubin. However, in jaundice caused by galactosemia, breastfeeding is contraindicated.[9]

Since bilirubin is a cell toxin, concern arises when TSB elevates to levels with the potential to cause permanent damage. The brain and brain cells, if destroyed by bilirubin deposits, do not regenerate.[81] *Bilirubin encephalophathy,* or *kernicterus,* has a mortality rate of 50%, and survivors usually are burdened with severe problems including cerebral palsy, hearing loss, paralysis of upward gaze, and intellectual and other handicaps.[78] While full-scale kernicterus is rare, there has been an increase in reported cases in the last two decades.[81] There is also concern that mild effects of bilirubin on the brain may be manifested clinically in later life with symptoms such as incoordination, hypertonicity, and mental retardation, or perhaps learning disabilities.[9]

Hyperbilirubinemia and Breastfeeding

Jaundice in the breastfed infant has been divided into types, early or late, based on the age of onset (Table 7.11). It is important to differentiate between the two to establish effective prevention and treatment. Early onset of elevated unconjugated bilirubin unexplained by other pathologic factors is associated with inadequate feeding and is called "breastfeeding jaundice" or more precisely, "breast-nonfeeding jaundice."[84] Late onset (after day 5) prolonged elevation of unconjugated bilirubin associated with the ingestion of breast milk is called breast milk jaundice.[84,87]

Breast-Nonfeeding Jaundice   The optimally breastfed infant initiates breastfeeding in the first hours, followed by at least 10 to 12 breastfeeds per day for the first 1 to 2 weeks without any water or other food supplementation, and with good positioning that assures effective milk transfer to the infant, and weight loss of less than 8% of birth weight.[88] Differences in bilrubin levels between adequately breastfed infants and formula-fed infants have not been found to be significant.[84] In contrast, infants who nurse infrequently or inefficiently ingest fewer calories and lose more weight than formula-fed infants are at risk for elevated bilirubin levels. Suboptimal breastfeeding can delay passage of meconium and reduce fecal weight, increasing the enterohepatic circulation of bilirubin. In addition, reduced milk intake produces a state of partial starvation in the infant, which further increases the intestinal absorption of bilirubin.[88] Delay in initiation of breastfeeding beyond

Table 7.11   Comparison of early and late jaundice associated with hyperbilirubinemia while breastfeeding

	Early Jaundice	Late Jaundice
Onset	Occurs 2–5 days of age	Occurs 5–10 days of age
Duration	Transient: 10 days	Persists >1 month
Incidence	More common with first child; approximately 60% of U.S. newborns become clinically jaundiced	All children of a given mother
Feeding	Infrequent feeds Receiving water or dextrose water	Milk volume not a problem May have abundant milk supply No supplements
Stools	Stools delayed and infrequent	Normal stooling
Total Serum Bilirubin	Peaks <15 mg/dL None or phototherapy	May be >20 mg/dL Treatment: phototherapy Discontinue breastfeeding temporarily Rarely, exchange transfusion.
Associations	Low Apgar scores, water or dextrose water supplement, prematurity	None identified

SOURCE: Adapted from *Breastfeeding: A Guide for the Medical Profession*, 6th ed., by R. A. Lawrence and R. M. Lawrence, copyright 2005, with permission from Elsevier.

the first hour of life, and administration of water to infants either before initiation of breastfeeding or in addition to breastfeeding significantly reduce the frequency of breastfeeding and increase bilirubin concentrations. Excessive hyperbilirubinemia causes lethargy and poor feeding in some infants, further reducing breastfeeding frequency, and duration and milk production, and feeding a vicious cycle of increasing bilirubin levels.

It is now well established that early onset breastfeeding jaundice results from reduced volume of milk transfer to the infant limiting caloric intake and producing a state of partial starvation and weight loss equivalent of the adult disorder known as starvation jaundice. Lawrence outlines treatment guidelines (Table 7.12) aimed at treating the actual cause—that is, failed breastfeeding or inadequate stooling or underfeeding.[9] The goal is to evaluate the breastfeeding for frequency, length of suckling, and apparent supply of milk, and then adjust the breastfeeding to

Table 7.12 Management outline for early jaundice while breastfeeding (breast-nonfeeding jaundice)

1. Monitor all infants for initial stooling. Stimulate stool if no stool in 24 hours.
2. Initiate breastfeeding early and frequently. Frequent, short feeding is more effective than infrequent, prolonged feeding, although total time may be the same.
3. Discourage water, dextrose water, or formula supplements.
4. Monitor weight, voidings, and stooling in association with breastfeeding pattern.
5. When bilirubin level approaches 15 mg/dL, stimulate stooling, augment feeds, stimulate breast milk production with pumping, and use phototherapy if this aggressive approach fails and bilirubin exceeds 20 mg/dL.
6. Be aware that no evidence suggests early jaundice is associated with "an abnormality" of the breast milk, so withdrawing breast milk as a trial is only indicated if jaundice persists longer than 6 days or rises above 20 mg/dL or the mother has a history of a previously affected infant.

SOURCE: Reprinted from *Breastfeeding: A Guide for the Medical Profession*, 6th ed., by R. A. Lawrence and R. M. Lawrence, copyright 2005, with permission from Elsevier.

solve the problem.[88] If stooling is an issue, the infant should be stimulated to stool. If starvation is the problem, the infant should receive temporary supplemental feeding by cup or bottle while the milk supply is being increased by better breastfeeding techniques.

Early discharge of infants from the hospital at less than 72 hours of life has raised concerns about the ability to evaluate breastfeeding and the opportunity to evaluate infants for jaundice.[78,81] Formal observation of breastfeeding with evaluation of effectiveness of breastfeeding and milk transfer at regular intervals throughout the first days of life can identify breastfeeding problems sufficiently early to ensure correction of problems. The AAP strongly recommends that all breastfed infants be evaluated by a trained observer within 2 or 3 days after discharge from the hospital.[78] Particular attention must be paid to infants who are <38 weeks gestation and infants at moderate or high risk for severe hyperbilirubinemia (Table 7.10).[78]

Breast Milk Jaundice Syndrome   In contrast to physiological newborn jaundice, which peaks in the third day and then begins to drop, breast milk jaundice syndrome becomes apparent after the third day, and bilirubin levels may peak any time from the seventh to the tenth day, with untreated cases being reported to peak as late as the fifteenth day.[9] In breast milk jaundice syndrome, no correlation exists with weight loss or gain, and stools are normal. Initially

breast milk jaundice syndrome was thought to be an unusual and distinct type of newborn jaundice affecting 1% of all breastfed neonates. More recent research reports demonstrated that at least one-third of all breastfed infants are clinically jaundiced in the third week of life and that two-thirds have significant unconjugated hyperbilirubinemia in the third week in contrast to the absence of hyperbilirubinemia in the third week in full-term artificially fed infants.[88] What was once believed to be a clinical disorder is now recognized as a normally occurring extension of physiolgic jaundice of the newborn.[87,88] However, at this time there is insufficient evidence to support the popular theory that breast milk jaundice may provide protective effects for newborns by the antioxidant effects of bilirubin, compensating for the relative deficiency of endogenous antioxidants in newborns.[84]

The undisputed cause of breast milk jaundice syndrome is unresolved. It is believed to be caused by a combination of factors: a substance in most mothers' milk that increases intestinal absorption of bilirubin and individual variations in the infant's ability to process bilirubin.[9,84] To treat severe elevations in unconjugated bilirubin in breast milk jaundice syndrome, the AAP Clinical Practice Guidelines for Hyperbilirubinemia Management[78] are applied with the goal of promptly lowering TSB bilirubin levels substantially limits set for the infant's age and level of risk. To establish the diagnosis of breast milk jaundice firmly when the bilirubin level is above 16 mg/dL for more than 24 hours, a short, temporary interruption of breastfeeding (12 to 24 hours) while monitoring bilirubin levels is recommended.[9,84]

The belief that severe hyperbilirubinemia in breastfed infants cannot result in kernicterus is erroneous and dangerous; 98% of the 105 cases of kernicterus in the U.S. Kernicterus Registry were breastfed infants.[89] Kernicterus, while rare, can develop in otherwise healthy full-term breastfed newborns or breastfed infants with sepsis. Breastfed infants need to be followed closely, supported effectively, and evaluated appropriately in order to avoid rare cases of severe hyperbilirubinemia and kernicterus.[78]

Interrelationships between Breast-nonfeeding Jaundice and Breast Milk Jaundice Syndrome   While breast-non-feeding jaundice and breast milk jaundice are two separate entities, they can have an interactive effect on each other. Infants with breast milk jaundice who manifest higher levels of bilirubin in the second and third weeks of life, often over 15 mg/dl, have been noted to have had relatively high serum bilirubin concentrations during the first 3 to 5 days of life due to either breast-nonfeeding jaundice, hemolysis, or unknown etiology.[88] Gartner postulates that these early, elevated bilirubin levels may produce an enlarged bilirubin pool. Then the ingestion of mature milk and a consequent enhancement of the enterohepatic circulation may enlarge the pool even further.[88]

Prevention and Treatment for Severe Jaundice

The American Academy of Pediatrics guidelines for the management of hyperbilirubinemia in healthy term newborns include a detailed algorithm for the management of jaundice in the newborn nursery and guidelines for initiating phototherapy.[78] Phototherapy involves placing the newborn under special fluorescent lights that, like sunlight, assist in removing jaundice from the skin. The light is absorbed by the bilirubin, changing it to a water-soluble product, which can then be eliminated without having to be conjugated by the liver.

Historically, treatment for jaundice in American hospitals involved phototherapy and discontinuing breastfeeding either permanently or until the bilirubin levels were acceptable. In addition, many health professions believed that newborn infants would become dehydrated if they were not supplemented with water or formula during the first days of breastfeeding. Recent research shows these practices are counterproductive.[78,84] The benefits of early and frequent breastfeeding in the first days of life to prevention of hyperbilirubinemia through maintaining hydration and stimulating the passage of stool are now well documented. The passage of stool in the newborn is important because there are 450 mg of bilirubin in the intestinal tract *meconium* of the average newborn.[9] To avoid reabsorption of bilirubin from the gut into the serum, passing meconium in the stool is critical. The current AAP hyperbilirubinemia management guidelines discourage the interruption of breastfeeding in healthy term newborns and encourages continued and frequent breastfeeding (at least 8 to 12 times every 24 hours). The AAP recommends against routine supplementation of nondehydrated breastfed infants with water or dextrose water as this practice will not prevent hyperbilirubinemia or decrease total serum bilirubin levels.[78]

Meconium Dark green mucilaginous material in the intestine of the full-term fetus.

Information for Parents

Health professionals should convey a balanced approach when communicating with parents about jaundice. Parents need to know that most breastfed infants will become jaundiced and that the overwhelming majority of cases will be benign. Only a small fraction of these infants is at risk for developing extreme hyperbilirubinemia and kernicterus. However, parents also need to fully understand the serious consequences of extremely elevated bilirubin levels and should have their infant evaluated by a health professional if jaundice develops. Health professionals need to understand that feelings of guilt are common among mothers of jaundiced infants as many mothers feel that they caused the jaundice by breastfeeding.[90] By providing accurate information and encouragement to breastfeed, health professionals have great impact on whether a mother continues breastfeeding after her experience with neonatal jaundice.

Breastfeeding Multiples

Since 1980, the birthrate of twins has increased by 59%, and the birthrate of higher-order multiples (triplets and more) has increased by over 400%. In the United States multiples currently represent over 3% of live births. The benefits of breastfeeding to mother and infant are multiplied with twins and higher-order multiples, who often are born at risk.[92] History and numerous case reports[93–95] provide ample evidence that an individual mother can provide adequate nourishment for more than one infant. In seventeenth-century France, wet nurses in foundling homes fed 3 to 6 infants, who were often of differing ages with different daily requirements.[9,96] Breastfeeding initiation rates of nearly 70% have been reported by surveys of members of Mothers of Super Twins (MOST), Parents of Multiple Births Association (POMBA) of Canada, and Double Talk, a newsletter for parents of multiples. Some mothers of triplets and quadruplets have fully breastfed their babies.[93]

Frequency and effectiveness of breastfeeding are the keys to building a plentiful milk supply. The more often a baby nurses, the more milk there will be.[9,92] Mothers who exclusively breastfeed twins or triplets can produce 2 to 3 kg/day, although this involves nursing an average of 15 or more times per day.[97] The main obstacle to nursing multiples is not usually the milk supply, but time and fatigue of the mother. Parents of twins and higher-order multiples need support in four major areas: organization, feeding, individualization, and stress management.[92]

Mothers of twins or higher-order multiples often face special challenges in the establishment of lactation after birth. Approximately 60% of twins and 90% of higher-order multiples in the United States are born at less than 37 weeks gestation, and premature multiples often experience medical complications that potentially interfere with breastfeeding.[99] Breastfeeding initiation may take place in the neonatal intensive care unit, usually because of prematurity and low birth weight. Mothers of multiples may be coping with the effects of a more physically demanding pregnancy and birth or complications of pregnancy. Mothers may experience exaggerated postpartum sleep deprivation related to round-the-clock care of two or more newborns, concern for sick newborns, or staggered infant discharge, which results in time divided between infants at home and in the hospital. In addition, every aspect of breastfeeding management is affected by the dynamics that multiple newborns create.[9,92] In a recent study, breastfeeding initiation rates as high as 73% were found among full-term multiples compared with initiation rates of only 57% in preterm multiples. Breastfeeding duration was also lower in preterm multiples than in term multiples (12 weeks versus 24 weeks).[99]

Health care professionals can help the mothers of multiples face the many challenges of breastfeeding by offering consistent, informed, individualized care and support in the hospital and after discharge.[97] Knowledge-able care providers can help parents distinguish between multiples-specific issues, normal variations in individual infant's breastfeeding abilities and patterns, and actual breastfeeding problems. Mothers need information on when and how to initiate simultaneous feedings, practical tips for managing nighttime nursing and fatigue, and how to assure herself that her babies are getting ample nourishment.[92] Parents need to be informed of resources for parenting multiples and for receiving support for breastfeeding in their community (see Resources at the end of chapter).

A well-defined plan for the health care of the lactating woman that includes screening for nutritional problems and providing dietary guidance is also important.[55] Mothers should be encouraged to drink to satisfy their thirst, to eat nutritious foods, and to sleep when the babies sleep. As in singleton nursing, women nursing multiples should be encouraged to obtain their nutrients from a well-balanced, varied diet rather than from vitamin–mineral supplements.

Infant Allergies

Protection from *allergic diseases* is one of the most important benefits of breastfeeding. There is substantial evidence that exclusive breastfeeding for at least 4 months can protect against ectopic dermatitis and wheezing illnesses in children up to age 6.[100,101] Elimination of major food allergens from the diet of breastfeeding mothers of infants at high risk for atopic disease was previously recommended to delay or prevent some food allergy and atopic dermatitis. However recent expert reviews concluded that there is no strong evidence to support this recommendation.[100,102] A recent well-controlled study confirmed that the presence of food proteins in human milk is common, but this can be highly variable between women consuming the same challenge (dose) food.[102] A number of other studies measuring proteins in human milk following challenge doses of eggs, cow's milk, gluten, and wheat protein also found variable response between women.[101,102]

The development of infant *food allergy* is influenced by genetic risk for allergy, duration of breastfeeding, time for introduction of other foods, maternal cigarette smoking during pregnancy and parental smoking, air pollution, exposure to infectious disease, and by maternal diet and immune systems.[101] Several mechanisms (Table 7.13) are thought to contribute to the protective effect of breastfeeding. There is preliminary evidence that increasing the breastfeeding mother's dietary omega-3 (n-3) polyunsaturated fatty acids may offer protection from some childhood allergies.[102]

Table 7.13 Possible reasons for allergy preventive effects of breastfeeding[101,102]

- Low content of allergens
- Transfer of maternal immunity
- Long-chain fatty acids and IGA in breast milk protect against inflammation and infections
- Regulation of infant immunity
- Influence on gut microbial flora

Common pediatric food allergens include:

- Cow's milk
- Wheat
- Eggs
- Peanuts
- Soybeans
- Tree nuts (e.g., almond, Brazil nut, walnut, hazelnut)

Infants with a positive family history of allergies should be exclusively breastfed for at least 4 to 6 months with continuance of breastfeeding for as long as possible.[9,16,101] Advice to breastfeeding mothers with a family history of allergies regarding elimination of common allergens in their own diets should be individualized. Although several studies have documented the presence of food allergens—particularly milk and peanut protein—in breast milk sufficient to induce food reactions in infants, the amounts are variable. The American Academy of Pediatrics specifically suggests restricting peanuts and tree nuts in the maternal diet of high-risk infants.[16] If there is no history of allergy to a specific food in the mother's or father's family, avoiding a food because it is a potential allergen is an unnecessary precaution. Only if there is a family history of allergies, or if a baby shows allergic symptoms should a mother consider avoiding certain foods. If a mother avoids certain foods, care must be taken to ensure that her diet remains nutritionally adequate.[103]

Allergic Diseases   Conditions resulting from hypersensitivity to a physical or chemical agent.

Food Allergy (Hypersensitivity) Abnormal or exaggerated immunologic response, usually immunoglobulin E (IgE) mediated, to a specific food protein.

Food Intolerance   An adverse reaction involving digestion or metabolism but not the immune system.

Food Intolerance

Although infants may have sensitivities to certain foods, there is no scientific basis for the concern about gassy foods, such as cabbage or legumes, causing gas in the breastfed baby. In mothers, the normal intestinal flora produces gas from the action on fiber in the intestinal tract. Neither the fiber nor the gas is absorbed from the intestinal tract, and neither enters the mother's milk. Likewise, the acid content of the maternal diet does not affect the breast milk because it does not change the pH of the maternal plasma.

The role of diet in infantile colic, which affects up to 28% of infants in the first months of life,[104] is controversial. While breastfeeding is not protective against infantile colic, several studies report a reduction in persistent crying after elimination of cow's milk and other food proteins from the maternal diet. In a recent randomized, controlled trial, a low-allergen maternal diet was associated with a reduction in distressed behavior among breastfed infants with colic.[104] The mothers excluded cow's milk, eggs, peanuts, tree nuts, wheat, soy, and fish from their diet.

Characteristic essential oils in foods such as garlic and spices may pass into the milk, and an occasional infant objects to their presence. Studies by Mennella and Beauchamp confirm that the diet of lactating women alters the sensory qualities of her milk.[105,106] Extensive clinical experience also suggests that some infants are sensitive to certain foods in the mother's diet. According to Lawrence, garlic, onions, cabbage, turnips, broccoli, beans, rhubarb, apricots, or prunes may be bothersome to some infants, making them colicky for 24 hours.[9] In the summer, a heavy diet of melon, peaches, and other fresh fruits may cause colic and diarrhea in the infant. Red pepper has been reported to cause dermatitis in the breastfed infant within an hour of milk ingestion.[107] Contrary to popular belief, chocolate rarely causes problems and can be consumed in moderation without causing colic, diarrhea, or constipation in most infants.[9]

If a mother suspects that her baby reacts to a specific food, it may be helpful for her to keep a record of foods eaten, along with notes on the baby's symptoms or behavior. If highly allergic or sensitive, infants may react to foods their mothers have eaten within minutes, although symptoms generally show up between 4 and 24 hours after exposure. While symptoms will improve in most infants after the offending food has been removed from the mother's diet for 5 to 7 days, it may take 2 weeks to totally eliminate all traces of the offending substance from both the mother and baby.[108] See Case Study 7.2.

Near-Term Infants

Infants born near 37 weeks may have subtle immaturity that makes establishing breastfeeding difficult and places, the infant at risk for insufficient milk intake, hypoglycemia, jaundice, and poor weight gain.[109] Infants may have cardiorespiratory instability especially in upright position, poor temperature control, lower glycogen and fat stores to prevent hypoglycemia, an immature immune system, and the suck–swallow coordination may be poorly coordinated and result in poor latch-on and milk transfer. The main emphasis in postpartum care should be on building and maintaining the milk supply and feeding the infant. If an infant is not sucking vigorously, mothers of near-term infants should

pump after each feeding attempt, or at least every 3 hours, to build the milk supply. A specific feeding plan including the plan to wake a sleepy infant every 2–4 hours is recommended to avoid the near-term breastfeeding cascade (Illustration 7.1). Mothers and their near-term infants should have a feeding assessment by a trained lactation professional who can provide intervention for positioning, potential suck problems, and other breastfeeding issues. Weighing an infant before and after feedings may be useful to determine milk transfer. Discharge planning should include follow-up care, home visits, and lactation counseling as needed. Close follow-up should be continued until the infant is gaining weight and the mother is comfortable.[109]

Human Milk and Preterm Infants

"Human milk is the preferred feeding for all infants, including premature and sick newborns, with rare exceptions. The ultimate decision on feeding of the infant is the mother's. Pediatricians should provide parents with complete, current information on the benefits and methods of breastfeeding to ensure that the feeding decision is a fully informed one. When direct breastfeeding is not possible, expressed human milk, fortified when necessary for the premature infant, should be provided."

AAP[110]

The benefits of breastfeeding may be most visible among preterm infants who are born immature and without adequate stores of nutrients. The nutritional benefits include ease of protein digestion, fat absorption, and improved lactose digestion.[110–113] The known health and developmental benefits include better visual acuity, greater motor and mental development at 1.5 years of age, greater verbal intelligence quotient at 7–8 years of age, and a lower incidence of serious infectious disease including necrotizing enterocolitis and sepsis even among infants who also receive some human milk substitutes.[112–114] Lower weight and length gains and poorer bone mineralization has been reported for preterm infants fed human milk.[115]

The composition of milk from women who deliver preterm infants is higher in protein, slightly lower in lactose, and higher in energy content (58–70 Kcal/oz) compared to the milk of women who deliver full-term infants (approximately 62 Kcal/oz). Once growth is established, the nutritional needs of the preterm infant exceed the content of human milk for protein, calcium, phosphorus, magnesium, sodium, copper, zinc, riboflavin, pyridoxine, folic acid, and vitamins C, D, E, and K.[16] Human milk fortifiers are available that provide additional protein, minerals, and vitamins.[16] Infants who receive fortified human milk do not need additional supplementation unless a specific nutritional problem is identified. The health benefits, particularly reduction in sepsis and necrotizing

Case Study 7.2

Breastfeeding Premature Infants

At 30 weeks gestation, 35-year-old Stacey delivers twin boys: baby Andrew is 2 pounds, 9 ounces and 13.5 inches long; baby Mark is 1 pound, 13 ounces and 14 inches long. Stacey had a difficult pregnancy that included severe nausea and vomiting, heartburn, preeclampsia, and preterm labor.

Stacey is very committed to breastfeeding and was able to use an electronic breast pump approximately 18 hours after delivery. She pumps successfully every 3–4 hours or 7–8 times daily in order to establish her milk supply. At the end of the first week, Stacey is pumping about 14–16 oz daily and by 3½ weeks she is pumping 2–2.5 oz per breast at each pumping. Stacey wakes at night to pump when she is engorged. She has placed the pumping equipment by the bed and become adept at pumping, getting out of bed only to put the milk in the refrigerator. At 2 weeks postpartum, Stacey experiences a plugged duct and has difficulty emptying the right breast for 2 days.

The twin boys have suffered the usual preterm difficulties with breathing, apnea, and bradycardia. Initially the twins are tube-fed. As their condition improves, baby Andrew is first put to breast 3 weeks after his birth, and baby Mark several weeks later. Baby Mark has more difficulty learning to latch on and suck and is growing more slowly than is his brother. Multiple interventions are used to achieve breastfeeding success. On advice from a lactation consultant, a nipple shield helps baby Mark latch on. In response to slow weight gain in baby Mark, the lactation consultant recommends that the baby receive hindmilk, which is often higher in fat and calories.

Mark and Andrew are released from the hospital a day after their due date. Mom continues using the nipple shield for several weeks with Mark, trying without the nipple shield every few days. After 3 weeks at home, baby Mark is able to latch on without the nipple shield. The twins are fed human milk and up to 3 bottles of premature infant formula for the first 2 months at home. The babies take feedings equally well from bottle or breast.

Questions

1. Describe the appropriate steps that a mother of a premature infant should take to establish a good milk supply by pumping.
2. What are the causes of plugged ducts?
3. What are the treatments for plugged ducts?
4. Describe how a mother would go about feeding a baby only hindmilk.

---

enterocolitis, outweigh the slightly lower rate of weight gain and length gain observed among preterm infants fed fortified human milk compared to those who receive a human milk substitute for premature infants.[115]

Early feeding seems to be important for preterm infants (when medically appropriate). Early feeding may be important to the ability to digest[111] and to the development of the infant's digestive system.[116] Often, milk must be expressed and stored for preterm infants. Establishing a milk supply early seems to be important to the mother's ability to maintain a milk supply that will meet her infant's demands after several weeks (see Chapter 6, Can

Women Make Enough Milk?). A woman who is pumping less than 750 mL of milk by 2 weeks may need additional support to establish a milk supply that will meet her infant's needs beyond the first month.

Despite the health benefits of feeding human milk to preterm infants, the incidence and duration of breastfeeding of preterm infants in the United States is about 30% lower than rates in full-term infants.[113] The challenges of feeding low-birth-weight infants include provision of adequate calorie and nutrient intake, establishing and maintaining an adequate milk supply, and transitioning from gavage feeding to feeding from the breast.[113] Strategies to improve

Near-Term Breastfeeding Cascade

Illustration 7.1 Near-term infant breastfeeding cascade.

SOURCE: Reprinted with permission from N. E. Wight, "Breastfeeding the borderline (near-term) preterm infant," in Pediatrics Annals, 32:329–336. Copyright © 2003 Slack Inc.

breastfeeding rates in this vulnerable population include providing the parents with information necessary to make an informed decision to breastfeed; assisting the mother with the establishment and maintenance of a milk supply; ensuring correct breast milk management (storage and handling) techniques; providing skin-to-skin (kangaroo) care and opportunities for non-nutritive sucking at the breast; managing the transition to the breast; measuring milk transfer; preparing the infant and family for discharge; and providing appropriate followup care.[117] Counseling mothers of very low-birth-weight infants on the benefits of breast milk and providing breast-feeding support increases breastfeeding initiation and breast milk feeding without increasing maternal anxiety and stress.[118]

> **HIV**   Human immunodeficiency virus.

Medical Contraindications to Breastfeeding

Few medical problems in the mother or baby are absolute contraindications to breastfeeding (Table 7.14). There are very few infectious pathogens that pose a risk to the newborn that outweighs the potential benefits of breastfeeding.[119] Even infants with metabolic disorders such as pheynylketonuria can continue to breastfeed in combination with a specialized formula to meet calorie and protein needs. When mothers or infants have medical or other problems that cause a poor suck or other feeding problems, early identification and appropriate support from a lactation consultant is necessary for successful breast-feeding. In some cases, pumping milk may be necessary to maintain a supply of milk while problems with the infant suck are addressed and corrected. Whenever a medical situation presents a potential risk for breastfeeding infants, the theoretical risk must be carefully measured against the projected benefits of breastfeeding.[120]

Breastfeeding and HIV Infection

Every year approximately 750,000 children worldwide become infected with *HIV*, mostly through mother-to-child transmission during pregnancy, delivery, or breastfeeding.[121] The transmission of HIV type 1 from mother to child through breastfeeding is well documented. Reports of transmission rates range between 5 and 20%, with prolonged breastfeeding doubling the rate to 35–40%.[121] Factors contributing to these variable rates include strain of HIV, maternal illness, immune status and viral load, duration of breastfeeding (timing of transmission), primary infection of the mother during the breastfeeding period, exclusive breastfeeding versus mixed feeding, mastitis, maternal vitamin deficiencies (A, C, E, or B vitamins), and the availability of antiretroviral therapy.[119–123]

In developed countries, where safe and affordable breast milk substitutes are available, HIV-infected women should be counseled strongly not to breastfeed.[124] The U.S. Department of Health and Human Services's Blueprint for Action on Breastfeeding states that "HIV infected women in the United States should not breastfeed or provide their breast milk for the nutrition of their own or other infants because of the risk of HIV transmission to the child" (p. 12).[74]

The choice for women with HIV in developing countries is not so clear. In most cases, breast milk substitutes are not affordable to families or to government-sponsored public health programs, and they pose a serious health risk to infants both with and without HIV.[121,123] Women and infants in developing countries who are most at risk for HIV also face poor water quality and sanitation and are at high risk for diarrheal diseases and other infections. Boiling of water for decontamination presents another obstacle as fuel to boil water is either difficult to obtain or expensive. A WHO-sponsored meta-analysis documented a sixfold increase in mortality due to diarrheal disease in the first 6 months of life for infants who were not breast-fed when compared with breastfed infants and twice the risk of pneumonia deaths.[125] In certain populations, the benefits of breastfeeding may outweigh the risks of HIV transmission. Breastfeeding is thought to be responsible for about 300,000 HIV infections per year, while UNICEF estimates that not breastfeeding is responsible for 1.5 million child deaths per year.[126] A mathematical modeling exercise that looked at the risks of infant morbidity and mortality by feeding choices suggests that exclusive breastfeeding for 6 months would be the best option in locations where infant mortality is over 40 per 1000.[126] Locations in most of sub-Saharan Africa fall into this

Table 7.14 Summary of medical contraindications to breastfeeding in the United States

Problem	OK to Breastfeed in U.S.?	Condition
Infectious Diseases		
Acute infectious disease	Yes	Respiratory, reproductive, gastrointestional infection
HIV	No	HIV positive in developed countries
Active tuberculosis	Yes	After mother has received 2 or more weeks of treatment
Hepatitis		
A	Yes	As soon as mother receives gamma globulin
B	Yes	After infant receives HGIB; first dose of hepatitis B vaccine should be given before hospital discharge
C	Yes	If no co-infections (e.g., HIV)
Veneral warts	Yes	
Herpes viruses		
Cytomegalovirus	Yes	
Herpes simplex	Yes	Except if lesion on breast
Varicella-zoster (chickenpox)	Yes	As soon as mother becomes noninfectious
Epstein-Barr	Yes	
Toxoplasmosis	Yes	
Mastitis	Yes	
Lyme disease	Yes	As soon as mother initiates treatment
HTLV-I	No	
Over-the-Counter/Prescription Drugs and Street Drugs		
Antimetabolites	No	
Radiopharmaceuticals		
Diagnostic dose	Yes	After radioactive compound has cleared mother's plasma
Therapeutic dose	No	
Drugs of abuse	No	Exceptions: cigarettes, alcohol
Other medications	Yes	Drug-by-drug assessment
Environmental Contaminants		
Herbicides	Usually	Exposure unlikely (except workers heavily exposed to dioxins)
Pesticides		
DDT, DDE	Usually	Exposure unlikely
PCBx, PBBs	Usually	Levels in milk very low
Cyclodiene pesticides	Usually	Exposure unlikely
Heavy metals		
Lead	Yes	Unless maternal level >40 mg/dL
Mercury	Yes	Unless mother symptomatic and levels measurable in breast milk
Cadmium	Usually	Exposure unlikely
Radionuclides	Yes	Risk greater to bottle-fed infants
Metabolic Disorders		
Galactosemia	No	
Pheynylketonuria	Yes	Human milk supplemented with phyenylalanine-free formula

SOURCE: Adapted from *Breastfeeding: A Guide for the Medical Profession*, 6th ed., by R. A. Lawrence and R. M. Lawrence, copyright 2005, with permission from Elsevier.

category. Recent studies have provided strong evidence that breastfeeding does not pose any mortality or other health risks to the HIV-infected mother.[126]

Originally, the World Health Organization (WHO) recommended that in developing countries or areas where the risk of infant mortality from infection is great, breastfeeding is recommend even if the mother has **AIDS**.[127] This policy was clarified at a meeting of the WHO, the Children's Charity UNICEF, and UNAIDS in Geneva in October 2000. The policy now states: "When replacement feeding is acceptable, feasible, affordable, sustainable and safe, avoidance of all breastfeeding by HIV-infected mothers is recommended. Otherwise, exclusive breastfeeding is recommended during the first months of life."[128] A recent review of the advantages and disadvantages of replacement feeding options concluded that microbicide treatment of breast milk shows the greatest promise. This option has been shown to be broad spectrum, quick, inexpensive, capable

AIDS   Acquired immunodeficiency syndrome.

of preserving the milk's nutritional and immune status and does not require a heat source.[129]

Detailed instructions on counseling HIV-infected women are available from WHO. All women should be encouraged to know their HIV status and seek early prenatal care. Women need to be aware of the risks of HIV transmission during pregnancy and lactation.

Human Milk Collection and Storage

"Human milk is the most appropriate food for infants, and is also used as medical therapy for older children and adults with certain medical conditions. Human milk has a long history and proven track record both as nutrition and therapy."

Human Milk Banking Association of North America

The appropriate collection and storage of human milk is important whether the milk is for the mother's own infant or to be donated. All of the collection containers and tubing used should be cleaned by dishwasher or sterilized by boiling. Hand pumps, electric handheld pumps, hospital-grade electric breast pumps, and manual expression can be used to extract the milk. The American Academy of Breastfeeding Medicine has published evidence-based guidelines for the collection and storage of human milk for home and human milk banking.[130] Table 7.15 presents current recommendations for milk storage for home use.

Milk Banking

Human milk banks provide human milk to infants who cannot be breastfed by their mother. Premature and sick infants are most likely to receive banked milk. A woman can donate milk once or on a continuing basis if her supply exceeds the demands of her infant. There is a long history of providing human milk to infants by persons other than the biological mother.[131] Wet nurses were the main source of human milk until the early 1900s for infants not fed by their biological mothers. Milk banks began in Europe and followed in the United States. Some neonatal intensive care units had informal milk banks until the 1980s. As a result of the human immunodeficiency virus, the resurgence of tuberculosis, and risks related to donors who might abuse drugs, human milk banks are now scarce in North America, but because of recognition of the importance of human milk, demand is increasing. A network of milk banks meets and shares information through the Human Milk Banking Association of North America (HMBANA). A copy of the association's guidelines for milk storage is available for a fee.

Human milk donors are chosen by their health profile. Women are carefully screened before they can donate extra milk to milk banks. Milk banks that belong to the Milk Banking Association of North America require telephone screening, a written health and lifestyle history, and verification of the health of the mother and baby by the health care provider of each. Blood samples are tested for heptitis B, hepititis C, HIV, HTLV, and syphilis by the milk bank. Women are not accepted if they are acutely ill, have had a blood transfusion or an organ transplant within a year, drink more than 2 ounces of liquor daily, regularly use medications or megavitamins, smoke, or use street drugs. Additionally, women who eat no animal products must take vitamin supplements with B_{12} to be eligible to donate.

Human milk is carefully pasteurized to kill any potential pathogens while preserving the nutrients and active immune properties of the milk. The North American

Table 7.15 Guidelines for storage of human milk for home use

Breast Milk	Room Temperature	Refrigerator	Freezer
Freshly expressed into a closed container	6–8 hr @ 78° F or lower	3–5 days @ 39° F or lower	2 weeks in freezer compartment inside refrigerator 3–6 months in freezer section of a refrigerator with separate door 6–12 months in deep freezer @ 0° F
Previously frozen—thawed in refrigerator but not warmed or used	4 hr or less	Store in refrigerator 24 hr	Do not refreeze
Thawed outside refrigerator in warm water	For completion of feeding	Hold for 4 hr or until next feeding	Do not refreeze
Infant has begun feeding	Use only for completion of feeding and then discard	Discard	Discard

SOURCE: Reprinted from *Breastfeeding: A Guide for the Medical Profession*, 6th ed., by R. A. Lawrence and R. M. Lawrence, copyright 2005, with permission from Elsevier.

Human Milk Banking Association communicates closely with the Food and Drug Administration to follow guidelines for use of human tissues and fluids. Human milk for milk banks is stored frozen to preserve the immunologic and nutritional components. Rigid plastic (polypropylene) containers are recommended for keeping the milk composition stable. White blood cells stick to glass, but not to plastic containers.[132]

A prescription from a physician or a hospital is needed to order milk for an infant from one of the North American Milk Banking Association milk banks. Costs are approximately $3.50 per ounce before shipping charges, significantly more than the cost of human milk substitutes.[131] Some insurance companies and Medicaid programs cover the fees when it is demonstrated that donor milk is the most appropriate therapy for a specific patient.

Model Programs

Breastfeeding Promotion in Physicians' Office Practices (BPPOP)

The American Academy of Pediatrics (AAP) receives funding support from the Maternal and Child Health Bureau (MCHB), USDHHS and the AAP Friends of the Children Fund for this innovative program designed to boost breastfeeding promotion and support in underserved populations. Initiated in 1997, BPPOP's original mission was to improve the ability of AAP members to support new mothers and their breastfeeding infants and to encourage pediatricians to collaborate with others to develop breastfeeding promotion programs. Pediatricians enrolling in the program received a resource kit of educational materials and other strategies to more effectively promote, support, and manage breastfeeding with all families in their practice. In addition, pediatricians were provided technical assistance by telephone and e-mail regarding breastfeeding concerns from AAP staff and were encouraged to participate in community and regional collaborative action groups. After over 700 pediatricians nationwide joined the program, BPPOP was expanded in 2002 (BPPOP II) to include

Getty

obstetricians, family physicians, and other health care providers and to specifically target office practices working with racially and ethnically diverse populations. A speaker's kit and materials targeting underserved populations were added to the resource kit, along with the newest strategies and opportunities for multidisciplinary networking for community breastfeeding promotion. Physicians joining the program complete a self-assessment questionnaire at the beginning and end of the program and measure the impact of breastfeeding promotion efforts by tracking the breastfeeding initiation and duration rates within their practices.

BPPOP II concluded in 2004. Gaps in the program were identified—including the need for focused training related to breastfeeding support and management. It also became evident that the need for this training should occur well before physicians are in practice, preferably in residency and medical school. To address this shortcoming, the BPPOP program entered its third phase (BPPOP III), which aims to educate pediatric, obstetric/gynecologic, and family medicine residents. This will be done through a pilot-tested breastfeeding residency curriculum that will be distributed to residency programs in need, especially those that are composed of racially and ethnically diverse populations of residents and patients.

The Rush Mothers' Milk Club

The Rush Mothers' Milk Club at Rush Presbyterian–St. Luke's Medical Center is an evidence-based program of breastfeeding interventions for the neonatal intensive-care unit (NICU).[133] The program uses a team approach to feeding very low-birth-weight infants their own mothers' milk (OMM). The infants' mothers work in partnership with neonatologists, neonatal nurse practitioners, bedside nurses, and other health care professionals to ensure that the latest research is applied to an infant's OMM feeding plan. Interventions to sustain lactation for program participants have evolved from the evidence about barriers to providing OMM to NICU infants. They include preventing and treating low milk volume, achieving adequate infant growth on OMM feedings, and making the transition to at-breast-feedings in the NICU and postdischarge periods. Additionally, research about the effectiveness of peer support to sustain lactation has been incorporated into the program.[133]

Major program components include: (1) providing information for mothers to make an informed decision, (2) providing access to a hospital-grade electric breast pump, (3) providing skin-to-skin care and suckling at the empty breast as practice for the newborn, (4) babies feeding at the breast as soon as they are able to suck and swallow effectively, and (5) nursery staff helping the mother prepare for breastfeeding after discharge (www.rush.edu/patients/children/publications/notes/preemies.html). The club serves as a place for mothers to discuss their goals and concerns about breastfeeding. Mothers learn the value of their milk to their high-risk baby from the Special Care Nursery Staff and from the Rush Mothers' Milk Club. The mothers learn to measure the amount of fat and calories in their own milk and learn to capture the highest calorie portion of milk, which is usually produced during the last 10 minutes of pumping. To create a bond between the mother and baby, mothers are encouraged to use the breast pumps at the baby's bedside. Family members and friends are also encouraged to participate in the weekly Mothers' Milk Club meetings to learn the importance of breastfeeding to the high-risk infant.

The success of the Rush Mothers' Milk Club is measured by its breastfeeding initiation rates. Between 95 and 97% of all mothers who deliver high-risk infants at Rush Presbyterian–St. Luke's Medical Center begin to nurse, compared with national rates for high-risk infants of only 30 to 40%. A group of low-income African American mothers who delivered babies below 1500 grams had OMM initiation rates of 63.4%, the highest reported rates for this population in the nation.[133] The program has evidence that these high rates of initiation are due to two primary interventions: the clarity of the message that the mothers received about the importance of OMM from health care providers and their immediate access to electric breast pump rental.

Key Points

1. The majority of mothers and infants do not experience significant problems with breastfeeding. Many of the more common problems can be prevented through prenatal breastfeeding education and from informed, consistent, and individualized care and support from health professionals both in the hospital and after discharge.

2. Most medications (prescription or over-the-counter) and herbal supplements taken by the mother are excreted in her breast milk and should not be ingested until the risks to the infant are established. For most maternal conditions, required drug therapy choices are available that will not cause harm to the breastfeeding infant; recommending that a mother discontinue breastfeeding to take a medication is rarely required.

3. The level of alcohol in breast milk matches the maternal plasma alcohol levels at the time of the infant feeding; a nursing woman who drinks occasionally can limit her infant's exposure to alcohol by timing breastfeeding in relation to her drinking.

4. Regardless of feeding choice (human milk or human milk substitute), maternal smoking presents significant health risks for infants.

5. While low levels of environmental pollutants are present in most human milk, their impact on the well-being of the mother and infant is unknown. The World Health Organization and other scientific groups state that the advantages of breastfeeding far outweigh the potential risks from environmental pollutants and recommend breastfeeding in all but extreme circumstances.

6. A thorough understanding of the normal and abnormal patterns and mechanisms of jaundice (hyperbilirubinemia) in the newborn period is important for all health professionals to prevent toxicity from excessive jaundice and for protecting and ensuring successful breastfeeding. Early and frequent breastfeeding (at least 8 to 12 times in 24 hours) in the first days of life help prevent hyperbilirubinemia through maintaining infant hydration and stimulating the passage of stool. The AAP recommends against routine supplementation of nondehydrated breastfed infants with water or dextrose water as this practice will not prevent jaundice.

7. Twins and other multiples can be successfully breastfed without supplementation.

8. Exclusive breastfeeding for at least 4 months can protect against ectopic dermatitis and wheezing illnesses in children up to age 6. Advice to breastfeeding mothers with a family history of allergies regarding elimination of common allergens in their own diet should be individualized. If there is no history of allergy to a specific food in the mother's or father's family, avoiding a food because it is a potential allergen is an unnecessary precaution.

9. Human milk is the preferred food for all premature and sick newborns with rare exceptions.

10. In developed countries, where safe and affordable breast milk substitutes are available, HIV-infected women should be counseled strongly not to breastfeed to prevent mother to child transmission of HIV through breast milk.

11. Health professionals should provide breastfeeding mothers with current evidence-based guidelines for the collection (through hand pumps, electric handheld pumps, hospital-grade electric breast pumps, or manual expression) and storage of human milk for home use or human milk banking.

12. In most situations the medical problems of the mother or infant can be managed without discontinuing breastfeeding. Any medical decision to limit breastfeeding must be justified by the fact that the risk to the infant clearly outweighs the benefits of breastfeeding.

Resources

The Academy of Breastfeeding Medicine
Contains evidence-based protocols.
Website: www.bfmed.org

Best Start Social Marketing
Website: www.Bestartinc.org

The Human Milk Banking Association of North America, Inc.
Represents all of the North American human milk banks that collect, pasteurize, and distribute donated mother's milk.
Website: www.hmbana.org

International Lactation Consultants Association
Website: www.ilca.org

La Leche League
Website: www.Lalecheleague.org

C. Martin and N. F. Krebs
The Nursing Mother's Problem Solver. New York: Fireside Publishing, 2000.

Resources for the parents of multiples
La Leche League International
http://www.llli.org/NB/NBmultiples.html
Granada, K. K. *Mothering Multiples: Breastfeeding and Caring for Twins and More!!!* (Revised Edition). Schaumburg, IL: La Leche League International, 1999.
Website: www.tripletconnection.org

References

1. Morland-Schultz K, Hill, PD. Prevention of and therapies for nipple pain: a systematic review. J Obstet Gynecol Neonatal Nurs. 2005;34:428–37.

2. Mass S. Breast pain: engorgement, nipple pain and mastitis. Clin Obstet Gynecol. 2004;47:676–82.

3. Inch S. Breastfeeding problems: prevention and management. Community Pract. 2006;79:165–67.

4. Smith JW, Tully MR. Midwifery management of breastfeeding: using the evidence. J Midwifery Womens Health. 2001;46:423–38.

5. Martin C, Krebs NF. The nursing mother's problem solver. New York: Simon and Schuster, 2000.

6. Wight NE. Management of common breastfeeding issues. Pediatr Clin North Am. 2001;48:321–44.

7. Leung AK, Sauve RS (2005). Breast is best for babies. J Natl Med Assoc. 2005;97:1010–19.

8. Matheson I et al. Bacteriological findings and clinical symptoms in relation to clinical outcome in puerperal mastitis. Acta Obstet Gynecol Scand. 1988;67:723–26.

9. Lawrence RA. Breastfeeding: a guide for the medical profession. 6th ed. St Louis: Mosby, Inc., 2005.

10. Hale TW. Maternal medications during breastfeeding. Clin Obstet Gynecol. 2004;47:696–711.

11. Lagoy CT et al. Medication use during pregnancy and lactation: an urgent call for public health action. J Womens Health. 2005;14:104–09.

12. Berlin CM, Briggs GG. Drugs and chemicals in human milk. Semin Fetal Neonatal Med. 2005;10:149–59

13. Begg EJ et al. Studying drugs in human milk: time to unify the approach. J Hum Lact. 2002;18:323–32.

14. Ito S. Drug therapy for breast-feeding women. N Engl J Med. 2000;343:118–26.

15. Transfer of drugs and other chemicals into human milk. Pediatrics. 2001,108:776–89.

16. Committee on Nutrition, American Academy of Pediatrics. Pediatric Nutrition Handbook. First Edition. Elk Grove Village, IL: American Academy of Pediatrics; 2004.

17. Chung AM, Reed MD, Blumer JL. Antibiotics and breast-feeding: a critical review of the literature. Paediatr Drugs. 2002;4:817–37.

18. Guthmann RA, Bang J, Nashelsky J. Combined oral contraceptives for mothers who are breastfeeding. Am Fam Physician. 2005;72:1303–4.

19. Diaz S. Contraceptive implants and lactation. Contraception. 2002;65:39–46.

20. Nice FJ, Snyder JL, Kotansky BC. Breastfeeding and over-the-counter medications. J Hum Lact. 2000;16:319–31.

21. Briggs GG, Freeman RE, Yaffee SJ. Drugs in pregnancy and lactation. Baltimore: Williams & Wilkins; 2001.

22. Hale TW. Medications and mothers milk. Amarillo, TX: Pharmasoft Medical Publishing; 2004.

23. Department of Child and Adolescent Health. Breastfeeding and Maternal Medication: Recommendations for Drugs in the Eleventh WHO Model List of Essential Drugs, 2003. UNICEF World Health Organization, 2003.

24. The complete German Commission E monographs: therapeutic guide to herbal medicines. Boston: Integrative Medicine Communications; 1998.

25. Hardy ML. Herbs of special interest to women. J Am Pharm Assoc (Wash.). 2000;40:234–42.

26. Scott CR, Jacobson HA. Selection of international nutritional & herbal remedies

for breastfeeding concerns. Midwifery Today. 2005;38–39.

27. Kopec K. Herbal medications and breastfeeding. J Hum Lact. 1999;15:157–61.

28. Humphrey SL, McKenna DJ. Herbs and breastfeeding. Breastfeeding Abstracts. 1997;17:11–2.

29. Conover E, Buehler BA. Use of herbal agents by breastfeeding women may affect infants. Pediatr Ann. 2004;33:235–40.

30. Woolf AD. Herbal remedies and children: do they work? are they harmful? Pediatrics. 2003;112:240–46.

31. Humphrey S. Sage advice on herbs and breastfeeding. LEAVEN. 1998;34:43–47.

32. Leung A, Foster S. Encyclopedia of common natural ingredients used in foods, drugs, and cosmetics. New York: John Wiley & Sons; 1996.

33. Cunningham E, Hansen K. Question of the month: where can I get information on evaluating herbal supplements? J Am Diet Assoc. 1999;99:1240.

34. The American Botanical Council. The ABC Clinical Guide to Herbs. American Botanical Council, 2005.

35. Cartwright MM. Herbal use during pregnancy and lactation: a need for caution. The Digest (Summer), 1–3. 2001. American Dietetic Association Public Health/Community Nutrition Practice Group.

36. Klier CM et al. St. John's wort (Hypericum perforatum) and breastfeeding: plasma and breast milk concentrations of hyperforin for 5 mothers and 2 infants. J Clin Psychiatry. 2006;67(2):305–9.

37. Lee A et al. The safety of St. John's wort (Hypericum perforatum) during breastfeeding. J Clin Psychiatry. 2003;64(8):966–68.

38. Gabay MP. Galactogogues: medications that induce lactation. J Hum Lact. 2002;18:274–79.

39. Betzold CM (2004). Galactagogues. J Midwifery Womens Health. 2004;49:151–54.

40. Snowden HM, Renfrew MJ, Woolridge MW. Treatments for breast engorgement during lactation. Cochrane Database Syst Rev. 2003;CD000046.

41. Roemheld-Hamm B. Chasteberry. Am Fam Physician. 2005;72:821–24.

42. Foote J, Rengers B. Medicinal herb use during pregnancy and lactation. The Perinatal Nutrition Report. 1998;5:1–2.

43. Mennella J. Alcohol's effect on lactation. Alcohol Res Health. 2001;25:230–34.

44. Giglia R, Binns C. Alcohol and lactation: a systematic review. Nutrition and Dietetics. 2006;63:103.

45. Ho E et al. Alcohol and breast- feeding: calculation of time to zero level in milk. Biol Neonate. 2001;80:219–22.

46. Mennella JA, Pepino MY, Teff KL. Acute alcohol consumption disrupts the hormonal milieu of lactating women. J Clin Endocrinol Metab. 2005;90:1979–85.

47. Mennella JA. Regulation of milk intake after exposure to alcohol in mothers' milk. Alcohol Clin Exp Res. 2001;25:590–93.

48. Mennella JA, Gerrish CJ. Effects of exposure to alcohol in mother's milk on infant sleep. Pediatrics. 1998;101:E2.

49. Mennella JA, Garcia Comez PL. Sleep disturbances after acute exposure to alcohol in mother's milk. Alcohol Int Biomed J. 2001;25:153–58.

50. Gunzerath L et al. National Institute on Alcohol Abuse and Alcoholism report on moderate drinking. Alcohol Clin Exp Res. 2004;28:829–47.

51. Little RE et al. Maternal alcohol use during breast-feeding and infant mental and motor development at one year. N Engl J Med. 1989;321:425–30.

52. Little RE. Maternal use of alcohol and breastfed infants. N Engl J Med. 1990;322:339.

53. Little RE, Northstone K, Golding J. The ALSPAC Team. Alcohol, breastfeeding, and development at 18 months. Pediatrics. 2002;109:1–6.

54. Mohrbacher N, Stock J. The Breast-feeding Answer Book. Schaumburg, IL: La Leche League International; 2003.

55. Subcommittee on Nutrition During Lactation, Committee on Nutritional Status During Pregnancy and Lactation. Food and Nutrition Board Institute of Medicine National Academy of Sciences. Nutrition during Lactation: Summary, Conclusions, and Recommendations. Washington DC: National Academy Press; 1991.

56. Howard CR, Lawrence RA. Xenobiotics and breastfeeding. Pediatr Clin North Am. 2001;48:485–504.

57. Dempsey DA, Benowitz NL. Risks and benefits of nicotine to aid smoking cessation in pregnancy. Drug Saf. 2001;24:277–322.

58. Amir LH, Donath SM. Does maternal smoking have a negative physiological effect on breastfeeding? The epidemiological evidence. Birth. 2002;29:112–23.

59. Ilett KF et al. Use of nicotine patches in breast-feeding mothers: transfer of nicotine and cotinine into human milk. Clin Pharmacol Ther. 2003;74:516–24.

60. Dahlstrom A et al. Nicotine and cotinine concentrations in the nursing mother and her infant. Acta Paediatr Scand. 1990;79:142–47.

61. Gaffney KF (2006). Postpartum smoking relapse and becoming a mother. J Nurs Scholarsh. 2006;38:26–30.

62. Nickerson K. Environmental contaminants in breast milk. J Midwifery Womens Health. 2006;51:26–34.

63. Molyneux A. Nicotine replacement therapy. British Medical Journal. 2006;328:454–56.

64. Fernandez-Ruiz J et al. Cannabinoids and gene expression during brain development. Neurotox Res. 2004;6:389–401.

65. Astley SJ, Little RE. Maternal marijuana use during lactation and infant development at one year. Neurotoxicol Teratol. 1990;12:161–68.

66. Rivera-Calimlim L. The significance of drugs in breast milk. Pharmacokinetic considerations. Clin Perinatol. 1987;14:51–70.

67. Nehlig A, Debry G. Consequences on the newborn of chronic maternal consumption of coffee during gestation and lactation: a review. J Am Coll Nutr. 1994;13:6–21.

68. World Health Organization. Safe food: crucial for child development. Available from: www.who.int/ceh/publications/en/poster15new.pdf accessed 10/06.

69. Massart F et al. Human breast milk and xenoestrogen exposure: a possible impact on human health. J Perinatol. 2005;25:282–88.

70. Condon M (2005). Breast is best, but it could be better: what is in breast milk that should not be? Pediatr Nurs. 2005;31:333–38.

71. Wang RY et al. (2005). Human milk research for answering questions about human health. J Toxicol Environ Health. 2005;68:1771–1801.

72. Food and Nutrition Board, Committee on the Evaluation of the Addition of Ingredients New to Infant Formula. Infant formula: Evaluating the safety of new ingredients. Institute of Medicine of the National Academies. Washington, DC: National Academy Press; 2004.

73. Schreiber JS. Parents worried about breast milk contamination. What is best for baby?

74. U.S. Department of Health and Human Services. HHS Blueprint for action on breastfeeding 2000. Department of Health and Human Services, Office of Women's Health.

75. World Health Organization. Report of the WHO working group on the assessment of health risks for human infants from exposure to PCDDs, PCDFs, PCBs. Chemosphere. 1998;57:1627–43.

76. Pronczuk J et al. Global perspectives in breast milk contamination: infectious and toxic hazards. Environ Health Perspect. 2002;110:A349–A351.

77. LaKind JS, Berlin CM, Naiman DQ. Infant exposure to chemicals in breast milk in the United States: what we need to learn from a breast milk monitoring program. Environ Health Perspect. 2001;109:75–88.

78. American Academy of Pediatrics Subcommittee on Hyperbilirubinemia. Management of hyperbilirubinemia in the newborn infant 35 or more weeks of gestation. Pediatrics. 2004;114:297–316.

79. Cohen SM. Jaundice in the full-term newborn Pediatr Nurs. 2006;32:202–8.

80. Stokowski LA. Early recognition of neonatal jaundice and kernicterus. Adv Neonatal Care. 2002;2:101–14.

81. Watchko JF. Vigintiphobia revisited. Pediatrics. 2005;115:1747–53.

82. Geiger AM, Petitti DB, Yao JF. Rehospitalisation for neonatal jaundice: risk factors and outcomes. Paediatr Perinat Epidemiol. 2001;15:352–58.

83. Joint Commission on Accreditation of Healthcare Organizations. Kernicterus threatens healthy newborns. Sentinel Event Alert. April 2001 Issue 18. Joint Commission on Accreditation of Healthcare Organizations. 2001.

84. Gartner LM, Herschel M. Jaundice and breastfeeding. Pediatr Clin North Am. 2001;48:389–99.

85. Reiser DJ. Neonatal jaundice: physiologic variation or pathologic process. Crit Care Nurs Clin North Am. 2004;16:257–69.

86. Reiser DJ. Hyperbilirubinemia. AWHONN. Lifelines. 2001;5:55–61.

87. Neonatal jaundice and kernicterus. Pediatrics. 2001;108:763–65.

88. Gartner LM. Breastfeeding and jaundice. J Perinatol. 2001;21 Suppl 1:S25–S29.

89. Gourley GR. Breast-feeding, neonatal jaundice and kernicterus. Semin Neonatol. 2002;7:135–41.

90. Willis SK, Hannon PR, Scrimshaw SC. The impact of the maternal experience with a jaundiced newborn on the breastfeeding relationship. J Fam Pract. 2002;51:465.

91. Geraghty SR et al. Breast milk feeding rates of mothers of multiples compared to mothers of singletons. Ambul Pediatr. 2004;4:226–31.

92. Flidel-Rimon O, Shinwell ES. Breastfeeding multiples. Semin Neonatol. 2002;7:231–39.

93. Gromada KK, Spangler AK. Breastfeeding twins and higher-order multiples. J Obstet Gynecol Neonatal Nurs. 1998;27:441–49.

94. Gromada KK. Breastfeeding more than one: multiples and tandem breastfeeding.

NAACOGS. Clin Issu Perinat Womens Health Nurs. 1992;3:656–66.

95. Auer C, Gromada KK. A case report of breastfeeding quadruplets: factors perceived as affecting breastfeeding. J Hum Lact. 1998;14:135–41.

96. Palmer G. The Politics of Breastfeeding. 1993; 2nd ed. 32. London, Pandora Press/ Harper Collins.

97. Saint L, Maggiore P, Hartmann PE. Yield and nutrient content of milk in eight women breast-feeding twins and one woman breast-feeding triplets. Br J Nutr. 1986;56:49–58.

98. Hattori R, Hattori H. Breastfeeding twins: guidelines for success. Birth. 1999; 26:37–42.

99. Geraghty SR, Khoury JC, Kalkwarf HJ. Comparison of feeding among multiple birth infants. Twin Res. 2004;7:542–47.

100. Zeiger RS, Friedman NJ. The relationship of breastfeeding to the development of atopic disorders. Nestle Nutr Workshop Ser Pediatr Program. 2006;57:93–105.

101. O'Connell EJ. Pediatric allergy: a brief review of risk factors associated with developing allergic disease in childhood. Ann Allergy Asthma Immunol. 2003;90:53–8.

102. Palmer DJ, Makrides M. Diet of lactating women and allergic reactions in their infants. Curr Opin Clin Nutr Metab Care. 2006;9:284–88.

103. Mofidi S. Nutritional management of pediatric food hypersensitivity. Pediatrics. 2003;111:1645–53.

104. Hill DJ et al. Effect of a low-allergen maternal diet on colic among breastfed infants: a randomized, controlled trial. Pediatrics. 2005;116:e709–e715.

105. Mennella JA, Beauchamp GK. The effects of repeated exposure to garlic-flavored milk on the nursling's behavior. Pediatr Res. 1993;34:805–8.

106. Mennella JA, Beauchamp GK. Maternal diet alters the sensory qualities of human milk and the nursling's behavior. Pediatrics. 1991;88:737–44.

107. Cooper RL, Cooper MM. Red pepper-induced dermatitis in breast-fed infants. Dermatology. 1996;193:61–2.

108. Zeretzke K. Allergies and the breast-feeding family. New Beginnings. 1998; 15(4):100.

109. Wight NE. Breastfeeding the borderline (near-term) preterm infant. Pediatr Ann. 2003;32:329–36.

110. American Academy of Pediatrics Work Group on Breastfeeding. Breastfeeding and the use of human milk. Pediatrics. 1997;100: 1035–39.

111. Shulman RJ et al. Early feeding, feeding tolerance, and lactase activity in preterm infants. J Pediatr. 1998;133:645–49.

112. Spatz DL. State of the science: use of human milk and breast-feeding for vulnerable infants. J Perinat Neonatal Nurs. 2006; 20:51–55.

113. Callen J, Pinelli J. A review of the literature examining the benefits and challenges, incidence and duration, and barriers to breastfeeding in preterm infants Adv Neonatal Care. 2005;5:72–88.

114. Rodriguez NA, Miracle DJ, Meier PP. Sharing the science on human milk feedings with mothers of very-low-birth-weight infants. JOCNN. 2005;34:109–119.

115. Schanler RJ, Shulman RJ, Lau C. Feeding strategies for premature infants: beneficial outcomes of feeding fortified human milk versus preterm formula. Pediatrics. 1999;103:1150–57.

116. Schanler RJ. Overview: the clinical perspective. J Nutr. 2000;130:417S–9S.

117. Spatz DL. Ten steps for promoting and protecting breastfeeding for vulnerable infants. J Perinat Neonatal Nurs. 2004;18: 385–96.

118. Sisk PM et al. Lactation counseling for mothers of very low birth weight infants: effect on maternal anxiety and infant intake of human milk. Pediatrics. 2006;117:67–75.

119. Lawrence RM, Lawrence RA. Breast milk and infection. Clin Perinatol. 2004;31: 501–28.

120. Lawrence RM, Lawrence RA. Given the benefits of breastfeeding, what contraindications exist? Pediatr Clin North Am. 2001;48: 235–51.

121. Newell ML. Current issues in the prevention of mother-to-child transmission of HIV-1 infection. Trans R Soc Trop Med Hyg. 2006;100:1–5.

122. Dorosko SM. Vitamin A, mastitis, and mother-to-child transmission of HIV-1 through breast-feeding: current information

and gaps in knowledge. Nutr Rev. 2005; 63:332–46.

123. Jackson DJ et al. HIV and infant feeding: issues in developed and developing countries. J Obstet Gynecol Neonatal Nurs. 2003;32:117–27.

124. American Academy of Pediatrics, Section on Breastfeeding. Breastfeeding and the use of human milk. Pediatrics. 2005;115:496–506.

125. Effect of breastfeeding on infant and child mortality due to infectious diseases in less developed countries: a pooled analysis. WHO Collaborative Study Team on the Role of Breastfeeding on the Prevention of Infant Mortality. Lancet. 2000;355: 451–55.

126. Coutsoudis A. Breastfeeding and the HIV positive mother: the debate continues. Early Hum Dev. 2005;81:87–93.

127. World Health Organization. HIV and Infant Feeding. A review of HIV transmission through Brestfeeding: World Health Organization. 1998. Geneva, Switzerland, World Health Organization.

128. WHO Technical Consultation on Behalf of the UNFPA/UNICEF/WHO/UNAIDS Inter-Agency Task Team on Mother-to-Child Transmission of HIV. New data on the prevention of mother-to-child transmission of HIV and their policy implications: Conclusions and recommendations. Geneva. Oct 13 2000. Geneva, World Health Organization.

129. Hartmann SU, Berlin CM, Howett MK. Alternative modified infant-feeding practices to prevent postnatal transmission of human immunodeficiency virus type 1 through breast milk: past, present, and future. J Hum Lact. 2006;22:75–88.

130. Academy of Breastfeeding Medicine. Protocol for the Storage of Human Milk. www.bfmed.org accessed 11/06.

131. Jones F. History of North American donor milk banking: one hundred years of progress. J Hum Lact. 2003;19: 313–18.

132. Paxson CI. Survival of human milk leukocytes. J Pediatr. 1979;94:61–64.

133. Meier PP et al. The Rush Mother's Milk Club: breastfeeding intereventions for mothers with very low birth weight infants. JOGNN. 2004;33:164–74.

> "A babe is fed with milk
> and praise."
> Charles Lamb, *The First Tooth*

Chapter 8

Infant Nutrition

Chapter Outline

- Introduction
- Assessing Newborn Health
- Infant Development
- Energy and Nutrient Needs
- Physical Growth Assessment
- Feeding in Early Infancy
- Development of Infant Feeding Skills
- Nutrition Guidance
- Common Nutritional Problems and Concerns
- Cross-Cultural Considerations
- Vegetarian Diets
- Nutrition Intervention for Risk Reduction

Digital Stock

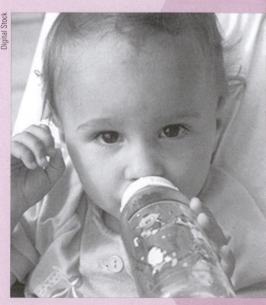

Digital Stock

Photo Disc

Prepared by **Janet Sugarman Isaacs**

Key Nutrition Concepts

1 The dynamic growth experienced in infancy is the most rapid of any age. Inadequate nutrition in infancy, however, leads to consequences that may be lifelong, harming both future growth and future development.

2 Progression in feeding skills expresses important developmental steps in infancy that signal growth and nutrition status.

3 Nutrient requirements of term newborns have to be modified for preterm infants. Knowing the needs of sick and small newborns results in greater understanding of the complex nutritional needs of all newborns and infants.

4 Changing feeding practices, such as the care of infants outside the home and the early introduction of foods, markedly affects nutritional status of infants.

Introduction

This chapter is about healthy *full-term infants* born at 37 weeks of gestation or later, and healthy *preterm infants* born at 34 weeks or later. These newborns are expected to have typical growth and development.

Full-Term Infants   Infants born between 37 and 42 weeks of gestation.

Preterm Infants   Infants born before 37 weeks of gestation.

Infant Mortality   Death that occurs within the first year of life.

The term *normal* is not used much in this chapter, because its opposite, *abnormal,* is an emotionally laden term, particularly when describing babies to their parents. *Typical* is used in place of *normal* when possible.

This chapter discusses how nutrition is an important contributor to the complex development of infants. Both biological and environmental factors interact during infant growth and development. Models about the interaction of biological and environmental factors are often incomplete. They are not always adequate for describing complex interactions, such as how mealtime stimulates language development and how food preferences develop during infancy. The complexity of infant development contributes to our individuality later on.

The Healthy People 2010 objectives involving infants concern reducing infant mortality, preterm birth rates, incidence of spina bifida and neural tube defects, fetal alcohol syndrome, and other birth defects (see also Chapter 9 for discussion of these topics). Progress is slow as tracked by the Centers for Disease Control and Health Promotion as more infants are surviving with conditions previously considered fatal, who require prolonged use of high-technology health care.

Assessing Newborn Health

Birth Weight as an Outcome

The weight of a newborn is a key measure of health status during pregnancy. The average gestation for a full-term infant is 40 weeks, with a range from 37 to 42 weeks. Full-term infants usually weigh 2500–3800 grams (5.5 to 8.5 lb) and are 47–54 centimeters (18.5–21.5 inches) in length.[1] There were over 4 million births in the United States in 2002, and 88% of these newborns were full term.[2] Infants with normal birth weights are less likely to require intensive care and are usually healthy in the long run. Conversely, preterm births, regardless of birth weight, are those born at 37 weeks or less. Preterm means incomplete development has taken place.[3]

Infant Mortality

Worldwide *infant mortality* rates rank the United States at twenty-seventh place, far worse than many other countries.[3] The reasons for the high rate of mortality are many, but the prevalence of low birth weight is a major factor. In 2002, 7.8% of live births in the United States were low birth weight, or less than 2500 grams (5.5 lb).[2] The three leading causes of infant mortality in 2002 were congenital malformations, preterm births, and sudden infant death syndrome.[3] The higher incidence of infant mortality, low birth weight, and preterm births in African American infants is of particular concern. Table 8.1 shows infant deaths in the United States based on race of the mother.[3] The basis for racial disparity and preterm birth is a major focus in federal initiatives to combat infant deaths in the United States. Despite efforts to lower neonatal and infant deaths, rates are still too high.

Combating Infant Mortality

Efforts to improve newborn health are under way on many levels. In the United States, improved access to specialized care for mothers and infants has been credited in

Table 8.1 United States infant mortality rates by race of the mother[3]

	Deaths per 1000 Live Births
All newborns	6.8
White	5.7
Black	13.5
Asian/Pacific Islander	4.8
American Indian or Alaska Native	8.7
Hispanic/Mexican	5.5

part for a decline in the infant mortality rate.[2] This is a multifaceted problem, however, affected by:

- Social and economic status of the families and women
- Access to health care
- Medical interventions
- Teenage pregnancy rates
- Availability of abortion services
- Failure to prevent preterm and low-birth-weight births

Resources have been concentrated on the proportion of newborns identified at risk. Some of these resources are major payers of health services, such as Medicaid and the Child Health Initiatives Program (CHIP).[4] The following concepts underscore the commitment of resources for infants:

- Recognition that birth weight is important for long-term health outcomes.
- Prevention and treatment of complications for at-risk infants are investments for the future.

The emphasis on prevention is seen in various programs. The Early Periodic Screening, Detection, and Treatment Program (*EPSDT*) is a major source of preventive and routine care for infants in low-income families. Immunizations during infancy are another example of a prevention approach.

Nutrition is included in some national prevention programs. The Special Supplemental Food Program for Women, Infants, and Children (WIC) and the Centers for Disease Control (CDC) collaborate to track infant growth as a part of the Nutrition Surveillance Program.[5] The Bright Futures program promotes and improves the health, education, and well-being of infants and children. Nutrition is one component of the program's guidelines about common issues and concerns. Bright Futures is an example of a comprehensive approach to health supervision by a collaboration of government and professional groups.[6]

Standard Newborn Growth Assessment

Newborn health status is assessed by various indicators of growth and development taken right after birth. Indicators include birth weight, length, and head circumference for gestational age. The designation "small for gestational age" (SGA)—also called "small for dates," *"intrauterine growth retardation (IUGR),"* or "intrauterine growth restriction"—means the newborn's birth weight falls below the 10th percentile of weight for gestational age.[7] Infants above the 90th percentile are considered large for gestational age (LGA). Those in-between are appropriate for gestational age (AGA).[8]

Infant Development

Monitoring infants' nutritional status requires an understanding of their overall development. Full-term newborns have a wider range of abilities than previously recognized; they hear and move in response to familiar sounds, such as the mother's voice.[9] Newborns demonstrate four states of arousal, ranging from sleeping to fully alert, and responsiveness differs in part on their state of arousal.[9] Recognizing the state of arousal is a part of nursing successfully.

Organs and systems developed during gestation continue to increase in size and complexity during infancy. The newborn's central nervous system is immature; that is, the neurons in the brain are less organized compared to those of the older infant. As a result, the newborn gives inconsistent or subtle cues of hunger and other needs, compared to the cues given later. The fact that newborns can *root, suckle,* and coordinate swallowing and breathing within hours of birth shows that feeding is directed by reflexes and the central nervous system.[9,10] Newborn *reflexes* are protective for them. These reflexes fade as they are replaced by purposeful movements during the first few months of life.[9] Table 8.2 on the next page lists major reflexes in newborns.[9-11]

> **EPSDT**   The Early Periodic Screening, Detection, and Treatment Program is a part of Medicaid and provides routine checkups for low-income families.
>
> **Intrauterine Growth Retardation (IUGR)**   Fetal undergrowth from any cause, resulting in a disproportionality in weight, length, or weight-for-length percentiles for gestational age. Sometimes called *intrauterine growth restriction.*
>
> **Reflex**   An automatic (unlearned) response that is triggered by a specific stimulus.
>
> **Rooting Reflex**   Action that occurs if one cheek is touched, resulting in the infant's head turning toward that cheek and the infant opening his mouth.
>
> **Suckle**   A reflexive movement of the tongue moving forward and backward; earliest feeding skill.

Motor Development

Motor development reflects an infant's ability to control voluntary muscle movement. There are several models for describing infant development, but none provide a complete description and explanation of the rapid advances in motor skills achieved during infancy.[9,11] Illustration 8.1 on page 222 depicts motor development during the first 15 months.[7] It is a great source of pride for parents when their baby first rolls over or sits up. The development of muscle control is top down, meaning head control is the start, and last comes lower legs.[10] Muscle development is also from central to peripheral, meaning the infant learns to control the shoulder and arm muscles before muscles in the hands.[10,12] Motor development influences both the ability of the infant to feed and the amount of calories expended in the activity.[13] An example of how motor development affects feeding is the ability to sit in a high chair. Only when an infant has achieved the motor development of head control and sitting balance and certain

Table 8.2   Major reflexes found in newborns

Name	Response	Significance
Babinski	A baby's toes fan out when the sole of the foot is stroked from heel to toe.	Perhaps a remnant of evolution
Blink	A baby's eyes close in response to bright light or loud noise.	Protects the eyes
Moro	A baby throws its arms out and then inward (as if embracing) in response to loud noise or when its head falls.	May help a baby cling to its mother
Palmar	A baby grasps an object placed in the palm of its hand.	Percursor to voluntary grasping
Rooting	When a baby's cheek is stroked, it turns its head toward the cheek that was stroked and opens its mouth.	Helps a baby find the nipple
Stepping	A baby who is held upright by an adult and is then moved forward begins to step rhythmically.	Precursor to voluntary walking
Sucking	A baby sucks when an object is placed in its mouth.	Permits feeding
Withdrawal	A baby withdraws its foot when the sole is pricked with a pin.	Protects a baby from unpleasant stimulation

SOURCE: Robert V. Kail and John C. Cavanaugh, *Human Development: A Lifespan View*. 2nd ed. Belmont, CA: Wadsworth/Thomson Learning, 2000, p. 99.[9]

Based on Shirley, 1931, and Bayley, 1969.

Illustration 8.1   Gross motor skills.

SOURCE: Robert V. Kail and John C. Cavanaugh, *Human Development: A Lifespan View*. 2nd ed. Belmont, CA: Wadsworth/Thomson Learning, 2000, p. 99.[9]

reflexes have disappeared can oral feeding with a spoon be achieved.[12] The development of motor skills slowly increases the caloric needs of infants over time because increased activity requires more energy.[13] Infants who crawl expend more calories in physical activity than do younger infants who cannot roll over.

Critical Periods

The concept of critical period is based on a fixed time period during which certain behaviors emerge. This theory of development suggests that there is a time period, or window of development, during which certain skills must be learned in order for subsequent learning to occur.[11] A critical period for the development of oral feeding may explain some later feeding problems in infancy.[12] In the typical healthy newborn, the mouth is a source of pleasure and exploring, an important form of early learning. When a newborn has a prolonged period of respiratory support, for example, the baby may not associate mouth sensations with pleasure, but rather with discomfort. Under such circumstances the critical period for associating mouth sensations with pleasure and exploration may have been missed. After discharge, such an infant may be a reluctant feeder and have difficulty learning to enjoy food from a spoon.

Cognitive Development

The concept of biological and environmental systems interacting is seen in Illustration 8.2 on page 224 showing *sensorimotor* development.[9] These skills influence feeding in important ways. For example, the stage during which infants are very sensitive to food texture is also when their speech skills are emerging.[9] The interaction between the environment and stimulating the senses in the developing brain is now seen as structuring the nervous system long term.[14] The latest research suggests that access to adequate calories and protein may not be sufficient for maximizing brain maturation if the social and emotional growth of the infant is not stimulated simultaneously.[14] Cognitive development is also subjected to genetic controls, which turn genes on and off in different time frames and sites within the body.[15] Specific vitamins being needed at specific time frames of development is an example of the interaction of genetics and environment.

Digestive System Development

"Now good digestion wait on appetite, and health on both."

William Shakespeare, *Macbeth*

A healthy digestive system is necessary for successful feeding. Parents may worry about gastrointestinal problems, in part because of misinformation about nutrition in infancy.[16] For example, an infant with soft, loose stools may be considered by the parents to have diarrhea if they do not know that this is typical for breastfed infants.[17] Another common example is parents being concerned that their infant's gastrointestinal discomfort may interfere with weight gain, even though growth usually progresses well. It takes up to 6 months for the infant gastrointestinal tract to mature, and the time required varies considerably from one individual to another.[6,18]

During the third trimester the fetus swallows amniotic fluid, and this stimulates the lining of the intestine to grow and mature.[7] At birth, the healthy newborn's digestive system is sufficiently mature to digest fats, protein, and simple sugars and to absorb fats and amino acids. Although healthy newborns do not have the same levels of digestive enzymes or rate of stomach emptying as older infants, the gut is functional at birth.[7]

After birth and through the early infancy period, the coordination of peristalsis within the gastrointestinal tract improves. Maturation of peristalsis and rate of passage are associated with some forms of gastrointestinal discomfort in infants.[18] Infants often have conditions that reflect the immaturity of the gut, such as colic, *gastroesophageal reflux (GER),* unexplained diarrhea, and constipation.[6,8] Such conditions do not interfere with the ability of the intestinal villa to absorb nutrients, and typically do not hinder growth. Other factors influence the rate of food passage through the colon and the gastrointestinal discomfort seen in infants. These include:

- *Osmolarity* of foods or liquids (which affects how much water is in the intestine)
- Colon bacterial flora
- Water and fluid balance in the body[7]

> **Sensorimotor**   An early learning system in which the infant's senses and motor skills provide input to the central nervous system.
>
> **Gastroesophageal Reflux (GER)** Movement of the stomach contents backward into the esophagus, due to stomach muscle contractions. The condition may require treatment depending on its duration and degree. Also known as *gastroesophageal reflux disease (GERD).*
>
> **Osmolarity**   Measure of the number of particles in a solution, which predicts the tendency of the particles to move from high to low concentration. Osmolarity is a factor in many systems, such as in fluid and electrolyte balance.

Parenting

"A babe in the house is a well-spring of pleasure."

Martin Farquhar Tupper, *On Education*

Even though the newborn can breast- or bottle-feed after birth, skills of new parents develop slowly. The parents' ability to recognize and respond to infant cues of hunger and satiety improve over time. New parents have to learn the temperament of their infants. Temperament has a biological basis and includes the infant's style or patterns of behavior.[9] Temperament includes the infant's emotional reactions to new situations, activity level, and sociability. The fit between the infant's temperament and that of the

SUBSTAGES DURING THE SENSORIMOTOR STAGE OF DEVELOPMENT

Substage	Age (months)	Accomplishment	Example	
1	0–1	Reflexes become coordinated.	Sucking a nipple	
2	1–4	Primary circular reactions appear—an infant's first learned reactions to the world.	Thumb sucking	
3	4–8	Secondary circular reactions emerge, allowing infants to explore the world of objects.	Shaking a toy to hear a rattle	
4	8–12	Means–end sequencing of schemes is seen, marking the onset of intentional behavior.	Moving an obstacle to reach a toy	
5	12–18	Tertiary circular reactions develop, allowing children to experiment.	Shaking different toys to hear the sounds they make	
6	18–24	Symbolic processing is revealed in language, gestures, and pretend play.	Eating pretend food with a pretend fork	

Illustration 8.2 Sensorimotor stage of development.

SOURCE: Robert V. Kail and John C. Cavanaugh, *Human Development: A Lifespan View*. 2nd ed. Belmont, CA: Wadsworth/Thomson Learning, 2000, p. 128.[9]

parents can increase or decrease feeding problems. For example, new parents may take a while to recognize that the newborn is more comfortable nursing at one noise level in the home than at another. Infants who are 6 months of age or older are better able to let parents know their needs and temperament. Conflicts in temperament may escalate as time progresses; they may become a factor in failure to thrive or other growth and feeding problems in later years.[12,18]

Energy and Nutrient Needs

The 2005 Dietary Guidelines for Americans address the needs of children aged 2 and older—not infants. Recommendations for infants are largely based on information from the Dietary Reference Intakes (DRI) and professional groups such as the National Academy of Medicine and the American Academy of Pediatrics.[13,19–21]

Caloric Needs

The caloric needs of typical infants are higher per pound of body weight than at any other time of life. The range in caloric requirements for individual infants is broad, ranging from 80 to 120 calories per kg (2.2 lb) body weight.[13]

The average caloric need of infants in the first 6 months of life is 108 cal per kg body weight, based on growth in breastfed infants.[13] From 6 to 12 months of age, the average caloric need is 98 cal/kg.[13] Factors that account for the range of caloric needs of infants include the following:

- Weight
- Growth rate
- Sleep/wake cycle
- Temperature and climate
- Physical activity
- Metabolic response to food
- Health status and recovery from illness

Current recommendations for infants of 108 and 98 cal are considered about 15% too high, based on new study results.[13] However, too few energy expenditure studies have been done on infants to reach a consensus about changing energy requirements.

Protein Needs

Recommended protein intake from birth up to 6 months averages 2.2 g of protein per kg of body weight, and from 6 to 12 months the need is for 1.6 g of protein per kg.[13]

Protein needs of individual infants vary with the same factors listed for calorie needs. Protein needs are influenced more directly by body composition than calorie needs are, because metabolically active muscles require more protein for maintenance.

Most young infants who breastfeed or consume the recommended amounts of infant formula meet protein needs without added foods. Infants may exceed their protein needs based on the DRI when they consume more formula than recommended for age and when protein sources such as baby cereal are added to infant formula.[22] Inadequate or excessive protein intake can result for infants who are offered formula that is not made correctly, such as when less or more water is used than appropriate in preparation. Essential amino acids required by healthy infants are constant across the first year of life.

Fats

There is no specific recommended intake level of fats for infants. Fat restriction is not recommended. Breast milk provides 55% of its calories from fat, and this percentage reflects an adequate intake of fat by infants.[17] The main source of fat in most infant diets is breast milk or formula. Cholesterol intake should not be limited in infancy because infants have a high need for it and its related metabolites in gonad and brain development.[23] The percentage of fat in the diet drops after the infant has baby foods on a spoon added, since most baby foods are low in fat. Infants need fat, which is a concentrated source of calories, to support their high need for calories. Fat requirements in infancy are complicated by the differences in digestion and transport of fats based on fatty acid chain length.[23] *Short-* and *medium-chain fats* such as those in breast milk are more readily utilized than *long-chain fats*, such as in some infant formulas. Long-chain fatty acids are the most common type in food, but they are more difficult for young infants to utilize. Examples of long-chain fatty acids are C16–C18 and include palmitic (C16:0), stearic (C18:1), and linoleic (C18:2) acids.[23,24]

Infants use fats to supply energy to the liver, brain, and muscles including the heart. The fact that infants have high caloric needs compared to those for older children means that infants use fats more regularly for generating energy. Young infants cannot tolerate fasting for long because it quickly uses up both carbohydrate and fat energy sources. This effect of fasting explains in part why infants cannot sleep through the night. In rare cases, some infants cannot metabolize fats due to a genetic condition that blocks specific enzymes needed to generate energy. Such infants may get sick suddenly; a few have been identified with this rare condition of fat metabolism only after dying of what appeared to be sudden infant death syndrome (SIDS).[25]

Fats in food provide the two essential fatty acids, linoleic and alpha-linolenic acid. Essential fatty acids are substrates for hormones, steroids, endocrine, and neuroactive compounds in the developing brain.[23] The long-chain polyunsaturated fatty acids docosahexaenoic acid (DHA) and eicosapentaenoic acid (EPA) are derived to some extent from an essential fatty acid.[25] Full-term breast-fed babies do not need supplemental fat components or essential fatty acids.[26]

Metabolic Rate, Calories, Fats, and Protein—How Do They All Tie Together?

The metabolic rate of infants is the highest of any period after birth.[7] The higher rate is primarily related to infants' rapid growth rate and the high proportion of infant weight that is made up of muscle. The usual body fuel for metabolism is glucose. When sufficient glucose is available, growth is likely to proceed. When glucose from carbohydrates is limited, amino acids will be converted into glucose and used for energy and therefore are made unavailable for growth. The conversion of amino acids into glucose is a more dynamic process in infants as compared to adults. The breakdown of amino acids for use as energy occurs during illness in adults, but it can occur daily in fast-growing infants. Circulating amino acids in the blood from ingested foods will be used for glucose production, and if these are not sufficient, the body will release amino acids from muscles. This process of breaking down body protein to generate energy is known as catabolism. If catabolism goes on too long, it will slow or stop growth in infants. The precise site of all this metabolic activity is inside organs such as the liver, and in *mitochondria* within cells. If carbohydrates are not provided in sufficient amounts, growth will plateau because ingested protein and fats will be used for meeting energy needs.

> **Short-Chain Fats**   Carbon molecules that provide fatty acids less than 6 carbons long as products of energy generation from fat breakdown inside cells. Short-chain fatty acids are not usually found in foods.
>
> **Medium-Chain Fats**   Carbon molecules that provide fatty acids with 6–10 carbons, again not typically found in foods.
>
> **Long-Chain Fats**   Carbon molecules that provide fatty acids with 12 or more carbons, which are commonly found in foods.
>
> **Mitochondria**   Intracellular unit in which fatty acid breakdown takes place and many enzyme systems for energy production inside cells are regulated.

Other Nutrients and Non-nutrients

Fluoride   The DRI for fluoride is 0.1 mg daily for infants less than 6 months of age, and 0.5 mg daily for 7- to 12-month-olds.[27] Fluoride is incorporated into the enamel of forming teeth, including those not yet erupted. Dental caries in early childhood are more frequent if an infant does not meet the DRI for fluoride. If an infant has more fluoride than recommended, tooth discoloration may result later. Community water fluoridation is safe for breastfeeding women and for infants.[6] Most infants who live in areas with fluoridated water do not require another source of fluoride. Fluoride is low in breast milk.[27] In areas in which fluoridated water is not available, prescribed fluoride is recommended for breastfed infants. If families routinely purchase bottled water, they should select water that has fluoride added.[6]

Vitamin D   Vitamin D, or preformed forms of vitamin D such as cholecalciferol, is required for bone mineralization with calcium.[27] Vitamin D supplementation is recommended for exclusively breastfed infants to obtain a minimum intake of 200 IU per day starting at 2 months of age. The identification of babies with rickets has resulted in this recommendation for breastfed infants from the American Academy of Pediatrics (AAP).[20] Vitamin D is not supplied by human milk in sufficient amounts, but it is added to infant formulas. Vitamin D is discussed in Chapter 6, page 163. When outside, parents are encouraged to apply ultraviolet-light skin-protective lotions to their children's exposed skin, which block vitamin D formation from sunlight.

Sodium   Sodium is a major component of extracellular fluid and an important regulator of fluid balance. Estimated minimum requirements for sodium are 120 mg for 0- to 5-month-olds and 200 mg for 6- to 12-month-olds.[19] Breast milk's content of sodium was used as the basis for setting the sodium requirements for infants, and infant formula is supplemented with sodium to match the amount in breast milk. Typical infants do not have difficulty maintaining body fluids and electrolytes, even though they may not show thirst as a separate signal from hunger. Young infants do not sweat as much as older children, so losses from sweating are not major losses for infants. Illnesses such as diarrhea or vomiting cause the loss of sodium and water and increase the risk of dehydration. Infants do not need salt added to foods to maintain adequate sodium intake.

Fiber   Although there are dietary fiber recommendations for toddlers and children, there are no dietary fiber recommendations for infants.[28] Commercial and home-made baby foods are generally not significant sources of dietary fiber because preparation methods reduce dietary fiber.[28] However, fiber-containing foods such as fruits, vegetables, and grains are appropriate foods for older infants.[29]

Lead   Although lead is not a nutrient, it can be associated with iron and calcium status during infancy. Elevated blood lead levels can be toxic to the developing brain, interfere with calcium and iron absorption, and bring about slowed growth and shorter stature.[30] Infants may inadvertently be exposed to environmental sources of lead. Lead may be a contaminant in water

from lead pipes, particularly if the house was built before 1950. Older homes may contain lead-based paints that taste sweet to infants. Screening for lead poisoning is recommended starting at 9 to 12 months of age.[6,29] If siblings have been found to have lead poisoning, screening for lead may be started at 6 months. Infancy is not the peak age for lead poisoning, but infants can be exposed if their parents work with lead-containing products. For example, if the father is a truck driver who uses leaded gasoline, his work clothes may have lead dust on them. If these clothes are mixed with the rest of the household laundry, or children play in the laundry room where lead dust has settled, an infant in the home may become exposed.

Physical Growth Assessment

Tracking growth in length and in weight helps identify health problems early, preventing or minimizing slowing of the growth rate. Parents understand that a sign of health is growth of their babies. By the time children reach school age, most families have a wall in their homes that proudly displays many height measurements over time. Healthy newborns double their birth weight by age 4–6 months and triple it by 1 year.[1] Growth reflects nutritional adequacy, health status, and economic and other environmental influences on the family. There is a wide range of growth attainment considered normal, however, and healthy babies may follow different patterns of growth. Often, healthy infants have short periods when their weight gain is slower or faster than at other times. Slight variations in growth rate can result from illness, teething, inappropriate feeding position, or family disruption. The overall growth pattern is important, and each assessment is compared to the whole picture. Table 8.3 shows typical growth rates during the first year of life.

Accurate assessment of growth and interpretation of growth rates are important components of health care for infants (Illustration 8.3). Accuracy requires calibrated scales, a recumbent-length measurement board with an attached right-angled headpiece, and a nonstretch tape for measurement of head circumference. Table 8.4 shows how to avoid common errors that interfere with accurate measurements. Makers of measuring equipment recommend checking their accuracy periodically, such as once a month. Calibration of measuring equipment is carried out by using standard weights (or lengths) to confirm accuracy and precision over the range that the equipment measures.

Standard techniques should be used to measure growth; these require practice and consistency. Equipment needed to measure infant growth is different from

Table 8.3 Typical gains in weight and height for age in infancy[31]

Age	Weight Gain grams	Weight Gain grams (pounds)	Length Gain mm	Length Gain mm (inches)
	Per Day	Per Month	Per Day	Per Month
0–3 months	20–30	600–900 (1.3–2)	1	30 (1.2)
3–6 months	15–21	450–630 (1–1.4)	0.68	20 (0.8)
6–12 months	10–13	300–390 (0.7–0.9)	0.47	14 (0.6)

Janet Sugarman Isaacs

Illustration 8.3 Infant being measured on length board and scale.

equipment for measuring children and adults. The scale bed must be long enough to allow the infant to lie down. Length is measured with the infant lying down with a head and foot board at right angles to the firm surface. Positioning the baby quickly and carefully is a skill needed for accurate measurement of recumbent length. In weighing, clothing, hair ornaments, and how much the baby jiggles the scale are examples of factors that could add error to measurements.

Table 8.4 Measuring growth accurately in infants

<u>To Avoid Measurement Errors</u>

- Use measuring equipment that was calibrated recently.
- Confirm that the scale is on zero before starting.
- Make sure the infant is not holding or wearing anything that adds weight or length.
- Confirm the position of the infant for length measurements:

    Head position—the infant's eyes are looking straight up and the head is in midline, touching the head board.

    Neither hips nor knees are bent.

    Heel is measured with foot flat against the foot board.

- Head circumference measure is at the widest part of the head.

<u>To Avoid Growth Plotting Errors</u>

- Calculate the age accurately in months after confirming the date of birth.
- Confirm plotting on the metric scale if kilograms were measured, not the pound scale.
- Confirm that the plotted weight and length are marked well enough to read easily without being so large as to change percentiles.

Interpretation of Growth Data

The National Center for Chronic Disease Prevention and Health Promotion (CDC) 2000 infant growth charts are based on five national surveys and represent a larger sample of infants than that used for previous growth charts.[1] The infant growth charts are based on infants weighed nude on calibrated scales. The charts take into account differences in the growth patterns for formula-fed compared to breastfed infants, regardless of race or ethnic background.[1] Various growth charts are provided in Appendix A. Growth charts for 0- to 36-month-olds consist of a prepared graph for each gender, showing:

- Weight for age
- Length for age

- Weight for length
- Head circumference for age

The more times a baby has been measured and the growth plotted, the more likely the growth trend will be clear, in spite of minor errors. Measures over time can identify a change in rate of weight or length gain and the need for intervention. Growth is so fast during infancy that it may be easier to determine growth problems during infancy than later. Every month in infancy there is an increase in both weight and length, which is not expected in older children. Warning signs of growth difficulties are lack of weight or height gain; plateau in weight, length, or head circumference for more than 1 month; or a drop in weight without regain within a few weeks.[7] Head circumference increases as a result of brain growth. If head circumference is not increasing typically, neither weight nor height increases are likely to track on standard growth percentiles. In the rare circumstance that an infant has a rapid increase in head circumference, this is not a sign of good nutrition or normal growth, but it may signal a condition that requires immediate attention to protect brain development. The rate of weight gain during infancy is not necessarily predictive of future growth patterns after infancy, nor a risk for long-term overweight, compared to the weight-gain pattern later in childhood.[31]

Feeding in Early Infancy

"Food is the first enjoyment of life."

Lin Yutan, *The Importance of Living*

Breast Milk and Formula

The American Academy of Pediatrics and the American Dietetic Association recommend exclusive breastfeeding for the first 6 months of life and continuation of breastfeeding for the second 6 months as optimum nutrition in infancy.[32,33] Infants who are born preterm benefit from breastfeeding, too. Encouragement of breastfeeding right after birth, before the mother's milk supply is available, is an example of a birthing practice that is endorsed.[34] Other recommended practices include teaching safe handling and storage of expressed human milk.[34] The nutrient composition of breast milk is presented in Chapter 6. For young infants less than 4–6 months old, no other liquids or foods are recommended in addition to breast milk and formula.[6] Recommendations for formula intake of young infants are shown in Table 8.5.

The growth rate and health status of an infant are better indicators of the adequacy of the baby's intake than is the volume of breast milk or formula. Infant formulas for full-term newborns are typically 20 cal/fl oz when

Table 8.5	Typical daily volumes for young infants not being breastfed
Age of Infant	Typical Intake of Formula per Day (24 hours)
Birth to 1 month	16–20 fl oz per day, 8–12 feedings/day, 1–2.5 fl oz per feeding
1 to 2 months	18–26 fl oz per day, 8–10 feedings/day, 2–4 fl oz per feeding
2 to 3 months	22–30 fl oz per day, 6–8 feedings/day, 3–5 fl oz per feeding
3 to 4 months	24–32 fl oz per day, 4–6 feedings/day, 4–8 fl oz per feeding

prepared as directed. Formulas for infants born prematurely provide 22 or 24 cal/fl oz. Some health providers recommend further increasing the caloric density of formula for some preterm infants, but such recommendations are not appropriate for most infants. Most infants can be quite flexible in accepting formula, lukewarm or cold, or changes in formula brands.

Table 8.6 gives an overview of the composition of commercially available infant formulas compared to breast milk, and Table 8.7 compares various formulas to one another.[35–38] Some formulas have been developed for common conditions of healthy infants, such as gastroesophageal reflux (GER) or frequent diarrhea. The specialty formula market appears to be growing, such as follow-up formulas, hypoallergenic formulas, and "organic" formulas. Selenium and nucleotides for preterm infants are examples of recent formula additives.[25] Table 8.7 shows how little different types of commonly used infant formulas vary in key nutrients, as they are all based on the same nutrient guidelines.[22]

Cow's Milk during Infancy

The American Academy of Pediatrics recommends that whole cow's milk, skim milk, and reduced-fat milks not be used in infancy.[39,40] Iron-deficiency anemia has been linked to early introduction of whole cow's milk. Low iron availability may come about as a result of gastrointestinal blood loss, low absorption of other minerals (calcium and phosphorus), or the lack of other iron-rich foods in the diet.[41] Studies on infants who were 7.5 months of age confirmed earlier findings that blood loss with whole cow's milk is more likely if the infant had been breastfed earlier rather than fed infant formula.[41] In 1997, 39% of infants were fed whole cow's milk at or before age 12 months.[41] The high cost of infant formulas may result

Table 8.6 How infant formulas are modified compared to breast milk

Macronutrients	Breast Milk	Cow's Milk–Based Formula	Soybean-Based Formula
Protein	7% of calories	9–12%	11–13%
Carbohydrates	38% of calories	41–43%	39–45%
Fats	55% of calories	48–50%	45–49%

Other ways infant formulas are modified compared to breast milk

What Is Modified	How It Is Modified	Examples from Two Major Manufacturers
Calorie level	Increase in calories from 20 calories/fl oz to 22 or 24 calories/fl oz (for preterm infants).	EnfaCare Lipil is 22 calories/fl oz. Similac with Iron 24 is 24 calories/fl oz.
Form of protein	Protein is broken down to short amino acid fragments (hydrolyzed protein) or into single amino acids. Source of protein changed.	Similac Neosure Advance has amino acids. Enfamil Nutramigen has hydrolyzed milk protein. Prosobee has hydrolyzed soy protein in place of milk-based protein.
Type of sugar	Lactose is replaced by other sugars, such as sucrose or glucose polymers from various carbohydrate sources.	Enfamil LactoFree has lactose replaced by corn syrup solids (which provides glucose). Prosobee has carbohydrates from corn syrup solids. Neither has sucrose or lactose.
Type of fat	Long-chain fatty acids partially replaced with medium-chain fatty acids (MCT) and source of fat changed.	Pregestimil has about half of the long-chain fats, replaced by a mixture of vegetable oils. Enfamil Nutramigen has no MCT oil, but has vegetable oils in place of animal-based fats.
Allergy/intolerance	Replacement of milk-based protein with protein from soybeans or replacement of whole proteins with amino acid fragments or single amino acids.	Similac Isomil and Enfamil Prosobee have milk protein replaced by soy protein.
Micronutrients	Increased calcium and phosphorus concentration for preterm infants. Decreased minerals related to renal function. Added essential fatty acids (see above). Lower supplemental iron.	Enfamil PrematureLipil Similac PM 60/40 is modified in calcium, phosphorus, and is low in iron. Similac Special Care Advance 24 is a low-iron formula sold only to hospitals for preterm infants. Enfamil Low Iron and Similac Low Iron have lower levels of iron than the standard formula.
Thickness	Added rice or fiber for gastrointestinal problems.	Similac Isomil D.F. (DF = diarrhea free) for short-term use; it has added fiber from soy. Enfamil A.R. has added rice.
Age of infant	Target age 0–12 months Target age 9–24 months	Similac Isomil Advance Similac Isomil 2

in families selecting cow's milk for older infants who are not breastfed.

Development of Infant Feeding Skills

Infants are born with reflexes that prepare them to feed successfully. As noted earlier, these reflexes include rooting, mouthing, head turning, gagging, swallowing, and coordinating breathing and swallowing.[9] Infants are also born with food-intake regulation mechanisms that adjust over time with development of the infant.[14] In early infancy, self-regulation of feeding is mediated by the pleasure of the sensation of fullness. Inherent preferences are in place for a sweet taste, which is also a pleasurable sensation.[26] After the first 4–6 weeks, reflexes fade and infants learn to purposely signal wants and needs. However, it is not until much later—about age 3—that children can verbalize that they are hungry. In-between the reflexes fading and the child speaking, appetite and food intake are regulated by biological and environmental factors interacting with one another.

Table 8.7 Key nutrients in popular infant formulas per 100 calories as served (about 5 fl oz)

Major Nutrients	Units	Ross Similac Advance	Mead-Johnson Enfamil Lipil	Nestle Good Start DHA+ARA
Protein	g	2.07	2.1	2.2
Fat	g	5.4	5.3	5.1
DHA*	mg	8	17	17
Linoleic acid	mg	1000	860	900
Carbohydrates	g	10.8	10.9	11.2
Thiamine B$_1$	micrograms	100	80	100
Vitamin B$_{12}$	micrograms	0.25	0.3	0.33
Calcium	mg	78	78	64
Iron	mg	1.8	1.8	1.5
Zinc	mg	0.75	1	0.8

*docosahexaenoic acid

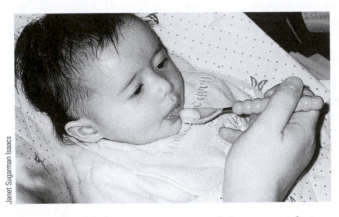

Janet Sugarman Isaacs

Illustration 8.4 Infant reaching with her tongue for a spoon.

Table 8.8 shows infant developmental milestones and readiness for feeding skills.[9,10,12] The interaction of biology and environment prevails here, too. For example, depression in a caregiver may be an underestimated variable in the development of infant feeding. Maternal depression may bring about a lower level of interaction between the parent and infant during feeding, reducing the number or volume of feedings and increasing the risk of slower weight gain.[40] Media influences and changes in social practices also affect how babies are fed. Examples are cultural and ethnic perceptions of breastfeeding and the availability of quality child care for infants.

Several models help assess readiness for a breastfed infant to begin eating from a spoon at around 4–6 months. The developmental model is based on looking for signs of readiness, such as being able to move the tongue from side to side without moving the head.[10,12] The infant must be able to keep her head upright and sit with little support before initiating spoon-feeding. Models based on

chronological age as well as those based on cues from the infant are considered outdated. Most infants adapt to a variety of feeding regimens, and various feeding practices can be healthy for them. The parents' ability to read the infant's cues of hunger, satiation, tiredness, and discomfort influence feeding-skill progression. The cues infants give may include:

- Watching the food being opened in anticipation of eating
- Tight fists or reaching for the spoon as a sign of hunger
- Showing irritation if the feeding pace is too slow or if the feeder temporarily stops
- Starting to play with the food or spoon as the infant begins to get full
- Slowing the pace of eating, or turning away from food when they want to end the meal
- Stopping eating or spitting out food when they have had enough to eat

Infants relate positive and pleasurable attributes to satisfaction of their hunger as part of a successful feeding experience. If there are long episodes of pain from gastroesophageal reflux or constipation, these can become the basis for later feeding problems as the association of eating and pleasure is replaced by an association of eating and discomfort.[18] An infant who makes the association between eating and discomfort is likely to be seen as an irritable baby. This may set up a cycle of the infant being difficult to calm and the parent being frustrated. If this cycle is not replaced by the more positive association of eating and pleasure, the feeding difficulty in infancy may later be characterized by pickiness, food refusals, and difficult mealtime behavior in an older child.[10] A negative association of pain and eating may persist, and appear as a behavioral problem at mealtimes.

Table 8.7 (continued)

Wyeth Bright** Beginnings Ultra	Wyeth Bright** Beginnings Organic	Wyeth Bright** Beginnings Soy	Ross Isomil Advance (Soy)	Mead-Johnson Prosobee (Soy)	Nestle Good Start DHA+ARA Soy
2.2	2.2	2.7	2.45	2.5	2.5
5.3	5.3	5.3	5.465	5.3	5.1
19	19	19	8	17	17
500	750	750	1000	860	920
10.6	10.6	10.4	10.3	10.6	11.1
100	100	101	60	80	60
0.20	0.2	0.3	0.45	0.3	0.3
63	63	90	105	105	105
1.8	1.8	1.8	1.8	1.8	1.8
0.8	0.75	0.75	0.75	1.2	0.9

**often renamed as various grocery store brands

Table 8.8 Development of infant feeding skills

Chronological Age	Developmental Milestone	Feeding Skills
Birth to 1 month	Vision is blurry; hears clearly. Head is oversized for muscle strength of the neck and upper body.	Suckling and sucking reflexes. Frequent feedings of 8–12 per 24 hr. Only thin liquids tolerated.
1–3 months	Cannot separate movement of tongue from head movements. Head control emerges. Smiles and laughs. Puts hands together.	Volume increases up to 6–8 fl oz per feeding, so number of feedings per day drops to 4–8 per 24 hr. Sucking pattern allows thin liquids to be easily swallowed. Learns to recognize bottle (if bottle-fed).
4–6 months	Able to move tongue from side to side. Working on sitting balance with stable sitting emerging. Drooling is uncontrolled. Disappearance of newborn reflexes allows more voluntary movements. Teething and eruption of upper and lower central incisors.	Interest in munching, biting, and new tastes. Cannot easily swallow lumpy foods, but pureed foods swallowed. 6–8 fl oz per feeding and 4–5 feedings per day (may be variable if breastfeeding). Holds bottle (if bottle-fed).
7–9 months	Hand use emerges, with pincer grasp and ability to release. Stable independent sitting. Crawling on hands and knees. Starting to use sounds, may say "mama" and "dada."	Self-feeding with hands emerges. Munching and biting emerges. Indicates hunger and fullness clearly. Prefers bottle, but little loss from a held, open cup.
10–12 months	Can pull to stand; standing alone emerges. Enjoys making sounds as if words. Can pick up small objects, such as a raisin. Can bang toys together with two hands. Has consistent routines about bedtime, diaper changing. Usually does not drool anymore.	Likes self-feeding with hands. Spoon self-feeding emerges. Drinks from an open cup as well as from a bottle. Uses upper and lower lip to clear food off a spoon. Enjoys chopped or easily chewed food or foods with lumps. Sitting position for eating. Enjoys table foods even if some baby foods still used.

Introduction of Solid Foods

Infants begin with food offered on a spoon in a small portion size of 1–2 tablespoons for a meal, with one or two meals per day. The purpose of offering food on a spoon to infants at 4–6 months is to stimulate mouth muscle development, and less so for nutritional needs, which are met from breast milk. Watching a baby learn how to eat from a spoon is fun for new parents. If the baby has achieved the developmental milestones in Table 8.8, it may take him only a few days of practice to start spoon-feeding and to learn to consume 1 tablespoon of semisoft food as a meal. Spoon-feeding is really two new experiences for a baby: a spoon is not soft and warm in the mouth like the breast, and whatever food is selected does not feel the same as breast milk on the tongue. At first the baby tries to suck food on a spoon like it was a liquid, so some food will come out of the mouth.

Babies respond strongly to new tastes or smells, regardless of the first food. Introducing a baby to food on a spoon includes these recommendations (in addition to those in the following section on infant feeding position):

- Time the first spoon-feeding experiences for when the baby is not overly tired or hungry, but active and playful.
- Offer a small spoon with a shallow bowl. The temperature of the spoon may have to be considered if it can conduct hot or cold readily.
- Give the baby time to open his or her mouth and extend the tongue toward the food. If the baby cannot extend the tongue farther out than the lower lip, the baby is not ready for spoon-feeding.
- Place the bowl of the spoon on the tongue with slight downward pressure toward the front of the mouth. Touching the back of the tongue may elicit a gag response.
- The spoon should be almost level. It is not a good practice to scrape the food off the spoon with the baby's gums by tilting the spoon handle up too high. The baby's chin should be slightly down to protect the airway.
- The pace of eating should be based on watching for the baby to swallow. Rushing will increase the risk of choking and of the infant having an unpleasant experience.
- First meals may be small in volume—only 5 or 6 baby spoons—and last about 10 minutes, based on the baby's interest.

After mastering the new skill of eating from a spoon, babies quickly teach their parents how to feed them by indicating the rate of eating they like. Common mistakes happen if the person feeding the baby does not read the signs that the baby is giving.

Weaning Discontinuation of breastfeeding or bottle-feeding and substitution of food for breast milk or infant formula.

The Importance of Infant Feeding Position

Positioning infants for feeding with a bottle and for eating from a spoon are important because improper positioning is associated with choking, discomfort while eating, and ear infections.[12] The semiupright feeding position as exemplified in car seats or infant carriers is recommended for the first few months.[10] Unsafe feeding positions, such as propping a bottle or placing the baby on a pillow, increase the risk for choking and overfeeding. The recommended sleeping position for young infants is lying on the back without elevating the head on a pillow. This position is not recommended for feeding, which reinforces why feeding an infant in bed with a bottle is not generally recommended.[6,10]

Spoon-feeding also has a recommended infant feeding position. The infant can better control his mouth and head in a seated position with good support for the back and feet. The person offering the spoon should sit directly in front of the infant and make eye contact without requiring the baby to turn his head.[10,12] A high chair is an appropriate feeding chair when the infant can sit without assistance. The infant should be kept in a sitting position by use of a seat belt so that the hips and legs are at 90 degrees. This position assists the infant's balance and digestion. If the infant is sliding out from under the tray of the high chair with the hips forward, the stomach is under more pressure, and spitting up is more likely.

Some apparently healthy infants show resistance to learning feeding skills or react to the introduction of foods in an unusual manner. These problems in early feeding experiences are sometimes warning signs of more general health or developmental difficulties.[18] They may signal emerging problems that cannot be diagnosed until later. Families who call attention to early feeding problems may assist their infants in the long run by having the problem recognized earlier. For example, some infants who start and stop feeding frequently, but then do not feed for several minutes in a row, may later be discovered to have heart problems. The coordination of eating and breathing may have been the basis for the starting and stopping. Some infants are very reluctant feeders and are later diagnosed with a milk protein intolerance. Case Study 8.1 describes a baby who refuses to eat when her mother thinks she must be hungry.

Preparing for Drinking from a Cup

The process of *weaning* starts in infancy, and usually is completed in toddlerhood. The recommended age for weaning the infant from the breast or from a bottle to drinking from a cup is from 12 to 24 months.[12] Breastfed infants may make the transition to drinking from a cup without ever having any liquids in a bottle. If breastfeeding is continued as recommended for the first year of life, introducing a cup for water and for juices after 6 months is recommended, near the time that foods are offered on a spoon. By the time weaning from breastfeeding is planned, the 1-year-old infant will be skilled enough at drinking from a cup to meet fluid needs without a bottle.

Infants who are not exclusively breastfed, or are breastfed for less than 12 months, need to have additional fluids offered in a bottle because their ability to meet their fluid needs by drinking from a cup are not sufficiently developed. Developmental readiness for a cup begins at 6–8 months.[12] Eight-month-old infants enjoy trying to mimic

Case Study 8.1

Baby Samantha Will Not Eat

Samantha is a healthy 8-month-old girl who lives with her mother, Kathy, her father, and her older sister, who is almost 3 years old. Both parents now work full-time, and both children attend day care full-time. Kathy nursed Samantha exclusively before she returned to work and built up a supply of frozen breast milk. She nurses her twice per day now, early in the morning and before Samantha goes to sleep. Samantha gets breast milk offered in bottles at day care. Samantha is reported by the day care staff to be a good baby. However, when Kathy picks her up after work, Samantha wants to be held and will not sit in her high chair or eat dinner. She cries if she is not held. Samantha's sister wants to eat as soon at they get home. Kathy has so much to do at home after work, she finds it difficult to hold Samantha at such a busy time. Kathy thinks that Samantha must be hungry and that she would be less irritable if she ate her dinner.

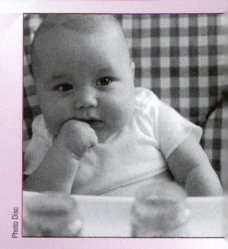

Photo Disc

Questions

1. What signs is baby Samantha giving to show that she needs comforting rather than hunger?
2. How might Kathy change her routine to give baby Samantha more attention and meet the needs of her older daughter?
3. At 8 months, is Samantha too young to overeat out of emotional needs?
4. Should Kathy stop or continue breastfeeding to improve Samantha's eating?

---

drinking from open cups that they see in the home. However, the ability to elevate the tongue and control the liquid emerges later, at closer to a year. The 10- to 12-month-old infant enjoys drinking from a held cup and trying to hold his own cup, even though the main feeding method is the breast or bottle. Infants are likely to decrease total intake of calories from breast milk or infant formula if served in a cup, because they are less efficient in the mouth skills needed. At first the typical portion size of fluid from a cup is 1 to 2 ounces. The infant who is weaned too soon may plateau in weight because of decreased total calorie intake. The drop in total fluids consumed may result in constipation. Changing from a bottle to a covered sippy cup with a small spout is not the same developmental step as weaning to an open cup.[12] The mouth skills needed in controlling liquids with the tongue are more advanced with an open cup. The skills learned in drinking from an open cup also encourage speech development.

Food Texture and Development

> "They say fingers were made before forks and hands before knives."
>
> Jonathan Swift

Weaning is not complete until the caloric intake from breast milk is provided from foods and beverages. Infants advance from swallowing only fluids to pureed soupy foods at

4–6 months.[6,12] Before that they can move liquids only from the front to the back of the mouth. The mouth is exquisitely sensitive to texture. If food with soft lumps is presented too soon, it causes an unpleasant sensation of choking. When infants are 4–6 months old, they can move their tongues from side to side. At 6–8 months, they are ready for foods with a lumpy but soft texture to elicit munching and jaw movements,[12] these movements simulate chewing. By 8–10 months, infants are able to chew and swallow soft mashed foods without choking. It is important to offer infants foods that do not require much chewing, because infants do not develop mature chewing skills until they are toddlers.

First Foods

The first food generally recommended for infants at 4–6 months is baby cereal, such as iron-fortified cereal mixed with water or breast milk. Rice cereal is a common first food because it is easily digested and *hypoallergenic*. When to add baby cereal or other food to an infant's diet is determined not only by developmental milestones as recommended, but by other reasons, such as these:

Hypoallergenic Foods or products that have a low risk of promoting food or other allergies.

- Some parents add baby foods because they think this will make the baby sleep longer. This practice is neither recommended nor effective for most infants. This common belief may result in introduction of baby cereal before the infant has developed the skills to eat from a spoon.

- Some families are instructed by pediatricians to add dry rice cereal as part of the treatment for gastrointestinal problems, because it tends to thicken infant formula.

Fruits and vegetables, such as pears, applesauce, or carrots, are also sometimes first foods for infants. What are considered healthy first foods for infants varies in different cultures and ethnic groups. Regardless of what foods are offered first, the timing and spacing of new foods can be used to identify any negative reactions. Common recommendations for parents of 6-month-olds are to add only one new food at a time and to offer it over 2 or 3 days. There are specific recommendations regarding the timing and spacing of foods known to trigger food allergies in families with histories of this problem (discussed later in the "Food Allergies and Intolerances" section).

Commercial baby foods are not required by infants. Parents can prepare baby foods at home using a blender or food processor, or by mashing with a fork. Care must be taken, however, to provide a soupy texture and to avoid contamination of home-prepared baby food by bacteria on food or from unsanitary storage methods. The nutrient content of home-prepared baby foods can vary widely depending on how they are prepared and stored. Adding salt and sugar to home-prepared baby foods are examples of variables that can decrease nutritional quality. The advantage of home-prepared baby foods is that a wider variety of foods may be introduced that are likely to be a part of the diet later. Additionally, money not spent on commercial baby foods may be significant savings for some families.

Commercial baby foods are commonly selected because of their sanitation and convenience. Several are listed in Table 8.9.[28] Moreover, families who pack food for day care or who travel with infants find commercial baby foods convenient. Parents have a lot of choices to make in selecting baby foods. Selection should be based on the nutritional needs of the infant, not on what is available in

Table 8.9 Commercial baby foods[28]

Type	Portion Sizes	Target Age/Skills
Single-ingredient pureed fruits and vegetables, 25–70 cal/jar	71 g (2.5 oz) jar	4–6 months, introduced on a spoon, with portion size of 2–3 Tbsp/meal
Powdered cereals, 60 cal/serving	Dry: ½ oz or 4 Tbsp/serving, mixed with water, juice, breast milk, formula, or other liquids	4–8 months
Jarred cereal mixed with fruits, 90 cal/jar	"Wet:" 113 g (4 oz) jar	4–8 months
Juices, fruit and vegetables, 60–100 cal/bottle with added vitamin C, calcium	4 and 6 fl oz bottles	4–10 months
Fruits and vegetables, pureed textures, added ingredients, such as tapioca and mixtures	113 g (4 oz) jar, two jars per meal	4–8 months, no munching or food intolerances
Meat mixtures, containing 3–4 g protein/jar, 50–70 cal/jar	113 g (4 oz) jar, one jar per meal	6–9 months, no munching or food intolerances
Desserts, 0–2 g protein/jar and 80–100 cal/jar	113 g (4 oz) jar	6–9 months, no munching
Fruits and vegetables with mixed textures, 0–4 g protein/jar, 70–160 cal/jar	170 g (6 oz) jar, one jar or 12 Tbsp/meal	9–12 months, side-to-side tongue movements and munching
Meat-based dinners, with mixed textures, 3–5 g protein/jar and 90–130 cal/jar	170 g (6 oz) jar, one jar or 12 Tbsp/meal	9–12 months, side-to-side tongue movements and munching
Meats with textures, 10–11 g protein/jar and 70–100 cal/jar	71 g (2.5 oz) jars	9–12 months, side-to-side tongue movements and munching
High-texture baby foods: diced fruits, vegetables, and meats; dinners have pieces as lumps	Dices 71 or 128 g jars (2.5 or 4.5 oz), dinners 170 g (6 oz) jars	Finger foods (to be picked up), requires self-feeding with hands and spoon, chewing
Baked products, such as zwieback toast and biter biscuits	One zwieback toast 7 g, one biter biscuit 11 g	10–12 months, requires biting and munching, limited chewing

local stores and the eating habits of the adult shoppers. Examples of baby foods that may reflect shopper's selections rather than baby needs are fruits with added tapioca or baby food desserts and snack foods. They are not recommended for most infants. Jar and plastic tub serving sizes of baby foods are based on industry standards, not necessarily recommended portion sizes. Portion sizes for infants should be based on appetite. Finishing an opened container of baby food may encourage overeating if parents do not pay attention to signs from the infant.

Many foods eaten by other family members are appropriate foods for infants who are 9–12 months of age. Examples are regular applesauce, yogurt, soft-cooked green beans, mashed potatoes, cooked hot cereal, and Cheerios.

Inappropriate and Unsafe Food Choices

New parents may inadvertently select foods for infants based on their own likes and dislikes, rather than on the infant's needs. Such choices are problematic when they increase the risk of choking. Here are some examples of unsafe foods for infants:[6,10,12]

- Popcorn
- Peanuts
- Raisins, whole grapes
- Uncut stringy meats
- Gum and gummy-textured candies
- Hard candy, jelly beans
- Hot dog pieces
- Hard raw fruits or vegetables, such as apples, green beans.

Some foods present a choking risk for infants because of their lower chewing skills. Underchewed pieces of food can obstruct the infant's airway because voluntary coughing and clearing the throat are skills not yet learned.[10]

Moreover, the infant may not be able to clear food from the roof of the mouth. A sticky food such as peanut butter against the hard palate may fall to the back of the mouth and present a risk of choking. Foods that do not easily stay together, such as potato chips, also can cause choking. A chip breaks apart in the mouth, but little pieces may stay crunchy. Small pieces may present to the back of the mouth before the infant can use the tongue to move the pieces to the sides and initiate munching.

Water

Breast milk or formula generally provides adequate water for healthy infants for the first 4–6 months.[6,17] Infant drooling generally does not increase the need for water. In hot, humid climates, infants have increased needs for water; but water should not replace breast milk or formula. Added water can be used to meet fluid needs, but not caloric and nutrient needs. All forms of fluids contribute to meeting the infant's water needs. Often parents are reluctant to say they offered their infants sips from their own glasses containing soft drinks or drinks containing caffeine or alcohol. This may be important information to include in a food intake record, especially if the contents are not recommended for babies. The replacement of an infant formula with a less nutritionally rich alternative such as juice, "sports drinks," cola, or tea has been found to be a contributor to lower-quality diets for infants.[42]

Water needs of infants are a concern because dehydration is such a common response to illness in infancy. The infant has limited ability to signal thirst, especially when sick. Vomiting and diarrhea result in dehydration more rapidly in infants than in older children, with symptoms that are more difficult to interpret.[10,43] Replacement of electrolytes has been the basis for a variety of over-the-counter fluid replacement products, such as Pedialyte, "sports drinks," and Gatorade, that are marketed to parents. These products contain some glucose (dextrose) along with sodium, potassium, and water. The amount of glucose provides significantly lower calories than do breast milk or formula—usually 3 cal per fl oz compared to 20 cal per fl oz. Such liquids can be overused, and they may result in weight loss even in healthy infants. Juice is not needed to meet the fluid needs of infants. The American Academy of Pediatrics recommends that juices need not be introduced into the diet before 6 months of age, and never at bedtime.[42] There is a recommendation for juice after age 1, but not before. If juice is offered to an infant over 6 months, it is recommended to be offered in a cup and not in a bottle. Infancy may be the time in which a habit of excessive juice intake starts, so limiting juice volume is a way to avoid problems later.

How Much Food Is Enough for Infants?

Parents improve in understanding infant feeding behavior as the infant–parent interaction matures from early to late infancy. During early infancy while the sleep/wake cycle of the infant is irregular, it is common for new parents to interpret all discomfort as signs of hunger. The infant's ability to self-calm is a developmental step that plays out differently with different temperaments and parenting styles.[44] Infants who are quite sensitive to what is happening around them are likely to be viewed as irritable and hungry if they cry frequently.[45] In contrast, infants who sleep through usual household noises and are less reactive to their immediate environment are likely to be offered food fewer times per day. As a result of different responses to temperament, a pattern of excessive or inadequate food and formula intakes may result.

The following is an example of excessive intake for a 3-month-old not being breastfed. Total formula intake: 33 fl oz (seven bottles per day, ranging from 3 to 5 fl oz per bottle), offered at 7:30 a.m., 11:00 a.m., 12:45 p.m., 2:30 p.m., 5:30 p.m., 8:45 p.m., and 11:30 p.m., one serving of baby food applesauce, fed on a spoon at 9:00 a.m.

This is overfeeding because excessive formula is being offered, along with premature offering of spooned food. The frequency of the bottles being offered suggest that the parents are interpreting the baby's signs of discomfort as hunger when she may have other needs, such as for being held, changed, or calmed by movement or touch. Overfeeding is less likely with breastfeeding.

In the first few months, the oral need to suck is easily confused with hunger by new parents. The typical forward and backward tongue movements of the infant's first attempts to eat from a spoon may seem to be a rejection of food by new parents.[12] The infant appears to spit out the food, but this is a sign of learning to swallow and not necessarily a taste preference. The same food that appears to be rejected will be accepted as the infant learns to move the food from the front to the back of the mouth.[46] It may appear to a parent that the infant does not like a food if he or she appears to choke. This choking response is more likely based on the position of the spoon on the tongue.[10,12] The mouth is very sensitive, particularly toward the back. If the bowl of the spoon is too far back, it will cause a gagging reaction, regardless of the taste of the food.[47]

How Infants Learn Food Preferences

Infants learn food preferences largely based on their experiences with food. Breastfed infants may be exposed to a wider variety of tastes within breast milk than are infants offered formula.[48] The different foods that the breastfeeding mother eats may result in some flavor compounds being transmitted to the nursing infant.[49] Studies on infants in the age range of 4–7 months showed that acceptance of new foods was more rapid than acceptance of new foods after the first year of life.[46] In the 1920s and 1930s, pediatrician Clara Davis conducted studies on the self-selection of food by weaning infants.[50] Dr. Davis was able to demonstrate that older infants are able to select and consume amounts of food needed to sustain normal growth.[50] Her classic studies were interpreted by later generations as supporting the concept that infants and children will instinctively select a well-balanced diet. However, these studies were subject to misperception because careful attention was not paid to the original methods, in which only nutritious and unsweetened foods were available.[51]

Food preferences of infants are largely learned, but genetic predisposition toward sweet tastes and against bitter foods may modify food preferences. Food preferences developed in infancy set the stage for lifelong food habits. The development of trust and security for an infant are crucial, but this need not be linked to overfeeding the infant.[14] Recognizing an infant's specific needs and responding to them appropriately is important. If offered only a limited variety of foods with little interaction during the meal, infants may learn to refuse to eat as a method of gaining attention. For example, 10-month-old infants enjoy throwing food on the floor just to have someone bring

them more. They may enjoy the sound of banging a spoon on a high chair more than eating. Infants who learn to get attention by not eating are likely to manipulate the behavior of adults even more successfully as toddlers.

Nutrition Guidance

Nutrition guidance materials have been developed for parents from many sources, such as the WIC program, makers of infant foods, and professionals such as those in the Bright Future in Practice initiative.[6,28,35] The need for nutrition education was demonstrated in a study of mothers and pregnant women, which showed that misunderstanding about infant nutrition was common.[16] Infant feeding recommendations from nutrition education materials are sampled in Table 8.10.

Infants and Exercise

The exercise and fitness benefits for adults do not apply to infants. Providing a stimulating environment is recommended, so infants can explore and move as a part of their developmental milestones. The American Academy on Pediatrics Committee on Sports Medicine policy statement recommends that structured infant exercise programs should not be promoted as being therapeutically beneficial for healthy infants.[52] Infants do not have the strength or reflexes to protect themselves, and their bones are more easily broken than those of older children and adults.

Supplements for Infants

Specific supplements are recommended for breastfed infants in the United States and Canada, under certain circumstances:

- Fluoride supplements are recommended if the family lives in a place that does not provide fluoridated water.
- If breast milk is the only form of nutrition after 4–6 months, fluoride is recommended.[32]
- Elemental iron (at 3 mg per kg body weight of the infant) may be prescribed if the mother was anemic during pregnancy.[6]
- Vitamin B_{12} may be prescribed if the mother is a vegan.[48,53]
- Vitamin D supplements may be needed if the infant is not exposed to adequate amounts of sunlight.

Supplements may also be prescribed for infants who were born early at low birth weight. They may need vitamins A and E and iron due to low stores of these nutrients usually accumulated late in pregnancy. A liquid multivitamin and mineral with fluoride is a common prescription for the healthy premature baby, regardless of breastfeeding status.[6]

Table 8.10 Infant feeding recommendations

Topics	Nutrition Education Sample Content
Appropriate use of infant formula (if not breastfeeding)	Mixing instructions for diluting concentrated formula, keeping formula sanitary by refrigeration, and monitoring how long offered formula is left out. Feeding positions for the infant and bottle, and burping the infant during feedings.
Baby food and sanitation	Serving sizes for infants of different ages, refrigeration and sanitation for opened jars of baby food, problems from mixing different baby foods.
Preparing baby foods at home	Avoiding spices, salt, and pepper in baby foods. Using safe food-handling techniques in preparing and storing servings.
Prevention of dental caries	Recommendations for bedtime and nap time to avoid sugary liquids pooling in the mouth. Identifying liquids that may promote dental caries.
Feeding position	Feeding positions for starting food on a spoon. How to tell if the infant's high chair is safe for feeding.
Signs of hunger and fullness	Identifying early signs compared to later signs of hunger. How infants of various ages communicate at mealtime. Reinforcing and rewarding infant signs of hunger and fullness.
Preventing accidents and injury	Checking temperature of baby foods and liquids. Use of appropriate car seats and safety belts.
Spitting up—when to be concerned	Typical feeding behaviors in young infants. Signs of overfeeding. Spitting up and signs of illness. Discussing signs and symptoms with health providers.

Common Nutritional Problems and Concerns

Common nutritional problems during infancy are failure to thrive, colic, iron-deficiency anemia, constipation, dental caries, and food allergies. Parents often overestimate the association between eating and these common health problems. Parents should discuss their concerns with the infant's health care providers.

Failure to Thrive

Failure to thrive (FTT) is a diagnosis that can be made during infancy or later. Various terms are used to refine FTT, such as *organic* (meaning a diagnosed medical illness is the basis), *nonorganic* (meaning not based on a medical diagnosis), and mixed type. Although growth failure may be brought about by a variety of medical and social conditions, FTT is primarily used to describe conditions in which a calorie deficit is suspected.[8,54] FTT is an emotionally loaded diagnosis for parents, because the term implies that someone failed. Examples of nonorganic or environmental factors are maternal depression, mental illness, alcohol or drug abuse in the home, feeding delegated to siblings or others unable to respond to the infant, and overdilution of the formula. The relationship of FTT to poverty has been well documented.[55] Examples of organic reasons for FTT commonly found are untreated GER, chronic illnesses such as ear infection or respiratory illness, and *developmental disabilities*. (The connection between FTT and developmental disabilities is further

discussed in Chapter 9.)[56] If there is a medical basis for expecting that the infant will not fit standard growth projections, FTT is not an appropriate term to use. For example, growth for an infant born with IUGR should be based on this medical history and related testing near birth. If this infant at 11 months of age is taken to a new health care setting, FTT may be suspected because of the infant's small size—unless the true cause, IUGR, is revealed.

Table 8.11 provides an example of an assessment that can be used to determine if FTT is present. The assessment of FTT depends on tracking growth. Once FTT is suspected, review of medical records often indicates that growth measurements have been taken in a variety of health care settings with different equipment and personnel, at times when the infant was both well and sick. These records may produce an irregular growth pattern that is difficult to interpret.

Nutrition Intervention for Failure to Thrive

Failure to thrive may be a basis for referral to a registered dietitian. Correction of FTT usually is not as

Failure to Thrive (FTT) Condition of inadequate weight or height gain thought to result from a caloric deficit, whether or not the cause can be identified as a health problem.

Organic Failure to Thrive Inadequate weight or height gain resulting from a health problem, such as iron-deficiency anemia or a cardiac or genetic disease.

Nonorganic Failure to Thrive Inadequate weight or height gain without an identifiable biological cause, so that an environmental cause is suspected.

Developmental Disabilities General term used to group specific diagnoses together that limit daily living and functioning and occur before age 21.

Table 8.11 Complete nutritional assessment of an infant to rule out failure to thrive

- Review records or weight, length, head circumference, fetal or maternal risk factors such as rate of weight gain during pregnancy, newborn screening results, Apgar scores, and physical exams after birth.
- Interpret all available growth records from provders, WIC, and emergency room visits.
- Interpret current growth measurements and indicators of body composition, such as fat measurements.
- Review of family structure, education, and social supports with attention to access to food and formula (if not breastfed).
- Analyze and interpret current food and fluid intakes as reported by the primary caregiver(s).
- Rule out a biological basis for FTT from available records and laboratory results.
- Observe and interpret parent–child interactions, feeding duration, and the feeding skills of the infant.

Colic A condition marked by a sudden onset of irritability, fussiness, or crying in a young infant between 2 weeks and 3 months of age who is otherwise growing and healthy.

simple as just feeding the baby, but increasing caloric and protein intake is the first step.[49,50] The registered dietitian's role is to assess the growth and nutritional adequacy, establish a care plan, and provide follow-up as part of a team approach. She may work with other specialists concerning medical or psychological aspects. Nutrition interventions may establish caloric and protein intake goals and a feeding schedule to assure adequate nutrition is being provided.[35] Other interventions may include:

- Gaining agreement from the caregivers about how and when intake and weight monitoring will be done
- Enrolling the infant in an early intervention program in the local area
- Arranging for transportation or solving other barriers to follow-up care
- Assessing social supports to assure a constant supply of food and formula (if used)
- Assisting the family in advocating for the infant within the health care delivery system, such as locating a local pediatrician and getting prescriptions filled

FTT is one reason that social service agencies become involved with families. Most new parents handle stress without hurting their infants, but a few react in ways that result in infants presenting with FTT or worse. After investigation, FTT may be determined to be a form of

child abuse, as a result of neglect. Some infants diagnosed with FTT become at risk for child abuse and need foster care if the home situation does not improve.[45]

Colic

Colic is the sudden onset of irritability, fussiness, or crying in a young infant.[57] Parents usually think that the infant has abdominal pain. Episodes may have a pattern of onset at the same time of day, for about the same duration every day, with all symptoms disappearing by the third or fourth month. The association of colic symptoms, gastrointestinal upset, and infant feeding practices has been studied, but no definitive cause has been shown.[57]

The response to colic is often to change baby formulas if the infant is not breastfeeding, although the change in formulas usually does not change the pattern of colic. Recommendations to relieve colic may include rocking, swaddling, bathing, or other ways of calming the infant, positioning the baby well for eating, or burping to relieve gas.[6,35] One theory about why infants have colic points to the mother's diet while breastfeeding, particularly her consumption of milk or specific foods such as onions. Identifying the origin of colic requires more research.

Iron-Deficiency Anemia

Iron deficiency in infants is less common than iron deficiency in toddlers. Iron reserves in full-term infants reflect the prenatal iron stores of the mother.[58] Women with iron-deficiency anemia during pregnancy pass on less iron to the fetus, a condition that may increase risk of anemia during infancy. Infants who have iron deficiency may be exposed to other risk factors to their overall development, such as low birth weight, elevated lead levels, or generalized undernutrition.[58] Family income at or below the poverty level is also a risk for iron deficiency.[58,59] Research in infants who have long-term and severe iron deficiency anemia suggests inadequate iron contributes to long-term learning delays from its role in central nervous system development.[15,29] Treatment of iron-deficiency anemia in infancy is generally by prescribed oral elemental iron administered as a liquid.[6,59]

Breastfed infants may be prescribed elemental oral iron and also receive iron through iron-fortified baby cereal after 4–6 months of age. For infants who are not breastfed, a usual source of iron for formula-fed babies is iron-fortified infant formula. Iron from this source improves iron status measured during the first year and is well accepted.[53] In a randomized study of healthy infants, those who received iron-fortified infant formula had by the end of the first year significantly improved biochemical measures of iron status compared to those who received infant formula without added iron. However, there were no differences in the developmental scores of the two groups by 15 months of age.[54]

The level of iron in iron-fortified formula has been 15 mg per liter, or 11.5 mg per quart, based on the RDA of 6 mg of iron for infants up to 6 months and 10 mg for infants from 6 to 12 months.[18] New infant formulas with a lower level of supplementation are also marketed in part as a result of gastrointestinal side effects that have been attributed to iron added to formula.[43] The "low-iron" formula has 8 mg iron per liter, or 4.5 mg of iron per quart.[35] Some manufacturers are not recommending the low-iron formula beyond 4 months of age, because it would not meet the RDA for iron.

Diarrhea and Constipation

Diarrhea and constipation may be attributed to dietary components such as breast milk or use of an iron supplement. Parents think that diarrhea and constipation are related to the infant's diet and want to change the diet or feeding plan to lower gastrointestinal upset. In fact, diarrhea can result from viral and bacterial infections, food intolerance, or changes in fluid intake.[43] Typically, young infants have more stools per day than do older infants, and have them soon after oral intake.[55] The number of stools varies widely from two per day to six per day, decreasing as the infant matures.[55] Parents of breastfed infants generally do not have concerns about constipation, because the infant's stool is generally soft. Infants fed soybean-based infant formulas may have more constipation than those fed a cow's milk–based formula. Recommendations for avoiding constipation are to assure that the infant is getting sufficient fluids and to avoid medications unless prescribed for the infant. Some parents use prune or other juices that have a laxative effect for an older infant, but there is a risk of creating a fluid imbalance and subsequent diarrhea.[43] Foods with high dietary fiber are generally not recommended for infants with constipation because many sources such as whole wheat crackers or apples with peels present a choking hazard and are not recommended for infants.

The cause of diarrhea during infancy may or may not be identifiable. Diarrhea in an infant may become a serious problem if the infant becomes dehydrated or less responsive.[43] Most infants have 1 or 2 days of loose stools without weight loss or signs of illness, such as after getting immunizations. General recommendations are to continue to feed the usual diet during diarrhea.[6,35] Breast milk does not cause diarrhea. During a bout of diarrhea, continuing adequate intake of fluids such as breast milk or infant formula is generally sufficient to prevent dehydration.

Prevention of Baby-Bottle Caries and Ear Infections

Baby-bottle caries are found in children older than 1 year, but are initiated by feeding practices during infancy. Infants have high oral needs, which means they love to suck and to explore by putting things in their mouths.

They derive comfort from sucking and may relax or fall asleep while sucking. The use of a bottle containing formula, juice, or other high-carbohydrate foods to calm a baby enough to sleep, however, may set her up for dental caries.[56] During sleep the infant swallows less, allowing the contents of the bottle to pool in the mouth. These pools of formula or juice create a rich environment for the bacteria that cause tooth decay to proliferate, increasing the risk for tooth decay.

Risk for ear infections is also correlated with excessive use of a baby bottle as a bedtime practice, as a result of the feeding position.[56] The shorter and more vertical tubes in the ears of infants are under different pressure during the process of sucking from a bottle.[35] If the infant is feeding by lying down while drinking, the liquid does not fully drain from the ear tubes. The buildup of liquid in the tubes increases the risk of ear infections. In a study of over 200 infants, pacifiers and bottle-feeding were correlated with greater prevalence of ear infections.[60] Infants who were breastfed did not have as high a rate of ear infections.[60]

Here are some good feeding practices to limit baby-bottle caries and ear infections related to baby bottles:

- Limit the use of a bottle as part of a bedtime ritual.
- Offer juices in a cup, not a bottle.
- Put only water in a bottle if offered for sleep.
- Examine and clean emerging baby teeth to prevent caries from developing.

Food Allergies and Intolerances

The prevalence of true food allergies is higher in younger than in older children. About 6–8% of children under 4 years of age have allergies that started in infancy.[61,62]

An infant may develop a food allergy to the protein in a cow's milk–based formula over time. Often such a problem follows a gastrointestinal illness. When the infant is well, protein is broken down during digestion so that absorption in the small intestine is of groups of two or three amino acids linked together. After an illness, small patches of irritated or inflamed intestinal lining may allow protein fragments of larger lengths of amino acids to be absorbed. Such fragments are hypothesized to trigger a reaction as if a foreign protein had invaded, setting up a local immune or inflammatory response.[63] This absorption of intact protein fragments is the basis for allergic reactions. When this happens with cow's-milk protein, it is likely that soy-based formulas will also cause the same allergic reaction.

The most common allergic reactions are respiratory and skin symptoms, such as wheezing or skin rashes.[62] Food allergies are confirmed by specific laboratory tests after infancy.[57] True allergies can present as an array of reactions building up over time, so that it may take several years for the initiating cause to be identified.

Food intolerances are frequently suspected in infants. Families may consider skin rashes, upper airway congestion, diarrhea, and other forms of gastrointestinal upset to be food allergies, but often they are not.[57] As in the case of food allergies, infants are not usually given all the tests for food intolerance, but instead are treated for symptoms. The infant with suspected protein intolerance may be changed to a specialized formula composed of *hydrolyzed protein*.[58] Because the protein of a hydrolyzed formula is already broken down, it does not trigger the same response as intact protein fragments do. A family with a known allergy or intolerance may lower the risk of the allergy occurring in their infant by breastfeeding, and by postponing into the second or third year introduction of allergy-causing foods such as wheat, eggs, and peanut butter. It is important for families not to overly restrict such foods thought to cause allergies from an infant's diet unless required. If many foods are being avoided, there may be consequences such as decreasing the nutritional adequacy of the diet and reinforcing behaviors of rejecting foods and limiting variety. Allergy and intolerance symptoms are more common in response to nonfood items such as grasses and dust, so many different sources of symptoms must be considered.

> **Hydrolyzed Protein Formula**   Formula that contains enzymatically digested protein, or single amino acids, rather than protein as it naturally occurs in foods.
>
> **Lactose**   A form of sugar or carbohydrate composed of galactose and glucose.

Lactose Intolerance

Lactose intolerance is a food intolerance in infancy characterized by cramps, nausea, and pain, and by alternating diarrhea and constipation. Infants who are breastfeeding may develop lactose intolerance, because breast milk has lactose in it.[17] Gastrointestinal infections may temporarily cause lactose intolerance because the irritated area of the intestine interferes with lactose breakdown.[26] The ability to digest lactose generally returns shortly after the illness subsides. *Lactose* is in all dairy products, so it is in cow's milk–based infant formulas.[17,35–38] Lactose-free infant formulas are soybean based or based on lactose-free cow's milk. Lactose intolerance is less common during infancy than at older ages in groups that are susceptible to it. An infant who was fed a lactose-free formula is likely to be able to eat dairy products later. Because dairy products are such an important source of calcium, introducing foods with low lactose is recommended for older infants who appeared to be lactose intolerant when younger.

Cross-Cultural Considerations

Commercial baby foods reflect the bias of the dominant American culture. There is little ethnic diversity in baby foods—no collards or Mexican beans. Many successful avenues to nourish a healthy infant are available, and room ought to be made for different cultural patterns in the development of feeding practices. Some cultural practices are clearly unsafe and must be discouraged, such as a mother prechewing meats for a baby. Cultural practices that support the development of competence in parents can be encouraged, even if not part of the dominant culture. Examples of practices that may reflect cultural choices are swaddling an infant, or having an infant sleep in the parent's bed or in a room with a certain temperature. Practices based on family traditions may be forms of social support for new parents. Only if new parents have not considered the safety of the infant or have little knowledge of other, safer alternatives should cultural practices be discouraged. For example, it may be a cultural practice to offer meats to adults but not to infants. Parents should be informed that older infants may safely eat meats that are cut up or soft cooked to avoid causing choking. Some cultures consider meat-based soups as infant foods.

Cultural considerations may affect the family's willingness to participate in assistance programs such as WIC or early intervention programs. The dignity of the family unit, including extended relatives, has to be considered when counseling families about infant feeding practices. Food-based cultural patterns may be part of a religious tradition, so sensitivity to the family unit would include recognizing and understanding such beliefs.

Vegetarian Diets

Studies have found that infants who receive well-planned vegetarian diets grow normally.[59] The most restrictive diets, vegan and macrobiotic diets, have been associated with slower growth rates in infancy, particularly if infants do not receive enough breast milk.[48,59] Breastfed vegan infants should receive supplements containing vitamin D, vitamin B_{12}, and possibly iron and zinc.[59] The composition of the breast milk from vegan mothers may differ in small ways from standard breast milk.[48] An example is in the ratio of types of fat, although the total fat in breast milk from vegan mothers is the same. Impacts of these differences on the health of infants are generally not known.

Vegetarian diets range from adequate to inadequate, depending on the degree to which the diet is restricted, just as diets for omnivores range from adequate to inadequate.[59] Either food sources, such as fortified infant cereals and soymilk, or supplements can be used to assure adequate vitamin and mineral intake. Vegetarian families who avoid all products of animal origin, including milk and eggs, require carefully selected fortified foods or a higher degree of supplementation for their infants.[59] Periodic assessments of dietary intake, growth, and health status can be used to monitor the infant fed a vegetarian diet. Vegetarian infants have similar risk for developing food allergies from soy products, wheat, and nuts as do other infants.

Nutrition Intervention for Risk Reduction

The Early Head Start program is an example of a federal program that is focused on preventing and reducing risks to infant development.[6] The Early Head Start program was developed to work with infants and their families, especially new families at risk due to drug abuse, infants with disabilities, or teenage mothers. Nutrition services are among a wide range of services typically offered in an Early Head Start program. Other services may include home-based early childhood education, case management, and mental health support services, as well as health and socialization services. The Early Head Start program assists families in coordinating WIC participation with food stamps, routine well-baby visits, and day care, as needed.

Model Program: Newborn Screening and Expanded Newborn Screening

In the United States and many other countries, all new-borns are screened for rare conditions that may cause disability or death. Such screening, from a small dried blood spot, was initiated in the 1960s after early treatment of phenylketonuria (PKU) was shown to prevent later mental retardation in young children.[24] Most states screen for three to six different conditions, such as PKU, *galactosemia, hypothyroidism,* and sickle-cell disease.[24] New technology has resulted in expanded newborn screening, so some states are now testing for as many as 57 different conditions from the same dried-blood test.[31] Many of the disorders that can be detected by expanded newborn screening are treated by diet. Dietary treatment avoids the substance that has a metabolic block and replaces other dietary components that are usually provided in the foods that are avoided.[24] Expanded infant screening for genetic disorders is likely to continue to expand nutrition knowledge overall.

Galactosemia A rare genetic condition of carbohydrate metabolism in which a blocked or inactive enzyme does not allow breakdown of galactose, causing serious illness in infancy.

Hypothyroidism Condition in which thyroid hormone is not produced in sufficient quantities, interfering with growth and mental development if untreated in infants.

Key Points

1. Infants born full-term and preterm infants born between 34 weeks and 38 weeks of gestation are the same in their milestones of growth, development, and feeding in the first year of life.
2. In addition to access to adequate nutrition, environmental and societal factors have been credited with decreasing infant mortality.
3. The ability of infants to feed and eat is based on developmental skills that show readiness for the next step; parents learn to read the signals of readiness from their infants over time.
4. Energy and nutrient needs of infants are modulated by individual differences in sleep/wake cycle, exposure to temperatures and state of health, among other factors.
5. The priority is energy needs first; protein and carbohydrates will be converted to meet energy needs if sufficient calories are not consumed, slowing growth over time.
6. Limiting micronutrients are Vitamin D and fluoride in some environments, so supplementation is recommended.
7. Growth as weight, length, and head circumference accretion is monitored and interpreted over the first year.
8. Introduction of solid foods is also a developmental stage for parents in learning to read signs of hunger, fullness, and food preferences in their infants and to know safe food choices.
9. Common nutrition problems in the first year such as failure to thrive, colic, and prevention of iron deficiency anemia and baby bottle caries are usually solved by combining parent educational, nutritional, and medical approaches.
10. Infants can thrive with many different cultural and parenting styles; nutrition guidance for new parents is available from reliable sources in every community.

Resources

American Academy of Pediatrics
Reliable and credible sources of pediatric medical expertise in position papers, with sections for consumers and health care providers.
Website: www.pediatrics.org

American Dietetic Association
Consumer and provider information includes child nutrition and health and access to credible resources.
Website: www.eatright.org

Food Allergy Network

Reliable and scientifically based information for families with diagnosed allergies. This site routinely includes recipes.
Website: www.foodallergy.org

Health/Infants

Consumer and marketing information about infant growth and development, based on information available to mass media.
Website: www.cnn.com/health/indepth.health/infants

National Center for Growth Statistics (source of growth charts)

Site for obtaining growth charts and guidelines for their use.
Website: www.cdc.gov/nchs

Canadian Perinatal Surveillance System

The Public Health Agency of Canada produced this good source of growth charts for various gestational ages.
Website: www.phac-aspc.gc.ca/rhs-ssg/bwga-pnag

References

1. National Center for Health Statistics: NCHS growth curves for children 0–19 years. U.S. Vital and Health Statistics, Health Resources Administration. Washington, DC: U.S. Government Printing Office; 2000.

2. U.S. Vital and Health Statistics. Births: preliminary data for October 2005. National Vital Statistics Report. 2006;54:1–7.

3. U.S. Vital and Health Statistics. Infant mortality statistics from the 2003 period linked birth/infant death data set. National Vital Statistics Report. 2006;54:1–30.

4. United States Department of Health and Human Services Centers for Medicare and Medicaid Services; www.cms.hhs.gov/schip, accessed 10/06.

5. United States Department of Health and Human Services National Center for Chronic Disease Prevention and Health Promotion; www.cdc.gov/nccdphp/dnpa/pednss.htm, accessed 10/06.

6. Story M, Holt K, Sofka D, eds. Bright Futures in practice: nutrition. Arlington, VA: National Center for Education and Child Health; 2000:25–56, 266–70.

7. Engle WA, Boyle DW. Newborn care: perinatal care: delivery room management and transitional care. In Osborn et al, eds. Pediatrics. St. Louis, MO: Mosby; 2005: 1250–38.

8. Kleinman RE, ed. Pediatric nutrition handbook, 5th ed. Chicago, IL: American Academy of Pediatrics; 2004:2–22.

9. Kail RV, Cavanaugh JC. Human development: a lifespan view. 2nd ed. Belmont, CA: Wadsworth/Thomson Learning; 2000:83–121.

10. Nardella MT, Owens-Kuehner A. Feeding and eating. In Lucas, BL.Children with special health care needs Nutrition Care Handbook. Chicago, IL: American Dietetic Association; 2004:59–87.

11. Santrock JW. Life-span development. 7th ed. New York: McGraw Hill; 1999: 124–9.

12. Shelton TL, Dobbins TR, Neal JM. Principles of child development and developmental assessment. In: Osborn et al, eds. Pediatrics. St. Louis, MO: Mosby; 2005:41–57.

13. Institute of Medicine Food and Nutrition Board. Dietary Reference Intakes for energy, carbohydrate, fiber, fat, fatty acids, cholesterol, protein, and amino acids. Washington, DC: National Academy Press; 2002.

14. Regalado M, Halfon N. Developmental and behavioral surveillance and promotion of parenting skills. In Osborn et al., eds. Pediatrics. St. Louis, MO: Mosby; 2005:224–33.

15. Batshaw ML. Chromosomes and heredity. In: Batshaw ML ed. Children with disabilities. 6th ed. Baltimore, MD: Paul H. Brookes; 2007:3–26, 770.

16. Hobbie C, Baker S, Bayerl C. Parental understanding of basic infant nutrition: misinformed feeding choices. J Pediatr Health Care. 2000;14:26–31.

17. Lawrence RA, Lawrence RM. Breastfeeding: a Guide for the medical profession 6th ed. Philadelphia, PA: Mosby; 2005: 105–170.

18. Kleinman RE, ed. Pediatric nutrition handbook, 5th ed. Chicago, IL: American Academy of Pediatrics; 2004:425–42.

19. Trumbo P et al. Dietary reference intakes; vitamin A, vitamin K, arsenic, boron, chromium, copper, iodine, iron, manganese, molybdenum, nickel, silicon, vanadium and zinc. J Am Diet Assoc. 2001;101:294–301.

20. American Academy of Pediatrics. Prevention of rickets and vitamin D deficiency: new guidelines for vitamin D intake. Pediatrics. 2003;111:908–10.

21. Report of the Scientific Committee on Food on the Revision of Essential Requirements of Infant Formulae and Follow-Up Formulae, European Commission Health and Consumer Protection Directorate-General. Adopted April 4, 2003; Brussels, Belgium.

22. Wright CM, Waterston AJR. Relationships between paediatricians and infant formula milk companies. Archives of Disease in Childhood. 2006;91;383–85.

23. Clandinin MT et al. Growth and development of preterm infants fed infant formulas containing docosahexaenoic acid and arachidonic acid. J Pediatr. 2005; 146:461–68.

24. Birch EE et al. A randomized controlled trial of long-chain polyunsaturated fatty acid supplementation of formula in term infants after weaning at 6 wks of age. American Journal of Clinical Nutrition. 2002; 75:570–80.

25. Joanna Briggs Institute. Early childhood pacifier use in relation to breastfeeding, SIDS, infection and dental malocclusion. Nurs Stand. 2006;20:52–55.

26. Carver JD. Advances in nutritional modifications of infant formulas. Am J Clin Nutr. 2003;77(suppl):1550S–4S.

27. Yates AA, Schlicker SA, Suitor CW. Dietary reference intakes; the new basis for recommendations for calcium and related nutrients, B vitamins, and choline. J Am Diet Assoc. 1998;98:699–706.

28. Hampl JS, Betts NM, Benes BA. The "age+5" rule: comparisons of dietary fiber intake among 1–4-year-old children. J Am Diet Assoc. 1998;98:1418–23.

29. Infant foods as listed www.ars.usda.gov/ food search, accessed 10/06.

30. Ballew C et al. Blood lead concentration and children's anthropometric dimensions in the Third National Health and Nutrition Examination Study (NHANES III), 1988–1994. J Pediatr. 1999;134:623–30.

31. American College of Medical Genetics. Newborn screening: toward a uniform screening panel and system. 2005 Final Report. http://mchb.hrsa.gov/screening, accessed 10/06.

32. American Academy of Pediatrics Work Group on Breastfeeding. Breastfeeding and the use of human milk. Pediatrics. 1997; 100:1035–39.

33. American Dietetic Association. Position of the American Dietetic Association: promotion and support of breast-feeding. J Am Diet Assoc. 1997;97:662.

34. Kleinman RE, ed. Pediatric nutrition handbook, 5th ed. Chicago, IL: American Academy of Pediatrics; 2004:23–54.

35. Ross products handbook. http://rpdcon40. ross.com/pn/PediatricProducts.NSF/ web_Ross.com_XML_PediatricNutrition, accessed 10/05.

36. Wyeth infant formula. www.bright beginnings.com/products/bright_beginnings. asp, accessed 10/06.

37. Nestle Good Start infant formulas. www. verybestbaby.com/GoodStart/Default.asp, accessed 10/06.

38. American Dietetic Association. Pediatric manual of clinical dietetics. N Nevin-Folino, ed., 2003; appendix 1.

39. American Academy of Pediatrics. Committee on Nutrition (RE9251). Policy Statement. The use of whole cow's milk in infancy. Pediatrics. 1992; 89:1105–09.

40. Kleinman RE, ed. Pediatric nutrition handbook, 5th ed. Chicago, IL: American Academy of Pediatrics; 2004;108–11.

41. Ziegler EE et al. Cow's milk and intestinal blood loss in late infancy. J Pediatrics. 1999;135:720–26.

42. American Academy of Pediatrics. The use and misuse of fruit juice in pediatrics. Pediatrics. 2001;107:1210–13.

43. Isaacs JS. Fluid and bowel problems. In Lucas BL, ed. Children with special health care needs nutrition care handbook. Chicago, IL: American Dietetic Association; 2004:103–18.

44. Tervo R. Parent's reports predict their child's developmental problems. Clinical Pediatrics. 2005;44:601–11.

45. Cadzow SP, Armstrong KL, Fraser JA. Stressed parents with infants: reassessing physical abuse risk factors. Child Abuse & Neglect. 1999;23:845–53.

46. Birch LL et al. Infants' consumption of a new food enhances acceptance of similar foods. Appetite. 1998;30:283–95.

47. Sandritter T. Gastroesophageal reflux disease in infants and children. J Pediatr Health Care. 2003;17(4):198–205.

48. Kleinman RE, ed. Pediatric nutrition handbook, 5th ed. Chicago, IL: American Academy of Pediatrics; 2004:459–70.

49. Mennella JA, Beauchamg GK. Maternal diet alters the sensory qualities of human milk and the nurslings' behavior. Pediatrics. 1991;88:737–44.

50. Davis CM. Self-selection of diet by newly weaned infants: an experimental study. Am J Diseases of Children. 1928;36:651–79.

51. Story M, Brown JE. Do young children instinctively know what to eat? New England J Med. 1998;103–6.

52. American Academy of Pediatrics. Policy Statement (RE8132). Infant exercise programs. Pediatrics. 1988;82:800. Reaffirmed 11/94.

53. Mangels AR, Messina V. Considerations in planning vegan diets: infants. J Am Diet Assoc. 2001;101:670–77.

54. Khoshoo V, Reifen R. Use of energy dense formula for treating infants with nonorganic failure to thrive. Eur J Clin Nutr (England). 2002;56(9):921–24.

55. Goldsmith E. Nonorganic failure to thrive. MCN Am J Matern Child Nurs. 2001;26(4):221.

56. Needell B, Barth RP. Infants entering foster care compared to other infants using birth status. Child Abuse & Neglect. 1998;22:1179–87.

57. Metcalf T et al. Simethicone in the treatment of infant colic: a randomized, placebo-controlled multicenter trial. Pediatrics. 1994;94:29–34.

58. Recommendations to prevent and control iron deficiency in the United States. MNWR April 3, 1998;47.

59. Kleinman RE, ed. Pediatric nutrition handbook, 5th ed. Chicago, IL: American Academy of Pediatrics; 2004:299–312.

60. Jackson JM, Mourino AP. Pacifier use and otitis media in infants twelve months of age or younger. Pediatric Dentistry. 1999;21:255–60.

61. Wood RA. Prospects for the prevention of allergy: a losing battle or a battle still worth fighting? Arch Pediatr Adolesc Med 2006;160:552–4 and Arch Pediatr Adolesc Med. 2006;160:502–7.

62. Falci KJ, Gombas KL, Elliot EL. Food allergen awareness: an FDA priority. Food Safety Magazine. 2001; Feb–March.

63. Kleinman RE, ed. Pediatric nutrition handbook, 5th ed. Chicago, IL: American Academy of Pediatrics; 2004:593–608.

Chapter 9

Infant Nutrition:
Conditions and Interventions

Chapter Outline

- Introduction
- Infants at Risk
- Energy and Nutrient Needs
- Growth
- Nutrition for Infants with Special Health Care Needs
- Severe Preterm Birth and Nutrition
- Infants with Congenital Anomalies and Chronic Illness
- Feeding Problems
- Nutrition Interventions
- Nutrition Services

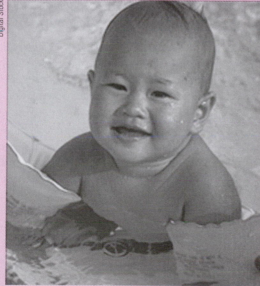

Prepared by **Janet Sugarman Isaacs**

Key Nutrition Concepts

1   Infants who are born preterm or who are sick early in life often require nutritional assessment and interventions that ensure that they are meeting their nutritional needs for growth and development.

2   Early nutrition services and other interventions can improve long-term health and growth among infants born with a variety of conditions.

3   The number of infants requiring specialized nutrition and health care is increasing due to the improved survival rates of small and sick newborns.

Introduction

Most infants are born healthy and then grow and develop in the usual manner. This chapter addresses the nutritional needs of infants who have health problems before or shortly after birth and are at risk for health or developmental difficulties. Within the first year of life, most infants have minor illnesses that do not interfere with growth and development. However, infants who were sick or small as neonates are likely to have conditions that may change the course of growth or development. *Children with special health care needs* is a broad term that includes the infants discussed in this chapter. As in Chapter 8, this chapter models sensitive communication with families by avoiding the word *normal* and, by implication, *abnormal* when referring to infants with special health care needs. Similarly, the designation *normal growth* is replaced by the phrase *standard growth* when referring to the CDC growth charts. Language such as "She is below normal on the growth charts" is replaced with "family-friendly" language, such as "She is the weight of a typical 6-month-old." Most families use the word *premature* (or *prematurity*) comfortably, but *preterm birth* is the conventional usage in maternity care. Both terms refer to infants born before 37 weeks of gestation, and both are used in this chapter.

Children with Special Health Care Needs   A federal category of services for infants, children, and adolescents with, or at risk for, physical or developmental disability, or with a chronic medical condition caused by or associated with genetic/metabolic disorders, birth defects, prematurity, trauma, infection, or perinatal exposure to drugs.

Low-Birth-Weight Infant (LBW)   An infant weighing <2500 g or <5 lb 8 oz at birth.

Very Low-Birth-Weight Infant (VLBW)   An infant weighing <1500 g or <3 lb 5 oz at birth.

Extremely Low-Birth-Weight Infant (ELBW)   An infant weighing <1000 g or 2 lb 3 oz at birth.

Neonatal Death   Death that occurs in the period from the day of birth through the first 28 days of life.

Perinatal Death   Death occurring at or after 20 weeks of gestation and through the first 28 days of life.

Down Syndrome   Condition in which three copies of chromosome 21 occur, resulting in lower muscle strength, lower intelligence, and greater risk for overweight.

Seizures   Condition in which electrical nerve transmission in the brain is disrupted, resulting in periods of loss of function that vary in severity.

Infants at Risk

The overall U.S. infant mortality rate decreased 45% between 1980 and 2002, but it has plateaued in this new century.[1] The health care system has been more successful at saving ill infants than in preventing preterm birth, low birth weight, or chronic conditions. The costs of preterm birth are not only the longer stay for care after birth, but costs over the first year of life are more than doubled from higher rates of physician and hospital visits.[2] The number of infants requiring nutritional services is increasing in large measure because of advances in neonatal intensive care. Small preterm infants who did not survive in the past are now being "saved." These are *low-birth-weight infants, very low-birth-weight infants*, and *extremely low birth-weight infants* who require the most intensive resources to support life and account for many **neonatal** and **perinatal deaths**. The smallest living newborns, who weigh 501–600 grams (1 lb 2 oz to 1 lb 5 oz), have a 31% chance of survival at birth.[2] This birth-weight range corresponds to about 23 weeks of gestation in the second trimester of pregnancy. Infants with birth weights of 901–1000 grams (2 lb to 2 lb 3 oz) are in the 29-week range of gestation and have an 88% change of survival.[2] Infants with genetic disorders, malformations, or birth complications have also benefited from advances in treatment and are less likely to die in infancy. However, they are much more likely to have chronic conditions, with increased need for medical, nutritional, and educational services later.

Regardless of what condition is involved, these nutrition questions are likely:

- How is the baby growing?
- Is the diet providing all required nutrients?
- How is the infant being fed?

In-depth nutrition assessments make sure nutrition is not limiting an infant's growth and development. Such assessments are needed by three main groups of infants:

- Infants born before 34 weeks of gestation. Preterm infants are born at less than 37 weeks of gestation, but generally only those born before 34 weeks of gestation have higher nutritional risks.
- Infants born with consequences of abnormal development during pregnancy, such as infants born with heart malformations as a result of the heart not forming correctly or exposure to toxins during gestation. Exposure to alcohol during gestation may interfere with brain formation and result in permanent changes in brain function. This second category includes infants with genetic syndromes, such as **Down syndrome**.
- Infants at risk for chronic health problems. Risks may come from the treatment needed to save their lives, or from the home environment that the baby enters. Examples of conditions that increase risks are **seizures** or cocaine withdrawal

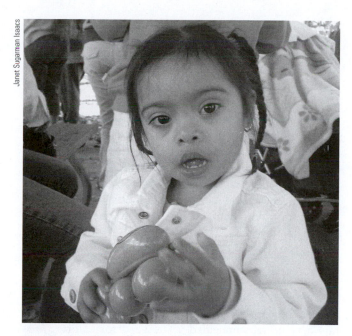

Janet Sugarman Isaacs

Illustration 9.1 Infant girl with Down syndrome after her heart surgery.

symptoms. Long-term consequences, such as later learning problems, may not be known for years.

Families of Infants with Special Health Care Needs

Every parent's wish is that his or her baby will be healthy. When parents find out their newborn has medical problems, they grieve for the loss of the perfect child of their dreams. The emotional impact of having a sick newborn overwhelms many parents, and providers of services for these families must be sensitive to parents' emotional needs. Coping styles of various family members vary, even if they are well prepared. It may take over a year for some family members to understand how the baby is doing and to adjust to the special needs of the child. For conditions with long-term consequences, the first year may not be long enough for parents to see that their infant is developing differently from other babies.

Energy and Nutrient Needs

Nutrient requirements for infants with health conditions are based on the recommendations for healthy infants.[3] Specific nutrients may be adjusted higher or lower based on the health condition involved. Adjusting the diet to changing conditions and close monitoring of growth and development may result in changing recommendations quite frequently in the first year. Scientific frontiers of medicine, genetics, nutrition, and technology interact in caring for sick infants. Nutrition requirements are not known for every condition, and individuals respond at their own pace of growth and development, so many nutrition

recommendations are based on the best judgment under the circumstances.

Energy Needs

For infants with special health care requirements, RDA may not be appropriate, since they are based on the needs of generally healthy infants.[3] DRI are used instead, so that the estimated energy requirements (EER) and adequate intakes (AI) can be considered as appropriate to the circumstances. For infants with special health care needs, caloric needs may be the same, less, or more than the DRI for infants (570 kcal/day).[3] Some conditions in newborns that have caloric needs based on the DRI are cleft lip and palate or phenylketonuria (PKU). The more common situation is that caloric needs are increased. Caloric needs of sick infants can be estimated with measurements such as *indirect calorimetry*.[4] Machines that measure indirect calorimetry are often available in intensive care nurseries. Estimated energy needs can change, however,

> **Indirect Calorimetry** Measurement of energy requirements based on oxygen consumption and carbon dioxide release.

with medications, activity, health conditions, and growth. Such estimates show that sick and small infants vary in their caloric needs more widely than expected, and that caloric deficits in preterm infants may be more common than previously known.[5] Extra calories are needed in circumstances such as the following:

- Infections
- Fever
- Difficulty breathing
- Temperature regulation
- Recovery from surgery and complications

Infants who are born preterm at less than 34 weeks of gestation particularly need higher energy intakes. The American Academy of Pediatrics suggests that premature infants need 120 cal/kg.[3] Intakes for recovering premature babies may be even higher, and the range of caloric need can be wide.[5,6] The European Society for Gastroenterology and Nutrition gives a caloric intake range of 95–165 cal/kg.[7] The amount of calories needed to gain 15 grams per day is recommended.[5] Infants born with VLBW or ELBW may still be weak and have difficulty feeding when they come home. Their higher calorie needs may be more difficult to meet given the small volume that they can consume. Over time, the recovering infant may increase intake so that even 180 cal/kg per day may be consumed.

Some infants need less energy than the RDA. These are infants born with smaller muscles or lower activity as a result of the inability to move certain muscles. An infant who has Down syndrome or one with repaired spina bifida needs fewer calories than the DRI of 108 cal/kg body weight.[3] Too many calories would interfere with her efforts to crawl; her weak muscles would have more body weight to move.

Protein Requirements

As noted for energy needs, protein requirements of infants with special health care needs may be higher, lower, or the same as other infants. Protein requirements based on the DRI of 1.52 grams of protein per kg body weight are recommended if the condition does not affect growth or digestion.[3] Protein recommendations are sufficient if total calories are high enough to meet energy needs.[8] The concept of protein sparing is important in fast-growing infants. If enough energy is available from glucose generated from foods containing fats and carbohydrates, then the amino acids generated from protein-containing foods are spared; that is, the amino acids are available for growth. If glucose from foods is not sufficient for meeting energy needs, however, amino acids from digestion of protein foods will be used to meet energy needs, and therefore less will be available for growth. When total caloric intake is low, protein-rich foods become an energy source. In this circumstance, providing the DRI for protein may be inadequate and result in slow growth. With preterm infants a sign of inadequate protein may be slow head growth, which is an indicator of brain growth.[9]

Conditions that could slow growth may require higher protein levels than the DRI for protein for infants. Higher protein recommendations are common in early infancy for conditions such as recovery from surgery or LBW. Protein intakes of 3.0–3.5 g/kg are appropriate for premature and recovering infants.[8] For recovering from some complications of ELBW, high protein intakes—as much as 4 g/kg—appear safe with adequate fluids, and without kidney problems.[8] The importance of protein to growing neonates is hard to overemphasize. Protein deficits in preterm infants have been shown even when high protein is provided, depending on how small and sick they are and how soon after birth the infant is being fed.[5]

MCT Oil   A liquid form of dietary fat used to boost calories; composed of medium-chain triglycerides.

Catch-Up Growth   Period of time shortly after a slow growth period when the rate of weight and height gains is likely to be faster than expected for age and gender.

Protein recommendations lower than the DRI are unusual in infancy. Infants with lower muscle activity as a result of smaller-sized muscles generally need lower protein. One example is Down syndrome. Conditions that lower physical activity and movement are often not identified until the infant is old enough to be moving around, closer to the end of the first year. During infancy, muscle tone is known to change over time, so muscle coordination and movement problems are usually confirmed later.

Form of Protein   Many illnesses interfere with the functioning of the gastrointestinal tract and digestion, even though newborns are born with intact enzymes for protein digestion. Protein and fat digestion depend on liver and pancreatic enzymes for intestinal absorption. However, many conditions associated with preterm birth and illness stress the liver and reduce its ability to function, causing changes in protein and fat digestion. Sick infants may require forms of protein in which amino acids are in short chains, such as in hydrolyzed protein, or single amino acids.[9] Other examples of infants needing protein that has been broken down are those with metabolic disorders.[10] Total protein may be limited, and partially replaced by mixtures of specific amino acids. For infants with PKU, meats and dairy product intake have to be limited because they contain too much of the amino acid phenylalanine.

Fats

Infants need a high-fat diet compared to older people, because fats provide energy. Up to 55% of calories from fats may be recommended.[9] The need for calories provided by fats is especially important in sick or recovering infants. Low-fat diets are generally not recommended for infants. Conditions that require limiting fats for infants are uncommon. One example is very sick infants who require heart-lung bypass machines as a part of major surgery. Such infants are given low-fat diets after age 2, without a specific level of fat restriction.[8] Fats are more difficult to absorb for infants with VLBW or ELBW because they require pancreatic and liver enzymes.[9] These enzyme systems may also be impaired in sick infants. Naturally occurring long-chain fats in breast milk may be supplemented with medium-chain fatty acids for sick infants. Medium-chain triglycerides do not require bile for absorption, so they are preferred.[9] Making sufficient bile for digesting long-chain fats requires healthy livers, like those in full-term infants, that are not likely in preterm births. *MCT oil* can be added to ensure calories from fats are available. Additionally, the essential fatty acids—alpha-linolenic and linoleic acid, as well as docosahexaenoic acid (DHA) and arachidonic acid (AA)—are provided in breast milk, human-milk fortifier, or special formulas.[11]

Vitamins and Minerals

DRIs for vitamins and minerals are appropriate for many infants with health conditions because recommendations are set with a safety margin.[12] However, DRIs are based on growth of typical infants, not those in which *catch-up growth* is required.[5] Vitamin and mineral requirements are affected by various health conditions, particularly those involving digestion. Prescribed medications may increase the turnover of specific vitamins.[13] Some infants with special health care needs have restrictions in volume consumed or activity that increase or decrease needs for specific vitamins or minerals. For example, limited volume of liquids may rule out vitamin-rich juices in the diet of infants with breathing problems.

High-potency vitamin and mineral supplements are usually prescribed for sick or recovering infants. Calcium is a potentially limiting nutrient in sick infants because calcium imbalance and *hypocalcemia* occur with a variety of conditions.[9] Iron, vitmain B_{12}, vitamin D, and fluoride are limiting in some specific situations.[8,13] Even after infants who were LBW or VLBW are eating well, vitamin and mineral requirements may be higher than the DRI, depending on specific health conditions. After preterm birth and discharge to the home, deficiencies in copper, zinc, and vitamin D are rare; they may be checked by blood tests.[14] The early signs of rickets as seen by X-ray are considered a sign of needing more vitamin D, above the 400 IU daily recommendation for term and preterm infants.[14]

Human milk fortifiers are used to boost calories as well as to provide additional vitamins and minerals in some infants in neonatal intensive care units.[14] In order to meet the higher requirements of specific vitamins and minerals for VLBW infants, such products can be added to breast milk. Such products are used under specific conditions, such as when an infant can only tolerate a low volume per feeding. They are intended to bridge the gap between breast milk and the extra needs of a VLBW infant.[8] Major ingredients are vitamins A, D, and C, and minerals such as calcium, phosphorus, sodium, and chloride.[9] Iron is not provided in human milk fortifiers, and is prescribed as needed.[9]

Some infant formula products also provide vitamins and minerals in concentrated amounts in formulas for premature infants. Such formulas are supplemented with extra calcium, phosphorus, copper, and zinc compared to standard formulas.[8] (The high levels of vitamins and minerals in premature infant formulas are shown in Table 9.2 later in this chapter). For some conditions, vitamins are used not only as dietary components, but as pharmaceuticals. An example is a condition in which vitamin B_{12} is injected as a part of therapy for a rare genetic disorder of protein metabolism.[10]

Growth

Growth in infancy is usually a reassuring sign that sufficient calories and nutrients are provided. The Centers for Disease Control's 2000 growth charts are a good starting point for monitoring of growth in infants with health risks.[15] The first goal of nutritional care is to maintain growth for age and gender. Later, this approach may be modified if there is a growth pattern typical for a specific condition that is identified after the first year. A steady accretion of weight or height is a sign of adequate growth, even if gains are not at the typical rate. Plateaus in weight or height, or weight gain followed by weight loss, are signs of inadequate growth. Growth may be assessed reliably using each infant as his or her own control regardless of

health conditions. As noted in Chapter 8, the methods of assessing growth require consistency and accuracy in order to make sure growth is interpreted correctly. Errors such as confusing pounds and kilograms in plotting interfere with interpretation no matter what growth chart is being used.

> **Hypocalcemia** Condition in which body pools of calcium are unbalanced, and low levels are measured in blood as a part of a generalized reaction to illnesses.

Usually, providing sufficient calories and nutrients results in good growth, but not in all cases. Sometimes slow growth is a symptom of an underlying condition, rather than a sign of inadequate nutrition. For example, infants who are born with genetic forms of kidney disease are short even when they consume adequate diets during the first year. Refinements in the usual methods and interpretation of growth are needed in conditions known to influence growth and development. These include:

- Using growth charts for specific diagnoses, such as Down syndrome growth charts.[16] (A list of specialty growth charts is included in Chapter 11.)
- Biochemical indicators of tissue stores of nutrients such as iron or protein, and of electrolytes such as potassium and sodium.
- Indicators of body composition, such as body fat measurement. These can be used to show caloric intake is not limiting growth because fat stores are adequate.
- Special attention to indicators of brain growth, such as measuring head circumference, may be helpful to explain short stature or other unusual growth patterns.
- Using treatment guidelines or published protocols, including disorder-specific weight-gain graphs and recommendations instead of standard growth charts.
- Medications that change weight gain, appetite, or body composition. Side effects of medications can explain rapid changes in weight.

Growth in Preterm Infants

Growth charts developed by the CDC can be used to assess growth progress of preterm infants with birth weights over 2500 grams.[15] The body composition of infants born preterm is not the same as that of term infants, in part because these infants have missed part of the third trimester, when fat is added rapidly.[11] In fact, body fat buildup is a late sign of recovery from preterm delivery. Body composition at various gestational ages is used to adjust growth expectations based on age. Treatment of the infant's medical condition may also affect growth expectations; for instance, fluid accumulation may artificially increase weight. As a result of such considerations, VLBW and ELBW infants are not represented by the standard growth charts.[15] Growth for newborns with birth weights between 501 and 1501 grams can be tracked by the Neonatal Research Network Growth Observational Study Research Network, sponsored by the National Institutes of Health, National Institute of Child Health and Human Development (NICHD). Its website (https://neonatal.rti.org/birth_curves/dsp_BirthCurves.cfm) has a VLBW Postnatal Growth Chart component that projects growth from whatever birth weight, birth length, or birth head circumference is entered. The management of preterm birth is so rapidly changing that recently published growth charts may not reflect growth patterns achieved with current practices.[13]

Growth charts for preterm infants are needed in hospitals for determining the age of gestation at birth, and for tracking growth until discharge. Standard growth charts often replace them when the child goes home. Growth charts for preterm births have been updated to cover more of fetal life, starting at 22 weeks of gestation and ending usually at 10 weeks after the expected 40 weeks of gestation.[17] An example set is called the "IHDP Growth Charts"; these are based on the *Infant Health and Development Program (IHDP)*.[17] This large research program has created four charts, two for girls and two for boys, at two different birth weights. One set, called "LBW Premature," is for infants with birth weights in the range of 1501–2500 grams; the other is "VLBW Premature," and is for use with infants who weighed 1500 grams or less and were less than 38 weeks of gestation.[17] Each chart has:

> **Infant Health and Development Program (IHDP)**   Growth charts with percentiles for VLBW (<1500-g birth weight) and LBW (<2500-g birth weight).

- Head circumference for gestation-adjusted age
- Weight for gestation-adjusted age
- Length for gestation-adjusted age

Canadian premature growth charts (see Resources at the end of this chapter) and U.S. National Institutes of Health Child Health and Human Development Neonatal Research Network growth charts are also used routinely.[18] All preterm growth charts show that head circumference is a main indicator of growth approaching that in term infants; both weight and length lag at discharge even when 40 weeks gestation age is reached.[18]

Correction for Gestational Age

Gestation-adjusted age is calculated by subtracting gestational age at birth from 40 weeks (the length of a full-term pregnancy). The resulting number of weeks is divided by four to obtain months. The result in months is then subtracted from the infant's current age. For example, if an infant was born at 30 weeks gestation, she is 10 weeks early. This equals 2.5 months preterm. When she is 3 months old, her gestation-adjusted age is 2 weeks, or 0.5 months. This age of 0.5 months would be used in plotting her growth on the IHDP growth chart as part of assessing her growth and development.

Does Intrauterine Growth Predict Growth Outside?

"Apples don't fall from a pear tree."

French saying

The answer is yes, no, and maybe! Fetal monitoring during pregnancy and in-depth knowledge about the development of various organ systems provides a clear pattern of growth at various gestational ages. However, many factors during and after pregnancy are known to affect growth rate. In summary, these are:

- Intrauterine environment, particularly the adequacy of the placenta in delivering nutrients; the presence of toxins such as viruses, alcohol, or maternal medications; or the depletion of a needed substance, such as folic acid
- Fetal-origin errors in cell migration or formation of organs, whether or not a cause is known. Various nutrients, such as vitamin A, have been implicated in such errors
- Unknown factors that cause preterm birth, such as environmental toxins in air pollution

Knowledge gained from studies of gene-nutrient interactions will identify additional genetic and nutritional factors influencing fetal development and improve preventative and treatment strategies.

Research is emerging that shows some infants were born prematurely due to conditions originating during the intrauterine period.[6] As discussed in Chapter 8, SGA and intrauterine growth retardation are terms used to describe infants who are smaller than expected at birth. SGA is the more general term because it is based on the population of infants of the same gestational age.[6] Both predict higher medical risks and need for close growth monitoring.

If the intrauterine insult was early in gestation, body weight, length, and head size (brain size) and parent–child interaction are affected.[9,13] There has been a change in the number and size of fetal cells. The abnormal fetal growth pattern may persist despite adequate medical and nutrition support after birth. Examples of conditions causing early insults are infants born after cocaine or alcohol exposure, both of which are associated with IUGR and preterm birth.[18] Later exposure in the second or early part of the third trimester may result in preservation of head size and body length, but low weight. Some genetic conditions characterized by small size are not diagnosed until childhood, but have IUGR noted in the medical history.[16] Long-term IUGR is linked with higher risk for attention and hyperactivity disorders during childhood, even when maternal factors such as smoking are considered.[16]

Intrauterine growth may not predict growth for some infants whose birth removes them from adverse exposure within the intrauterine environment. Examples of this situation are maternal uncontrolled diabetes, smoking, phenylketonuria, or maternal seizures treated by medications. In such cases the rate of growth after birth may be improved and normalized during the first year of life.[9] In general, the earlier the exposure to the toxin, the worse the effects on later growth.[21] Sometimes marijuana, alcohol, tobacco, and crack/cocaine have been used at various times during pregnancy, so growth impacts may be based on amount, timing, and interactions of toxins. The most important risks for later growth may be neonates born with smaller head circumferences.

A large category that changes the rate of intrauterine growth is early medical treatment. If the intrauterine growth was fine, preterm birth or its complications may slow or plateau growth early in infancy. This may mask whether or not typical growth goals are appropriate all during infancy. Various studies have measured whether infants recover from early growth problems.[19–21] A group of ELBW infants who were followed into adolescence and did not have major disabilities were shorter in stature and lower in weight than others of the same age and gender, suggesting growth is affected long term.[20] Another study of infants who were short for age and were provided nutritional supplements for 2 years found that the infants were still short for age at 11 and 12 years.[21] Whether or not adequate nutrition is provided early in life of course depends on how *adequate nutrition* is defined. Advocates of increasing current nutritional recommendations for preterm infants expect growth problems could be lessened if higher levels of nutrition are provided, regardless of other factors.[5]

Conditions that affect growth may be time limited. Catch-up growth may be seen during recovery in many infants, resulting in changing growth interpretation. Surgery, when required for heart conditions, may delay growth for a short time. Growth may slow and then catch up with respiratory illnesses that resolve with medications. Usually increased access to adequate nutrition improves growth. Only close monitoring over time may show signs of catch-up growth early, such as an increase in fat stores or length.

The amount of time needed for catch-up growth for premature infants differs based on gestational age at birth and subsequent complications.[22] Clinical conventions are to provide 1 year for catch-up growth for infants born 32 weeks or later, and 3 years for catch-up growth for VLBW or ELBW infants. Catch-up growth after low birth weight is usually encouraged, but rapid growth in infancy may be raising risks for later chronic conditions, such as cardiovascular disease and diabetes if the rate of weight gain is excessive.[23] It is likely such infants will have growth percentiles in the lower end of the normal range during childhood.

Interpretation of Growth

Hospital discharge after preterm birth may be based on a pattern of weight gain, such as 20–30 grams per day.[14] Strong emphasis is placed on growth as a sign of improving health in monitoring small and sick infants after discharge, but complications make this difficult to achieve. An example of a common diagnosis in preterm babies and difficulty in interpretation of growth is the lung condition bronchopulmonary dysplasia (BPD). Rates of growth among infants with BPD are different from those in full-term infants and preterm infants who do not have BPD. While infants with BPD are recovering, the growth pattern of these infants is affected for the entire first year of life.[19] The reasons for the slower growth are higher nutritional requirements, changes in endocrine and pulmonary systems, and perhaps interaction among these systems.

Microcephaly   Small head size for age and gender as measured by centimeters (or inches) of head circumference.

Macrocephaly   Large head size for age and gender as measured by centimeters (or inches) of head circumference.

Growth rate changes are closely associated with the frequency of illness, hospitalizations, and medical history.[14] Conditions acquired as a result of preterm birth that make growth difficult to interpret include:

- Symptoms related to intestinal absorption that can temporarily or permanently change nutrient requirements.
- *Microcephaly* (small head size) or *macrocephaly* (large head size) compared to other growth indicators may be a sign that growth may be affected as a result of neurological consequences. Both large and small head size in infancy can affect muscle mass, body composition, and subsequent growth.
- Variable rates of recovery and growth are seen for many infants. Infants are as different from each other as the rest of us, but these differences are hard to see soon after birth.

Nutrition for Infants with Special Health Care Needs

Infants who are small or sick near birth may have major growing and feeding problems.[23] Infancy is such a vulnerable time of life that most health conditions occurring this early interfere with growth and development. Over time, most of these problems resolve, although some become chronic, and a few result in death. Nutrition plays an important part in preventing illness, maintaining health, and treating conditions in infancy. Nutrition tends to become more important over time to maintain growth if conditions are chronic and impact feeding (Illustration 9.2).[22] Table 9.1 on the next page shows nutrition problems in infants with special health care needs. Nutrition assessment documents

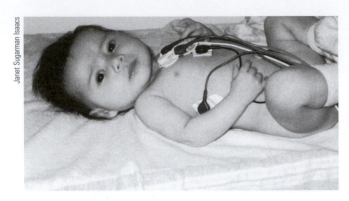

Illustration 9.2 Baby girl wearing a heart monitor at home.

these concerns, and nutrition services are provided based on the assessment.

Nutrition Risks to Development

Many health conditions change the infant's rate of development. *Developmental delay* describes the interaction of a chronic condition with development. The terms *children with special health care needs* and *developmental delay* are general terms used to allow nutritional, medical, and developmental services to be provided for infants.

Developmental delay is used to describe a wide range of symptoms that reflect slow development. Symptoms that relate to nutrition are common. These symptoms include infants who are growing slower than expected for age, or have difficulty in feeding, such as refusing food from a spoon by 8 months of age. An example is a 2-month-old girl who does not breastfeed for more than a few minutes per side. At first this may appear to be a problem of breastfeeding position or frequency. By 4 months, weight gain is slower than expected, so now growth concerns and feeding concerns are interacting; it is not clear if they are separate or related problems. These concerns are sufficient to request an evaluation for eligibility for intervention services. Several months later, after various services have been put in place, it may still be unclear if these nutrition problems are from development, a health condition such as a heart murmur, or the interaction of both. In any case, the girl fits

> **Developmental Delay** Conditions represented by at least a 25% delay by standard evaluation in one or more areas of development, such as gross or fine motor, cognitive, communication, social, or emotional development.

Table 9.1 Nutrition concerns in infants with special health care needs	
Growth	Slow rate of weight gain
	Fast rate of weight gain
	Slow rate of gain in length
	Disproportionate rate of weight to height gain
	Unusual growth pattern with plateau in weight or length gain
	Altered body composition that decreases or increases muscle size or activity
	Altered brain size that decreases or increases muscle size or activity
	Altered size of organs or skeleton, such as an enlarged liver or shortened leg length
Nutritional adequacy	Calorie needs are higher or lower
	Nutrient requirements higher or lower overall
	Specific nutrients, such as protein or sodium, are required in higher or lower amounts
	Vitamins, minerals, or cofactors (such as carnitine) are required in higher or lower amounts
Feeding	Disruption of the delivery of nutrients as a result of:
	• Structure or functioning of the mouth or oral cavity
	• Structure or functioning of the gastrointestinal tract, including diarrhea, vomiting, and constipation
	• Appetite suppression by constipation or medications
	• Disrupted interaction of the infant with the parent, such as infant cues being so subtle that parent responses are delayed
	• Posture or position that promotes or interferes during meal times
	• Timing of nursing, meals, and snacks throughout the day
	• Inappropriate food choices or methods of preparation
	• Interruptions in adequate shelter for feeding and sleeping
	• Instructions were unclear or too complicated for the parent to follow

the category of a child with a special health care need. This allows the family to benefit from nutritional, medical, and developmental interventions, without requiring a specific diagnosis. Infants generally are not old enough to have a specific diagnosis related to development, such as mental retardation or *autism*.[23]

Down syndrome is an example of a condition in which developmental delay is noted in infancy. Down syndrome prevalence is about 13 cases per 10,000 live births.[1] Nutrition concerns with infants who have Down syndrome are feeding difficulties related to weak muscles in the face, and overall; high risk of overweight; and constipation. Heart and intestinal conditions are more common in infants with Down syndrome, so their nutrient needs may be increased if surgery is required. This is also an example of a chronic condition in which nutrition problems such as overweight increase over time if prevention and maintaining health are not addressed. Growth requires close monitoring to identify and prevent overweight starting in infancy. Infants with Down syndrome love to suck and have things in their mouths so much that it is easy to overfeed them. Development of movement occurs at a slower rate, with lower physical activity, which also can contribute to overweight. Giving parents their own copy of Down syndrome growth charts for infants is recommended after the diagnosis is confirmed.[16] These charts may be helpful in recognizing typical growth and preventing overweight early. These special growth charts are available from places that serve children with Down syndrome, such as developmental or genetics clinics in major medical centers.[16]

Not all children with developmental delay in infancy have developmental disabilities later. For example, an infant with breathing problems may be slower to grow and to crawl as a result of his higher caloric needs during the first year of life. Such an infant may show developmental delay in motor skills, but by age 3 he will have improved in overall health and have caught up to others in motor skills. He would not have a developmental disability. Other examples are infants from high-risk pregnancies, such as those born large for gestational age as a result of poorly controlled gestational diabetes. Many infants require short stays in intensive care units for glucose regulation; some may have long-term risks for their development. Some infants with developmental delays continue to have slower development over time. After infancy, when standard testing and evaluation can be performed, the term *developmental delay* may be replaced with a more specific type of medical or developmental diagnosis.[24]

Severe Preterm Birth and Nutrition

The yearly incidence of VLBW in the United States is approximately 60,000 infants—about the same as the population of a small city, such as Iowa City, Iowa.

Infants with birth weights near 1500 grams (3 lb 4 oz) have gestational ages from 28 to 32 weeks and a survival rate of almost 90%.[25] Each infant requires immediate intensive care hospitalization; and if they survive, these infants continue to have high nutritional needs throughout infancy. ELBW infants weigh less than 1500 grams and have gestational ages ranging from 23 to 28 weeks. Despite advances in the care of such infants, disability such as delayed development is a common outcome of ELBW.[26] Attention and learning problems in school age children are at higher rates than in children who were born at term, although the majority of children who were ELBW do not have disabilities.[21] Some outcome studies are demonstrating lifelong consequences of low birth weight, such as impacting later employment as adults.[27]

Nutrition problems resulting from VLBW and ELBW preterm births are addressed as they present. The initial problem after birth is that the newborn cannot nurse like a full-term infant, and most require respiratory support to breathe. Getting adequate calories and nutrients into the preterm infant requires *nutrition support*, usually first *parenteral feeding* and then *enteral feeding* methods.[9,27] Feeding problems of preterm infants are discussed later in this chapter. All newborns have high metabolic rates, and they will use fat stores and protein in tissues and muscles to meet glucose needs if consumed calories and nutrients are not sufficient. This happens sooner in infants than in adults.[4,5] Providing sufficient calories and nutrients to meet requirements and preserve ingested protein and calories for growth is the goal, but it may be difficult and take more time than expected in sick and recovering infants.

Autism Condition of deficits in communication and social interaction with onset generally before age 3, in which mealtime behavior and eating problems occur along with other behavioral and sensory problems.

Nutrition Support Provision of nutrients by methods other than eating regular foods or drinking regular beverages, such as directly accessing the stomach by tube or placing nutrients into the bloodstream.

Parenteral Feeding Delivery of nutrients directly to the bloodstream.

Enteral Feeding Method of delivering nutrients directly to the digestive system, in contrast to methods that bypass the digestive system.

How Sick Babies Are Fed

Gastrointestinal upset is a response to many conditions in newborns, whether the intestines are the initial problem or not. VLBW, ELBW, and sick infants are especially vulnerable to problems related to the gastrointestinal tract. Such problems directly affect how calories and nutrients are provided, as well as the composition of the diet. For example, if a newborn gets an infection, an early sign may be inflammation of the intestine. As a response, the method of feeding the infant has to be adjusted. Inflamed or damaged areas may slow or interrupt typical intestinal muscle movements, resulting in signs of increasing illness.[9,13] Blood loss from the intestines is a sign of *necrotizing enterocolitis (NEC),* a serious condition in the neonate. When this occurs, oral feeding is stopped and replaced by parenteral nutrition.

Many gastrointestinal conditions interfere with infant feeding, such as gastroesophogeal reflux, constipation, spitting up, and vomiting. In small and sick newborns, these gastrointestinal conditions may represent slow or uncoordinated movements of the intestinal muscles.[9] These conditions do not rule out enteral feeding, which stimulates the intestines and keeps them healthy. Feeding methods are selected based on the length of time before it is expected the baby can nurse or feed without help. Gavage feedings may be used. These are slow feedings sent from the mouth or nose into the stomach though a tube. Infants who are too weak to breastfeed may be offered the comfort of the breast or pacifier along with gavage feeding. *Oral-gastric (OG) feeding* is also used. Other enteral methods are *transpyloric feeding (TP), gastrostomy feeding,* and *jejunostomy feeding.*[9] These methods are used when nutrition support is expected to be needed for several months.[23]

> **Necrotizing Enterocolitis (NEC)** Condition with inflammation or damage to a section of the intestine, with a grading from mild to severe.
>
> **Oral-Gastric (OG) Feeding** A form of enteral nutrition support for delivering nutrition by tube placement from the mouth to the stomach.
>
> **Transpyloric Feeding (TP)** Form of enteral nutrition support for delivering nutrition by tube placement from the nose or mouth into the upper part of the small intestine.
>
> **Gastrostomy Feeding** Form of enteral nutrition support for delivering nutrition by tube placement directly into the stomach, bypassing the mouth through a surgical procedure that creates an opening through the abdominal wall and stomach.
>
> **Jejunostomy Feeding** Form of enteral nutrition support for delivering nutrition by tube placement directly into the upper part of the small intestine.

Food Safety Preterm babies with immature immunological systems are prone to infection, so every effort is made to assure their feedings do not become contaminated. The rate of feeding preterm infants is often much slower than that for full-term infants, and formula or breast milk is at room temperature for a longer time. Contamination of feeding equipment increases with time; consequently, hospitals have policies requiring them to change the feedings often, such as every 4 hours.[28–30] Hospitals avoid using powdered formulas as much as possible since liquid formulas can be heat-treated to higher temperatures.[28] Frozen and then thawed breast milk is tracked so that it is given to the correct baby (usually too weak to suck) inside set time frames.

What to Feed Preterm Infants

Breast milk is the recommended source of nutrition for preterm infants. Colostrum and breast milk are produced even when the mother delivers very early.[31] Preterm human milk has increased protein content compared to term milk.[31] Hospital protocols and policies for having mothers pump and freeze breast milk for later use by their preterm infants are highly recommended.[32] Staff training to encourage new mothers at home to rest enough and pump enough to stimulate breast milk production is also recommended.[33] Barriers to breastfeeding small and sick newborns are partially based on their abilities, on how sick they are, and on the care system. Promoting breast milk for preterm infants is recommended as hospital policy by the American Academy of Pediatrics, but hospitals differ in practices.[31] Medical conditions in the infant may undermine breastfeeding. Infants are generally able to nurse successfully at about 37 weeks of gestation. Prior to that age, they may benefit from being put to the breast to stimulate non-nutritive sucking, or sucking that does not deliver milk to be swallowed. There are a few conditions in which human milk is unsafe for preterm or sick infants. Breastfeeding is contraindicated when breast milk contains harmful medications, street drugs, viruses, or other infective agents, or when the infant has a specific type of gastrointestinal tract malformation or inborn errors of metabolism.[31]

Depending on the infant's birth weight and health status, breast milk may be insufficient in nutrients unless supplemented by human milk fortifier and/or other sources of calories, such as MCT oil. If not fed modified breast milk or nursing, the infant's source of nutrition may be cow's milk or soybean-based formulas.[32] Whey as the predominant form of protein from cow's milk is recommended because its amino acids profile is closer to that of human milk.[8]

Infant formulas for preterm infants are available for home use after hospital discharge, if breast milk is not available. They provide the higher calories and nutrient levels that small infants need compared to term infants.[34] Standard formula that is 20 calories per fluid ounce can also be used for preterm infants, modified in a manner similar to breast milk to boost calories and nutrients. High-calorie formulas, such as 28 cal/fl oz, may be appropriate for some infants. Such high-calorie formulas are not routinely used because they have high osmolarity, which may affect fluid and electrolyte balance. Table 9.2 shows a comparison of premature and standard formulas.[34] If the infant easily fatigues or is too weak to suck enough volume, 22 or 24 cal/fl oz formulas may be recommended.[9] The sources to add extra calories and nutrients are selected based on the infant's gastrointestinal tolerance and volume requirements. They may include MCT oil; polycose; rice baby cereal; and rarely, human milk fortifier. Routine nutritional assessment of the infant's growth tracks the diet's effectiveness in providing adequate nutrients and calories.

Preterm Infants and Feeding

VLBW or ELBW infants usually progress at their own rate of development regarding feeding skills. The goal is the

Table 9.2 Selected nutrient composition of term and preterm formulas

Nutrients	20 Cal/Fl Oz	22 Cal/Fl Oz	24 Cal/Fl Oz
Protein	2.1 g	2.8 g	3 g
Linoleic acid	860 mg	950 mg	1060 mg
Vitamin A	300 IU	450 IU	1250 IU
Vitamin D	60 IU	80 IU	270 IU
Vitamin E	2 IU	4 IU	6.3 IU
Thiamin (B_1)	80 mcg	200 mcg	200 mcg
Riboflavin(B_2)	140 mcg	200 mcg	300 mcg
Vitamin B_6	60 mcg	100 mcg	150 mcg
VitaminB_{12}	0.3 mcg	0.3 mcg	0.25 mcg
Niacin	1000 mcg	2000 mcg	4000 mcg
Folic acid	16 mcg	26 mcg	35 mcg
Pantothenic acid	500 mcg	850 mcg	1200 mcg
Biotin	3 mcg	6 mcg	4 mcg
Vitamin C	12 mg	16 mg	20 mg
Inositol	6 mg	30 mg	17 mg
Calcium	78 mg	120 mg	165 mg
Copper	75 mcg	120 mcg	125 mcg

Table 9.3 Preterm and term infant feeding differences

Preterm Infant	Term Infant
Central nervous system does not signal hunger	Signals hunger; has supportive newborn feeding reflexes
Unstable feeding position, such as a forward head position	Stable and facilitating feeding position from newborn reflexes
Oral hypersensitivity	Readily accepts food by mouth

same as for all infants—to achieve good nutritional status, as indicated by growth and feeding skills progression. Most families enjoy feeding their infants and experience few long-term feeding problems. Infants who were born preterm, however, may be hard to feed. There are several reasons for this:[32]

- Fatigue. Low levels of arousal of weak or sick infants may lessen feeding duration.
- Low tolerance of volume. Abdominal distention due to feeding may result in changes in breathing and heart rate, so that the infant stops feeding.
- "Disorganized feeding" may result from the infant having experienced defensive and unpleasant reactions to feedings or procedures, so anything coming to the mouth causes a stress reaction rather than a pleasurable reaction.[9]

Regardless of the associated conditions, certain feeding characteristics of preterm infants are distinct from those of term infants (Table 9.3). Most recovering infants improve in their feeding abilities with time. Anxiety decreases as parents become more comfortable with caring for their infant at home. The underlying reflexes that associate pleasure with feeding reemerge, and the interaction of the infant with the feeder improves. There is a lot of room for hope in the feeding process after discharge.

Major advances in understanding the nutritional needs of preterm babies have come about from working with smaller and smaller infants. Table 9.4 presents an example of a typical diet for a premature baby who was not breastfed and had a 3-month hospital stay after birth. The infant has gastroesophageal reflux and prescribed medications that are included in his feeding instructions. This example shows that providing an adequate diet is an important part of the infant's growth and development and his recovery from preterm birth complications (see Case Study 9.1 on the next page).

Table 9.4 Example diet for a VLBW infant at 8 months of age, with a corrected age of 4.5 months (weight 4.5 kg, 5–25% on VLBW Premature Boys growth chart)

Food and Formula	Feeding Instructions
Five feedings per day of formula, each 5 fl oz, High-calorie formula (24 cal/fl oz) with 2 Tbsp rice cereal added	Provide support for semi-reclining feeding position during and up to 30 minutes after feeding
Medications for stomach added to 2 fl oz of 50% diluted apple juice two times per day	Encourage use of pacifier for comfort between feedings
Two meals of pureed baby foods fed with spoon, total intake one 2-oz jar	Offer bottles every 3 hours except overnight if no signs of hunger before then
Liquid vitamin/mineral supplement	Keep scheduled appointments for WIC; weigh in at MD office, bringing diet log book

Case Study 9.1

Photo Disc

Premature Birth in an At-Risk Family

A baby named Eric is born at 30 weeks of gestation, appropriate for gestational age, at 1.4 kilograms (3 lb). Like his mother, he tests positive for cocaine. Eric receives routine intensive care services, which include ruling out sepsis, and is given head ultrasound studies. By 33 weeks he is being fed OG and has only transient respiratory difficulties. Prior to discharge at 37 weeks, he appears to be developing normally. He drinks 22 fl oz formula/day at the rate of about 1.5–2 fl oz/feeding, with 10 feedings per day. He is placed in foster care with an experienced foster mother who has two older children.

Eric's custody is reconsidered when his biological mother expresses interest when he is about 9 months of age. Shortly after that, his mother's parental rights are terminated based on criminal charges. His foster mother reports that he has been a colicky baby, and he has had at least three ear infections during his first year of life. He is enrolled in an *early intervention program* based on his prematurity and intrauterine drug exposure. Eric's initial developmental testing was within normal limits at 6 months.

Eric's foster mother expresses concern about his intervals of periodic crying, during which he accepts neither soothing nor a bottle. Eric is diagnosed with gastroesophageal reflux (GER) and slow gastric emptying, and he is treated medically starting at 8 months. He is slow to accept foods on a spoon, with gagging and spitting up. His growth on the IHDP LBW chart is near the 50th percentile for weight and height. His head circumference is at the 5th percentile. Eric does not sit up unassisted until 8.5 months, but he turns over from stomach to back and vice versa easily. He is sent to a genetic specialist because he appears to have some facial features consistent with fetal alcohol syndrome, such as low-set ears, wide nasal bridge, and thin upper lip. The diagnosis is not confirmed because he is too young, but reported as a possibility to be reevaluated after he is older and can be given developmental tests. If this diagnosis were confirmed, standard growth charts would not fit Eric, because short stature and low weight are part of the diagnosis, even when adequate nutrition is provided. In infancy Eric's growth was within normal limits after correction for prematurity, with the same trend of a lower head circumference.

Eric's foster mother expresses an interest in adoption when he is almost 1 year old. She is pleased that he needs no medication and is growing well. His early intervention services are continued based on his at-risk status, because no specific 25% delay has been documented at 1 year. (Later, at 34 months of age, he is diagnosed with mixed developmental delay based on cognitive and speech delays.) Eric is adopted by the foster family.

Early Intervention Program
Educational intervention for the development of children from birth up to 3 years of age.

Questions

1. Did Eric's early birth account for his slow growth later?
2. Did nutrition affect when Eric's developmental delay probably started? Note that it was diagnosed at about age 3.
3. What are the signs that Eric can outgrow his problems?

Infants with Congenital Anomalies and Chronic Illness

Infants who are not preterm but require neonatal intensive care may be at risk for chronic illness. About half of babies in neonatal intensive care units have normal birth weights and experience lower mortality than LBW infants. They tend to have a higher rate of *congenital anomalies* (22%) and often require rehospitalization.[35] These infants need more nutrition services than typical infants because their

Congenital Anomaly Condition evident in a newborn that is diagnosed at or near birth, usually as a genetic or chronic condition, such as spina bifida or cleft lip and palate.

growth and feeding development requires close monitoring and intervention.

Congenital anomalies are recorded in the Birth Defects Monitoring Program (BDMP).[36] The Centers for Disease Control publishes prevalence data based on states and hospitals that participate voluntarily in surveillance programs. The United States keeps prevalence data not just for infants but also for children of various ages. *Infant mortality attributable to birth defects (IMBD)* accounts for approximately 2 deaths per 1000 live births.[36] Whereas infant mortality is decreasing in the United States, the proportion of infants who die from IMBD is increasing.[36] The major types of birth defects associated with death are first, heart malformations and next, central nervous system defects.[36] Examples of central nervous system congenital anomalies are spina bifida and *anencephaly.* Since folic acid has been added as a supplement in grains and flours, rates of spina bifida and related conditions have declined 26%.[37]

Infants with congenital anomalies, genetic syndromes, and malformations fit in the category of children with special health care needs, so they are eligible for a wide range of medical, nutritional, and educational services to maximize their growth and development.[25] All babies born with such conditions have risks to maintaining good nutrition status as they receive treatment. The nutrition concerns are growth, adequacy of the diet in providing required nutrients, and feeding development, as shown in Table 9.1. Nutrition services range from temporary to long term, and they are as diverse as the many different types of conditions involved. Several examples of disorders with minor and major nutritional consequences follow.

Major nutritional impacts are exemplified by disorders that involve the gastrointestinal tract. Infants with *diaphragmatic hernia* or *tracheoesophageal atresia* cannot safely eat by mouth and require nutrition support and several surgeries during infancy.[32,36] Diaphragmatic hernia occurs in 1 in 4000 live births due to failure of the diaphragm to form completely. Tracheoesophageal atresia occurs in 1 in 4500 live births due to an error in development of the trachea. These examples of conditions treated by neonatal surgery and intensive care have been credited with lower infant mortality for such congenital anomalies.[36] Both conditions change the motility of the gastrointestinal tract, so sufficient calories, nutrients to maintain growth, and oral feeding are important parts of the treatment plan.[32] Such infants miss the windows of development when oral feeding is pleasurable and may have residual feeding problems, such as disliking eating by mouth, well into early childhood. Financing such intensive health care and maintaining the child's normal social and emotional development are major issues for the family. The families of such infants have many specialty health care providers, and their complex financial and emotional reactions can also influence the

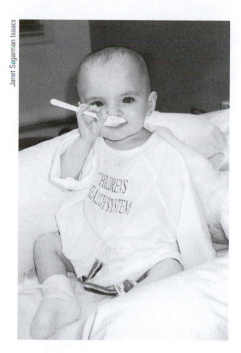

Janet Sugarman Isaacs

Illustration 9.3 Infant in hospital being encouraged to eat with a spoon.

infant's development. Both of these conditions eventually result in children being able to eat like everyone else (Illustration 9.3).

Common examples of congenital anomalies are those for infants with *cleft lip and palate.* Major feeding difficulties occur before and after corrective surgeries, and they sometimes interfere with growth in infancy and early childhood.[38,39] Assistance in feeding by registered dietitians as part of a team approach is needed, because hearing, speech, and language problems are associated with cleft lip and palate. Positions for eating are adapted, and use of special feeding devices for the infant's cleft are needed. Cleft lip and palate may occur alone or as a part of various rare genetic conditions, so growth and feeding problems should also be assessed after corrective surgery.[38]

Infant Mortality Attributable to Birth Defects (IMBD) Category used in tracking infant deaths in which specific diagnoses have a high mortality.

Anencephaly Condition initiated early in gestation of the central nervous system in which the brain is not formed correctly, resulting in neonatal death.

Diaphragmatic Hernia Displacement of the intestines up into the lung area due to incomplete formation of the diaphragm in utero.

Tracheoesophageal Atresia Incomplete connection between the esophagus and the stomach in utero, resulting in a shortened esophagus.

Cleft Lip and Palate Condition in which the upper lip and roof of the mouth are not formed completely and are surgically corrected, resulting in feeding, speaking, and hearing difficulties in childhood.

Infants with Genetic Disorders

Infants diagnosed with genetic disorders near birth are a small subset of infants with congenital anomalies or chronic conditions. They also fit in the category of infants with special health care needs. The expanded use of prenatal genetic tests has the consequence that some families know ahead of time that the baby has a specific condition at birth. When the condition is treated by a special diet, nutritional therapy may begin right after birth. The mother can plan ahead to breastfeed by pumping and freezing her milk. The newborn is fed the appropriate diet while confirmatory testing proceeds. Once the newborn's testing confirms the prenatal test results, therapy can set the amounts of breast milk to be mixed with a special formula at home.

The number of genetic disorders that can be identified in newborns is increasing rapidly, particularly through expanded infant screening programs. Population-based prevalence estimates keep decreasing, such as 1 in 800 live births compared to 1 in 2500–5000 live births.[36] The nutritional implication of expanded genetic screening and diagnosis is that more newborns require special diets immediately. Infants with rare genetic conditions such as galactosemia or *maple syrup urine disease* need the diet started within days of birth—waiting even 1 week can result in irreversible brain damage or death.[10] The infants who are identified in newborn screening with metabolic or genetic conditions usually require special formulas. During the diagnostic process, relatives of the affected newborn may be identified with later onset, milder forms of the same disorder. Genetic centers or inborn errors of metabolism clinics are notified when a newborn screening result needs follow-up. Immediate action is taken to locate the family and confirm the diagnosis. In such circumstances, newborn screening results in early diagnosis and avoids a costly stay in the hospital intensive care unit. For example, an infant with galactosemia, if not picked up by the initial abnormal screening result, is likely to have been hospitalized with possible sepsis or liver problems by the time a second screen is to be collected. If the baby with galactosemia receives supportive measures, such as sugar solutions, and then soy-based infant formulas without galactose, recovery is usually rapid. Some of the disorders picked up by newborn screening do not make the baby sick in early infancy, but later. It is difficult for parents of a healthy-appearing newborn to be told that a special diet is needed to prevent illness later. An example of a condition that can be picked up by newborn screening before illness presents is cystic fibrosis. Although the prevalence of spina bifida and anencephaly has decreased since supplementation of the food supply with folic acid, genetic forms of these disorders have also been more

Maple Syrup Urine Disease   Rare genetic condition of protein metabolism in which breakdown by-products build up in blood and urine, causing coma and death if untreated.

DiGeorge Syndrome   Condition in which chromosome 22 has a small deletion, resulting in a wide range of heart, speech, and learning difficulties.

clearly identified and account for 50–70% of these defects.[37]

This concept of increased use of genetic tests is exemplified by the condition called *DiGeorge syndrome.* This test may be ordered for any infant with a heart defect. DiGeorge syndrome is a condition in which a small piece of chromosome 22 is deleted.[40] Recent incidence estimates place it at 1 in every 4000 births.[40] This makes DiGeorge relatively common for a genetic condition, more common than PKU or cystic fibrosis, and second only to Down syndrome as a cause of mental retardation. Infants with DiGeorge syndrome may have a wide range of conditions affecting the heart, immune system, and calcium balance, and later speech and learning problems. Only when the genetic probe became available was it understood that three separate disorders involved the same deletion. As a result, the incidence of DiGeorge syndrome was underreported before the probe was available. Nutrition services may be required based on short stature, heart malformations, heart surgery, and resulting feeding problems (see Case Study 9.2).

Feeding Problems

Infants who were born preterm or have chronic health problems tend to be more irritable and less able to signal their wants and needs compared to healthy infants. Feeding difficulties are reported in 40–45% of families with VLBW infants.[32] Children with developmental disabilities have more frequent feeding problems, as high as 70%, that may or may not be identified in infancy.[6] Table 9.5, on page 260, gives signs of feeding difficulties that can be caught early.

By the time feeding problems require interventions to prevent further growth and developmental problems, families and infants may be frustrated from their feeding experiences. Infants who are difficult to feed are at risk for failure to thrive (FTT), child abuse, and neglect.[23] Infant feeding guidelines for term infants are appropriate for many preterm infants if they were healthy and had gestational ages such as 35 weeks. Preterm infants who were VLBW or ELBW need infant feeding guidelines based on their adjusted gestational age. As an example, the recommendation for adding food on a spoon at 4–6 months would be adjusted to 6–8 months for an infant who was born at 32 weeks of gestation. Even with this adjustment, feeding problems are common because preterm infants may be extra sensitive. The emphasis on weight gain and catch-up growth inadvertently results in overfeeding and signs of gastrointestinal discomfort, such as spitting up. Some preterm infants by late infancy have learned to get attention by devices such as refusing to eat, dropping food off the high chair, or throwing a cup.

Case Study 9.2

Noah's Cardiac and Genetic Condition

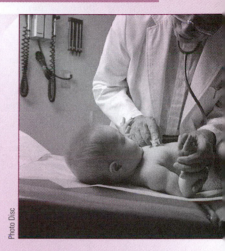

Photo Disc

A baby named Noah requires open heart surgery, which results in a diagnosis of DiGeorge syndrome. His mother is successful at expressing her milk and maintaining her breast milk supply by using the breast pump provided by the medical center during the hospitalization. Pumping and freezing the breast milk for the baby is important to the family. Noah is a reluctant eater by breast or bottle, with intake usually only 1 or 2 ounces per feeding, when allowed to feed. He is too weak to feed throughout most of the hospitalization.

At discharge the family is referred to a local early intervention program, WIC, Supplemental Social Insurance (SSI), and the state program for children with special needs. Specialty clinic and local follow-up appointments are made. Feeding difficulties concern the family, and both a lactation consultant and a registered dietitian are involved at home.

Noah nurses frequently, but briefly, due to fatigue. Growth is slower than expected, but the cardiologist does not think Noah's slow weight gain is a result of the cardiac problem. Noah tolerates only small volumes, even after his family implements the recommendations from the lactation consultant. The consultant recommends offering breast milk in a bottle so that its caloric density can be increased by added rice cereal and MCT oil. The family perceives this recommendation as undermining the mother's effort to breastfeed, and they are unwilling to try it. Noah likes to nurse, but for such a short time that the richer hindmilk may not be available. Over time, the family offers food on a spoon and continues nursing. Weight and height gain do not fit the expected rate, but appear to be fairly consistent.

The family increasingly enjoys parenting, and thinks Noah is a beautiful infant. They have not contacted WIC or the early intervention program. They regard the baby as getting better after surgery. They consider his small size a result of heart surgery. Many people they meet assume he is a premature baby, but his small size is not as much a matter of concern to his parents as it is to health providers.

This case example demonstrates inadequate growth rate, feeding problems, and questionable adequacy of the diet. The genetic syndrome, heart condition, nutritional inadequacy of the baby's diet, and the impact of stress and coping on the mother's milk supply all could explain his slow growth. Growth expectations are unclear because there is no growth chart for this genetic syndrome, or for infants with cardiac anomalies. The standard growth chart is the only one available, but it may not be appropriate to predict future growth.

Important parts of Noah's growth assessment include that his fat stores are good, showing he has access to enough calories, and that his head circumference percentile is low, which suggests that his brain is not growing at the typical rate. This could be due to neurological damage during surgery or afterward, or to the underlying genetic syndrome. It is probably not due to inadequate nutrition in early infancy, because the body tries to preserve brain growth, but there is no way to rule that out. Noah's feeding problems are subtle signs of his developmental delay, although this is clear only in hindsight.

Questions

1. How does breastfeeding benefit Noah?
2. How do you know what is good growth in such a case, when the standard growth chart may not fit?
3. Why doesn't the family want to have WIC or early intervention services, although they are eligible?

Table 9.5 Signs of feeding problems in infants

In Early Infancy (under 6 months of age)

- Baby has a weak suck and cannot make a seal on the nipple; breast milk or formula runs out of the mouth on whatever side is lower, with obvious fatigue after a few minutes of sucking.
- Baby appears to be hungry all the time due to low volume consume-per feeding, and/or time between feedings does not appear to increase from one month to the next.
- Extended feeding times are seen, with the baby napping during the feeding despite efforts to keep the baby interested in the feeding.
- The mother is not sure that the baby is swallowing, although she is appearing to suck.

In Later Infancy (over 6 months of age)

- The baby cannot maintain good head control while being fed from a spoon.
- The baby resists spoon-feeding by not opening her mouth when food is offered.
- The baby drinks from a bottle but does not accept baby foods after trying repeatedly.
- The baby resists anything in the mouth except a bottle, breast, nipple, or pacifier.
- The baby does not explore the mouth with fingers or try to mouth toys.
- The baby resists lumpy and textured foods; may turn face away or push food away.
- The baby does not give signs to the parents that clearly indicate hunger or fullness.

Nutrition Interventions

When feeding problems are identified in infancy, interventions are required to assure growth and development. Interventions may include any or all of these:

- Assess growth more frequently or in more depth, such as by measuring body fat stores to identify a change in rate of weight or length gain. This would include head growth measurement.
- Monitor the infant's intake of all liquids and foods by a diet analysis to document that enough calories and nutrients are being consumed. The infant's intake may be variable due to illness, congestion, or medications that lower the appetite.
- Change the frequency or volume of feedings as needed to meet calorie and nutrient needs.
- Adjust the timing of nursing, snacks, or meals as needed to fit medication or sleeping schedules.
- Assess the infant's feeding position and support as needed. This may be important if the infant cannot sit without support.
- Change the diet composition to improve the nutrient density, so that the infant has to expend less effort to meet energy or nutrient needs.
- Provide parent education or support services as needed, so that the feeding environment is positive and low in stress.
- Observe the interaction of the infant and mother (or whoever is routinely feeding the infant) at home or in a developmental program to make sure that signs of hunger and comfort result in a positive feeding experience for the pair.

- Adjust routine nutrition guidelines to the developmental abilities of the infant even if different from the chronological age or gestation-corrected age.

Often, attempts to improve the feeding experience are successful in meeting the infant's calorie and nutrient needs. However, when calorie and nutrient needs are higher than usual, additional steps are needed to make sure that the diet is enriched. Table 9.6 shows some of the special formulas that may be used by infants who have feeding problems or chronic conditions that increase their nutrient requirements.

Nutrition Services

Infants who were born preterm or with special health care needs have access to more nutrition services than other infants. The following programs are sources of nutrition services or finances to pay for nutrition services:[41]

- Federal disability programs
- Individuals with Disabilities Education Act (IDEA), Part C
- Early Head Start
- WIC
- State funding from the Maternal and Child Health (MCH) Block Grant

Infants with disabling conditions are eligible for Supplemental Social Insurance (SSI), a federal program within the Social Security Administration.[41] SSI provides the family with a disability check and access to health insurance if its income meets federal guidelines.

Table 9.6 Examples of infant formula for special needs

Condition	Example of Special Infant Formula
Pulmonary problems such as bronchopulmonary dysplasia or cardiac defect	Breast milk or standard infant formula with polycose and MCT oil to provide 28 cal/fl oz (high calories in a low volume)
Phenylketonuria (genetic disorder of protein metabolism)	Mixture of amino acids, carbohydrates, fats, vitamins, and minerals without the amino acid phenylalanine
Maple syrup urine disease (genetic disorder of protein metabolism)	Mixture of amino acids, carbohydrates, fats, vitamins, and minerals without the amino acids leucine, isoleucine, and valine
VLBW infant who required surgery after necrotizing enterocolitis	Mixture of amino acids, carbohydrates, fats, vitamins, and minerals
Gastroesophageal reflux and swallowing problem	Standard infant formula with baby rice cereal (increased thickness is to lower risk of choking and vomiting)
Chronic renal failure (hereditary kidney disease)	Concentrated natural protein, fats, and carbohydrates providing 40 cal/fl oz

Nutrition services are part of educational programs in IDEA, including services for children 0–2 years of age.[47] Early Head Start programs enroll infants with special health care needs. Infants with a nutrition-related diagnosis such as PKU or a cardiac problem would be eligible. Early Head Start staff are trained to feed infants special diets.

Each state has to designate a portion of the federal MCH Block Grant for children with special health care needs.[41] Services differ from state to state, but all states provide care for infants with chronic conditions. Following are some examples of how nutrition services are provided:

- Specialty clinic services, such as having a nutrition consultant attend a cystic fibrosis clinic

- Contractual services for providing special formulas or therapy for groups of patients who need more nutrition care than usually provided

- Visiting at schools or programs to conduct nutrition assessments or coordinate follow-up recommendations, such as making sure that a specific diet is being offered or that mealtime behavior is being monitored

- Transporting teams of specialists to rural or isolated areas for direct care

- Development and distribution of nutrition education materials for staff training

Every state also has a program funded by MCH to identify and advocate for children with special needs, such as the Developmental Disabilities Council.[41]

Key Points

1. The number of infants at risk for having special health care needs is increasing even though survival rates of preterm infants are improving over decades.

2. Nutrition guidance for infants with special health care needs has to be adjusted to fit their individual energy and nutrient needs.

3. Infants born severely preterm temporarily require modified forms of protein, fats, carbohydrates, vitamins, and minerals that are not in breast milk or typical infants formulas.

4. Growth in infants with special health care needs reflects nutritional intake and many other factors, such as intrauterine environment, developmental delay, and underlying medical conditions.

5. Infants with severe preterm birth or congenital anomalies can be fed directly into the stomach or blood stream when oral feeding is not safe.

6. Feeding and eating difficulties are common in infants who require intensive medical care, so nutrition services track growth patterns closely.

7. Educational and developmental services for infants at risk for special health care needs are encouraged; such services include nutrition experts and assistance in feeding and eating problems.

8. Among the many genetic conditions identified in infancy, some require special formulas and close nutrition management to promote normal development.

Resources

Emory University Pediatrics Department

This website includes information on preterm births and risks for providers and parents, with sections on nutrition in various health conditions.

Website: www.emory.edu/PEDS

National Association of Developmental Disabilities Councils

This website includes public policy and advocacy resources for providers and parents all over the United States concerning various disabilities, including services for infants.

Website: www.naddc.org

National Early Childhood Technical Assistance Center

This website provides publications and other resources about programs for infants and children. Staff training materials are also available for those who are working with infants and young children.

Website: www.nectas.unc.edu

Neonatology on the Web

This website provides resources for health care professionals, including practice guidelines and consensus statements.

Website: www.neonatology.org

Preemieparents.com

This website lists sites for recommended reading and parent resources.

Website: www.Preemieparents.com

Gaining and Growing: Assuring Nutritional Care of Preterm Infants

This site provides solid information and case studies about nutrition and feeding problem interventions for outpatient follow-up of infants who had been born early. The case study approach gives practical suggestions, "red flags," and progress over time.

Website: www.depts.washington.edu/growing/Cases

United Cerebral Palsy Associations

This website provides links to service sites in the United States.

Website: www.ucp.org

References

1. Centers for Disease Control and Prevention. Improved national prevalence estimates for 18 selected major birth defects United States, 1999–2001. MMWR 2005;54: 1301–5.

2. Cuevas KD et al. The Cost of prematurity: hospital charges at birth and frequency of rehospitalizations and acute care visits over the first year of life: a comparison by gestational age and birth weight. American Journal of Nursing. 2005; 105:56–64.

3. Institute of Medicine Food and Nutrition Board. Dietary Reference Intakes for energy, carbohydrate, fiber, fat, fatty acids, cholesterol, protein, and amino acids. National Academy Press; 2002.

4. Stallings VA. Resting energy expenditure. In: Altschuler SM, Liacouras CA, eds. Clinical pediatric gastroenterology. Philadelphia: Harcourt Brace; 1998:607–11.

5. Brine E, Ernst JA. Total parenteral nutrition for premature infants. Newborn and Infant Nursing Reviews. 2004;4:133–55.

6. Ponce NA et al. Preterm birth: the interaction of traffic-related air pollution with economic hardship in Los Angeles neighborhoods. American Journal of Epidemiology. 2005;162:140–48.

7. Committee on Nutrition of the Preterm Infant. European Society of Paediatic Gastroenterology and Nutrition in Nutrition and feeding of the preterm infant. Oxford, UK: Blackwell Scientific Publications; 1987.

8. Hall RT, Carroll RE. Infant feeding. Pediatrics in Review. 2000;21:191–200.

9. Rais-Bahrami K et al. Premature and small-for-dates infants. In: Batshaw ML, ed. Children with disabilities. 6th ed. Baltimore, MD: Paul H. Brookes; 2007:85–103.

10. American College of Medical Genetics. Newborn screening: toward a uniform screening panel and system. 2005 Final Report. http://mchb.hrsa.gov/screening, accessed 10/06.

11. Linnet KM et al. Gestational age, birth weight, and the risk of hyperkinetic disorder. Archives of Disease in Childhood. 2006; 91:655–60.

12. Trumbo P et al. Dietary reference intakes; vitamin A, vitamin K, arsenic, boron, chromium, copper, iodine, iron, manganese, molybdenum, nickel, silicon, vanadium, and zinc. J Am Diet Assoc. 2001;101:294–301.

13. Keilty B, Freund M. Caregiver-child interaction in infants and toddlers born extremely premature. Journal of Pediatric Nursing. 2005;20:181–89.

14. York J, Devoe M. Health issues in survivors of prematurity. South Med J. 2002;95(9):969–76.

15. National Center for Health Statistics: NCHS growth curves for children 0–19 years. U.S. Vital and Health Statistics, Health Resources Administration. Washington, DC: U.S. Government Printing Office; 2000.

16. Growth references: third trimester to adulthood. 2nd ed. Clinton SC: Greenwood Genetic Center; 1998.

17. The Infant Health and Development Program: enhancing the outcomes of lowbirth-weight, premature infants. JAMA 1990;263(22):3035–42.

18. Fenton TR. A new growth chart for preterm babies: Babson and Benda's chart updated with recent data and a new format, http://www.biomedcentral.com/1471–2431/3/13, accessed 10/06.

19. Huysman WA et al. Growth and body composition in preterm infants with bronchopulmonary dysplasia. Arch Dis Child Fetal Neonatal Ed. 2003; 88:F46–F51.

20. Peralta-Carcelen M et al. Growth of adolescents who were born at extremely low birth weight without major disability. Pediatrics. 2000;136:633–40.

21. Hintz SR et al. Changes in Neurodevelopmental Outcomes at 18 to 22 Months Corrected Age Among Infants of Less Than 25 Weeks' Gestational Age Born in 1993–1999. Obstetrical & Gynecological Survey. 2005;60:714–15.

22. Weaver L. Rapid growth in infancy: balancing the interests of the child. Journal of Pediatric Gastroenterology & Nutrition. 2006;43:428–32.

23. Eicher PS. Feeding. In: Batshaw, ML ed. Children with disabilities. 5th ed. Baltimore, MD: Paul H. Brookes; 2002:549–66.

24. Persson B et al. Screening for infants with developmental deficits and/or autism. Journal of Pediatric Nursing. 2006;21: 313–24.

25. U.S. Vital and Health Statistics. Births: preliminary data for October 2005. National Vital Statistics Report. 2006;54:1–7.

26. Wood NS et al. Neurological and developmental disability after extremely preterm birth. N Eng J Med. 2000;343:378–84.

27. Strauss R. Adult functional outcome of those born small for gestational age: twenty-six- year follow-up of the 1970 British birth cohort. JAMA. 2000;283:625–32.

28. Matlow A et al. Microbial contamination of enteral feed administration sets in a pediatric institution. Amer J Infect Contr. 2003;51:49–53.

29. American Dietetic Association. Infant feedings: guidelines for preparation of formula and breast milk in health care facilities, 2003.

30. Mudd SH et al. Infantile hypermethioninemia and hyperhomocysteinemia due to high methionine intake: a diagnostic trap. Molecular Genetics and Metabolism. 2003;79:6–16.

31. American Academy on Pediatrics. Breastfeeding and the use of human milk. (RE9729) Pediatrics. 1997;100:1035–39.

32. Lawrence RA, Lawrence RM. Breastfeeding: a Guide for the medical profession, 6th ed. Philadelphia, PA: Mosby; 2005:479–513.

33. Elliott S, Reimer C. Postdischarge telephone follow-up program for breastfeeding preterm infants discharged from a special care nursery. Neonat Net. 1998;17:41–45.

34. Mead-Johnson Nutritionals product information website; available at http://www.meadjohnson.com/professional/products/enfamillipilwithiron.htm, accessed 10/06.

35. Gray JE et al. Normal birth weight intensive care unit survivors: outcome assessment. Pediatrics. 1996;97:832–38.

36. Lee PJ, Cook P. Frequency of metabolic disorders: more than one needle in the haystack. Archives of Disease in Childhood. 2006;91:879–80.

37. Centers for Disease Control and Prevention. Spina bifida and anencephaly before and after folic acid mandate—United States, 1995–1996 and 1999–2000. MMWR 2004;53:362–65.

38. Lee J et al. Height and weight achievement in cleft lip and palate. Arch Dis Child. 1997;76:70–72.

39. Milerad J et al. Associated malformations in infants with cleft lip and palate: a prospective, population-based study. Pediatrics 1997;100:180–86.

40. Bassett AS et al. Clinical features of 78 adults with 22q11 deletion syndrome. American Journal of Medical Genetics. 2005;138A:307–13.

41. General information about disabilities. National information Center for Children and Youth with Disabilities; available at www.nichcy.org/general.htm, accessed 10/06.

Chapter 10
Toddler and Preschooler Nutrition

Chapter Outline

- Introduction
- Tracking Toddler and Preschooler Health
- Normal Growth and Development
- Physiological and Cognitive Development
- Energy and Nutrient Needs
- Common Nutrition Problems
- Prevention of Nutrition-Related Disorders
- Dietary and Physical Activity Recommendations
- Nutrition Intervention for Risk Reduction
- Public Food and Nutrition Programs

Photo Disc

Photo Disc

Jupiter Images

Prepared by **Nancy H. Wooldridge**

Key Nutrition Concepts

1 Children continue to grow and develop physically, cognitively, and emotionally during the toddler and preschool age years, adding many new skills rapidly with time.

2 Learning to enjoy new foods and developing feeding skills are important components of this period of increasing independence and exploration.

3 Children have an innate ability to self-regulate food intake. Parents and caretakers need to provide children nutritious foods and let children decide how much to eat.

4 Parents and caretakers have tremendous influence on children's development of appropriate eating, physical activity, and other health behaviors and habits formed during the toddler and preschool years. These lessons are mainly transferred by example.

Introduction

This chapter describes the growth and development of toddlers and preschool age children and their relationships to nutrition and the establishment of eating patterns. Growth during the toddler and preschool age years is slower than in infancy but steady. This slowing of *growth velocity* is reflected in a decreased appetite; yet, young children need adequate calories and nutrients to meet their nutritional needs. The eating and health habits established at this early stage of life may impact food habits and subsequent health later in life. The development of new skills and increasing independence mark the toddler and preschool stages. Learning about and accepting new foods, developing feeding skills, and establishing healthy food preferences and eating habits are important aspects of this stage of development.

Growth Velocity The rate of growth over time.

Toddlers Children between the ages of 1 and 3 years.

Gross Motor Skills Development and use of large muscle groups as exhibited by walking alone, running, walking up stairs, riding a tricycle, hopping, and skipping.

Fine Motor Skills Development and use of smaller muscle groups demonstrated by stacking objects, scribbling, and copying a circle or square.

Preschool-Age Children Children between the ages of 3 and 5 years, who are not yet attending kindergarten.

Definitions of the Life-Cycle Stage

Toddlers are generally defined as children between the ages of 1 and 3 years. This stage of development is characterized by a rapid increase in *gross* and *fine motor skills* with subsequent increases in independence, exploration of the environment, and language skills. *Preschool-age children* are between 3 and 5 years of age. Characteristics of this stage of development include increasing autonomy; experiencing broader social circumstances, such as attending preschool or staying with friends and relatives; increasing language skills; and expanding their ability to control behavior.

Importance of Nutrition

Adequate intake of energy and nutrients is necessary for toddlers and preschool-age children to achieve their full growth and developmental potential. Undernutrition during these years impairs children's cognitive development as well as their ability to explore their environments.[1] Long-term effects of undernutrition, such as failure to thrive and cognitive impairment, may be prevented or reduced with adequate nutrition and environmental support.

Tracking Toddler and Preschooler Health

There are approximately 12 million children under age 3 years and 11.8 million children between the ages of 3 and 5 years in the United States today.[2] The poverty rate for children under 6 years of age in 2004 was 21%.[2] Ten percent of children under age 6 years had no parent in the labor force in 2004. When evaluating young children's nutritional status and offering nutrition education to parents, it is important to consider children's home environments. Establishing healthy eating habits may not be high on a family's priority list when the home environment is one of poverty and food insecurity. Disparities in child health indicators, including those of nutrition status in this age group, exist among races.

Healthy People 2010

Healthy People 2010—objectives for the nation for improvements in health status by the year 2010—includes a number of objectives that directly relate to toddlers and preschoolers.[3] These are listed in Table 10.1.

Normal Growth and Development

An infant's birth weight triples in the first 12 months of life, but growth velocity slows thereafter until the adolescent growth spurt. On average, toddlers gain 8 ounces (0.23 kg)/month and 0.4 inches (1 cm) of height/month, while preschoolers gain 4.4 pounds (2 kg) and 2.75 inches (7 cm) per year.[4] This decrease in rate of growth is accompanied by a reduced appetite and food intake in toddlers and preschoolers. A common complaint of parents of children this age is that their children have much less appetite and a lower interest in food or eating compared to their appetite and food intake during infancy. Parents need to be reassured that a decrease in appetite is part of normal growth and development for children in this age group.

Table 10.1   Healthy People 2010 objectives related to toddlers and preschool age children[3]

	Healthy People 2010 Objective	Baseline	Target
19.4:	Reduce growth retardation among low-income children under age 5	6%	5%
19.5:	Increase the proportion of persons aged 2 years and older who consume at least two daily servings of fruit	28%	75%
19.6:	Increase the proportion of persons aged 2 years and older who consume at least three daily servings of vegetables, with at least one-third being dark green or deep yellow vegetables	3%	50%
19.7:	Increase the proportion of persons aged 2 years and older who consume at least six daily servings of grain products, with at least three being whole grains	7%	50%
19.8:	Increase the proportion of persons aged 2 years and older who consume less than 10% of calories from saturated fat	36%	75%
19.9:	Increase the proportion of persons aged 2 years and older who consume no more than 30% of calories from fat	33%	75%
19.10:	Increase the proportion of persons aged 2 years and older who consume 2400 mg or less of sodium daily	21%	65%
19.11:	Increase the proportion of persons aged 2 years and older who meet dietary recommendations for calcium	46%	75%
19.12:	Reduce iron deficiency among young children	4%	1%

In monitoring a child's physical growth, it is important for children to be accurately weighed and measured at periodic intervals. Toddlers less than 2 years of age should be weighed without clothing or a diaper. The *recumbent length* of toddlers should be measured on a length board with a fixed head board and moveable foot board. Proper measurement of recumbent length requires two adults—one at the child's head making sure the crown of the head is placed firmly against the head board, and the other making sure that the child's legs are fully extended and placing the foot board at the child's heels. Proper positioning for measurement of a child's recumbent length is shown in Illustration 10.1. Preschool-age children should be weighed and measured without shoes and in lightweight clothing. Calibrated scales should be used and a height board should be used for measuring *stature*. Illustrations 10.2 and 10.3 on the next page further demonstrate the proper techniques for weighing and measuring young children. It is important that both weight and height be plotted on the appropriate growth charts, such as the 2000 CDC growth charts discussed next.

The 2000 CDC Growth Charts

A full set of the "CDC Growth Charts: United States" can be found in Appendix A.[5] Illustration 10.4 on page 269 shows an example of one of the charts, which depicts the growth of a healthy child. Charts are gender specific and are available for birth to 36 months and for 2 to 20 years. With these charts, the health care professional can plot and monitor weight-for-age, length- or stature-for-age, head circumference-for-age, weight-for-length, weight-for-stature, and BMI-for-age. There is overlap between the two sets of growth charts for children between 24–36 months of age. If the child's recumbent length is measured, then the birth–36-month growth chart is the appropriate one to use. If the child over 2 years of age is measured standing, the 2–20-year-old growth chart is the correct choice. Children's growth usually "tracks" within a fairly steady percentile range. It is important to monitor a child's growth over time and to identify any deviations in growth.

Recumbent Length   Measurement of length while the child is lying down. Recumbent length is used to measure toddlers <24 months of age, and those between 24 and 36 months who are unable to stand unassisted.

Stature   Standing height.

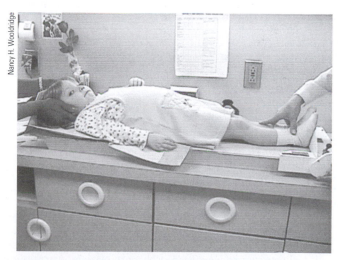

Nancy H. Wooldridge

Illustration 10.1   Measuring the recumbent length of a toddler is a two-person job!

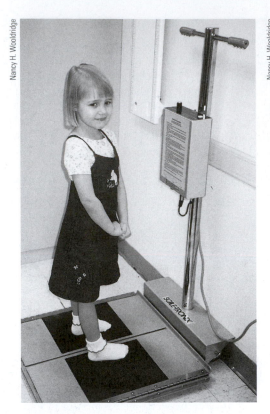

Illustration 10.2 Young child being weighed.

Illustration 10.3 Measuring the stature of a preschool-age child.

toddler and preschool stage of development. Although the curves for weight-for-age and length- or stature-for-age continue to increase during the toddler and preschool-age years, the slope of the curve is not as steep as during the first year of life.

Common Problems with Measuring and Plotting Growth Data

Growth in young children that is measured or plotted incorrectly can lead to errors in health status assessment. Standard procedures should be followed, calibrated and appropriate equipment should be used, and plotting should be double-checked—including checking the age of the child—to avoid such errors. Choosing the appropriate growth chart based on how the child was measured (recumbent length versus stature) and on the child's gender is as important as using the current growth charts.

It is the pattern of growth that is important to assess rather than any one single measurement. A weight measurement without a length or stature measurement doesn't indicate how appropriate the weight is for the child's length or stature.

Body mass index, or BMI, provides a guideline for assessing underweight and overweight in children and adults. Body mass index is predictive of body fat for children over 2 years of age. A BMI of 85th percentile or greater but less than the 95th percentile indicates risk of being overweight, and a BMI of 95th percentile or greater indicates overweight.[8] A BMI less than 5th percentile indicates underweight. BMI fluctuates throughout childhood. BMI increases in infancy; it decreases during preschool years, hitting its lowest point at approximately 4–6 years of age; and then increases into adulthood. Because of this normal fluctuation of BMI, the only way to know if a child's BMI is within a normal range is to plot BMI-for-age on the appropriate growth curve. In pediatrics, the goal is to strive for a BMI-for-age in the normal range and not a specific BMI range as is the goal in adults.

Growth charts visually aid parents by demonstrating the expected slowing of the growth velocity during the

Body Mass Index An index that correlates with total body fat content or percent body fat, and is an acceptable measure of adiposity or body fatness in children and adults.[6,7] It is calculated by dividing weight in kilograms by the square of height in meters (kg/m²).

Physiological and Cognitive Development

Toddlers

An explosion in the development of new skills occurs during the toddler years. Most children begin to walk independently at about their first birthday. At first the walking is more like a "toddle" with a wide-based gait.[4] After practicing for several months, the toddler achieves greater steadiness and soon will be able to stop, turn, and stoop without falling over. Gross motor skills, such as sitting on a small chair and climbing on furniture, develop rapidly at this age; and with practice, great improvements in balance and agility take place. At about 15 months, children can crawl up stairs; by about 18 months, they can run stiffly. Most toddlers can walk up and down stairs one step at a time by 24 months, and jump in place. At about 30 months, children have advanced to going up stairs alternating their feet. By 36 months of age, children are ready for tricycles.

Children become increasingly mobile and independent with improvements in gross motor skills. Toddlers are fascinated with these newfound skills, showing a readiness to put these skills into practice and to develop new skills. However, toddlers have no sense of dangerous situations. At this age,

Illustration 10.4 Birth to 36 months: Girls length-for-age and weight-for-age percentiles.[5]

toddlers often "orbit" around their parents, like planets, moving away, looking back, moving farther, and then returning.

From a socialization standpoint, the child moves from being primarily self-centered to being more interactive. The toddler now possesses the ability to explore the environment and to develop new relationships. Fears of certain situations, such as separation, darkness, loud sounds, wind, rain, and lightning commonly emerge during this period as the child learns to deal with changes in the environment. Children develop rituals in their daily activities in an attempt to deal with these fears.

Social development also involves imitating others, such as parents, caretakers, siblings, and peers, during this time. The child in this stage begins to learn about the family's cultural customs, including those related to meals and food.

Dramatic development of language skills occurs from 18 to 24 months. Once a child realizes that words can stand for things, his vocabulary erupts from 10 to 15 words at 18 months to 100 or more words at 2 years of age. The toddler soon begins combining words to make simple sentences. By 36 months, the child uses three-word sentences.[4]

An important social change for toddlers is increased determination to express their own will. This expression often comes in the form of negativism and the beginning of temper tantrums, which give this stage of development its label of "the terrible twos." With an increase in motor development coupled with an increasing quest for independence, toddlers try to do more and more things, pushing their capabilities to the limit. Thus the toddler can become easily frustrated and negative. The child seeks more independence and at the same time needs the parents and caretakers for security and reassurance. Toddler behavior uncannily parallels the same type of behavior commonly seen in adolescents!

Development of Feeding Skills in Toddlers Many babies begin to wean from the breast or the bottle at about

children are especially vulnerable to accidental injuries and ingestion of harmful substances. In fact, the leading cause of death among young children is unintentional injuries.[3] Parents and caregivers have to constantly watch over toddlers, preferably in environments made "child safe."

Cognitive Development in Toddlers With the toddler's newly acquired physical skills, exploring the environment accelerates, and exerting their newfound independence becomes very important to them. Toddlers now have the power to control the distance between themselves and their parents. *Nelson's Textbook of Pediatrics*[4] describes how

9 to 10 months of age, when their solid food intake increases, and they learn to drink from a cup.[9] Parents need to pay attention to cues of readiness for weaning, such as disinterest in breastfeeding or bottle feeding. The time it takes to wean is variable and depends on both the child and the mother. Weaning will be easier for those babies who adapt well to change. Weaning is a sign of the toddler's growing independence and is usually complete by 12 to 14 months of age, although the age varies from child to child.

Gross and fine motor development during the toddler years enhances children's ability to chew foods of different textures and to self-feed. Between 12 and 18 months, toddlers are able to move the tongue from side to side (or laterally) and learn to chew food with rotary, rather than just up and down, movements. Toddlers can now handle chopped or soft table food.

At about 12 months, children have a refined pincer grasp that enables them to pick up small objects, such as cooked peas and carrots, and put them into their mouths. Children will be able to use a spoon around this age, but not very well. At 18 to 24 months, toddlers are able to use the tongue to clean the lips and have well-developed rotary chew movements. Now the toddler can handle meats, raw fruits and vegetables, and multiple textures of food.

A strong need for independence in self-feeding emerges during the toddler age. "I do it" and "No, no, no!" are commonly heard phrases in households where toddlers reside. As toddlers busily practice their newly found skills, they become easily distracted. Parents need to realize that their toddler's sometimes-fierce independence is part of normal growth and development and represents an ongoing process of separation from dependency on the parents and caretakers.

Increasing fine motor and visual motor coordination skills allow toddlers to use cups and spoons more effectively. Although toddlers' skill with a spoon increases during the second year, they prefer to eat with their hands. Initial attempts at self-feeding are inevitably messy, as Illustration 10.5 depicts, but represent an important stage of development. It is important that parents and caretakers keep distractions, such as television, to a minimum during mealtimes, and allow their toddlers to practice self-feeding skills and to experience new foods and textures. The child derives pleasure in self-feeding and exploring new tastes. Learning to self-feed allows the child to develop mastery of an important part of everyday life.

Adult supervision of eating is imperative due to the high risk of choking on foods at this age. Toddlers should always be seated during meals and snacks, preferably in a high chair or booster seat with the family, and not allowed to "eat on the run." Foods that may cause choking, such as hard candy, popcorn, nuts, whole grapes, and hot dogs, should not be served to children less than 2 years of age.[9]

Feeding Behaviors of Toddlers   The toddler's need for rituals, a hallmark of this stage of development, may be

Nancy H. Wooldridge

Illustration 10.5 Toddler enjoying mealtime!

linked to the development of food jags. Many toddlers demonstrate strong food preferences and dislikes. They can go through prolonged periods of refusing a particular food or foods they previously liked. The intensity of the refusal or the negative attitude toward a particular food will be influenced by the child's temperament (see Temperament Differences on page 273). To circumvent food jags, parents can serve new foods along with familiar foods. New foods are better accepted if they are served when the child is hungry and if she sees other members of the family eating these foods. Eventually, toddlers' natural curiosity will get the best of them. Toddlers are great imitators, which includes imitating the eating behavior of others.

Mealtime is an opportunity for toddlers to practice newly acquired language and social skills and to develop a positive self-image. It is not the time for battles over food or "force feedings." Establishing the habit of eating breakfast is an important part of healthy eating behaviors. Family mealtime provides an opportunity for parents and caretakers to model healthy eating behaviors for the young child.

Appetite and Food Intake in Toddlers   Parents need to be reminded that toddlers naturally have a decreased interest in food because of slowing growth, and a corresponding decrease in appetite. Besides, with all of their newfound gross and fine motor skills, they have places to go and new environments to explore! It is a part of normal growth and development for toddlers to have a decreased interest in food and to be easily distracted at mealtime.

Toddlers need toddler-sized portions. One rule of thumb for serving size is 1 tablespoon of food per year of age. So, a serving for a 2-year-old child would be about 2 tablespoons. It is better to give the child a small portion and allow him to ask for more than to serve large portions. Parents often overestimate portion sizes needed by their

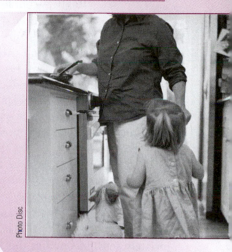

Photo Disc

Case Study 10.1

Making Meal Time Pleasant

Lindsey, a 24-month-old little girl, lives with her parents. She stays at a child care center during the week while both of her parents are at work. On the weekends, her parents enjoy their time with Lindsey, although a lot of their time is spent running errands and catching up on household chores. Partly to appease Lindsey, her parents allow her to have as much of her favorite beverage, apple juice, from a "sippy" cup as she wants between meals. Lindsey also has free access to snacks such as crackers, slices of cheese, and cookies. When the family sits down to have a meal together, Lindsey plays with her food and usually doesn't eat much. She tells her parents that she doesn't like the food being served and wants "something else." She soon becomes fussy and wants to get down from her booster seat. To try to keep her at the table with them, her parents turn on the television or play Lindsey's favorite cartoons. If that does not quiet Lindsey, her mother offers to prepare another food item of Lindsey's choice. Mealtime has become an unpleasant experience for the family.

Lindsey's parents need to stop allowing Lindsey free access to the apple juice and snack foods between meals. Lindsey needs snacks, and foods such as the cheese and crackers are appropriate snack foods. But she should not be allowed to "graze" throughout the day. It is no wonder that she is not hungry at mealtime! If she is allowed to build up an appetite for her meals and snacks, she will be more interested in eating and in the food being served. Lindsey's parents have the right idea in trying to promote the family eating together, but they need to eliminate the distraction of watching TV and cartoons during mealtime. And Lindsey's mother needs to stop preparing special items for Lindsey. Lindsey's parents can serve her favorite basic foods along with a small portion of a new food. By employing these suggestions, the family's mealtimes should become much more pleasant.

Questions

1. Identify some of the inappropriate eating habits that Lindsey's parents have allowed her to develop.
2. Considering her stage of development, what advice would you give Lindsey's parents in their attempts to increase the number of foods she will eat?
3. In what types of food-preparation activities would it be appropriate for Lindsey's parents to have her participate? Why is this important?
4. What suggestions do you have for snack food items for Lindsey?
5. Would you advise Lindsey's parents to give her a daily multivitamin supplement?
6. What advice would you give the family regarding physical activity?

young child, which may contribute to labeling the child as a "picky" eater. Because toddlers can't eat a large amount of food at one time, snacks are vital in meeting the child's nutritional needs. It is important that toddlers not be allowed to "graze" throughout the day on sweetened beverages and foods such as cookies and chips. These foods can "kill" their limited appetite for basic foods at meal and snack times. In considering the toddler's need for rituals and limit setting, parents and caretakers need to establish regular but flexible meal and snack times, allowing enough time between meals and snacks for the toddler to get hungry.

Preschool-Age Children

Preschool-age children continue to expand their gross and fine motor capabilities. At age 4, the child can hop, jump on one foot, and climb well. The child can ride a tricycle, or a bicycle with training wheels, and can throw a ball overhand.[4]

Cognitive Development of Preschool-Age Children
Magical thinking and egocentrism characterize the preschool period.[4] Egocentrism does not mean that the

child is selfish, but rather that the child is not able to accept another's point of view. The child is beginning to interact with a widening circle of adults and peers. During the preschool years, children gradually move from primarily relying on external behavioral limits, such as those demanded by parents and caregivers, to learning to limit behavior internally. This transition is a prerequisite to functioning in a school classroom.[4] Also during this time, children's play starts to become more cooperative, such as building a tower of blocks together. Toward the end of the preschool years, children move to more organized group play, such as playing tag or "house."

Control is a central issue for preschool children. They will test their parents' limits and still resort to temper tantrums to get their way. Temper tantrums generally peak between the ages of 2 and 4 years.[4] The child's challenge is to separate, and the parent's challenge is to appropriately set limits and at the same time to let go, another parallel with adolescence. Parents need to strike an appropriate balance for setting limits. Too tightly controlled limits can undermine the child's sense of initiative and cause him or her to act out, whereas loose limits can cause the child to feel anxious and that no one is in control.

Language develops rapidly during the preschool years and is an important indicator of both cognitive and emotional development. Between ages 2 and 5, children's vocabularies increase from 50 to 100 words to more than 2000 words, and their language progresses from two- to three-word sentences to complete sentences.[4]

Development of Feeding Skills in Preschool-Age Children
The preschool-age child can use a fork and a spoon and uses a cup well. Cutting and spreading with a knife may need some refinement. Children should be seated comfortably at the table for all meals and snacks. Eating is not as messy a process during the preschool years as it was during toddlerhood. Spills still do occur, but they are not intentional. Foods that cause choking in young children should be modified to make them safer, such as cutting grapes in half lengthwise and cutting hot dogs in quarters lengthwise and then cutting into small bites. Adult supervision during mealtime is still important.

Feeding Behaviors of Preschool-Age Children
As during the toddler years, parents of preschool-age children need to be reminded that the child's rate of growth continues to be relatively slow, with a relatively small appetite and food intake. Growth occurs in "spurts" during childhood. Appetite and food intake increase in advance of a growth spurt, causing children to add some weight that will be used for the upcoming spurt in height. Therefore, the appetite of a preschool-age child can be quite variable.

Preschool-age children want to be helpful and to please their parents and caretakers. This characteristic makes the preschool years a good time to teach children about foods, food selection, and preparation by involving them in simple

---

> **Table 10.2 Meal preparation activities for young children[10]**
>
> Let children select and help prepare a whole grain side dish.
>
> Let children help shop for, clean, peel, or cut up vegetables and fruits, depending on their age.
>
> Let children decide on the dinner vegetable or what goes into salads.

food-related activities. For instance, outings to a farmers' market can introduce children to a variety of fresh vegetables and fruit. Allowing children to be involved in meal-related activities, such as those listed in Table 10.2 from MyPyramid for Kids, can be quite instructive.[10] Families of preschool-age children need to continue to be encouraged to eat together, like the family in Illustration 10.6.

Innate Ability to Control Energy Intake
An important principle of nutrition for young children, and one with direct application to child feeding, is children's ability to self-regulate food intake. If allowed to decide when to eat and when to stop eating without outside interference, children eat as much as they need.[11,12] Children have an innate ability to adjust their caloric intake to meet energy needs. The preschool-age child's intake may fluctuate widely from meal to meal and day to day. But over a week's time, the young child's intake remains relatively stable.[13] Parents who try to interfere with the child's ability to self-regulate intake by forcing the child to "clean her plate" or using food as a reward are asking the child to overeat or undereat.

Although children can self-regulate caloric intake, no inborn mechanisms direct them to select and consume a well-balanced diet.[14] Children learn healthful eating habits.[15] Parents give up some control over what their preschool child eats if the child spends more time away from home in a child care center or with extended family members. Preschool children continue to learn about food and food habits by observing their parents, caretakers, peers, and siblings, and they begin to be influenced by what they see on television. Their own food habits and food preferences are established at this time.

Appetite and Food Intake of Preschoolers
Parents of preschool-age children often describe their children's appetite as being "picky." One reason a child may want the same foods all of the time is because familiar foods may be comforting to the child. Another reason is that the child may be trying to exert control over this aspect of her life. The child's eating and food selection can easily become a battleground between parent and child; this scenario should be avoided. Some practical suggestions for parents and caretakers of children this age include serving child-sized portions and serving the food in an attractive way. Young children

Illustration 10.6 Sharing family meals is an important aspect of development in young children.

often do not like their foods to touch or to be mixed together, such as in casseroles or salads. They typically do not like strongly flavored vegetables and other foods, or spicy foods, at this young age. Just as with toddlers, parents of preschool-age children should not allow their children to eat and drink indiscriminately between meals and snacks. This behavior often "kills" the appetite at mealtime. Children should not be forced to stay at the table until they have eaten a certain amount of food as determined by the parent.

Temperament Differences

> "Better is a dinner of herbs where love is, than a fatted ox and hatred with it."
>
> Proverbs 15:17

Temperament is defined as the behavioral style of the child, or the "how" of behavior. This definition was derived from analysis of data from the New York Longitudinal Study, begun in 1956 by Chess and Thomas.[16] These investigators defined three temperamental clusters: the "easy" child (about 40% of children), the "difficult" child (10%), and the "slow-to-warm-up" child (15%). The remaining children, classified as "intermediate-low" or "intermediate-high," demonstrated a mixture of behaviors but gravitated to one end of the spectrum.[16]

Children's temperaments affect feeding and mealtime behavior. The "easy" child is regular in function, adapts easily to regular schedules, and tries and accepts new foods readily. The "difficult" child on the other hand, is characterized by irregularity in function and slow adaptability.

This child is more reluctant to accept new foods and can be negative about them. The "slow-to-warm-up" child exhibits slow adaptability and negative responses to many new foods with mild intensity. With repeated exposures to new foods, this child can learn to accept them over time with limited complaining.[16]

The "goodness of fit" between the temperaments of the child and the parent or caretaker can influence feeding and eating experiences.[16] A mismatch can result in conflict over eating and food. Parents and caretakers need to be aware of the child's temperament when attempting to meet nutritional needs. The difficult or slow-to-warm-up child may pose special challenges that need to be addressed by gradually exposing the child to new foods and not hurrying him or her to accept them.[16]

Food Preference Development, Appetite, and Satiety

Food preference development and regulation of food intake have been studied extensively by Leann Birch and associates.[17,18] It is clear that children's food preferences do determine what foods they consume. Children naturally prefer sweet and slightly salty tastes and generally reject sour and bitter tastes. These preferences appear to be unlearned and present in the newborn period. Children eat foods that are familiar to them, a fact that emphasizes the importance environment plays in the development of food preferences. Children tend to reject new foods but may learn to accept a new food with repeated exposures to it. It may, however, take eight to ten exposures to a new food before it is accepted. Children who are raised in an environment where all members of the family eat a variety of foods are more likely to eat a variety of foods. One study showed that 5-year-old girls' fruit and vegetable intakes were related to their parents' fruit and vegetable intakes.[19]

Children also appear to have preferences for foods that are energy dense due to high levels of sugar and fat.[17,18] This preference may develop because children associate eating energy-dense foods with pleasant feelings of satiety, or because these types of foods may be associated with special social occasions such as birthday parties. The context in which foods are offered to a child influences the child's food preferences. Foods served on a limited basis but used as a reward become highly desirable. Restricting a young child's access to a palatable food may actually promote the desirability and intake of that food.[20] Coercing or forcing children to eat foods can have a long-term negative impact on their preference for these foods.[17–19]

Media Influence   Young children are also influenced by media. One study of advertisements during programming aimed specifically at toddlers and preschool-aged children on three different networks found that more than half of all food advertisements were aimed specifically at children, and the majority of these advertisements were for fast food chains or sweetened cereals. The ads associated the advertised product with fun and/or excitement and energy. Fast food ads seemed to focus on building brand recognition through the use of licensed characters, logos, and slogans and were less likely to show food during the ads.[21]

Appetite, Satiety   Children's energy intake regulation has been studied by giving children *preloads* of food or beverage of varying energy content followed by self-selected meals. In one such study, children ages 3 to 5 years were given either a low-energy preload beverage made with aspartame (Nutrasweet), a low-calorie sugar substitute, or a high-energy preload beverage made with sucrose. Fat and protein content of the preloads did not differ. Children were then allowed to self-select their lunches. Children who had the low-calorie beverage before lunch consumed more calories at lunch, while those who had the higher-calorie beverage consumed fewer calories. These results indicate that young children are able to adjust caloric intake based on caloric need.[18,22] Similar studies were conducted in 2- to 5-year-olds using foods with dietary fat or olestra, a nonenergy fat substitute. Results indicate that children compensated for the lower level of calories in food when olestra was substituted for dietary fat.

The preloading protocol just described was also used to study children's responsiveness to caloric content of foods in

Preloads   Beverages or food such as yogurt in which the energy/macronutrient content has been varied by the use of various carbohydrate and fat sources. The preload is given before a meal or snack and subsequent intake is monitored. This study design has been employed by Birch et al. in their studies of appetite, satiety, and food preferences in young children.[18]

the presence or absence of common feeding advice from adults. In one group, teachers were trained to minimize their control over how much the children ate. In the other group, teachers were trained to focus the children on external factors to control their intake, such as rewarding the children for finishing the portions served to them or encouraging them to eat because "it was time to eat." Results of this investigation show that when the adults focused the children on external cues for eating, children lost their ability to regulate food intake based on calories. It appears children's innate ability to regulate caloric intake can be altered by child-feeding practices that focus on external cues rather than the child's own hunger and satiety signals.[22]

The effects of portion size on children's intakes were compared between classes of 3-year-old and 5-year-old children. The children were served either a small, medium, or large portion of macaroni and cheese along with standard amounts of other foods in their usual lunchtime setting. Analysis of amount of food eaten showed that portion size did not affect the younger children's intakes; their intakes remained constant despite the amount of food served to them. In contrast, the 5-year-old children's intakes increased significantly with the larger portion sizes. The researchers conclude that by 5 years of age, children are influenced by the size of portions served to them, another external factor that influences intake.[23] These investigators also raise the question as to what effect large portion sizes have on overeating and, consequently, on the development of childhood overweight.

Another study of 5-year-old girls and their parents looked at the effects of parents' restriction of palatable foods on their children's consumption of these foods. After a self-selected standard lunch, these 5-year-old girls were given free access to snack foods, such as ice cream, potato chips, fruit-chew candy, and chocolate bars. The daughters of parents who reported restricting access to snack foods indicated to the investigators that they ate "too much" of the snack foods and also reported negative emotions about eating the snack foods. Parents' restriction of foods actually promoted the consumption of these foods by their young daughters and, of even more concern, the daughters reported feeling bad about eating these "forbidden" foods.[24] A related study found a lower self-concept in 5-year-old girls with high body weight.[25] Daughters of parents who restricted access to food and expressed concern about their daughter's weight status tend to have negative self-evaluations.[25] Mothers in particular seem to have more influence over their young daughters' beliefs about food and dieting.[26]

Satter describes the optimal "feeding relationship" as one in which parents and caretakers are responsible for what children are offered to eat and the environment in which the food is served, while children are responsible for how much they eat or even whether they eat at a particular meal or snack. According to Satter, if this feeding relationship is respected, then feeding and potential weight problems can be prevented.[27] Parenting includes influencing

Table 10.3 Practical applications of child feeding research[18]

- Parents should respond appropriately to children's hunger and satiety signals.
- Parents should focus on the long-term goal of developing healthy self-controls of eating in children, and look beyond their concerns regarding composition and quantity of foods children consume or fears that children may eat too much and become overweight.
- Parents should not attempt to control children's food intakes by attaching contingencies ("No dessert until you finish your rutabagas.") and coercive practices ("Clean your plate. Children in Bangladesh are starving.").
- Parents should be cautioned not to severely restrict "junk foods," foods high in fat and sugar, as that may make these foods even more desirable to the child.
- Parental influence should be positively focused on the child developing food preferences and selection patterns of a variety of foods consistent with a healthy diet. Parental modeling of eating a varied diet at family mealtime will have a strong influence on children.
- Children have an unlearned preference for sweet and slightly salty tastes; they tend to dislike bitter, sour, and spicy foods.
- Children tend to be wary of new foods and tastes, and it may take repeated exposures to new foods before they are accepted.
- Children need to be served appropriate child-sized servings of food.
- Child feeding experiences should take place in secure, happy, and positive environments with adult supervision.
- Children should never be forced to eat anything.

what is served to children and the environment in which it is served, at home and in child care settings as well.

What implications does all this research have on child feeding practices? Based on the results of these studies, it appears that by late preschool age, children are more responsive to external cues than to their innate ability to self-regulate intake. Table 10.3 sums up the practical applications of Birch's work.[18] The importance of appropriate parenting skills in helping children learn to self-regulate food intake and possibly avoid problems with obesity is echoed by a panel of obesity experts.[28] Birch's research also reinforces the important role that parents and caretakers play in modeling healthy eating behaviors for young children.

Energy and Nutrient Needs

Dietary Reference Intakes (DRIs) were developed from 1997 through 2004. Information about the various DRI publications can be found on the National Academy Press website (see Resources at the end of this chapter). Published DRI tables are provided on the inside front cover of this book.

Energy Needs

DRIs have been established for the energy needs of young children.[29] The formula for Estimated Energy Requirements (EER) for children ages 13–35 months is (89 × weight of infant [kg] − 100) + 20 (Kcal for energy deposition). For example, a healthy 24-month-old girl who weighs 12 kg would have an EER of: (89 × 12 kg − 100) + 20 = 988 kilocalories. Beginning at age 3, the DRI equations for estimating energy requirements are based on a child's gender, age, height, weight, and physical activity level (PAL). Categories of activity are defined in terms of walking equivalence. Table 10.4 on the next page depicts estimated energy requirements for reference boys and girls at selected ages. Energy needs of toddlers and preschool-age children reflect the slowing of the growth velocity of children in this age group.

Protein

The DRIs for protein for the toddler/preschool age groups can be found in Table 10.5.[29] *Recommended Dietary Allowances (RDAs)* have been established for protein. These recommendation are easily met with typical American diets as well as with vegetarian diets. Adequate energy intake to meet an individual child's needs has a protein-sparing effect; that is, with adequate energy intake, protein is used for growth and tissue repair rather than for energy. Ingestion of high-quality protein, such as milk and other animal products, lowers the amount of total protein needed in the diet to provide the essential amino acids.

Dietary Reference Intakes (DRIs) Quantitative estimates of nutrient intakes, used as reference values for assessing the diets of healthy people. DRIs include Recommended Dietary Allowances (RDAs), Adequate Intakes (AI), Tolerable Upper Intake Level (UL), and Estimated Average Requirement (EAR).

Recommended Dietary Allowances (RDAs) The average daily dietary intake levels sufficient to meet the nutrient requirements of nearly all (97% to 98%) healthy individuals in a population group. RDAs serve as goals for individuals.

Vitamins and Minerals

DRIs for vitamins and minerals have been established for the toddler and preschool-age child. Analysis of data from NHANES I, II, and III and Continuing Survey of Food Intake by Individuals (CSFII) indicate that children's average intakes of most nutrients meet or exceed the recommendations.[30,31] Most children from birth to 5 years are meeting the targeted levels of consumption of most nutrients, except for iron, calcium, and zinc. The DRIs for these key nutrients are listed in Table 10.6.[32,33]

Table 10.4 Estimated energy requirements (in Kcals) for reference boys and girls at selected ages and varying physical activity levels (PAL)[29]

Age/Gender	Reference Weight (kg [lbs])	Reference Height (m [in])	Sedentary PAL (Kcal/d)	Low Active PAL (Kcal/d)	Active PAL (Kcal/d)	Very Active PAL (Kcal/d)
3-year-old boy	14.3 (31.5)	0.95 (37.4)	1162	1324	1485	1683
4-year-old boy	16.2 (35.7)	1.02 (40.2)	1215	1390	1566	1783
5-year-old boy	18.4 (40.5)	1.09 (42.9)	1275	1466	1658	1894
3-year-old girl	13.9 (30.6)	0.94 (37.0)	1080	1243	1395	1649
4-year-old girl	15.8 (34.8)	1.01 (39.8)	1133	1310	1475	1750
5-year-old girl	17.9 (39.4)	1.08 (42.5)	1189	1379	1557	1854

Table 10.5 Dietary Reference Intakes for protein[29]

Age	RDA* g/kg/d
1–3 years	1.1 g/kg/d or 13 g/day*
4–8 years	0.95 g/kg/d or 19 g/day*

*RDA based on average weight for age (reference individual).

Table 10.6 Dietary Reference Intakes for key nutrients for toddlers and preschoolers[32,33]

Age	Recommended Dietary Allowances Iron (mg/d)	Zinc (mg/d)	Adequate Intake Calcium (mg/d)
1–3 years	7	3	500
4–8 years	10	5	800

Table 10.7 Progression of iron deficiency

Common Nutrition Problems

Iron-Deficiency Anemia

Iron deficiency and iron-deficiency *anemia* are prevalent nutrition problems among young children in the United States, although the prevalence is decreasing. A rapid growth rate coupled with frequently inadequate intake of dietary iron places toddlers, especially 9- to 18-month-olds, at the highest risk for iron deficiency.[34] According to the 1999–2000 National Health and Nutrition Examination Survey (NHANES 1999–2000), 7% of toddlers ages 1 to 2 years are iron deficient (down from 9% in NHANES III); of these, 2% have iron-deficiency anemia (down from 3% in NHANES III).[35] In numbers, these percentages translate to approximately 700,000 toddlers with iron deficiency, and of these, 240,000 have iron-deficiency anemia.[35] The full impact of this nutrition problem is profound. Iron-deficiency anemia in young children appears to cause long-term delays in cognitive development and behavioral disturbances.[1,34]

Table 10.7 depicts the progressing signs of iron deficiency. Iron deficiency can be defined as absent bone marrow iron stores, an increase in hemoglobin concentration of <1.0 g/dL after treatment with iron, or other abnormal lab values, such as serum ferritin concentration.[34] The definition of iron-deficiency anemia is less than the 5th percentile of the distribution of *hemoglobin* concentration or *hematocrit* in a healthy reference population. Age- and sex-specific cutoff values for anemia are derived from NHANES III data. For children 1 to 2 years of age, the diagnosis of anemia would be made if the hemoglobin concentration is <11.0 g/dL and hematocrit <32.9%. For children ages 2 to 5 years, a hemoglobin value <11.1 g/dL or hematocrit <33.0% is diagnostic of iron-deficiency anemia.

Not all anemias are due to iron deficiency. Other causes of anemia include other nutritional deficiencies such as folate or vitamin B$_{12}$, chronic inflammation, or recent or current infection.[34]

Anemia A reduction below normal in the number of red blood cells per cubic mm in the quantity of hemoglobin, or in the volume of packed red cells per 100 ml of blood. This reduction occurs when the balance between blood loss and blood production is disturbed.

Hemoglobin A protein that is the oxygen-carrying component of red blood cells. A decrease in hemoglobin concentration in red blood cells is a late indicator of iron deficiency.

Hematocrit An indicator of the proportion of whole blood occupied by red blood cells. A decrease in hematocrit is a late indicator of iron deficiency.

One Healthy People 2010 objective is to reduce iron deficiency in children ages 1 to 2 years from 9% to 5% and in children ages 3 to 4 years from 4% to 1%.[3] Part of reaching this goal will mean reducing or eliminating disparities in iron deficiency by race and family income level. The prevalence of iron deficiency is higher in African Americans than in white children (10% versus 8% for children ages 1 to 2 years) and is highest in Mexican American children (17% of children ages 1 to 2 years).[3] Children of families with incomes <130% of the poverty threshold have a higher incidence of iron deficiency than those with a higher income (12% versus 7%).

Preventing Iron Deficiency

The Centers for Disease Control have published recommendations for preventing iron deficiency in the United States.[34] It is recommended that children 1 to 5 years of age drink no more than 24 ounces of cow's milk, goat's milk, or soy milk each day because of the low iron content of these milks. Larger intakes may displace high-iron foods. For detecting iron deficiency, it is recommended that children at high risk for iron deficiency, such as low-income children and migrant and recently arrived refugee children, be tested for iron deficiency between the ages of 9 and 12 months, 6 months later, and then annually from ages 2 to 5 years. For children who are not at high risk for iron deficiency, selective screening of children at risk only is recommended. Children at risk include those who have a low-iron diet, consume more than 24 ounces of milk per day, have a limited access to food because of poverty or neglect, and who have special health care needs, such as an inborn error of metabolism or chronic illness.

Nutrition Intervention for Iron-Deficiency Anemia

Treatment of iron-deficiency anemia includes supplementation with iron drops at a dose of 3 mg/kg per day, counseling of parents or caretakers about diets that prevent iron deficiency, and repeat screening in 4 weeks. An increase of >1 g/dL in hemoglobin concentration, or >3% in hematocrit, within 4 weeks of initiation of treatment confirms the diagnosis of iron deficiency. If the anemia is responsive to treatment, dietary counseling should be reinforced, and the iron treatment should be continued for 2 months. At that time, the hemoglobin and hematocrit should be rechecked, and the child should be reassessed in 6 months. If the hemoglobin and hematocrit do not increase after 4 weeks of iron treatment, further diagnostic tests are needed. Iron status will not improve with iron supplements if the cause of the anemia is not directly related to a need for iron.[34]

Dental Caries

Approximately one in five children aged 2 to 4 years has decay in the primary or permanent teeth.[3] One of the Healthy People 2010 objectives is to reduce the proportion of children and adolescents who have dental caries experience in their primary or permanent teeth.[3] A primary cause of dental decay is habitual use of a bottle with milk or fruit juice at bedtime or throughout the day. Prolonged exposure of the teeth to these fluids can produce *baby-bottle tooth decay.* Upper front teeth are most severely affected by decay, which is where fluids pool when toddlers fall asleep while drinking from a bottle. Toddlers with baby-bottle tooth decay are at increased risk for caries in the permanent teeth.[36] The incidence of baby-bottle tooth decay is highest among Hispanic, American Indian, and Alaska Native children, and among children whose parents have less than a high school education.[3]

> **Baby-Bottle Tooth Decay** Dental caries in young children caused by being put to bed with a bottle or allowed to suck from a bottle for extended periods of time. Also called "baby or nursing-bottle dental caries."
>
> **Fluorosis** Permanent white or brownish staining of the enamel of teeth caused by excessive ingestion of fluoride before teeth have erupted.[36]

Food sources of carbohydrates such as milk and fruit juice can have direct effects on dental caries development because *Streptococcus mutans,* the main type of bacteria that cause tooth decay, use carbohydrates for food. Bacteria present in the mouth excrete acid that causes the tooth decay.[36] Consequently, the more often and longer teeth are exposed to carbohydrates, the more the environment in the mouth is conducive to the development of tooth decay. Foods containing carbohydrates that stick to the surface of the teeth, such as sticky candy like caramel, are strong caries promoters. Rinsing the mouth with water or brushing teeth to get rid of the carbohydrate stuck to teeth reduces caries formation. Young children allowed to "graze" or indiscriminately eat or drink throughout the day likely expose their teeth to carbohydrates longer, which encourages bacteria proliferation and tooth decay. Crunchy foods such as carrot sticks and apple slices, when age appropriate, are good choices for snacks because they are less likely to promote tooth decay than sticky candies.

Fluoride

Children need a source of fluoride in the diet. If the water supply is not adequately fluoridated, then a fluoride supplement is recommended. The American Dental Association, the American Academy of Pediatrics, and the American Academy of Pediatric Dentistry have devised a fluoride supplementation schedule, which is based on the child's age and the fluoride content of the local water supply.[37] Children ages 6 months to 3 years need 0.25 mg of fluoride per day if their local water supply has <0.3 ppm of fluoride. Children 3 to 6 years of age need 0.5 mg fluoride per day if their water supply has <0.3 ppm, but only 0.25 mg fluoride per day if the local water has 0.3 to 0.6 ppm of fluoride.[37] Excessive fluoride supplementation, consumption of toothpaste with fluoride, and natural water supplies high in fluoride can cause *fluorosis.*

Although otherwise harmless, fluoriosis produces permanent staining of the enamel of teeth, particularly permanent teeth. Because of the risk of fluorosis, fluoride supplements are only available by prescription. Few foods contain much fluoride but fluoridated water used in beverages and food preparation does provide fluoride.

Constipation

Constipation, or hard and dry stools associated with painful bowel movements, is a common problem of young children. Sometimes "stool holding" develops when the child does not completely empty the rectum, which can lead to chronic overdistension so that eventually the child is retaining a large fecal mass.[38] Then having a bowel movement can become painful to the child, which leads to more "stool holding," and a vicious cycle ensues. A pediatrician should manage the treatment of "stool holding."

Diets providing adequate total or dietary fiber for age (see fiber recommendations on page 283) guard against constipation. Some of the best food sources of dietary fiber for toddlers and preschoolers are whole grain breads and cereals, legumes, and fruits and vegetables appropriate for age. Too much fiber should be avoided, however. Young children easily develop diarrhea from high amounts, and high-fiber foods may displace other energy-dense foods and may decrease the bioavailability of some minerals, such as iron and calcium.

Lead Poisoning

Approximately 2.2% of children 1 to 5 years of age have high blood lead levels, exceeding 10 mcg/dL.[39] According to the latest surveillance data, the number of children with high blood lead levels continues to decline throughout the United States.[3,39] Young children are particularly at risk for developing high levels of lead because, in exploring their environment, they enjoy putting things into their mouths. Depending on their surroundings, some of these objects may be high in lead. There are racial and ethnic disparities in children with high lead levels with higher rates found in black and Mexican American children than in white children living in comparable housing.[40]

High blood lead levels affect the functioning of many tissues in the body, including the brain, blood, and kidneys. Low-level exposure to lead is associated with decreases in IQ and behavioral problems, and elevated blood lead levels may decrease growth in young children.[41]

Children living in housing built before 1950 are at increased risk of high lead levels because lead-based paint may have been used on these houses. Lead-based paint chips taste sweet, tempting children to consume them. As the age of housing decreases nationally, so does the incidence of high lead levels in children.[3] Lead can enter the food supply through lead-soldered water pipes, contaminated water supplies, and from certain canned goods from other countries that contain lead-solder seals. Nonfood items containing lead

Food Security Access at all times to a sufficient supply of safe, nutritious foods.

include dirt, lead weights, and other objects. Lead is also found in ceramic glazes, pewter, and in some folk remedies. One study of a small group of toddlers, 18 to 36 months of age, who lived in an urban area with potentially high lead levels, found a correlation between the amount of lead wiped from the children's hands and the level of lead in their food.[42] These results emphasize the importance of hand washing when preparing and consuming food. Damage caused by lead exposure may begin during pregnancy as lead is transported across the placenta to the fetus.

Iron deficiency in preschool-age children increases the risk of elevated lead levels.[43] With iron deficiency and a reduction in iron availability, binding receptors for iron are made available for lead transport. Also, with iron deficiency, iron uptake receptors in the small intestine are available for lead absorption. Adequate dietary calcium intake appears to protect against high blood lead levels by decreasing absorption of lead.

The CDC published guidelines for screening children for lead poisoning in 1997.[44] The American Academy of Pediatrics endorsed these guidelines in a policy statement published in 1998.[45] Lead screening should be considered at 9–12 months of age and again around 24 months of age. Local and state health departments have established recommendations for universal or targeted screening. Universal screening is recommended for communities with >27% of the housing built before 1950 or where data is insufficient to know the prevalence of elevated blood lead levels. Targeted screening is recommended in communities where <27% of the housing was built before 1950 or where <12% of children have blood lead levels >10 mcg/dl.[9,39,44,45] Besides the age of the housing, other risk factors for high blood lead levels include living in poverty and having a sibling or playmate who has had high blood lead levels. Most states require lead screeening of children who receive public assistance services such as Medicaid and WIC.[9,39,44,45]

To summarize, eliminating sources of lead in the child's environment is the most important step toward eliminating elevated blood lead levels in children. In addition, preventing iron deficiency and promoting a well-balanced diet, which includes good sources of calcium, help to prevent this problem in young children.

Food Security

One of the Healthy People 2010 objectives is to increase *food security* among U.S. households to 94% from a baseline of 88%.[3] In 1998, nearly 10 million people, more than one-third of them children, lived in households in which at least some members experienced hunger due to the lack of food.[46] According to the Household Food Security Report, the level of households having uncertain food availability has remained relatively constant since 1995. Food insecurity is more likely to exist among American Indian or Alaska Native, African American, and Hispanic or Latino people than among whites; in households with children, particularly those headed by single women; and in lower-income-level households (<130% of poverty threshold).

Food security is particularly important for young children because of their high nutrient needs for growth and development. Young children are a vulnerable group because they must depend on their parents and caretakers to supply them with adequate access to food. It appears that children who are hungry and have multiple experiences with food insufficiency are more likely to exhibit behavioral, emotional, and academic problems as compared to other children who do not experience hunger repeatedly.[47]

Food Safety

Young children are especially vulnerable to food poisoning because they can become ill from smaller doses of organisms. Key foodborne pathogens include *Campylobacter* species and *Salmonella* species, which are the most frequently reported foodborne illnesses in the United States, and the emerging pathogens, *E. coli 0157:H7* and *Listeria monocytogenes*.[3] The highest rate of *Campylobacter* species infections is seen in children under age 1.[3] *Campylobacter* is transmitted by handling raw poultry, eating undercooked poultry, drinking raw milk or nonchlorinated water, or handling infected animal or human feces.[48] The most common cause of salmonella food poisoning is consumption of foods containing undercooked or raw eggs, such as raw cookie dough containing eggs. Children less than 10 years of age account for a disproportionate percent of cases of *E. coli*. It is a serious disease and can cause bloody diarrhea and **hemolytic uremic syndrome (HUS)**. Outbreaks of *E. coli* have been associated with ingestion of contaminated, undercooked hamburger meat, unpasteurized apple cider and juice, and unpasteurized milk. Employing proper food storage and preparation techniques at home, in child care centers, and in retail food establishments is essential for decreasing the incidence of foodborne illnesses in young children. Contamination of food products can occur at any point along the way from production to consumption. Therefore, risk reduction and controls can be targeted at various steps in food processing. One food safety education program called FightBAC™, was developed by the Partnership for Food Safety Education, a partnership of industry, state and consumer organizations, and government agencies including the CDC and EPA. FightBAC™ has four food safety practice messages:[3,49]

- Clean: Wash hands and surfaces often.
- Separate: Don't cross-contaminate.
- Cook: Cook to proper temperatues.
- Chill: Refrigerate promptly.

Under the Dietary Guideline on Food Safety, there is a key recommendation for young children. The guidelines recommend that young children do not eat or drink raw (unpasteurized) milk or any products made from unapasteurized milk, raw or undercooked meat and poultry, raw or undercooked fish or shellfish, unpasteurized juices, and raw sprouts.[50] The Dietary Guidelines offer additional strategies for keeping food safe. Child care workers as well as family members and other caretakers of children need to be well educated in food safety issues.

The U.S. Environmental Protection Agency is in the process of evaluating all existing standards for pesticides.[3] A major objective of this evaluation is to ensure that the current levels of pesticides in the food supply and drinking water are safe for young children.

Hemolytic Uremic Syndrome (HUS) A serious, sometimes fatal complication associated with illness caused by *E. coli 0157:H7*, which occurs primarily in children under the age of 10 years. HUS is characterized by renal failure, *hemolytic anemia*, and a severe decrease in *platelet* count.[3]

Hemolytic Anemia Anemia caused by shortened survival of mature red blood cells and inability of the bone marrow to compensate for the decreased life span.

Platelets A component of the blood that plays an important role in blood coagulation.

Overweight Body mass index for age at or above the 95th percentile.

Obesity BMI-for-age greater than the 95th percentile with excess fat stores as evidenced by increased triceps skinfold measurements above the 85th percentile.[51,52]

Prevention of Nutrition-Related Disorders

The prevalence of *overweight* and *obesity* among children, adolescents, and adults in the United States has increased and represents a major public health problem. High-energy, high-fat diets coupled with sedentary lifestyles are thought to be major contributors to the increase in weight. Cardiovascular disease, a major cause of death and morbidity in the United States today, is also thought to be influenced by diets and sedentary lifestyles. Food habits, preferences, and behaviors established during the toddler and preschool ages logically influence dietary habits later in life and subsequent health status. Behaviors associated with risk factors for cardiovascular disease, including dietary habits, physcial activity behaviors, and the use of tobacco can be acquired in childhood.[53] The American Heart Association strongly advocates that primary prevention of atherosclerotic disease begin in childhood.[53] Families are encouraged to adopt dietary and exercise patterns that promote a healthy lifestyle.

Overweight and Obesity in Toddlers and Preschoolers

According to the NHANES 2003–2004 data, 13.9% of children ages 2 to 5 years are overweight, defined as having a BMI greater than or equal to the 95th percentile. This is up 3.3% from the NHANES 2001–2002 survey. Another 26.2% of children ages 2–5 years are at risk for overweight with BMIs greater than or equal to the

85th percentile but less than the 95th percentile. There was a significant increase in the prevalence of overweight in children over 1999–2000, 2001–2002, and 2003–2004. No significant difference in prevalence of overweight was found between male and female children, but differences do exist by race/ethnicity. The prevalence of overweight in Mexican American male children was significantly greater than in non-Hispanic white male children. Mexican American and non-Hispanic black female children were significantly more likely to be overweight than non-Hispanic white female children.[54]

During the preschool years, a decrease in body mass index (BMI), or weight-for-height squared [wt(kg)/ht(m)^2], is a normal part of growth and development. BMI usually reaches its lowest point at approximately 4 to 6 years of age and then increases gradually in the period called **adiposity rebound** or **BMI rebound**. Early adiposity rebound in children increases the risk of adult obesity.[55] An annual increase in BMI of three to four units may reflect a rapid increase in body fat for most children.[28]

Obesity is a multifaceted problem that is difficult to treat, making prevention the preferred approach. A committee of obesity experts has recommended that children with a BMI greater than or equal to the 85th percentile on the CDC growth charts with complications, such as hypertension or gallbladder disease, or with a BMI greater than or equal to the 95th percentile, be evaluated and possibly treated for obesity.[28] Children less than 2 years of age who fall into this category should be evaluated by a pediatric obesity specialist. Obtaining a **triceps skinfold** measurement and comparing it to standards will further validate that the child has excess fat stores rather than increased lean muscle mass. Standards for triceps skinfold measurements include those derived from data of the National Health and Nutrition Examination Survey (NHANES) I, which are age and gender specific.[57]

Maintaining weight while gaining height can be the best treatment for obese children between the ages of 2 and 7. This approach allows the obese child to "grow into his or her weight" and to lower BMI. The Dietary Guideline on Weight Management includes a key recommendation for overweight children that the rate of body weight gain be reduced, allowing growth and development. It is further recommended that a health care provider be consulted before placing a child on a weight-reduction diet.[50] However, if the child is already exhibiting secondary complications of obesity, such as mild hypertension or high cholesterol or triglyceride levels, gradual weight loss may be indicated.[28] This is also supported by a key recommendation of the Dietary Guideline on Weight Management for overweight children with chronic diseases and/or on medication. The recomendation is that a health care provider be consulted about weight-loss strategies prior to starting a weight-reduction program to ensure appropriate management of other health conditions.[50] Reducing weight at this young age is tricky because sufficient nutrients must be provided for children to reach their full height potential and to remain healthy.

Prevention and Treatment of Overweight and Obesity

Prevention is the best approach for overweight and obesity. Parenting techniques, such as finding reasons to praise the child's behavior, but never using food as a reward, foster the development of healthy eating behaviors in children and help children to self-regulate food intake. (See Food Preference Development, Appetite, and Satiety, page 273.) When weight control for a young child is warranted, the expert committee on obesity evaluation and treatment recommends a general approach, including family education and involvement.[28] Examples of behavior changes, which a family can incorporate into their lifestyle, include increasing physical activity, offering nutrient-dense and not calorie-dense snacks, and focusing on behavior changes rather than weight changes.

Nutrition and Prevention of Cardiovascular Disease in Toddlers and Preschoolers

Heart disease is the number one cause of death in the United States today.[3] A leading risk factor for cardiovascular disease, which includes diseases of the heart and blood vessels, is elevated levels of **LDL cholesterol**. Children with **familial hyperlipidemias** and obese children can have high levels of LDL cholesterol. High intakes of saturated fat, **trans fatty acids,** and, to a lesser extent, dietary cholesterol elevate LDL-cholesterol levels in children and adults alike. Other nutrition-related risk factors for cardiovascular disease include high triglyceride

---

Adiposity or BMI Rebound   A normal increase in body mass index that occurs after BMI declines and reaches its lowest point at 4 to 6 years of age.[56]

Triceps Skinfold   A measurement of a double layer of skin and fat tissue on the back of the upper arm. It is an index of body fatness and measured by skinfold calipers. The measurement is taken on the back of the arm midway between the shoulder and the elbow.

Heart Disease   The leading cause of death and a common cause of illness and disability in the United States. Coronary heart disease is the principal form of heart disease and is caused by buildup of cholesterol deposits in the coronary arteries, which feed the heart.

LDL Cholesterol   Low-density lipoprotein cholesterol, the lipid most associated with atherosclerotic disease. Diets high in saturated fat, trans fatty acids, and dietary cholesterol have been shown to increase LDL-cholesterol levels.

Familial Hyperlipidemia   A condition that runs in families and results in high levels of serum cholesterol and other lipids.

Trans Fatty Acids   Fatty acids that have unusual shapes resulting from the hydrogenation of polyunsaturated fatty acids. Trans fatty acids also occur naturally in small amounts in foods such as dairy products and beef.

levels and high body mass index.[53] Fatty streaks, which can be precursors to the buildup of fat deposits in blood vessels, have been found in the arteries of young children. Some experts believe that these streaks can represent the beginning of *atheroscleorsis* and cardiovascular disease.[53]

The American Heart Association (AHA) has published guidelines for the primary prevention of atherosclerotic cardiovascular disease beginning in childhood.[53] In these guidelines, the AHA recommends that all children be screened for risk factors of developing future cardiovascular disease. Furthermore, the AHA recommends that all children have an overall healthy eating pattern while maintaining an appropriate body weight, desirable lipid profile, and desirable blood pressure. Avoidance of smoking is encouraged as well as daily physical activity and the reduction of sedentary time.[53]

The American Heart Association has published dietary recommendations for children, which have been endorsed by the American Academy of Pediatrics and which are consistent with the U.S. Dietary Guidelines (see Dietary Guidelines, page 282).[50,58] For children ages 2 to 3 years, 30–35% of total energy from fat is recommended. For children 4 years of age or older, the recommendation is 25–35% of total energy from fat.[58] Studies have shown that such dietary restrictions are safe and effective in reducing risk factors in childhood. In one study, the Special Turku Risk Intervention Program (STRIP), a randomized dietary intervention trial was begun at about 7 months of age.[59] When the children were 5 years of age, the intervention children had lower intakes of saturated fat and cholesterol than the control children. The intervention boys had significantly lower mean serum cholesterol values than the control boys between 13 and 60 months of age.[59] No differences in growth between the intervention and control groups were shown.[59]

According to the new DRIs,[29] the acceptable macronutrient distribution ranges (AMDRs) for fat are 30–40% for children 1 to 3 years and 25–35% for children 4 to 18 years. No specific recommendations for total fat per day in the diets of young children have been made. Adequate intake levels for the essential fatty acids, linoleic acid and alpha-linolenic acid, have been determined.[29]

Dietary recommendations are different for children who are at increased risk of developing premature cardiovascular disease because a parent has heart disease or because of familial hyperlipidemias. These children need periodic screening for blood cholesterol levels and close follow-up. If LDL-cholesterol levels are high, restriction of total calories from saturated fat to less than 7% and of dietary cholesterol to no more than 200 mg per day is recommended. These children need to be closely monitored by a physician and registered dietitian.[53]

Vitamin and Mineral Supplements

Children who consume a variety of basic foods can meet all of their nutrient needs without vitamin or mineral supplements. Eating a diet of a variety of foods is the preferred way to get needed nutrients because foods contain many other substances, such as phytochemicals, in addition to nutrients that benefit health.

The American Academy of Pediatrics recommends vitamin and mineral supplementation for children who are at high risk of developing or have one or more nutrient deficiencies.[37] Children at risk of nutrient deficiency include:

> **Atherosclerosis**   A type of hardening of the arteries in which cholesterol is deposited in the arteries. These deposits narrow the coronary arteries and may reduce the flow of blood to the heart.
>
> **Tolerable Upper Intake Levels**   Highest level of daily nutrient intake that is likely to pose no risk of adverse health effects to almost all individuals in the general population; gives levels of intake that may result in adverse effects if exceeded on a regular basis.

- Children from deprived families or who suffer from abuse or neglect
- Children with anorexia or poor appetite and poor eating habits
- Children who consume a "fad diet" or only a few types of food
- Children who consume a vegetarian diet without dairy products

Despite these recommendations, data from the NHANES III indicate that children 1 to 5 years of age are major users of supplements.[60] Approximately one in two 3-year-olds in the United States are given a vitamin and mineral supplement by their parents.[61] Characteristics of mothers who give their children supplements include non-Hispanic white, older, more years of education, married, have health insurance, receive care from a private health care provider, have greater household income, and took supplements themselves during pregnancy. Children most likely to receive a supplement are those at low risk of developing nutrient deficiencies—children who would most likely benefit from a supplement are less likely to receive them.

If given to children, vitamin and mineral supplement doses should not exceed the DRI for age. Parents and caretakers should be warned against giving high amounts of vitamins and minerals to children, particularly vitamins A (retinol) and D. The *Tolerable Upper Intake Levels* shown in the DRI tables should serve as a guide to excessive levels of nutrient intake from fortified foods and supplements.

Herbal Supplements

The use of herbal remedies for various disorders is increasing in the United States today as is the use of complementary and alternative medicine practices in general. Parents and caretakers who take herbs are likely to give these products to their children. Few definitive studies exist on the effectiveness of these substances in preventing disease and promoting health in adults, much less in children. In spite of the lack of scientific evidence, anecdotal reports of benefits abound. However, some reports have linked herbal preparations to adverse effects.[62] Information on herb use should be obtained during the nutrition assessment of a child to rule out herbs as a source of health problems. At the current time, no regulation of, or consistency in, the composition of the products can lead to uncertain results. Children given various herbs are the "test subjects" in these uncontrolled studies. Parents should be advised of the potential risks of herbal therapies and the need for close monitoring of their child if they choose to give herbs to their child. The National Institutes of Health's (NIH) National Center for Complementary and Alternative Medicine's (NCCAM) website is listed in the Resources section at the end of this chapter and provides reports on the known safety and effectiveness of various herbal remedies and alternative medical practices.

Dietary and Physical Activity Recommendations

"Children aged 2 to 11 years should achieve optimal physical and cognitive development, attain a healthy weight, enjoy food, and reduce the risk of chronic disease through appropriate eating habits and participation in regular physical activity."[63]

The American Dietetic Association

Taking into consideration the energy and nutrient needs of young children and the common nutritional problems and concerns of this age group, it is easy to understand the importance that underlies dietary recommendations for toddlers and preschoolers. A primary recommendation is that young children eat a variety of foods. This recommendation is more easily achieved if healthful food preferences and eating habits are acquired during the early years. Food preferences in conjunction with food availability form the foundation of the child's diet. Limited food selection, therefore, will influence the adequacy of the child's diet by decreasing variety. Parents and caretakers cannot expect a child to "do as I say, but not as I do." Nutrition education aimed at the adults in the child's life becomes as important, if not more so, as nutrition education directed to the child.

Dietary recommendations have been developed and disseminated by the federal government and professional organizations. Two sets of guidelines for young children's diets are available: the Dietary Guidelines for Americans and the Food Guide Pyramid.[10,50] Recommendations for caloric and nutrient intake are represented in the DRIs.

Dietary Guidelines

Dietary Guidelines for Americans 2005 (discussed in Chapter 1), include some key recommendations for specific population groups, including children. The guidelines emphasize that children be offered a variety of foods including grain products, at least half of which should be whole grains, vegetables and whole fruits, and low-fat dairy products.[50] The guidelines also recommend reducing the amount of added sugar to children's diets. Children 2 to 8 years should drink 2 cups per day of fat-free or low-fat milk or equivalent milk products. In regard to fats, the guidelines recommend keeping total fat intake between 30–35% of calories for children 2 to 3 years of age and between 25–35% of calories for children and adolescents 4 to 18 years of age. Most fats in children's diets should come from sources of polyunsaturated and monunsaturated fatty acids, such as fish, nuts, and vegetable oils. Some fish may contain higher levels of mercury that may harm a young child's developing nervous system. The Food and Drug Administration (FDA) and the Environmental Protection Agency (EPA) advise that young children eat fish and shellfish that are lower in mercury. Information about mercury content of fish in a specific area can be obtained from the FDA (see Resources).

The guidelines also recommend that beans, lean meats, and poultry be added as appropriate for the child. Foods high in fat and sugar, such as candy, cookies, and cakes, should be limited in children's diets. The Dietary Guidelines also emphasize the importance of parents modeling this type of diet for their children or "to do as I say and as I do." The guidelines emphasize the importance of physical activity. Parents are advised to encourage their children to engage in at least 60 minutes of physical activity on most, preferably all, days of the week and to limit the time spent in sedentary activities, such as TV watching and computer game playing, that replace physical activity.[50] It is important that parents model for their children a lifestyle that includes a varied diet and regular physical activity. There are also some key recommendations regarding weight management for overweight children (see Overweight and Obesity in Toddlers and Preschoolers, page 279) and for food safety (see Food Safety, page 279).

As mentioned previously, the American Heart Association dietary recommendations are consistent with the U.S. Dietary Guidelines. The AHA places special emphasis on adequate intakes of omega-3 fatty acids and recommends introducing and regularly serving fish as an entrée to children. The recommendations also emphasize physical activity and balancing intake with physical activity.

Food Guide Pyramid

The USDA has developed a MyPyramid for Kids (Illustration 10.7) targeted at young children.[10] The color-coded

Illustration 10.7 MyPyramid for Kids.[10]

Table 10.8 MyPyramid for Kids Tips for Families[10]	
Eat Right	**Exercise**
1. Make half your grains whole.	1. Set a good example.
2. Vary your veggies.	2. Take the President's Challenge as a family.
3. Focus on fruits.	3. Establish a routine.
4. Get your calcium-rich foods.	4. Have an activity party.
5. Go lean with protein.	5. Set up a home gym.
6. Change your oil.	6. Move it!
7. Don't sugarcoat it.	7. Give activity gifts.

pyramid encourages children to consume a variety of foods, to make wise food choices from each of the food groups, and to limit foods high in fat and sugars. To emphasize the importance of physical activity, illustrations of children being active are depicted around the pyramid. MyPyramid for Kids Tips for Families can be found in Table 10.8. Illustration 10.8 depicts a MyPyramid Steps to a Healthier You food guide chart for a 4-year-old boy who is physically active for at least 60 minutes every day.

Recommendations for Intake of Iron, Fiber, Fat, and Calcium

Adequate iron intake is necessary to prevent iron deficiency and iron deficiency anemia in toddlers and preschoolers. Appropriate fiber intake is needed to prevent constipation

and may provide long-term disease prevention. Fat is an important source of calories, essential fatty acids, and fat-soluble vitamins in young children's diets. Adequate calcium intake is important for children to achieve peak bone mass, and yet about 20% of children do not meet the DRIs for calcium.[3]

Iron   Adequate iron intake is important in this age group to prevent iron deficiency. Good sources of dietary iron can be found in Chapter 1, page 27. Meats, which are good sources of iron, can be ground or chopped to make them easier for toddlers to chew. Fortified breakfast cereals and dried beans and peas are also good sources of iron.

"Toddler" milks, or iron-fortified commercial formulas for toddlers, are available. Healthy children who consume a variety of foods, and whose milk intake is less than 24 ounces daily, obtain adequate iron without these special products. Other commercial beverages being marketed to parents include formulas that were originally designed for children with illnesses or who had to be fed complete nutrition through a feeding tube. Such special products are expensive and are unnecessary for healthy children. It would be better for parents of healthy children to spend their food dollars on a variety of healthy foods rather than on these special products.

Fiber   Ample dietary fiber intake has been associated with the prevention of heart disease, certain cancers, diabetes, and hypertension in adults. Whether fiber helps prevent these problems as young children become adults is not known, but it is clear that fiber in a child's diet helps prevent constipation and is part of a healthy diet. Too much fiber in a child's diet can be detrimental because high-fiber diets have the potential of reducing the energy density of the diet, which could impact growth.[64] High-fiber diets could also impact the bioavailbilty of some minerals, such as iron and calcium.

The recommendations for total fiber intake based on the DRIs can be found in Table 10.9.[29] Total fiber is the sum of dietary fiber and functional fiber. Earlier recommendations were based on dietary fiber alone. Including fruits, vegetables, and whole grain breads and cereal products in the diet can increase the dietary fiber intake of children.

Table 10.9 Adequate intake of total fiber for children[29]	
1–3 years of age	19g day of total fiber
4–8 years of age	25g day of total fiber

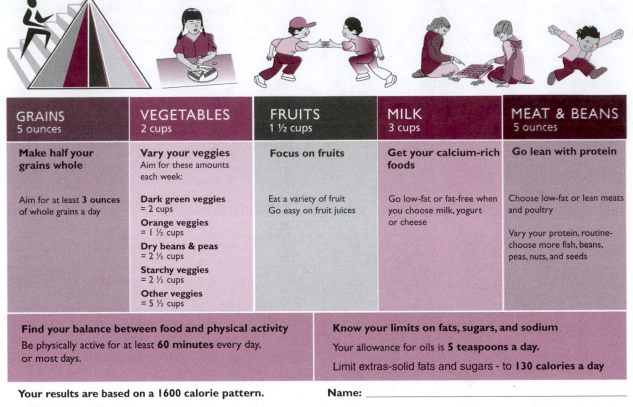

GRAINS 5 ounces	VEGETABLES 2 cups	FRUITS 1 ½ cups	MILK 3 cups	MEAT & BEANS 5 ounces
Make half your grains whole Aim for at least **3 ounces** of whole grains a day	**Vary your veggies** Aim for these amounts each week: **Dark green veggies** = 2 cups **Orange veggies** = 1 ½ cups **Dry beans & peas** = 2 ½ cups **Starchy veggies** = 2 ½ cups **Other veggies** = 5 ½ cups	**Focus on fruits** Eat a variety of fruit Go easy on fruit juices	**Get your calcium-rich foods** Go low-fat or fat-free when you choose milk, yogurt or cheese	**Go lean with protein** Choose low-fat or lean meats and poultry Vary your protein, routine- choose more fish, beans, peas, nuts, and seeds

Find your balance between food and physical activity

Be physically active for at least **60 minutes** every day, or most days.

Know your limits on fats, sugars, and sodium

Your allowance for oils is **5 teaspoons a day.**

Limit extras-solid fats and sugars - to **130 calories a day**

Your results are based on a 1600 calorie pattern.   Name: _____

This calorie level is only an estimate of your needs. Monitor your body weight to see if you need to adjust your calorie intake.

Illustration 10.8 MyPyramid Steps to a Healthier You—Individual plan for a 4-year-old boy who is physically active for at least 60 minutes every day.[10]

Those who meet the recommendation consume more high- and low-fiber breads and cereals, fruits, vegetables, legumes, nuts, and seeds than those who do not. Children with adequate fiber intake tend to have lower intakes of fat and cholesterol, and higher intakes of dietary fiber, vitamins A and E, folate, magnesium, and iron than those children who have low dietary fiber intakes.[65]

Fat   An appropriate amount of fat in a young child's diet can be achieved by employing the principles of the Dietary Guidelines and the Food Guide Pyramid that promote a diet of whole grain breads and cereals, beans and peas, fruits and vegetables, low-fat dairy products after 2 years of age, and lean meats.[10,50] Foods high in fat are used sparingly, especially foods high in saturated fat and trans fatty acids. However, an appropriate amount of dietary fat is necessary to meet children's needs for calories, essential fatty acids, and fat-soluble vitamins. As discussed in Chapter 1, good sources of the essential fatty acid linoleic acid are peanut, canola, corn, safflower, and other vegetable oils. Flaxseed, soy, and canola oils are good sources of the essential fatty acid alpha-linolenic acid.

It is important to include sources of fat-soluble vitamins in the diets of young children. Good sources of vitamin A

include whole eggs and dairy products. Sources of vitamin D include exposure to sunlight and vitamin D–fortified milk. Corn, soybean, and safflower oils are excellent sources of vitamin E. Vitamin K is widely distributed in both animal and plant foods.

Calcium   Adequate calcium intake in childhood affects peak bone mass. A high peak bone mass is thought to be protective against osteoporosis and fractures later in life.[66] Yet, many children do not consume adequate calcium. An estimated 21% of children 2 to 8 years of age consume less than their DRI for calcium.[3] The recommendations for daily calcium intake in the DRI table is 500 mg/day for children ages 1 to 3 years, and 800 mg/day for children ages 4 to 8 years. An important aspect of adequate calcium intake in toddlers and preschoolers is the development of eating patterns that will lead to adequate calcium intake later in childhood.[66]

Dietary sources of calcium are listed in Chapter 1, page 26. Dairy products are good sources of calcium, as are canned fish with soft bones such as sardines, dark green leafy vegetables such as kale and bok choy, tofu made with calcium, and calcium-fortified foods and beverages such as calcium-fortified orange juice. Nonfat and low-fat dairy products are low in saturated fat while still providing a good source of calcium.

Fluids

Healthy toddlers and preschoolers will consume enough fluid through beverages, foods, and sips and glasses of water to meet their needs. Fluid requirements increase with fever, vomiting, diarrhea, and when children are in hot, dry, or humid environments.

Consumption of milk has decreased among young children since the late 1970s, but consumption of carbonated soft drinks has increased by about the same amount. Since that time, consumption of noncitrus juices has increased almost threefold.[67] According to food consumption surveys, young children consume large amounts of sweetened beverages, including fruit juice, soft drinks, and sweetened iced tea to the detriment of the overall nutritional balance of their diet and oral health. Approximately 50% of children ages 2 to 5 years consume soft drinks.[68] Children with high consumption of regular soft drinks (more than 9 ounces per day) consume more calories and less milk and fruit juice than those with lower consumptions of regular soft drinks. Water is a good but underused "thirst quencher" for toddlers and preschoolers, as long as milk (2 cups) and fruit juice consumption (1 cup) is part of their regular diet. Parents and caretakers can offer children water to drink between meals and snacks.

Recommended versus Actual Food Intake

Several national surveys have examined food and nutrient intakes of young children. Data is available from the USDA's Continuing Survey of Food Intakes by Individuals (CSFII) 1994 to 1996 and the Supplemental Children's Survey conducted in 1998. In these national surveys, dietary data for children less than 6 years of age is provided by an adult caretaker.

According to the Food and Nutrient Intakes by Children report,[67] young children meet their energy needs. Mean percentages of total energy from carbohydrate, protein, total fat, saturated fat, and cholesterol intake in the diets of toddlers and preschoolers are shown in Table 10.10. Toddlers who participated in the 2002 Feeding Infants and Toddlers Study (FITS) exceeded estimated energy requirements by 31%.[69]

The total fat intake of about 32% of total calories is within the target range for this age child. Three- to five-year-olds have sodium intakes of 2600 mg per day, which slightly exceed the sodium recommendation of 2300 mg per day.

In general, young children consume more than enough protein and fat. Mean vitamin and mineral intakes of young children exceeded the recommended levels except for vitamin E and zinc.[67] In a longitudinal study of the nutrient and food intakes of preschool children ages 24 to 60 months, mean intakes of zinc, folic acid, and vitamins D and E were consistently below the recommended levels.[70] Low intakes of zinc, vitamin E, and iron were found in toddlers ages 12 to 18 months, a time of dietary transition.[71] Vitamin E intakes less than the estimated average requirement was found in 58% of toddlers who participated in the 2002 Feeding Infants and Toddlers Study (FITS).[69] The means for nutrient intakes often hide problems at the extremes, however. They fail to indicate the percentage of children with low nutrient intakes less than 66% of the recommended levels, and children with high nutrient intakes that exceed the Tolerable Upper Intake Levels.

According to the USDA report Food and Nutrient Intakes by Children, about 25% of children ages 4 to 8 years consumed fast food. The diets of children who ate fast food were found to be higher in total energy, total fat, total carbohydrate, added sugars, and sugar-sweetened beverages, and to have less fiber, less milk, and fewer fruits and nonstarchy vegetables than the diets of children who did not eat fast food.[72] Using this same data set, it was found that 11% of children ages 2 to 3 years and 12% of children ages 4 to 5 years consumed greater than 25% of total energy from added sugar. Increased added sugar consumption was associated with decreased nutrient and food group intakes and increased percentage of children not meeting the DRIs.[73] According to the FITS Study, by 19 to 24 months, 62% of toddlers consumed a baked dessert, 20% consumed candy, and 44% consumed a sweetened beverage.[74]

According to the USDA report Food and Nutrient Intakes by Children,[67] 89% of 1- to 2-year-old children ate snacks, and for these children snacks contributed 24% of calories and 23% of total fat intake for the day. Similarly, 88% of 3- to 5-year-old children consumed snacks, which contributed 22% of total energy and 20% of total fat. These data indicate the important contribution of snacks to children's overall nutrient intakes.

Children's portions sizes have remained constant over the years except for meat portions, which have decreased. This stability in portion sizes of young children over time reinforces the hypothesis that young children are capable of self-regulating energy intake. Portion sizes were positively

Table 10.10 Mean percentages of total calories from carbohydrate, protein, total fat, saturated fatty acids, and cholesterol intake[67]

Age	Carbohydrate (%)	Protein (%)	Total fat (%)	Saturated Fatty Acids (%)	Cholesterol (mg/d)
1–2 years	54	15	32	13	189
3–5 years	55	14	32	12	197

related to both body weight percentiles and energy intake. It seems that young children self-regulate energy intake by adjusting portion size.[75]

Cross-Cultural Considerations

When working with families from various cultures, it is important to learn as much as possible about the culture's food-related beliefs and practices. Ask the parents and caretakers about their experiences with food, including foods used for special occasions. It is also helpful to know whether foods are used for home remedies or to promote certain aspects of health. Cultural beliefs influence many child feeding practices, such as what foods are best for young children, which cause digestive upsets, or which help relieve illnesses. Perhaps one of the best-known examples is the use of chicken soup to cure what ails you! It is important for the health care provider to build on the cultural practices and to reinforce those positive practices while attempting to affect change in those practices that could be harmful to the young child.

A series of booklets entitled *Ethnic and Regional Food Practices* is available from the American Dietetic Association. This series addresses food practices, customs, and holiday foods of various ethnic groups. They provide examples for incorporating traditional foods of various groups of people into dietary recommendations. For example, peanut or polyunsaturated oils are recommended to Chinese Americans for stir-frying instead of the more traditional use of lard or chicken fat. Ordering information for this series can be found in Resources.

Vegetarian Diets

Young children can grow and develop normally on vegetarian or vegan diets, as long as their dietary patterns are intelligently planned. Vegetarian diets are rich in fruits, vegetables, and whole grains, the consumption of which is encouraged for the general population. However, young children in particular need some energy-dense foods to reduce the total amount of food required. The amount of vegetarian foods needed to meet nutrient needs may be more food than young children can eat. Young children need to eat several times a day to meet their energy needs because their stomachs cannot hold a lot of food at one time.

Children who are fed *vegan* and *macrobiotic diets* tend to have lower rates of growth, although still within the normal ranges, during the first 5 years of life compared to children given a mixed diet.[76] Strict vegan diets, which exclude all foods of animal origin, may be deficient in vitamins B_{12} and D, zinc, and omega-3 fatty acids, and may also be low in calcium, unless fortified foods are consumed.

Vegan Diet   *The most restrictive of vegetarian diets, allowing only plant foods.*

Macrobiotic Diet   *This diet falls between semivegetarian and vegan diets and include, foods such as brown rice, other grains, and vegetables, fish, dried beans, spices, and fruits.*

Protein needs are usually met if the diet is adequate in energy and a variety of foods are included.[77] Children on vegan diets should receive vitamin B_{12} supplements or consume fortified breakfast cereals, textured soy protein, or soy milk fortified with vitamin B_{12}. The vitamin B_{12} status of children following vegetarian and vegan diets should be monitored on a regular basis as vitamin B_{12}–deficiency may cause vitamin B_{12}–deficiency anemia. Iron-deficiency anemia is an infrequent problem among children consuming a vegetarian diet.

Vitamin D adequacy can be met by diet or by sun exposure. Good sources of vitamin D for children include fortified soy milk, fortified breakfast cereals, and fortified margarines. Zinc is found in foods of animal origin. Plant sources of zinc include legumes, nuts, and whole grains. Vegetable products are also lacking in omega-3 fatty acids. Therefore, including a source of these fatty acids, such as canola or soybean oils, is advisable.[76]

Foods containing phytates, such as unrefined cereals, may interfere with calcium absorption. So if the child's diet contains a lot of unrefined cereals, higher calcium intakes may be needed.[76] Good sources of calcium for children on strict vegetarian diets include fortified soy milk, calcium-fortified orange juice, tofu processed with calcium, and certain vegetables, such as broccoli and kale.[9] Supplements may be necessary for some children with inadequate intakes that are not remedied by dietary means.

Guidelines recommended for vegetarian diets for young children have been developed and are given here:[9]

- Allow the child to eat several times a day (i.e., three meals and two to three snacks).
- Avoid serving the child bran and an excessive amount of bulky foods, such as bran muffins and raw fruits and vegetables.
- Include in the diet some sources of energy-dense foods such as cheese and avocado.
- Include enough fat (at least 30% of total calories) and a source of omega-3 fatty acids, such as canola or soybean oils.
- Include sources of vitamin B_{12}, vitamin D, and calcium in the diet, or supplement if required.

Tables 10.11 and 10.12 provide suggested daily food guides for lacto-ovovegetarians and vegans.[9]

Child Care Nutrition Standards

"All child care programs should achieve recommended standards for meeting children's nutrition and nutrition education needs in a safe, sanitary, supportive environment that promotes healthy growth and development."[78]

The American Dietetic Association

An estimated 23 million children in the United States require child care while their parents work, making foods children

Table 10.11 Suggested daily food guide for lacto-ovovegetarians at various intake levels[9]

Food Groups	Servings per Day, by Age and Daily Caloric Intake	
	1–2 Years (1300 Kcal)	3–6 Years (1800 Kcal)
Breads, grains, cereal	5	5
Legumes	½	1
Vegetables	2	3
Fruits	3	3
Nuts, seeds	½	½
Milk, yogurt, cheese	3	3
Eggs (limit 3/week)	½	½
Fats, oils (added)	2	4
Sugar (added teaspoons)	3	4

Table 10.12 Suggested daily food guide for vegan children at various intake levels[9]

Food Groups	Servings per Day, by Age and Daily Caloric Intake	
	1–2 Years (1300 Kcal)	3–6 Years (1800 Kcal)
Breads, grains, cereal	5	6
Legumes	½	1
Vegetables, dark-green leafy	1	1
Vegetables, other	1	2
Fruits	2	4
Nuts, seeds	½	½
Milk alternatives	3	3
Fats, oils (added)	2	4
Sugar (added teaspoons)	3	4

eat away from home a major contribution to their overall intake. Young children eat away from home often: 37% of 1- to 2-year-olds, and 52% of 3- to 5-year-olds eat away from home daily. The most common places for away-from-home meals in children 5 years of age and younger are fast food restaurants, day care centers, and friends' houses.[67]

Nutrition standards for child care services exist and specify minimum requirements for amounts and types of foods to include in meals and snacks and food service safety procedures.[78,79] These standards also address nutrition learning experiences and education for children, staff, and parents as well as the physical and emotional environment in which meals and snacks are served. It is recommended that children in part-day programs (4 to 7 hours per day) receive food that provides at least one-third of their daily calorie and nutrient needs in at least one meal and two snacks or two meals and one snack. A child in a full-day program (8 hours or more) should receive foods that meet one-half to two-thirds of the child's daily needs based on the DRIs in at least two meals and two snacks or three snacks and one meal. Food should be offered at intervals of not less than 2 hours and not more than 3 hours and should be consistent with the Dietary Guidelines for Americans.[50]

Physical Activity Recommendations

Inactivity is thought to be a major contributor to the increasing prevalence of obesity. The Dietary Guidelines for Americans 2005 recommend that young children engage in play activity for at least 60 minutes on most, preferably all, days of the week.[50] Some suggested activities from MyPryamid for Kids[10] include:

- Taking a nature walk
- Riding a tricycle or bicycle
- Walking, skipping, or running
- Most important—having fun while being active!

This advice is reinforced by the American Academy of Pediatrics (AAP).[80] The AAP suggests that toddlers, under the supervision of an adult caregiver, engage in activities such as walking in the neighborhood, park, or zoo, and in free play outdoors. For the preschool-age child, the AAP lists appropriate activities as running, swimming, tumbling, throwing, and catching, under adult supervision. No television viewing is recommended for children less than 2 years of age while screen time should be limited to less than 2 hours per day for all other age groups.[80]

Parents are encouraged to set a good example for their children by being physically active themselves. Parents are also encouraged to limit the time that they allow their children to watch television and play computer games.

Nutrition Intervention for Risk Reduction

Model Program

"Bright Futures is a vision, a philosophy, a set of expert guidelines, and a practical developmental approach to providing health supervision for children of all ages, from birth through adolescence."[81]

Bright Futures in Practice: Nutrition is an example of a model program for nutrition intervention for risk reduction.[9] This guide is a component of the larger project *Bright Futures Guidelines for Health Supervision of Infants, Children, and Adolescents.*[81]

The purpose of Bright Futures is to foster trusting relationships between the child, health professionals, the family, and the community to promote optimal health for the child.[81] Bright Futures guidelines are developmentally based and address the physical, mental, cognitive, and social development of infants, children, adolescents, and their families.

Many different "user-friendly" materials and tools are available from this program to assist in the implementation of the guidelines. In addition to the *Bright Futures Guidelines for Health Supervision,* implementation guides have been published for oral health, general nutrition, physical activity, mental health, and one for families. *Bright Futures in Practice: Nutrition* is based on three critical principles:[9]

1. Nutrition must be integrated into the lives of infants, children, adolescents, and families.
2. Good nutrition requires balance.
3. An element of joy enhances nutrition, health, and well-being.

The program is based on the premise that optimal nutrition for children be approached from the development of the child and put in the context of the environment in which the child lives.[9] It emphasizes the development of healthy eating and physical activity behaviors. Nutrition supervision guidelines are given for each age group, and within each broad age group, interview questions, screening and assessment, and nutrition counseling topics are provided. It lists desired outcomes for the child and discusses the role of the family, as well as frequently asked questions. For example, by utilizing the guidelines, health care providers will be able to provide anticipatory guidance to parents of toddlers for the proper advancements of their diets based on growth and development. The implementation guide also addresses special topics related to pediatric nutrition including oral health, vegetarian eating practices, iron-deficiency anemia, and obesity. Useful information is included in the appendix such as nutrition questionnaires for the various age groups. *Bright Futures in Practice: Nutrition* is a valuable resource to anyone who is interested in promoting healthy eating and physical activity behaviors in children. Ordering information for Bright Futures materials is available on their website, which is listed in Resources.

Public Food and Nutrition Programs

Young children and their families can benefit from a number of federally sponsored food and nutrition programs. Four example programs are presented here.

WIC

The Special Supplemental Nutrition Program for Women, Infants, and Children,[82] previously described in Chapter 8, is administered by the Food and Nutrition Service of the U.S. Department of Agriculture (USDA). It is one of the most successful federally funded nutrition programs in the United States. Participation in WIC services improves the growth, iron status, and the quality of dietary intake of nutritionally at-risk infants and children up to age 5 years.[83] The WIC program is cost effective in that each dollar invested in the program saves up to $3 in health care costs.[82]

As in infancy, children must live in a low-income household, 185% or less of the federal poverty level, and be at "nutrition risk" to be eligible for WIC services. "Nutrition risk" means a child has a medical or dietary-based condition that places the child at increased risk. Such conditions include iron-deficiency anemia, underweight, overweight, a chronic illness, such as cystic fibrosis, or the child consumes an inadequate diet.[82]

Children receive nutrition assistance, education, and follow-up services by specially trained registered dietitians and nutritionists. Vouchers for food items such as milk, juice, eggs, cheese, peanut butter, and fortified cereals are given to eligible families. These vouchers are exchanged for the food items at authorized retailers.

WIC's Farmers' Market Nutrition Program

The Farmers' Market Nutrition Program is a special seasonal program for WIC participants. This program provides vouchers for the purchase of locally grown produce at farmers' markets. The program is designed to help low-income families increase their consumption of fresh fruits and vegetables.

Head Start and Early Head Start

Administered by the U.S. Department of Health and Human Services, Head Start and Early Head Start are comprehensive child development programs, serving children from birth to 5 years of age, pregnant women, and their families. Nearly one million U.S. children participate in this program. The overall goal is to increase the readiness for school of children from low-income families. A range of individualized, culturally appropriate services are provided through Head Start and related agencies working in education and early childhood development; medical, dental, and mental health services; nutrition services; and parent education.[84] About three in four Head Start families have incomes <$12,000 annually. More specific information about Early Head Start can be found in Chapter 9.

Food Stamps

The Food Stamp Program, administered by the USDA, is designed to help adults in low-income households buy food. In 2005, about 25.7 million people living in 11.2 million households received food stamps in the United States. About half of the food stamp recipients were children. The monetary amount of food vouchers provided to an eligible household depends on the number of people in it and the income of the household. Income eligibility criteria for this and a number of other federal programs can be found at the USDA's Food and Nutrition Service website: www.fns.usda.gov/fsp. The average monthly amount of benefits received through the Food Stamp Program in 2006 was $209 per household, enough to help families and individuals pay for a portion of the food they need. Each state must develop a food stamp nutrition education plan based on federal guidance.[85] Participation in the Food Stamp Program is associated with increased intakes of a number of nutrients.[83]

Key Points

1. Periodic and accurate measurements of a young child's growth are important indicators of a child's nutritional status.

2. The types of foods offered to children and methods of feeding are based on an individual child's growth and development.

3. Food habits are learned behaviors.

4. Common nutritional problems such as iron-deficiency anemia, dental caries, and constipation in a healthy child can be addressed through adjustments in the child's diet.

5. The prevalence of overweight and risk of over-weight is increasing even among young children in the United States.

6. The U.S. Dietary Guidelines recommends that children eat a variety of foods and increase physical activity.

7. The MyPyramid for Kids reinforces eating a variety of foods and increasing physical activity.

8. Surveys of food intake indicate that most children meet their nutritional needs, but that their intakes exceed recommendations for energy, sodium, total fat, and saturated fat.

9. A well-planned vegetarian diet can meet the needs of a growing child.

10. Public food and nutrition programs are important resources to many young children in the United States.

Resources

American Academy of Pediatrics

The American Academy of Pediatrics website contains consumer information on current news topics affecting the pediatric population. Policy statements can be found on this page and consumer publications can be ordered through the Bookstore.
Website: www.aap.org

American Dietetic Association

Besides information for members of the American Dietetic Association, this website contains information for consumers on various topics of interest. There is a "Daily Tip" and a feature article, plus consumer information on topics such as food safety and healthy lifestyles.
Website: www.eatright.org

Bright Futures

The Bright Futures publications are developmentally based guidelines for health supervision and address the physical, mental, cognitive, and social development of infants, children, adolescents, and their families. Three implementation guides have also been published to date addressing oral health, general nutrition, and physical activity.
Website: www.brightfutures.org

Centers for Disease Control and Prevention, National Center for Health Statistics

CDC Growth Charts: United States. May 30, 2000.
The CDC website provides background information on the growth charts. Also, individual growth charts can be downloaded and printed.
Website: www.cdc.gov/nchs/about/major/nhanes/growthcharts/charts.htm

Diabetes Care and Education, Dietetic Practice Group of the American Dietetic Association

Ethnic and Regional Food Practices, A Series. Chicago, IL: The American Dietetic Association, 1994–1999.
This series of booklets addresses food practices, customs, and holiday foods of various ethnic groups. Booklets describing the following ethnic groups are available: Alaska Native, Chinese American, Filipino American, Hmong American, Jewish, and Navajo.

FitSource: A Web Directory for Providers

This resource, produced by the Child Care Bureau, contains a wide variety of tools for incorporating nutrition and physical activity into child care and after-school programs. It is designed for use by program administrators, directors, technical assistance providers, and others interested in promoting proper nutrition and physical activity.
Website: www.nccic.caliber.com/fitsource/index.cfm

Food and Drug Administration

Call FDA's food information line toll-free at 1-888-SAFE-FOOD or visit **www.cfsan.fda.gov/~dms/admehg3.html.**
The FDA provides information about the safety of the food supply, including information about the mercury content of fish and shellfish in certain areas.

Graves DE, Suitor CW

Celebrating Diversity: Approaching Families Through Their Food. 2nd ed. Arlington, VA: National Center for Education in Maternal and Child Health, 1998.
The purpose of this publication is to assist health professionals in learning to communicate effectively with a diverse clientele.

Topics covered in the book include using food to create common ground, changing food patterns, how food choices are made, communicating with clients and families, and working within the community.

Green M, Palfrey JS (Eds)

Bright Futures: Guidelines for Health Supervision of Infants, Children, and Adolescents. 2nd ed. Arlington, VA: National Center for Education in Maternal and Child Health, 2000. This Bright Futures publication contains developmentally based guidelines for health supervision and addresses the physical, mental, cognitive, and social developments of infants, children, adolescents, and their families.

MyPyramid.gov

As described in Chapter 1, MyPyramid.gov is Internet-based and resource-filled. As the name suggests, the MyPyramid for Kids is designed for children and includes such resources as a Kids Coloring Page, a Kids Worksheet, and Classroom materials.
Website: www.MyPyramid.gov

National Academy Press, Dietary Reference Intakes

The website of the National Academy Press, the publisher for The National Academies, contains descriptions of available books. Over 2000 books are available on-line free of charge. The current Dietary Reference Intakes are also available on this website and can be read online free of charge.
Website: www.nap.edu

National Center for Complementary and Alternative Medicine, National Institutes of Health

The National Center for Complementary and Alternative Medicine is dedicated to science-based information on complementary and alternative healing practices. This website provides information for consumers and practitioners as well as information about related news and events.
Website: www.nccam.nih.gov/nccam

National Network for Child Care

This website, hosted by Iowa State University Extension, provides articles, resources, and links on a variety of topics of interest to professionals and families who care for children and youth. Topics include child development, nutrition, and health and safety.
Website: www.nncc.org

Partnership for Food Safety Education

This partnership was formed in 1997 for the purpose of educating the public about safe food handling to reduce foodborne illnesses. This website promotes the Partnership's four food safety practices to the educator, consumer, and the media.
Website: www.fightbac.org

Patrick K, Spear BA, Holt K, and Sofka D.

Bright Futures in Practice: Physical Activity. Arlington, VA: National Center for Education in Maternal and Child Health, 2001.

This colorful and user-friendly spiral-bound book is one of the implementation guides of the Bright Futures publications. Developmentally appropriate activities are presented for infants, children, and adolescents. Special issues and concerns are addressed, such as physical activity for the child with a chronic condition such as asthma or diabetes mellitus.

USDA, Center for Nutrition Policy and Promotion

Tips for Using the Food Guide Pyramid for Young Children 2 to 6 Years Old. Government Printing Office, 202/512–1800, Stock Number 001–00004665–9.

This website provides information on USDA materials including the Dietary Guidelines for Americans and the MyPyramid for kids.
Website: www.cnpp.usda.gov/Resources.htm

References

1. Center on Hunger, Poverty and Nutrition Policy. Statement on the link between nutrition and cognitive development in children. Tufts University, School of Nutrition Science and Policy; 1998.

2. The Annie E. Casey Foundation. 2006 kids ccount data book. Baltimore, MD: The Annie E. Casey Foundation, 2006.

3. U.S. Department of Health and Human Services. Healthy People 2010 (conference edition, in two volumes). Washington, DC: U.S. Department of Health and Human Services, January 2000.

4. Behrman RE, Kliegman RM, Jenson HB, eds. Nelson's textbook of pediatrics, 17th ed. Philadelphia, PA: WB Saunders Co; 2004.

5. Centers for Disease Control and Prevention, National Center for Health Statistics. CDC growth charts: United States. Available from www.cdc.gov/ nchs/about/major/ nhanes/growthcharts/charts.htm, accessed 05/00.

6. Pietrobelli A et al. Body mass index as a measure of adiposity among children and adolescents: a validation study. J Pediatr. 1998;132: 204–10.

7. Dietz WH, Robinson TN. Use of the body mass index (BMI) as a measure of overweight in children and adolescents. J Pediatr. 1998;132:191–93.

8. Troiano RP, Flegal KM. Overweight children and adolescents: description, epidemiology, and demographics. Pediatrics. 1998; 101:497–504.

9. Story M, Holt K, Sofka D, eds. Bright futures in practice: nutrition. Arlington, VA: National Center for Education in Maternal and Child Health; 2000.

10. USDA, MyPyramidforkids. www.MyPyramid.gov.

11. Satter E. The feeding relationship: problems and interventions. J Pediatr. 1990;117:181–89.

12. Birch LL. Children's food acceptance patterns. Nutr Today. 1996;31:234–40.

13. Birch LL et al. The variability of young children's energy intake. N Eng J Med. 1991;324:232–35.

14. Story M, Brown JE. Do young children instinctively know what to eat? The studies of Clara Davis revisited. New Eng J Med. 1987;316:103–6.

15. Van den Bree MBM et al. Genetic and environmental influences on eating patterns of twins ages >50 years. Am J Clin. Nutr. 1999;70:456–65.

16. Chess S, Thomas A. Dynamics of individual behavioral development. In: Levine MD, Carey WB, Crocker AC, eds. Developmental-behavioral pediatrics. Philadelphia, PA: WB Saunders Co.; 1992:84–94.

17. Birch LL, Fisher JO. Development of eating behaviors among children and adolescents. Pedatrics. 1998;101:539–49(S).

18. Birch LL, Fisher JA. Appetite and eating behavior in children. Pediatr Clinic N Amer. 1995;42:931–53.

19. Fisher JO et al. Parental influences on young girls' fruit and vegetable, micronutrient, and fat intakes. J Amer Diet Assoc. 2002;102:58–64.

20. Fisher JO, Birch LL. Restricting access to palatable foods affects children's behavioral response, food selection, and intake. Am J Clin Nutr. 1999;69:1264–72.

21. Connor SM. Food-related advertising on preschool television: building brand recognition in young viewers. Pediatrics. 2006; 118:1478–85.

22. Birch LL, Fisher JO. Food intake regulation in children, fat and sugar substitutes and intake. Ann NY Acad Sci. 1997;819: 194–220.

23. Rolls BJ, Engell D, Birch LL. Serving portion size influences 5-year-old but not 3-year-old children's food intake. J Amer Diet Assoc. 2000;100:232–34.

24. Fisher JO, Birch LL. Parents' restrictive feeding practices are associated with young girls' negative self-evaluation of eating. J Amer Diet Assoc. 2000;100:1341–46.

25. Davison KK, Birch LL. Weight status, parent reaction, and self-concept in five-year-old girls. Pediatrics. 2001;107: 46–53.

26. Abramovitz BA, Birch LL. Five-year-old girls' ideas about dieting are predicted by their mothers' dieting. J Amer Diet Assoc. 2000;100:1157–63.

27. Satter E. Feeding dynamics: helping children to eat well. J Pediatr Health Care. 1995;9:178–84.

28. Barlow SE, Dietz WH. Obesity evaluation and treatment: expert committee recommendations. Pediatrics [serial on-line] 1998;102: E3. Available from www. pediatrics.org/cgi/content/full/102/3/e29, accessed 07-00.

29. Institute of Medicine, Food and Nutrition Board. Dietary Reference Intakes for energy, carbohydrate, fiber, fat, protein and amino acids. Washington, DC: National Academy Press; 2002/2005.

30. Kennedy E, Goldberg J. What are American children eating? implications for public policy. Nutr Rev. 1995;53:111–26.

31. Kennedy E, Powell R. Changing eating patterns of American children: a view from 1996. J Amer Coll of Nutr. 1997;16:524–29.

32. Institute of Medicine, Food and Nutrition Board. Dietary reference intakes for calcium, phosphorus, magnesium, vitamin D, and fluoride. Washington, DC: National Academy Press; 1997.

33. Institute of Medicine, Food and Nutrition Board. Dietary reference intakes for vitamin A, vitamin K, arsenic, boron, chromium, copper, iodine, iron, manganese, molybdenum, nickel, silicon, vanadium, and zinc. Washington, DC: National Academy Press; 2001.

34. Centers for Disease Control and Prevention. Recommendations to prevent and control iron deficiency in the United States. Morbidity and Mortality Weekly Report. April 3, 1998;47:No. RR-03.

35. Centers for Disease Control and Prevention. Iron deficiency—United States, 1999–2000. Morbidity and Mortality Weekly Report. October 11, 2002;51:No. 40.

36. Casamassimo P. Bright futures in practice; oral health. Arlington, VA: National Center for Education in Maternal and Child Health; 1996.

37. American Academy of Pediatrics, Committee on Nutrition. Pediatric nutrition handbook. 5th ed. Kleinman RE (ed.) Elk Grove Village, IL: American Academy of Pediatrics; 2004.

38. McClung HJ, Boyne L, Heitlinger L. Constipation and dietary fiber intake in children. Pediatrics. 1995;96:999–1001(S).

39. Meyer PA et al. Surveillance for elevated blood lead levels among children—United States, 1997–2001. In: Surveillance Summaries, Septemeber 12, 2003. MMWR 2003;52 (No. SS-10), 1–21.

40. Centers for Disease Control. National Center for Environmental Health. Childhood Lead Poisoning. Available from www.cdc.gov/nceh/lead/factsheets/childhoodlead.htm

41. Ballew C, et al. Blood lead concentration and children's anthropometric dimensions in the Third National Health and Nutrition Examination Survey (NHANES III), 1988–1994. J Pediatr. 1999;134:623–30.

42. Stanek K, et al. Lead consumption of 18- to 36-month-old children as determined from duplicate diet collections: nutrient intakes, blood lead levels, and effects on growth. J Amer Diet Assoc. 1998;98:155–58.

43. Mushak P, Crocetti AF. Lead and nutrition. Nutr Today. 1996;31:12–17.

44. Centers for Disease Control and Prevention. Screening young children for lead poisoning: guidance for state and local public health officials. Atlanta, GA: Centers for Disease Control and Prevention, 1997. Also available at www.cdc.gov/nceh/programs/lead/guide/1997/guide97.htm.

45. American Academy of Pediatrics. Screening for elevated blood lead levels. Pedaitrics. 1998;101:1072–78.

46. Household food security in the United States 1995–1998, advance report—summary. Available from www.fns.usda.gov/oane/MENU/Published/FSP/FILES/fsecsum.htm.

47. Kleinman RE, et al. Hunger in children in the United States: potential behavioral and emotional correlates. Pediatrics [serial on-line] 1998; 101:E3. Available from www.pediatrics.org/cgi/content/full/101/1/e3, accessed 7/00.

48. National Institute of Allergy and Infectious Diseases. National Insitutes of Health. U.S. Department of Health and Human Services. Foodborne Diseases. Available from www.niaid.nih.gov/factsheets/foodbornedis.htm

49. Partnership for Food Safety education. FightBAC™. Available from: www.fightbac.org

50. U.S. Department of Health and Human Services and U.S. Department of Agriculture.

Dietary Guidelines for Americans, 2005. 6th Edition. Washington, DC: U.S. Government Printing Office, January 2005.

51. Must A, Dallal GE, Dietz WH. Reference data for obesity: 85th and 95th percentiles of body mass index (wt/ht^2) and triceps skinfold thickness. Am J Clin Nutr. 1991;53:839–46.

52. Dwyer JT, Stone EJ, Yang M. Prevalence of marked overweight and obesity in a multiethnic pediatric population: findings from the Child and Adolescent Trial for Cardiovascular Health (CATCH) study. J Amer Diet Assoc. 2000;100:1149–56.

53. Kavey RW et al. American Heart Association guidelines for primary prevention of atherosclerotic cardiovascular disease beginning in childhood. Circulation. 2003;107:1562–66.

54. Ogden CL et al. Prevalence of overweight and obesity in the United States, 1999–2004. JAMA 2006;295:1549–55.

55. Whitaker RC et al. Early adiposity rebound and the risk of adult obesity. Pediatrics [serial on-line] 1998;101:E5. Available from www.pediatrics.org/cgi/content/full/101/3/e5, accessed 07/00.

56. Dietz WH, Gortmaker SL. Preventing obesity in children and adolescents. Annu Rev Public Health. 2001;22:337–53.

57. Frisancho AR. New norms of upper limb fat and muscle areas for assessment of nutritional status. Am J Clin Nutr. 1981;34: 2540–45.

58. Gidding SS et al. Dietary recommendations for children and adolescents: a guide for pracitioners. Consensus statement from the American Heart Association. Circulation. 2005; 112:2061–75.

59. Rask-Nissila L et al. Prospective, randomized, infancy-onset trial of the effects of a low-saturated-fat, low-cholesterol diet on serum lipids and lipoproteins before school age: the special turku coronary risk factor intervention project (STRIP). Circulation 2000; 102: 1477–83.

60. Vital & Health Statistics—Series 11: Data from the National Health Survey. 1999;244:1–14.

61. Yu SM, Kogan MD, Gergen P. Vitamin-mineral supplement use among preschool children in the United States. Pediatrics [serial on-line] 1997;100:E4. Available from www.pediatrics.org/ cgi/content/full/100/5/e4, accessed 07/00.

62. Buck ML, Michael RS. Talking with families about herbal products. J Pediatr. 2000;136:673–78.

63. American Dietetic Association. Position of the American Dietetic Association: dietary guidance for healthy children aged 2 to 11 years. J Amer Diet Assoc. 2004; 104:660–77.

64. A summary of conference recommendations on dietary fiber in childhood: Conference on Dietary Fiber in Childhood, New York, May 24, 1994. Pediatrics. 1995;96:1023–28.

65. Hampl JS, Betts NM, Benes BA. The "age+5" rule: comparisons of dietary fiber intake among 4- to 10-year-old children. J Amer Diet Assoc. 1998;98:1418–23.

66. American Academy of Pediatrics, Committee on Nutrition. Calcium requirements of infants, children, and adolescents. Pediatrics. 1999;104:1152–7.

67. U.S. Department of Agriculture, Agricultural Research Service. 1999. Food and nutrient intakes by children 1994–96, 1998. ARS Food Surveys Research Group, available on the "Products" page at www.barc. usda.gov/bhnrc/foodsurvey/home.htm, accessed 07/00.

68. Harnack L, Stang J, Story M. Soft drink consumption among US children and adolescents: nutritional consequences. J Amer Diet Assoc. 1999;99:436–41.

69. Devaney B et al. Nutrient intaktes of infants and toddlers. J Am Diet Assoc. 2004;104:S15–21.

70. Skinner JD et al. Longitudinal study of nutrient and food intakes of white preschool children aged 24 to 60 months. J Amer Diet Assoc. 1999;99:1514–21.

71. Picciano MF et al. Nutritional guidance is needed during dietary transition in early childhood. Pediatrics. 2000;106:109–14.

72. Bowman SA et al. Effects of fast-food consumption on energy intake and diet quality among children in a national household survey. Pediatrics. 2004;113: 112–18.

73. Kranz S et al. Adverse effect of high added sugar consumption on dietary intake in American preschoolers. J Pediatr. 2005; 146:105–11.

74. Fox MK et al. Feeding infants and toddlers study: what foods are infants and toddlers eating? J Am Diet Assoc. 2004;104:S22–30.

75. McConaby KL et al. Food portions are positively related to energy intake and body weight in early childhood. J Pediatr. 2002;140:340–47.

76. Sanders TAB. Vegetarian diets and children. Pediatr Clinic N Amer. 1995; 42:955–65.

77. American Dietetic Association. Position of the American Dietetic Association and Dietetians of Canada: vegetarian diets. J Amer Diet Assoc. 2003;103:748–65.

78. American Dietetic Association. Position of the American Dietetic Association: benchmarks for nutrition programs in child-care settings. J Amer Diet Assoc. 2005;105:979–86.

79. American Public Health Association, American Academy of Pediatrics. Caring for our children, national health and safety performance standards: guidelines for out-of-home child care programs; 1992.

80. American Academy of Pediatrics, Council on Sports Medicine and Fitness and Council on School Health. Pediatrics. Active healthy living: prevention of childhood obesity through increased physical activity. 2006;117:1834–42.

81. Green M, Palfrey JS, eds. Bright futures: guidelines for health supervision of infants, children, and adolescents, 2nd ed. Arlington, VA: National Center for Education in Maternal and Child Health; 2000.

82. U.S. Department of Agriculture. WIC. Available from www.usda.gov.

83. Rose D, Habicht JP, Devaney B. Household participation in the food stamp and WIC programs increase the nutrient intakes of preschool children. J Nutr. 1998;128:548–55.

84. U.S. Department of Health and Human Services. Head Start. Available from www. dhhs.gov.

85. U.S. Department of Agriculture. Food Stamps Program. Available from www. fns.usda.gov/fsp.

Chapter 11

Toddler and Preschooler Nutrition:
Conditions and Interventions

Chapter Outline

- Introduction
- Who Are Children with Special Health Care Needs?
- Nutrition Needs of Toddlers and Preschoolers with Chronic Conditions
- Growth Assessment
- Feeding Problems
- Nutrition-Related Conditions
- Food Allergies and Intolerance
- Dietary Supplements and Herbal Remedies
- Sources of Nutrition Services

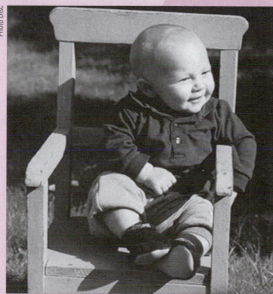

Prepared by **Janet Sugarman Isaacs**

Key Nutrition Concepts

1 Nutrition problems in young children with special health care needs include underweight, overweight, feeding difficulties, and higher nutrient needs as a result of chronic health conditions.

2 Feeding difficulties in preschoolers and toddlers appear as food refusals, picky appetites, and concerns about growth.

3 Nutrition services for toddlers and preschoolers with chronic health problems are provided in various settings, including schools and other educational programs and specialty clinics.

4 Toddlers and preschoolers at risk for chronic conditions have the same nutritional problems, concerns, and needs as other children.

Introduction

Most toddlers and preschoolers are healthy and develop as expected. This chapter discusses children who do not fit the typical pattern, *children with special health care needs* associated with a *chronic condition* or disability, or children who are at risk. Sometimes no diagnosis has been made, and yet parents, health care providers, or preschool teachers have a nagging feeling that something is not right about how the child is growing and developing. This chapter covers nutrition needs and services for young children with food allergies, breathing or *pulmonary* problems, feeding and growth problems, or developmental delays, and those at risk for needing nutrition support. The Healthy People 2010 objectives (in Chapter 10, Table 10.1) are not appropriate for children with special health care needs if their underlying health care condition affects any of the nutrients involved. Similarly the 2005 U.S. Dietary Guidelines are not customized to the underlying conditions of children with special health care needs.

> **Children with Special Health Care Needs**   A general term for infants and children with, or at risk for, physical or developmental disabilities, or chronic medical conditions from genetic or metabolic disorders, birth defects, premature births, trauma, infection, or prenatal exposure to drugs.
>
> **Chronic Condition**   Disorder of health or development that is the usual state for an individual and unlikely to change, although secondary conditions may result over time.
>
> **Pulmonary**   Related to the lungs and their movement of air for exchange of carbon dioxide and oxygen.
>
> **Early Intervention Services**   Federally mandated evaluation and therapy services for children in the age range from birth to 3 years under the Individuals with Disabilities Education Act.

Who Are Children with Special Health Care Needs?

The child who does not see, hear, or walk is easily recognized as having a chronic condition. It can be difficult and expensive, however, to identify some other children with special health care needs. Criteria for labeling chronic conditions in children vary from state to state. More than 40 different federal definitions describe the term *disability*.[1] Criteria used for identifying disabilities in adults do not fit children, because the criteria are related to a person's ability to work or perform household chores. *Chronic condition* and *disability* mean the same thing in referring to toddlers and preschoolers. Prevalence estimates for disabilities range from 5% to 31% of children.[1,2] Whatever the number, nutrition problems are common in children with disabilities. Up to 90% of children with disabilities have some type of nutritional problem.[3]

Toddlers and preschoolers with chronic illnesses are entitled to the same services as older people with chronic illnesses are, with additional help. They are covered by the Americans with Disabilities Act, the Social Security Disability Program, Supplemental Social Security Insurance (SSI) Program, and services for families without health insurance coverage.[4] Additional help comes from educational regulations ensuring that all children with disabilities have a free, appropriate public education. Nutrition services are funded within education regulations in the Individuals with Disabilities Education Act (IDEA).[4] Most children start school at age 5 or 6 years, but children at risk or who have special needs may attend well before that, as soon as the need is identified. The sooner special educational, nutritional, and health care interventions are started, the better for the overall development of the child. Parents of a typical child choose and pay for day care or a preschool program. For a child with special health care needs, day care or educational programs are selected based on nutrition and other types of therapy provided by state and federal resources. Nutrition services can be provided to young children within special educational programs and services both as preschoolers (3 to 5 years old) and from birth up to 3 years old.[4,5] Services have to be culturally appropriate for various ethnic groups, reflecting food preferences, religious beliefs, and sensitivity to dress and language; otherwise, they are likely to be rejected.

Eligibility for services does not require a specific diagnosis. *Early intervention services* are based on the following:[4,5]

- Developmental delays in one or more of the following areas: cognitive, physical, language and speech, psychosocial, or self-help skills
- A physical or mental condition with a high probability of delay, such as Down syndrome
- At risk medically or environmentally for substantial developmental delay if services are not provided

A number of chronic conditions are suspected, but not obvious, in the first year of life. The diagnosis often becomes clear in the toddler and preschool years, however. Standardized developmental screening, evaluation, and testing for these ages show more reliability than they do for infants. Parents who were told about possible disabilities during infancy move beyond coping by denial or disbelief and are willing to seek out services. The tendency is to resist labeling a young child with a diagnosis, so some suspected conditions are not confirmed until school age if the delay in diagnosis will not harm the child.

Nutrition Needs of Toddlers and Preschoolers with Chronic Conditions

Toddlers or preschoolers with chronic health conditions are at risk for the same nutrition-related problems and concerns as other children.[6] Consequently, every attempt should be made to meet their overall nutritional needs and to assure normal growth and development. The DRIs for toddlers and preschoolers provide a good starting point for setting protein, vitamin, and mineral needs for children with chronic conditions.[7] (DRI tables are located on the inside front cover of this text.) The recommendations for typical children concerning dietary fiber, prevention of lead poisoning, and iron-deficiency anemia apply to children at risk or already diagnosed with special health care needs.[5] However, some specific conditions require adjustments to the general guidelines. The DRI for dietary fiber for children may be too low for some conditions and too high for others.[8] Children with sickle-cell disease have more specific blood iron and lead testing than indicated by the usual guidelines. Iron-rich foods to increase their iron stores may not be appropriate when iron also comes with blood transfusions. Consequently, nutritional needs must be customized to the child.[6,8]

Chronic conditions may result in poor appetites, although there are increased caloric needs.[8] Table 11.1 gives examples of conditions in which caloric needs may be high or low.[6,8] Each child must be assessed to confirm caloric needs. A child may have an interval of needing additional calories based on the course of the chronic condition. Changes in caloric needs may explain why both underweight or overweight are more common in children with chronic conditions than in other children.[6] Overweight and obesity are common in Down syndrome and spina bifida in part because lower caloric needs are due to low muscle mass, lower mobility, and short stature.[6] Overall health status is worsened by excessive body weight, so matching caloric intake to needs is important no matter how difficult. Underweight results in part from the chronic illness and its treatment. Children with chronic illnesses may be more likely to experience weight loss with any illness. Underweight children with a chronic condition may or may not benefit from food choices for weight gain. In underweight children, it is inappropriate

Table 11.1	Chronic conditions generally associated with high and low caloric needs	
Higher Caloric Need Conditions	Lower Caloric Need Conditions
Cystic fibrosis	Down syndrome
Renal disease	Spina bifida
Ambulatory children with *diplegia*	Nonambulatory children with diplegia
Pediatric AIDS	*Prader-Willi syndrome*
Bronchopulmonary dysplasia (BPD)	Nonambulatory children with short stature

to make some of the usual recommendations, such as reducing fat intake.[5]

Recommendations regarding food intake, vitamin and mineral supplementation, and mealtime behaviors also should be customized to the individual child. Children who are frequently sick or have low energy levels and appetites may dislike eating foods that are hard to chew or take a long time to eat. Some food-intake problems related to chronic illness may result from the children's behavior. It is age appropriate for children to express their food likes and dislikes, insist on their independence, and go on food jags. It can be difficult but important to distinguish between food-intake problems related to the chronic condition and those related to "growing up" in the toddler and preschool years.

Cystic Fibrosis Condition in which a genetically changed chromosome 7 interferes with all the exocrine functions in the body, but particularly pulmonary complications, causing chronic illness.

Diplegia Condition in which the part of the brain controlling movement of the legs is damaged, interfering with muscle control and ambulation.

Pediatric AIDS Acquired immunodeficiency syndrome in which infection-fighting abilities of the body are destroyed by a virus.

Prader-Willi Syndrome Condition in which partial deletion of chromosome 15 interferes with control of appetite, muscle development, and cognition.

Growth Assessment

Most toddlers and preschoolers with chronic conditions are provided an assessment of nutritional status as a first step in determining whether more intensive levels of nutrition services are needed. The need for nutrition services is identified by answers to these sorts of questions:

- Is the child's growth on track?
- Is his or her diet adequate?
- Are the child's feeding or eating skills appropriate for the child's age?
- Does the diagnosis affect nutritional needs?

A variety of nutrition screening tools exist for assessing the nutritional status of children with chronic conditions.[6,8] Such tools are useful for children at risk as well as those already diagnosed with conditions such as asthma,

HIV infection, allergies, and cerebral palsy. After assessment, nutrition intervention services provide methods to improve nutritional status. Several conditions that require nutrition intervention services include failure to thrive, celiac disease, breathing problems, and muscle coordination problems.

Children with special health care needs often have conditions that affect growth even when adequate nutrients are provided. In such cases the 2000 CDC growth charts require interpretation based on the child's previous growth pattern.[9] If a thin and small-appearing child has adequate fat stores, adding calories may be harmful; it is important to recognize the growth pattern as healthy for that child. Trying to add calories in such a case may promote overweight in the form of excess fat stores. Growth patterns in children with special health care needs are also affected by some prescribed medications, particularly those that change body composition, such as steroids and growth hormone.[10,11]

Specific growth charts developed for chronic conditions, when available, are preferred. For children up to age 38 months who are born low birth weight (LBW) or very low birth weight (VLBW), the Infant Health and Development Program (IHDP) growth percentile charts are appropriate.[12] Correction for prematurity, as discussed in Chapter 9, makes the charts useful for preterm babies as well as toddlers. For a child born 3 months early, for example, the IHDP growth charts could be used at a chronological age of 41 months. Plotting on both the 2000 CDC growth charts and the IHDP chart documents catch-up growth. When catch-up growth happens, the child's growth pattern crosses channels on the growth chart, for weight as well as for length.

Special health care providers commonly use a special head growth chart.[13] It provides head circumference percentiles from birth to age 18 years and is used to determine whether head growth falls within normal limits or indicates a neurological condition, such as *Rett syndrome*.[13] Rett

syndrome is a rare disorder, characterized by a reduced rate of head growth beginning in the toddler years (see Illustration 11.1).[14] Later, over time, the rate of weight and height accretion slows in girls with Rett syndrome.[14] Decreased rate of head growth in toddlers and preschoolers may be indicative of problems from infancy, such as prematurity, or consequences of infection such as *meningitis*. Some clinics plot head circumference on the back of the CDC 2000 growth chart (up to age 3 years) as well as on the special head circumference growth chart.

Feeding Problems

Children with special health care needs have many of the same developmental feeding issues as other children, such as using food to control their parents' behavior at mealtime and going on food jags. Feeding problems that are part of underlying health conditions may emerge in the toddler and preschool years on top of the usual feeding difficulties and require extra attention. Some feeding problems during the toddler years are typical in children who are later diagnosed with a chronic condition. Examples of such conditions include gastroesophageal reflux, asthma (pulmonary problems in general), developmental delay, cerebral palsy, attention deficit hyperactivity disorder, and autism.[3,8] These children as toddlers tend to

Illustration 11.1 Nellhaus head circumference growth chart plotted for girl with Rett syndrome.

Table 11.2 Food choices of a 2.5-year-old child with suspected developmental delay

Likes	Dislikes
3 packets instant Cream of Wheat with added sugar and margarine (refused offered apple slices)	Hamburger meats, or any other kind of meat
Macaroni and cheese (refused offered sandwich with lettuce and bologna)	Green beans or any kind of vegetables
Banana, with peel removed (refused other cut-up fruits offered)	Vegetable soup
Pudding, only chocolate	Salads of all kinds
Cheese puffs (refused corn chips)	Casseroles or any mixtures of foods
Juices of all kinds, in a sippy cup	Milk, and milk with any flavoring added

display signs of feeding problems, such as low interest in eating, long mealtimes (over 30 minutes), preferring liquids over solids, and food refusals. Children at risk for developmental delay often prove more difficult to feed as toddlers and preschoolers.[3] The child may drink liquids excessively, or eat foods usually preferred by younger children (see Case Study 11.1 on the next page). Recognizing that the child needs to be treated as younger than current chronological age may be a necessary step. Offering the child food textures that he or she can eat successfully within a monotonous diet, or continuing to offer a bottle, may be appropriate choices in these circumstances.

Table 11.2 shows an example of the likes and dislikes of a 2.5-year-old child. The child likes only a few foods that are not especially nutritious. Usual recommendations are to add variety to the child's diet and to assure intake of meats, milk, and vegetables. This recommendation is appropriate for a typical child, but this child's eating pattern suggests a feeding problem. The soft textures and mild tastes of preferred foods characterize a child closer to 1 year of age, and eating the foods the child dislikes require higher oral skills. An evaluation of the child's overall level of functioning will likely indicate a developmental delay of the child's feeding skills.

Behavioral Feeding Problems

"Every mouth prefers its own soup."

Sephardic saying

Mealtime behavioral problems and food refusals are common in children with behavioral and attention disorders. These concerns often bring parents to nutrition experts for solutions. Behavioral disorders that affect nutritional status are autism and attention-focusing problems, such as ***attention deficit hyperactivity disorder (ADHD)***. ADHD may be suspected during the preschool years, but it is primarily treated during school years.[15] (ADHD is discussed further in Chapter 13.) Table 11.3 shows the intake of a 2-year-old with a feeding problem resulting in a self-restricted diet typical of autism. The child refuses to eat

many foods and is rigid in what he will eat. He does not respond to feeling hungry, as other 2-year-old children do. When he is not given foods he likes, he refuses to eat at all and has temper tantrums during which he can injure himself. His self-restricted diet is a part of the condition, which affects how he senses everything in his environment. He prefers to drink rather than eat foods, so a high proportion of his total calories come from one type of drink. Interventions to improve the diet for this child may include a complete vitamin and mineral supplement, and adding one new food by offering it many times (15–20 times) over 1 or 2 months. Nutrition interventions should be incorporated into this child's overall treatment plan, provided within a special education program. See Case Study 11.1 on the next page.

Attention Deficit Hyperactivity Disorder (ADHD) Condition characterized by low impulse control and short attention span, with and without a high level of overall activity.

Excessive Fluid Intake

The food intake noted in Table 11.3 highlights a common issue related to excessive fluids. Many young children prefer to drink rather than eat solid foods, especially when they are not feeling well. Families of chronically ill children

Table 11.3 Dietary intake of 2-year-old child with suspected autism

Dry Fruit Loops cereal
10 fl oz calcium-supplemented orange juice drink
Chicken fingers from a specific fast-food restaurant
French fries
10 fl oz calcium-supplemented orange juice drink
Waverly crackers
Pringles potato chips
10 fl oz calcium-supplemented orange juice drink
Oatmeal cake
10 fl oz calcium-supplemented orange juice drink

Case Study 11.1

A Picky Eater

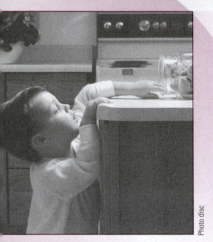

Photo disc

Greg is a well-groomed boy almost 3 years old. He has been growing as expected, but he does not talk. He can walk and move about well, but he prefers to play alone. Favorite foods are juices in his sippy cup, which he likes to carry around; macaroni and cheese; white bread without crusts; mashed potato; Honeycomb cereal; and crackers. Greg cries and throws food that he does not like, such as hamburgers, fruits, most vegetables, and any food combinations. He will periodically eat cheese pizza, scrambled eggs, and applesauce.

His mother tries talking to the pediatrician about his picky appetite, but the pediatrician reassures her that Greg will eat when he is hungry and not to worry. Greg's mother is frustrated that he is so difficult to take out to eat because of the tantrums he throws in restaurants and friends' homes. He sometimes eats a large portion of a food he likes. Most of the time, he is satisfied just drinking juices all day from his sippy cup, and he is rarely interested in eating when others eat. He is able to eat with a spoon, but he does not like to touch foods with his hands. Greg has been referred for speech therapy, but his therapy does not address his eating. His medical history shows that he was born full-term and has had three ear infections but no major illnesses.

Nutrition assessment shows that Greg is consuming adequate calories at 1350 calories/day, or 85 cal/kg. His diet is excessive in vitamin C and B vitamins, with adequate protein at the RDA for his age. His sources of protein are mainly his starchy foods of bread, crackers, and dry cereal.

Questions

1. What are the signs that Greg's feeding problem may be related to his speech?
2. Because Greg is growing well and meeting his calorie needs, why not just wait for him to mature to accept other foods?
3. Was his pediatrician wrong to say that Greg will eat when he is hungry?

---

tend to offer juices and lower-nutrient beverages in an effort to achieve growth when eating is difficult. The American Academy of Pediatrics recommendation to limit juice intake to 4 to 6 fluid ounces per day for ages 1–6 years applies to all children.[16] Calcium-fortified juices may be appropriate if other sources of calcium are limited, but these juices can also be overconsumed. In a child who already has gastrointestinal problems, it may not be clear if problems are caused by excessive juice intake.[16] Excess juice resulting in a pattern of low milk intake has been documented to result in smaller stature and lower bone density.[17] In young children who may be less active due to chronic conditions, the negative impact may be larger.

Feeding Problems and Food Safety

Toddlers and preschoolers with chronic conditions are at greater risk for food-contamination problems. Some feeding problems result in prolonged needs for soft, easy-to-eat food textures well past the age when baby foods are eaten. Forkmashing or blending foods may invite bacterial contamination or spoilage over time.[6] Similarly, complete nutritional supplements and formulas are subject to contamination, particularly in the tubing used to give them. How often tubing and devices to deliver formulas are changed can be a food-safety issue. Some families, aware of the high cost of such devices, tend to use them longer than recommended.[18] Microbial contamination of powdered dry formulas has been found, so pasturized canned liquid formulas are recommended.[18]

Feeding Problems from Disabilities Involving Neuromuscular Control

Children who have feeding problems related to muscle control of swallowing or control of the mouth or upper body may choke or cough while eating or refuse foods that require chewing.[3] These types of feeding problems result from conditions such as cerebral palsy or other *neuromuscular disorders* and genetic disorders such as Down syndrome. These signs of feeding and swallowing problems in toddlers or preschoolers generally appear more severe than the reactions of infants who are learning how to munch and chew foods.[8] The decrease in appetite expected in toddlers and preschoolers may be pronounced in children who find eating difficult and unpleasant. These feeding problems may require further study to make sure eating is safe for a child, and not related to frequent illness such as bronchitis or pneumonia.

A child with *hypotonia* or *hypertonia* in the upper body may experience difficulty sitting for a meal and self-feeding with a spoon.[8] If these feeding problems are not resolved by providing therapy in early intervention programs or schools, children are likely to resist eating over time. They may then need a form of nutrition support, such as placement of a *gastrostomy*.

Nutrition-Related Conditions

Failure to Thrive

Failure to thrive (FTT) is a condition in which a caloric deficit is suspected.[19] FTT has a slightly different basis in toddlers and preschoolers, who may have grown adequately during the first year. Their decrease in growth rate occurs at the age when appetite typically decreases and control issues at mealtime are expected, making identifying the cause of FTT more difficult. Generally FTT is suspected when a child's growth declines more than 2 growth percentiles, placing him near or below the lowest percentile in weight-for-age, weight-for-length, and/or BMI. FTT may result from a complex interplay of medical and environmental factors, such as the following:[4,6]

- Digestive problems such as gastrointestinal reflux or celiac disease
- Asthma or breathing problems
- Neurological conditions such as seizures
- Pediatric AIDS

Children who have chronic illnesses or were born preterm have a higher risk of FTT as a result of abuse or *medical neglect*.[19] They have greater needs than other children do, and they may be more irritable and demanding, which places them at risk. Often a specific nutrient or group of nutrients are suspected of being inadequate in the diet of children with FTT, when the more appropriate emphasis should be placed on energy and protein. Copper and zinc in the blood of toddlers with failure to thrive was reported to be the same as age-matched

controls, although protein intake was lower.[20]

Recovery from FTT can include catch-up growth, which is an acceleration in growth rate for age.[3] If calories are provided at a higher level than for a typical child of the same age, catch-up growth is likely (see Illustration 11.2). The length of time needed for catch-up growth varies, but some weight gain should be documented within a few weeks. For example, recovery from FTT for one 3-year-old was a gain of 6 pounds—more weight than typical to gain in 1 year—within the first 3 months of living in a new home.

Toddler Diarrhea and Celiac Disease

Toddlers are likely to develop diarrhea. The condition is called toddler diarrhea, in which otherwise healthy growing children have diarrhea so often that their parents bring them for a checkup.[3] Testing shows no intestinal damage and normal blood levels, without FTT or weight loss. The dietary culprit is likely to be excessive intake of juices that contain sucrose or sorbitol. The diarrhea results from excess water being pulled into the intestine, so limiting juices intake may be recommended.[16]

Celiac disease occurs in people who are sensitive to gluten, a component of wheat, rye, and barley. It has a prevalence of 1 in 3000 people within certain ethnic groups, such as those of Middle Eastern or Irish ancestry.[21] Symptoms of diarrhea and other digestive problems usually develop by 2 years of age. Confirmation of the condition is based on testing blood for the antibodies to gluten. Dietary management requires complete restriction of any foods with gluten. This list includes everything made with flour, such as bread and pasta, as well as foods with wheat, barley, or rye as an additive.[21] The allowed foods include rice, soy, corn, and potato flours. Oats are gluten-free, but may be contaminated with gluten from wheat mixed in. Meats, fruits, and vegetables are not restricted, but many processed foods use wheat flour for thickening. After instituting dietary restrictions, the intestinal damage heals, and the digestive symptoms disappear. The parents of preschoolers with celiac disease learn to be expert readers of food labels because intestinal damage recurs if gluten is eaten by mistake.

Neuromuscular Disorders   Conditions of the nervous system characterized by difficulty with voluntary or involuntary control of muscle movement.

Hypotonia   Condition characterized by high muscle tone, stiffness, or spasticity.

Hypertonia   Condition characterized by low muscle tone, floppiness, or muscle weakness.

Gastrostomy   Form of enteral nutrition support for delivering nutrition by tube directly into the stomach, bypassing the mouth through a surgical procedure that creates an opening through the abdominal wall and stomach.

Medical Neglect   Failure of parent or caretaker to seek, obtain, and follow through with a complete diagnostic study or medical, dental, or mental health treatment for a health problem, symptom, or condition that, if untreated, could become severe enough to present a danger to the child.

Birth to 36 months: Girls
Length-for-age and Weight-for-age percentiles

NAME _____

RECORD # _____

Revised April 20, 2001.
SOURCE: Developed by the National Center for Health Statistics in collaboration with
the National Center for Chronic Disease Prevention and Health Promotion (2000).
http://www.cdc.gov/growthcharts

Illustration 11.2 Growth chart for a girl with failure to thrive before and after intervention.

Autism

The toddler and preschool years are when behavioral signs of what may later be found to be autism are noted by families. No scientifically proven diet is now recommended for prevention or treatment of autism.[22] A gluten-free and casein-free diet is well known to families who educate themselves on the Internet and through autism support

Spastic Quadriplegia A form of cerebral palsy in which brain damage interferes with voluntary muscle control in both arms and legs.

resources. Many products are marketed for the diet. Milk substitutes to avoid casein may or may not meet the child's need for calcium, vitamin D, protein, or other nutrients. Parents may choose to restrict gluten in the child with autism who has been found not to have celiac disease. Until the major studies now under way about autism release their findings, dietary recommendations for autism are the same as for any other child of the same age who has feeding problems.

Muscle Coordination Problems and Cerebral Palsy

Toddlers and preschoolers at risk for or confirmed with cerebral palsy need nutrition assessments that include body composition indexes, such as fat stores.[6,23] Nutrition interventions are then based on these findings, and they may include encouraging weight gain if body fat stores are low. If the child appears thin as a result of small muscle size and not low fat stores, weight gain is not needed. Growth tracking may be based on spastic quadriplegia growth charts for some toddlers or preschoolers.[24] Part of the growth assessment for a preschooler with cerebral palsy may include an estimate of caloric needs for activity, which may be higher or lower than expected. A girl may expend higher energy in her efforts to coordinate walking while receiving physical therapy 3 days per week at school. Her activity may be lower if she is in a wheelchair most of the time.

Feeding assessment for a child with severe cerebral palsy (also called *spastic quadriplegia*) may be necessary as part of the overall nutritional assessment.[8] The assessment may include an observation of eating to determine any restrictions in the type of foods that the child can eat, and whether coordinating muscles for chewing, swallowing, and/or using a spoon or fork are working well. Table 11.4 on page 302 provides a food-intake record for a 4-year-old girl with spastic quadriplegia who does not walk and is

Case Study 11.2

Early Intervention Services for a Boy at Risk for Nutrition Support

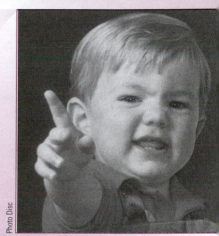

Photo Disc

Robert is 2.3 years old and in an early intervention program. He is eligible based on his preterm birth at 30 weeks gestation. His premature birth was related to exposure to an intrauterine infection. All in the family agree that he is small, but their main concern is that he is difficult to feed. He cries and refuses to eat when offered meals, even those with his favorite foods.

The registered dietitian who consults at the early intervention program meets with the family, assesses Robert, and reviews his medical records. Nutrition services are first planned to boost calories to stimulate weight gain. Observing Robert being fed by his mother is part of the nutrition services. Other therapists at the early intervention center are involved in making sure that Robert is positioned well to eat, so that he is sitting up without extra effort. The nutritionist and occupational therapist are concerned that Robert is choking so easily, and they talk to the family about contacting his pediatrician. They send a fax to the pediatrician's office recommending tests to study Robert's swallowing.

Robert does not attend the early intervention program for the next 3 weeks. The tests demonstrate that he is aspirating some of his liquids into his lungs, so oral feeding is unsafe. He requires a gastrostomy for feeding and is hospitalized for surgery. His parents learn how to feed him through the gastrostomy.

When Robert returns to the early intervention program, his pediatrician asks the early intervention staff to monitor his weight and to reinforce the discharge feeding instructions with the family. Nutrition services provided in the early intervention program are changed from working on Robert's oral feeding to monitoring and documenting his growth as adjustments are made in his gastrostomy feeding schedule. Over the next 6 months, Robert gains weight. He starts being more interactive with the staff at the early intervention center and makes some developmental progress in his walking and speaking. He is still a small child, but his improved nutritional status is confirmed by his adequate body fat measurements.

His ability to return to eating by mouth will be reassessed later in the year.

Questions

1. What are the signs that Robert needs gastrostomy feeding?
2. Could the gastrostomy placement have been prevented if Robert had gained weight?
3. Can Robert enjoy life if he cannot eat?

receiving nutritional services for weight gain. Her meal pattern was adjusted because she tires easily while eating. She does not like to eat too much at a time, and refuses to be fed by another person (which is appropriate for her age). She can chew foods such as fresh apple, but then is too tired to eat something else. She eats a larger portion if the food is soft and does not require her to work so hard. She has gained weight at a slow rate, and her fat stores are low. The first plan is to use regular foods that are easy for her to eat to meet her nutritional needs, including cooked rather than fresh vegetables and fruits, and to avoid hard-to-chew foods, such as roast beef or corn on the cob. If she does not

gain weight by eating foods such as those suggested in Table 11.4, she may need nutritional supplementation to assure her nutritional needs are met within her feeding limitations. See Case Study 11.2.

Pulmonary Problems

Breathing conditions are examples of common problems in children with special health care needs with major nutritional consequences. Breathing problems increase nutritional needs, lower interest in eating, and can slow growth rate.[25] (See Case Study 11.2.) Infants who were born preterm are espe-

Table 11.4 Meal pattern and recommended foods for an underweight girl with feeding problems as a result of weakness

Meal Pattern: Small, Frequent Meals and Snacks to Prevent Tiredness at Meals	Recommended Foods That Are Easy to Chew, with Small Portions
Breakfast at home	**Breakfast:** ½ c oatmeal with added soft fruit, margarine, and brown sugar
Mid-morning snack	
Lunch (at preschool)	**Snacks:** 1 slice deli meat with 6 fl oz whole milk with Carnation Instant Breakfast added
Afternoon snack (at preschool)	
After-school snack	½ c soft-cooked sliced apples with added margarine
Dinner	cake-type cookie (frosting allowed)
Bedtime snack	**Dinner:** ½ c mashed potato with added margarine
	3 Tbsp meat loaf
	3 Tbsp soft-cooked carrots with added margarine
	Bedtime snack: chocolate cake with frosting and 4 fl oz whole milk

- Interrupted sleep and fatigue make hunger and fullness cues harder to interpret.

By the preschool years, the impact of BPD on slowing the rate of weight gain is usually clear. Exposure to common respiratory illnesses, which are minor in typical children, can require a trip back to the hospital for some children with BPD. Increased frequency of infections adds another limitation to catch-up growth. Neither the CDC growth chart nor the IHDP preterm growth chart may be helpful in predicting the child's growth pattern, but periods of good health are usually accompanied by a increase in weight gain and appetite.

Dietary recommendations for toddlers with BPD are similar to those for children with weakness (see Table 11.4). Small, frequent meals with foods that are concentrated sources of calories are needed. Easy-to-eat foods may still be recommended so that fatigue from meals is low. If the toddler with breathing problems does not gain weight as a result of dietary recommendations such as those in Table 11.4, the next step will be to add complete nutritional supplements to meet the higher caloric needs. The supplements, such as Pediasure, are also a source of vitamins and minerals.

Bronchopulmonary Dysplasia (BPD) Condition in which the underdeveloped lungs in a preterm infant are damaged so that breathing requires extra effort.

Asthma Condition in which the lungs are unable to exchange air due to lack of expansion of air sacs. It can result in a chronic illness and sometimes unconsciousness and death if not treated.

Work of Breathing (WOB) A common term used to express extra respiratory effort in a variety of pulmonary conditions.

Mental Retardation Substantially below average intelligence and problems in adapting to the environment, which emerge before age 18 years.

cially likely as toddlers to have breathing problems. Up to 80% of 1000-gram infants can develop chronic lung disease.[25] Examples of pulmonary diseases or chronic lung disease are *bronchopulmonary dysplasia (BPD)* and *asthma*. Asthma is self-reported in 58 of every 1000 children under 5 years of age.[26] Asthma results in more emergency room visits for children under 5 years—at 121 visits per 10,000 people—than it does in older children with asthma.[26] Asthma does not necessarily require nutrition services, but some children have asthma as a result of food allergies.[27]

Toddlers and preschoolers with BPD have a positive long-term prognosis because new lung tissue can grow until about 8 years of age.[25] Toddlers and preschoolers with serious breathing problems generally need extra caloric intake due to the extra energy expended in breathing. Increased *work of breathing (WOB)* occurs with different pulmonary conditions and generally leads to low interest in feeding, partially as a result of tiredness.[26] Feeding difficulties have several causes in a toddler treated for BPD:[25]

- The normal progression of feeding skills is interrupted.
- Medications and their side effects contribute to high nutrition needs.

Developmental Delay and Evaluations

Developmental delay may be suspected when specific nutrients are consumed in inadequate or excessive amounts. Iron deficiency and lead toxicity are risk factors for developmental problems.[5,6] Developmental evaluations are recommended for young children who have been sick for a long time and isolated from other children. Standardized testing aids in finding a definitive diagnosis and appropriate educational programs. Developmental delay is a specific diagnosis that may be replaced by *mental retardation* when the child is 6 or 7 years old.[4] Changes in growth rate are typical in children with developmental delay.[23] Short stature is common and part of the unusual growth pattern that often prompts referrals for genetic testing.[28] The evaluation of growth from a genetic expert may include more in-depth analyses, such as measurements of hand and foot size and bone age.[28] Genetic syndromes also can be associated with unusually fast growth. Soto's syndrome is a rare disorder in which the child is tall and large, but has delayed development.[28]

Food Allergies and Intolerance

True food allergies are estimated to be present in 2% to 8% of children.[27] Food allergies are usually identified in toddlers and preschoolers because allergy testing in infancy is not useful due to the incomplete development of the immune system. True food allergies can result in life-threatening episodes of *anaphylaxis*.[27] Examples of food allergies that may result in anaphylaxis for some children include the following:[27]

- Milk
- Eggs
- Wheat
- Peanuts
- Walnuts
- Soy
- Fish

Cow's milk protein allergy rarely persists into the toddler and preschool years. However, when cow's milk protein allergy does persist, symptoms in the toddler and preschool years may appear as more general allergy symptoms, such as asthma or skin rashes.[29] A high incidence of other food allergies are present in a child with confirmed cow's milk protein allergy, with, for example, 35% reacting also to oranges or 47% reacting also to soy milk.

Strict and complete avoidance of the food that causes the allergy is required. This abstinence includes all settings, such as eating nothing prepared at bake sales when the food ingredients are unknown. If the preschool child is on an extensively restricted diet, the quality of the diet may not meet all her nutritional needs. Such restrictions are also likely to result in mealtime behavioral problems. The parents may become overprotective, or the child may quickly learn to use restricted foods to get a parent's concern and attention. Diagnosed food allergies can greatly affect the family. For children at risk for anaphylaxis, parents and caregivers should be given instruction in emergency lifesaving procedures and use of an injectible form of epinephrine.[27]

Dietary Supplements and Herbal Remedies

"It is better to take food into the mouth than to take worries into the heart."

Yiddish saying

Families who are concerned that something may be wrong with their young children may be attracted to health and nutritional claims targeted and packaged for adults. The family that is having difficulty finding effective treatment for a child is most at risk for inappropriate or ineffective alternative products. Parent coalitions and advocacy groups are excellent sources of networking for families, but they can also be sources of nutritional claims for products and dietary regimens that have no scientific testing behind them. Down syndrome, for example, is a disorder for which nutritional supplementation has been marketed to parents. No specific nutrients, combinations of nutrients, or herbal remedies have been shown to improve the intellectual functioning of individuals with Down syndrome.[30,31] The National Down Syndrome Society cautions parents about the ineffectiveness of nutrient and herbal supplements to discourage their use, but interest continues.[30,32] What is really being marketed is hope, which families always want and need.

> **Anaphylaxis** Sudden onset of a reaction with mild to severe symptoms, including a decrease in ability to breathe, which may be severe enough to cause a coma.

Constipation remedies are examples of over-the-counter products used often for children with special health care needs. Constipation is a common condition in children with various neuromuscular conditions in which muscles are weak.[6] Parents tend to try over-the-counter remedies, dietary methods, and home remedies for constipation management. The effectiveness of dietary fiber may be low when muscle weakness is an underlying problem.[3] Both overtreatment and undertreatment can get the child in trouble by worsening the constipation problem. A young child recently died as a result of poisoning by a laxative product administered at a higher dose than recommended.[33] Effective prescription medications for constipation management are available, but the family has to bring the problem to the attention of the health care provider. Encouraging the family to discuss the problem with the physician before trying over-the-counter products is important for many children.

Sources of Nutrition Services

Infants and toddlers who have chronic conditions are served by a variety of resources. Registered dietitians who have training in pediatrics are qualified to provide services to toddlers and preschool children with chronic conditions.

Programs in which nutrition care may be accessed include the following:[34]

- State programs for children with special health care needs
- Early intervention programs (age 0 up to 36 months)
- Early childhood education programs (IDEA, ages 3–5 years)
- Head Start; regular program or special needs category (ages 3–5 years)
- Early Head Start; regular program or special needs category (0 up to 36 months)
- WIC
- Low birth weight follow-up programs
- Child care feeding programs (USDA)

These programs are described in Chapters 8 and 10. Efforts to increase program accessibility come from state and federal governmental offices, toll-free outreach services, and websites. Specific outreach programs to locate toddlers and preschoolers at risk are funded in each state, under names such as "Child Find."[4] Because every child at risk is eligible for a screening, contacting a neighborhood public school is a good starting place to locate services, even if the child is not old enough to attend the school.

Key Points

1. Toddlers and preschoolers with special health care needs may require medication for their underlying condition that may interfere with their growth, appetite, and meal pattern. Early intervention programs and early childhood education programs include adjusting the timing of meals and snacks.

2. Nutritional, educational, and developmental providers for toddlers and preschoolers demonstrate how parents are to advocate for their children's special health care needs later on in schools and job sites.

3. Families of children with special health care needs are targets of invalid nutritional claims by those hoping to sell nutritional supplements. Over-the-counter nutrition-related products, such as vitamin or constipation remedies for adults may be dangerous for such young children.

4. Toddlers and preschoolers with special health care needs are quite varied in their nutrition needs, but the basic concepts of supporting growth, typical feeding skill development, and meeting nutrition needs for age and activity still apply.

5. Failure to thrive is often the reason children with special health care needs enter medical, educational, and developmental services. Such cases of failure to thrive cannot be corrected with additional energy— as it is in children without special health care needs. Unusual growth patterns can be signs of conditions that is not directly related to nutrition.

6. Examples of conditions likely to appear in toddlers and preschoolers with special health needs are autism, Rett syndrome, spastic quadriplegia, asthma, developmental delay, and true food allergies.

Resources

Federal Interagency Coordinating Council Site for Families with Children with Disabilities

Identifies, by state and city, resources for finding local intervention programs.
Website: www.fed-icc.org

Federation for Children with Special Needs

Includes support for families of children with special needs.
Website: www.fcsn.org

Food Allergy

A credible source of recommendations for preventing food allergy reactions; the newsletter provides recipes to avoid foods that cause reactions.
Website: www.foodallergy.org

National Information Center for Children and Youth with Disabilities

A useful site for parents and providers who are looking for intervention services for children with special needs. It is targeted mainly toward educational programs.
Website: www.NICHCY.org

National Organization for Rare Diseases (NORD)

A credible source for parents and providers of information and resources about rare "orphan" diseases.
Website: www.rarediseases.org

Quackwatch

Includes information on dietary supplements and products that are claimed to benefit health and nutrition.
Website: www.quackwatch.com

Studies to Advance Autism Research and Treatment (STAART) Network

The U.S. National Institutes of Health (NIH) has clinical studies on autism to define the characteristics of different subtypes and possible treatments. Nutrition in toddlers who show signs of developmental delay and dietary changes as a treatment for autism are examples from the eight STAART sites. One study, "Diet and Behavior in Young Children with Autism," is testing if a gluten- and casein-free diet has specific benefits for children with autism in the preschool-age group. This research is needed since many families with a child with autism use Internet advocacy sites as their source of nutrition information, as this therapy is not accepted by most health care providers.

Pacific West Maternal and Child Health Distance Learning Network

Personnel who can work in early intervention programs and schools on nutrition problems are limited in part because qualified nutrition providers are not familiar with children with special health care needs, Web-based distance learning resources for training about nutrition needs for children with special health care needs are vital tools for recruiting and advancing nutrition providers. This training network was originated due to the distances from Hawaii and Alaska to the mainland, but it is now used internationally.
Website: **http://washington.edu/pwdlearn**

References

1. Westbrook LE, Silver EJ, Stein, REK. Implications for estimates of disability in children: a comparison of definitional components. Pediatrics. 1998;101:1025–30.

2. Centers for Disease Control and Prevention. Mental health in the United States: Health care and well being of children with chronic emotional, behavioral, or developmental problems—United States, 2001. MMWR:2005:985–88.

3. Staiano A. Food refusal in toddlers with chronic diseases. J Ped Gastro Nutr. 2003;37:225–27.

4. National Information Center for Children and Youth with Disabilities. General information about disabilities. Available at www.nichcy.org/general.htm, accessed 10/06.

5. Story M, Holt K, Sofka D, eds. Bright futures in practice: nutrition. Arlington, VA: National Center for Education in Maternal and Child Health; 2000:266–70.

6. Isaacs JS. The role of the nutritionist in life care planning. In: Riddick-Grisham S, ed. Pediatric life care planning and case management. Boca Raton, FL:CRC Press; 2004:325–38.

7. Trumbo P et al. Dietary reference intakes: vitamin A, vitamin K, arsenic, boron, chromium, copper, iodine, iron, manganese, molybdenum, nickel, silicon, vanadium, and zinc. J Amer Diet Assoc. 2001;101:294–301.

8. Kleinman RE, ed. Pediatric nutrition handbook, 5th ed. Chicago IL: American Academy of Pediatrics; 2004:425–42.

9. National Center for Health Statistics. NCHS growth curves for children 0–19 years. U.S. Vital and Health Statistics, Health Resources Administration. Washington, DC: U.S. Government Printing Office; 2000.

10. Anneren G et al. Normalized growth velocity in children with Down's syndrome during growth hormone therapy. J Intel Disabil Res. 1993;371:381–87.

11. Wollmann HA et al. Reference values for height and weight in Prader-Willi syndrome based on 315 patients. Eur J Pediatr. 1998; 157:634–42.

12. The Infant Health and Development Program: enhancing the outcomes of low birth weight, premature infants. JAMA. 1990;263(22):3035–42.

13. Nellhaus G. Composite international and interracial graphs. Pediatrics 1968; 41:106–14.

14. Isaacs JS et al. Eating difficulties in girls with Rett syndrome compared with other developmental disabilities. J Amer Diet Assoc. 2003;103(2):224–30.

15. American Academy of Pediatrics clinical practice guideline: diagnosis and evaluation of the child with attention deficit/hyperactivity disorder. Pediatrics. 2000;105:1158–70.

16. American Academy of Pediatrics. The use and misuse of fruit juice in pediatrics. Pediatrics. 2001;107:1210–13.

17. Black RE et al. Children who avoid drinking cow milk have low dietary calcium intakes and poor bone health. Am J Clin Nutr. 2002;76:675–80.

18. Bott L. Contamination of gastrostomy feeding systems in children in a home-based enteral nutrition program. J Pediatr Gastroentol. 2001;33:266–70.

19. Orelove FP, Hollahan DJ, Myles KT. Maltreatment of children with disabilities: training needs for a collaborative response. Child Abuse & Neglect. 2000;24:185–94.

20. Berkovitch M et al. Copper and zinc blood levels among children with nonorganic failure to thrive. J Clin Nutr. 2003; Apr; 22(2):183–86.

21. Jennings JSR, Howdle PD. New developments in celiac disease. Curr Opin Gastroenterol. 2003;19:118–29.

22. Cunningham E, Marcason W. Is there any research to support a gluten- and caseinfree

diet for a child that is diagnosed with autism? J Amer Diet Assoc. 2001;101:222.

23. Minns RA. Neurological disorders. In: Kelnar CJH et al., eds. Growth disorders. Chapman and Hall; 1998:447–70.

24. Krick J et al. Pattern of growth in children with cerebral palsy. J Amer Diet Assoc. 1996;96:680–85.

25. Kleinman RE, ed. Pediatric nutrition handbook, 5th ed. Chicago IL: American Academy of Pediatrics; 2004:757–72.

26. Centers for Disease Control and Prevention. Asthma prevalence and control charactistics by race/ethnicity—United States, 2002. MMWR. 2004;53:145–48.

27. Falci KJ, Gombas KL, Elliot EL. Food allergen awareness: an FDA priority. Food Safety Magazine. 2001:Feb–March.

28. American Academy of Pediatrics Policy Statement: Quality early education and child care from birth to kindergarten. Pediatrics. 2005;114:187–91.

29. Bishop JM, Hill DJ, Hosking CS. Natural history of cow milk allergy: clinical outcome. J Pediatrics. 1990;116:862–67.

30. Position statement on vitamin-related therapies. National Down Syndrome Society; 1997. www.ndss.org/content.cfm, accessed 10/06.

31. Trissler RJ. Folic acid and Down syndrome. J Amer Diet Assoc. 2000;159.

32. Statement on nutritional supplements and piracetam for children with Down syndrome. American College of Medical Genetics; 1996. www.acmg.net/resources/policies/pol-006.asp, accessed 10/06.

33. McGuire J, Kulkarni M, Baden H. Fatal hypermagnesemia in a child treated with megavitamin/megamineral therapy. J Pediatr. 2000;105:318.

34. Maternal and Child Health Bureau, Health and Human Services. Available at http://mchb.hrsa.gov/html/drte.html, accessed 10/06.

> "Men are but children
> of a larger growth"
> John Dryden

Chapter 12
Child and Preadolescent Nutrition

Chapter Outline

- Introduction
- Tracking Child and Preadolescent Health
- Normal Growth and Development
- Physiological and Cognitive Development of School-Age Children
- Energy and Nutrient Needs of School-Age Children
- Common Nutrition Problems
- Prevention of Nutrition-Related Disorders in School-Age Children
- Dietary Recommendations
- Physical Activity Recommendations
- Nutrition Intervention for Risk Reduction
- Public Food and Nutrition Programs

Ann Hoffman

Photo Disc

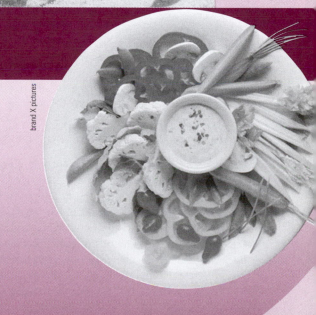

brand X pictures

Prepared by **Nancy H. Wooldridge**

Key Nutrition Concepts

1 Children continue to grow and develop physically, cognitively, and emotionally during the middle childhood and preadolescent years in preparation for the physical and emotional changes of adolescence.

2 Children continue to develop eating and physical activity behaviors that affect their current and future states of health.

3 Although children's families continue to exert the most influence over their eating and physical activity habits, external influences, such as teachers, coaches, peers, and the media, begin to have more impact on children's health habits.

4 With increasing independence, children begin to eat more meals and snacks away from home and need to be equipped to make good food choices.

Introduction

This chapter focuses on the growth and development of school-age and preadolescent children and their relationships to nutritional status. Children continue to grow physically at a steady rate during this period, but development from a cognitive, emotional, and social standpoint is tremendous. This period in a child's life is preparation for the physical and emotional demands of the adolescent growth spurt. Having family members, teachers, and others in their lives who model healthy eating and physical activity behaviors will better equip children for making good choices during adolescence and later in life.

Definitions of the Life Cycle Stage

Middle childhood is a term that generally describes children between the ages of 5 and 10 years. This stage of growth and development is also referred to as school-age, and the two terms are used interchangeably in this chapter. *Preadolescence* is generally defined as ages 9 to 11 years for girls and ages 10 to 12 years for boys. School-age is also used to describe preadolescence.

Middle Childhood   Children between the ages of 5 and 10 years; also referred to as school-age.

Preadolescence   The stage of development immediately preceding adolescence; 9 to 11 years of age for girls and 10 to 12 years of age for boys.

Importance of Nutrition

Adequate nutrition continues to play an important role during the school-age years in assuring that children reach their full potential for growth, development, and health. Nutrition problems can still occur during this age, such as iron-deficiency anemia, undernutrition, and dental caries. Regarding weight, both ends of the spectrum are seen during this age. The prevalence of obesity is increasing, but the beginnings of eating disorders can also be detected in some school-age and preadolescent children. Therefore, adequate nutrition and the establishment of healthy eating behaviors can help to prevent immediate health problems as well as promote a healthy lifestyle, which in turn may reduce the risk of the child developing a chronic condition, such as obesity, type 2 diabetes, and/or cardiovascular disease later in life.[1] Adequate nutrition, especially eating breakfast, has been associated with improved academic performance in school and reduced tardiness and absences.[2] Meeting energy and nutrient needs, addressing common nutrition problems, and preventing nutrition-related disorders while establishing healthy eating and physical activity habits will be discussed later in this chapter.

Tracking Child and Preadolescent Health

The statistics regarding the environments in which many children are growing up are alarming. In 2004, approximately 8% of U.S. children lived in extreme poverty, where the household income was below 50% of the poverty level, and 40% of children lived in low-income families, where the household income was below 200% of the poverty level.[3] Eleven percent of children did not have health insurance in 2003. Additional statistics include:

- 33% of children had no parent in the household who had a full-time, year-round job in 2004.
- 31% of families with children were headed by a single parent in 2004.
- 35% of children lived in female-headed families receiving child support in 2003.

The environment in which a child lives affects the child's health and education. In 2005, 38% of fouth grade students scored below basic reading level.[3] Lack of transportation is a significant limitation for many families. In the discussions that follow regarding nutrition during childhood, the recommendations must always be considered in the context of the individual child's environment.

Disparities in nutrition status indicators exist among the races. For example:

- The prevalence of overweight in Mexican American male children is significantly greater than in non-Hispanic white male children.[4]
- The prevalence of overweight in Mexican American and non-Hispanic black female children is significantly greater than in non-Hispanic white female children.[4]
- Minorities have higher percentages of total calories from dietary fat.[5]

Healthy People 2010

A number of objectives in the Healthy People 2010 document are specific to children's health and well-being. Table 12.1 lists the specific Healthy People 2010 objectives

Table 12.1 Healthy People 2010 Objectives related to school-age children[5]

	Healthy People 2010 Objective	Baseline %	Target %
19.3:	Reduce the proportion of children and adolescents who are overweight or obese	11.0	5.0
19.5:	Increase the proportion of persons aged 2 years and older who consume at least two daily servings of fruit	28.0	7.0
19.6:	Increase the proportion of persons aged 2 years and older who consume at least three daily servings of vegetables, with at least one-third being dark green or deep yellow vegetables	3.0	50.0
19.7:	Increase the proportion of persons aged 2 years and older who consume at least six daily servings of grain products, with at least three being whole grains	7.0	50.0
19.8:	Increase the proportion of persons aged 2 years and older who consume less than 10% of calories from saturated fat	36.0	75.0
19.9:	Increase the proportion of persons aged 2 years and older who consume no more than 30% of calories from fat	33.0	75.0
19.10:	Increase the proportion of persons aged 2 years and older who consume 2400 mg or less of sodium daily	21.0	65.0
19.11:	Increase the proportion of persons aged 2 years and older who meet dietary recommendations for calcium	46.0	75.0
19.15:	Increase the proportion of children and adolescents aged 6 to 19 years whose intake of meals and snacks at schools contributes proportionally to good overall dietary quality (developmental objective)		
22.8:	Increase the proportion of the nation's public and private schools that require daily physical education for all students (middle and junior high schools).	6.4	9.4
22.14:	Increase the proportion of trips made by walking	31.0	50.0
22.15:	Increase the proportion of trips made by bicycling	2.4	5.0

that are pertinent to a discussion of middle childhood and preadolescence.

Normal Growth and Development

During the school-age years, the child's growth is steady, but the growth velocity is not as great as it was during infancy or as great as it will be during adolescence. The average annual growth during the school years is 7 pounds (3–3.5 kg) in weight and 2.5 inches (6 cm) in height.[6] Children of this age continue to have spurts of growth that usually coincide with periods of increased appetite and intake. During periods of slower growth, the child's appetite and intake will decrease. Parents should not be overly concerned with this variability in appetite and intake in their school-age children.

Periodic monitoring of growth continues to be important in order to identify any deviations in the child's growth pattern. Children should continue to be weighed on calibrated scales without shoes and in lightweight clothing. The child's stature or standing height should be measured without shoes and utilizing a height board (see Illustration 10.3 in Chapter 10). A height board consists of a nonstretchable tape on a flat surface like a wall with a moveable right-angle head board. The child's heels should be up against the wall or flat surface; and the child should be instructed to stand tall, looking straight ahead with arms by the side, during the measurement. Both weight and height should be plotted on the appropriate 2000 CDC growth charts, discussed next.

The 2000 CDC Growth Charts

The "CDC Growth Charts: United States," found in Appendix A, are excellent tools for monitoring the growth of a child.[7] The growth charts, which are pertinent to the school-age child, are weight-for-age, stature-for-age, and body mass index (BMI)-for-age for boys and girls. The growth charts are based on data from cycles 2 and 3 of the National Health and Examination Survey (NHES) and the National Health and Nutrition Examination Surveys (NHANES) I, II, and III. However, weight data for children greater than 6 years of age who participated in NHANES III, were not included because there was a known higher prevalence of overweight for these ages. Incorporating this data into the growth charts would reflect an unhealthy standard.[8] Gender-specific BMI-for-age greater than or equal to the 95th percentile defines overweight, and BMI-for-age values greater than or equal to the 85th but less than the 95th percentiles identify children at risk for becoming overweight.

Illustrations 12.1 and 12.2 depict the growth of a healthy child. A chart for weight-for-stature up to a height of 122 cm (or 48 in) is also available for the younger school-age child. As with the toddler and preschooler, it is the child's pattern of growth over time that is important rather than any single measurement. The tracking of BMI-for-age is an important screening tool for overweight as well as undernutrition. Making sure to use the correct age of the child when plotting on the growth charts and using the most current growth curves will help to avoid errors.

Physiological and Cognitive Development of School-Age Children

Physiological Development

During middle childhood, muscular strength, motor coordination, and stamina increase progressively.[6] Children are able to perform more complex pattern movements, therefore affording them opportunities to participate in activities such as dance, sports, gymnastics, and other physical activities.

During the early childhood years, percent body fat reaches a minimum of 16% in females and 13% in males. Percent body fat then increases in preparation for the adolescent growth spurt. This increase in percent body fat, which usually occurs on average at 6.0–6.3 years of age, is called *adiposity rebound* or *BMI rebound* and is reflected in the BMI-for-age growth charts.[9] The increase in percent body fat with puberty is earlier and greater in females than in males (19% for females versus 14% for males). During middle childhood, boys have more lean body mass per centimeter of height than girls. These differences in body composition become more pronounced during adolescence.[1] It is important to understand that BMI is not constant throughout childhood. Plotting BMI-for-age on the growth charts is the only way to know if a child's BMI is outside the normal range for his age. The goal for children is not to strive for a certain range of BMI values as it is in adults, but rather to have a BMI for age percentile within the normal range.

With the increase in body fat, preadolescents, especially girls, may be concerned that they are becoming overweight. Parents need to be aware that an increase in body fat during this stage is part of normal growth and development. Parents need to be able to reassure their child that these changes are most likely not permanent; parents also need to be careful not to reinforce a preoccupation with weight

Illustration 12.1 2 to 20 years: Girls stature-for-age and weight-for-age percentiles.[7]

and size. Boys may become concerned about developing muscle mass and need to understand that they will not be able to increase their muscle mass until middle adolescence (see Chapter 14).[1]

Cognitive Development

The major developmental achievement during middle childhood is self-efficacy, the knowledge of what to do and the ability to do it. During the school-age years, children move from a preoperational period of development to one of "concrete operations."[6] This stage is characterized by being able to focus on several aspects of a situation at the same time; being able to have more rational cause/effect reasoning; being able to classify, reclassify, and generalize; and a decrease in egocentrism, which allows the child the ability to see another's point of view. Schoolwork becomes increasingly complex as the child gets older. School-age children also enjoy playing strategy games, displaying growing cognitive and language development.

During this stage, the child is developing a sense of self. Children become increasingly independent and are learning their roles in the family, at school, and in the community.[1] Peer relationships become increasingly important, and children begin to separate from their own families by spending the night at a friend's or relative's house. More and more time is spent watching television and playing video games. Older children may be able to walk or ride a bicycle to a neighborhood store and purchase snack items. Thus, influences outside the home environment play an increasing role in all aspects of the child's life.

Development of Feeding Skills

With increased motor coordination, school-age children develop increased feeding skills. During childhood, the child masters the use of eating utensils, can be involved in simple food preparation, and can be assigned chores related to mealtime such as setting the table. By performing these tasks, the child learns to contribute to the family, which boosts developing self-esteem. Complexities of the tasks can be increased as the child grows older. At the same time, the child can be learning about different food, simple food preparation, and some basic nutrition facts.

Eating Behaviors Parents and older siblings continue to have the most influence on a child's attitudes toward food and food choices during middle childhood and preadolescence. The eating behaviors and cultural food practices and preferences of parents will influence the child's food likes and dislikes. The feeding relationship between parent and child, as described in Chapter 10, still applies to the school-age child. Parents are responsible for the food environment in the home,

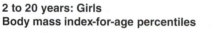

2 to 20 years: Girls
Body mass index-for-age percentiles

SOURCE: Developed by the National Center for Health Statistics in collaboration with the National Center for Chronic Disease Prevention and Health Promotion (2000). http://www.cdc.gov/growthcharts

Illustration 12.2 2 to 20 years: Girls body mass index-for-age percentiles.[7]

what foods are available, and when they are served. The child is responsible for how much she eats.[10] Parents need to continue to be positive role models for their children in terms of healthy eating behaviors. They also need to provide the necessary guidance so that the child will be able to make healthy food choices when away from home.

Families should try to eat meals together. When children are involved in school-related activities, eating together is often difficult for families to achieve because of the family members' hectic schedules. But eating together as a family should be encouraged as a goal, allowing time for conversation (see Illustration 12.3). Excessive reprimanding and arguments should be avoided during mealtime.

One study of 9- to 14-year-old children of participants of the Nurses' Health Study II found a positive relationship between families eating dinner together and the overall quality of the children's diets.[11] Children who ate dinner with their families had higher energy intakes as well as higher intakes of nutrients such as fiber, calcium, folate, iron, and vitamins B_6, B_{12}, C, and E. These children also reported eating more fruits and vegetables, eating less fried foods when away from home, and drinking fewer soft drinks. The percentage of children who reported eating family dinner decreased with the age of the child. So, a higher percentage of 9-year-olds than 14-year-olds ate dinner with the family, indicating that family dinner becomes more of a challenge as children get older.

School-age children spend more and more time away from home, which is an important part of normal growth and development. Peer influence becomes greater as the child's world expands beyond the family. The increased peer influence extends to attitudes toward foods and food choices. Children may suddenly request a new food or refuse a previous favorite food, based on recommendations from a peer.

Teachers and coaches have an increasing influence on the child's attitudes toward food and eating behaviors. Nutrition should be part of the health curriculum, and what is learned in the classroom should be reinforced by foods available in the school cafeteria. Vending machines present in school as a source of extra funding can also reinforce good nutrition with appropriate choices or be a source of high-fat, high-sugar foods and beverages.

In their expanding world, children come under the influence of the media. Children want to try foods they see advertised on television. One study that analyzed the commercials aired during Saturday morning television programming found that 56.5% of all advertisements were for food.[12] Of these, 43.6% were classified in the fats, oils, and sweets food groups, which is not in line with the recommendations of the U.S. Department of Agriculture's MyPyramid (see Chapter 1).[12,13] Fast food establishments, with their playgrounds and give-aways, are also attractive to children. With more and more children having access to the Internet, food companies are finding new ways to market their products to children. The new forms of marketing include "advergames," online games that feature the company's product or brand character; "viral marketing" in which children are encouraged to send e-mails to their friends about a product; television advertising online, which blurs the lines between advertising and entertainment; and "advercation" a combination of advertising and education. The impact of online marketing on children's food choices needs further study.[14]

Snacks continue to contribute significantly to a child's daily intake. During middle childhood, children cannot consume large amounts of food at one time and therefore

Illustration 12.3 A family enjoying mealtime together.

need snacks to meet their nutrient needs. Many children prepare their own breakfasts or after-school snacks. These children need to have a variety of foods available to them, be equipped with nutrition education for making their own food choices, and have some age-appropriate knowledge and skill in food preparation—assuming, of course, that the family has adequate access to food.

Body Image/Excessive Dieting   Food preference development, appetite, and satiety in young children were thoroughly discussed in Chapter 10. Researchers have described the innate ability of young children to internally control their energy intake and their responsiveness to energy density. The internal controls can be altered by external factors such as child-feeding practices. Studies in 9- to 10-year-old children found that these older children were not as responsive to energy density as young preschool-age children.[15] External factors such as the time of day, the presence of other people, and the availability of good food begin to override the internal controls of hunger and satiety as children get older.

Birch and associates, who have performed extensive research in the area of the development of food preferences and appetite control in children, have also examined the relationships among children's adiposity, child-feeding practices, and children's responsiveness to energy density.[16] These researchers found that children of parents who imposed authoritarian controls on their children's eating were less likely to be responsive to energy density. In other words, these children were not able to listen to internal cues in energy regulation. In girls, regulation of energy intake was inversely related to their adiposity. Heavier girls were less likely to be able to regulate their intake based on internal cues. Parents who had difficulty controlling their own intakes seemed to impose more restrictions on their children. A study of mothers and their 5-year-old daughters found that this transfer of "restrictive" eating practices may begin as early as the preschool age.[17] The more the mother is concerned with her own weight and with the risk of her daughter becoming overweight, the more likely she is to employ restrictive child-feeding practices. These researchers hypothesize that the chronic dieting and dietary restraints, which are commonly seen in adolescent girls and young women, may have their beginnings in the early regulation of energy intake and may be related to the amount of parental control exerted over the child's eating.[16]

Young girls seem to have a preoccupation with weight and size at an early age. With the normal increase in body mass index or body fatness in preadolescence, many girls and their mothers may interpret this phenomenon of normal growth and development as a sign that the child is developing a weight problem. By imposing controls and restrictions over their daughters' intakes, mothers may actually be promoting the intake of the forbidden or restricted foods.[18] Similar results were found in 5-year-old girls whose parents restricted palatable snack foods.[19] Not only did parental

restriction promote the consumption of these forbidden foods by the young girls, but these children reported feeling badly about eating these foods. Early "dieting" may actually be a risk factor for the development of obesity.[20] Dieting, which imposes restrictions, is similar to controlling child-feeding practices, which restrict children's intake. Both methods ignore internal cues of hunger and satiety. Not only do these types of child-feeding practices contribute to the onset of obesity and possibly a nutritionally inferior diet, but they may also be contributing to the beginnings of eating disorders. Eating disorders are discussed in more detail in Chapter 14.

Many studies of ethnic differences in body image and body size preferences have been conducted, hypothesizing that there are ethnic and gender differences in these parameters. The research conducted to date is not conclusive. A recent study of men and women of four ethnicities/race—black, Hispanic, Asian, and white—was conducted to examine body image and body size assessments, while controlling for age, body weights, and education level as a measure of socioeconomic status.[21] This study found that ethnicity alone did not influence the preference for body shapes or tolerance for obesity. In working with individual families, however, it is important to try to assess their health beliefs and their preference for body size, which may impact their readiness for nutrition counseling.

Energy and Nutrient Needs of School-Age Children

Dietary Reference Intakes (DRIs) have been developed (1997–2002) . DRI tables are provided on the inside front cover of this book. Children need a variety of foods that provide enough energy, protein, carbohydrate, fat, vitamins, and minerals for optimal growth and development.[1]

Energy Needs

Energy needs of school-age children reflect the slow but steady growth rate during this stage of development. Energy needs of an individual child are dependent on the child's activity level and body size. Equations for estimating energy requirements have been developed as part of the Dietary Reference Intakes, based on a child's gender, age, height, weight, and physical activity level (PAL).[22] Estimated energy expenditure (EER) has been defined as total energy expenditure plus kilocalories for energy deposition. Categories of activity are defined in terms of walking equivalence. For example, an 8-year-old girl who weighs 56.4 pounds (25.6 kg) and is 50.4 inches (128 cm) tall will require 1360 kilocalories per day if sedentary, 1593 kcal/day if she is low active, 1810 kcal/day if active, and 2173 kcal/day if very active. Energy allowances based on body weight are lower for school-age children than for toddlers and preschoolers. The decrease in

Table 12.2 Dietary Reference Intakes for key nutrients for school-age children[25,26]

	Recommended Daily Allowances		Adequate Intake
Age	Iron (mg/d)	Zinc (mg/d)	Calcium (mg/d)
4–8 years	10	5	800
9–13 years	8	8	1300

the energy requirement per kilogram of body weight is a reflection of slowing growth rate.

Protein

Based on the DRIs, the recommended protein intake for school-age children is 0.95 gram of protein/kg body weight/day for 4- to 13-year old girls and boys.[22] School-age children can meet this recommendation by consuming diets that follow the MyPyramid recommendations.[13] Vegetarian diets are also appropriate for school-age children if they provide sufficient energy, complementary protein foods, a variety of foods, and adequate levels of intake of vitamins and minerals.[1] By meeting an individual child's energy needs, protein is spared for tissue repair and growth.

Vitamins and Minerals

Dietary Reference Intakes (DRIs) for vitamins and minerals have been established for the school-age and preadolescent child. Analysis of data from NHANES I, II, and III and the Continuing Survey of Food Intake by Individuals (CSFII) indicates that children's mean intakes of most nutrients meet or exceed the recommendations. Still, certain subsets of children do not meet their needs for key nutrients such as iron and zinc, which are important for growth; and calcium, needed to achieve peak bone mass.[23,24] According to NHANES III data, calcium intakes are declining in 6- to 11-year-old children.[5] Dietary Reference Intakes for these key nutrients are listed in Table 12.2.[25,26]

Common Nutrition Problems

During the last century, common nutrition problems have shifted from problems of nutrient deficiencies to problems of excess nutrition, such as energy, fat, and salt. During middle chidhood, some children still experience problems such as iron deficiency anemia and dental caries, especially with easy accessibility to high-sugar foods. These nutrition problems are addressed here, followed by a thorough discussion of prevention of nutrition-related disorders.

Iron Deficiency

Iron deficiency is not as common a problem in middle childhood as it is in the toddler age group. According to the NHANES 1999–2000 survey data, 4% of 6- to 11-year-old children were found to be iron deficient, compared to 7% of toddlers.[27] Although the prevalence of iron-deficiency anemia is decreasing, these rates are still above the 2010 national health objectives.

Age- and gender-specific cutoff values for anemia are based on the 5th percentile of hemoglobin and hematocrit for age from NHANES III. For children 5 to under 8 years of age, the diagnosis of anemia is made if the hemoglobin concentration is <11.5 g/dL and hematocrit <34.5%. For children from 8 to under 12 years of age, a hemoglobin value <11.9 g/dL or hematocrit <35.4% is diagnostic of iron-deficiency anemia.[27,28]

Dental Caries

Approximately one in two children age 6 to 8 years have decay in their primary or permanent teeth.[5] The amount of time that children's teeth are exposed to carbohydrates influences the risk of dental caries or tooth decay. (See the explanation of the carogenic process in Chapter 10.) Complex carbohydrates such as fruits, vegetables, and grains are better choices than simple sugars, such as soft drinks and candy, in relation to oral health and nutrition. Sticky carbohydrate-containing foods, such as raisins and gummy candy, are strong caries promoters. Fats and proteins may have a protective effect on enamel. So choosing snacks that are combinations of carbohydrates, proteins, and fats may decrease the risk of developing dental caries. Having regular meal and snack times versus continually snacking throughout the day is also beneficial. Rinsing the mouth after eating—or better yet, brushing the teeth regularly—also decreases the development of cavities.[29] It is important that the school-age child continue to have a source of fluoride, either from the water supply or through supplementation. Details about fluoride supplementation were reported in Chapter 10.

During middle childhood, children lose their primary or baby teeth and begin to get their permanent teeth. If several teeth are missing, children may experience difficulty in chewing some foods such as meat. Also, the orthodontic appliances commonly worn by school-age children may interfere with the child's ability to eat certain foods. Modifying food, such as by chopping meat or slicing fresh fruit, can help.[1]

Prevention of Nutrition-Related Disorders in School-Age Children

The prevalence of overweight among children is increasing at an alarming rate. Increases in prevalence of overweight are present in the adult population and in populations of other countries, indicating that social and environmental factors may be having an impact. Despite the increase in the prevalence of overweight, analysis of dietary data from NHANES I, II, and III indicates no corresponding increase in energy intake among children over the years. This finding suggests that physical inactivity may be a significant contributing factor to the increased prevalence of overweight.[30] The problem of increasing overweight in the United States needs to be addressed from a public health perspective.[8] Furthermore, children who are overweight are at increased risk for developing risk factors for chronic conditions, such as cardiovascular disease and type 2 diabetes mellitus.[31]

Overweight and Obesity in School-Age Children

Prevalence According to the NHANES 2003–2004 data, aproximately 18.8% of children ages 6 through 11 years are overweight, with BMIs-for-age greater than or equal to the 95th percentile; an additional 37.2% are considered to be at risk for overweight, with BMIs greater than or equal to the 85th but less than the 95th percentiles.[4] These percentages have increased significantly since the NHANES III study (1988–1994), when 11% of children ages 6 through 11 years were overweight, and 14% were considered to be at risk for overweight. According to the 2003–2004 data, the proportion of children ages 6 through 11 years who are at risk for becoming overweight ranges from about 34.5% for non-Hispanic black males to approximately 47.9% for Mexican American males. Using data from 1999–2000, 2001–2002, and 2003–2004, the prevalence of overweight in Mexican American male children was significantly greater than in non-Hispanic white male children. Prevalence of overweight among non-Hispanic white male children did not differ significantly from non-Hispanic black male children. Mexican American and non-Hispanic black female children were significantly more likely to be overweight compared with non-Hispanic white female children.[4]

The prevalence of overweight among children has increased over time. In fact, an intrasurvey increase of about 2 to 6 percentage points occurred for most of the gender, age, and racial-ethnic groups during the six years of the NHANES III survey.[8] As mentioned in the description of the growth charts, BMI data for children older than 6 years of age were not included in the revised growth charts because of the known increased prevalence of overweight for these ages in NHANES III. Inclusion of this data in the revised growth charts would reflect a heavier population and would not be a healthy standard. Further analysis shows that the heaviest, or obese, children are getting heavier.

Characteristics of Overweight Children Overweight children are usually taller, have advanced *bone ages*, and experience sexual maturity at an earlier age than

their nonoverweight peers. From a psychosocial standpoint, overweight children look older than they are, and

> **Bone Age** Bone maturation; correlates well with stage of pubertal development.

often adults expect them to behave as if they were older. Health consequences of obesity, such as hyperlipidemia, higher concentrations of liver enzymes, hypertension, and abnormal glucose tolerance, occur with increased frequency in obese children than in children of normal weight.[32] Analysis of data from the Bogalusa Heart Study, a community-based study of adverse risk factors in early life in a biracial population, confirms an increase in chronic disease risk factors with increasing BMI-for-age.[31] Increasing insulin levels show the strongest association with increasing BMI-for-age. Additionally, overweight children are more likely to have more than one chronic disease risk factor.

Type 2 diabetes mellitus, typically considered to be a disease of adults, is increasing in children and adolescents in the United States today, with up to 85% of affected children being either overweight or obese at diagnosis.[33] According to the recommendations of a panel of experts in diabetes in children, any child who is overweight, which is defined by this group as having a BMI above the 85th percentile, and who has other risk factors, should be monitored for type 2 diabetes beginning at age 10 or at puberty. Other risk factors include a family history of type 2 diabetes, belonging to certain race and ethnic groups, including African American, Hispanic American, Asian and South Pacific Islanders, and Native American, and having signs of insulin resistance.[33]

It is still unclear what effect an early onset of obesity in childhood has on the risk of adult morbidity and mortality.[32] But consequences of obesity and the precursors of adult disease do occur in obese children. More studies have been performed on the relationship between obesity during adolescence and the risks of obesity in adulthood than have been performed on the relationship between childhood obesity and obesity and Type 2 diabetes in adulthood (see Chapter 14).[34]

Predictors of Childhood Obesity Dietz[9] describes critical periods in childhood for the development of obesity: gestation and early infancy, the period of adiposity rebound, and adolescence. "Adiposity rebound" (or rebound in BMI) is the normal increase in body mass index, which occurs after BMI declines and reaches its lowest point, at about 4 to 6 years of age, and is reflected in the BMI-for-age growth chart. Studies suggest that the age at which adiposity rebound occurs may have a significant effect on the amount of body fat that the child will have during adolescence and into adulthood. Early adiposity rebound is defined as beginning before 5.5 years of age, while the average age of adiposity rebound is 6.0–6.3 years. Adiposity rebound after age 7 is considered late. Studies have shown that adolescents and adults who as children had an early adiposity rebound have higher BMI and *subscapular skinfold thicknesses* than those subjects who had an average or late adiposity rebound.

> **Subscapular Skinfold Thickness**
> A skinfold measurement that can be used with other skinfold measurements to estimate percent body fat; the measurement is taken with skinfold calipers just below the inner angle of the scapula or shoulder blade.
>
> **Waist-to-Hip Ratio** The ratio of the waist circumference, measured at its narrowest, and the hip circumference, measured where it is widest. This ratio is an easy way to measure body fat distribution, with a higher ratio indicative of an abdominal fat pattern. A high waist-to-hip ratio is associated with a high risk of chronic disease.
>
> **Resting Energy Expenditure** The amount of energy needed by the body in a state of rest.

Several possible mechanisms may explain the relationship between adiposity rebound and subsequent obesity.[35] The period of adiposity rebound may be when children are beginning to express learned behaviors related to food intake and activity. Early adiposity rebound may be related to infants who were exposed to gestational diabetes during fetal development and consequently have high birth weights. Although more study is needed, the conclusion is that preventive efforts need to focus on these developmental stages.[9]

Another predictor of childhood obesity is the child's home environment. Children from birth to 8 years were followed over a 6-year period as part of the National Longitudinal Survey of Youth.[36] The associations between the home environment and socioeconomic factors and the development of childhood obesity were examined. Maternal obesity was found to be the most significant predictor of childhood obesity, followed by low family income and lower cognitive stimulation.

Parental obesity is associated with an increased risk of obesity in children.[37] In one study, parental obesity doubled the risk of adult obesity for both obese and nonobese children less than 10 years of age. An analysis of data from NHANES III indicated a higher percentage of overweight youth who had one obese parent as compared to those children who had no obese parent. The percentage of overweight youth increased further if both parents were obese.[38] The connection between parental obesity and obesity in children is likely due to genetic as well as environmental factors.[37]

Effects of Television Viewing Time on the Incidence of Overweight

Analysis of data collected during cycles II and III of the National Health Examination Survey (NHES) revealed significant associations between the time spent watching television and the prevalence of obesity in children and adolescents.[39] A dose–response relationship was detected. For each additional hour of television viewed in the 12- to 17-year-old group, the prevalence of obesity increased by 2%.

A strong dose–response relationship between TV viewing time and the prevalence of overweight was found in the National Longitudinal Survey of Labor Market Experience, Youth Cohort (NLSY). This study consists of a nationally representative sample of youths aged 10 to 15 years.[40] The odds of having a BMI above the 85th percentile for age

and gender are significantly greater for those youths who view more than 5 hours of television per day as compared to those who watch 2 or fewer hours of television daily. These odds remain the same when adjustments are made for confounding variables such as previous overweight of the child, maternal overweight, socioeconomic status, household structure, ethnicity, and child aptitude test scores. Approximately 33% of the youth report watching more than 5 hours of television per day, while only 11% watch 2 or fewer hours of daily television, which is a Healthy People 2010 objective.[5,40]

According to NHANES III data, children aged 11 through 13 years have the highest rates of daily television viewing.[41] Children, both males and females, who watch 4 or more hours of television daily have greater body fat and BMI than those who watch less television.[41,42] A school-based intervention program aimed at reducing third- and fourth-grade children's television, videotape, and video game use was shown to be effective in reducing television viewing and meals eaten in front of the television. In addition, decreases in BMI, triceps skinfold, waist circumference, and *waist-to-hip ratio* were seen.[43]

The proposed mechanisms by which television viewing contributes to obesity include reduced energy expenditure by displacing physical activity and increased dietary intake by eating during viewing or as a result of food advertising.[43] Analysis of NHANES III data showed a positive correlation between intake and number of hours of television watched.[42] One study found that energy expenditure during television viewing was actually significantly lower than *resting energy expenditure* in 15 obese children and 16 normal-weight children who ranged in age from 8 to 12 years.[44] Based on these findings, it is hypothesized that television viewing does contribute to the prevalence of obesity, and that treatment for childhood obesity should include a reduction in the number of hours spent watching television and videos and playing video and computer games.

One of the Healthy People 2010 objectives is to increase the proportion of children and adolescents who view television 2 or fewer hours per day from 60% to 75%. In 2005, this proportion was reported to be 63% for students in grades 9 to 12.[45] Related data analyzed by race and ethnicity, gender, and family income level are depicted in Table 12.3.[5,45]

Prevention of Overweight and Obesity

"An ounce of prevention is worth a pound of cure."

Recognizing the increase in the prevalence of childhood overweight and its associated chronic health problems, the American Academy of Pediatrics (AAP) has issued a policy statement on the prevention of pediatric overweight and obesity.[46] The policy statements advocates for (1) early recognition, using BMI-for-age as a screening tool, and

Table 12.3 Percentage of children and adolescents viewing television 2 or fewer hours per day by race/ethnicity, gender, and family income level[5,45]

Children and Adolescents Aged 8 to 16 Years	Television 2 or Fewer Hours Per Day	
	1988–1994	2005
Race and Ethnicity		
Mexican American	53%	54%
Black or African American	42	36
White	65	71
Gender		
Female	64	64
Male	54	62
Family Income Level		
Poor	53	not available
Near poor	54	not available
Middle/high income	64	not available

providing anticipatory and appropriate guidance regarding healthy eating and physical activity; and (2) advocacy for opportunities for physical activity, improvements in foods available to children, research, and third-party reimbursement for treatment of overweight.[46] To further emphasize the role of increased physical activity in the prevention of childhood obesity, the Council on Sports Medicine and Fitness and Council on School Health of the AAP released another policy statement recommending that increased physical activity for children can be encouraged, monitored, and advocated for by pediatric health care providers and public health officials.[47]

Treatment of Overweight and Obesity Recommendations of an expert committee on obesity evaluation and treatment for children are described in Chapter 10 and depicted in Illustration 12.4.[48] The committee recommends that children and adolescents with BMIs greater than or equal to the 95th percentile for age and gender should have an in-depth medical assessment. For children older than 7 years of age, prolonged weight maintenance is an appropriate goal if their BMI is greater than or equal to the 85th but less than the 95th percentile and if they do not have any secondary complications of obesity. However, for children whose BMI is in the range just described and who have a nonacute secondary complication of obesity, such as mild hypertension or hyperlipidemia—or whose BMI is at or above the 95th percentile—weight loss is recommended. Reducing sedentary behaviors and increasing physical activity are important components of obesity treatment for children.[49] Rather than focusing on attaining an ideal body weight, treatment should focus on changing unhealthy lifestyle behaviors and maintaining healthy behaviors.[48] See Case Study 12.1 on the next page.

An evidence-based analysis of intervention literature showed positive effects of multicomponent, family-based programs for children between the ages of 5 and 12 years. Recommended components include: parent training, dietary counseling/nutrition education, physical activity and addressing sedentary behaviors, and behavioral counseling.[50]

Some potential consequences of a weight-loss program in childhood are a slowing of linear growth and the

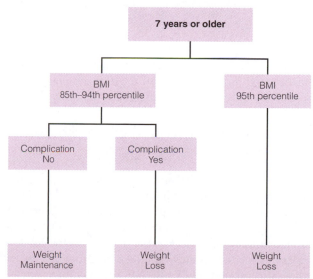

Illustration 12.4 Recommendations for weight goals.

Case Study 12.1

Pediatric Overweight

Photo Disc

Seven-year-old Timothy's mother takes him to his pediatrician for a checkup. His weight of 77 pounds, plotted at the 95th percentile, and his height of 52 inches are between the 75th and 90th percentiles for his age. His body mass index of 20 kg/m² is plotted at just over the 95th percentile for his age. His growth percentiles have been increasing over the last several years.

Timothy's mother expresses concern to the pediatrician about her son's weight. His older and younger brothers are both thinner than Timothy. Timothy's mother is obese, but his father is a normal weight for height. Timothy is in the second grade. He rides the school bus to and from school. He participates in the School Lunch Program at his school, but his parents gave him extra money in case he wants to buy some additional à la carte food items from the cafeteria or items from the vending machines. After school, Timothy and his brothers stay in their home with a babysitter until one of their parents returns home from work. Timothy usually watches TV or plays video games after school. His parents leave snack foods—chips, cookies, and sodas—in the house for their sons to have after school. His mother usually prepares their evening meal, which consists of a meat, starch, vegetables, and a dessert item. After dinner, Timothy does his homework and then usually watches more TV with his parents. He usually has a dish of ice cream before going to bed.

Questions

1. What is your assessment of Timothy's body size based on his weight-for-age, height-for-age, and BMI-for-age percentiles?

2. What suggestions do you have for Timothy's parents about improving his eating habits?

3. What suggestions do you have for Timothy's parents for increasing his physical activity level?

4. Is it significant that Timothy's mother also has a weight problem?

---

beginnings of eating disorders. To reduce the risks associated with weight loss in childhood, the program must ensure nutritional adequacy of the diet, a nonjudgmental approach, and attention to the child's emotional state.[48]

Nutrition and Prevention of Cardiovascular Disease in School-Age Children

In the new DRIs, no recommendations for total grams of fat per day in the diet of children have been made.[22] Studies have shown that as long as enough energy is provided for growth, no effect of fat intake on growth has been found. In addition, the evidence is still insufficient to be able to define the optimal fat intake for promoting growth while also preventing obesity and other chronic diseases. According to the DRIs, the Acceptable Macronutrient Distribution Ranges (AMDRs) for fat is 25 to 35% of energy for children 4 to 18 years of age.[22]

The new DRIs do stress the importance of including sources of linoleic acid (omega-6 fatty acid) and alpha-linolenic acid (omega-3 fatty acid). Adequate intake levels for these essential fatty acids have been determined and can be found in Table 12.4. Sources of linoleic acid include vegetable oils, seeds, nuts, and whole-grain breads

Table 12.4 Adequate intake of linoleic acid and alpha-linolenic acid[22]*

Gender and Age	Linoleic Acid g/day	Alpha-Linolenic Acid g/day
Children 4–8 years	10	0.9
Boys 9–13 years	12	1.2
Girls 9–13 years	10	1.0

*See Chapter 1 for a list of food sources.

and cereals. Fish, as well as flaxseed, soy, and canola oils, are good sources of alpha-linolenic acid. (See Chapter 1 for a complete list of food sources of linoleic acid and alpha-linolenic acid.) A diet that emphasizes fruits, vegetables, low-fat dairy products, whole grain breads and cereals, nuts, seeds, fish, and lean meats is recommended for promoting nutrition and preventing cardiovascular disease in school-age children.[22]

The American Heart Association and the American Academy of Pediatrics have jointly issued guidelines for cardiovascular health promotion in all children and adolescents.[51,52] For children over 2 years of age, these guidelines recommend limiting foods high in saturated fats (<10 percent of total calories per day), cholesterol (<300 mg per day), and trans fatty acids.[51,52]

As discussed in Chapter 10, children with hyperlipidemias require further dietary restrictions to help control LDL cholesterol. For these children, restricting saturated fat to less than 7% of total calories and cholesterol to 200 mg per day is recommended.[51,52] Restricting trans fatty acids to a level as low as possible is also important. Increasing soluble fiber intake, emphasizing weight management and physical activity, and follow-up by a registered dietitian are also treatment recommendations.[51,52]

Dietary Supplements

> "For most healthy children who eat a variety of foods, experts generally agree that dietary supplements beyond a daily multivitamin and mineral supplement are not necessary, let alone safe or effective."[53]

Children who are healthy and consume a diet of a variety of foods do not require a vitamin and mineral supplement to meet their nutrient needs. The American Academy of Pediatrics recommends vitamin and mineral supplementation for children who are at high risk of developing, or have one or more, nutrient deficiencies.[54] (See Chapter 10 for a list of children at risk for nutrient deficiency.)

If vitamin and mineral supplements are given to school-age children, the supplement should not exceed the Dietary Reference Intakes for age. Parents should be warned against giving amounts of vitamins and minerals that exceed the Tolerable Upper Intake levels designated in the DRI tables.

It is not clear to what extent herbal supplements are given to school-age children. Herbal supplements are used in some cultures as home remedies. It is important to obtain this information from parents and caretakers as part of the child's health history. The use of herbal supplements, botanicals, and vitamin/mineral supplements may be a more prevalent practice by parents of children with special health care needs (see Chapters 11 and 13).

Dietary Recommendations

The basic dietary recommendation for school-age and preadolescent children is to eat a diet of a variety of foods, which is why it remains so important throughout these school years for children to have a variety of foods available to them. The available food environment will affect children's food choices. Parents and other adult role models need to continue to model appropriate eating behaviors for children.

Dietary recommendations, as outlined by the USDA in the Dietary Guidelines for Americans and the MyPyramid, apply to school-age children as well as to other segments of the population.[13,55] Professional organizations, such as the American Dietetic Association, have also published positions on dietary guidance for healthy children, supporting the federal guidelines.[56]

Recommendations for Intake of Iron, Fiber, Fat, and Calcium

Adequate iron nutrition is still important during middle childhood and preadolescence to prevent iron deficiency anemia and its consequences. According to food consumption surveys, children are not eating the recommended amounts of fiber in their diets. Children are exceeding the recommendations of total calories from fat and saturated fat. Calcium requirements increase during the preadolescent years, but calcium intake decreases with age.

Iron   Although iron deficiency is not as prevalent during the school-age years as it was during the toddler and preschool-age years, adequate intake of iron is still important. The inclusion of iron-rich foods—such as meats, fortified breakfast cereals, and dry beans and peas—in children's diets is important. A good vitamin C source, such as orange juice, will enhance the absorption of iron. (See Chapter 1 for a more complete list of high-iron foods.)

Fiber   As reported in Chapter 10, many health effects of fiber intake have been identified—including prevention of chronic disease in adulthood, such as heart disease, certain cancers, diabetes, and hypertension. The new recommendations for *total fiber* intake based on the DRIs can be found in Table 12.5. Total fiber is the sum of *dietary fiber* and *functional fiber.* Earlier recommendations were based on dietary fiber.[22]

To increase the dietary fiber in children's diets, parents and caretakers can begin by increasing the amount of fresh fruits and vegetables

Total Fiber   Sum of dietary fiber and functional fiber.

Dietary Fiber   Complex carbohydrates and *lignins* naturally occurring and found mainly in the plant cell wall. Dietary fiber cannot be broken down by human digestive enzymes.

Lignin   Noncarbohydrate polymer that contributes to dietary fiber.

Functional Fiber   Nondigestible carbohydrates including plant, animal, or commercially produced sources, that have beneficial effects in humans.

Table 12.5 Adequate intake of total fiber[22]

Gender and Age	Total fiber, g/day
Children 4–8 years	25
Boys 9–13 years	31
Girls 9–13 years	26

and whole grain breads and cereals being offered. High-fiber fruits, such as apples with peels, have about 3 grams per serving, while fruit juices are low in fiber. High-fiber vegetables, such as broccoli, have about 2.5 grams per serving. Whole grain breads, cereals, and brown rice have about 2.5 grams per serving. High-fiber cereals, such as bran flakes and raisin bran, have about 8 to 10 grams per serving. Served alone, these high-fiber cereals may not be well accepted by young children; but they can be mixed with other cereals or used in recipes for food items such as muffins. Dried beans and peas are also excellent sources of fiber, providing 4 to 7 grams of fiber per 1/2 cup serving.[57]

Fat   Food intakes that follow the recommendations of the Dietary Guidelines for Americans and the MyPyramid provide an appropriate amount of fat for school-age and preadolescent children.[13,55] Healthy diets include whole grain breads and cereals, beans and peas, fruits and vegetables, low-fat dairy products, and lean meats, fish, and poultry. Foods high in fat, especially those high in saturated fat and trans fatty acids, should be kept to a minimum. However, an appropriate amount of dietary fat is necessary to meet children's needs for calories, essential fatty acids, and fat-soluble vitamins.

Calcium   The recommendations for adequate daily intakes of calcium are 800 milligrams for children aged 4 to 8 years and 1300 milligrams for children 9 through 18 years.[25] The higher recommendation for older children reflects the fact that most bone formation occurs during puberty. Adequate calcium intake during this time is necessary to achieve peak bone formation, which may prevent osteoporosis later in life.[58]

Good sources of calcium are listed in Chapter 1. It is difficult to meet the higher recommendations of calcium without the inclusion of dairy products, preferably low-fat dairy products. For those individuals who are lactose intolerant, lactose-reduced dairy products are available. Calcium-fortified foods such as fruit juice and soy milk are also available for children such as those on a vegan diet. For children whose calcium intake is inadequate, calcium supplements need to be given under the guidance of a physician or registered dietitian.

Fluids   It is of particular importance for school-age children to drink enough fluids to prevent dehydration during periods of exercise and during participation in sports, because children are at risk for dehydration and heat-related stress. Preadolescent children need to be more careful about staying hydrated than do adults and adolescents, for several reasons.[59] Children sweat less, and they get hotter during exercise. Some sports, such as football and hockey, require special protective gear that may prevent the body from being able to cool off. Children should never deprive themselves of food or water in order to meet a certain weight category, such as in wrestling.

Adults who are supervising children's physical activities need to make sure that children drink fluids before, during, and after exercise. The thirst mechanism may not work as well during exercise, and children may not realize that they need fluids. Cold water is the best fluid for children. However, children may be more likely to drink more fluids if they are flavored. Sports drinks, which contain 4 to 8% carbohydrate and diluted fruit juice, are appropriate for children. Children should not be given soft drinks or undiluted juice, because the high carbohydrate load is too high to be hydrating and could cause stomach cramps, nausea, and diarrhea.[59]

Soft Drinks   Approximately 32% of school-age children consume up to 8.9 ounces of soft drinks per day, 32% consume greater than or equal to 9 ounces per day; only 36% are nonconsumers of soft drinks.[60] School-age children consume more soft drinks than preschool-age children do, but not as much as adolescents—indicating an increase in consumption with age. Energy intake increases with increased consumption of nondiet soft drinks. Children with high consumption of regular soft drinks (more than 9 ounces per day) consume less milk and fruit juice than do those with lower consumptions of regular soft drinks. According to analysis of NHANES III data, overweight children have a higher proportion of their energy intake from soft drinks than non-overweight children do.[30] A study of 548 ethnically diverse school-age children, average age 11.7 years, showed that BMI and the frequency of high BMI greater than 95th percentile increased along with increased consumption of sugar-sweetened beverages.[61] These investigators controlled for anthropometric, demographic, dietary, and lifestyle variables. Soft drinks can contribute significantly to children's overall calorie intake, while contributing little to the overall nutritional value of their diets and displacing more nutritious foods. A study of 30 children 6 to 13 years of age found that sweetened drink consumption displaced milk from children's diets and resulted in intakes that were lower in protein, calcium, magnesium, phosporus, and vitamin A but higher in calories.[62] According to USDA food consumption data, children who ate fast food consumed more carbonated non-diet beverages and less milk than children who did not eat fast food.[63] Soft drinks can also contribute to a child's caffeine intake, providing 35–50 mg caffeine per 12-ounce serving.[64] Diet soft drinks do not provide sugar, and the aspartame content of diet sodas do not appear to pose a risk to healthy children.[65] Soft drinks in

Table 12.6 Mean percentages of food energy from carbohydrate, protein, total fat, saturated fatty acids, and cholesterol intake of 6- to 11-year-old children[66]

Gender and Age	Carbohydrate (%)	Protein (%)	Total Fat (%)	Saturated Fatty Acids (%)	Cholesterol (mg/d)
Males:					
6–9 years	55	14	33	12	225
6–11 years	55	14	33	12	232
Females:					
6–9 years	55	14	32	12	190
6–11 years	55	14	33	12	199

excess are not recommended for school-aged children because they provide empty calories and promote tooth decay.

Recommended versus Actual Food Intake

The composition of children's diets, based on data from the Food and Nutrient Intakes by Children report, can be found in Table 12.6.[66] This table depicts the mean percentages of total energy from carbohydrate, protein, total fat, and saturated fatty acids, and cholesterol intake for 6- to 9-year-old and 6- to 11-year-old males and females. Fat and saturated fat intakes have decreased slightly since the NHANES III survey, which found a total fat intake of 33.8% and a saturated fat intake of 12.6% of total calories among 6- to 11-year-old boys and a total fat intake of 33.6% and a saturated fat intake of 12.3% among 6- to 11-year-old girls.[30] According to these data, both boys and girls are exceeding the recommendation for total calories from saturated fat of less than 10%. Cholesterol intake is well below the recommendation of 300 mg per day. Analysis of NHANES III data shows that the percentage of energy from fat is higher for black and Mexican American girls and black boys than for white girls and boys.[67] These differences are seen by 6 to 9 years of age in black and Mexican American girls and by 10 to 13 years of age in black boys.

As in the toddler and preschool-age group, mean vitamin and mineral intakes for school-aged children exceeded the recommendations except for vitamin E and zinc.[66] Table 12.7 depicts a further analysis of children's diets in relation to dietary fiber, sodium, and caffeine intake. School-age children's caffeine intake has risen dramatically for both males and females from an average daily intake of 12.7 milligrams during the preschool years. Higher caffeine consumption is seen in the 6- to 11-year-old group versus the 6- to 9-year-old group, indicating an increased consumption of caffeine with age. This coincides with an increase in soft drink consumption. Based on the Food and Nutrient Intake by Children report, 30.3% of U.S. children reported consuming fast food on a typical day. Children who ate fast food consumed more total energy, more total fat, more total carbohydrates,

more added sugars, more sugar-sweetened beverages, less fiber, less milk, and fewer fruits and non starchy vegetables than children who did not eat fast food.[63]

Tables 12.8 and 12.9 on the next page depict the mean percentages of nutrient intake contributed by foods eaten at snacks and foods eaten away from home for 1 day. These figures further illustrate the important contributions that snacks and meals eaten away from home make to the total daily food intake of children, especially in terms of total calories, total fat, and saturated fat. Analysis of food consumption data indicates that snacking among children has increased over the years, and the contribution of snacks to energy intake has increased from 20% in 1977 to 25% in 1996.[68] When eating away from home, school-age children most often eat at the school cafeteria, followed by someone else's house, and fast food restaurants.[66]

According to the NHANES III data, children ages 6 to 11 years obtain about 20% of their total energy intake from beverages, with milk, soft drinks, and juice drinks being the largest contributors.[30] Drinking whole milk contributes significantly to children's saturated fat intakes.

A measure of diet quality is provided by the Healthy Eating Index (HEI; available at www.usda.gov/cnpp), which is part of the Continuing Survey of Food Intake by Individuals.[69] The HEI measures the degree to which a person's diet conforms to the MyPyramid, measures total fat and saturated fat as a percent of total energy, total cholesterol, and sodium intake; and indicates the variety of foods in a person's diet. According to the most recently available HEI data, only 8% of children aged 7 to 12 years had a good diet, while 79% needed to improve their diets, and 13% had a poor diet. The average HEI score for children aged 7 to 12 years was 64.1 out of 100, indicating that their diets needed improvement.[69]

Table 12.7 Mean dietary fiber, sodium, and caffeine intake of 6- to 11-year-old children[66]

Gender and Age	Dietary Fiber (g)	Sodium (mg)	Caffeine (mg)
Males:			
6–9 years	13	3195	23
6–11 years	14	3264	25
Females:			
6–9 years	12	2764	19
6–11 years	12	2839	23

Table 12.8 Mean percentages of nutrient intake contributed by foods eaten at snacks for 1 day[66]

Gender and Age	Individuals Eating Snacks (%)	Food Energy (%)	Total Fat (%)	Saturated Fatty Acids (%)
Males:				
6–9 years	83	21	19	19
6–11years	83	21	20	20
Females:				
6–9 years	84	21	20	20
6–11 years	82	20	19	19

Table 12.9 Mean percentage of nutrient intake contributed by foods obtained and eaten away from home for 1 day[66]

Gender and Age	Individuals Eating Snacks (%)	Food Energy (%)	Total Fat (%)	Saturated Fatty Acids (%)
Males:				
6–9 years	65	26	27	28
6–11 years	65	26	28	28
Females:				
6–9 years	64	29	30	31
6–11 years	66	30	31	32

Cross-Cultural Considerations

Healthy People 2010 has as one of its major goals the elimination of health disparities among different segments of the population.[5] The reasons for the health disparities are complex but may include genetic variations, environmental factors, and health behaviors, including diet. Access to community-based, culturally competent, linguistically appropriate preventive health care is needed to eliminate these disparities.[5]

A unique characteristic of every ethnic group in America is its culturally based foods and food habits.[70] As discussed earlier, children learn food habits within the context of their family's culture. It is important for a health care professional to try to learn as much as possible about the foods and diets of the ethnic groups served, including where food is purchased and how it is prepared. The next step is to evaluate the diet within the context of the culture. Which foods or food habits have positive health benefits and should be encouraged? Which food behaviors have harmful effects on health and should be limited or modified? For example, in working with the Latino population, it is important to first of all establish the country of origin. Latino immigrants may be from Mexico, Central America, South America, or the Carribean. Food habits are unique for each of these ethnic groups. For example, Central Americans eat a lot of legumes, rice, and corn. Fruits and vegetables are also included in the diet. So these dietary practices form the basis of a healthy diet. However, lard is the most commonly used fat. Encouraging Central Americans to use a vegetable oil instead of lard is an example of a modification of a food practice to make it healthier.

As mentioned in Chapter 10, a series of booklets entitled *Ethnic and Regional Food Practices* is available from the American Dietetic Association. These booklets provide examples for incorporating traditional foods of various ethnic groups into dietary recommendations. Ordering information for this series can be found in the Resource section at the end of this chapter.

Vegetarian Diets

Young children who are consuming vegetarian diets are usually following their parents' eating practices. Preadolescents, on the other hand, may choose to follow a vegetarian diet independently of the family, motivated by concerns about animal welfare, ecology, and the environment.[1] A vegetarian diet is a socially acceptable way to reduce caloric intake and may be adopted by adolescents with eating disorders (see Chapter 14). A Vegetarian Food Guide Pyramid with suggested number of servings from different food groups has been developed.[71] Key nutrients in planning vegetarian diets for children include adequate calories, protein, calcium, zinc, iron, omega-3 fatty acids, vitamin B_{12}, riboflavin, and vitamin D.[72] Suggested daily food guides for lacto-ovo vegetarians and for vegans can be found in Tables 12.10 and 12.11.[1]

Physical Activity Recommendations

Physical activity has many proven health benefits including prevention of coronary heart disease. Physical activity is one of the health behaviors that is important to establish in childhood to increase the chances of a physically active lifestyle that will continue into adolescence and adulthood. With the increased prevalence of childhood obesity, increasing physical activity and decreasing sedentary behaviors become important factors in controlling childhood overweight.[42]

Table 12.10 Suggested daily food guide for lacto-ovovegetarians at various intake levels[1]

Food Groups	Servings per Day, by Age and Daily Caloric Intake	
	7–10 Years (2000 Kcal)	11+ Years (2200–2800 Kcal)
Breads, grains, cereal	6	9–11
Legumes	1	2–3
Vegetables	3–5	4–5
Fruits	4	4
Nuts, seeds	1	1
Milk, yogurt, cheese	3	3
Eggs (limit 3/week)	½	½
Fats, oils (added)	4	4–6
Sugar (added teaspoons)	6	6–9

Table 12.11 Suggested daily food guide for vegan children and adolescents at various intake levels[1]

Food Groups	Servings per Day, by Age and Daily Caloric Intake	
	7–10 Years (2000 Kcal)	11+ Years (2200–2800 Kcal)
Breads, grains, cereal	7	10–12
Legumes	1	2–3
Vegetables, dark-green leafy	1	1
Vegetables, other	2	3–4
Fruits	5	4–6
Nuts, seeds	1	1
Milk alternatives (fortified with calcium and vitamins B_{12} and D)	3	3
Fats, oils (added)	4	4–6
Sugar (added teaspoons)	6	6–9

Recommendations versus Actual Activity

It is recommended that children engage in at least 60 minutes of physical activity every day.[55] Parents are encouraged to:

- Set a good example by being physically active themselves and joining their children in physical activity.
- Encourage children to be physically active at home, at school, and with friends.
- Limit television watching, computer games, and other inactive forms of play by alternating with periods of physical activity.

Additionally, physical activity and daily physical education should be encouraged at schools and during after-school care programs. But presently, only about 17% of middle and junior high schools require daily physical activity for all students.[5] Healthy People 2010 objectives include increasing the proportion of trips that school-age children make by walking and by bicycling. As of 2001, only about 36% of children and adolescents ages 5 to 15 years take walking trips to school less than 1 mile, and only about 2% of children and adolescents ages 5 to 15 years take bicycle trips to school less than 2 miles.[45] In order to meet these goals, communities need to assure safe places for children to walk and ride their bicycles. Bicycle safety measures, such as wearing a helmet, need to be employed. Communities can also offer youth sports and recreation programs that are developmentally appropriate and fun for all young people. To achieve such a community environment, partnerships need to be established among federal, state, and local governments, nongovernment organizations, and private entities. The Centers for Disease Control and Prevention (CDC) have proposed strategies for promoting physical activity for children in family, school, and community settings.[73]

Determinants of Physical Activity

It is important to understand children's physical activity patterns and determinants of physical activity, so that vulnerable groups can be identified and appropriate intervention programs designed. Potential determinants of physical activity behaviors among children include physiological, environmental, psychological, social, and demographic factors.[74] Childhood physical activity has been difficult to assess and to track into adulthood. Many of the studies have identified correlates of physical activity behavior rather than predictors. The determinants of childhood physical activity are probably multidimensional and interrelated. More work needs to be done in this area, but some generalities resulting from existing studies are listed here:

- Girls are less active than boys.
- Physical activity decreases with age.
- Seasonal and climate differences are seen in children's activity levels.
- Physical education in schools has decreased.

School and neighborhood safety is an important issue in promoting physical activity. In addition, parents have direct and indirect effects on children's physical activity levels.

Organized Sports

Many school-age and preadolescent children participate in organized sport activities, through schools or other community organizations. An analysis of NHANES III data indicates that children who participate in team sports and exercise programs are less likely to be overweight as compared to nonparticipants.[38] The American Academy of Pediatrics (AAP) recommends that children who are involved in sports be encouraged to participate in a variety of different activities. The proper use of safety equipment, such as helmets, pads, mouth guards, and goggles should be encouraged. The AAP warns against intensive, specialized training for children. The AAP's recommendations include:

- Regularly monitoring the child athlete's physical condition and development
- Preventing stress or overuse injuries, with the child's physician and coach working together
- Identifying and addressing eating disorders
- Instructing families, coaches, and child athletes about recognizing and preventing heat injury.[75]

Recommendations for sports participation by children stress the importance of proper hydration. As noted earlier, children are ar greater risk for dehydration and heat-related stresses than are adults.

Nutrition Intervention for Risk Reduction

"It is the position of The American Dietetic Association, the Society for Nutrition Education, and the American School Food Service Association that comprehensive nutrition services must be provided to all of the nation's preschool through grade twelve students. These nutrition services shall be integrated with a coordinated, comprehensive school health program and implemented through a school nutrition policy...."[76]

Nutrition Education

Eating a healthy diet and participating in physical activity are important components of a healthy lifestyle that may prevent chronic disease in childhood and into adolescence and adulthood. School age is a prime time for learning about healthy lifestyles and incorporating them into daily behaviors. Schools can provide an appropriate environment for nutrition education and learning healthy lifestyle behaviors. Nutrition education studies have been conducted in school settings as well as outside of schools. Some of these programs have been knowledge-based nutrition education programs, with the focus on improving the knowledge, skills, and attitudes of children in regard to food and nutrition issues.[77] Other nutrition education programs have been more behaviorally focused, emphasizing disease risk

---

Table 12.12 Recommendations for school health programs promoting healthy eating[78]

1. **Policy:** Adopt a coordinated school nutrition policy that promotes healthy eating through classroom lessons and a supportive school environment.
2. **Curriculum for nutrition education:** Implement nutrition education from preschool through secondary school as part of a sequential, comprehensive school health education curriculum designed to help students adopt healthy eating behaviors.
3. **Instruction for students:** Provide nutrition education through developmentally appropriate, culturally relevant, fun, participatory activities that involve social learning strategies.
4. **Integration of school food service and nutrition education:** Coordinate school food service with nutrition education and with other components of the comprehensive school health program to reinforce messages on healthy eating.
5. **Training for school staff:** Provide staff involved in nutrition education with adequate preservice and ongoing in-service training that focuses on teaching strategies for behavioral change.
6. **Family and community involvement:** Involve family members and the community in supporting and reinforcing nutrition education.
7. **Program evaluation:** Regularly evaluate the effectiveness of the school health program in promoting healthy eating, and change the program as appropriate to increase its effectiveness.

---

reduction as well as enhancing health. The CDC has published "Guidelines for School Health Programs to Promote Lifelong Healthy Eating"; see Table 12.12.[78]

Nutrition Integrity in Schools

"The schools and the community have a shared responsibility to provide all students with access to high-quality foods and school-based nutrition services as an integral part of the total education program. Educational goals, including the nutrition goals of the National School Lunch Program and the School Breakfast Program, should be supported and extended through school district wellness policies that create overall school environments that promote access to healthful school meals and physical activity and provide learning experiences that enable students to develop lifelong healthful eating habits"[79]

The American Dietetic Association

Nutrition integrity in schools is defined as ensuring that all foods available to children in schools are consistent with the U.S. Dietary Guidelines for Americans and the Dietary Reference Intakes.[5,55] School nutrition programs are vital to reinforcing healthy eating habits in school-age children.

Sound nutrition policies need the support of the community and school environments, and must involve students in order to be successful. Preparing community leaders for involvement in policy development is one of the nutrition integrity core concepts.[79] Training food service personnel, teachers, administrators, and parents is an integral part of this process. The school environment must be one that supports healthy eating and exercise patterns. Foods sold from vending machines and snack bars often do not support healthy eating and may undermine sound nutrition programs. However, in some schools, vending machine proceeds are important sources of revenue for underfunded schools. Some schools have *pouring rights* contracts with soft drink companies and receive a percentage of the profits.[76] Many schools sell à la carte items in addition to standard school lunches to increase revenue. One study showed that seventh-grade students in schools with à la carte items ate more fat and fewer fruits and vegetables than did students at schools without an à la carte program.[80] In the same study, for each snack vending machine, students' mean intake of fruit servings declined by 11%. It is against USDA regulations to sell *competitive foods* of minimal nutritional value. However, it is not against USDA regulations to sell these foods to students at times other than mealtimes or in other areas of the school, outside of food service areas.

Adequate time allotted for meals is another important component of a sound nutrition program. Students can be involved in a nutrition advisory council, providing feedback about menu preferences and meal environment, and serving as a communication link with other students.

The American Academy of Pediatrics Committee on School Health has issued a policy statement about nutrition concerns regarding soft drink consumption in schools. The AAP advocates for the elimination of sweetened beverages in schools and the replacement with beverages such as real fruit and vegetable juices, water, and low-fat white or flavored milk. Vended food or drink contracts are discouraged, and the health and nutritional interests of students should form the foundation of nutritional policies in schools.[81]

The School Health Index (SHI) for Physical Activity, Healthy Eating and a Tobacco-free Lifestyle is a self-assessment and planning tool for schools, which is offered by the National Center for Chronic Disease Prevention and Health Promotion, Centers for Disease Control and Prevention.[82] The SHI helps schools:

- Identify strengths and weaknesses in health promotion policies and strategies.
- Develop an action plan for improving student health.
- Involve all stakeholders, including teachers, parents, students, and the community in improving school policies and programs.[82]

The SHI has eight different modules for elementary schools and middle and high schools in the self-assessment,

Illustration 12.5 SHI figure.[82]

which correspond to the eight components of a coordinated school health program as depicted in Illustration 12.5. Each module consists of a score card, a questionnaire with guidance for arriving at a score, planning questions, and recommendations for implementation. Illustration 12.6 on the following page is an example of items on the Nutrition Services score card.

Pouring Rights   Contracts between schools and soft drink companies whereby the schools receive a percentage of the profits of soft drink sales in exchange for the school offering only that soft drink company's products on the school campus.

Competitive Foods   Foods sold to children in food service areas during meal times that compete with the federal meal programs.

Model Programs

The 5 A Day for Better Health program models public–private partnership to enhance community nutrition education and to affect change in eating behaviors. In 1991, the National Cancer Institute (NCI), acknowledging the strong association of increased fruit and vegetable consumption with decreased risk of certain cancers, launched the 5 A Day for Better Health program. NCI partnered with the Produce for Better Health Foundation, a nonprofit organization that represents the produce industry, for this campaign. Industry participants included supermarkets, suppliers, commodity groups, and food service operations.[83]

The 5 A Day program includes retail, media, community, and research components. Supermarkets provide information to consumers at the retail level. The NCI and Produce for Better Health Program work together to develop a media campaign. At the community level, health, educational, agricultural, and voluntary agencies work together, sometimes forming coalitions, to reach consumers at the local level. For the research component, NCI funded

School Heath Index – Elementary School

Module 4: Nutrition Services

Score Card (photocopy before using)

Instructions
1. Carefully read and discuss the Module 4 Questionnaire, which contains questions and scoring descriptions for each item listed on this Score Card.
2. Circle the most appropriate score for each item.
3. After all questions have been scored, calculate the overall Module Score and complete the Module 4 Planning Questions located at the end of this module.

		Fully in place	Partially in place	Under Development	Not in place
4.1	Breakfast and lunch programs	3	2	1	0
4.2	Variety of foods in school meals	3	2	1	0
4.3	Low-fat and skim milk available	3	2	1	0
4.4	Meals include appealing, low-fat items	3	2	1	0
4.5	A la carte offerings include appealing, low-fat items	3	2	1	0
4.6	Sites outside the cafeteria include appealing, low-fat items	3	2	1	0
4.7	Food purchasing and preparation practices to reduce fat content	3	2	1	0
4.8	Promote healthy cafeteria selections	3	2	1	0
4.9	Clean, safe, pleasant cafeteria	3	2	1	0
4.10	Preparedness for food emergencies	3	2	1	0
4.11	Collaboration between food service staff and teachers	3	2	1	0
4.12	Degree and certification for food service manager	3	2	1	0
4.13	Professional development for food service manager	3	2	1	0

Column Totals: For each column, add up the numbers that are circled and enter the sum in this row

Total points: Add the four sums above and enter the total to the right.

Module Score = (Total points/39) x 100   %

Illustration 12.6 School Health Index Nutrition Services Score Card for Elementary Schools.[82]

nine studies to develop, implement, and evaluate interventions in specific communities to increase the consumption of fruits and vegetables. The results of one of these funded studies, based in elementary schools, are presented here. It is a good example of what can be accomplished through nutrition education and community involvement of parents, teachers, school food service personnel, and industry.

Model Program: High 5 Alabama

The purpose of this study was to evaluate the effectiveness of a school-based dietary intervention program in increasing the fruit and vegetable consumption among fourth-graders.[84] Twenty-eight elementary schools in the Birmingham, Alabama, metropolitan area were paired within three school districts based on ethnic composition and the proportion of students receiving free or reduced-price meals through the National School Lunch Program. One school in each pair was randomly assigned to an intervention group or a usual care control group. Assessments were completed at baseline (at the end of third grade), after year 1 (at the end of fourth grade), and after year 2 (at the end of fifth grade).

The intervention consisted of three components: classroom, parent, and food service. The classroom component of the intervention included 14 lessons, taught biweekly by trained curriculum coordinators with assistance from the regular classroom teachers. The parent component consisted of an overview during a kickoff meeting and completion of seven homework assignments by the parent and the child. Parents were also asked to encourage and support behavior change in their children. The food service component consisted of food service managers and workers receiving half-day training by High 5 nutritionists in purchasing, preparing, and promoting fruit and vegetables within the High 5 guidelines. Data analyzed included 24-hour recalls from the students, cafeteria observations, psychosocial measures, and parent measures.[84]

Results indicate that mean daily consumption of fruit and vegetables was higher at year 1 follow-up (3.96 vs 2.28) and year 2 follow-up (3.2 vs. 2.21) for the intervention group as compared to the controls. At year 1 follow-up, the mean daily consumption of fruits and vegetables was higher for the intervention parents as compared to control parents, but no difference was found at year 2 follow-up. The intervention was found to be effective in subsamples suggesting that the program can be used with boys and girls; African American and European Americans; low-, middle-, and higher-income families; and parents of low, medium, and high educational levels. The intervention was found to be effective in changing the fruit and vegetable consumption of fourth-grade students. Future studies are recommended to enhance the effectiveness of the intervention in changing parents' consumption patterns and to test the effectiveness of the intervention when delivered by the regular classroom teachers.

Public Food and Nutrition Programs

"It is the position of the American Dietetic Association that all children and adolescents, regardless of age, sex, socioeconomic status, racial diversity, ethnic diversity, or linguistic diversity, or health status, should have access to food and nutrition programs that ensure the availability of a safe and adequate food supply that promotes optimal physical, cognitive, social, and emotional growth and development. Appropriate food and nutrition programs include food assistance and meal programs, nutrition education initiatives, and nutrition screening and assessment followed by appropriate nutrition intervention and anticipatory guidance to promote optimal nutrition status"[85]

The American Dietetic Association

Child Nutrition Programs, which have had a federal legislative basis since 1946, contribute significantly to the food intake of school-age children. The purpose of the child nutrition programs is to provide nutritious meals to all children. These programs can also reinforce nutrition education, which takes place in the classroom. Increasing the proportion of children and adolescents ages 6 to 19 years whose overall dietary quality is enhanced by meals and snacks at schools is addressed in one of the Healthy People 2010 objectives.[5] Beginning with the school year 2006–2007, schools are required to develop a wellness plan.

Child nutrition programs include the National School Lunch Program, School Breakfast Program, Child and Adult Care Food Program, Summer Food Service Program, Special Milk Program, Commodity Assistance for Child Nutrition Programs, Special Supplemental Food Program for Women, Infants, and Children, Nutrition Education and Training Program, and the National Food Service Management Institute.[86] Descriptions of several of these programs follows.

The National School Lunch Program

The federal government provides financial assistance to schools participating in the National School Lunch Program (NSLP) through cash reimbursements for all lunches served, with additional cash for lunches served to needy children,

and through commodities.[86,87] Schools must meet five major requirements in order to participate in the NSLP:

1. Lunches must be based on nutritional standards.
2. Children who are unable to pay for lunches must receive lunches for free or at a reduced price, with no discriminating between paying and nonpaying children.
3. The programs operate on a nonprofit basis.
4. The programs must be accountable.
5. Schools must participate in the *commodity program*.

School lunches must provide one-third of the Dietary Reference Intakes for the age/grade group of children being served for energy, protein, calcium, iron, vitamin A, and vitamin C and must be consistent with the most recent version of the U.S. Dietary Guidelines for Americans when analyzed over a week's time.[55,86,87] Special emphasis is placed on serving a variety of foods and having menus that contain a variety of fruits and vegetables, low-fat dairy products, and lean meats. In addition, food school service personnel must make food safety a priority. These programs must also meet the needs of children with disabilities and special health care needs (see Chapter 13). Although not federally mandated, schools are encouraged to allow adequate time for children to eat their lunches. Schools receive payments from the federal government based on the number of meals served by category, paid, free, or reduced-price.

> **Commodity Program** A USDA program in which food products are sent to schools for use in the Child Nutrition Programs. Commodities are usually acquired for farm price support and surplus removal reasons.[79]

Schools participating in the National School Lunch Program can choose from among four menu planning approaches to plan school lunches:[88]

1. *Traditional Food-Based Menu-Planning Approach* Schools using this approach must plan menus with specific component and quantity requirements by offering five food items from four food components (meat/meat alternate, vegetables and/or fruits, grains/breads, and milk). Minimum portion sizes are established by ages and grade groups. Table 12.13 on the next page shows the meal pattern for lunches.
2. *Enhanced Food-Based Menu-Planning Approach* A variation of the traditional food-based menu-planning approach, this approach increases calories from low-fat food sources and increases the weekly servings of vegetables and fruits and grains/breads, in order to meet the Dietary Guidelines. The five food components are retained.
3. *Nutrient Standard Menu-Planning Approach* This is a computer-based menu-planning system that analyzes the specific nutrient content of menus, using approved computer software. This system allows more flexibility in planning, at the same time assuring that the nutrient standards are being met.

Table 12.13 Traditional food-based menu planning approach–meal pattern for lunches[88]

Food Components and Food Items	Minimum Quantities Ages 9 and Older Grades 4–12	Recommended Quantities Ages 12 and Older Grades 7–12	Sample Menu Minimum Quantities for Ages 9 and Older
Milk (as a beverage)	8 fl oz	8 fl oz	8 fl oz low-fat milk
Meat or Meat Alternate (quantity of the edible portion as served)			
Lean meat, poultry, or fish	2 oz	3 oz	2-oz hamburger patty
Alternate Protein Products	2 oz	3 oz	
Cheese	2 oz	3 oz	
Large egg	1	½ c	
Cooked dry beans or peas	½ c	¾ c	
Peanut butter or other nut or seed butters	4 Tbsp	6 Tbsp	
Yogurt, plain or flavored, unsweetened or sweetened	8 oz or 1 c	12 oz or 1½ c	
The following may be used to meet no more than 50% of the requirement and must be used in combination with any of the above: peanuts, soy nuts, tree nuts, or seeds, as listed in program guidance, or an equivalent quantity of any combination of the above meat/meat alternate (1 oz of nuts/seeds = 1 oz of cooked lean meat, poultry, or fish)	1 oz = 50%	1½ oz = 50%	
Vegetable or Fruit			
2 or more servings of vegetables, fruits, or both	¾ c	¾ c	Lettuce, tomato, Carrot/raisin salad Fresh apple
Grains/Breads (servings per week)			
Must be enriched or whole grain. One serving is 1 slice of bread or an equivalent serving of biscuits, rolls, etc., or ½ c of cooked rice, macaroni, noodles, other pasta products, or cereal grains.	8 servings per week, (For the purposes of this table, a week = 5 days.) minimum of 1 serving per day	10 servings per week, minimum of 1 serving per day	Enriched hamburger bun Condiments

4. *Assisted Nutrient Standard Menu-Planning Approach* This variation of the nutrient standard menu planning approach is for schools that do not have the necessary technical resources to conduct the nutrient analyses. This approach allows schools to have an outside source plan and analyze the menus based on the school's needs and preferences.

An additional provision of the National School Lunch Program allows states and school districts to use any reasonable approach to menu planning as long as the method is reviewed and approved by the state agency or USDA.

School Breakfast Program

First authorized as a pilot program in 1966, the School Breakfast Program is a voluntary federal program. Many state legislatures have mandated breakfast programs for their districts, especially in schools serving needy populations.[87] In general, the NSLP rules also apply to the School Breakfast Program. School breakfasts must provide one-fourth of the Dietary Reference Intakes for the children being served, based on age or grade groups, and comply with the U.S. Dietary Guidelines for Americans when analyzed over a week's time. Table 12.14 shows the traditional, food-based meal pattern for breakfast. It is a special challenge for schools to allow enough time for

Table 12.14 Traditional food-based menu planning approach—meal pattern for breakfasts[88]

Food Components and Food Items	Grades K–12	Sample Menu
Milk (fluid)		
As a beverage, on cereal, or both	8 fl oz	8 fl oz low-fat milk
Juice/Fruit/Vegetable		
Fruit and/or vegetable; or full-strength fruit juice or vegetable juice	½ c	½ c orange juice
Select one serving from each of the following components, two from one component, or an equivalent combination.		
Grains/Breads		
Whole grain or enriched bread	1 slice	1 slice enriched toast, butter, jelly
Whole grain or enriched biscuit, roll, muffin, etc.	1 serving	
Whole grain, enriched or fortified cereal	¾ c or 1 oz	¾ c raisin bran cereal
Meat or Meat Alternates	1 oz	
Meat/poultry or fish	1 oz	
Alternate protein products	1 oz	
Cheese	1 oz	
Large egg	½	
Peanut butter or other nut or seed butters	2 Tbsp	
Cooked dry beans and peas	4 Tbsp	
Nuts and/or seeds (as listed in program guidance)	1 oz	
Yogurt, plain or flavored, unsweetened or sweetened	4 oz or ½ c	

school breakfasts before school when most of the participating children arrive at about the same time. Currently, universal breakfast programs in elementary schools are being tested as pilot programs.

Summer Food Service Program

The Summer Food Service Program provides meals to children from needy areas when school is not in session. Schools, local government agencies, or other public and private nonprofit agencies operate these programs. The federal government provides financial assistance to these programs for providing meals in areas where 50% or more of the participating children are from families whose incomes are lower than 185% of the poverty level.[87] The Summer Food Service Program is an important source of food for many children from food-insecure families.

Team Nutrition

Team Nutrition is a program of the USDA's Food and Nutrition Service. The program is aimed at improving children's lifelong eating and physical activity habits through application of information in the Dietary Guidelines for Americans and MyPyramid.[13,55,89] Team Nutrition is a partnership of public and private organizations, including private sector companies, nonprofit organizations, and advocacy groups that are interested in improving the health of the nation's children. Team Nutrition is an excellent example of a program that addresses the establishment of healthy eating and physical activity patterns for children on multiple fronts.

Team Nutrition operates through three behavior-oriented strategies:

- Provide training and technical assistance for Child Nutrition food service professionals to help them serve meals that meet nutrition standards while tasting and looking good.
- Provide integrated nutrition education for children and their parents with the goal of establishing healthy food and physical activity choices as part of a healthy lifestyle.
- Provide support for healthy eating and physical activity by involving community partners, including school administrators and other school and community partners.

Six communication channels are utilized: (1) food service initiatives; (2) classroom activities; (3) schoolwide events; (4) home activities; (5) community programs and events; and (6) media events and coverage.[86] Schools are recruited to become Team Nutrition schools with the benefit of receiving a resource kit of materials. Additional information is available on the Team Nutrition website at www.fns.usda.gov/tn. The site includes activities for educators, parents, and students.[89]

Key Points

1. School-aged and preadolescent children continue to grow at a slow, steady rate.

2. Monitoring BMI-for-age percentiles is important for screening for over- or underweight.

3. Family mealtimes should be encouraged as there is a positive relationship between families eating together and the overall quality of the child's diet.

4. A child's food choices are being influenced by peers, teachers, coaches, the media, and the Internet.

5. The prevalance of overweight and the risk of overweight continue to increase in the school-aged and preadolescent age group.

6. Complications of overweight in children and adolescents, such as type 2 diabetes mellitus, are increasing.

7. Sedentary lifestyles and limited physical activity are contributing factors to the increase in childhood overweight.

8. School-age and preadolescent children are encouraged to eat a variety of foods and increase physical activity as outlined by the U.S. Dietary Guidelines for Americans and the MyPyramid.

9. Sweetened soft drink consumption is associated with increased calorie consumption and poorer diet quality.

10. Schools play an important community role in promoting healthy nutrition and physical activity patterns for children and adolescents.

Resources

The American Dietetic Association, Diabetes Care and Education Dietetic Practice Group
Ethnic and Regional Food Practices A Series. Chicago, IL: The American Dietetic Association, 1994–1999.
This series of booklets addresses food practices, customs, and holiday foods of various ethnic groups.

Bright Futures
The Bright Futues publications are developmentally based guidelines for health supervision, and address the physical, mental, cognitive, and social development of infants, children, and adolescents and their families.
Website: www.brightfutures.org

The Center for Health and Health Care in Schools
This Center provides sample "best practices" school policies on healthy eating; it also provides questions for parents and interested community members to ask in assessing school nutrition policies.
Website: www.healthinschools.org

Centers for Disease Control and Prevention, National Center for Health Statistics
The CDC website provides background information on the growth charts. Also, individual growth charts can be downloaded and printed from this web page.
Website: www.cdc.gov/nchs/about/major/nhanes/growthcharts/charts.htm

Child Nutrition Programs
The Child Nutrition Program website provides information on all of its programs including the National School Lunch Program, School Breakfast Program, Special Milk Program, Summer Food Service Program, and Child and Adult Care Food Program.
Website: www.fns.udsa.gov/cnd

Coalition for Healthy and Active Kids
This coalition's mission is the education of parents, children, schools, and communites about the importance of physical fitness and nutrition education in trying to reverse the childhood obestiy trend.
Website: www.chaausa.org

Dietary Reference Intakes: National Academy Press
Over 2,000 books are available online at this site, free of charge, including the current Dietary Reference Intakes.
Website: www.nap.edu

Federal Trade Commission, Bureau of Consumer Protection
See *"Promotions for Kids" Dietary Supplements Leave Sour Taste.* This article explores what is known about the use of dietary supplements in the pediatric population and includes a list of pointers for parents.
Website: www.ftc.gov/bcp/conline/features/kidsupp.htm

Fit, Healthy, and Ready to Learn: A School Health Policy Guide
This in-depth guide includes policies to encourage physical activity, encourage healthy eating, and discourage tobacco use. The policy to encourage healthy eating addresses nutrition education, the food service program, and other food choices at school.
Website: www.nasbe.org/healthyschools/fithealthy.mgi

Graves DE, Suitor CW.
Celebrating Diversity: Approaching Families Through Their Food. 2nd ed. Arlington, VA: National Center for Education in Maternal and Child Health, 1998. The purpose of this publication is to assist health professionals in learning to communicate effectively with a diverse clientele.

Green M, Palfrey JS. (eds).
Bright Futures: Guidelines for Health Supervision of Infants, Children, and Adolescents. 2nd ed. Arlington, VA: National Center for Education in Maternal and Child Health, 2000. This publication addresses the physical, mental, cognitive, and social development of infants, children, adolescents, and their families

Healthy Edge 2000
This continuing education program of the American School Food Service Association provides training for school food service personnel and nutrition professionals in nutrition education and in offering healthy meals that are appealing to children while meeting their nutritional needs.
Website: www.asfsa.org/continuinged/profdev/he2k

Healthy Habits for Healthy Kids

A joint project of the American Dietetic Association and WellPoint, this program provides guidance to health professionals in counseling families about childhood overweight.
Website: www.eatright.org

Institute of Medicine Committee on Prevention of Obesity in Children and Youth

Preventing Childhood Obesity: Health in the Balance. This report provides a comprehensive national strategy that recommends specific actions for families, schools, industry, communities, and government.
Website: www.iom.edu

Kidnetic

This colorful and interactive website is geared to teach older school-age children about healthy eating and exercise habits. It is supported by the International Food Information Council (IFIC) Foundation.
Website: www.kidnetic.com

Kids Count

This website provides data on critical issues affecting at-risk children and their families. The publication *Kids Count* by the Annie E. Casey Foundation, is also available online at this Web address.
Website: www.kidscount.org

KidsHealth

This website, supported by the Nemours Foundation, has sections for parents, kids (older school-age), and teens. It addresses all types of health concerns, including healthy eating and exercise.
Website: www.KidsHealth.org

Kittler PG, Sucher KP.

Food and Culture. 3rd ed. Belmont, CA: Wadsworth/Thomson Learning, 2001. This book covers culturally related food and nutrition topics. The traditional food habits of key ethnic, religious, and regional groups are comprehensively reviewed.

Information on traditional health beliefs and practices is also discussed.

National Center for Complementary and Alternative Medicine, National Institutes of Health

This website provides science-based information for consumers and practitioners as well as information about related news and events.
Website: www.nccam.nih.gov/nccam

Patrick K, Spear BA, Holt K, and Sofka D.

Bright Futures in Practice: Physical Activity. Arlington, VA: National Center for Education in Maternal and Child Health, 2001. Developmentally appropriate activities are presented for infants, children, and adolescents.

Promoting Better Health for Young People Through Physical Activity and Sports

Report to the President from the Secretary of Health and Human Services and the Secretary of Education. This report outlines 10 strategies to promote health through lifelong participation in physical activity and sports.
Website: www.cdc.gov/nccdphp/dash/presphysactrpt

Team Nutrition

This website provides information on the USDA's Team Nutrition program, which is designed to help implement the Dietary Guidelines in Child Nutrition Programs.
Website: www.fns.usda.gov/tn

U.S. Department of Health and Human Services, Maternal and Child Health Bureau

"Overweight and Physical Activity Among Children: A Portrait of States and the Nation 2005." This chartbook presents national and state-level data on the prevalence of overweight in children and adolescents (ages 10–17) within the context of family sructure, poverty level, parental health and habits, and community surroundings.
Websites: www.mchb.hrsa.gov and www.cdc.gov/nchs/slaits.htm

References

1. Story M, Holt K, Sofka D, eds. Bright futures in practice: nutrition. Arlington, VA: National Center for Education in Maternal and Child Health; 2002.

2. Meyer AF et al. School breakfast program and school performance. Am J Dis Child. 1989;143:1234–9.

3. The Annie E. Casey Foundation. 2006 Kids Count Data Book. Baltimore, MD: The Annie E. Casey Foundation; 2006.

4. Ogden CL et al. Prevalence of overweight and obesity in the United States, 1999–2004. JAMA. 2006; 295:1549–55.

5. U.S. Department of Health and Human Services. Healthy People 2010 (Conference Edition, in two volumes). Washington, DC; January 2000.

6. Behrman RE, Kliegman RM, Jenson HB, eds. Nelson textbook of pediatrics, 17th ed. Philadelphia: WB Saunders Co; 2004.

7. Centers for Disease Control and Prevention, National Center for Health Statistics. CDC growth charts: United States. Available at http://www.cdc.gov/nchs/about/major/nhanes/growthcharts/charts.htm, accessed 4/04.

8. Troiano RP, Flegal KM. Overweight children and adolescents: description, epidemiology, and demographics. Pediatrics. 1998;101:497–504.

9. Dietz WH. Critical periods in childhood for the development of obesity. Am J Clin Nutr. 1994;59:955–9.

10. Satter E. Feeding dynamics: helping children to eat well. J Pediatr Health Care. 1995;9:178–84.

11. Gillman MW et al. Family dinner and diet quality among older children and adolescents. Arch Fam Med. 2000;9:235–40.

12. Kotz K, Story M. Food advertisements during children's Saturday morning television

programming: are they consistent with dietary recommendations? J Amer Diet Assoc. 1994;94:1296–1300.

13. US Department of Agriculture. MyPyramid. www.MyPyramid.gov.

14. Moore E and the Kaiser Family Foundation. It's child play: advergaming and the online marketing of food to children. Menlo Park, CA: Henry J. Kaiser Family Foundation, 2006.

15. Birch LL, Fisher JO. Food intake regulation in children, fat and sugar substitutes and intake. Ann NY Acad of Sci. 1997; 819:194–220.

16. Birch LL, Fisher JA. Appetite and eating behavior in children. Pediatr Clinic N Amer. 1995;42:931–53.

17. Birch LL, Fisher JO. Mothers' child-feeding practices influence daughters' eating and weight. Amer J Clin Nutr. 2000;71: 1054–61.

18. Birch LL. Psychological influences on the childhood diet. J Nutr. 1998;128:407S–10S.

19. Fisher JO, Birch LL. Parents' restrictive feeding practices are associated with young girls' negative self evaluation of eating. J Amer Diet Assoc. 2000;100:1341–6.

20. Birch LL, Fisher JO. Development of eating behaviors among children and adolescents. Pediatrics. 1998;101:539–49(s).

21. Cachelin F et al. Does ethnicity influence body-size preference? a comparison of body image and body size. Obesity Research. 2002;10:158–66.

22. Institute of Medicine, Food and Nutrition Board. Dietary Reference Intakes for energy, carbohydrate, fiber, fat, protein, and amino acids. Washington, DC: National Academy Press; 2002.

23. Kennedy E, Goldberg J. What are American children eating? Implications for public policy. Nutr Rev. 1995;53:111–26.

24. Kennedy E, Powell R. Changing eating patterns of American children: a view from 1996. J Amer Coll Nutr. 1997;16:524–9.

25. Institute of Medicine, Food and Nutrition Board. Dietary Reference Intakes for calcium, phosphorus, magnesium, vitamin D, and fluoride. Washington, DC: National Academy Press; 1997.

26. Institute of Medicine, Food and Nutrition Board. Dietary Reference Intakes for vitamin A, vitamin K, arsenic, boron, chromium, copper, iodine, iron, manganese, molybdenum, nickel, silicon, vanadium, and zinc. Washington, DC: National Academy Press; 2001.

27. Centers for Disease Control and Prevention. Iron deficiency. United States, 1999–2000. MMWR. 2002;51:897–9.

28. Centers for Disease Control and Prevention. Recommendations to prevent and control iron deficiency in the United States. Morbidity and Mortality Weekly Report. April 3, 1998;47:RR-03.

29. Casamassimo P. Bright futures in practice: oral health. Arlington, VA: National Center for Education in Maternal and Child Health; 1996.

30. Troiano RP et al. Energy and fat intakes of children and adolescents in the United States: data from the National Health and Nutrition Examination Surveys. Am J Clin Nutr. 2000;72 (suppl):1343S–53S.

31. Freedman DS et al. The relation of overweight to cardiovascular risk factors among children and adolescents: The Bogalusa Heart Study. Pediatrics. 1999;103:1175–82.

32. Dietz WH. Health consequences of obesity in youth: childhood predictors of adult disease. Pediatrics. 1998;101:518S–25S.

33. American Diabetes Association. Type 2 diabetes in children and adolescents. Pediatrics. 2000;105:671–80.

34. Dietz WH. Childhood weight affects adult morbidty and mortality. J Nutr. 1998;128:411S–14S.

35. Dietz WH, Gortmaker SL. Preventing obesity in children and adolescents. Annu Rev Public Health. 2001;22:337–53.

36. Strauss RS, Knight J. Influence of the home environment on the development of obesity in children. Pediatrics [serial online] 1999;103:e85. Available from http://www.pediatrics.org/cgi/content/full/103/1/e85.

37. Whitaker RC et al. Predicting obesity in young adulthood from childhood and parental obesity. New Eng J Med. 1997;337:869–73.

38. Dowda M, Ainsworth BE, Addy CL. Environmental influences, physical activity, and weight status in 8- to 16-year-olds. Arch Pediatr Adolesc Med. 2001;155:711–7.

39. Dietz WH, Gortmaker SL. Do we fatten our children at the television set? obesity and television viewing in children and adolescents. Pediatrics. 1985;75:807–12.

40. Gortmaker SL et al. Television viewing as a cause of increasing obesity among children in the United States, 1986–1990. Arch Pediatr Adolesc Med. 1996;150:356–62.

41. Andersen RE et al. Relationship of physical activity and television watching with body weight and level of fatness among children: results from the Third National Health and Nutrition Examination Survey. JAMA. 1998;279:938–42.

42. Crespo CJ et al. Television watching, energy intake, and obesity in US children. Arch Pediatr Adolesc Med. 2001;155:360–5.

43. Robinson TN. Reducing children's television viewing to prevent obesity: a randomized controlled trial. JAMA. 1999;282:1561–67.

44. Klesges RC, Shelton ML, Klesges LM. Effects of television on metabolic rate: potential implications for childhood obesity. Pediatrics. 1993;91:281–6.

45. Centers for Disease Control. http://wonder.cdc.gov/data2010/focus.htm, accessed 10/06.

46. American Academy of Pediatrics. Committee on Nutrition: Prevention of pediatric overweight and obesity. Pediatrics 2003;112:424–30.

47. American Academy of Pediatrics Council on Sports Medicine and Fitness and Council on School Health. Active heathly living: prevention of childhood obesity through increased physical activity. Pediatrics. 2006;117:1834–42.

48. Barlow SE, Dietz WH. Obesity evaluation and treatment: expert committee recommendations. Pediatrics [serial online] 1998;102:e29. Available from http://www.pediatrics.org/cgi/content/full/102/3/e29, July 1, 2000.

49. Epstein LH et al. Decreasing sedentary behaviors in treating pediatric obesity. Arch Pediatr Adolesc Med. 2000;154:220–6.

50. American Dietetic Association. Position of the American Dietetic Association: individual-, family-, school-, and community-based interventions for pediatric overweight. J Am Diet Assoc. 2006;106:925–45.

51. Kavey RW et al. American Heart Association guidelines for primary prevention of atherosclerotic cardiovascular disease beginning in childhood. Circulation. 2003;107:1562–6.

52. Kavey RW et al. American Heart Association guidelines for primary prevention of atherosclerotic cardiovascular disease beginning in childhood. J Pediatr. 2003;142:368–72.

53. Federal update. J Amer Diet Assoc. 2000;100:877.

54. American Academy of Pediatrics, Committee on Nutrition. Pediatric nutrition handbook. 5th Ed. Kleinman RE (ed.) Elk Grove Village, IL: American Academy of Pediatrics; 2004.

55. U.S. Department of Health and Human Services and U.S. Department of Agriculture. Dietary Guidelines for Americans, 2005, 6th ed. Washington, DC:U.S. Government Printing Office; January 2005.

56. American Dietetic Association. Position of the American Dietetic Association: dietary guidance for healthy children aged 2 to 11 years. J Amer Diet Assoc. 2004;104:660–77.

57. Dwyer JT. Dietary fiber for children: how much? Pediatrics. 1995;96:1019S–22S.

58. American Academy of Pediatrics, Committee on Nutrition. Calcium requirements of infants, children, and adolescents. Pediatrics. 1999;104:1152–7.

59. Jennings DS, Steen SN. Play hard eat right, a parents' guide to sports nutrition for children. Minneapolis, MN: Chronimed Publishing; 1995.

60. Harnack L, Stang J, Story M. Soft drink consumption among US children and adolescents: nutritional consequences. J Amer Diet Assoc. 1999;99:436–41.

61. Ludwig DS, Peterson KE, Gortmaker SL. Relation between consumption of sugar-sweetened drinks and childhood obesity: a prospective, observational analysis. Lancet. 2001;357:505–8.

62. Mrdjenovic G, Levitsky DA. Nutritional and energtic consequences of sweetened drink consumption in 6- to 13-year-old children. J Pediatr. 2003;142:604–10.

63. Bowman SA et al. Effects of fast-food consumption on energy intake and diet quality among children in a national household survey. Pediatrics. 2004;113:112–18.

64. Pennington JATs. Bowes and Church's food values of portions commonly used, 18th ed. Philadelphia: Lippincott Williams Wilkins; 2005.

65. American Dietetic Association. Position of the American Dietetic Assocaition: use of nutritive and nonnutritive sweeteners. J Amer Diet Assoc. 2004;104:255–75.

66. U.S. Department of Agriculture, Agricultural Research Service. 1999 Food and nutrient intakes by children 1994–1996, 1998. ARS Food Surveys Research Group.

Available on "products" page at www.barc.usda.gov/bhnrc/foodsurvey/home.htm, accessed 4/04.

67. Winkleby MA et al. Ethnic variation in cardiovascular disease risk factors among children and young adults: findings from the Third National Health and Nutrition Examination Survey, 1988–1994. JAMA. 1999; 281:1006–13.

68. Jahns L, Siega-Riz AM, Popkins BM. The increasing prevalence of snacking among US children from 1977 to 1996. J Pediatr. 2001;138:493–8.

69. Federal Interagency Forum on Child and Family Statistics. America's children: key national indicators of well-being 2002. Federal Interagency Forum on Child and Family Statistics, Washington, DC: U.S. Government Printing Office.

70. Kittler PG, Sucher KP. Food and culture, 3rd ed. Belmont, CA: Wadsworth/Thomson Learning; 2001.

71. Messina V, Melina V, Mangels AR. A new food guide for North American vegetarians. J Amer Diet Assoc. 2003;103:771–5.

72. Messina V, Mangel AR. Considerations in planning vegan diets: children. J Amer Diet Assoc. 2001;101:661–9.

73. Department of Health and Human Services and Department of Education. Promoting better health for young people through physical activity and sports: a report to the president from the secretary of Health and Human Services and the secretary of

Education. Silver Spring, MD: Centers for Disease Control and Prevention; 2000. Available from www.cdc.gov/nccdphp/dash/presphysactrpt.

74. Kohl HW, Hobbs KE. Development of physical activity behaviors among children and adolescents. Pediatrics. 1998;101: 549–54.

75. American Academy of Pediatrics, Committee on Sports Medicine and Fitness. Intensive training and sports specialization in young athletes. Pediatrics. 2000;106:154–7.

76. American Dietetic Association. Position of ADA, SNE, and ASFSA: school nutrition services: an essential component of comprehensive school health programs. J Amer Diet Assoc. 2003;103:505–14.

77. Lytle L. Nutrition education for school-aged children. J Nutr Ed. 1995;27:298–311.

78. Centers for Disease Control. Guidelines for school health programs to promote lifelong healthy eating. J Sch Health. 1997; 67:9–26.

79. American Dietetic Association. Position of The American Dietetic Association: local support for nutrition integrity in schools. J Amer Diet Assoc. 2006;106:122–33.

80. Kubic MY et al. The association of the school food environment with dietary behaviors of young adolescents. Am J Pub Health. 2003;93:1168–72.

81. American Academy of Pediatrics Committee on School Health. Soft drinks in schools. Pediatrics. 2004;113:152–4.

82. Centers for Disease Control and Prevention. School Health Index for physical activity, healthy eating and a tobacco-free lifestyle: a self-assessment and planning guide. elementary school version. Atlanta, GA:2002. Available at www.cdc.gov/nccdphp/dash/SHI, accessed 4/04.

83. Havas S, Heimendingeer J, Reynolds K. 5 a day for better health: a new research initiative. J Amer Diet Assoc. 1994;94:32–6.

84. Reynolds KD et al. Increasing the fruit and vegetable consumption of fourth-graders: results from the high 5 project. Prev Med. 2000;30:309–19.

85. American Dietetic Association. Position of the American Dietetic Association: child and adolescent food and nutrition programs. J Amer Diet Assoc. 2006; 106:1467–75.

86. 7CFR Parts 210 and 220. Child Nutrition Programs; School Meal Initiative for Healthy Children; Final Regulation. June 1995.

87. Martin J. Overview of federal child nutrition legislation. In: Martin J, Conklin MT, eds. Managing child nutrition programs leadership for excellence. Gaithersburg, MD: Aspen Publishers; 1999.

88. U.S. Department of Agriculture, Food and Nutrition Service. FNS-303, 1998; also available online at www.fns.usda.gov/fns, accessed 4/04.

89. U.S. Department of Agriculture, Food and Nutrition Service. Team Nutrition. Available at www.fns.usda.fov/tn, accessed 4/04.

Chapter 13

Child and Preadolescent Nutrition:
Conditions and Interventions

Chapter Outline

- Introduction
- "Children Are Children First"—What Does That Mean?
- Nutritional Requirements of Children with Special Health Care Needs
- Growth Assessment
- Nutrition Recommendations
- Eating and Feeding Problems in Children with Special Health Care Needs
- Dietary Supplements and Herbal Remedies
- Sources of Nutrition Services

Prepared by **Janet Sugarman Isaacs**

Key Nutrition Concepts

1   Children are children first, even if they have conditions that affect their growth and nutritional requirements.

2   Common nutrition problems in children with chronic conditions are underweight and overweight, and difficulties in eating enough to meet higher nutrient requirements.

3   Children who have special health conditions receive more intensive nutrition services in schools and health care settings than other children do.

4   Family meal patterns and routines affect nutrition for children, so providing adequate support for families improves the nutritional status of children with chronic health conditions.

Introduction

Nutrition services need to be part of the goal to help a child reach his or her full potential. This chapter discusses the nutrition needs of children with chronic conditions, such as cystic fibrosis, diabetes mellitus, cerebral palsy, phenylketonuria (PKU), and behavioral disorders. Nutrition recommendations are based on those for children generally, but they may be modified by the condition and its consequences on growth, nutrient requirements, and/or eating abilities. Other factors such as activity that increase or decrease caloric needs are also discussed. Expanding school and community resources include nutrition services for children with special health care needs or those with developmental disabilities. Advocates for those with disabilities prefer "people-first language," which this chapter models. This convention names the person first and then the condition. An example is "a girl with *Down syndrome*," rather than the "Down's girl." Advocates for those with disabilities also prefer the word *disabilities* rather than *handicapped*. (The word *handicap* comes from the practice of using a cap to beg.)

"Children Are Children First"—What Does That Mean?

Children with special health care needs are children first, even if their conditions change their nutrition, medical, and social needs. Children are expected to become more independent in making food choices, assisting with meal preparation, and participating at mealtime with other family members. These same expectations are appropriate for children with special health care needs. For example, the child with spina bifida should be encouraged to make a salad or set the table. Modifications may be needed to help the child be successful, such as storing plates and utensils on low shelves, or lowering counter heights to accommodate a wheelchair at a kitchen sink. The cognitive developmental gains of childhood and participation in meal preparation and mealtime are the same for children with chronic conditions.

This concept has been acknowledged in schools through federal legislation in the Individuals with Disabilities Education Act (IDEA). This law requires the least restrictive environment and is resulting in inclusive settings for more children with disabilities.[1] This concept of inclusion has major ramifications for how children receive all types of services, such as schools providing alternative foods as required in the main cafeteria with all the other children. As a result of inclusion, children in wheelchairs or with Down syndrome spend time in regular classrooms with others of the same age. Nutritional problems related to food refusals, mealtime behavior, or special diets are being addressed in the neighborhood school in the regular classroom as often as possible.

This same concept of treating the child as a child first is recommended at home too. A special diet is part of an overall treatment plan that incorporates normal developmental steps of childhood. Children with diabetes or PKU do not benefit from being treated in a special manner at mealtime. As soon as possible, children are taught to take responsibility for making food choices consistent with their diet plans. Consistency and structure in the home support a child's normal development. This structure includes regular meal and snack times, and the child accepting increasing levels of responsibility in preparing foods. These approaches lower the chance of the child being overprotected or manipulating adults because of her illness or its treatment.

General nutrition guidelines for children, school nutrition educational materials, and nutrition prevention strategies may or may not be applicable to children with specific conditions. Many nutrition education curricula appropriately target the goals of preventing overweight, lowering fat, and increasing fruit and vegetable intake. These curricula provide appropriate education for the most part to children with conditions such as diabetes and Down syndrome. Conditions in which slow weight gain and underweight are common, such as cerebral palsy, may not fit such curricula.[2] Nutrition education may not address such children's need for high-fat foods as part of high caloric needs. Another example of how general nutrition recommendations may not fit all children concerns the widely used MyPyramid.[3] A child with PKU cannot ever have protein-rich meats or dairy products.[4] When this pyramid is discussed and encouraged, the child with PKU may feel isolated and confused about whether his diet is healthy. When possible, children with special health care needs are encouraged to participate in school nutrition education programs with modifications as needed.

"Train a child in the way he should go, and when he is old, he will not depart from it."

Proverbs

Nutritional Requirements of Children with Special Health Care Needs

Caloric and protein needs are lower on a body weight basis in childhood than during the preadolescence, toddler, and preschool years.[5] Children with special health care needs have a wide range of nutritional requirements and more variability than other children based on these factors:[6,7]

- Low caloric intake may be appropriate with small muscle size.
- High protein is needed with high protein losses, such as skin breakdown.
- High fluid volume is needed with frequent losses from vomiting or diarrhea.
- High fiber may be needed for chronic constipation management.
- Long-term use of prescribed medications may increase or decrease vitamin or mineral requirements, or change the balance of vitamins and minerals needed as a result of medication side effects.
- Routine illness is more likely to result in hospitalization or resurgence of symptoms of the underlying disorder.

These factors may replace the U.S. Dietary Guidelines and the Healthy People 2010 objectives for most children with special health care needs.

Energy Needs

Children with special health care needs may need more, less, or the same caloric intakes as other children of the same age. Energy needs are amazingly complex in children, let alone those with special needs. Under ideal conditions the caloric needs are measured by indirect calorimetry, but usually they are estimated using standard calculations that cannot take into account the specific conditions involved.[8] Machines that measure indirect calorimetry are becoming available, but they still can give only estimates of energy needs at rest, without considerations of energy needed for activity and growth. Conditions that slow growth or decrease muscle size generally result in lower caloric needs.[6,9,10] Caloric needs in a child with Prader-Willi syndrome may be only 66% of the caloric needs of a child of the same age and gender.[6]

Other factors that change energy requirements are related to activity level and frequency of illnesses.[5,6] Children with a chronic condition are encouraged to participate in age-appropriate sports activities. Conditions in which activity may be especially beneficial include diabetes and mild cerebral palsy. The level of activity of children with chronic conditions may be higher or lower than activity in other children. Children who are very active may appear thin as a result of low caloric intake. Children with autism and attention deficit hyperactivity disorder are generally more active than other children, and/or they may sleep less.[11,12] Such a range in level of activity is addressed as a part of a thorough nutrition assessment. Questions such as "Is the child receiving physical therapy one or three times per week?" and "How much time does the child use a walker compared to a wheelchair?" are examples of how activity can be assessed in determining caloric needs.

Protein Needs

Protein needs also can be higher, lower, or the same as those for other children, based on the condition. Healing burns and cystic fibrosis are examples of disorders with high protein needs—at 150% of the DRI.[13] Conditions such as PKU and other protein-based inborn errors of metabolism require greatly reduced amounts of natural protein in the diet.[4] Children with diabetes mellitus do not have modified protein needs.[14] The importance of protein for wound healing and for maintaining a healthy immune system makes protein requirements key for various conditions with frequent illnesses or surgeries. For example, a child with cerebral palsy who is scheduled to have hip surgery would have protein needs evaluated in a complete nutritional assessment. Higher protein may be recommended for wound healing and for skin breakdown while in a cast after surgery.

Other Nutrients

DRIs are good starting places to assess the need for vitamins and minerals in chronic conditions.[15] (DRIs are listed inside the back cover of this textbook.) As in all children, if the diet provides sufficient foods to meet the needs for protein, fats, and carbohydrates, it is likely the vitamin and mineral needs are also met. However, children with chronic conditions may have more difficulty meeting the DRI for vitamins and minerals as a result of these considerations:[15,16]

- Eating or feeding problems may restrict intake of foods requiring chewing, such as meats, so that certain minerals may be low in the diet.
- Prescribed medications and their side effects can increase turnover for specific nutrients, raising the recommended amount needed.
- Food refusals are common with recurrent illness, so total intake may be more variable day to day than in other children of the same age.

- Treatment of the condition necessitates specific dietary restrictions, so that vitamins and minerals usually provided in restricted foods have to be supplemented.

Nutrients such as calcium that are low in the general population of children are also problem nutrients for children with chronic conditions.[17] The American Academy of Pediatrics statement on calcium applies to children with chronic conditions. This statement recommends good-quality food sources of calcium for all children. Food sources of calcium may avoid lead contamination that has been reported in a variety of over-the-counter calcium supplements.[18] Taking high levels of supplemental calcium does not clearly benefit children with chronic conditions.[17]

Growth Assessment

The Centers for Disease Control (CDC) 2000 growth charts are a good starting place for assessing the growth of any child. Identifying children at risk for overweight and preventing long-term cardiovascular risks are important purposes underlying growth assessments of children. Such concerns may or may not apply to children with chronic conditions, but a nutrition assessment can tailor nutrition goals to specific conditions. Families dealing with conditions that shorten life, such as cystic fibrosis, would not benefit from including overweight and its long-term risks as part of the child's nutrition assessment.

Cerebral Spinal Atrophy Condition in which muscle control declines over time as a result of nerve loss, causing death in childhood.

Secondary Condition Common consequence of a condition, which may or may not be preventable over time.

Seizures Condition in which electrical nerve transmission in the brain is disrupted, resulting in periods of loss of function that vary in severity.

Scoliosis Condition in which the vertebral bones in the back show a side-to-side curve, resulting in a shorter stature than expected if the back were straight.

Neuromuscular Term pertaining to the central nervous system's control of muscle coordination and movement.

Families of children with severe disabilities who have wheelchairs may be apprehensive about growth as a goal for the child. They may have long-term concerns about caring for the child at home. Families may not want the child to grow at the usual rate when activities such as lifting him or her from the bathtub or out of a wheelchair may become more difficult. Also, children with rare degenerative conditions such as *cerebral spinal atrophy* have such major decreases in muscle size that growth may not occur.[10]

Most children with chronic conditions do grow, and assessing growth is an important component of nutrition services. If the child's condition is known to change the rate of weight or height gain—either slowing or accelerating it—the following signs need attention regardless of what growth chart is used:

- A plateau in weight
- A pattern of gain and then weight loss

- Not regaining weight lost during an illness
- A pattern of unexplained and unintentional weight gain

Growth Assessment and Interpretation in Children with Chronic Conditions

Factors that affect growth assessment and interpretation in childhood may not have been detectable earlier in younger children. These factors are the age of onset of the condition, *secondary conditions*, and activity. The child's age when the condition started may influence whether CDC growth charts are applicable. Early onset is more likely than later onset to affect growth in conditions such as *seizures*.[10] If the seizures started in middle childhood, the standard growth chart may be appropriate because the child's growth pattern is already established. Onset of seizures in the neonatal period may reflect more severe brain damage, which markedly slows growth rate. Then the child's own growth record over time would be the best indicator of future growth.

Toddlers and preschoolers with cerebral palsy usually do not develop secondary conditions until childhood or later. *Scoliosis* is a secondary condition that interferes with accurate measurement of stature.[6] It may develop as a result of muscle incoordination and weakness in some forms of cerebral palsy in preadolescence. If a child with cerebral palsy has stature measurements that plateau or decline, it may be a result of cerebral palsy, scoliosis, lack of adequate nutritional intake, or a combination of these three factors.[10] Nutritional interventions cannot prevent scoliosis, although nutritional consequences of its treatment may arise. Children may be provided custom-fitted back braces, so weight gain means the brace needs to be replaced. Children with scoliosis braces also may become less active because the brace restricts some types of movement. If scoliosis surgery is performed, the child may become slightly taller immediately, again showing that stature measurements have to be interpreted with care.

Body Composition and Growth

Children with special health care needs may or may not be typically proportioned in muscle size, bone structure, and fat stores. Some children in good nutritional status may plot at or below the lowest percentile on a standard growth chart for height.[19] In fact, low-percentile heights are usual for a child with Down syndrome if growth is plotted on the CDC chart rather than the special growth chart for Down syndrome.[20] Short stature, low muscle tone, and low weight compared to age-matched peers are not attributed to caloric intake. They characterize the natural consequences of the *neuromuscular* changes within Down syndrome. Similarly, a child with low muscle size could have a low weight and short stature. It would be unfair to assume that the child's diet is inadequate because of the low weight. A thorough assessment that includes body composition is necessary. For example, a thin-appearing child needs to have body fat stores measured before diet recommendations are made. If body fat stores are fine, adding calories is more likely to contribute to overweight.

Children with small muscle size will have lower weights than those with regular-sized muscles.[8,10] Conditions with altered muscle size may be described using terms such as *hypotonia* or *hypertonia*. Examples include cerebral palsy, Down syndrome, and spina bifida.[10,20] Not all muscles are affected. For example, some children with spina bifida have larger muscle size in the upper body and smaller muscle size in the lower body. Variation in size of muscles may make growth interpretation more difficult. Any assessment must address risks for overweight, such as body mass index (BMI) and adiposity rebound. Standard interpretation may suggest a risk of overweight, but it may not accurately reflect that short stature is part of the child's condition. By standard interpretation, every child with Down syndrome or spina bifida could be overweight. (See Case Study 13.1, page 340). For now, no established BMI tables cover specific conditions or the appropriate time for adiposity rebound.

Measuring body fat is another indicator of body composition. Skinfold fat measurements and their interpretation have to be based on consistent and repeatable standard methods.[21,22] Measuring fat stores in children is not like measuring fat stores in adults, because of the changes in body composition that come with age and growth. Calculated formulas and methods for body composition for children are not the same as for adults.[8,23] Estimates of body composition for children with chronic conditions may be based on smaller sample sizes than for other children, but still such information is helpful. Identified low fat stores trigger recommendations to boost calories.

In-depth growth assessment may include head circumference measurement for all ages, with plotting and interpretation based on the Nellhaus head circumference growth chart, as discussed in Chapter 9.[24] Head circumference is important because children with unusually small heads have smaller brains, a characteristic associated with short stature. Even with adequate diet and no documented eating problems, children with various genetic disorders tend to be shorter than age-matched peers.[20,25]

Special Growth Charts   Special growth charts have been published for a variety of genetic conditions.[20] Table 13.1 includes examples of these special growth charts. The number of children reported in such growth charts is not as large, nor as representative, as the CDC 2000 growth charts. Special growth charts are revised often, based on new information emerging about the natural course of rare conditions. Some special growth charts are based on only the most severe forms of the condition, such as for children living in residential care. Many chronic conditions

Table 13.1 Examples of specialty growth charts[2,20]

Conditions with Special Growth Charts	Comment
Achrondroplasia	Form of dwarfism
Down syndrome Trisomy13 Trisomy 18	Short stature, variable weight
Fragile X syndrome	Short stature, primary in males
Prader-Willi syndrome	Short stature, overweight
Rubinstein-Tabyi syndrome	Short stature
Sickle-cell disease	Short stature
Turner syndrome	Short stature
Spastic quadriplegia	Short stature, low weight
Marfan syndrome	Tall stature

do not have special growth charts because they present with a wide range of severity. Conditions without a specialty growth chart, which may or may not match the standard growth charts, include the following:

- *Juvenile rheumatoid arthritis*
- Cystic fibrosis
- Rett syndrome
- Spina bifida
- Seizures
- Diabetes

> **Juvenile Rheumatoid Arthritis**
> Condition in which joints become enlarged and painful as a result of the immune system; generally occurs in children or teens.

Nutrition Recommendations

Children with chronic conditions require nutrition assessments to determine whether they are meeting their nutrient and caloric needs, whether eating problems such as food refusals or mealtime behavior are interfering with meeting nutritional needs, and whether growth is on target for their age and gender. Then nutrition interventions are provided based on the assessment. The goal is for the child to maintain good nutritional status, and to prevent nutrition-related problems from being superimposed on the primary condition. (See Case Study 13.1 on the following page.)

Children with special health care needs benefit from the same nutritional recommendations other children do, particularly in general areas such as dietary fiber or appropriate use of soft drinks. However, children with special health care needs may require particular formulas and nutrition support not needed for most children. Most children develop feeding skills during the toddler and preschool years; by childhood, abilities and/or disabilities that limit self-feeding and using utensils may require more aggressive support. Nutritional supports common for children are enteral supplements, when oral feeding of regular foods is insufficient in quality or amount to

Case Study 13.1

Photo Disc

Adjusting Caloric Intake for a Child with Spina Bifida

Sam is a third-grader in regular classes at his public school. He uses a wheelchair all the time and can transfer from his wheelchair to a chair by himself. He is on a toileting schedule at school with the assistance of a nurse. He participates in modified physical education as part of his physical therapy treatment. He likes to eat with his friends at school. His mother tries to make him cut back at the evening meal and has stopped buying some of his favorite snacks. He is mad at his mother because he likes his snacks after school when he is bored.

Nutrition assessment from Sam's last visit at the spina bifida clinic at the local hospital showed that he was overweight by measuring his fat stores. Because he cannot stand, his stature was estimated by measuring his length lying down and comparing it with his last length measurement. Standard methods could not be used to measure him, which limits the interpretation of his growth using the CDC growth chart. The chart showed Sam at the 75th percentile in weight for his age, which is not overweight for his age. His rate of weight gain of 8 pounds per year, typical for a boy his age, is too fast for his low level of activity. His estimated calorie needs are 1100 per day due to low activity and short stature, or about two-thirds of the caloric needs of others his age. Sam says he does not care about his size or being overweight. His mother is quite concerned that she would not be able to assist him if he fell or needed to be lifted.

Recommendations: The nutritionist at the clinic completes a school lunch prescription to reduce Sam's caloric intake from 650 calories to 350 calories per lunch. His meal pattern is adjusted to two meals (breakfast and lunch) and two snacks per day at home, which better fits his low caloric needs. Sam is allowed to choose his favorite snack foods to replace his evening meal. Giving him choices about his snacks increases Sam's sense of being in control and lowers the instances of his expressing anger at his mother about snack foods. The clinic nutritionist calls the school to review Sam's level of activity and confirm that the lunch changes are being implemented. The physical therapist at school has found after-school swimming lessons and recommends them to Sam's mother as a way to increase his activity and socialization. To motivate Sam to pay attention to his eating and weight gain, his teacher and his mother set up a monthly nonfood reward for him if he does not gain any weight. The effectiveness of the plan to cut Sam's caloric intake and increase his activity will be assessed at his next clinic visit, when he will be weighed and have his fat stores measured.

Questions

1. Since Sam does not care about his size or being overweight, why is a diet plan necessary?
2. What are the risks from Sam's weight since he is only at the 75th percentile for his age on the standard growth chart?
3. Will he grow taller when he goes though puberty and be able to eat more calories each day?

maintain health and to assure growth is not limited. Table 13.2 provides a list of commonly used complete nutritional supplements and examples of their use. Children under 10 years of age are generally provided a pediatric formula, but adult formulas may be used for children.

Methods of Meeting Nutritional Requirements

Children with special health care needs who cannot meet their nutrient requirements from regular foods may receive complete nutritional supplements in addition to

Table 13.2 Examples of nutritional supplements and formula for children[26]

Formula	Comments
Pediatric versions of complete nutritional supplements, such as Pediasure	Generally recommended for children under 10 years of age; can be used for gastrostomy or oral nutrition support.
Adult complete nutritional supplements, such as Ensure	Generally 1 calorie per milliliter is recommended for children.
Enrichment of beverages, such as Carnation Instant Breakfast added to milk	Requires that milk is tolerated.
Predigested formula with amino acids and medium-chain fatty acids, such as Peptamen Junior	For conditions in which intestinal absorption may be impaired.
Special formulas for inborn errors of metabolism (PKU), such as Phenex-2	Usually a powder that is mixed as a beverage, but other forms such as bars and capsules are available.
High-calorie booster for cystic fibrosis, such as Scandishake	Generally 2.5 calories per milliliter to concentrate calories in small volume.

Table 13.3 Example of a feeding and eating schedule for an 8-year-old who eats by mouth and by gastrostomy

Daily Schedule	Comments
6:30 a.m. Night feeding pump turned off 7:15 a.m. Breakfast: refused 8:00 a.m. Bus to school 11:30 a.m. School lunch offered and about half is eaten: ½ chicken sandwich, all of french fries, with ⅓ pint of whole-milk 3:30 p.m. After-school snack at home of 4 oz pudding cup, two plain cookies, and 4 fl oz orange drink 6:30 p.m. Evening meal at home: ½ cup mashed potato, 6 fl oz whole milk, refused vegetable and meat 8:30 p.m. Bedtime	Overnight feeding by gastrostomy runs from 9:30 p.m. until 6:30 a.m., providing about 3 fl oz per hour, so no hunger in the morning is common. Child has slow eating pace and is easily distracted by school lunchroom sounds. Mealtime behavior at home includes many attempts to leave the table, with prompting to eat from parents. Parents hook up night feeding pump while the child is sleeping.

meals or for partial or complete replacement for meals. The first choice is that required supplements are drunk or eaten in the usual way. If this method does not work out, complete nutritional supplements can be administered by placement of a feeding *gastrostomy*.[6] Gastrostomy feeding may be required in children with kidney diseases, some forms of cancer, and severe forms of cerebral palsy and cystic fibrosis.[26–28] Many families experience difficulty accepting a gastrostomy for meeting nutritional requirements because feeding is such an important aspect of parenting.[29] Aside from emotional aspects, insurance coverage and financial questions of paying for formulas fed by gastrostomies are major concerns for some families.

Children fed by gastrostomy can have many different schedules, such as eating orally during school and being fed by gastrostomy overnight. Table 13.3 gives an example of a feeding plan that includes gastrostomy feeding and oral feeding. If medications are required, they can be given through the gastrostomy also. For example, for children with pediatric AIDS who require many medications during the day, compliance with taking the drugs improved after gastrostomies were placed. The parents spend less time trying to administer the medications to their children, and some children's health improved as a result of taking all of the required medications. Another example is a child

who cannot safely drink liquids as a result of cerebral palsy. The child could have fluids given by gastrostomy, but eat solids foods by mouth. Children with gastrostomies can swim, bathe, and do any activities they could do before the gastrostomy was placed.

Most formula fed by gastrostomy can be consumed as beverages. Even regular foods can be blended in a recipe for gastrostomy feeding for some children. Such "home brews" for gastrostomy feeding have to be carefully monitored because they are a rich medium for bacterial contamination. Part of the decision-making process about use of special formulas and gastrostomy feeding often hinges on prior rejection of other feeding methods. When possible, gastrostomy feeding is planned as a temporary measure, with a return to oral eating later. For example, a child who has a gastrostomy because of a kidney condition may have a kidney transplant that allows removal of the gastrostomy after recovery.

Other nutrition supplements fed by gastrostomy have specific components that are unusual in beverages because they have such a strange taste. For example, formulas that contain individual amino acids generally are accepted only by those who have had them from infancy, as in the formulas for PKU. If a child required a new formula with amino acids, gastrostomy feeding would be more successful than oral feeding in most cases.

Vitamin and Mineral Supplements for Chronic Conditions   Children's complete vitamin and mineral supplements are recommended for a variety of chronic conditions to assure that the DRIs for essential nutrients are provided. However most over-the-counter supplements are in the form of chewable tablets, so children who cannot chew well may require a liquid form of vitamins and minerals.

The composition of vitamin and mineral supplements may be important because some have added ingredients not recommended for certain chronic conditions. Examples are vitamin and mineral brands with added carbohydrates, which are not allowed on a *ketogenic diet*, or those made with an artificial sweetener containing phenylalanine not recommended for children with PKU. (The ketogenic diet and PKU are discussed later in this chapter.)

> **Ketogenic Diet**   High-fat, low-carbohydrate meal plan in which ketones are made from metabolic pathways used in converting fat as a source of energy.
>
> **Galactosemia**   A rare genetic condition of carbohydrate metabolism in which a blocked or inactive enzyme does not allow breakdown of galactose. It can cause serious illness if not identified and treated soon after birth.
>
> **Osteoporosis**   Condition in which low bone density or weak bone structure leads to an increased risk of bone fracture.

The underlying diagnosis can make specific nutrients so important in the diet that they may be prescribed as pharmaceuticals. Cystic fibrosis treatment (discussed later in this chapter) requires fat-soluble vitamin supplements due to poor intestinal absorption of these nutrients. Vitamins A, D, E, and K are needed in cystic fibrosis. Vitamin B_{12} injections are needed for some protein-based inborn errors of metabolism.[4,13] Vitamin C may be prescribed above the DRI for some children with spina bifida who have frequent bladder infections.[6] The high dose of vitamin C functions as a medication rather than as a nutrient in this instance.

Use of excessive levels of vitamins and minerals can be risky, especially for underweight children. For example, a child may be counseled to add Carnation Instant Breakfast, while another provider adds a complete nutritional supplement and is unaware that the child takes a chewable children's vitamin/mineral tablet too. Determination of the total intake of supplemented nutrients is part of a nutrition assessment. All medications, prescribed and over the counter, are identified within the nutrition and medical care plans.

Children with chronic conditions that limit activity or require medications that affect bone growth need special attention regarding calcium and vitamin D requirements.[17] *Galactosemia*, in which dairy products are eliminated from the diet, and cerebral palsy are two examples of conditions affecting calcium. Some children with these conditions are like older women with *osteoporosis* in that their calcium may move out of bones faster than it goes in. Providing additional calcium, phosphorous, and vitamin D may be recommended. Selecting calcium supplements that can be taken for years or decades by children raises concerns about the lead content found in some calcium supplements.[18]

Fluids

Guidelines for fluids for all children are appropriate. Particular considerations for children with special health care needs are high fluid losses—for example, from drooling as a result of cerebral palsy—or behaviors that result in low fluid intakes.[6] Because constipation is common in children with neuromuscular disorders, adequate fluids are often stressed as part of a bowel management program. Children with limited ability to talk may have more difficulty indicating thirst. Many chronic health conditions carry higher risks for dehydration due to side effects of prescribed medications. A chronic condition generally does not change the fluid requirement when the child is well.

Eating and Feeding Problems in Children with Special Health Care Needs

Eating and feeding problems are diagnosed when children have difficulty accepting foods, chewing them safely, or ingesting sufficient foods and beverages to meet their nutritional requirements. About 70% of children with developmental delays have feeding difficulties, independent of whether neuromuscular problems have been identified.[6] Examples of these feeding problems include the following situations:

- Self-feeding skills are lower than the child's chronological age, requiring assistance and supervision to assure adequate intake.
- Meals take so long or so much food is lost in the process of eating that the actual food intake is too low.
- The condition requires adjustment in the timing of meals and snacks at home and at school.

In children who do not have developmental delay, the impact of chronic conditions on eating may include behavioral problems at mealtimes, conflicts about control over food choices, and variability in appetite. Families of children with chronic conditions may focus on mealtimes and foods as methods of coping with their concerns about the child's future. For example, families may be overprotective and restrict a child from eating at friends' homes, when such activities may be appropriate for social development.

Specific Disorders

Cystic Fibrosis   Cystic fibrosis (CF) is one of the most common lethal genetic disorders, with an incidence of 1 in 1500–2000 live births.[13] It is highest among Caucasians, with an incidence of 1 in 17,000 live births for African Americans. The CF gene is located on the long arm of chromosome 7, and it has many different genetic versions. The most common genetic mutation characterizes 67% of the cases. CF affects all the exocrine functions in the body, with lung complications often causing death. Its major nutrition-related consequence is malabsorption of various nutrients due to the lack of pancreatic enzymes. This can result in a slower rate of weight and height gain, and higher energy needs due to chronic lung infections. Children with cystic fibrosis are likely to develop malnutrition as the condition progresses. Intensive nutrition interventions may be required to meet higher caloric needs.[26,27]

Nutrition interventions for CF include monitoring growth, assessing dietary intake, and increasing calories and protein by two to four times the usual recommendations to compensate for malabsorption. Every time a child with CF eats a meal or snack, he must take pills containing enzymes. Frequent eating and large, calorie-dense meals are encouraged. Gastrostomy feeding at night to boost calories is sometimes required. Vitamin and mineral supplementation, particularly fat-soluble vitamins, is a part of daily management. Children with CF are at risk for developing diabetes because the pancreas is a target organ of CF damage.[13] In recent years children diagnosed with CF have achieved longer life expectancies, and many have survived into the young adult years.

Nutrition experts working with children who have CF struggle to balance the children's high nutrient needs and frequent illnesses. Many children with CF have slow growth and are lower in weight and shorter than expected. Even with nutrition support, decline in pulmonary function over time continues. Some children with CF have lung transplants if they meet strict eligibility requirements. CF is on the leading edge of gene therapy research, giving hope to families with young children.

Diabetes Mellitus   Diabetes mellitus is a disorder of *insulin* regulation in which dietary management is crucial. The hormone insulin may be underproduced or mistimed in its release, and/or peripheral tissues may become insensitive to it.[30,31] Both type 1 and type 2 diabetes mellitus are increasing in children, for reasons that are not well understood.[31] Diabetes incidence estimates vary widely in children based on age and ethnicity. Children of Pima Indians in Arizona have a high incidence of type 2 diabetes, at 22 per 1000 children as compared to 7.2 cases per 100,000 in children seen in Ohio.[14] In children, diabetes mellitus type 1 is more common than type 2. Type 1 diabetes is related to immune function and results in virtually no insulin production. For type 2 diabetes, which is more common in older children, teens, and adults, some insulin may be produced in the body. Because it is closely associated with overweight, type 2 diabetes may partially be managed by weight loss.[30,31] Consequences of poorly controlled diabetes in children are the same as in adults: risk of heart, eye, and kidney damage and premature death. Major changes in diabetes management for adults are based on 1994 guidelines that recommend increased monitoring of blood glucose and diet flexibility.[14] These concepts are also appropriate for children with diabetes, although some specific blood-monitoring levels for adults are not appropriate for children. Children with

diabetes type 1 are more likely to have both high and low blood sugars, not just high blood sugars.

Treatment for diabetes is regulation of the timing and composition of meals and exercise, along with insulin injections or medications.[14] A third-grader with type 1 diabetes is likely to require an insulin injection once or twice per day, with oversight and modification of school breakfast, school lunch, and snack time based on physical activity in school and after-school activities. If the child is invited to a birthday party, the timing of meals and snacks can be adjusted to allow the child to attend the party and eat most of the foods there. Common colds, or foods a child refuses to eat, can cause wide variation in blood sugar, contributing to irritability, sleepiness, or difficulty with schoolwork.

In the summertime, many localities organize summer camps for children with diabetes; diet education and controlled access to a diabetic diet are provided along with the usual camp activities. Such disease-specific camps are good for breaking the social isolation that children experience when they feel they are the only ones required to follow special diets.

Seizures   Seizures are uncontrolled electrical disturbances in the brain. Epilepsy and seizures are the same disorder. Seizures in children are a relatively common condition, with an incidence of 3.5 per 1000 children.[32,33] Seizure activity has a range of outward signs, from uncontrollable jerking of the whole body to mild blinking. Currently, no known nutrients bring on seizures. Children who have seizures are usually treated by medications that prevent them. After some types of seizures, the child may have a period of semiconsciousness called a *postictal state* and appear to be sleeping, but he or she is difficult to wake.[32,33] Feeding or eating during the postictal state is not recommended, because the child may choke. Some children have long enough postictal states to miss meals. Then adding other eating times is needed to make up for the lost calories and nutrients.

> **Insulin**   Hormone usually produced in the pancreas to regulate movement of glucose from the bloodstream into cells within organs and muscles.
>
> **Postictal State**   Time after a seizure of altered consciousness; appears like a deep sleep.
>
> **Ketones**   Small 2-carbon chemicals generated by breakdown of fatty acids for energy.

When seizures are controlled by medications, growth usually continues at the rate typical for that child. Dietary consequences of controlled seizures are primarily related to drug–nutrient side effects, such as change in hunger or sleepiness. Some drugs should be taken without food, and others may be offered with snacks or meals. Most drugs have to be taken on a strict schedule and are not stopped without medical supervision.

Some children have uncontrolled seizures that may cause further brain damage over time. For reasons that remain unknown, seizures decrease when brain metabolism is switched from the usual fuel, glucose, to *ketones*

Table 13.4 Sample day's menu for a ketogenic diet for a 7-year-old treated for seizures; portion sizes are prescribed, and parents are taught to measure foods

Breakfast
Beverage: heavy whipping cream (4 fl oz)
Scrambled egg with bacon and mushrooms

Lunch
Heavy whipping cream (4 fl oz)
Hard-boiled egg mixed with mayonnaise
2 Tbsp green beans with butter

After-school snack
Carbohydrate-free multivitamin and mineral pill
Sugar-free Popsicle
Diet soda

Dinner
Heavy whipping cream (4 fl oz)
Black olives (3)
Sugar-free gelatin topped with whipped cream
Slice of full-fat ham
Slice of tomato

Snack
Walnuts (3)
Diet soda

Illustration 13.1 Boy with CP in a walker.

from fat metabolism.[33] Some specialty clinics administer the *ketogenic diet* for uncontrolled seizures. The ketogenic diet severely limits carbohydrates and increases calories from fat. An example intake on a ketogenic diet is given in Table 13.4. The diet is adequate in calories and protein. Vitamins and minerals have to be added as supplements because the allowed food sources of carbohydrates are not sufficient to meet vitamin and mineral requirements. The ketogenic diet may allow seizure medications to be reduced or eliminated over time. However, many difficulties, such as measuring growth, blood glucose, and ketones in urine, lie in monitoring the body's reaction to such severe carbohydrate restriction. Growth during the time on a ketogenic diet may be different from that seen in the child's previous pattern. Some children improve in both weight and height when seizure activity declines. The ketogenic diet is so high in fat that some children gain weight faster than expected. The diet is generally recommended for 2 years, if it shows demonstrated effectiveness.

Cerebral Palsy Cerebral palsy (CP) is one of the most common conditions in children with severe disabilities (Illustration 13.1). Overall incidence of cerebral palsy is about 1.4–2.4 per 1000 children.[34] *Cerebral palsy* is a general term well understood by the public; it covers a broad range of conditions resulting from brain damage. Causes of CP all involve damage to the brain early in life, either

before or after birth. The initial site of brain damage does not progress, but progression of secondary effects occurs over time. Secondary effects may include contractures, scoliosis, gastroesophageal reflux, and constipation.[6,10] Many children with CP have constipation because coordinated muscle movements are part of bowel emptying, including the muscles in and over the intestines. Muscle coordination problems most easily seen in movements of the arms and legs may occur in muscles all over the body, including the abdominal muscles that assist in bowel evacuation.[34]

The form of cerebral palsy that presents the most nutrition problems is spastic quadriplegia, involving all limbs.[2] Most children with spastic quadriplegia appear thin, but this appearance may be a result of brain damage or muscle size. Children with cerebral palsy often have other forms of brain damage as well: 39–44% have mental retardation; 26–36% have seizures, 14–18% have severe visual impairment. Causes of CP are unknown for more than one-third of affected children, and the condition may or may not be related to preterm birth. About half the time after a preterm birth, a basis for CP can be identified from the perinatal period.[34] The prevalence of CP in children born with very low birth weight has been increasing, but this trend may be a consequence of the overall increase in survivors of severe prematurity.[6] Children with CP can enjoy many activities, attend school, and later contribute to society. Persons with CP display a wide range of abilities. As indicated in the growth chart in Illustration 13.2, children with spastic quadriplegia grow; but their growth is slower than others, with or without gastrostomy feeding.[2]

Nutritional consequences of spastic quadriplegia are slow weight gain and other growth concerns, difficulty

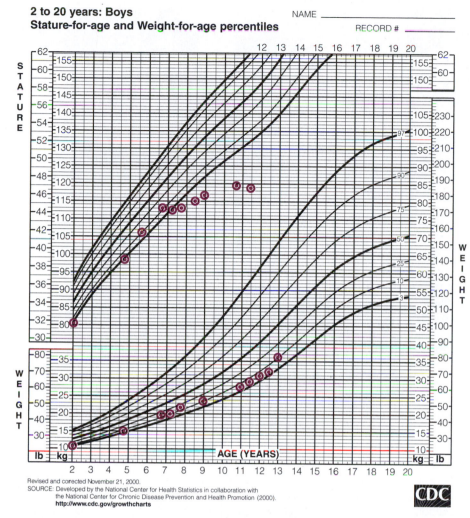

2 to 20 years: Boys
Stature-for-age and Weight-for-age percentiles

NAME _____

RECORD # _____

Revised and corrected November 21, 2000.
SOURCE: Developed by the National Center for Health Statistics in collaboration with
the National Center for Chronic Disease Prevention and Health Promotion (2000).
http://www.cdc.gov/growthcharts

CDC

Illustration 13.2 Growth chart for gastrostomy feeding for a boy with spastic quadriplegia and scoliosis.

and sitting, and those in the jaw, tongue, and lips and used in swallowing, may contribute to feeding and eating problems.[34]

Nutrition experts who provide services for children with CP monitor their growth and then make recommendations for food choices that fit the children's abilities for eating, for nutritional supplements if food and beverages are not providing sufficient nutrients, and for nutrition support if needed. Nutrition interventions may include the following:

- Stimulating oral feeding
- Promoting healthy eating at school
- Adjusting menus and timing of meals and snacks at home or school for meeting nutrient needs from foods that minimize fatigue during meals
- Assessing and adjusting the child's diet over time
- Using adapted self-feeding utensils or other types of feeding equipment

Phenylketonuria (PKU)   PKU is an inborn error of protein metabolism with a prevalence of 1 in 12,000 live births.[4] The only treatment is lifelong dietary management, in which more than 80% of protein intake from foods and beverages is replaced by protein in which the amino acid phenylalanine has been removed. The enzyme that uses phenylalanine as a substrate is either not working at all, or only partially active, in a person with PKU. Treatment reduces intake of this amino acid to the minimum amount needed as an essential amino acid. This strategy limits toxic breakdown products of accumulated phenylalanine, which the body has difficulty clearing. How excess phenylalanine causes mental retardation is not known. The PKU diet is required throughout life (Illustration 13.3). If foods with protein are consumed in too high amounts, PKU slowly becomes a degenera-

> **Athetosis**   Uncontrolled movements of the large muscle groups as a result of damage to the central nervous system.

tive disease affecting the brain at whatever age the treatment is stopped. A woman with PKU has to continue strict dietary adherence because high levels of phenylalanine affect every pregnancy, even if the infant did not inherit PKU.

with feeding and eating, and changes in body composition. No specific vitamins or minerals are known to correct CP. Problem nutrients are likely to be those related to bone density, calcium, and vitamin D, or nutrients needed in higher amounts as a result of medication side effects. Recommendations for caloric needs are difficult to determine, even with an in-depth growth assessment. Children with small or weak muscles have lower caloric needs because they are less active as a result of little voluntary muscle control. In contrast, types of CP characterized by increased uncontrolled movement require extra calories as a result of a higher activity level. *Athetosis* is an example of this less common form of CP, in which increased energy needs have been documented.[35] Altered body composition affects many aspects of the child's nutrition and eating abilities.[7] Eating or feeding problems may appear in the forms of spilling food, long mealtimes, fatigue at mealtime, and/or requiring assistance to eat. Difficulty in controlling muscles such as those in the neck and back, used in head position

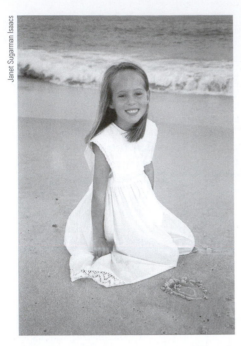

Illustration 13.3 This girl does not appear to have a chronic illness, but she has PKU.

Sugars, oils and sweets
Limted amounts

Fruits and vegetables
5–6 servings

Low-protein
substitute
foods
for bread,
pasta,
cereal

Formula or bar without phenylalanine 2–3 servings

Illustration 13.4

SOURCE: © 2000 SHS North America. Reprinted with permission.

When their diets are managed correctly, children with PKU appear to be eating meals providing less food than the meals of other children. The diet is adequate in all vitamins, minerals, protein, fats, and calories, but more nutrients are in liquid rather than solid forms. Foods to be avoided completely are protein-rich foods such as meats, eggs, regular dairy products, peanuts, and soybeans in all forms. Allowed natural sources of protein are limited amounts of regular crackers, potato chips, rice, and potatoes. Many fruits and vegetables are encouraged, if offered without added sources of protein. Some foods that are high in fats and sugars and generally low in natural protein, such as fried vegetables or candy canes, are safe for children with PKU. Illustration 13.4 shows a MyPyramid with the base of the pyramid the special medical food or formula that replaces the protein in foods.

The phenylalanine-deficient protein is generally served as a liquid, called a medical food or formula. The vitamins and minerals required to meet the RDA are in the phenylalanine-deficient protein powder. If the child does not drink enough of the PKU formula, foods that the child eats to meet her vitamin, mineral, and calorie needs will elevate the blood phenylalanine. Table 13.5 is a dietary recall from a child in good control of PKU. The phenylalanine-deficient protein also is available as bars and pills. It can be expensive to buy substitute low-protein alternative foods, such as low-protein pizza crusts, low-protein cheese, and low-protein baking mixes. Successful compliance requires use of low-protein foods to allow variety in the diet, such as low-protein pasta.

Table 13.5 Dietary recall for a 5-year-old child with well-controlled PKU

Breakfast
2 slices low-protein bread with jelly and margarine
6 cut-up orange pieces
8 fl oz PKU formula

Lunch
½ c fruit cocktail in heavy syrup
1 c tossed salad (lettuce, tomato, celery, cucumber only) with 2 Tbsp ranch dressing
17 french fries with ketchup
6 fl oz apple juice

After-school snack
½ c microwave popcorn
8 fl oz PKU formula

Dinner
pickle spears (dill, 3 wedges)
1 c low-protein imitation rice containing 1.5 tb margarine
½ c grilled onions, green peppers, and mushrooms (on rice)
1 c canned peaches in heavy syrup
8 fl oz PKU formula

Snack
Skittles candy (small snack size)
4 fl oz apple juice

Attention Deficit Hyperactivity Disorder   Attention deficit disorder and attention deficit hyperactivity (ADHD) disorder are the most common *neurobehavioral* conditions in children. The incidence of ADHD is estimated at 3% to 5% of school-age children.[36,37] Children suspected of having ADHD may have a chaotic meal and snack pattern and the inability to stay seated for a meal. They may be given fewer opportunities to use kitchen appliances and get their own snacks due to impulsiveness. Theories about specific foods or nutrients causing ADHD have not been proven scientifically, but high interest in nutrition as a cause and treatment continues. The sale of herbal medicines and nutritional supplements to families with children with ADHD is common. One large survey found that 64% of children with ADHD had tried at least one type of alternative therapy and that 13% of children took some sort of multivitamin supplement.[38] Megavitamins were found ineffective in treating ADHD in a double-blind crossover trial, but such information has not stopped claims made for them or marketing and sales.[38] Often health care providers, including nutritionists, are not told about the use of these supplements by families.

Recommended procedures to confirm ADHD include at least two sites completing observation checklists about behavior. Treatment with the following two approaches has been most effective:

1. A structured behavioral approach that may also include mental health counseling and support, such as parenting classes
2. A prescribed *psychostimulant* medication; examples are Ritalin or Adderal

Nutritional concerns in ADHD include medication side effects that decrease appetite, maintaining growth while being medicated, and mealtime behavior. Low appetite as a result of treatment of ADHD is quite variable; it depends on the timing of the medications compared to meals, dosage, and medication schedule and on how long the child has had to adjust to the medication. Less interference with appetite and growth is likely if the child does not take the medication during school holidays.

Regardless of the child's dosage schedule, ADHD medication peak activity is aimed for school hours, which includes school lunch. Nutrition interventions for children on psychostimulant medications call for timing meals and snacks around the medication's action peaks. For example, adding a large bedtime snack when the medication's effects are low is a typical recommendation. Monitoring weight and height carefully over time helps identify growth plateaus. Education for the school's lunchroom supervisors and teachers may be helpful to deal with food refusals and mealtime behavior for the child with ADHD.[37]

Pediatric HIV   Most children with HIV were infected at around the time of birth. HIV in children under age 13 is classified differently than it is in adults, by age-based categories as well as level of immunosuppression from the virus. Only in the last few years have affected infants lived long enough to benefit from the combination highly active antiretroviral therapy that became available in the mid-1990s.[39] As experience has accumulated with these potent medications, growth of children with low viral loads has improved, and opportunistic infections have occurred at a lower rate.

Nutrition is an important component of HIV management. Failure to thrive is common in infected infants regardless of drug treatment, due to the dampening effect of antiretroviral therapy on appetite and food intake. For children too young to be in charge of their own medication and eating schedules, educating the family and arranging support may be part of the nutrition intervention. Other nutrition concerns include food-related infection-control measures, assuring access to complete nutritional supplements, and referrals to food banks. If weight-gain and medication compliance problems are unresolved, gastrostomy placement for medications and supplemental feedings may be needed.[39]

Working with children with HIV is complicated and demanding, and dietary approaches have to be customized to the behavioral and developmental realities of each child. For example, an 11-year-old girl is being treated for HIV and its related illnesses. Her diet prescription includes a high-protein/high-calorie diet, with one complete vitamin/mineral supplement daily and three meals and three snacks. Her family members are to check her weight at home weekly and call in if they observe weight loss and low appetite. The girl takes four kinds of HIV-related medications, totaling 17 capsules per day. Two medications have no food-related restrictions, one is best taken with food at two different times per day, and one is best taken on an empty stomach (30 minutes before a meal or 2 hours after a meal). She also gets an injection every other week to strengthen her blood counts.

Neurobehavioral   Pertains to control of behavior by the nervous system.

Psychostimulant   Classification of medication that acts on the brain to improve mental or emotional behavior.

Dietary Supplements and Herbal Remedies

Families with children in a lengthy process of diagnosis—where the diagnosis does not lead to a definite treatment and when expense, insurance coverage, and administrative problems tax their ability to cope—are more likely to seek alternative therapies. Some of these alternatives have questionable effectiveness and perhaps are even harmful. No herbal remedies or nutritional supplements have been found effective to prevent or treat the conditions covered in this chapter; however, nutritional claims abound for various chronic conditions. Families hear from one another about micronutrients—such as magnesium, zinc, and B_6—sold with various combinations of amino acids for Down syndrome and autism.[40] Restrictive diets, such as avoiding dairy products or gluten, have been researched for one condition and then extrapolated for another. Sports drinks and high-protein products marketed for athletes may attract families with children who have difficulty gaining weight.

Strategies to counter unscientific nutritional claims for various products include the following:

- Recognize the benefits of support for families, such as advocacy groups.
- Improve communication with health care providers, so that families ask more questions about nutrition claims of alternative treatments.
- To give them some control over decision making for their children, give families reliable information, such as scientific literature or fact sheets, without endorsing any claim.

Sources of Nutrition Services

Children with chronic conditions that interfere with their ability to function may be eligible for Supplemental Social Insurance (SSI). Low-income families are eligible for SSI depending on the child's condition. Examples of conditions usually qualifying for SSI are chromosomal disorders; mental retardation; and severe forms of seizures, cerebral palsy, and CF. A child with treated PKU is generally not eligible for SSI because treatments prevent decline in learning abilities. Also, the Americans with Disabilities Act applies to all ages. It requires, for example, that school cafeteria lines accommodate wheelchairs.

USDA Child Nutrition Program

The U.S. Department of Agriculture Child Nutrition Program, as described in previous chapters, requires that school breakfast and lunch menus be modified for children with diagnosis-specific diets or changes in the texture of foods. Parents who want their children to participate in the Child Nutrition Program cannot be charged an additional fee for providing a special diet for the child. A registered dietitian or another health provider completes a prescription ordering special breakfasts or lunches. Examples of diet prescription orders are a reduced-calorie school lunch and breakfast, a pureed diet, or a nutrient-modified diet, such as a PKU diet (see Case Study 13.2). If families do not want to participate in the Child Nutrition Program and prefer to pack lunches, they can change their decision any time during the school year. Formulas administered by gastrostomy are not required to be supplied by the Child Nutrition Program.

Maternal and Child Health Block Program of the U.S. Department of Health and Human Services (HHS)

Every state has a designated portion of federal funding for children with special health care needs.[41,42] A wide range of services can be provided based on state planning as reported back to HHS. Nutrition services may be in specialty clinics or county health departments; or they can be contracted with for providing care, assuring resources such as formulas, foods, and nutrition education. Nutrition experts work with children in various settings, including schools, early intervention programs, homes, clinics, and facilities. Also, a program in every state identifies and advocates for children with special needs. An example is the Developmental Disabilities Council.

Public School Regulations: 504 Accommodation and IDEA

Two sets of regulations guide how schools provide nutrition services in addition to the Child Nutrition Program. Nutrition services in schools are generally more available to younger children than they are to older children. Children in regular education have different access to services than do children eligible for special educational services. See Case Study 13.2.

504 Accommodation The USDA Child Nutrition Program requires that elementary and junior and senior high school breakfast and lunch fit individual special diets when the child's physician orders them. This is called the "504 accommodation" when the child has a regular curriculum. If the child requires special education services, this requirement becomes included in their individualized education plan. For a child with MSUD, diabetes, or another chronic condition, the Child Nutrition Program school district director works with the child's hospital-based dietitian who prescribes the diet, if a parent requests that the child receive school meals. Child Nutrition Program regulations do not allow the parents of the child to be charged any additional cost for the modified school lunch. The child's dietary requirements also have to be met for classroom birthday parties or when food is used in classroom projects. These rules do not apply in private school, although the Americans with Disabilities Act may protect children from their special dietary needs being the basis for discrimination against them.

Individuals with Disabilities Education Act Children eligible for special education are covered by regulations within the Individuals with Disabilities Education Act (IDEA).[1] It requires each child to have an individualized education plan (IEP) that may include nutrition-related

Dealing with Food Allergies in School Settings

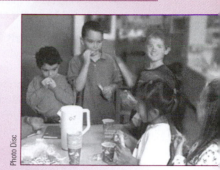

Judy is to start regular kindergarten now. When she was 2 years old, she was diagnosed with a peanut food allergy after many episodes of asthma and hives. Her health has improved as a preschooler with avoidance of peanuts in all forms. The family has carefully watched what she eats. However, at age 4 she had an episode of breathing difficulty that required an emergency room visit. This incident makes the family quite concerned about Judy's eating at school. She is generally not allowed to go to friends' homes to play; friends come to her house so the family can watch out for her. She has been instructed not to take any food from anyone. She has not been in day care or preschool, so starting school is a big step for the family.

Judy's mother meets with the school staff to discuss plans for Judy at school to avoid exposure to peanuts in any form. The family does not want to participate in the school lunch program. Her mother plans to pack Judy a lunch from home, although most children eat food provided at school. Judy's mother obtains a letter from her pediatrician about the peanut allergy and gives copies to the school nurse, teacher, and principal. With the school staff, Judy's mother discusses snack time for Judy and her eating at the cafeteria, which periodically serves food cooked in peanut oil, or food containing peanuts. The snack at kindergarten is provided by parents, based on a rotation schedule. It is usually milk or juice with cookies or fruit. A plan is put in place for the teacher to check the snack foods and offer a replacement snack provided by Judy's mother if she is unsure whether that day's snack contains peanuts.

The school working-group has written out a 504 accommodation plan for Judy's peanut allergy. It includes making sure that tables where children are eating peanut-containing foods are washed well, and posting signs in the cafeteria with Judy's picture to make sure she does not inadvertently get peanut-containing food from another child or in a food activity.

After Judy has been in school for 1 month, the family meets with the school group. In that month, two episodes have resulted in what may have been hives, and her family is worried that Judy is not adjusting well. At snack time Judy does not recognize some of the foods; she refuses to eat a snack most days. She appears hungry after school at home. Her mother says she would like to send to school a snack that she knows her daughter will eat, and she wants to attend school during snack time to make sure Judy is not being teased.

Questions

1. Why is it the school's job to check for peanuts when other parents are sending snacks?
2. The parents seem overprotective. Can the teacher transfer Judy to another classroom?
3. What are the chances that Judy will outgrow the peanut allergy by next year?

---

goals and objectives as needed. The school staff must involve the parent in developing the IEP. For some diagnoses, it would be appropriate for a nutritionist to attend the IEP meeting to make sure the teacher, teacher's aide, and other staff understand what the child needs. Nutrition, eating, and feeding problems may be a part of that plan, and apply to food offered in the classroom as well as that served in the regular school cafeteria. An example of IEP goals and objectives can be found in Table 13.6. For this particular plan, the child's education includes learning to eat by mouth with prompting and assistance. Nutritional supplements may

be purchased as part of an education intervention called for in the child's IEP.

Nutrition Intervention Model Program

The Maternal and Child Health Bureau (MCH) is a part of the Department of Health and Human Services and funds nutrition services for chronically ill children.[41,42] MCH develops and promotes model programs by funding competitive grants that emphasize training health care providers, including nutrition experts. Model programs that are targeted for children with special health care

Table 13.6 Example of nutrition objectives in an individualized education plan for an 8-year-old boy with limited oral feeding skills

1. In three of five trials JR will hold food on the spoon as he moves it to his mouth without hand-over-hand assistance of his aides during three meals per week.
2. JR will point to what he wants to eat with his left hand in three trials after two prompts per meal three days each week.
3. JR will cooperate in having his gastrostomy site checked at feedings by pulling up his shirt three days in a row of each week.

needs are necessary because most health care providers are not comfortable caring for children with rare conditions. Families of children with special health care needs often have difficulty locating nutrition experts and other providers who are familiar with their complex nutritional, medical, and educational needs. Training programs vary in length from short, intensive courses to year-long traineeships. Topics vary from nutrition for infants receiving intensive care services to nutrition problems of adolescence, such as warning signs of anorexia nervosa. Examples of such federal grant programs are the Pediatric Pulmonary Centers and Leadership Education in Neurodevelopmental Disabilities.[42] Funding nutrition faculty is one of the provisions of these grants.

Key Points

1. Children with chronic health conditions still want to fit in with everyone else and be treated like others their age. Paying too much attention to the special health care need may not help the individual become independent over time.
2. Nutrition requirements have to be customized to the individual: guidelines for healthy children may not be appropriate.
3. Energy needs (calories) are based on the child's activity as affected by the underlying condition; it may be higher or lower than for others of the same age.
4. Feeding and eating problems are likely to interfere with appetite and meal patterns for conditions that require medications that have side-effects.
5. The goal for meeting nutritional needs is to eat by mouth, if this is enjoyable and safe for the individual. An alternative may be to add complete nutritional supplements taken orally or by overnight feedings.
6. Vitamins and minerals are needed for some conditions at levels that are higher or lower than usually recommended.
7. Dietary management of cystic fibrosis and diabetes mellitus are examples of lifelong treatments that are combined with other medical approaches for children with special health care needs.
8. Cerebral palsy is one of the more severe conditions in which growth and eating are impacted; alternatives to using the standard growth chart and eating guidelines are necessary based on the abilities of the individual.
9. Children with special health care needs attend school like everyone else, but they may need to have a modified school lunch or eat different foods than others based on their underlying condition.
10. Nutrition and medical providers are encouraged to help families find in their community the educational and developmental services suitable for their children.

Resources

Ability On-Line Support Network
Connects young people with disabilities and chronic illnesses to peers and mentors with and without disabilities.
Website: www.ablelink.org

American Diabetes Association
Allows searches about diabetes in children around the world and provides professional and consumer publications.
Website: www.diabetes.org

Cystic Fibrosis Foundation
This large organization has information about research, services, and policy related to cystic fibrosis, including nutritional products and recommendations.
Website: www.cff.org

Exceptional Parent
This magazine is an excellent resource for a wide variety of conditions, and includes tips for parents on how to work with care providers and educators.
Website: www.eparent.com

Ketogenic Diet
This website is maintained by the Packard Children's Hospital at Stanford, which provides credible information for parents and providers about the ketogenic diet and places that use it.
Website: www.stanford.edu/group/ketodiet

National Down Syndrome Society
Includes information about its policy regarding nutritional products and directs parents to local resources for working with schools.
Website: www.ndss.org

References

1. General information about disabilities. National Information Center for Children and Youth with Disabilities. Washington, DC: U.S. Government Printing Office. Available at www.Nichcy.org/general.htm, accessed 10/06.

2. Krick J, et al. Pattern of growth in children with cerebral palsy. J Am Diet Assoc. 1996;96:680–85.

3. The U.S. Department of Agriculture's MyPyramid Food Guide. Available at: www.mypyramid.gov, accessed 10/06.

4. Isaacs JS, Zand D. Single-Gene autosomal recessive disorders and Prader-Willi syndrome: an update for food and nutrition professionals. American Dietetic Association; 2007;3.

5. Institute of Medicine Food and Nutrition Board. Dietary reference intakes for energy, carbohydrate, fiber, fat, fatty acids, cholesterol, protein, and amino acids; 2002.

6. Isaacs JS. Non-oral feeding. In: Lucas B, ed. Children with special health care needs: nutrition care handbook. Chicago IL: American Dietetic Association; 2004.

7. Kleinman RE, ed. Pediatric nutrition handbook, 5th ed. Chicago IL: American Academy of Pediatrics; 2004:629–42.

8. Stallings VA. Resting energy expenditure. In: Altschuler SM, Liacouras CA. Clinical pediatric gastroenterology. Philadelphia: Harcourt Brace; 1998:607–11.

9. L'Alleman D et al. Cardiovascular risk factors improve during three years of growth hormone therapy in Prader-Willi syndrome. Eur J Pediatr. 2000;159:836–42.

10. Minns RA. Neurological disorders. In: Kelnar CJH et al., eds. Growth disorders. London: Chapman and Hall Medical; Philadelphia, PA; 1998:447–70.

11. Schertz M et al. Predictors of weight loss in children with attention deficit hyperactivity disorder treated with stimulant medication. Pediatrics. 1996;98:763–69.

12. Berney TP. Autism—an evolving concept. British J of Psychiatry. 2000;176:20–5.

13. Dodge JA, Turck D. Cystic fibrosis: nutritional consequences and management. Best Practice & Research in Clinical Gastroenterology. 2006;20:531–46.

14. Lefebvre DE et al. Dietary proteins as environmental modifiers of type 1 diabetes mellitus. Annual Review of Nutrition. 2006;26:175–202.

15. Trumbo P et al. Dietary reference intakes; vitamin A, vitamin K, arsenic, boron, chromium, copper, iodine, iron, manganese, molybdenum, nickel, silicon, vanadium, and zinc. J Am Diet Assoc. 2001;101:294–301.

16. American Academy of Pediatrics. Alternative routes of drug administration—advantages and disadvantages (subject review). Pediatrics. 1997;100:143–52.

17. American Academy of Pediatrics. Calcium requirements of infants, children, and adolescents (RE9904). Pediatrics. 1999;1152–57.

18. Ross EA, Szabo NJ, Tabbett IR. Lead content of calcium supplements. J Am Med Assoc. 2000;284:1425–29.

19. National Center for Health Statistics. NCHS growth curves for children 0–19 years. U.S. Vital and Health Statistics, Health Resources Administration. Washington, DC: U.S. Government Printing Office; 2000.

20. Growth references: third trimester to adulthood, 2nd ed. Clinton, SC: Greenwood Genetic Center; 1998.

21. Frisancho AR. Triceps skinfold and upper arm muscle size norms for assessment of nutritional status. Am J Clin Nutr. 1974; 27:1052–157.

22. Fomon S et al. Body composition of reference children from birth to age 10 years. American J Clin Nutr. 1982;35:1169–75.

23. Nassis GP. Sidossis LS. Methods for assessing body composition, cardiovascular and metabolic function in children and adolescents: implications for exercise studies. Current Opinion in Clinical Nutrition & Metabolic Care. 2006;5:560–67.

24. Nellhaus G. Composite international and interracial graphs. Pediatrics. 1968; 41:106–14.

25. Batshaw ML. Chromosomes and heredity. In: Batshaw ML, ed. Children with disabilities, 6th ed. Baltimore, MD: Paul H. Brookes; 2007:3–26,770.

26. Steinkamp G, Von der Hardt H. Improvement of nutritional status and lung function after long-term nocturnal gastrostomy feedings in cystic fibrosis. J Pediatrics. 1994;121:244–49.

27. Walkowiak J, Przyslawski J. Five-year prospective analysis of dietary intake and clinical status in malnourished cystic fibrosis patients. J Hum Nutr Diet. 2003; 16:225–31.

28. Sullivan PB et al. Does gastrostomy tube feeding in children with cerebral palsy increase the risk of respiratory morbidity? Arch Dis Child. 2006;91:478–82.

29. Shingadia D et al. Gastrostomy tube insertion for improvement of adherence to highly active antiretroviral therapy in pediatric patients with human immunodeficiency virus. Pediatrics. 2000;105:1–5. Available at www.pediatrics.org/cgi/content/full/105/6/e80, accessed 10/06.

30. American Diabetes Association. Type 2 diabetes in children and adolescents. Pediatrics. 2000;671–80.

31. Kleinman RE, ed. Pediatric nutrition handbook, 5th ed. Chicago IL: American Academy of Pediatrics; 2004:491–514.

32. Kleinman RE, ed. Pediatric nutrition handbook, 5th ed. Chicago IL: American Academy of Pediatrics; 2004:773–88.

33. Tallian KB, Nahata MC, Tsao C-Y. Role of the ketogenic diet in children with intractable seizures. Ann Pharmacotherapy. 1998;32:349–56.

34. Pellegrino L. Cerebral palsy. In: Batshaw ML, ed. Children with disabilities, 5th ed. Baltimore, MD: Paul H. Brookes; 2002: 443–66.

35. Johnson RK et al. Athetosis increases resting metabolic rate in adults with cerebral palsy. J Amer Diet Assoc. 1996;96:145–48.

36. Gray RF, Indurkhya A, McCormick MC. Prevalence, stability, and predictors of clinically significant behavior problems in low birth weight children at 3, 5, and 8 years of age. Pediatrics. 2004;114:736–43.

37. Zachor DA et al. Effects of long-term psychostimulant medication on growth of children with ADHD. Research in Developmental Disabilities. 2006;27:162–74.

38. Cala S, Crismon ML, Baumgartner J. A survey of herbal use in children with attention-deficit-hyperactivity disorder or depression. Pharmacotherapy. 2003; 23:222–30.

39. Kleinman RE, ed. Pediatric nutrition handbook, 5th ed. Chicago IL: American Academy of Pediatrics; 2004:671–88.

40. National Down Syndrome Society public position statements vitamin-related therapies. Available at www.ndss.org, accessed 10/2006.

41. Story M, Holt K, Sufka D, eds. Bright futures in practice: nutrition. Arlington, VA: National Center for Education and Child Health; 2000: 266–70.

42. Maternal and Child Health Bureau, Health and Human Services. Available at http://mchb.hrsa.gov/html/drte.html, accessed 10/06.

"The willingness of adolescents to try out new behaviors creates a unique opportunity for nutrition education and health promotion. Adolescence is an especially important time in the life cycle for nutrition education since dietary habits adopted during this period are likely to persist into adulthood."

Jamie Stang

Chapter 14

Adolescent Nutrition

Chapter Outline

- Introduction
- Nutritional Needs in a Time of Change
- Normal Physical Growth and Development
- Normal Psychosocial Development
- Health and Eating-Related Behaviors during Adolescence
- Energy and Nutrient Requirements of Adolescents
- Nutrition Screening, Assessment, and Intervention
- Physical Activity and Sports
- Promoting Healthy Eating and Physical Activity Behaviors

Prepared by **Jamie Stang** with the assistance of **Shira Feldman** and **Mary Story**

Key Nutrition Concepts

1 Nutrition needs should be determined by the degree of sexual maturation and biological maturity (biological age) instead of by chronological age.

2 Unhealthy eating behaviors common among adolescents include frequent dieting, meal skipping, use of unhealthy dieting practices, and frequent consumption of foods high in fat and sugar, such as fast foods, soft drinks, and savory snacks.

3 Concrete thinking and abstract reasoning abilities do not develop fully until late adolescence or early adulthood; therefore, education efforts need to be highly specific and based on concrete principles.

4 Adolescent eating behaviors are influenced by a variety of factors including peer influences, parental modeling, food availability, food preferences, cost, convenience, personal and cultural beliefs, mass media, and body image.

5 Family meals decline during adolescence, but they are an important factor in improving the nutritional quality of adolescents' diets.

6 A small number of adolescents are meeting nutritional requirements for fruits, vegetables, whole grains, and calcium intake. At the same time many adolescents exceed daily energy requirements.

7 Nutrition messages for adolescents need to focus on what is important to their lives. Focusing on the present and how good nutrition can positively impact appearance, sports performance, or academic performance is likely to have greater impact than focusing on long-term disease prevention.

8 Calcium intake is important during adolescence for development of peak bone mass, but most adolescents consume less than half of the recommended intake. Participation in physical activity during adolescence also plays a role in the development of bone mass.

Introduction

Adolescence is defined as the period of life between 11 and 21 years of age. It is a time of profound biological, emotional, social, and cognitive changes during which a child develops into an adult. Physical, emotional, and cognitive maturity is accomplished during adolescence (Illustration 14.1). Many adults view adolescence as a tumultuous, irrational phase that children must go through. However, this view does disservice to its developmental importance. The tasks of adolescence, not unlike those experienced during the toddler years, include the development of a personal identity and a unique value system separate from parents and other family members, a struggle for personal independence accompanied by the need for

Illustration 14.1 Average ages of pubertal, cognitive, and psychosocial maturation.

SOURCE: Reprinted with permission from Johnson, RL. Adolescent Growth and Development. In: Hoffman A, Greydanus D, eds. Adolescent Medicine. New York: McGraw Hill; 1988.

economic and emotional family support, and the adjustment to a new body that has changed in shape, size, and physiological capacity. When the seemingly irrational behaviors of adolescents are reframed as essential endeavors and viewed in light of these developmental tasks, adolescence can and should be viewed as a unique, positive, and integral part of human development.

Nutritional Needs in a Time of Change

The biological, psychosocial, and cognitive changes associated with adolescence have direct effects on nutritional status. The dramatic physical growth and development experienced by adolescents significantly increases their needs for energy, protein, vitamins, and minerals. However, the struggle for independence that characterizes adolescent psychosocial development often leads to the development of health-compromising eating behaviors, such as excessive dieting, meal skipping, use of unconventional nutritional and nonnutritional supplements, and the adoption of fad diets. These disparate situations create a great challenge for health care professionals. The challenging behaviors of adolescents can become opportunities for change at a time during which adult health behaviors are being formed. The search for personal identity and independence among adolescents can lead to positive, health-enhancing behaviors such as adoption of healthful eating practices, participation in competitive and noncompetitive physical activities, and an overall interest in developing a healthy lifestyle. These interests and behaviors provide a good foundation on which nutrition education can build.

This chapter provides an overview of normal biological and psychosocial growth and development among adolescents and how these experiences affect the nutrient needs and eating behaviors of teens. Common concerns related to adolescent nutrition and effective methods for educating and counseling teens are also discussed.

Normal Physical Growth and Development

Early adolescence encompasses the occurrence of *puberty*, the physical transformation of a child into a young adult. The biological changes that occur during puberty include sexual maturation, increases in height and weight, accumulation of skeletal mass, and changes in body composition. Even though the sequence of these events during puberty is consistent among adolescents, the age of onset, duration, and tempo of these events vary a great deal between and within individuals. Thus, the physical appearance of adolescents of the same chronological age covers a wide range. These variations directly affect the nutrition requirements of adolescents. A 14-year-old male who has already experienced rapid linear growth and muscular development will have noticeably different energy and nutrient needs than a 14-year-old male peer who has not yet entered puberty. For this reason, sexual maturation (or biological age) should be used to assess biological growth and development and the individual nutritional needs of adolescents rather than chronological age.

Sexual Maturation Rating (SMR), also known as "Tanner Stages," is a scale of *secondary sexual characteristics* that allows health professionals to assess the degree of pubertal maturation among adolescents, regardless of chronological age (Table 14.1). SMR is based on breast development and the appearance of pubic hair among females, and on testicular and penile development and the appearance of pubic hair among males.[1] SMR stage 1 corresponds with prepubertal growth and development, while stages 2 through 5 denote the occurrence of puberty. At SMR stage 5, sexual maturation has concluded. Sexual maturation correlates highly with linear growth, changes in weight and body composition, and hormonal changes.[1]

The onset of *menses* and changes in height relative to the development of secondary sexual characteristics that occur in females during puberty are shown in Illustration 14.2. Among females, the first signs of puberty are the development of breast buds and sparse, fine pubic hair occurring on average between 8 to 13 years

Puberty The time frame during which the body matures from that of a child to that of a young adult.

Secondary Sexual Characteristics Physiological changes that signal puberty, including enlargement of the testes, penis, and breasts and the development of pubic and facial hair.

Menses The process of menstruation.

Table 14.1 Sexual maturity rating for girls and boys

Girls

Stage	Breast Development	Pubic Hair Growth
1	Prepubertal; nipple elevation only	Prepubertal; no pubic hair
2	Small, raised breast bud	Sparse growth of hair along labia
3	General enlargement of raising of breast and areola	Pigmentation, coarsening, and curling, with an increase in amount
4	Further enlargement with projection of areola and nipple as secondary mound	Hair resembles adult type, but not spread to medial thighs
5	Mature, adult contour, with areola in same contour as breast, and only nipple projecting	Adult type and quantity, spread to medial thighs

Boys

Stage	Genital Development	Pubic Hair Growth
1	Prepubertal; no change in size or proportion of testes, scrotum, and penis from early childhood	Prepubertal; no pubic hair
2	Enlargement of scrotum and testes; reddening and change in texture in skin of scrotum; little or no penis enlargement	Sparse growth of hair at base of penis
3	Increase first in length, then width of penis; growth of testes and scrotum	Darkening, coarsening, and curling; increase in amount
4	Enlargement of penis with growth in breadth and development of glands; further growth of testes and scrotum, darkening of scrotal skin	Hair resembles adult type, but not spread to medial thighs
5	Adult size and shape genitalia	Adult type and quantity, spread to medial thighs

SOURCE: Reprinted with permission. Tanner, JM. Growth at Adolescence. Oxford: Blackwell; 1962.

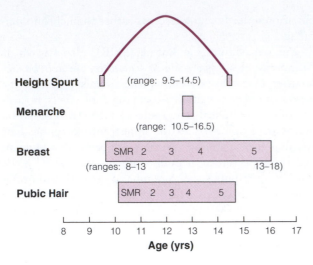

Illustration 14.2 Sequence of physiological changes during puberty in females.

SOURCE: From J. M. Tanner, Growth at Adolescence. Copyright © 1962 Blackwell Publishers. Reprinted with permission.

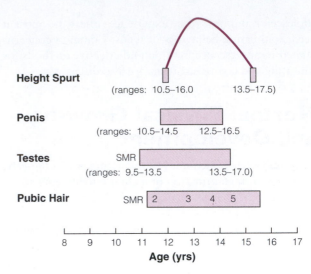

Illustration 14.3 Sequence of physiological changes during puberty in males.

SOURCE: From J. M. Tanner, Growth at Adolescence. Copyright © 1962 Blackwell Publishers. Reprinted with permission.

of age (SMR stage 2). *Menarche* occurs 2 to 4 years after the initial development of breast buds and pubic hair, most commonly during SMR stage 4. The average age of menarche is 12.4 years, but menarche can occur as early as 9 or 10 years or as late as 17 years of age. Menarche may be delayed in highly competitive athletes or in girls who severely restrict their caloric intake to limit body fat.

> **Menarche** The occurrence of the first menstrual cycle.
>
> **Testes** One of the two male reproductive glands located in the scrotum.

Ethnic and racial differences are evident in the initiation of sexual maturation among females.[2] A recent study found differences in the median age for pubic hair development among young females.[2] Non-Hispanic black females had developed pubic hair at 9.4 years, compared to 10.5 years for non-Hispanic white and 10.4 years for Mexican American females. The development of breast buds also varied by race, but to a lesser extent; they were noted among non-Hispanic black girls by 9.5 years, among Latina girls by 9.8 years, and among white females by 10.4 years of age. Pubic hair development appeared 2 years earlier among African American girls (average age 8.7 years) than among other adolescent females. The onset of menarche was similar among all racial and ethnic groups of females.

Onset of the linear growth spurt occurs most commonly during SMR stage 2 in females, beginning between the ages of 9.5 to 14.5 years in most females (Illustration 14.2). Peak velocity in linear growth occurs during the end of SMR stage 2 and during SMR stage 3, approximately 6 to 12 months prior to menarche. As much as 15% to 25% of final adult height will be gained during puberty, with an average increase in height of 9.8 inches (25 cm).[3] During the peak of the adolescent growth spurt, females gain approximately 3.5 inches (8–9 cm) a year.

The linear growth spurt lasts 24 to 26 months, ceasing by age 16 in most females. Some adolescent females experience small increments of growth past age 19 years, however. Linear growth may be delayed or slowed among females who severely restrict their caloric intake.

Enlargement of the *testes* and change in *scrotal* coloring are most often the first signs of puberty among males (Illustration 14.3), occurring between the ages of 10.5 and 14.5 years with 11.6 years being the average age. The development of pubic hair is also common during SMR stage 2. Testicular enlargement begins between the ages of 9.5 to 13.5 years in males (SMR 2 to 3) and concludes between the ages of 12.7 and 17 years. The average age of "spermarche" is approximately 14 years among males. Clearly, males show a great deal of variation in the chronological age at which sexual maturation takes place.

On average, peak velocity of linear growth among males occurs during SMR stage 4, coinciding with or just following testicular development and the appearance of faint facial hair. The peak velocity of linear growth occurs at 14.4 years of age, on average. At the peak of the growth spurt, adolescent males will increase their height by 2.8 to 4.8 inches (7 to 12 cm) a year. Linear growth will continue throughout adolescence, at a progressively slower rate, ceasing about 21 years of age.

Changes in Weight, Body Composition, and Skeletal Mass

As much as 50% of ideal adult body weight is gained during adolescence. Among females, peak weight gain follows the linear growth spurt by 3 to 6 months. During the peak velocity of weight change, which occurs at an average age of 12.5 years, girls will gain approximately

18.3 lb (8.3 kg) per year.[3] Weight gain slows around the time of menarche, but will continue into late adolescence. Adolescent females may gain as much as 14 lb (6.3 kg) during the latter half of adolescence. Peak accumulation of muscle mass occurs around or just after the onset of menses.

Body composition changes dramatically among females during puberty, with average lean body mass falling from 80% to 74% of body weight while average body fat increases from 16% to 27% at full maturity. Females experience a 44% increase in lean body mass and a 120% increase in body fat during puberty.[4] Adolescent females gain approximately 2.5 lb (1.14 kg) of body fat mass each year during puberty. Adolescent body fat levels peak among females between the ages of 15 to 16 years. Research by Rose Frisch suggests that a level of 17% body fat is required for menarche to occur and that 25% body fat is required for the development and maintenance of regular ovulatory cycles.[5] Normal changes in body fat mass can be mediated by excessive physical activity and/or severe caloric restriction.

Even though the accumulation of body fat by females is obviously a normal and physiologically necessary process, adolescent females often view it negatively. Weight dissatisfaction is common among adolescent females during and immediately following puberty, leading to potentially health-compromising behaviors such as excessive caloric restriction, chronic dieting, use of diet pills or laxatives, and, in some cases, the development of body image distortions and eating disorders.

Among males, peak weight gain coincides with the timing of peak linear growth and peak muscle mass accumulation.[3] During peak weight gain, adolescent males gain an average of 20 lb (9 kg) per year. Body fat decreases in males during adolescence, resulting in an average of approximately 12% by the end of puberty.

Almost half of adult peak bone mass is accrued during adolescence. By age 18, more than 90% of adult skeletal mass has been formed.[6] A variety of factors contribute to the accretion of bone mass, including genetics, hormonal changes, weight-bearing exercise, cigarette smoking, consumption of alcohol, and dietary intake of calcium, vitamin D, protein, phosphorus, boron, and iron. Because bone is comprised largely of calcium, phosphorus, and protein and because a great deal of bone mass is accrued during adolescence, adequate intakes of these nutrients are critical to support optimal bone growth and development.

Normal Psychosocial Development

During adolescence, an individual develops a sense of personal identity, a moral and ethical value system, feelings of self-esteem or self-worth, and a vision of occupational aspirations. Psychosocial development is most readily understood when it is divided into three periods: early

adolescence (11 to 14 years), middle adolescence (15 to 17 years), and late adolescence (18 to 21 years). Each period of psychosocial development is marked by the mastery of new emotional, cognitive, and social skills (Table 14.2).

During early adolescence, individuals begin to experience dramatic biological changes related to puberty. The development of body image and an increased awareness of sexuality are central psychosocial tasks during this period of adolescence. The dramatic changes in body shape and size can cause a great deal of ambivalence among adolescents, leading to the development of poor body image and eating disturbances if not addressed by family or health care professionals.

Peer influence is very strong during early adolescence. Young teens, conscious of their physical appearance and social behaviors, strive to "fit in" with their peer group. The need to fit in can affect nutritional intake among adolescents. Focus groups conducted with adolescent females revealed that situational factors, such as who students ate with and where they ate, were important factors in the food choices they made.[7] Consequently, teens express their ability and willingness to fit in with a group of peers by adopting food preferences and making food choices based on peer influences and by refuting family preferences and choices.

The wide chronological age range during which pubertal growth and development begins and proceeds can become a major source of personal dissatisfaction for many adolescents. Males considered to be "late bloomers" often feel inferior to their peers who mature earlier and may resort to the use of anabolic steroids and other supplements in an effort to increase linear growth and muscle development. Females who mature early have been found to have more eating problems and poorer body image than their later developing peers.[8] They are also more likely to initiate "grown-up" behaviors such as smoking, drinking alcohol, and engaging in sexual intercourse at an earlier age.[8,9] Education of young adolescents on normal variations in tempo and timing of growth and development can help to facilitate the development of a positive self-image and body image and may reduce the likelihood of early initiation of health-compromising behaviors.

Cognitively, early adolescence is a time dominated by concrete thinking, egocentrism, and impulsive behavior. Abstract reasoning abilities are not yet developed to a great extent in most adolescents, limiting their ability to understand complex health and nutrition issues. Young adolescents also lack the ability to see how their current behavior can affect their future health status or health-related behaviors.

Middle adolescence marks the development of emotional and social independence from family, especially parents. Conflicts over personal issues, including eating and physical activity behaviors, are heightened during midadolescence. Peer groups become more influential, and their influence on food choices peaks. Physical growth and development are mostly completed during this stage.

Table 14.2 Psychosocial processes and the substages of adolescent development

Substage	Emotional and Social Development	Cognitive Development
Early adolescence	• Alterations in body image secondary to dramatic changes in body shape and size • Increased awareness of sexuality • Strong need for social acceptance by peers • Strong sense of impulsivity	• Concrete thinking processes are dominant, often with limited abstract thought capacity
Middle adolescence	• Development of greater autonomy from parents and family • Continued need for peer acceptance, which may lead to risk-taking behaviors • Increased opportunities for employment outside of home, resulting in more decision making and the beginning of economic independence • Increased awareness of moral and social issues	• Development of abstract reasoning continues • May revert to concrete thinking when under stress
Late adolescence	• Further development of personal set of morals and values • Increased impulse control • Greater social, emotional, and economic independence from family • Reduced need for peer acceptance • Development of personal and vocational goals	• Abstract thought capacity fully develops

SOURCE: Adapted from Ingersoll GM. Psychological and Social Development. In: Textbook of Adolescent Medicine. Saunders Publishing; 1992.

Body image issues are still of concern, especially among males who are late to mature and females. Peer acceptance remains important, and the initiation of and participation in health compromising behaviors often occurs during this stage of development. Adolescents may believe they are invincible during this stage of development.

The emergence of abstract reasoning skills occurs rapidly during middle adolescence; however, these skills may not be applied to all areas of life. Adolescents will revert to concrete thinking skills if they feel overwhelmed or experience psychosocial stress. Teens begin to understand the relationship between current health-related behaviors and future health, even though their need to "fit in" may supplant this understanding.

Late adolescence is characterized by the development of a personal identity and individual moral beliefs. Physical growth and development is largely concluded, and body image issues are less prevalent. Older teens become more confident in their ability to handle increasingly sophisticated social situations, which is accompanied by reductions in impulsive behaviors and peer pressure. Adolescents become increasingly less economically and emotionally dependent on parents. Relationships with one individual become more influential than the need to fit in with a group of peers. Personal choice emerges.

Self-Efficacy The ability to make effective decisions and to take responsible action based upon one's own needs and desires.

Abstract thinking capabilities are realized during late adolescence, which assists teens in developing a sense of future goals and interests. Adolescents are now able to understand the perspectives of others and can fully perceive future consequences associated with current behaviors. This capability is especially important among adolescent females who plan to have children or who become pregnant.

Health and Eating-Related Behaviors during Adolescence

Eating patterns and behaviors of adolescents are influenced by many factors, including peer influences, parental modeling, food availability, food preferences, cost, convenience, personal and cultural beliefs, mass media, and body image. Illustration 14.4 presents a conceptual model of the many factors that influence eating behaviors of adolescents. The model depicts three interacting levels of influence that impact adolescent eating behaviors: personal or individual, environmental, and macrosystems. Personal factors that influence eating behavior include attitudes, beliefs, food preferences, *self-efficacy*, and biological changes. Environmental factors include the immediate social environment such as family,

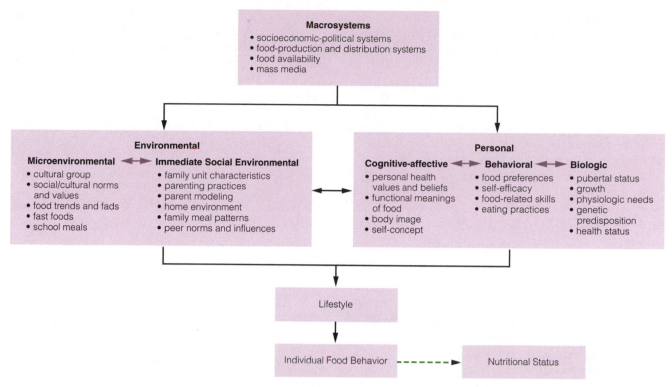

Illustration 14.4 Conceptual model for factors influencing eating behavior of adolescents.

SOURCE: Reprinted with permission. Story M, Alton I, Becoming a Woman: Nutrition in Adolescence. In: Krummel DA, Kris-Etherton PM, eds. Nutrition in Women's Health. Gaithersburg, Md: Aspen Publishers; 1996.

friends, and peer networks, and other factors such as school, fast food outlets, and social and cultural norms. Macrosystem factors, which include food availability, food production and distribution systems, and mass media and advertising, play a more distal and indirect role in determining food behaviors, yet can exert a powerful influence on eating behaviors. To improve the eating patterns of youth, nutrition interventions should be aimed at each of the three levels of influence. In addition to these influencing factors, socioeconomic status and racial/ethnic background also play a role in shaping adolescent eating patterns. Adolescents of a lower socioeconomic status and from minority backgrounds have been found to be more likely to engage in less healthy eating behaviors, which often leads to higher prevalence of overweight.[10]

Eating habits of adolescents are not static; they fluctuate throughout adolescence in relation to psychosocial and cognitive development. Longitudinal data of adolescent females suggest that even though body weight percentiles track throughout adolescence, little consistency guides the intakes of the energy, nutrients, vitamins, and minerals from early to late adolescence.[11] Health professionals must therefore refrain from jumping to conclusions about the dietary habits of adolescents (even if they have been evaluated for nutritional status at an earlier age) and take the time to assess the current dietary intake of the individual.

Adolescents lead busy lives. Many are involved in extracurricular sports or academic activities, others are employed, and many must care for younger children in a family for part of the day. These activities, combined with the increased need for social and peer contact and approval, and increasing academic demands as they proceed through school, leave little time for adolescents to sit down to eat a meal. Snacking and meal skipping are commonplace among adolescents. Almost all adolescents consume at least one snack per day, with a range of one to seven.[12,13] One study of adolescents found that during an average week, males ate 18.2 meals and 10.9 snacks, while females ate 16.9 meals and 9.9 snacks. Snacks account for 25% to 33% of daily energy intakes among adolescents.[13] National data suggest that the proportion of calories and nutrients from foods consumed as snacks has risen during the past decade.[13,14] The average size of snacks has remained approximately the same, but the number of snacks consumed has increased, which accounts for the increased caloric intake.[13] The increased calorie intake from snacks has paralleled the increase in consumption of foods outside of the home. Adolescents are consuming a larger proportion of calories from snacking away from home, often at fast food restaurants.[14]

Unfortunately, the food choices made by adolescents while snacking tend to be high in sugar, sodium, and fat, while relatively low in vitamins and minerals. Soft drinks are the most commonly chosen snacks for adolescent females and account for about 6% of total caloric intake.[13,14] This trend causes significant concern because high consumption of soft drinks may increase the risk for bone fractures over an individual's lifetime.[15] Practioners working with adolescents need to understand that snacking is a commonplace behavior among adolescents and work with adolescents to improve food choices. Due to the frequency of snacking, an improvement in food choices toward more nutrient-dense foods has the potential to positively impact eating behaviors of adolescents.

The occurrence of meal skipping increases as adolescents mature. Breakfast is the most commonly skipped meal; only 29% of adolescent females tend to eat breakfast daily.[16,17] Skipping breakfast can dramatically decrease intakes of energy, protein, fiber, calcium, and folate due to the absence of breakfast cereal or other nutrient-dense foods commonly consumed at breakfast. Lunch is skipped by almost one-quarter of adolescents.[16–18] As with breakfast, skipping lunch reduces intakes of energy, protein, and other nutrients. Adolescents who skip meals should be counseled on convenient, portable, and healthy food choices that can be taken with them and eaten as meals or snacks.

As adolescents mature, they spend less time with family and more time with their peer group. Eating away from home becomes prevalent: female adolescents, for example, eat almost one-third of their meals away from home and the average teen eats at a fast food restaurant twice a week. Fast food visits account for 31% of all food eaten away from home, and make up 83% of adolescent visits to restaurants.[16,18] Fast food restaurants and food courts are favorite eating places of teens for several reasons:

- They offer a social setting with an informal, comfortable atmosphere for adolescents.
- Fast foods are relatively inexpensive and offer socially acceptable choices.
- Fast foods can be eaten outside the restaurant, fitting into the busy schedules of adolescents.
- Service is fast and the limited offerings allow for quick decision making.
- Fast food restaurants employ many adolescents, increasing the social value of these restaurants.

Eating at fast food restaurants has direct bearing on the nutritional status of adolescents. Many fast foods are high in fat and low in fiber and nutrients. However, specific choices increase the nutrient content of fast food meals and decrease the fat content. Adolescents can be counseled to ask for juice, water, or milk instead of soft drinks, order small sandwiches instead of larger choices, choose a salad or baked potato as a side dish instead of french fries, order grilled items as opposed to fried sandwiches, and avoid "super sizing" meals even if it seems to offer a better economic deal. Fruit cups, pancakes, and fruit and yogurt parfaits are better choices for fast food breakfasts than breakfast sandwiches.

As adolescents spend more time with peers, participation in family meals tends to steadily decline.[19–21] This is an unfortunate occurrence as more frequent family meals are associated with improved dietary intake among adolescents, including higher intakes of grains, fruits, and vegetables, and decreased intake of soft drinks. Vitamin and mineral intakes including calcium, folate, fiber, iron, and vitamins A, C, E, B_6, and B_{12}, are higher among adolescent who frequently consume family meals.[19–22] A greater number of families that are eating together do so in front of the television, with a national sample of adolescents indicating that 64% of 11- to 18-year-olds had the TV on during meals.[23] Adolescents eating meals in front of the TV, even with family, may be more susceptible to food advertisements and product placements, which tend to be for higher-calorie and low-nutrient-dense foods.

Vegetarian Diets

The term *vegetarian* is used quite broadly and can consist of many different eating patterns. Table 14.3 lists the most common vegetarian diet patterns along with the foods most commonly excluded. Among low-literacy populations or those who do not speak English as a first language, vegetarian may be thought to mean that a person eats vegetables. Therefore, health professionals should ask adolescents to define what type of vegetarian diet they consume and to elicit a complete list of foods that are avoided.

The prevalence of vegetarianism among adolescents is small—approximately 1% of adolescents report consuming a vegetarian diet.[24] Adolescents adopt vegetarian eating plans for a variety of reasons, including cultural or religious beliefs, moral or environmental concerns, health beliefs, as a means to restrict calories and/or fat intake,

Table 14.3 Types of vegetarian diets and foods excluded

Type of Vegetarian Diet	Foods Excluded
Semi- or partial-vegetarian	Red meat
Lacto-ovo-vegetarian	Meat, poultry, fish, seafood
Lacto-vegetarian	Meat, poultry, fish, seafood, eggs
Vegan (total vegetarian)	Meat, poultry, fish, seafood, eggs, dairy products (may exclude honey)
Macrobiotic	Meat, poultry, eggs, dairy, seafood, fish (fish may be included in the diets of some macrobiotic vegetarians)

SOURCE: Reprinted with permission. Hadded E. Johnston P. Vegetarian Diets and Pregnant Teens. In: Story M, Stang J, eds. Nutrition and the Pregnant Adolescent: A Practical Reference Guide. Minneapolis. MN: Center for Leadership, Education, and Training in Maternal and Child Nutrition, University of Minnesota; 2000.

Case Study 14.1

Nutritional Issues and the Vegan Adolescent

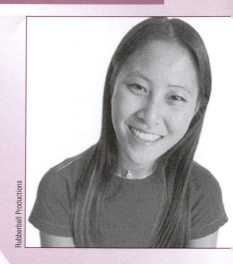

Rutherball Productions

Nikki is a 13-year-old female who is being seen by her family physician for her annual physical examination. Nikki's sexual maturation rating is estimated at between 2 and 3. Her body mass index (BMI) is plotted at just under the 20th percentile. During the visit, Nikki reports that she is following a strict vegan diet. She explains that she thinks it is a healthier way to eat and states that she avoids meat, poultry, fish, seafood, eggs, and dairy products.

Questions

1. How might a vegetarian diet affect an adolescent's growth and maturation?
2. Why is protein a concern in Nikki's diet?
3. Which vitamins and minerals may be deficient in Nikki's diet? For each vitamin or mineral, list possible food sources that are appropriate for a vegan diet.
4. Is it necessary to assess Nikki for the presence of an eating disorder? Explain your answer.

and as a means of exerting independence by adopting eating behaviors that differ from those of the teen's family. Regardless of the reason for consuming a vegetarian diet, the adolescent's diet should be thoroughly assessed for nutritional adequacy. As a rule, the more foods that are restricted in the diet, the more likely it is that nutritional deficiencies will result.

Vegetarian adolescents have been found to be shorter and leaner than omnivores during childhood and to enter puberty at a later age. On average, menarche occurs 6 months later in vegetarians than among omnivores.[25] After puberty, vegetarian adolescents are as tall or taller than omnivores and are generally leaner, although final adult height may be reached at a later age.[25,26]

Well-planned vegetarian diets can offer many health advantages to adolescents, such as a high intake of complex carbohydrates and relatively high intake of vitamins and minerals found in plant-based foods. Data suggest that vegetarian adolescents are twice as likely to consume fruits and vegetables, one-third as likely to consume sweets, and one-fourth as likely to consume salty snack foods once per day.[24] When well planned, vegetarian diets can provide adequate protein to promote growth and development among pubescent adolescents, particularly if small amounts of animal-derived foods, such as milk or cheese, are consumed at least two times per week. If vegetarian diets restrict intake of all animal-derived food products, however, careful attention must be paid to assure adequate intakes of protein, calcium, zinc, iron, and vitamins D, B_6, and B_{12}.[26] Supplements of vitamin B_{12} are often required among vegans. A suggested dietary food guide for adolescent vegetarians is listed in Table 14.4.

Table 14.4 Suggested daily food guide for lacto-ovo vegetarians and vegan adolescents at various intake levels

Food Groups	Servings per Day, by Daily Caloric Intake	
	Lacto-Ovo Vegetarians 11+ years (2200–2800 Kcal)	Vegans 11+ years (2200–2800 Kcal)
Breads, grains, cereal	9–11	10–12
Legumes	2–3	2–3
Vegetables	4–5	1
Fruits	4	3–4
Nuts, seeds	1	4–6
Milk, yogurt, cheese	3	1
Eggs (limit 3/week)	½	3
Fats, oils (added)	4–6	4–6
Sugar (added teaspoons)	6–9	6–9

SOURCE: Data used with permission from: Haddad EH. Development of a Vegetarian Food Guide. Am J Clin Nutr. 1994;1248S–1254S; and Story M, Holt K, Sofka D, eds. Bright Futures in Practice: Nutrition. Arlington, VA: National Center for Education in Maternal and Child Health; 2000.

Adolescents who consume vegan diets must also be assessed for adequacy of total fat and essential fatty acid intakes. Docosahexaenoic acid (DHA) is derived from alpha-linolenic acid. Although it is found in soy products, flaxseed, nuts, eggs, and canola oil, intake is very low in the

Table 14.5 Plant sources of alpha-linolenic acid

Food Source	Alpha-Linolenic Acid, g
Flaxseed, 2 Tbsp	4.3
Walnuts, 1 oz	1.9
Walnut oil, 1 Tbsp	1.5
Canola oil, 1 Tbsp	1.6
Soybean oil, 1 Tbsp	0.9
Soybeans, ½ c cooked	0.5
Tofu, ½ c	0.4

SOURCE: Reprinted with permission. Haddad E, Johnston P. Vegetarian Diets and Pregnant Teens. In: Story M, Stang J, eds. Nutrition and the Pregnant Adolescent: A Practical Reference Guide. Minneapolis. MN: Center for Leadership, Education, and Training in Maternal and Child Nutrition, University of Minnesota; 2000.

Table 14.6 Percentage of adolescents meeting the recommended number of MyPyramid servings for select food groups

	Male (%)	Female (%)
Dairy products	33	17
Fruits	23	27
Vegetables	35	34
Grains	64	48
Meat	50	17

SOURCE: Pyramid Serving Intakes in the U.S. 1999–2002, 1 Day. Community Nutrition Research Group. Beltsville Human Nutrition Research Center. Agricultural Research Service, USDA. Available online at www.ba.urs.usda.gov/enrg (accessed November 2, 2006).

diets of vegans. Diets that are low in fat may not supply an adequate ratio of linoleic acid to alpha-linolenic acid (5:1 to 10:1) in order to facilitate the metabolism of alpha-linolenic acid to DHA.[26] Therefore, particular attention should be paid to sources of fat in the diets of vegans and other vegetarians with low fat intakes. Some plant sources of alpha-linolenic acid are shown in Table 14.5.

Adolescents who consume a vegetarian diet, particularly if they report doing so for health- or weight-related reasons, should be carefully assessed for the presence of eating disorders, chronic dieting, and body image disturbances. Neumark-Sztainer and colleagues have found that vegetarian adolescents are somewhat more likely to report binge eating, almost twice as likely to report frequent or chronic dieting, four times more likely to report purging, and eight times more likely to report laxative use than nonvegetarian peers.[24] These results seem to stem from the fact that many individuals who are chronic dieters or who have eating disorders adopt a vegetarian diet as a means of self-denial or self-control rather than the vegetarian diet causing these behaviors. Adolescent vegetarians, especially adolescent males, have been found to be at high risk for engaging in unhealthy and often extreme weight loss behaviors.[27] Therefore, it is imperative that practioners working with adolescents who are following a vegetarian diet explore reasons for adopting a vegetarian diet as well as help adolescents follow a nutritionally balanced and healthy diet.

Dietary Intake and Adequacy Among Adolescents

Data on food intakes of U.S. adolescents suggest that many adolescents consume diets that do not match the Dietary Guidelines for Americans or the MyPyramid recommendations.[28,29] Inadequate consumption of dairy products, grain products, fruits, and vegetables is commonplace among adolescents (Table 14.6).

Few adolescents meet recommendations for fruit or vegetable consumption.[30] Data from NHANES suggests that 35% of adolescent males and 34% of adolescent females met recommendations for vegetable intake while 23% of males and 27% of females met recommendations for fruit intake.[30] Approximately 30% of adolescents consumed less than one serving of vegetables each day, with white potatoes making up half of the vegetables consumed by teens. Almost two-thirds (62%) of male and 57% of female teens consumed less than one serving of fruit per day. Fruit and vegetable intake by teens is not adequate to promote optimal health and reduce risk of chronic diseases.

Adequate intake of servings of grains were reported by 64% of teenage males and 48% of teen females; however, intake of whole grains was below recommended levels.[30] Intake of meat/meat alternatives was low among adolescents surveyed in NHANES.[30] Ten percent of males and 18% of females reported less than one serving per day with only 50% of males and 17% of females meeting recommended intakes. Intake of dairy products was especially low among adolescents. One-third of teen males and 17% of teen females met recommendations for dairy intake according to NHANES data.[30] More than one-third (39%) of female and 29% of male teens reported less than one intake of dairy per day.

It was reported that fat made up 32% of all energy consumed by adolescents.[30] Added sugars accounted for 21% of energy intake by teens. Male teens consumed 35 tsp of added sugar each day while female teens consumed 26 tsp per day. Clearly, adolescents do not consume diets that comply with the national nutrition recommendations or provide the recommended level of intakes for all food groups.

Energy and Nutrient Requirements of Adolescents

Increases in lean body mass, skeletal mass, and body fat that occur during puberty result in energy and nutrient needs that exceed those of any other point in life. Energy and nutrient requirements of adolescence correspond with the degree of physical maturation that has taken place. Unfortunately, few available data define optimal nutrient and energy intakes during adolescence. Most existing data is extrapolated from adult or child nutritional requirements. Recommended intakes of energy, protein, and some other nutrients are based on adequate growth as opposed to optimal physiological functioning. The Dietary Reference Intakes (DRIs) provide the best estimate of nutrient requirements for adolescents (Table 14.7). It should be

Table 14.7 Dietary reference intakes of selected nutrients for preadolescents and adolescents

Life-Stage Group	Calcium (mg/d)	Phosphorus (mg/d)	Magnesium (mg/d)	Vitamin D (mg/d)[a,b]	Fluoride (mg/d)	Thiamin (mg/d)	Riboflavin (mg/d)	Niacin (mg/d)[c]
Males								
9–13 years	1300*	**1250**	240	5*	2*	0.9	0.9	12
14–18 years	1300*	**1250**	410	5*	3*	1.2	1.3	16
19–30 years	1000*	**700**	400	5*	4*	1.2	1.3	16
Females								
9–13 years	1300*	**1250**	240	5*	2*	0.9	0.9	12
14–18 years	1300*	**1250**	360	5*	3*	1.0	1.0	14
19–30 years	1000*	**700**	310	5*	3*	1.1	1.1	14
Pregnancy								
<18 years	1300*	**1250**	400	5*	3*	1.4	1.4	18
19–30 years	1000*	**700**	350	5*	3*	1.4	1.4	18
Lactation								
<18 years	1300*	**1250**	360	5*	3*	1.4	1.6	17
19–30 years	1000*	**700**	310	5*	3*	1.4	1.6	17

Life-Stage Group	Vitamin B$_6$ (mg/d)	Folate, (mg/d)	Vitamin B$_{12}$ (mg/d)	Pantothenic Acid (mg/d)	Biotin (mg/d)	Choline (mg/d)[d]	Vitamin C (mg/d)	Vitamin E (mg/d)	Selenium (mg/d)
Males									
9–13 years	1.0	**300**	**1.8**	4*	20*	375*	**45**	**11**	**40**
14–18 years	1.3	**400**	**2.4**	5*	25*	550*	**75**	**15**	**55**
19–30 years	1.3	**400**	**2.4**	5*	30*	550*	**90**	**15**	**55**
Females									
9–13 years	1.0	**300**	**1.8**	4*	20*	375*	**45**	**11**	**40**
14–18 years	1.2	**400**[e]	**2.4**	5*	25*	400*	**65**	**15**	**55**
19–30 years	1.3	**400**[e]	**2.4**	5*	30*	425*	**75**	**15**	**55**
Pregnancy									
<18 years	1.9	**600**	**2.6**	6*	30*	450*	**80**	**15**	**60**
19–30 years	1.9	**600**	**2.6**	6*	30*	450*	**85**	**15**	**60**
Lactation									
<18 years	2.0	**500**	**2.8**	7*	35*	550*	**115**	**19**	**70**
19–30 years	2.0	**500**	**2.8**	7*	35*	550*	**120**	**19**	**70**

NOTE: This table presents Recommended Dietary Allowances (RDAs) in **bold type** and Adequate Intakes (AIs) in ordinary type followed by an asterisk (*). RDAs and AIs may both be used as goals for individual intake. RDAs are set to meet the needs of almost all (97–98%) individuals in a group. For healthy and breastfed infants, the AI is the mean intake. The AI for other life stage and gender groups is believed to cover needs of all individuals in the group, but lack of data or uncertainty in the data prevent being able to specify with confidence the percentage of individuals covered by this intake.

[a] As cholecalciferol. 1 mg cholecalciferol = 40 IU vitamin D.

[b] In the absence of adequate exposure to sunlight.

[c] As niacin equivalents (NE). 1 mg of niacin = 60 mg tryptophan; 0–6 months = preformed niacin (not NE).

[d] Although AIs have been set for choline, there are few data to assess whether a dietary supplement of choline is needed at all stages of the life cycle, and it may be that the choline requirement can be met by endogenous synthesis at some of these stages.

[e] In view of evidence linking folate intake with neural tube defects in the fetus, it is recommended that all women capable of becoming pregnant consume 400 mg from supplements or fortified food until their pregnancy is confirmed and they enter prenatal care, which ordinarily occurs after the end of the periconceptional period—the critical time for formation of the neural tube.

SOURCE: Food and Nutrition Board, Institute of Medicine, National Academies. Dietary Reference Intakes: Recommended Intakes for Individuals. Washington, DC: National Academy Press; 2000.

noted, however, that these nutrient recommendations are classified according to chronological age, as opposed to individual levels of biological development. Thus, health care professionals must use prudent professional judgment based on SMR status, and not solely on chronological age, when determining the nutrient needs of an adolescent.

Nutrient intakes of U.S. adolescents suggest that many adolescents consume inadequate amounts of vitamins and minerals; this trend is more pronounced in females than males. It is not surprising, given the fact that most adolescents do not consume diets that comply with the MyPyramid or the Dietary Guidelines for Americans. On average, adolescents consume diets inadequate in several vitamins and minerals, including folate; vitamins A, B_6, C, and E; and iron, zinc, magnesium, phosphorous, and calcium (Table 14.8).[31] Dietary fiber intake among adolescents is also low. Diets consumed by many teens exceed current recommendations for total and saturated

Table 14.8 Mean intakes of selected nutrients compared to DRIs, adolescent males and females

	9- to 13-Year-Old Males				14- to 18-Year-Old Males			
	Mean Intake	RDA/AI	EAR	% < EAR	Mean Intake	RDA/AI	EAR	% < EAR
Vitamin A	670	600	445	13	638	700	630	55
Thiamin	1.78	0.9	0.7	<3	1.96	1.2	1.0	<3
Riboflavin	2.51	0.9	0.8	<3	2.57	1.3	1.1	<3
Niacin	22.5	12	9	<3	27.0	16	12	<3
Vitamin B_6	1.87	1.0	0.8	<3	2.17	1.3	1.1	<3
Vitamin B_{12}	6.0	1.8	1.5	<3	6.69	2.4	2.0	<3
Folate	644	300	250	<3	683	400	330	4
Vitamin C	80.2	45	39	8	100.0	75	63	26
Vitamin E	6.0	11	9.0	97	7.3	15	12	>97
Calcium	1139	1300	NA	NA	1142	1300	NA	NA
Phosphorous	1431	1250	1055	9	1575	1250	1055	9
Magnesium	250	240	200	14	284	410	340	78
Iron	17.0	8	5.9	<3	19.1	11	7.7	<3
Zinc	13.0	8	7.0	<3	15.1	11	8.5	4
Sodium	3549	1500	NA	NA	2806	1500	NA	NA
Fiber	14.2	31	NA	NA	15.3	38	NA	NA

	9- to 13-Year-Old Females				14- to 18-Year-Old Females			
	Mean Intake	RDA/AI	EAR	% < EAR	Mean Intake	RDA/AI	EAR	% < EAR
Vitamin A	536	600	420	34	513	700	485	54
Thiamin	1.44	0.9	0.7	<3	1.4	1.0	0.9	12
Riboflavin	1.94	0.9	0.8	<3	1.80	1.0	0.9	6
Niacin	18.5	12	9	<3	18.6	14	11	6
Vitamin B_6	1.52	1.0	0.8	<3	1.48	1.2	1.0	16
Vitamin B_{12}	4.4	1.8	1.5	<3	4.16	2.4	2.0	8
Folate	512	300	250	<3	500	400	330	19
Vitamin C	81.0	45	39	9	75.6	65	56	42
Vitamin E	5.6	11	9	95	5.6	15	12	>97
Calcium	865	1300	NA	NA	804	1300	NA	NA
Phosphorous	1141	1250	1055	42	1099	1250	1055	49
Magnesium	215	240	200	44	206	360	300	91
Iron	13.7	8	5.7	<3	13.3	15	7.9	16
Zinc	9.8	8	7.0	10	9.5	9	7.3	26
Sodium	2806	1500	NA	NA	2799	1500	NA	NA
Fiber	12.3	26	NA	NA	11.7	26	NA	NA

Data from: Moshfegh A, Goldman J, Cleveland L. 2005. NHANES 2001–2002: Usual Nutrient Intakes from Food Compared to Dietary Reference Intakes. Accessed online at www.ars.usda.gov/foodsurvey, Oct 28, 2006.

fats, cholesterol, sodium, and sugar. Data on nutrient intakes of adolescents taken from *What We Eat in America 2001–2002* suggest that more than half of teens consume less than the Estimated Average Requirement (EAR) for vitamins A and E and magnesium. More than a quarter of adolescents consume less than the EAR for vitamin C, with more than a quarter of females also consuming less than the EAR for phosphorous and zinc as well.[32] More than one-third of females consume inadequate amounts of all of these nutrients.[31]

Based on growth and development of adolescents, as well as national findings on dietary intakes of foods and nutrients, adolescent diets should be assessed for adequacy of intake of vitamins, minerals, energy, protein, carbohydrates, and fiber. Nutrients of particular concern for teens are discussed in greater detail in the following sections.

Energy

Energy needs of adolescents are influenced by activity level, basal metabolic rate, and increased requirements to support pubertal growth and development. Basal metabolic rate (BMR) is closely associated with the amount of lean body mass of individuals. Because adolescent males experience greater increases in height, weight, and lean body mass, they have significantly higher caloric requirements than females. The estimated energy requirements for adolescents are listed in Table 14.9. Due to the great variability in the timing of growth and maturation among adolescents, the determination of energy needs based on velocity of growth will provide a better estimate than one based on chronological age.

The DRI for energy is based upon the assumption of a light to moderate activity level. Therefore, adolescents who participate in sports, those who are in training to increase muscle mass, and those who are more active than average may require additional energy to meet their individual needs. Conversely, adolescents who are not physically active or those who have chronic or handicapping

conditions that limit their mobility will require less energy to meet their needs. Physical activity has been found to decline throughout adolescence, with approximately 10% of adolescents involved in no physical activity.[33] Therefore, caloric needs of older adolescents who have completed puberty and are less active may be significantly lower than those of younger, active, still-growing adolescents.

Physical growth and development during puberty is sensitive to energy and nutrient intakes. When energy intakes fail to meet requirements, linear growth may be retarded and sexual maturation may be delayed. The standard way to gauge adequacy of energy intake is to assess height, weight, and body composition. If, over time, height as well as weight-for-height continuously fall within the same percentiles when plotted on gender-appropriate National Center for Health Statistics growth charts, it can be assumed that energy needs are being met. If percentile of weight-for-height measurements begin to fall or rise, a thorough assessment of energy intake should be done, and adjustments in energy intake should be made accordingly. The use of body fat measurements, such as triceps and subscapular skinfold measurements, can provide useful information when weight-for-height does not remain consistent. Remember, however, that transient increases and decreases in body fat are commonly noted among adolescents during puberty due to the variation in timing of increases in height, weight, and accumulation of body fat and lean body mass. Repeated measurements of weight, height, and body composition over a several-month period are needed to accurately assess adequacy of growth and development.

Protein

Protein needs of adolescents are influenced by the amount of protein required for maintenance of existing lean body mass, plus allowances for the amount required to accrue additional lean body mass during the adolescent growth spurt. The estimated protein need for adolescents is 0.85 g/kg body weight/day, slightly higher than that of

Table 14.9 Recommended intakes of macronutrients based on IOM daily recommended intakes

	Estimated Energy Requirements (Kcals)	Carbohydrate (g)	% of Daily Energy from Carbohydrate	Fiber (g)	% of Daily Energy from Fat	Linoleic acid (g)	alpha-linolenic acid (g)	Protein (g)	% of Daily Energy from Protein
Males									
9–13	2279	130	45–65	31	25–35	21	1.2	34	10–30
14–18	3152	130	45–65	38	25–35	16	1.6	52	10–30
Females									
9–13	2071	130	45–65	26	25–35	10	1.0	34	10–30
14–18	2368	130	45–65	26	25–35	11	1.1	46	10–30

adults. Because protein needs vary with the degree of growth and development, requirements based upon developmental age will be more accurate than absolute recommendations based upon chronological age.

Recommended protein intakes are shown in Table 14.9. Protein requirements are highest for females at 11 to 14 years, and for males at 15 to 18 years, when growth is at its peak. Similar to energy needs, estimation of protein needs based on timing of growth rather than chronological age is most accurate. As with energy, growth is affected by protein intakes. When protein intakes are consistently inadequate, reductions in linear growth, delays in sexual maturation, and reduced accumulation of lean body mass may be seen. Subgroups of adolescents may be at risk for marginal or low protein intakes, however, including those from food-insecure households, those who severely restrict calories, and those who consume semivegetarian or vegetarian diets, most notably vegans.

Carbohydrates

Carbohydrates provide the body's primary source of dietary energy. Carbohydrate-rich foods, such as fruit, vegetables, whole grains, and legumes are also the main source of dietary fiber. The recommended intake of carbohydrate among teens is 130 g/day or 45–65% of daily energy needs (Table 14.9). Sweeteners and added sugars provide approximately 21% of energy intake by teens. Males consume 35 tsp and female teens consume 26 tsp of added sugar per day.[30] Soft drinks, candy, baked goods, and other sweetened bevaerages are major sources of added sweeteners in the diets of adolescents.

Dietary Fiber

Dietary fiber is important for normal bowel function and may play a role in the prevention of chronic diseases such as certain cancers, coronary artery disease, and type 2 diabetes mellitus. Adequate fiber intake is also thought to reduce serum cholesterol levels, moderate blood sugar levels, and reduce the risk of obesity. The American Academy of Pediatrics (AAP) Committee on Nutrition has recommended that dietary fiber intakes among children and adolescents should be 0.5 grams per kilogram of body weight.[34] This corresponds to average fiber intakes of 15.5 to 34.5 grams per day among 10- to 18-year-old boys, and 16.0 to 28.5 grams per day among 10- to 18-year-old females. The AAP has further recommended that fiber intake not exceed 35 grams per day, as levels above this amount may reduce the bioavailability of some minerals. The DRIs set the recommended intake of dietary fiber for adolescent females at 26 g/day, for males <14 years of age at 31 g/day, and for older adolescent males at 38 g/day; these recommendations are slightly higher than the AAP guidelines for males and slightly lower than AAP recommendations for females.[36]

National data indicate that adolescent males consume 14.2 grams of fiber per day while adolescent females consume 12.3 g/day, well short of AAP and DRI recommendations.[32] During adolescence, fiber intake among males increases slightly with age while it decreases with age among females. The low intake of fruit and vegetables, combined with an average intake of less than one serving of whole grains per day among adolescents, are contributing factors affecting fiber intake among adolescents.[30]

Fat

The human body requires dietary fat and essential fatty acids for normal growth and development. Current recommendations suggest that children over the age of 2 years consume no more than 25–35% of calories from fat, with no more than 10% of calories derived from saturated fat.[35,36] Data on energy and macronutrient intakes among adolescents suggest that approximately 32% of total calories consumed are derived from fat.[30] Approximately two-thirds of teens meet the recommendations for total fat and saturated fat. National dietary guidelines also suggest that adolescents consume no more than 300 mg of dietary cholesterol per day.[35,36] The DRIs recommend specific intake of linoleic and alpha-linolenic acid to support optimal growth and development (Table 14.9).

Calcium

Achieving an adequate intake of calcium during adolescence is crucial to physical growth and development. Calcium is the main constituent of bone mass. Because about half of peak bone mass is accrued during adolescence, calcium intake is of great importance for the development of dense bone mass and the reduction of the lifetime risk of fractures and osteoporosis. Additionally, calcium need and absorption rates are higher during adolescence than any other time, except infancy.[6,37] Female adolescents appear to have the greatest capability to absorb calcium about the time of menarche, with calcium absorption rates decreasing from then on.[6,37] Calcium absorption rates in males also peak during early adolescence, a few years later than in females. Young adolescents have been found to retain up to four times as much calcium as young adults. By age 24 in females and 26 in males, calcium accretion in bone mass is almost nonexistent.[6,38] Clearly, an adequate intake of calcium is of paramount importance during adolescence.

The DRI for calcium for 9- to 18-year-olds is 1300 mg per day (Table 14.7). National data suggest that many adolescents, most notably females, do not consume the DRI for calcium. Adolescent females consume 865 mg calcium per day, while adolescent males have been found to consume about 1130 mg calcium each day.[32] These levels of dietary intake are not adequate to support the development of optimal bone mass. Supplements may be warranted for adolescents who do not consume adequate calcium from dietary sources.

Recent research suggests that adolescents are not able to meet daily calcium needs in diets that do not include dairy products without the use of calcium-fortified foods.[39] Adolescents increasingly consume their calcium in the form of fortified foods. Studies indicate that more than half of teens reported drinking calcium-fortified juices and 31% ate cereals fortified with calcium.[40,41] Soy beverages are another source of calcium consumed by youth that are being considered for inclusion in school meal programs.[42] The availability of calcium from soy beverages appears low, however, and the equivalency of soy versus dairy as a calcium source is highly debatable.[43] Other food sources of calcium must be carefully chosen when dairy intake is not adequate to meet daily needs.

The consumption of soft drinks by adolescents may displace the consumption of more nutrient-dense beverages, including milk and fortified juices. Studies have shown an inverse relationship between the intake of carbonated beverages and the intake of milk and juice.[44,45] Because milk and fortified juices are significant sources of calcium in the diets of adolescents, interventions aimed at reducing consumption of soft drinks may be warranted. Such interventions are especially important in light of the growing body of evidence that suggests that carbonated beverage consumption may increase the risk of bone fractures among children and adolescents.[15]

Calcium consumption drops as age increases among both male and female adolescents; however, males consume greater amounts of calcium at all ages than do females.[32] Calcium intakes among adolescents are highly correlated with energy intakes. When dietary calcium intake is adjusted for energy intake, no differences in calcium density of diets are found between males and females.[40] This fact suggests that females who restrict calories in an effort to control their weight are at particularly high risk for inadequate calcium intakes. Some variation in calcium intake follows race categories among females: Cuban, Asian, and black females consume less calcium on average than Mexican American, Puerto Rican, and white, non-Hispanic females.

A study of knowledge regarding calcium and bone health revealed some interesting findings. When adolescents were questioned about their knowledge of the health benefits of calcium, 92% knew that it was needed to strengthen bones, 60% knew that it was required for "good" teeth, and 60% realized that adolescence was a critical time for the development of peak bone mass. Only 19% of teens knew that the recommended intake of milk, milk products, or fortified soy milk was four servings per day.[40] Nutrition education and interventions that target calcium consumption by older children and young teens are needed.

In addition to the importance of calcium during adolescence, physical activity also plays a role in bone development during adolescence. Physical activity patterns and participation during adolescence have been shown to be a strong predictor of adult bone density. Participation in weight-bearing activities may lead to increased bone mineral density as compared to that accrued by more sedentary adolescents.[46–48]

Iron

The rapid rate of linear growth, the increase in blood volume, and the onset of menarche during adolescence increase a teen's need for iron. The DRIs for iron for male and female adolescents are shown in Table 14.7. These recommendations are based on the amount of dietary iron intake needed to maintain a suitable level of iron storage, with additional amounts of iron added to cover the rapid linear growth and onset of menstruation that occur in male and female adolescents, respectively. Note that even though DRIs are based on chronological age, the actual iron requirements of adolescents are based on sexual maturation level. Iron needs of an adolescent will be highest during the adolescent growth spurt in males, and after menarche in females.

The age-specific hemoglobin and hematocrit values used to determine iron deficiency anemia are listed in Table 14.10. Hemoglobin and hematocrit levels, although commonly used to screen for the presence of iron-deficiency anemia, are actually the last serum indicators of depleted iron stores to drop. More sensitive indicators of iron stores include *serum iron, plasma ferritin,* and *transferrin saturation.* These measures are expensive and not commonly used

Serum Iron, Plasma Ferritin, and **Transferrin Saturation** Measures of iron status obtained from blood plasma or serum samples.

Table 14.10 Hemoglobin and hematocrit cut-point values for iron-deficiency anemia in adolescents

Sex/Age[a]	Hemoglobin (<g/dL)	Hematocrit (<%)
Males and Females	Less Than:	Less Than:
8–12 years	11.9	35.4
Males		
12–15 years	12.5	37.3
15–18 years	13.3	39.7
18+ years	13.5	39.9
Females[b]		
12–15 years	11.8	35.7
15–18 years	12	35.9
18+ years	12	35.7

SOURCE: Abridged from Centers for Disease Control and Prevention. "Recommendations to Prevent and Control Iron Deficiency Anemia in the United States." Morb Mortal Wkly Rep. 2002;51(40):897–899.

[a]Age and sex-specific cutoff values for anemia are based on the 5th percentile from the third National Health and Nutrition Examination Survey.
[b]Nonpregnant and lactating adolescents.

in the traditional medical setting, however. Estimates of iron deficiency among adolescents are 9% of 12- to 15-year-old females, 5% of 12- to 16-year-old males, 11% of 15- to 19-year-old females, and 2% of 15- to 19-year-old males.[49] Interestingly, the prevalence of iron deficiency among adolescent males was five times higher in the 1999–2000 NHANES surveys than previous years, which is cause for concern. While iron deficiency occurs fairly frequently in all adolescents, iron-deficiency anemia occurs almost exclusively in females, with a prevalence of <1% of males and 2% among females.[49] Therefore, we must assume that although the prevalence of iron-deficiency anemia may be relatively low among adolescents, a larger proportion may have inadequate iron stores. Rates of iron deficiency and anemia are twice as high among black and Mexican American females compared to white females and are three times higher than the target prevalence goals set by Healthy People 2010.[50] This finding is particularly relevant among adolescents from low SES homes, because rates of iron deficiency tend to be higher in adolescents from low-income families.

The availability of dietary iron for absorption and utilization by the body varies by its form. The two types of dietary iron are *heme iron*, which is found in animal products, and *nonheme iron*, which is found in both animal and plant-based foods. Heme iron is highly bioavailable while nonheme iron is much less so. More than 80% of the iron consumed is in the form of nonheme iron.

> **Heme Iron** Iron contained within a protein portion of hemoglobin that is in the ferrous state.
>
> **Nonheme Iron** Iron contained within a protein of hemoglobin that is in the ferric state.

Bioavailability of nonheme iron can be enhanced by consuming it with heme sources of iron or vitamin C. This point is particularly salient for adolescents who avoid animal foods as a means of restricting calories and those who consume few animal-based foods (semivegetarian) or vegetarian diets for moral or cultural reasons.

Dietary intakes of iron are estimated at 17.0 mg/day among 9- to 13-year-old males, 19.1 mg/day among 14- to 18-year-old males, 13.7 mg/day among 9- to 13-year-old females, and 13.3 mg/day in 14- to 18-year-old females.[32] Data suggest that while <3% of adolescent males and young adolescent females consume less than the DRI for iron, the prevalence of very low iron intake is 16% among older adolescent females.[32] In light of the increasing frequency of iron deficiency among adolescents, nutrition education and counseling for teens to promote higher iron consumption is warranted.

Zinc

Zinc is particularly important during adolescence because of its role in the synthesis of RNA and protein, and its role as a cofactor in over 200 enzymes. The body's need for zinc, along with its ability to retain zinc, dramatically increases during the adolescent growth spurt. Zinc is required for sexual maturation to occur. Males who are zinc deficient experience growth failure and delayed sexual development. Zinc supplementation of both male and female zinc-deficient adolescents from developing countries often initiates accelerated growth and sexual development. Data on zinc nutrition of adolescents are limited, but evidence shows that serum zinc levels decline in response to the rapid growth and hormonal changes during adolescence.

The bioavailability of zinc from dietary sources is highly dependant upon the source of zinc. Zinc from animal sources is more bioavailable than plant-based sources. Undigestible fibers found in many plant-based sources of zinc can inhibit its absorption by the body. Zinc and iron compete for absorption, so elevated intakes of one can reduce the absorption of the other. Adolescents who take iron supplements may be at increased risk of developing mild zinc deficiency if iron intake is more than twice as high as that of zinc.

Dietary intakes of zinc range from 9.5 to 9.8 mg/day among adolescent females to 13.0 to 15.1 mg/day among males.[32] National surveys suggest that 75% of males and 81% of females consume less than the DRI for zinc, with 4% of males and 10–26% of females consuming less than the EAR.[32] Vegetarians, particularly vegans, and teens who do not consume many animal-derived products are at highest risk for low intakes of zinc.

Folate

Folate is an integral part of DNA, RNA, and protein synthesis. Thus, adolescents have increased requirements for folate during puberty. The DRI for folate is listed in Table 14.7. Folate in the form of folic acid is twice as bioavailable as other forms of folate. For this reason, dietary folate equivalents (DFEs) are used in the DRIs. One microgram of folic acid is equivalent to approximately 2 DFEs, while 1 microgram of other forms of folate is equivalent to 1 DFE. Folic acid is the form of folate added to fortified cereals, breads, and other refined grain products.

Severe folate deficiency results in the development of megaloblastic anemia, which is rare among adolescents. Evidence, however, indicates that a significant proportion of adolescents have inadequate folate status. Serum folate levels drop during adolescence among females as sexual maturation proceeds, suggesting that increased folate needs during growth and development are not being met. For this reason, sexual maturation level should be used to identify folate needs as opposed to chronological age.

Poor folate status among adolescent females also presents an issue related to reproduction. Studies show that adequate intakes of folate prior to pregnancy can reduce the incidence of spina bifida and selected other congenital anomalies and may reduce the risk of Down syndrome among offspring.[51] The protective effects of folate occur early in pregnancy, often before a woman knows she is pregnant. Thus, it is imperative that all women of reproductive age (15–44 years old) consume adequate folic acid, preferably through dietary sources, or if needed, through supplements.

National data suggest that many adolescents do not consume adequate amounts of folate. Median intakes of folate among adolescent males range from 644–683 DFEs/day whiles females consume 500–512 DFEs/day.[32] Four percent of male and 26% of female adolescents consume less than the EAR for folate.[32] Teens who skip breakfast or do not commonly consume orange juice and ready-to-eat cereals are at an increased risk for having a low consumption of folate.

Vitamin A

The DRI for vitamin A is shown in Table 14.7. Vitamin A deficiency is rare among adolescents in the United States; however, national studies have consistently shown low dietary intakes of this vitamin. It has been reported that 52% of teen males and 62% of teen females consume less than the DRI for vitamin A.[32] Intakes below the EAR were detected in 13% of young adolescent males, 55% of older adolescent males, 34% of young teen females, and 54% of older teen females.[32] Average intakes of vitamin A ranged from 638–670 retinol activity equivalents/day among males and 513–536 retional activity equivalents/day among females. The low intake of fruits, vegetables, and milk and dairy products by adolescents contributes to their less-than-optimal intake of vitamin A.

Vitamin E

Vitamin E is well known for its antioxidant properties, a role that becomes increasingly important as body mass expands during adolescence. The DRIs for vitamin E for adolescents are shown in Table 14.7. Few data are available on the vitamin E status of adolescents. National nutrition surveys suggest that dietary intakes of vitamin E are well below recommended levels, which may be indicative of poor vitamin E status. Almost all adolescents (≥97%) had vitamin E intakes below the EAR.[32] Median intakes among adolescent males ranged from 6.0–7.3 mg of tocopherol/day, with adolescent females consuming 5.6 mg of tocopherol/day—intakes that are approximately half of esitamted daily needs.

Vitamin C

Vitamin C is involved in the synthesis of collagen and other connective tissues. For this reason, vitamin C plays an important role during adolescent growth and development. The DRIs for 9- to 13-year-old and 14- to 18-year-old

adolescents are shown in Table 14.7. Vitamin C intakes are marginally adequate within the adolescent population. Median intakes are estimtaed to be 80–100 mg/day among teenage males and 76–81 mg/day among teenage females.[32] The prevalence of intakes below the EAR ranged from 8% in young adolescent males and 9% in young adolescent females to 26% of older adolescent males and 42% of older adolescent females.

Vitamin C acts as an antioxidant. Smoking increases the need for this antioxidant within the body because it consumes vitamin C in antioxidation reactions. Consequently, smoking results in reduced serum levels of vitamin C. Recommended levels of vitamin C intake are higher among smokers. On average, adolescents who use tobacco and other substances have poorer-quality diets and consume fewer fruit and vegetables, which are primary sources of vitamin C.[52]

Nutrition Screening, Assessment, and Intervention

The American Medical Association's Guidelines for Adolescent Preventive Services (GAPS) recommend that all adolescents receive annual health guidance related to healthy dietary habits and methods to achieve a healthy weight.[53] This health guidance begins by annually screening all adolescents for indicators of nutritional risk. Common concerns that should be investigated during nutrition screening include overweight, underweight, eating disorders, hyperlipidemia, hypertension, iron-deficiency anemia, food insecurity, and excessive intake of high-fat or high-sugar foods and beverages. Pregnant adolescents should also be assessed for adequacy of weight gain and compliance with prenatal vitamin–mineral supplement recommendations.

Nutrition screening should include an accurate measurement of height and weight, and calculation of BMI (body mass index). These data, plotted on age- and gender-appropriate National Center for Health Statistics 2000 growth charts, indicate the presence of any weight or other growth problems. Indicators of height and weight status are listed in Table 14.11. Teens below the 5th percentile of weight-for-height or BMI-for-age are considered to be underweight and should be referred for evaluation of metabolic disorders, chronic health conditions, or eating disorders. Adolescents with a BMI above the 85th percentile but below the 95th percentile are considered to be at risk for overweight. They should be referred for a full medical evaluation to determine the presence or absence of obesity-related complications. Teenagers with a BMI greater than 95th percentile are considered to be overweight and should also be referred for a medical evaluation. Referral to a weight management program specially designed to meet the needs of adolescents may also be warranted for overweight

Table 14.11 Indicators of height and weight status for adolescents

Indicator	Body Size Measure	Cutoff Values
Stunting (low height-for-age)	Height-for-age	<3rd percentile
Thinness (low BMI-for-age)	BMI-for-age	<5th percentile
At risk for overweight	BMI-for-age	>85th percentile, but <95th percentile
Overweight	BMI-for-age	>95th percentile

SOURCE: Compiled with permission from World Health Organization. Physical Status: The Use and Interpretation of Anthropometry. Report of a WHO expert committee. World Health Organization Technical Report Serties 854, 1995; and Himes J, Dietz W. Guidelines for Overweight in Adolescent Preventive Services: Recommendations from an Expert Committee. Am J Clin Nutr 59, 1994. In: Story M, Holt K, Sofka D, eds. Bright Futures in Practice: Nutrition. Arlington, VA: National Center Education in Maternal and Child Health: 2000.

adolescents who have completed physical growth (see Chapter 15 for additional information).

Nutrition screening should also include a brief dietary assessment. Food frequency questionnaires, 24-hour recalls, and food diaries or food records are all appropriate for use with adolescents. Table 14.12 lists the advantages and disadvantages of each dietary assessment method. Less formal dietary assessment questionnaires that target specific behaviors, such as consumption of savory snacks and high-sugar beverages, can also be used for initial nutrition screening. These "rate your plate" or rapid assessment questionnaires can be completed quickly and may be used to determine those adolescents in need of additional dietary assessment and nutrition counseling.

Nutrition risk indicators that may warrant further nutrition assessment and counseling are listed in Table 14.13. Adolescents who have a poor-quality diet characterized by an excessive intake of high-fat or high-sugar foods and

Table 14.12 Strengths and limitations of various dietary assessment methods used in clinical settings

	Strengths	Limitations	Applications
24-Hour Recall	• Does not require literacy • Relatively low respondent burden • Data may be directly entered into a dietary analysis program • May be conducted in person or over the telephone	• Dependent on respondent's memory • Relies on self-reported information • Requires skilled staff • Time consuming • Single recall does not represent usual intake	• Appropriate for most people as it does not require literacy • Useful for the assessment of intake of a variety of nutrients and assessment of meal patterning and food group intake • Useful counseling tool
Food Frequency	• Quick, easy, and affordable • May assess current as well as past diet • In a clinical setting, may be useful as a screening tool	• Does not provide valid estimates of absolute intake of individuals • Can't assess meal patterning • May not be appropriate for some population groups	• Does not provide valid estimates of absolute intake for individuals, thus of limited usefulness in clinical settings • May be useful as a screening tool; however, further development research is needed
Food Record	• Does not rely on memory • Food portions may be measured at the time of consumption • Multiple days of records provide valid measure of intake for most nutrients	• Recording foods eaten may influence what is eaten • Requires literacy • Relies on self-reported information • Requires skilled staff • Time consuming	• Appropriate for literate and motivated population groups • Useful for the assessment of intake of a variety of nutrients and assessment of meal patterning and food group intake • Useful counseling tool
Diet History	• Able to assess usual intake in a single interview • Appropriate for most people	• Relies on memory • Time consuming (60 to 90 minutes) • Requires skilled interviewer	• Appropriate for most people as it does not require literacy • Useful for assessing intake of nutrients, meal patterning, and food group intake • Useful counseling tool

SOURCE: Used with permission. Story M, Stang I, eds. Nutrition and the Pregnant Adolescent: A Practical Reference Guide. Minneapolis, MN: Center for Leadership, Education, and Training in Maternal and Child Nutrition, University of Minnesota; 2000.

Table 14.13 Key indicators of nutrition risk for adolescents

Indicators of Nutrition Risk	Relevance	Criteria for Further Screening and Assessment
FOOD CHOICES		
Consumes fewer than 2 servings fruit or fruit juice per day	Fruits and vegetables provide dietary fiber and several vitamins (such as A and C) and minerals. Low intake of fruits and vegetables is associated with an increased risk of many types of cancer. In females of childbearing age, low intake of folic acid is associated with an increased risk of giving birth to an infant with neural tube defects.	Assess the adolescent who is consuming less than 1 serving of fruit or fruit juice per day.
Consumes fewer than 3 servings of vegetables per day		Assess the adolescent who is consuming fewer than 2 servings of vegetables per day.
Consumes fewer than 6 servings of bread, cereal, pasta, rice, or other grains per day	Grain products provide complex carbohydrates, dietary fiber, vitamins, and minerals. Low intake of dietary fiber is associated with constipation and an increased risk of colon cancer.	Assess the adolescent who is consuming fewer than 3 servings of bread, cereal, pasta, rice, or other grains per day.
Consumes fewer than 3 servings of dairy products per day	Dairy products are a good source of protein, vitamins, and calcium and other minerals. Low intake of dairy products may reduce peak bone mass and contribute to later risk of osteoporosis.	Assess the adolescent who is consuming fewer than 2 servings of dairy products per day. Assess the adolescent who has a milk allergy or is lactose intolerant. Assess the adolescent who is consuming more than 20 ounces of soft drinks per day.
Consumes fewer than 2 servings of meat or meat alternatives (e.g., beans, eggs, nuts, seeds) per day	Protein-rich foods (e.g., meats, beans, dairy products) are good sources of B vitamins, iron, and zinc. Low intake of protein-rich foods may impair growth and increase the risk of iron-deficiency anemia and of delayed growth and sexual maturation. Low intake of meat or meat alternatives may indicate inadequate availability of these foods at home. Special attention should be paid to children and adolescents who follow a vegetarian diet.	Assess the adolescent who is consuming fewer than 2 servings of meat or meat alternatives per day or who consumes a vegan diet.
Has excessive intake of dietary fat	Excessive intake of total fat contributes to the risk of cardiovascular diseases and obesity and is associated with some cancers.	Assess the adolescent who has a family history of premature cardiovascular disease. Assess the adolescent who has a body mass index (BMI) greater than or equal to the 85th percentile.
EATING BEHAVIORS		
Exhibits poor appetite	A poor appetite may indicate depression, emotional stress, chronic disease, or an eating disorder.	Assess the adolescent if BMI is less than the 15th percentile or if weight loss has occurred. Assess if irregular menses or amenorrhea have occurred for 3 months or more. Assess for organic and psychiatric disease.
Consumes food from fast food restaurants three or more times per week	Excessive consumption of convenience foods and foods from fast food restaurants is associated with high fat, calorie, and sodium intakes, as well as low intake of certain vitamins and minerals.	Assess the adolescent who is at risk for overweight/obesity, or who has diabetes mellitus, hyperlipidemia, or other conditions requiring reduction in dietary fat.

continued

Table 14.13 Key indicators of nutrition risk for adolescents (continued)

Indicators of Nutrition Risk	Relevance	Criteria for Further Screening and Assessment
Skips breakfast, lunch, or dinner/supper three or more times per week	Meal skipping is associated with a low intake of energy and essential nutrients and, if it is a regular practice, could compromise growth and sexual development. Repeatedly skipping meals decreases the nutritional adequacy of the diet.	Assess the adolescent to ensure that meal skipping is not due to inadequate food resources or unhealthy weight-loss practices.
Consumes a vegetarian diet	Vegetarian diets can provide adequate nutrients and energy to support growth and development if well planned. Vegan diets may lack calcium, iron, and vitamins D and B$_{12}$. Low-fat vegetarian diets may be adopted by adolescents who have eating disorders.	Assess the adolescent who consumes fewer than 2 servings of meat alternatives per day. Assess the adolescent who consumes fewer than 3 servings of dairy products per day. Assess the adolescent who follows a low-fat vegetarian diet and experiences weight loss for eating disorder and adequacy of energy intake.
FOOD RESOURCES		
Has inadequate financial resources to buy food, insufficient access to food, or lack of access to cooking facilities	Poverty can result in hunger and compromised food quality and nutrition status. Inadequate dietary intake interferes with learning.	Assess the adolescent who is from a family with low income, is homeless, or is a runaway.
WEIGHT AND BODY IMAGE		
Practices unhealthy eating behaviors (e.g., chronic dieting, vomiting, and using laxatives, diuretics, or diet pills to lose weight)	Chronic dieting is associated with many health concerns (fatigue, impaired growth and sexual maturation, irritability, poor concentration, impulse to binge) and can lead to eating disorders. Frequent dieting in combination with purging is often associated with other health-compromising behaviors (substance use, suicidal behaviors). Purging is associated with serious medical complications.	Assess the adolescent for eating disorders. Assess for organic and psychiatric disease. Screen for distortion in body image and dysfunctional eating behavior, especially if adolescent desires weight loss, but BMI <85th percentile.
Is excessively concerned about body size or shape	Eating disorders are associated with significant health and psychological morbidity. Eighty-five percent of all cases of eating disorders begin during adolescence. The earlier adolescents are treated, the better their long-term prognosis.	Assess the adolescent for distorted body image and dysfunctional eating behaviors, especially if adolescent wants to lose weight, but BMI is less than the 85th percentile.
Has exhibited significant weight change in past 6 months	Significant weight change during the past 6 months may indicate stress, depression, organic disease, or an eating disorder.	Assess the adolescent to determine the cause of weight loss or weight gain (limited to too much access to food, poor appetite, meal skipping, eating disorder).
GROWTH		
Has BMI less than the 5th percentile	Thinness may indicate an eating disorder or poor nutrition.	Assess the adolescent for eating disorders. Assess for organic and psychiatric disease. Assess for inadequate food resources.
Has BMI greater than the 95th percentile	Obesity is associated with elevated cholesterol levels and elevated blood pressure. Obesity is an independent risk factor for cardiovascular disease and type 2 diabetes mellitus in adults. Overweight adolescents are more likely to be overweight adults and are at increased risk for health problems as adults.	Assess the adolescent who is overweight or at risk for becoming overweight (on the basis of present weight, weight-gain patterns, family weight history).

Table 14.13 Key indicators of nutrition risk for adolescents (continued)

Indicators of Nutrition Risk	Relevance	Criteria for Further Screening and Assessment
PHYSICAL ACTIVITY		
Is physically inactive: engages in physical activity fewer than 5 days per week	Lack of regular physical activity is associated with overweight, fatigue, and poor muscle tone in the short term and a greater risk of heart disease in the long term. Regular physical activity reduces the risk of cardiovascular disease, hypertension, colon cancer, and type 2 diabetes mellitus. Weight-bearing physical activity is essential for normal skeletal development during adolescence. Regular physical activity is necessary for maintaining normal muscle strength, joint structure, and joint function; contributes to psychological health and well-being; and facilitates weight reduction and weight maintenance throughout life.	Assess how much time the adolescent spends watching television/videotapes and playing computer games. Assess the adolescent's definition of physical activity.
Engages in excessive physical activity	Excessive physical activity (nearly every day or more than once a day) can be unhealthy and associated with menstrual irregularity, excessive weight loss, and malnutrition.	Assess the adolescent for eating disorders.
MEDICAL CONDITIONS		
Has chronic diseases or conditions	Medical conditions (diabetes mellitus, spina bifida, renal disease, hypertension, pregnancy, HIV infection/AIDS) have significant nutritional implications.	Assess adolescent's compliance with therapeutic dietary recommendations. Refer to dietitian if appropriate.
Has hyperlipidemia	Hyperlipidemia is a major cause of atherosclerosis and cardiovascular disease in adults.	Refer adolescent to a dietitian for cardiovascular nutrition assessment.
Has iron-deficiency anemia	Iron deficiency causes developmental delays and behavioral disturbances. Another consequence is increased lead absorption.	Screen adolescents if they have low iron intake, a history of iron-deficiency anemia, limited access to food because of poverty or neglect, special health care needs, or extensive menstrual or other blood losses. Screen annually.
Has dental caries	Eating habits have a direct impact on oral health. Calcium and vitamin D are vital for strong bones and teeth, and vitamin C is necessary for healthy gums. Frequent consumption of carbohydrate-rich foods (e.g., lollipops, soda) that stay in the mouth a long time may cause dental caries. Fluoride in water used for drinking and cooking as well as in toothpaste reduces the prevalence of dental caries.	Assess the adolescent's consumption of snacks and beverages that contain sugar, and assess snacking patterns. Assess the adolescent's access to fluoride (e.g., fluoridated water, fluoride tablets).
Is pregnant	Pregnancy increases the need for most nutrients.	Refer the adolescent to a dietitian for further assessment, education, and counseling as appropriate.
Is taking prescribed medication	Many medications interact with nutrients and can compromise nutrition status.	Assess potential interactions of prescription drugs (e.g., asthma medications, antibiotics) with nutrients.

continued

Table 14.13 Key indicators of nutrition risk for adolescents (continued)

Indicators of Nutrition Risk	Relevance	Criteria for Further Screening and Assessment
LIFESTYLE		
Engages in heavy alcohol, tobacco, and other drug use	Alcohol, tobacco, and other drug use can adversely affect nutrient intake and nutrition status.	Assess the adolescent further for inadequate dietary intake of energy and nutrients.
Uses dietary supplements	Dietary supplements (e.g., vitamin and mineral preparations) can be healthy additions to a diet, especially for pregnant and lactating women and for people with a history of iron-deficiency anemia; however, frequent use or high doses can have serious side effects. Adolescents who use supplements to "bulk up" may be tempted to experiment with anabolic steroids. Herbal supplements for weight loss can cause tachycardia and other side effects. They may also interact with over-the-counter prescription medications.	Assess the adolescent for the type of supplements used and dosages. Assess the adolescent for use of anabolic steroids and megadoses of other supplements.

SOURCE: Adapted from Story M, Holt K, Sofka D, eds. Bright Futures in Practice: Nutrition. Arlington, VA: National Center for Education in Maternal and Child Health; 2000.

beverages or meal skipping should be provided with nutrition counseling that provides concrete examples of ways to improve dietary intake. Adolescents who have been found to have a nutrition-related health risk, such as hyperlipidemia, hypertension, iron-deficiency anemia, overweight, or eating disorders should be referred for in-depth medical assessment and nutrition counseling. Pregnant adolescents may also benefit from in-depth nutrition assessment and counseling.

In-depth nutrition assessment should include a review of the full medical history, a review of psychosocial development, and evaluation of all laboratory data available. A complete and thorough dietary assessment should be performed, preferably using two dietary assessment methods. Most commonly, a food frequency questionnaire or a 3- to 7-day food record is combined with a 24-hour recall to provide accurate dietary intake data. Specific areas of nutrition concern can be identified during a complete nutrition assessment, and recommendations for nutrition education and counseling can be made accordingly.

Nutrition Education and Counseling

Providing nutrition education and counseling to teenagers requires a great deal of skill and a good understanding of normal adolescent physical and psychosocial development. When working with teens, it is important to treat them as individuals with unique needs and concerns. The initial component of the counseling session should involve getting to know the adolescent, including personal health or nutrition-related concerns. After establishing a rapport with the teen, the counselor should provide an overview of the events of the counseling session, including what specific nutrition topics will be discussed. Once again, the adolescent should be encouraged to add his or her own nutrition concerns to the list of topics to be discussed during the education session.

After agreeing to a list of topic areas to be covered during the nutrition education session, a complete nutrition assessment should be performed. Upon completion of the assessment, the counselor and teen should work together to establish goals for improving dietary intake and reducing nutrition risk.

It is important to involve the adolescent in decision-making processes during nutrition counseling. Allowing teens to provide input as to what aspects of their eating habits they think need to be changed and what changes they are willing to make accomplishes several important tasks during the counseling session. First, the importance of the adolescent in the decision-making process is stressed, and she or he is encouraged to become involved in personal decisions about health. Second, a good rapport established between the health professional and the adolescent may lead to greater interaction between both parties. Finally, behavior change is more likely when the adolescent has suggested ways to change, thus expressing a willingness to change.

One or two goals during a counseling session is a reasonable number to work toward. Setting too many goals reduces the probability that the adolescent can meet all of the goals and may seem overwhelming. For each goal set, several behavior-change strategies should be mutually agreed upon for meeting that goal. These strategies should be concrete in nature and instigated by the teen. The adolescent and the counselor should also work together to decide how to determine when a goal is met. The MyPyramid online system can be utilized by teens as a means of assessing how changes in food choices affect changes in nutrient intake across time. Frequent follow-up sessions also help to provide feedback and monitor progress toward individual goals.

Physical Activity and Sports

Regular physical activity leads to many health benefits. Physical activity improves aerobic endurance and muscular strength, may reduce the risk of developing obesity, and builds bone mass density.[46–48] Physical activity among adolescents is consistently related to higher levels of self-esteem and self-concept and lower levels of anxiety and stress. Thus, physical activity is associated with both physiological and psychological benefits, especially during adolescence, which offers opportunities to positively influence the adoption of lifelong activity patterns. Increasing physical activity among adolescents is an important goal because regular physical activity declines during adolescence and many American teens are inactive.

Physical activity is defined as any bodily movement produced by skeletal muscles that results in energy expenditure. This definition is distinguished from exercise, which is a subset of physical activity that is planned, structured, and repetitive and is done to improve or maintain physical fitness. Physical fitness is a set of attributes that are either health or skill related. The International Consensus Conference on Physical Activity Guidelines for Adolescents recommends that all adolescents be physically active daily, or nearly every day, as part of play, games, sports, work, transportation, recreation, physical education, or planned exercise. Further, it is recommended that adolescents engage in 60 minutes or more of physical activity on most days of the week.[28]

Despite common knowledge about the importance and benefits of physical activity, only 35% of U.S. adolescents met physical activity guidelines while 10% reported no moderate to vigorous physical activity.[33] Moreover, physical activity declines steadily throughout adolescence, especially among females. More males than females meet daily physical activity guidelines. Racial and ethnic differences are also noted in physical activity; white teens are more likely than black or Mexican American teens to meet physical activity targets.

Factors Affecting Physical Activity

Individual, social, and environmental factors are associated with physical activity among adolescents. Females are less active than males, and among adolescent females, blacks are less active than whites.[48] Individual factors positively associated with physical activity among young people include confidence in one's ability to engage in exercise (i.e., self-efficacy), perceptions of physical or sports competence, having positive attitudes toward physical activity, enjoying physical activity, and perceiving positive benefits associated with physical activity (i.e., excitement, fun, adventure, staying in shape, improved appearance, weight control, improving skills). Social factors associated with engaging in physical activity are peer and family support. Environmental factors associated with physical activity are having safe and convenient places to play, sports equipment, and transportation to sports or fitness programs.

Schools offer an ideal setting for promoting physical activity through physical education classes. About half of U.S. adolescents attend a physical education class at least once a week, but not all students are physically active for 20 minutes or more during these classes.[33] Community programs that offer noncompetetive activities where all teens get an opportunity to participate are essential because most physical activity among adolescents occurs outside of schools.

The Centers for Disease Control and Prevention published *Guidelines for School and Community Programs to Promote Lifelong Physical Activity Among Young People*, which provides a developmental framework for comprehensive school and community physical activity programs to be used by school districts, educators, health professionals, and policymakers.[48] The guidelines include recommendations (Table 14.14) to promote lifelong physical activity including school policies, physical and social environments that encourage and enable physical activity, developmentally appropriate physical education curricula and instruction, personnel training, family and community involvement, and program evaluation.

High levels of physical activity, combined with growth and development, increase adolescents' needs for energy, protein, and select vitamins and minerals. Participation in competitive sports often means an adolescent will participate in intense training and competition during an athletic season. If the athlete competes in several sports, energy and nutrient needs will remain relatively stable throughout the year. If an athlete participates in only one sport, however, energy and nutrient needs may fluctuate based on the timing of the sports season. Therefore, adolescents must be assessed for seasonal and yearly physical activity when energy and nutrient needs are determined.

Table 14.14 Recommendations for school health programs promoting healthy eating and physical activity

Healthy Eating	Physical Activity
• Adopt a coordinated school nutrition policy that promotes healthy eating through classroom lessons and a supportive school environment. • Implement nutrition education from preschool through secondary school as part of a sequential, comprehensive school health education curriculum designed to help students adopt healthy eating behaviors. • Provide nutrition education through developmentally appropriate, culturally relevant, fun participatory activities that involve social learning strategies. • Coordinate school food service with other components of the comprehensive school health program to reinforce messages on healthy eating. • Involve family members and the community in supporting and reinforcing nutrition education. • Evaluate regularly the effectiveness of the school health program in promoting healthy eating, and change the program as appropriate to increase the effectiveness.	• Establish policies that promote enjoyable lifelong physical activity among young people. • Provide physical and social environments that encourage and enable safe and enjoyable physical activity. • Implement physical education curricula and instruction that emphasize enjoyable participation in physical activity and that help students develop the knowledge, attitudes, motor skills, behavioral skills, and confidence needed to adopt and maintain physically active lifestyles. • Provide extracurricular physical activity programs. • Include parents and guardians in physical activity instruction and in extracurricular and community physical activity programs, and encourage them to support their children's participation in enjoyable physical activities. • Provide training for education, coaching, recreation, health care, and other school and community personnel that imparts the knowledge and skills needed to effectively promote enjoyable, lifelong physical activity among young people. • Assess physical activity patterns among young people, counsel them about physical activity, refer them to appropriate programs, and advocate for physical activity instruction and programs for young people. • Provide a range of developmentally appropriate community sports and recreation programs that are attractive to all young people. • Regularly evaluate school and community physical activity instruction, programs, and facilities.

SOURCE: Centers for Disease Control and Prevention. "Guidelines for School and Community Programs to Promote Lifelong Physical Activity among Young People, MMWR 45:1996; and Guidelines for School Health Programs to Promote Lifelong Healthy Eating," MMWR 45:1996.

The energy and nutrient needs of adolescent athletes vary widely. Many of the recommendations available are based on needs of young adult athletes or are extrapolated from usual nutrient needs of adolescents. The best method of assessing the nutrient needs of athletes is to begin with general dietary needs based on Sexual Maturation Rating (SMR), adding additional allowances based upon the unique needs of the individual and the intensity of physical activity they engage in. In order to assess individual nutrient needs, health care professionals must gather information such as:

• What sport(s) does the adolescent engage in, and what is the duration of the competition season?

• What is the level of competition of the adolescent? Is participation recreational, competitive, or highly competitive?

• What kind of training does the adolescent engage in? The method(s), intensity, and duration of training activities should be noted.

• Does the athlete typically sweat profusely or lose body weight during competition?

• Does the athlete follow a special diet or take supplements to improve athletic performance? Be sure to note the type, amount, and frequency of supplement use.

General energy and protein needs are shown in Table 14.8. These guidelines should provide the foundation for calculating protein and energy needs for athletes. Competitive athletes may require 500–1500 additional calories per day to meet their energy needs. Athletes and their parents should be encouraged to monitor weight stability. Any weight loss that is not transient (transient losses are often due to dehydration) signifies that the caloric intake is inadequate to support growth and development. A thorough assessment of energy and protein intakes, accompanied by measurements of body composition, should be taken when unexpected weight loss occurs. Protein should supply no more than 30% of calories in the diet. Groups at risk for inadequate intake would include athletes who follow vegetarian diets or restrict caloric intake. When the main sources of protein are plant-based, additional protein intake may be needed because plant-based sources of protein may be less bioavailable.

Dietary intakes of athletes should follow the MyPyramid recommendations, with the realization that the increased energy needs of athletes may require them to consume the upper limit of food group recommendations. Athletes should be encouraged to eat a pre-event meal at least 2 to 3 hours prior to exercise; eating too close to exercise may lead to indigestion and physical discomfort. Foods that are high in fat, high in protein, and high in dietary fiber should be avoided for at least 4 hours prior to exercise, because they take longer to digest and may cause physical discomfort during exercise. Protein and fat also displace complex carbohydrates, which are the most readily available source of energy during athletic events. Postevent meals should contain approximately 400–600 calories, and should be comprised of high-carbohydrate foods and adequate amounts of noncaffeinated fluids.

Calcium intakes have been shown to be below the DRIs in a significant proportion of adolescents, especially females. Athletes' increased risk for bone fractures makes adequate calcium intake extremely important. Data suggest that consumption of carbonated soft drinks by athletic females elevates their risk of developing bone fractures when compared to less active females who consume the same beverages.[15] Although the mechanism responsible for this tendency has not been identified, female adolescent athletes with low calcium consumption appear to be the highest risk group of all adolescents for bone fractures, and therefore, should make every effort to consume adequate calcium in their diets. Teen athletes who cannot or will not consume calcium from dietary sources should be counseled to take a daily calcium supplement that meets their daily needs.

Promoting Healthy Eating and Physical Activity Behaviors

Meeting the challenge of improving the nutritional health of teenagers requires the integrated efforts of teenagers, parents, educators, health care providers, schools, communities, the food industry, and policymakers all working together to create more opportunities for healthful eating.

Effective Nutrition Messages for Youth

Health professionals need to rethink how they frame messages to youth. Years ago, Leverton pointed out that too often teenagers have been given the message that good nutrition means "eating what you don't like because it's good for you."[54] Rather, they should be told to "eat well because it will help you in what you want to do and become." Teenagers are present-oriented and tend not to be concerned about how their eating will affect them in later years. However, they are concerned about immediate, socially relevant issues such as their physical appearance, achieving and maintaining a healthy weight, and having lots of energy. Many are also interested in optimizing sports performance. Even though adolescents need to be aware of the long-term risks of an unhealthy diet and benefits of a more healthful one, focusing on the short-term benefits will have more appeal to them.

Parent Involvement

Parents as well as teenagers should be targets for nutrition education because they fill the role of gatekeepers of foods and serve as role models for eating behavior. Even though parents may have little control over what their teenagers are eating outside the home, they have more control in the home environment. Teenagers tend to eat what is available and convenient. Parents can capitalize on this by stocking the kitchen with a variety of nutritious ready-to-eat foods and limiting the availability of high-sugar, high-fat foods within the home. Focus groups of parents of teenagers suggest that they have concerns over whether or not they should involve teens in choosing foods served at meals, prepare alternative foods for teens when they don't like what is served, or restrict intake of specific foods.[55] The use of different creative settings and outlets to deliver innovative nutrition education programs to parents—including work sites, places of worship, community centers, libraries, supermarkets, and restaurants—should be explored.

School Programs

School-based programs can play important roles in promoting lifelong healthy eating and physical activity. Efforts to promote physical activity and healthful eating should be part of a comprehensive, coordinated school health program and should include school health instruction (curriculum), school physical education, school food service, health services (screening and preventive counseling), school-site health promotion programs for faculty and staff, and integrated community efforts.[56] The Centers for Disease Control and Prevention published two complementary reports, *Guidelines for School Health Programs to Promote Lifelong Healthy Eating*[57] and *Guidelines for School and Community Programs to Promote Lifelong Physical Activity Among Young People,*[56] which provide a developmental framework for comprehensive school nutrition and physical activity programs to be used by school districts, educators, health professionals, and policymakers. The guidelines (Table 14.14) include recommendations to promote healthy eating and lifelong physical activity including school policies, physical and social environments that encourage and enable physical activity and healthy eating, developmentally appropriate nutrition and physical education curricula and instruction, personnel training, family and community involvement, and program evaluation. The CDC has also published the School Health Index which can be used to assess the strengths and weaknesses of schools.[58]

Nutrition Education   Two-thirds of state education departments require instruction on nutrition or dietary behavior; 82% of school districts and 87% of schools require instruction in nutrition at the high school level.[59] Data from the 2004 School Health Profile survey show that most nutrition instruction is offered as part of a required health education course.[60] An average of 5 hours of nutrition and dietary behavior instruction is provided to high school students. The nutrition topics taught in health education classes can vary widely. Only 32% of health education teachers were given time or opportunities for nutrition education training in a 2-year period, suggesting there is a need for health professionals to become more involved in teacher training and continuing education activities. Teacher training in basic nutrition and instructional, motivational, and behavioral change strategies increases the success of nutrition education curriculum. Training may be most effective if teachers have the opportunity to examine their own body image and assess their eating behaviors. Teacher training typically increases the time spent on teaching nutrition in the classroom.

By senior high, students are in the process of cognitive and social development changes that permit more advanced nutrition education concepts and activities. The ability for more abstract thinking coupled with the changing psychosocial terrain of young adolescents provides both a challenge and a unique opportunity for educators to offer new learning and teaching strategies to encourage them to make healthful food choices. Young adolescence is an ideal time to teach students how to assess their own behavior and set goals for change. As adolescents begin the social process of individuation, they become ready and eager to make their own decisions and show their individuality. Nutrition education often fails to take advantage of the social and cognitive transitions of adolescence to promote the adoption of more healthful behaviors. In addition, obstacles to implementing nutrition education programs persist, ranging from insufficient funding to teacher ambivalence to competition with other high-priority health concerns, such as HIV and substance use prevention.[57]

Because knowledge alone is inadequate when students must decide which foods to eat and how to deal with peer and social influences, as well as with a widely available supply of high-fat, high-sugar foods, the focus of nutrition education and teaching methods should be on behavior change strategies and skill acquisitions to make healthful

food decisions. Characteristics of teaching methods found to be most effective in school health education curricula include use of discovery learning; use of student learning stations, small work groups, and cooperative learning techniques; cross-age and peer teaching; positive approaches that emphasize the intrinsic value of good health; use of personal commitment to change and goal setting; and provision of opportunities to increase self-efficacy in modifying health behaviors. Most important, adolescents need to be given repeated opportunities to develop, demonstrate, practice, and master the skills needed to make informed decisions and cope with social influences. To be effective, programs also must take into account cultural factors as well as the developmental processes of adolescents.

Nutrition Environment of the School   The National School Lunch Program (NSLP), and School Breakfast Program (SBP) are federally sponsored nutrition programs administered by the Department of Agriculture (USDA), in conjunction with state and local education agencies. Youth from households with incomes between 130% and 185% of the poverty level receive meals at reduced rates; youth from households with incomes 130% of poverty and below receive meals free. The school lunch and breakfast programs can complement and reinforce what is learned in the classroom and serve as a learning laboratory for nutrition education. The synergistic interaction between the school lunch program and classroom learning should enhance the likelihood that adolescents will adopt healthful eating practices.

The school environment provides multiple food and nutrition activities and influences not only classroom nutrition education and school meals, but also food sold in vending machines, school stores, and snack bars; fund raising events; food rewards by teachers; corporate-sponsored nutrition education materials; and in-school advertising of food products.[61] The result can be inconsistent nutrition messages. The growing stream of commercial messages, food advertisements, and easy access to high-fat and high-sugar food products in school are at cross purposes and in direct conflict with the goals of nutrition education and may negate the efforts in the classroom and lunchroom to foster healthful eating practices.

Most schools offer students the opportunity to purchase foods that are not part of the NSLP or SBP, through the option of á la carte foods, canteens or school stores, or vending machines. Data collected as part of the School Health Policies and Programs Study (SHPPS) demonstrated that 46% of elementary schools, 64% of middle/junior high schools, and 76% of high schools offered pizza, hamburgers, or sandwiches as á la carte food items.[65] In addition, 30% of elementary schools, 46% of middle/junior high schools, and 62% of high schools offered french fried potatos.[59] This same study found that 20% of schools offered brand-name fast foods for sale.

Vending machines or school stores can be found in 43% of elementary and 74% of middle/junior high and

98% of high school corridors;[59] 69% of schools allow students to purchase foods and beverages from these sources during school lunch periods. The majority of foods in school vending machines and school stores are high-fat or high-sugar items such as snack chips, candy, and soft drinks. Healthy food choices or lower-fat alternatives are generally unavailable. In the promotion of a healthy nutrition environment, vending machines and school stores need to offer healthier choices and lower-fat alternatives.

Half of all school districts surveyed in the SHPPS reported having signed a contract with a soft drink company allowing them exclusive rights to sell their product within the school.[59] Almost 80% of these school districts received a percentage of the profits generated by beverage sales and almost one-third received other incentives such as the donation of sports equipment, bulletin boards, trophy cases or cash incentives.

A growing trend of commercialism and marketing in schools uses in-school advertising and corporate-sponsored education materials. A study by Consumers Union Education Services found that direct advertising in schools has mushroomed.[61] Almost half of schools districts that contract with soft drink companies allow the placement of advertisments for these products on school grounds (including playgrounds) and 35% allow advertising within the school building.[59] Examples include school bus advertising for soft drinks and fast food restaurants (2% of all school districts in the SHPPS), "free" textbook covers advertising candy, chips, and soft drinks; ads for high-sugar/high-fat products on wallboards and in hallways (7% of school districts in the SHPPS), in student publications such as newspapers and yearbooks, and on sports scoreboards; and product giveaways in coupons (6% of SHPPS schools). In addition, *Channel One,* the daily news program broadcast to millions of students in grades 6 through 12 in thousands of schools, has 2 minutes out of each daily 12-minute program devoted to paid commercials for products that include candy bars, snack chips, and soft drinks.

Local and district policy initiatives can be instrumental in creating a supportive and integrated school nutrition environment with consistent health-promoting messages. Serious consideration should be given to restricting the sale of foods high in fat and sugar in schools. Corporate-sponsored education materials and programs should be carefully evaluated for nutrition accuracy, objectivity, completeness, and noncommercialism. Schools should also consider the Consumer's Union recommendation for making schools ad-free zones, where young people can pursue learning without commercial influences and messages.[62] Schools should be an environment where healthful eating behavior is normative, modeled, and reinforced. To improve the health of adolescents in the United States, schools should strive for an integrated nutritionally supportive environment.

School Wellness Policies   In the Child Nutrition and WIC Reauthorization Act of 2004, the United States

Congress included a provision requiring that all school districts with a federally funded school meals program develop and implement wellness policies that address nutrition and physical activity by the beginning of the 2006–2007 school year (P.L. 108–265). At a minimum, each local policy should include:

- Goals for *nutrition education, physical activity and other school-based activities* that are designed to promote student wellness in a manner that the local educational agency determines is appropriate
- *Nutrition guidelines* selected by the local educational agency for all foods available on each school campus under the local educational agency during the school day with the objectives of promoting student health and reducing childhood obesity
- *Guidelines for reimbursable school meals*
- *A plan for measuring implementation* of the local wellness policy, including designation of one or more persons within the local educational agency or at each school, as appropriate, charged with operational responsibility for ensuring that each school fulfills the district's local wellness policy
- *Community involvement*—including parents, students, and representatives of the school food authority, the school board, school administrators, and the public—in the development of the school wellness policy.[63,64]

An analysis of local wellness polcies from the 100 largest school districts in the United States showed:

- 99% address school meal nutrition standards.
- 93% address nutrition standards for á la carte foods and beverages.
- 92% address nutrition standards for foods and beverages available in vending machines.
- 65% address nutrition standards/guidelines for fundraisers held during school hours.
- 63% address nutrition standards/guidelines for classroom celebrations or parties.
- 65% address nutrition standards/guidelines for teachers using foods as rewards in the classroom.
- 50% of school districts address a recess requirement for at least elementary grade.
- 96% require physical activity for at least some grade levels.
- 97% require nutrition education for at least some grade levels.
- 95% outlined a plan for implementation and evaluation, utilizing the superintendent, school nutrition director, or wellness policy task force as the entity responsible for monitoring the policy.[65]

Sample model wellness policies have been developed by the National Alliance for Nutrition and Activity, and more information can be found on their website at www.schoolwellnesspolicies.org. Additional information about school wellness policies can also be found in Resources sections.

Community Involvement in Nutritionally Supportive Environments

Promoting lifelong healthy eating and physical activity behaviors among adolescents requires attention to the multiple behavioral and environmental influences in a community. Adolescents are most likely to adopt healthy behaviors when they receive consistent messages through multiple channels (e.g., community, home, school, and the media) and from multiple sources (e.g., parents, peers, teachers, health professionals, and the media).

Most physical activity occurs outside the school setting, making community sports and recreation programs essential for promoting physical activity among young people. Healthy eating can be integrated into these efforts by providing nutritious snacks. Community coalitions or task forces can be established to assess community needs and to develop, implement, and evaluate physical activity and nutrition programs for young people. Few studies have reported on nutrition education or physical activity programs outside of the school setting.

Model Nutrition Program   One example of a model community-based nutrition intervention program is the California Adolescent Nutrition and Fitness (CANfit) Program.[66] Through competitive grants, the CANfit Program supports and empowers adolescent-serving, community-based organizations to develop and implement nutrition education and physical activity programs for ethnic adolescents from low-income communities. Using a capacity-building model, the CANfit Program attempts to change the community context by improving access to healthier food choices and safe, affordable physical activity opportunities and by enabling adolescents to have the decision-making skills and social support necessary for making healthy nutrition and fitness choices. Examples of CANfit grantees' projects include (1) development of 10-week curriculum for an after-school program for African American girls that focuses on self-esteem, body image, healthy eating, cooking, and physical activity (e.g., hip-hop dance); (2) a nutrition and physical activity program for adolescents and their parents attending Saturday Korean language schools in Los Angeles; and (3) Latino adolescents in a soccer league that worked with a local health department to train team coaches and parents in sports nutrition. Innovative programs using capacity-building models, such as the CANfit Program, can provide numerous benefits to other communities.

Key Points

1. Puberty may begin earlier among black girls when compared to white and Mexican American girls.

2. French fries account for 23% of all vegetables eaten by adolescents.

3. More than one-third of female adolescents choose diets that are inadequate in calcium, magnesium, iron, zinc, folate, and vitamins A, B_6, and E.

4. Half of peak bone mass is accrued during adolescence. The greatest capacity to absorb calcium occurs in early adolescence, when the body can absorb almost four times more calcium than during young adulthood. However, calcium consumption drops as age increases during adolescence.

Resources

American Medical Association Adolescent Health Online
   Links to health and nutrition-related websites specifically addressing adolescent health issues; information on key health issues that affect teens.
   Website: **www.ama-assn.org/adolhlth**

Bright Futures
   View and download Bright Futures publications including: *Bright Futures in Practice: Nutrition* and *Bright Futures in Practice: Physical Activity.*
   Website: **www.brightfutures.org**

Centers for Disease Control and Prevention Nutrition and Physical Activity Programs
   Facts on physical activity and nutrition among U.S. adults and youth, model programs, program guidelines; links to pediatric growth charts.
   Website: **www.cdc.gov/nccdphp/dnpa**

National Center for Education in Maternal and Child Health
   Search databases on adolescent health; view and download publications and reports.
   Website: **www.ncemch.org**

The Vegetarian Resource Group

Information on choosing a healthy vegetarian diet; links to other resources and websites.

Website: www.vrg.org/nutrition/teennutrition.htm

United States Department of Agriculture Food and Nutrition Information Center

Information on Dietary Guidelines for Americans, the MyPyramid, dietary supplements, food safety, the Healthy

School Meals Resource System, and other sources of nutrition information.

Website: www.nal.usda.gov/fnic

School Nutrition Association: www.schoolnutrition.org

National Alliance for Nutrition and Activity: www. schoolwellnesspolicies.org/

United States Department of Agriculture, Food and Nutrition Service, Team Nutrition: www.fns.usda.gov/tn/Default.htm

References

1. Tanner JM. Growth at adolescence. Oxford: Blackwell; 1962.

2. Sun SS et al. National estimates of the timing of sexual maturation and racial differences among US children. Pediatr. 2002; 110:911–19.

3. Barnes HV. Physical growth and development during puberty. Med Clin North Am. 1975;59:1305–17.

4. Frisch RE. Fatness, puberty, and fertility: the effects of nutrition and physical training on menarche and ovulation. In: Brooks-Gunn J, Peterson AC, eds. Girls at puberty: biological and psychosocial perspectives. New York: Plenum Press; 1983:29–49.

5. Frisch RE, McArthur JW. Menstrual cycles: fatness as a determinant of minimum weight for height necessary for their maintenance or onset. Science. 1974;185:949–51.

6. Matkovic V et al. Nutrition influences skeletal development from childhood to adulthood: a study of hip, spine and forearm in female adolescents. J Nutr. 2004; 134:701S–705S.

7. Neumark-Sztainer D, Story M, Perry C, Casey M. Factors influencing food choices of adolescents: findings from focus group discussions with adolescents. J Am Diet Assn. 1999;99(8):929–937.

8. Killen JD et al. Is puberty a risk factor for eating disorders? Am J Dis Child. 1992;146:323–25.

9. Neumark-Sztainer D et al. Covariations of eating behaviors with other health-related behaviors among adolescents. J Adol Health. 1997;20:450–458.

10. Delva J, O'Malley PM, Johnston LD. Racial/ethnic and socioeconimic status difference in overweight and health-related behaviors among American students: national trends 1986–2003. J Adolesc Health. 2006. Oct;39:536–45.

11. Cusatis DC et al. Longitudinal nutrient intake patterns of US adolescent women: the Penn State Young Women's Health Study. J Adolesc Health. 2000;26: 194–204.

12. American School Health Association, Association for the Advancement of Health Education, Society for Public Health Education. The National Adolescent Student Health Survey, a report on the health of America's youth. Oakland, CA: Third Party Publications; 1988:1–178.

13. Jahns L, Siega-Riz AM, Popkin BM. The increasing prevalence of snacking among US children 1977–1996. J Pediatr. 2001;138(4):493–498.

14. Nielsen SJ, Siega-Riz AM, Popkin BM. Trends in food locations and sources among adolescents and young adults. Prev Med. 2002;35(2):107–13.

15. Wyshak G. Teenaged girls, carbonated beverage consumption, and bone fractures. Arch Pediatr Adolesc Med. 2000;154: 610–13.

16. Nicklas TA et al. Impact of breakfast consumption on nutritional adequacy of the diets of young adults in Bogalusa, Louisiana: ethnic and gender contrasts. J Amer Diet Assoc. 1998;98:1432–38.

17. Siega-Riz AM, Carson T, Popkin B. Three squares or mostly snacks—what do teens really eat? A sociodemographic study of meal patterns. J Adolesc Health. 1998; 22:29–36.

18. Lin BH, Guthrie J, Blaylock J. The diets of America's children: influences of dining out, household characteristics, and nutrition knowledge: U.S. Department of Agriculture Economic Report Number 746 (AER-746); 1996.

19. Neumark-Sztainer D et al. Family meal patterns: associations with sociodemographic characteristics and improved dietary intake among adolescents. J Am Diet Assoc. 2003;103:317–22.

20. Gillman M et al. Family dinner and diet quality among older children and adolescents. Arch Fam Med. 2000;9:235–40.

21. Taveras E et al. Family dinner and adolescent overweight. Obesity Research. 2005; 13:900–6.

22. Videon T, Manning C. Influences on adolescent eating patterns: the importance of family meals. Journal of Adolescent Health. 2003;32:365–73.

23. Kaiser Family Foundation. Generation M: media in the lives of 8–18 year-olds. March 2005. Available at www.kff.org/ entmedia/7251.cfm, accessed 05/06.

24. Neumark-Sztainer D et al. Adolescent vegetarians: a behavioral profile of a school-based population in Minnesota. Arch Pediatr Adolesc Med. 1997;151:833–38.

25. Sabate J, Llorca MC, Sanchez A. Lower height of lacto-ovovegetarian girls at preadolescence: an indicator of physical maturation delay? J Amer Diet Assoc. 1992; 92:1263–64.

26. Johnston P, Haddad E. Vegetarian diets and pregnant teens. In: Story M, Stang J, eds. Nutrition and the pregnant adolescent: a practical reference guide. Minneapolis, MN: Leadership, Education and Training Program in Maternal and Child Nutrition, 2000;135–45.

27. Perry CL et al. Characteristics of vegetarian adolescents in a multiethnic urban population. J Adolesc Health. 2001; 29(6):406–16.

28. Dietary Guidelines Advisory Committee. Excerpt of the report of the Dietary Guidelines Advisory Committee on Dietary Guidelines for Americans, 2000.

29. USDA. Center for Nutrition Policy and Promotion. MyPyramid 2000.

30. Community Nutrition Research Group, Beltsville Human Nutrition Research Center, Agricultural Research Service, USDA. Pyramid serving intakes in the US 1999–2002, 1 day. Available at www.ba. urs.usda.gov/cnrg, accessed 11/06.

31. Stang J et al. Relationships between vitamin and mineral supplement use, dietary intake, and dietary adequacy among adolescents. J Amer Diet Assoc. 2000;100: 905–10.

32. Moshfegh A, Goldman J, Cleveland L. NHANES 2001–2002: usual nutrient intakes from food compared to dietary reference intakes. www.ars.usda.gov/foodsurvey, accessed 10/06.

33. Centers for Disease Control and Prevention. Youth risk behavior surveillance—surveillance summaries 2005. Morb Mortal Wkly Rep. June 9, 2006,55(No SS-5).

34. American Academy of Pediatrics Committee on Nutrition. Carbohydrate and dietary fiber. In: Barness L, ed. Pediatric nutrition handbook, 3rd ed. Elk Grove Village, IL: American Academy of Pediatrics; 1993;100–6.

35. National Heart, Lung, and Blood Institute, National Cholesterol Education Program. Report of the expert panel on blood cholesterol levels in children and adolescents. Bethesda, MD: National Institutes of Health; 1991.

36. Food and Nutrition Board, Institute of Medicine. Dietary Reference Intakes for energy, carbohydrate, fiber, fat, fatty acids, cholesterol, protein and amino acids. Washington, DC: National Academies Press; 2005.

37. Peacock M. Calcium absorption efficiency and calcium requirements in children and adolescents. Am J Clin Nutr. 1991;54: 261S–265S.

38. Hui SL, Johnston CC, Jr., Mazess RB. Bone mass in normal children and young adults. Growth. 1985;49:34–43.

39. Gao X et al. Meeting adequate intake for dietary calcium without dairy foods in adolescents aged 9 to 18 years (National Health and Nutrition Examination Survey 2001–2022). J Amer Diet Assn. 2006;106: 1759–65.

40. Harel Z et al. Adolescents and calcium: what they do and do not know and how much they consume. J Adolesc Health. 1998;22:225–28.

41. Albertson AM, Tobelmann RC, Marquart L. Estimated dietary calcium intake and food sources for adolescent females: 1980–92. J Adolesc Health. 1997;20:20–26.

42. Reilly JK et al. Acceptability of soymilk as a calcium-rich beverage in elementary school children. J Am Diet Assoc. 2006; 106:590–93.

43. Hearney RP, Rafferty K. The settling problem in calcium-fortified soybean drinks. J Amer Diet Assoc. 2006;106: 1753.

44. Harnack L, Stang J, Story M. Soft drink consumption among US children and adolescents: nutritional consequences. J Am Diet Assoc. 1999;99:436–41.

45. Cavadini C, Siega Riez AM, Popkin BM. US adolescents food intake trends from 1965 to 1996. Arch Dis Child. 2000;83:18–24.

46. Lloyd T et al. Adult female hip bone density reflects teenage sports exercise patterns, but not teenage calcium intake. Pediatrics. 2000;106:40–44.

47. Lloyd T et al. Modifable determinants of bone status in young women. Bone. 2002;30:416–21.

48. Wang MC et al. Diet in midpuberty and sedentary activity in prepuberty predict peak bone mass. Am J Clin Nutr. 2003; 77:495–503.

49. Centers for Disease Control and Prevention. Recommendations to prevent and control iron deficiency anemia in the United States. Morb Mortal Wkly Rep. 2002; 51(40):897–99.

50. U.S. Department of Health and Human Services. Healthy People 2010 (conference edition in two volumes). Washington, DC: January 2000.

51. Institute of Medicine. Nutrition during pregnancy: part I: weight gain; part III: nutrient supplements. Washington, DC: National Academy Press; 1990.

52. Neumark-Sztainer D et al. Correlates of inadequate fruit and vegetable consumption among adolescents. Prev Med. 1996;25: 497–505.

53. American Medical Association. Guidelines for adolescent preventive services. Chicago: American Medical Association, Department of Adolescent Health; 1992.

54. Leverton RM. The paradox of teen-age nutrition. J Amer Diet Assoc. 1968;53: 13–16.

55. O'Dougherty M, Story M, Lytle L. Food choices of young African-American and Latino adolescents: where do parents fit in? J Am Diet Assoc. 2006;106:1846–50.

56. Centers for Disease Control and Prevention. Guidelines for school and community programs to promote lifelong physical activity among young people. Morb Mortal Wkly Rep. 1997;46:1–36.

57. Centers for Disease Control and Prevention. Guidelines for school health programs to promote lifelong healthy eating. Morb Mortal Wkly Rep. 1996;45:1–37.

58. CDC. School health index. Atlanta, VA: US Dept of Health and Human Services, CDC; 2006.

59. Weschler H et al. Food service and foods and beverages available at school: results from the School Health Policies and Programs Study 2000. J School Health. 2001;71:313–24.

60. CDC. Secondary school health education related to nutrition and physical activity—selected sites, United States, 2004. MMWR. 2006;55(30):821–24.

61. Consumers Union Education Services. Captive kids: commercial pressures on kids at school. Yonkers, NY: Consumers Union of United States; 1995.

62. Consumers Union Education Services. Selling America's kids: commercial pressures on kids of the '90s. Yonkers, NY: Consumers Union of the United States; 1990.

63. School Nutrition Association. Local school wellness policies. Available at www.schoolnutrition.org, accessed 10/06.

64. United States Department of Agriculture, Food and Nutrition Service. Healthy schools: local wellness policy. Available at www.fns.usda.gov/tn/Healthy/wellnesspolicy.html, accessed 10/06.

65. School Nutrition Association. A foundation for the future: analysis of local wellness policies from the largest 100 districts. School Nutrition Association, 2006. Available at www.schoolnutrition.org, accessed 10/06.

66. Hinkle A. Community-based nutrition interventions: reaching adolescents from low-income communities. Ann NY Acad Sci. 1997;187:83–93.

Chapter 15

Adolescent Nutrition:
Conditions and Interventions

Chapter Outline

- Introduction
- Overweight and Obesity
- Supplement Use
- Nutrition for Adolescent Athletes
- Substance Use
- Iron-Deficiency Anemia
- Cardiovascular Disease
- Disordered Eating, Dieting, and Eating Disorders

Prepared by **Jamie Stang,** with assistance from **Rachel Kossover** and **Dianne Neumark-Sztainer**

Key Nutrition Concepts

1   Overweight adolescents are at increased risk for medical and psychosocial complications such as hypertension, hyperlipidemia, insulin resistance, type 2 diabetes mellitus, hypoventilation and orthopedic disorders, depression, and low self-esteem.

2   Competitive adolescent athletes require an additional 500–1500 Kcals per day to meet their energy needs. Additional protein may be required among adolescent athletes who are still growing.

3   Adolescents are concerned about their weight; a national study suggests that on any given day, approximately half of adolescent females and 15% of adolescent males were attempting weight loss.

Introduction

Multiple factors influence the nutritional needs and behaviors of adolescents. This chapter presents specific behaviors and nutrition concerns that affect significant numbers of adolescents, including overweight, participation in competitive sports, substance abuse, vegetarian diets, eating disorders, hypertension, and hyperlipidemia. Because overweight, sports participation, and eating disorders affect a larger group of adolescents than other conditions, they are presented in greater detail.

Overweight and Obesity

The increase in the prevalence of overweight among adolescents mirrors that of adults over the past two decades. Exact reasons for this increase have not been identified. Although genetics is known to contribute to the occurrence of overweight, and having one or more overweight parent(s) increases a teen's risk of developing obesity, it alone clearly cannot account for the dramatic increase in overweight during the past two decades.[1] Environmental factors, or interactions between genetic and environmental factors, are the most likely causes of the dramatic rise in overweight. Risk factors for the development of overweight among children and adolescents include having at least one overweight parent; coming from a low-income family; being the descendant of African American, Hispanic, or American Indian/Native Alaskan parents; and being diagnosed with a chronic or disabling condition that limits mobility.[1]

Inadequate levels of physical activity and consuming diets high in total calories and added sugars and fats are additional risk factors among a significant proportion of adolescents. These environmental factors increase the risk of developing overweight if an adolescent is genetically predisposed to obesity.

Weight status among adolescents should be assessed by calculating body mass index (BMI). BMI is calculated by dividing a person's weight (kg) by their height^2 (m^2). BMI values are compared to age- and gender-appropriate percentiles to determine the appropriateness of the individual's weight for height. Youth with BMI values greater than the 85th but lower than the 95th percentile are considered at risk for overweight; those with BMI values above the 95th percentile are considered overweight.[2] Growth curves based on BMIs for children and adolescents are available from the National Center for Health Statistics. An example of a BMI growth curve is shown in Illustration 15.1.

Illustration 15.1 CDC Growth Charts: United States.

Developed by the National Center for Health Statistics in collaboration with the National Center for Chronic Disease Prevention and Health Promotion; 2000.

Table 15.1 Prevalence of at-risk-for and overweight by race and gender among 12–19 year olds in NHANES 2003–04

	At-risk of Overweight	Overweight
Males		
White	38.7	19.1
Black	31.4	18.5
Mexican American	37.3	18.3
Male Total	36.8	18.3
Females		
White	30.4	15.4
Black	42.1	25.4
Mexican American	31.1	14.1
Female Total	31.7	16.4

SOURCE: All data taken from Ogden CL, Carroll MD, Curtin LR, McDowell MA, Tabak CJ, Flegal KM. Prevalence of Overweight and Obesity in the United States, 1999–2004. JAMA. 2006 Apr 5; 295(13):1549–55.

Data from the 2003–2004 NHANES suggest that 34.3% of U.S. students are at risk of overweight and 17.4% are overweight.[3] Table 15.1 provides prevalence estimates of overweight among adolescents in the United States, by gender and race/ethnicity. In general, the prevalence of being at risk for overweight among females is highest among black teens, while among males the prevalence is highest among white and Mexican American teens. The prevalence of overweight is similar among all adolescent males; among females black teens are more likely to be overweight than their peers. While American Indian students are not reported separately in the Youth Risk Behavior Survey (YRBS) or NHANES, regional data suggest that the prevalence of overweight among American Indian adolescent males is higher than other racial/ethnic groups.[4,5] Rates of overweight among American Indian adolescent females is higher than among white and Hispanic females but lower than among black females.

The persistence of overweight from childhood throughout adulthood has not been well quantified. Research suggests that the persistence of obesity from infancy to adulthood increases with age. More than 70% of overweight adolescents can be expected to remain overweight into adulthood.[6,7] Identification of overweight at an early age is important, as research data suggests that children with BMI above the 85th percentile are more likely than children with BMI below the 50th percentile to continue to gain weight and reach overweight status by adolescence.[1] The risk of persistence of obesity from childhood into adulthood increases if at least one parent is overweight.[7] Risk of persistence of overweight is also higher among the most overweight individuals, especially those whose weight is more than 180% of ideal weight.

Health Implications of Adolescent Overweight

A range of medical and psychosocial complications accompanies overweight among adolescents, including hypertension, dyslipidemia, insulin resistance, type 2 diabetes mellitus, sleep apnea and other hypoventilation disorders, orthopedic problems, hepatic diseases, body image disturbances, and lowered self-esteem.[8–10] Cardiorespiratory fitness is lower in male and female adolescents who are overweight than in those who are normal weight. Cardiorespiratory fitness is also lower in adolescents who have low levels of physical activity and high levels of sedentary behaviors.[11] Rates of type 2 diabetes mellitus are more common among overweight youth. In one study, a third of newly diagnosed diabetic patients under the age of 20 had type 2 diabetes mellitus; almost all of these patients had BMI values above the 90th percentile.[12]

All adolescents should be screened for appropriateness of weight-for-height on a yearly basis. Teens determined to be at-risk for overweight require an in-depth medical assessment to diagnose any obesity-related complications. Illustration 15.2 provides recommended screening and referral procedures for weight-for-height among adolescents.

Illustration 15.3 illustrates recommended weight goals based on BMI and age. Weight maintenance is recommended for adolescents who are at risk for overweight and have not yet completed puberty (unless medical complications are noted). Weight loss is recommended

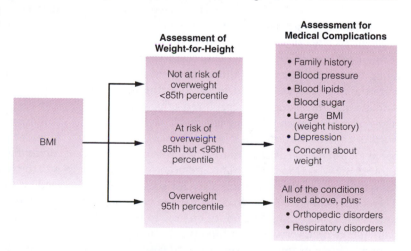

Illustration 15.2 Recommended overweight screening procedures.
Adapted with permission from Himes JH, Dietz WH. Guidelines for Overweight in Adolescent Preventive Services: Recommendations from an Expert Committee. The Expert Committee on Clinical Guidelines for Overweight in Adolescent Preventive Services. Am J Clin Nutr. 1994;59:307–16.

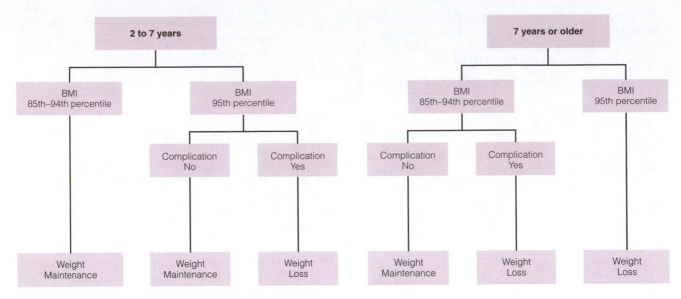

Illustration 15.3 Recommendations for weight goals.

SOURCE: Barlow SE, Dietz WH. Obesity Evaluation and Treatment: Expert Committee Recommendations. The Maternal and Child Health Bureau, Health Resources and Services Administration and the Department of Health and Human Services. Pediatrics. 1998;102:E29.

when medical complications are present in youths at risk for overweight, when the adolescent is determined to be overweight, and among older adolescents who have completed physical growth and development. Guidelines for treatment of overweight are presented in Table 15.2. Weight loss should be attempted only after the adolescent and his or her family show current weight can be maintained.

No specific physical activities or dietary regimens are recommended for adolescent weight management programs. Treatment should provide appropriate caloric intake and optimum nutrient intake for normal growth and development, and should help the adolescent develop and sustain healthful eating habits.[1,13] Reducing or eliminating the intake of "problem" foods such as savory snacks or high-sugar beverages is often the first dietary change recommended. Health care providers generally should not recommend calorie or fat gram counting to avoid encouraging "good and bad food" thinking and to try to minimize the risk of developing disordered eating behaviors. Adolescents more readily accept replacing high-fat or high-sugar foods with healthier substitutes and monitoring portion control, resulting in a better chance of long-term behavior changes. Fast foods are a way of life for many adolescents—fast food restaurants are prime employers of teens and are a prime gathering place for social activities. Therefore, asking teens to avoid fast food restaurants is an unrealistic goal. Counseling youth on how to balance out daily intake on days when fast food is consumed is useful, as are concrete examples of specific food choices that are lower in calories and added fats and sugars.

Ideas on how to decrease sedentary activities to less than 2 hours a day and on how to increase physical activity are beneficial.[1,13] Referrals to community centers, athletic clubs, local parks and recreation programs, and community education programs that offer fun, noncompetitive physical activities—such as yoga, tai chi, swimming, weight lifting, and bicycling or walking clubs—point adolescents in a beneficial direction. Examples of how to incorporate physical activity into daily living, such as walking or biking to school, may be beneficial for students who do not see themselves as athletically inclined.

For severely obese adolescents, as well as those with significant medical complications, rapid weight loss may be required. The use of very low-calorie diets or protein-sparing modified fasts should only be done under continuous medical supervision as these diets have been associated with many health risks including orthostatic hypotension, diarrhea, hyperuricemia, cholelithiasis, electrolyte imbalance, and reduced serum protein levels.[14] The use of these diets should not exceed 12 weeks in duration. The use of appetite suppressants and other weight loss medications among adolescents has been examined in only a few studies. As with intensive dietary therapies, these medications must be adminstered with intensive medical follow-up to minimize health risks. Side effects of weight-loss medications include insomnia, headache, hypertension, cardiac arrhythmia, depression, dizziness, edema, nausea, anxiety, steatorrhea, flatulence, fecal incontinence, blurred vision, and fat-soluble vitamin deficiencies.[14]

Bariatric surgery is performed on adolescents who are severely overweight and experience potentially life-threatening medical complications. Guidelines for the use of bariatric surgery among adolescents have been developed (Table 15.3).[15] In order to be considered as a candidate for bariatric surgery, adolescents must have a BMI of >40 with medical complications or a BMI of >50 without medical complications. In addition, teens

Table 15.2 General guidelines for weight management therapy

1. Early intervention is recommended, preferably before the child reaches the 95th percentile for BMI.

2. Parents should be informed of medical complications associated with childhood overweight, and all youth should be assessed for medical complications.

3. All family members and caregivers should be involved in the weight management program.

4. All family members should be assessed to determine their readiness to make behavior changes, and treatment should not begin until all family members are ready to adopt behavior changes.

5. Weight management programs should emphasize goals of improving eating and physical activity patterns as opposed to specific weight goals.

6. Programs should be skill based, with families taught to identify problem behaviors, monitor such behaviors, utilize behavior modification principles to address problem behaviors, and implement problem-solving skills when dealing with obstacles to behavior change.

7. Families should be involved in assessing current eating and activity patterns, deciding which behaviors need to be modified, setting goals for behavior changes, and determining how these goals will be achieved.

8. New behavior changes should not be instituted until previous changes have been accomplished and maintained.

9. Routine follow-up visits should be scheduled to monitor progress and prevent relapse to former eating and activity patterns.

SOURCE: Used with permission. Stang J, Holladay L. Treatment of Childhood Obesity Involves the Family. American Dietetic Association Public Health Community Nutrition Practice Group newsletter. The Digest; Spring 2000. Adapted from Barlow SE, Dietz WH. Obesity Evaluation and Treatment: Expert Committee Recommendations. Pediatrics. 1998;102:E29.

Table 15.3 Guidelines for consideration of bariatric surgery in adolescents

- Failure to obtain adequate weight loss after minimum of 6 months of intensive weight loss program participation
- SMR/Tanner at stage III or higher
- BMI ≥40 with medical complications or ≥50 without medical complications
- Participation in psychological and medical counseling before surgery with agreement to continue counseling after surgery
- Must have adequate support of family and a home environment conducive to long-term dietary change
- Capability to follow medical nutrition therapy protocol after surgery
- Agreement to prevent pregnancy for at least 1 year after surgery

Adapted from: American Pediatric Surgical Association Clinical Task Force on Bariatric Surgery 2004.

bariatric surgery.[16] In a recent study of 31 teens, 39% of patients undergoing bariatric surgery experienced complications, with 25% experiencing moderate complications, 13% experiencing severe complications, and 1 teen dying within a year of the surgery.[16] Of the 30 adolescents included in follow-up, 2 began regaining weight within the first year following surgery, with 1 adolescent regaining more than 50% of lost body weight.[16]

Supplementation with multivitamin–mineral preparations following bariatric surgery is imperative for adolescents. While all nutrients may require supplementation when food intake is severely limited following surgery, calcium, vitamin B_{12}, folate, and thiamin are especially important.[15,16] Iron status of menstruating females must be monitored closely as well, with supplementation provided as needed.

Supplement Use

Vitamin–Mineral Supplements

Supplements may be used by adolescents for a variety of reasons, including improving health, treating iron-deficiency anemia, increasing energy, building muscle, and losing weight. National data suggest that the prevalence of vitamin–mineral supplement use among teens in the United States ranges from 16% to 33%;[17] studies in Canada show a prevalence of vitamin/mineral use of 43%.[18] More than half of adolescents who report using vitamin–mineral supplements take them occasionally, with slightly less than half using them daily. Data on racial differences on adolescent supplement use are equivocal; one study suggests that white female adolescents are most likely to report using

should have completed the majority of their adolescent growth spurt before undergoing bariatric surgery in order to minimize potential side effects such as stunting of growth. Once the adolescent growth spurt is completed, nutrient needs are reduced so it is less likely that nutrient deficiencies will occur among adolescents who have limited food intakes following bariatric surgery. It is recommended that, like adults, adolescents attempt to lose weight through traditional avenues before undergoing bariatric surgery.[15]

The long-term success rate of bariatric surgery among teens is not well established. Research suggests that adolescents reduce their BMI values by 19 units and lose an average of 36–47% of their excess body fat following

Case Study 15.1

Photo Disc

Adolescent Nutrition, Growth and Development, and Obesity

Anna is a 12-year-old female being seen in the pediatric and adolescent medicine clinic for a routine physical examination for summer camp. She reports no medical concerns. Her mother reports concern over recent weight gain. Anna's last visit was approximately 5 months earlier when she was seen for an upper respiratory infection.

During the current visit, Anna's BMI value is calculated. Her hemoglobin and hematocrit levels are checked, and a blood pressure reading is taken. The primary health care provider also takes a medical history and determines Anna's sexual maturation rating (SMR). The following data are recorded:

Age: 12 years, 5.5 months
SMR: early stage 3
Menarche: not yet occurred
Height: 58 inches (1.47 meters)
Weight: 111 pounds (50.35 kilograms)

BMI: 23.3
Hematocrit: 35.1%
Hemoglobin: 11.9 g/dL
Blood Pressure: 121/77

Anna's BMI value is charted on the National Center for Health Statistics BMI reference growth chart for girls age 2–20 years, and is compared to previous data. On her previous medical visit, Anna's BMI was calculated to be 21.4. Her previous height was 56.0 inches (1.42 meters) and her previous weight was 95 pounds (43.09 kilograms).

Questions

1. What is the current BMI percentile for Anna, based on her age and gender?
2. What was her previous BMI percentile based on the BMI of 21.4 and age of 12.0 years?
3. What is her current classification for height-for-weight (BMI)? Normal, at-risk for overweight, or overweight?
4. Given her current weight classification based on BMI, what are the recommendations for weight management based on current recommendations?
   a. Maintain current weight and monitor food intake, physical activity levels, weight, height, growth, and development at 3- to 4-month intervals.
   b. Weight loss of 1–2 pounds per week.
   c. Weight gain of 1–2 pounds per week.
   d. No concerns or actions are identified at this time.
5. What do the SMR and the absence of menarche tell you about Anna's potential for linear growth and future weight gain?
6. Using the Centers for Disease Control and Prevention (CDC) guidelines, how would you classify Anna's blood pressure readings?
7. Using the CDC guidelines, how would you classify her hemoglobin and hematocrit levels?

The primary health care provider refers Anna to a nutritionist at the clinic for nutrition counseling and weight management. The nutritionist conducts a 24-hour recall of food intake and also asks about physical activity and sedentary activity. The results follow.

Food Intake in Past 24 Hours

Today:

7 a.m.	8 oz apple juice at home
10 a.m.	candy bar from school vending machine
12:15 p.m.	hamburger from school cafeteria and 12 oz of sweetened (nondiet) cola and 1 oz bag of potato chips from school vending machine
3:00 p.m.	tortilla chips with cheese sauce (approximately 3 oz chips and ½ c cheese sauce) and 12 oz nondiet cola at home

Case Study 15.1 (continued)

Previous evening:

6:30 p.m.	chicken rice casserole (approximately 2 cups), 8 oz apple juice, 2 crescent rolls with 1 Tbsp butter
8:30 p.m.	1 bag of butter flavored microwave popcorn (approximately 8 c)

Physical Activity in Past 24 Hours

Today:

7:45 to 2:45 p.m.	walked between classes for 1–2 minutes, six times that day
1:00 p.m.	35 minutes of physical education (Anna played basketball for only 15 minutes due to need to rotate students on teams); physical education is offered 2 days per week in winter semester only

Previous evening:

5:00 p.m.	walked from friend's house to home (approximately 3 blocks)

Sedentary Activity in Past 24 Hours

Today:

7:45 to 2:45 p.m.	sat in classroom for 35–40 minutes, five times that day
3:00 p.m.	on the computer in Internet chat rooms and e-mail for 40 minutes

Previous evening:

8:00 p.m.	watched television for 2 hours
10:45 p.m.	went to bed

The nutritionist talks with Anna about ways to improve her diet by eating more healthy, convenient breakfast foods such as sandwiches, cereal, peanut butter or cheese on whole grain crackers, and string cheese for breakfast. She also suggests replacing apple juice with calcium-fortified orange juice to increase folic acid, vitmain C, and calcium intakes. The nutritionist suggests lower fat, more nutrient-dense vending machine choices that Anna can make when she eats from vending machines at school, such as animal crackers, whole grain crackers with peanut butter or cheese, yogurt, pretzels, baked chips, trail mix, dried fruit, milk, and fruit juices. Anna is encouraged to choose items from the school cafeteria for lunch instead of vending machines, especially from the salad bar. She is also given a list of low-fat, high-nutrient snacks that are easy to make or quick to eat after school. Finally, the nutritionist discusses portion sizes based on the Food Guide Pyramid with Anna and provides guidance about meeting the recommended number of servings of fruits, vegetables, whole grains, and dairy products.

The nutritionist provides suggestions on how Anna can increase physical activity and reduce sedentary activity to help in weight maintenance. She provides a list of local community centers and health clubs that offer fun classes and activities that adolescents might enjoy after school and on weekends. She suggests that Anna take daily walks with her best friend or mother, and that she limit television and computer time to no more than 2–3 hours total per day.

The nutritionist recommends that Anna come in for a follow-up appointment in 3–4 weeks to see how she is progressing with physical activity and dietary changes. She also suggests that Anna make an appointment to see her primary care provider in 3–4 months for further evaluation of her height, weight, and physical development.

Questions

1. List two nutrients that are likely to be consumed in inadequate amounts in Anna's diet.
2. In which of the MyPyramid food group(s) is Anna's diet inadequate?
3. List two recommendations that you would suggest to Anna to improve her food intake. (Try to use concrete examples because she is a young adolescent.)
4. List two recommendations that you would suggest to Anna to improve her physical activity or reduce her sedentary activity. (Try to use concrete examples because she is a young adolescent.)

vitamin–mineral supplements, followed by white males, Mexican American females, black females, Mexican American males, and finally black males, while another study found no differences by race. Overall, approximately 28% of female adolescents and 24% of male adolescents use vitamin–mineral supplements.[17,19]

Approximately half of vitamin–mineral supplements consumed by adolescents are multivitamins without minerals, 34% are individual vitamins or minerals, 18% are multivitamins with minerals, and 17% are iron with vitamin C tablets. Among the individual nutrient supplements used, vitamin C is the most common, followed by calcium, iron, vitamin E, and B-vitamin complex.[19] Adolescents who take vitamin–mineral supplements tend to consume a more nutritionally adequate diet than those who don't.[19,20] Supplement users obtain a smaller proportion of calories from total and saturated fats, and more from carbohydrate compared to nonsupplement users. They also consume diets that tend to be more nutrient dense—higher in folate, calcium, iron, and vitamins E, C, and A—than these of nonusers.

Herbal Remedies

Few data are available to quantify the use of non-nutritional supplements such as herbs (including herbal weight-loss products) among adolescents. A small study of 78 Australian teens found that 18% has used herbal supplements, 5% had taken creatine and guarana, and 1% had used coenzyme Q.[21] A study of 353 teens from Canada found that 4.1% of adolescents used herbal weight-control products, 6% used energizers (e.g., bee pollen), 1.6% used L-carnitine, and 5.3% used creatine.[18]

The use of herbs and supplements by youth is highly controversial. Adolescents may take herbal supplements for several reasons including weight loss, treatment of attention deficit disorder, and to increase energy and stamina. Youth with special health care needs, such as autism spectrum disorders, attention deficit disorder, and cystic fibrosis, may use supplements more frequently than other adolescents. Studies are needed to determine exactly what types of herbal products are used by adolescents, because many herbs are known to have potentially dangerous side effects, and few recommendations are available to guide the use of herbs by children or adolescents.

Ergogenic Supplements Used by Teens

YRBS data suggest that 4% of adolescents report having used illegal steroids.[4] Steroid use is reported more frequently among male (5%) than female (3%) adolescents.[4] Hispanic males report the highest prevalence of steroid use (6%), followed by white males (5%) and black males (4%). Steroid use appears to decrease slightly with age, peaking during nineth or tenth grade. Supplements used by adolescent athletes include creatine; individual amino acids or protein powders; carnitine; anabolic-androgenic

steroids; anabolic steroid precursors, including dehydroepiandrosterone (DHEA) and androstenedione; beta-hydroxy-beta-methylbutyrate; growth hormone; and ephedra.[22,23] Steroids and other ergogenic supplements are taken orally, injected, or absorbed through transdermal patches.[22] They are most often used outside of the sport season to avoid detection of use. Steroids and ergogenic supplements are often taken in 1–3-month periods and are "stacked" so that the peak dose of one substance may overlap the introduction of another substance.[22] While the use of steroids and ergogenic aids is forbidden by national and NCAA regulations, few high school athletic programs test athletes for their use.[24]

Anabolic-androgenic steroids are controlled substances used to increase lean body mass and improve strength.[22] The use of these steroids has been linked to infertility, hypertension, physeal closure, depression, aggression, and increased risk of atherosclerosis.[22] DHEA and androstenedione are precursors of testosterone and estrogen. Androstenedione in also a controlled substance while DHEA is widely available as a supplement.[22] Naturally produced in the human body by the adrenal glands, DHEA levels fall in humans as age increases. Its reputed effects include reducing body fat, decreasing insulin resistance, increasing immune system function, increasing lean body mass, and decreasing risk of osteoporosis; however, no scientific evidence backs such claims.[22] As steroid precursors, andostenedione and DHEA may induce many of the same side effects as steroids such as irreversible gynecomastia (breast enlargement) and prostate enlargement among males and hirsuitism (facial hair) among females.[22,23] Growth hormone has been shown to decrease subcutaneous body fat.[22] Side effects of its use include physeal closure, hyperlipidemia, glucose intolerance, and myopathy.[22] Given the possibility of significant side effects in pubescent adolescents undergoing hormonal changes, steroids or their precursors and growth hormone should not be used by adolescents.

Creatine is sold as a nutritional supplement to increase lean body mass. Creatine, formed in the liver and kidney of the human body, can be obtained in more than adequate amounts from the consumption of meat. Studies of creatine in adults show mixed results.[25] It appears to be of no benefit to endurance athletes, and marginally beneficial during short-duration, anaerobic, strength-related sports.[22] Side effects of creatine use, which seem to be dose related, include abdominal pain and cramping, nausea, diarrhea, headache, dehydration, reduced renal function, increased tendency toward muscle strains, and muscle soreness.[22,25] No available data document the long-term health effects related to creatine use; however, chronic use may be associated with renal damage.[22,25]

Ephedra was sold as an over-the-counter supplement until 2004 when its sale was banned by the FDA. While it has been proven to increase metabolic rate, no known benefits on athletic performance have been documented.[22]

Ephedra was removed from the market due to side effects including cardiac arrhythmia, hypertension, increased risk of myocardial infarction and cerebral vascular accidents and, in extreme cases, death. Litigation to return ephedra to the marketplace in lower doses than previously sold is ongoing.[22]

Nutrition for Adolescent Athletes

More than half (54%) of U.S. adolescents report playing on one or more organized sports teams.[4] Participation is higher among male adolescents than female adolescents (62% versus 50%) and reduces with age. Among males, participation in organized sports is similar among adolescents of all racial and ethnic groups (62–65%). Among females, white females are more likely to participate than their peers (54% versus 45%).

Fluids and Hydration

Fluid intake is an important issue in sports nutrition for adolescents. Young adolescents and those who are prepubertal present a particular vulnerability to heat illnesses because their bodies do not regulate body temperature as well as older adolescents. Adolescents can become so mentally and physically involved in physical activities that they do not pay attention to physiological signals of fluid loss, such as excessive sweating and thirst. Some athletes commonly assume they do not need additional fluids if they are not actively moving all of the time during exercise. Other factors, such as ambient temperature and humidity levels and weight of equipment (helmets, padding, etc.) worn or utilized during exercise also play a role. For instance, hockey goalies do not skate for great distances during a match, yet they may lose 5 or more pounds of body weight due to the weight of the padding and equipment they wear. Therefore, all athletes should be counseled to regularly consume fluids, even if they do not feel thirsty.

Athletes should consume 6–8 oz of fluid prior to exercise, 4–6 oz every 15–20 minutes during physical activity, and at least 8 oz of fluid following exercise. Recommendations encourage athletes to weigh themselves periodically before and after exercise or competition to determine whether they have lost body weight. Each pound of body weight lost during an activity requires ingestion of 16 oz of fluid following the activity to maintain proper hydration. Athletes should drink no more than 16 oz of fluid each 30 minutes, however, to avoid potential side effects, such as nausea.

The type of fluid an athlete drinks is affected more by peer pressure and mass media than by actual physiological need. Sports drinks and energy drinks are very popular among teens, even those who do not participate in sports. Data on children suggest that even though water is an economical, easily available fluid, it may not provide optimal benefits for athletes who participate in physically intense events or those of great duration.[26] In such events, juice diluted at a ratio of 1:2 with water, or sports drinks that contain no more than 6% carbohydrate, may allow for better hydration and physical performance. Undiluted juices, fruit drinks, carbonated beverages, energy drinks, and sports drinks that contain more than 6% carbohydrate are not recommended during exercise because they may cause gastric discomfort. Their high carbohydrate content may also delay gastric emptying. Some carbonated soft drinks and many energy drinks contain significant amounts of caffeine, which promotes diuresis, reducing their effectiveness at rehydration.

Special Dietary Practices

Adolescent athletes may follow special diets or consume nutritional and non-nutritional supplements in an effort to improve physical performance and increase lean body mass. Even though data on the prevalence of supplements or special diets among adolescent athletes remain limited, data from surveys of collegiate athletes suggest that more than 10% of athletes use supplements.[23] Special diets that are noted among adolescent athletes include carbohydrate-loading regimens and high-protein diets. Distance runners and other endurance athletes traditionally used carbohydrate loading to improve the glycogen content of muscle. It involves the manipulation of training intensity and duration along with the carbohydrate content of meals to improve glycogen formation in muscle tissue. Carbohydrate loading is traditionally a weeklong process that begins with intense training 1 week prior to competition. For the first 3 days of a carbohydrate-loading week, athletes choose low-carbohydrate foods, but continue to exercise in an attempt to deplete muscle glycogen stores. During the 3 days prior to competition, athletes rest, or exercise minimally, while consuming a high-carbohydrate diet to promote glycogen formation and storage.

High-protein diets may take many forms for teen athletes. In general, athletes who follow high-protein diets may consume three to four times the recommended protein intake, accompanied by a relatively low intake of carbohydrate. High-protein diets should be discouraged among athletes for several reasons. First, many dietary protein sources are also sources of total and saturated fats, which may increase lifetime risk of coronary artery disease. Second, high protein and fat intakes delay digestion and absorption, limiting the amount of energy available for use during physical activity. Finally, more water is required for the breakdown of protein than either fat or carbohydrate due to the increased water loss that accompanies the excretion of nitrogen. This factor places an athlete at increased risk for dehydration, often accompanied by a decrease in physical performance. Adolescents should be reminded that the consumption of individual amino acids or high intakes of complete proteins do not increase physical performance—and may in fact decrease physical performance.

Substance Use

The use of substances, such as tobacco, alcohol, and recreational drugs, directly affects the nutritional status of adolescents. YRBS data suggest that 23% of adolescents smoke at least once per month and 13% smoke at least 1 cigarette each day.[4] Almost 11% of teens smoked >10 cigarettes per day. Rates of smoking are highest among white teens (26%), followed by Hispanic (22%) and black (13%) adolescents. Initiation of smoking among adolescents can begin as young as 9 years old, with 12–18 years of age being the highest risk age group.

Traditionally, male adolescents were more likely to smoke than females; however, that tendency is no longer true. Current data suggest that smoking rates in male and female adolescents are similar.[4] Twenty-seven percent of white females report smoking compared to 25% of white males. The lowest prevalence of smoking was reported among black females (12%). Males are more likely to use smokeless tobacco products, with 14% of males and 2% of female adolescents using these products.[4] Smokeless tobacco use is highest among white males (18%), followed by Hispanic males (9%) and black males (3%). Adolescents who use tobacco have been shown to have higher vitamin C requirements compared to peers who do not use tobacco.

Alcohol intake and substance use among adolescents increases with age. The 2005 YRBS suggested that by the twelfth grade, more than 80% of adolescents had tried alcohol.[4] Almost half (43%) of teens reported drinking at least 1 day in the past month; 26% reporting binge drinking (drinking five or more alcoholic drinks during one occasion) at least 1 day during the past month. Alcohol use is significantly higher among white and Hispanic youth compared to black youth. The consumption of alcohol may replace nutritional foods and beverages in the diet, compromising nutritional status. Thiamin and other B vitamin requirements may be higher among adolescents who frequently consume large quantities of alcohol.

According to YRBS data, illicit drug use is reported by a significant number of adolescents.[3] One in five adolescents reported using marijuana at least one time during the past month, 8% had tried cocaine, 12% had used inhalants, 9% had used hallucinonengs, 2.5% had used heroin, and 2% had used other injectible drugs. Six percent of teens reported using ecstacy and methamphetamines.[4] Across the board, illicit drug use was higher among males than females and was more prevalent among Hispanic youth than other races and ethnicities.

Much of the data on differences in dietary intake and nutrient needs between substance users and nonusers

Table 15.4 Potential effects of substance use on nutrition status

Appetite suppression
Reduced nutrient intake
Decreased nutrient bioavailability
Increased nutrient losses/malabsorption
Altered nutrient synthesis, activation, and utilization
Impaired nutrient metabolism and absorption
Increased nutrient destruction
Higher metabolic requirements of nutrients
Inadequate weight gain/weight loss
Iron deficiency anemia
Decreased financial resources for food

SOURCE: Reprinted with permission. Alton I. Substance Abuse During Pregnancy. In: Story M, Stang J, eds. Nutrition and the Pregnant Adolescent: A Practical Reference Guide. Minneapolis, MN: Center for Leadership, Education, and Training in Maternal and Child Nutrition, University of Minnesota; 2000.

comes from studies of adults. Data collected on adolescent substance users as part of the Minnesota Adolescent Health Survey and from high school students living in Israel provide some of the best data on dietary intakes among adolescent users. Adolescent substance users in these studies reported consuming diets that were less adequate than those of nonusers.[27,28] As a group they were more likely to skip breakfast, not eat a meal or snack at school, not eat three meals per day, report chronic dieting, and report purging than their non-substance-using peers. All of these behaviors suggest that teens who use substances may be at high risk of poor nutritional status.

Substance use can result in depleted stores of vitamins and minerals, including thiamin, vitamin C, and iron. Chronic ingestion of alcohol and drug use can result in a reduced appetite, leading to low dietary intakes of protein, energy, vitamins A and C, thiamin, calcium, iron, and fiber. Other adverse nutritional effects of substance use are listed in Table 15.4.

Iron-Deficiency Anemia

Iron-deficiency anemia is the most common nutritional deficiency noted among children and adolescents. Several risk factors are associated with its development among adolescents, including rapid growth, inadequate dietary intake of iron-rich foods or foods high in vitamin C, highly restrictive vegetarian diets, calorie-restricted diets, meal skipping, participation in strenuous or endurance sports, and heavy menstrual bleeding.[29,30] The effects of iron-deficiency anemia include delayed or impaired growth and development, fatigue, increased susceptibility to infection secondary to depressed immune system function, reductions in physical performance and endurance, and increased susceptibility to lead poisoning. Pregnant teens who are iron deficient in the early stages of gestation are at increased risk of preterm delivery and delivery of a low-birth-weight infant.

Assessment of iron-deficiency anemia compares individual hemoglobin and hematocrit levels to standard reference values. Table 15.5 lists the Centers for Disease Control and Prevention criteria for determining anemia, based on age and gender. Adjustments to these values must be made for individuals who live at altitudes greater than 3000 feet and for smokers. An adjustment of +0.3 g/dL is required for adolescents who smoke.[30] Because adolescent males are not at high risk for iron-deficiency anemia, they do not need to be screened unless they exhibit one or more of the risk criteria listed. All adolescent females should be screened every 5 years for anemia; those with one or more risk factors for anemia should be screened annually.

Treatment that follows a diagnosis of iron-deficiency anemia needs to include increased dietary intake of foods rich in iron and vitamin C as well as iron supplementation. Adolescents under the age of 12 should be supplemented with 60 mg of elemental iron per day, and teenagers over the age of 12 should receive 60 to 120 mg of elemental iron per day.[30] These recommendations spark some controversy, however, given their high doses of elemental iron. Adolescents often report gastrointestinal side effects from iron supplementation, such as constipation, nausea, and cramping. These side effects can be lessened by giving smaller doses of iron more frequently throughout the day and counseling the adolescent to take the iron supplement at meal times or with food sources of vitamin C. Calcium supplements, dairy products, coffee, tea, and high-fiber foods may decrease absorption of iron supplements; these foods should be avoided within 1 hour of taking an iron supplement.

Cardiovascular Disease

Hypertension

Criteria for the detection and diagnosis of hypertension are shown in Table 15.6. Adolescents are considered hypertensive if the average of three systolic and/or diastolic blood pressure readings exceed the 95th percentile, based on age, sex, and height.[31] Blood pressure levels for

Table 15.5 Maximum hemoglobin concentration and hematocrit values for iron-deficiency anemia[a]

Sex/Age[a]	Hemoglobin (<g/dL) Less Than:	Hematocrit (<%) Less Than:
8–12 years	11.9	35.4
Males		
12–15 years	12.5	37.3
15–18 years	13.3	39.7
18+ years	13.5	39.9
Females[b]		
12–15 years	11.8	35.7
15–18 years	12	35.9
18+ years	12	35.7

SOURCE: Abridged from Centers for Disease Control and Prevention. Recommendations to Prevent and Control Iron Deficiency Anemia in the United States. Morb Mortal Wkly Rep. 2002;51(40):897–899.

[a]Age and sex-specific cutoff values for anemia are based on the 5th percentile from the third National Health and Nutrition Examination Survey.
[b]Nonpregnant and lactating adolescents.

Table 15.6 Consensus statement guidelines for detection and diagnosis of hypertension and hyperlipidemia

	Guidelines
Hypertension	
Prehypertension	Systolic or diastolic blood pressure >90th percentile for age and gender or 120/80 mm Hg, whichever is less
Stage 1 Hypertension	Systolic or diastolic blood pressure >95th percentile for age and gender on 3 consecutive visits or 140/90 mm Hg, whichever is less
Stage 2 Hypertension	Systolic or diastolic blood pressure >99th percentile +5 mm Hg for age and gender or 160/110 mm Hg, whichever is less
Hyperlipidemia	
Total cholesterol, mg/dL	
borderline	≥170
abnormal	≥200
LDL cholesterol, mg/dL	
borderline	≥100
abnormal	≥130
HDL cholesterol, mg/dL	
abnormal	<40
Triglycerides, mg/dL	
abnormal	≥200

SOURCE: Adapted from American Heart Association. Dietary Recommendations for Children and Adolescents. A Guide for Practitioners. Consensus statement from the American Heart Association. Circulation. 2005;112:2061–2075.

Table 15.7 Blood pressure levels for the 90th and 95th percentiles of blood pressure for boys and girls, aged 10 to 17 years

| | | Systolic BP (mm Hg), by Height Percentile from Standard Growth Curves | | | | | | | | | | | | | | Diastolic BP (mm Hg), by Height Percentile from Standard Growth Curves | | | | | | | | | | | | | |
| | | Boys | | | | | | | Girls | | | | | | | Boys | | | | | | | Girls | | | | | | |
Age	BP Percentile*	5%	10%	25%	50%	75%	90%	95%	5%	10%	25%	50%	75%	90%	95%	5%	10%	25%	50%	75%	90%	95%	5%	10%	25%	50%	75%	90%	95%
10	90th	110	112	113	115	117	118	119	112	112	114	115	116	117	118	73	74	74	75	76	77	78	73	73	73	74	75	76	76
	95th	114	115	117	119	121	122	123	116	116	117	119	120	121	122	77	78	79	80	80	81	82	74	74	75	75	76	77	78
11	90th	112	113	115	117	119	120	121	114	114	116	117	118	119	120	74	74	75	76	77	78	78	74	74	75	75	76	77	77
	95th	116	117	119	121	123	124	125	118	118	119	121	122	123	124	78	79	79	80	81	82	83	78	78	79	79	80	81	81
12	90th	115	116	117	119	121	123	123	116	116	118	119	120	121	122	75	75	76	77	78	78	79	75	75	76	76	77	78	78
	95th	119	120	121	123	125	126	127	120	120	121	123	124	125	126	79	79	80	81	82	83	83	79	79	80	80	81	82	82
13	90th	117	118	120	122	124	125	126	118	118	119	121	122	123	124	75	76	76	77	78	79	80	76	76	77	78	78	79	80
	95th	121	122	124	126	128	129	130	121	122	123	125	126	127	128	79	80	81	82	83	83	84	80	80	81	82	82	83	84
14	90th	120	121	123	125	126	128	128	119	120	121	122	124	125	126	76	76	77	78	79	80	80	77	77	78	79	79	80	81
	95th	124	125	127	128	130	132	132	123	124	125	126	128	129	130	80	81	81	82	83	84	85	81	81	82	83	83	84	85
15	90th	123	124	125	127	129	131	131	121	121	122	124	125	126	127	77	77	78	79	80	81	81	78	78	79	79	80	81	82
	95th	127	128	129	131	133	134	135	124	125	126	128	129	130	131	81	82	83	83	84	85	86	82	82	83	83	84	85	86
16	90th	125	126	128	130	132	133	134	122	122	123	125	126	127	128	79	79	80	81	82	82	83	79	79	79	80	81	82	82
	95th	129	130	132	134	136	137	138	125	126	127	128	130	131	132	83	83	84	85	86	87	87	83	83	83	84	85	86	86
17	90th	128	129	131	133	134	136	136	122	123	124	125	126	128	128	81	81	82	83	84	85	85	79	79	79	80	81	82	82
	95th	132	133	135	136	138	140	140	126	126	127	129	130	131	132	85	85	86	87	88	89	89	83	83	83	84	85	86	86

SOURCE: Adapted from the National Heart, Lung, and Blood Institute, National High Blood Pressure Education Working Group on Hypertension Control in Children and Adolescents. Fourth Report on the Diagnosis, Evaluation, and Treatment of High Blood Pressure in Children and Adolescents. Bethesda, MN: National Institutes of Health; 2005.

*Blood pressure percentile determined by a single movement.

the 95th percentiles for males and females are shown in Table 15.7. Classifications of blood pressure based on the average of three readings are:

- Normal blood pressure: <90th percentile
- "prehypertensive:" >90th and <95th percentiles
- Stage 1 hypertension: >95th and <99th percentile + 5 mm Hg
- Stage 2 hypertension: >99th percentile + 5 mm Hg

Risk factors for hypertension among adolescents include a family history of hypertension, high dietary intake of sodium, overweight, hyperlipidemia, inactive lifestyle, and tobacco use.[31] Adolescents who display one or more of these risk factors should be routinely screened for hypertension. Nutrition counseling to decrease sodium intake, to limit fat intake to 30% or less of calories, and to consume adequate amounts of fruit, vegetables, whole grains, and low-fat dairy products should be provided when hypertension is diagnosed.[32] Table 15.8 outlines the dietary recommendations suggested for adolescents to promote health and reduce cardiovascular risk factors. Weight loss is recommended for adolescents who are hypertensive in the presence of overweight. If medications are prescribed, teens still must adhere to general dietary recommendations and should still be encouraged to reach and maintain a healthy weight for their height.

Hyperlipidemia

Approximately one in four adolescents in the United States has an elevated cholesterol level.[32] Table 15.6 provides the classification criteria for elevated cholesterol levels in children and adolescents. Risk factors for hypercholesterolemia include a family history of cardiovascular disease or high blood cholesterol levels, cigarette smoking, overweight, hypertension, diabetes mellitus, and low level of physical activity. Adolescents who exhibit these risk factors should be screened to determine causes of hyperlipidemia and should be referred for treatment as required.[32] Early intervention among adolescents who have high cholesterol levels may reduce their risk of coronary artery diseases later in life.

Dietary recommendations indicate that youth over the age of 5 years should obtain less than 35% of their calories from fat, with no more than 10% of calories derived from saturated fat.[32] Dietary cholesterol intakes of 300 mg per day or less have also been recommended. Counseling adolescents with hyperlipidemia to follow these guidelines can be challenging, given their frequent consumption of fast foods and their preferred food choices. Table 15.8 outlines American Heart Association recommendations for dietary intake to promote health and reduce cardiovascular disease risk. Suggested dietary changes should take into account the eating habits of adolescents and emphasize healthier food choices as opposed to restriction of favorite foods of teens. Health professionals can work with adolescents to make

Table 15.8 Dietary recommendations to promote health and prevent cardiovascular disease

	9–13 years	14–18 years
Calories†		
Female	1600 Kcal	1800 Kcal
Male	1800 Kcal	2200 Kcal
Fat	25–35% Kcal	25–35% Kcal
Milk/Dairy Products‡	3 c	3 c
Meat/Meat Alternatives	5 oz	
Female		5 oz
Male		6 oz
Fruits	1.5 c	
Female		1.5 c
Male		2 c
Vegetables		
Female	2 c	2.5 c
Male	2.5 c	3 c
Grains‖		
Female	5 oz	6 oz
Male	6 oz	7 oz

SOURCE: Abridged from: American Heart Association. Dietary Recommendations for Children and Adolescents. A Guide for Practitioners. Consensus statement from the American Heart Association. Circulation. 2005;112:2061–2075.

†Calorie estimates are based on a sedentary lifestyle. Increased physical activity will require additional calories: by 0–200 Kcal/d if moderately physically active; and by 200–400 Kcal/d if very physically active.
‡Milk listed is fat-free (except for children under the age of 2 years). If 1%, 2%, or whole-fat milk is substituted, this will utilize, for each cup, 19, 39, or 63 kcal of discretionary calories and add 2.6, 5.1, or 9.0 g of total fat, of which 1.3, 2.6, or 4.6 g are saturated fat.
‖Half of all grains should be whole grains.

healthier choices at fast food restaurants, to limit their portion sizes of high-fat food items such as high-fat snacks, and to consume adequate amounts of fruits, vegetables, grains, and low-fat dairy products.

Disordered Eating, Dieting, and Eating Disorders

The Continuum of Eating Concerns and Disorders

Eating concerns and disorders lie on a continuum ranging from mild dissatisfaction with one's body shape to serious eating disorders such as *anorexia nervosa*, *bulimia nervosa*, and *binge eating disorder*. Along the continuum,

between these endpoints, lie normative dieting behaviors and more severe disordered eating behaviors such as self-induced vomiting and binge eating (Illustration 15.4). Although engagement in anorexic behaviors and unhealthy dieting may not be frequent or intense enough to meet the formal criteria for being defined as an eating disorder, these behaviors may negatively impact health and may lead to the development of more severe eating disorders. All eating disorders present a serious public health concern in light of their prevalence and their potentially adverse effects on growth, psychosocial development, and physical health outcome.

Prevalence of Eating Disorders

An awareness of the prevalence of eating disorders is critical in effective planning for interventions aimed at their treatment and prevention. Conditions prevalent among youth warrant interventions that have the potential to reach large numbers of youth, such as community-based and school-based programs. The small percentage of the adolescent population affected by these types of conditions having severe health implications requires more intensive individual or small-group interventions. Estimates as to the prevalence of each of the eating disorders on the continuum are presented in Table 15.9.

Anorexia Nervosa

Anorexia nervosa and its impact on morbidity and mortality make it the most severe condition on the continuum of eating disorders. Among adolescent girls and young women, prevalence estimates of anorexia nervosa range from 0.2% to 1.0%.[33–35] Anorexia nervosa presents more frequently among females than among males; about 9 out of 10 individuals with anorexia nervosa are female. Only in recent years has attention been directed toward males with this condition; they may not be suspected of having anorexia nervosa and therefore may be diagnosed at later stages of the disease when treatment is more difficult.

Anorexia Nervosa An eating disorder characterized by extreme weight loss, poor body image, and irrational fears of weight gain and obesity.

Characteristics of anorexia nervosa include self-starvation and strong fears of being fat. An adolescent may begin with simple dieting behaviors due to social pressures to be thin or comments by others about an adolescent's weight. However, as these behaviors lead to weight loss, feelings of control, and comments from others about weight loss, anorexia nervosa develops, and the condition takes on a life of its own. Diagnostic criteria for anorexia nervosa are shown in Table 15.10. Key features of anorexia nervosa are refusal to maintain body weight over a minimal normal

Illustration 15.4 The continuum of weight-related concerns and disorders.

Table 15.9 Estimated prevalence and brief description of weight-related concerns/disorders among adolescents

Disorder	Estimated Prevalence
Anorexia nervosa	Approximately 0.2% to 1.0% of adolescent females and young women
Bulimia nervosa	Approximately 1% to 3% of adolescent females and young women
Binge eating	Estimated 30% of population currently dieting; 2% of general population
Disordered eating behaviors	Estimated 10% to 20% of adolescents although estimates vary
Dieting behaviors	Estimates vary and range from 44% of adolescent females, 15% adolescent males, to 50% to 60% of all adolescent females are attempting to lose weight
Body dissatisfaction	Estimates vary in accordance with type of measurement used and age, gender, and ethnicity of population: approximately 60% of girls and 35% of boys are not satisfied with their weight

Table 15.10 Diagnostic criteria for anorexia nervosa

- Refusal to maintain body weight at or above a minimally normal weight for age and height (e.g., weight loss leading to maintenance of body weight less than 85% of that expected; or failure to make expected weight gain during period of growth, leading to body weight less than 85% of that expected)
- Intense fear of gaining weight or becoming fat, even though underweight
- Disturbance in the way in which one's body weight or shape is experienced, undue influence of body weight or shape on self-evaluation, or denial of the seriousness of the current low body weight
- Amenorrhea in postmenarchal women; that is, the absence of at least three consecutive menstrual cycles (a woman is considered to have amenorrhea if her menstrual periods occur only following hormone-estrogen-administration)

Restricting Type

During the episode of anorexia nervosa, the person has not regularly engaged in binge eating or purging behavior (i.e., self-induced vomiting or the misuse of laxatives, diuretics, or enemas).

Binge-Eating/Purging Type

During the episode of anorexia nervosa, the person has regularly engaged in binge-eating or purging behavior (i.e., self-induced vomiting or the misuse of laxatives, diuretics, or enemas).

SOURCE: Reprinted with permission. Diagnostic and statistical manual of mental disorders, 4th ed. Washington, DC: American Psychiatric Association; 1994.

weight for age and height; intense fear of gaining weight or becoming fat, even though underweight; a distorted body image; and amenorrhea (in females).

The two subtypes of anorexia nervosa are restricting and nonrestricting. In the restricting subtype, the individual does not regularly engage in binge eating or purging behaviors. The nonrestricting subtype exhibits regular episodes of binge eating and purging behaviors. However, both subtypes present with a refusal to maintain a minimally normal body weight, which differentiates them from other types of eating disorders.

An estimated 10–15% of patients with anorexia nervosa die from their disease, although difficulties arise in assessing mortality rates from anorexia nervosa.[34,35] Reasons for fatality from anorexia include a weakened immune system due to undernutrition, gastric ruptures, cardiac arrhythmias, heart failure, and suicide. The adolescent or the family commonly denies the condition, which delays the diagnosis and treatment, resulting in a poorer prognosis for recovery. Early recognition of possible signs of anorexia nervosa and seeking out of professional help significantly affect the time and intensity of treatment and improve chances for a successful recovery. Recovery rates are estimated at 40–50% for individuals with anorexia nervosa.[34,35]

Bulimia Nervosa

Bulimia nervosa is an eating disorder characterized by the consumption of large amounts of food with subsequent purging by self-induced vomiting, laxative or diuretic abuse, enemas, and/or obsessive exercising. Whereas anorexia nervosa is characterized by severe weight loss, bulimia nervosa may show weight maintenance or extreme weight fluctuations due to alternating binges and fasts. In some individuals anorexia and bulimia nervosa overlap. Reliable estimates of bulimia nervosa range from 1.0% to 3.0%.[35] As with anorexia nervosa, about 90% of individuals with bulimia nervosa are female, probably due to the greater social pressures on women to be thin.

Diagnostic criteria for bulimia nervosa are shown in Table 15.11. Key features of bulimia nervosa include recurrent episodes of binge eating (rapid consumption of a

Bulimia Nervosa An eating disorder characterized by recurrent episodes of rapid, uncontrolled eating of large amounts of food in a short period of time. Episodes of binge eating are often followed by purging.

large amount of food in a discrete period of time); a feeling of lack of control over eating behavior during the binge; self-induced vomiting; use of laxatives or diuretics; strict dieting or fasting; vigorous exercise in order to prevent weight gain; and persistent overconcern with body shape and weight. People with bulimia nervosa can be overweight, underweight, or of average weight for their height and body frame. Bulimia nervosa may be proceeded by a history of dieting or restrictive eating, which are thought to contribute to the binge–purge cycle. Mortality for bulimia nervosa appears to be lower than for anorexia nervosa. Based on a review of the existing literature in this area, it has been estimated that approximately 5% of patients die of their disease, usually due to heart failure resulting from electrolyte abnormality or suicide.[34] Recovery rates for bulimia nervosa appear to be higher than for anorexia nervosa, with estimates of 50–60% for recovery.[34,35]

Binge-Eating Disorder

Binge-eating disorder (BED) is a condition in which one engages in eating large amounts of food and feels that these eating episodes are not within one's control.[35,36] BED is defined by recurrent episodes of binge eating at least 2 days a week for at least 6 months (Table 15.12). In addition, the person feels a subjective sense of a loss of control over binge eating, which is indicated by the presence of three of the following five criteria: eating rapidly, eating when

> **Binge-Eating Disorder**   An eating disorder characterized by periodic binge eating, which normally is not followed by vomiting or the use of laxatives. People must experience eating binges twice a week on average for over six months to qualify for this diagnosis.

Table 15.11   Diagnostic criteria for bulimia nervosa

A. *Recurrent episodes of binge eating.* An episode of binge eating is characterized by both of the following:

- eating, in a discrete period of time (e.g., within any 2-hour period), an amount of food that is definitely larger than most people would eat during a similar period of time and under similar circumstances.
- a sense of lack of control over eating during the episode (e.g., a feeling that one cannot stop eating or control what or how much one is eating).

B. Recurrent inappropriate compensatory behavior in order to prevent weight gain, such as self-induced vomiting; misuse of laxatives, diuretics, enemas, or other medications; fasting; or excessive exercise.

C. The binge eating and inappropriate compensatory behaviors both occur, on average, at least twice a week for 3 months.

D. Self-evaluation is unduly influenced by body shape and weight.

E. The disturbance does not occur exclusively during episodes of anorexia nervosa.

Purging Type: During the current episode of bulimia nervosa, the person regularly engages in self-induced vomiting or the misuse of laxatives, diuretics, or enemas.

Nonpurging Type: During the current episode of bulimia nervosa, the person has used other inappropriate compensatory behaviors, such as fasting or excessive exercise, but has not regularly engaged in self-induced vomiting or the misuse of laxatives, diuretics, or enemas.

SOURCE: Reprinted with permission. Diagnostic and Statistical Manual of Mental Disorders, 4th ed. Washington, DC: American Psychiatric Association; 1994.

Table 15.12   Diagnostic criteria for binge-eating disorder

A. *Recurrent episodes of binge eating.* An episode of binge eating is characterized by both of the following:

- eating, in a discrete period of time (e.g., within any 2-hour period), an amount of food that is definitely larger than most people would eat in a similar period of time and under similar circumstances.
- a sense of lack of control over eating during the episode (e.g., a feeling that one cannot stop eating or control what or how much one is eating).

B. The binge-eating episodes are associated with three (or more) of the following:

- eating much more rapidly than normal
- eating until feeling uncomfortably full
- eating large amounts of food when not feeling physically hungry
- eating alone because of being embarrassed by how much one is eating
- feeling disgusted with oneself, depressed, or guilty after overeating
- experiencing marked distress regarding binge eating
- occurring, on average, at least 2 days a week for 6 months

C. The method of determining frequency differs from that used for bulimia nervosa; future research should address whether the preferred method of setting a frequency threshold is counting the number of days on which binges occur or counting the number of episodes of binge eating.

D. The binge eating is not associated with the regular use of inappropriate compensatory behaviors (e.g., purging, fasting, excessive exercise) and does not occur exclusively during the course of anorexia nervosa or bulimia nervosa.

SOURCE: Reprinted with permission. Diagnostic and Statistical Manual of Mental Disorders, 4th ed. Washington, DC: American Psychiatric Association; 1994.

not physically hungry, eating when alone, eating until uncomfortably full, and feeling self-disgust about bingeing. BED differs from bulimia nervosa in that binge eating is not followed by compensatory behaviors such as self-induced vomiting, as occurs in bulimia nervosa. BED is more prevalent among overweight clinical populations (30%) than among community samples (5% of females and 3% of males).[35,36] Studies on adolescents that assess the prevalence of binge eating include few that document prevalence rates of BED. In a college student sample, the rate of BED was 2.6%. In contrast to other weight-related conditions, significant differences were not found between male and female students. Further study of the prevalence and etiology of BED among adolescents seems critical in light of the increasing rates of obesity among youth.

Disordered Eating Behaviors

Some adolescents engage in anorexic or bulimic behaviors, but with less frequency or intensity than required for a formal diagnosis of an eating disorder. Behaviors typically considered in this category include self-induced vomiting, laxative use, use of diet pills, fasting or extreme dieting, binge eating, and excessive physical activity. The heterogeneity of these behaviors makes it more difficult to estimate their prevalence. Data from the 2005 YRBS show that 12% of adolescents had gone for 24 hours or longer without eating (fasting) as an attempt to lose weight.[4] Fasting was more commonly reported among females (17%) than among males (8%). Eighteen percent of Hispanic and white females reported fasting to lose weight compared to 14% of black females, 9% of black males, 8% of white males, and 7% of Hispanic males. Six percent of students surveyed in YRBS reported using diet pills or other diet formulae to lose weight.[4] The use of diet pills was reported by 9% of white, 8% of Hispanic, and 5% of black females compared to 6% of Hispanic, 5% of black and 4% of white males. The use of vomiting and/or laxative use to lose weight was reported by 7% of Hispanic and white females, 4% of black females, 4% of Hispanic males, 3% of black males, and 2% of white males.[4]

The types of questions used to assess disordered eating behaviors may influence prevalence estimates and may account for the disagreement on how serious this issue is among youth. In a large national study of 6728 adolescents, disordered eating was reported by 13% of the girls and 7% of the boys.[37] In this study, disordered eating was assessed with the question: "Have you ever binged and purged (which is when you eat a lot of food and then make yourself throw up, vomit, or take something that makes you have diarrhea) or not?" In a large population-based study of Minnesota adolescents in grades 7 through 12, 12% of the girls reported that they had made themselves vomit for weight control purposes at least once in their lives, and 2% reported having used laxatives or diuretics for weight control purposes. Rates were lower among boys: 6% reported

vomiting and less than 1% reported laxative or diuretic use.[38] Binge eating (eating large amounts during a short period of time and feeling out of control while eating) was reported by 30% of the girls and 13% of the boys.[38] Based on research findings, it can be reasonably estimated that between 10% and 20% of adolescents have engaged in anorexic or bulimic behaviors. Disordered eating behaviors are often overlooked in overweight adolescents, but overweight adolescents have reported the use of unhealthy and extreme weight-control behaviors, which can include purging, laxative use, self-restriction, and excessive exercise.[39] Disordered eating behaviors such as self-induced vomiting and binge eating have serious implications for health and may be precursors to full-blown eating disorders. Therefore, interventions aimed at their prevention are essential.

Dieting Behaviors

Dieting behaviors among adolescents, and in particular among adolescent girls, tend to be alarmingly high. In a study of weight-control behaviors among 459 adolescents (12–17 years of age) from four regions of the United States, current weight-control behaviors were reported by 44% of the adolescent girls and 37% of the adolescent boys.[40] National data suggest that 62% of female and 30% of male adolescents have dieted in the past month to lose weight.[4] The national estimates are in concordance with a review of the literature that concluded that 50% to 60% of adolescent girls consider themselves overweight and have attempted to diet.[41] Of particular concern are the increasing rates of dieting behaviors among children and young adolescents.

Dieting was once considered to be a phenomenon of white, middle-class females. Current data suggest that dieting behaviors are more prevalent among Hispanic females (51%) than other racial and ethnic groups (46% of white and 39% of black females). More than one-third of Hispanic males (39%) report dieting compared to 29% of white and 24% of black males. The prevalence of dieting drops slightly with age, but remains a significant issue for adolescents of all ages, races, and ethnicities into adulthood.

Dieting and the use of unhealthy weight control behaviors may also place adolescents at increased likelihood of being overweight in the future. Neumark-Sztainer and colleagues found that over a 5-year period, adolescents who initially reported dieting or using unhealthy weight control behaviors were more likely to be overweight 5 years later than were peers who did not report using weight-control behaviors.[42] Effective nutrition messages aimed at adolescents should focus on making healthy lifestyle changes, rather than focusing on short-term dieting behaviors that are often difficult to sustain. Shifting focus toward long-term behavior changes is needed for prevention of both eating disorders as well as overweight.

Dieting behaviors among youth are of concern in that they are often used by youth who are not overweight. Furthermore, unhealthful dieting behaviors in which meals are skipped, energy intake is severely restricted, or food groups are lacking are common. Dieting behaviors have been found to be associated with inadequate intakes of essential nutrients. Restricting behaviors, leading adolescents to experience hunger or cravings for specific foods, may place them at risk for binge-eating episodes. Finally, dieting behaviors may be indicative of increased risk for the later development of eating disorders; research has found that the relative risk for dieters to develop an eating disorder was eight times higher than that for nondieters after a 1-year period.[43] Therefore, dieting should not be viewed as a normative and acceptable behavior, in particular among children and adolescents.

Body Dissatisfaction

During adolescence, body image and self-esteem tend to be closely intertwined; therefore, body image concerns should not be viewed as acceptable and normative components of adolescence. Furthermore, body dissatisfaction is probably the main contributing factor to dieting behaviors, disordered eating behaviors, and clinical eating disorders. Adolescents with low levels of body satisfacation

are at greater risk for using unhealthy weight-control behaviors, binge eating, and dieting as well as being more likely to participate in less physical activity.[44] Body dissatisfaction is reported by a high percentage of adolescents. In a study of 36,000 adolescents in Minnesota, Story and colleagues found that less than 40% of adolescent girls and about 65% of adolescent boys were satisfied with their weight.[45] Although actual weight status is directly associated with perceived weight status, a considerable number of teens who are not overweight perceived themselves as overweight.

In working with overweight youth who express body dissatisfaction, health counselors are challenged to help them improve their body image while simultaneously working toward weight control. All adolescents, including overweight youth, should be encouraged to appreciate the positive aspects of their bodies. Overweight adolescents may need help in accepting the fact that they may never achieve the thin ideal portrayed in the media, but they may strive toward a leaner and healthier body that is realistic for them. Tips for fostering a positive body image among adolescents, regardless of their weight, are shown in Table 15.13.

Table 15.13 Tips for fostering a positive body image among children and adolescents

Child or Adolescent	Parents	Health Professional
• Look in the mirror and focus on your positive features, not the negative ones.	• Demonstrate healthy eating behaviors, and avoid extreme eating behaviors.	• Discuss changes that occur during adolescence.
• Say something nice to your friends about how they look.	• Focus on non-appearance-related traits when discussing yourself and others.	• Assess weight concerns and body image.
• Think about your positive traits that are not related to appearance.	• Praise your child or adolescent for academic and other successes.	• If a child or adolescent has a distorted body image, explore causes and discuss potential consequences.
• Read magazines with a critical eye, and find out what photographers and computer graphic designers do to make models look the way they do.	• Analyze media messages with your child or adolescent.	• Discuss how the media negatively affects a child's or adolescent's body image.
• If you are overweight and want to lose weight, be realistic in your expectations and aim for gradual change.	• Demonstrate that you love your child or adolescent regardless of what he weighs.	• Discuss the normal variation in body sizes and shapes among children and adolescents.
• Realize that everyone has a unique size and shape.	• If your child or adolescent is overweight, don't criticize her or his appearance—offer support instead.	• Educate parents, physical education instructors, and coaches about realistic and healthy body weight.
• If you have questions about your size or weight, ask a health professional.	• Share with a health professional any concerns you have about your child's or adolescent's eating behaviors or body image.	• Emphasize the positive characteristics (appearance- and non-appearance related) of children and adolescents you see.
		• Take extra time with an overweight child or adolescent to discuss psychosocial concerns and weight-control options.
		• Refer children, adolescents, and parents with weight-control issues to a dietitian.

SOURCE: Story M, Holt K, Sofka D, eds. Bright Futures in Practice: Nutrition. Arlington, VA: National Center for Education in Maternal and Child Health; 2000.

Etiology of Eating Disorders

The etiology of eating disorders is multifactorial—that is, many factors contribute to their onset. Some of the major contributory factors include social norms emphasizing thinness, being teased about one's weight, familial relations (e.g., chaotic lifestyles, boundaries between family members, patterns of communication), physical and sexual abuse experiences, personal body shape and size, body image, and self-esteem. These factors do not operate in isolation but interact with each other to increase the adolescent's risk for engaging in potentially harmful eating and dieting behaviors. In considering etiological issues, it is essential to realize that different etiological pathways may lead to weight-related disorders in different adolescents. For some adolescents, family issues may be major factors, while for others social norms may be the key factors leading to the onset of a condition. Furthermore, different conditions tend to be influenced by different factors. That said, a number of factors play a major role across conditions and across individuals. Rosen and Neumark-Sztainer have grouped these potential contributory factors into socioenvironmental, personal, and behavioral domains.[46]

Socioenvironmental factors include:

- Sociocultural norms (regarding thinness, eating, food preparation, roles of women)
- Food availability (type of food, amount of food)
- Familial factors (communication patterns, parental expectations, boundaries, weight concerns and dieting behaviors of parents and siblings, family meals)
- Peer norms and behaviors (dieting behaviors, eating patterns, weight concerns)
- Abuse experiences (by family members, other adults; rape experiences)
- Media influences (images portrayed in the media, roles assigned to thin actors as compared to fat actors)

Personal factors include:

- Biological factors (genetic disposition, body mass index, age, gender, stage of development)
- Psychological factors (self-esteem, body image, drive for thinness, depression)
- Cognitive/affective factors (nutritional knowledge and attitudes, media internalization)

Behavioral factors include:

- Eating behaviors (meal patterns, fast food consumption, nutritional variety, bingeing)
- Dieting and other weight-management behaviors (dieting frequency, types of methods used, purging behaviors)
- Physical activity behaviors (TV viewing, sport involvement, daily activities)
- Coping behaviors (with dieting failures, with life frustrations)
- Skills (self-efficacy in dealing with harmful social norms, skills in food preparation, media advocacy skills)

An understanding of the etiology of eating disorders is essential to the development of effective interventions aimed at their treatment and prevention. An individual clinical setting needs to allow time to assess the factors leading to the onset of the condition for that particular adolescent. In developing prevention programs to reach larger groups of adolescents, it is more feasible to identify and address factors that may be contributing to the onset of weight-related behaviors and conditions for a broad sector of the targeted population. Although not all factors may be addressed within one intervention, it is important to be aware of the broad range of factors coming into play and the interactions between them.

Treating Eating Disorders

The complex etiology of eating disorders and their numerous potential psychosocial, physical, and behavioral consequences highlight the need for a multidisciplinary treatment approach. The health care team caring for an adolescent with an eating disorder will often include a physician, nutritionist, nurse, psychologist, and/or psychiatrist. The role of the nutritionist is paramount to the treatment of eating disorders at the stages of assessment, treatment, and maintenance. Initially an adolescent may be more willing to discuss his or her concerns with a nutritionist than with a psychologist. During treatment a major role of the nutritionist is to help the adolescent normalize eating patterns and to feel comfortable with these changes. For example, in the treatment of bulimia nervosa, some of the key goals of the nutritional care include the following:

- Establishing a regular pattern of nutritionally balanced meals and snacks
- Ensuring adequate but not excessive levels of energy intake with the goal of weight maintenance
- Ensuring adequate dietary fat and fiber intake to promote satiety
- Avoiding dieting behaviors and excessive exercise; gradually including formerly forbidden foods in the diet
- Keeping and reviewing dietary records; employing stimulus control strategies to control high-risk situations; and weighing at scheduled intervals only.[47]

For some adolescents, any denial of the condition or a lack of motivation for change make the work quite challenging. It is important for the nutritionist to work in close conjunction with other members of the health team to ensure that roles of different members of the team are clearly defined.

Preventing Eating Disorders

The high prevalence of eating disorders and their potentially harmful consequences point to a need for interventions aimed at their prevention. One of the most pressing current public health issues that needs to be addressed concerns the prevention of eating disorders described in previous sections and the prevention of obesity. Even the prevention of a small percentage of these conditions, at a population level, returns huge benefits in terms of reducing physical, emotional, and financial burdens.

In the development of interventions aimed at the prevention of eating disorders, it is essential to address factors that contribute to the onset of these conditions for a large proportion of the targeted population, factors that are potentially modifiable, and factors suitable for addressing within the designated setting. For example, media awareness and advocacy has been suggested as a suitable approach toward preventing eating concerns and disorders. Participants may learn about how the media influence one's body image and about techniques used within the media to improve the appearance of models, and then take action toward making changes in the media. This approach is suitable in that media influences, and the internalization of media messages, may contribute to weight concerns among a large sector of the adolescent population. These adolescent perceptions are potentially modifiable and suitable for addressing within clinical, community, and school-based settings where interventions may be implemented.

Any efforts toward prevention first must consider the target audience. An important question is whether to direct interventions to all adolescents or to adolescents at increased risk for eating disorders. Reasons for providing interventions for all adolescents include the high prevalence of eating concerns among adolescents, difficulties inherent in identifying and targeting high-risk individuals, and the advantages of developing positive social norms regarding eating issues within the peer group. Taking a more targeted approach offers the advantages of better use of limited resources, more intensive interventions, and interventions developed for specific high-risk groups (e.g., ballet dancers, youth with diabetes, or overweight girls). In order to most effective in preventing eating disorders, both types of interventions seem necessary; more general approaches address the issues of the general adolescent population while more refined approaches can better meet the needs of specific high-risk groups.

Prevention interventions may be implemented within clinical, community, and school-based settings. Clinical settings provide opportunities for identifying early signs of eating disorders and working toward their prevention. Questions about body concerns and eating patterns should be a part of routine health care visits and appropriate channels made available for discussing concerns. Key factors to be assessed in screening for eating disorders are shown in Table 15.14.

Table 15.14 Screening elements and warning signs for individuals with eating disorders

Screening	Warning Signs
Body image and weight history	• Distorted body image • Extreme dissatisfaction with body shape or size • Profound fear of gaining weight or becoming fat • Unexplained weight change or fluctuations greater than 10 lbs
Eating and related behaviors	• Very low caloric intake; avoidance of fatty foods • Poor appetite; frequent bloating • Difficulty eating in front of others • Chronic dieting despite not being overweight • Binge-eating episodes • Self-induced vomiting; laxative or diuretic use
Meal patterns	• Fasting or frequent meal skipping to lose weight • Erratic meal pattern with wide variations in caloric intake
Physical activity	• Participation in physical activity with weight or size requirement (e.g., gymnastics, wrestling, ballet) • Overtraining or "compulsive" attitude about physical activity
Psychosocial assessment	• Depression • Constant thoughts about food or weight • Pressure from others to be a certain shape or size • History of physical or sexual abuse or other traumatizing life event
Health history	• Secondary amenorrhea or irregular menses • Fainting episodes or frequent light-headedness • Constipation or diarrhea unexplained by other causes
Physical examination	• BMI <5th percentile • Varying heart rate, decreased blood pressure after arising suddenly • Hypothermia; cold intolerance • Loss of muscle mass • Tooth enamel demineralization

SOURCE: Used with permission. The Society for Nutrition Education. Adams LB, Shafer MB. Early Manifestations of Eating Disorders in Adolescents: Defining Those at Risk. Journal of Nutrition Education: 20, 1988; and American Medical Association. Perkins K, Ferrari N, Rosas A, et al. You Won't Know Unless You Ask: The Biopsychosocial Interview for Adolescents. Clinical Pediatrics: 36(2), 1997; and Guidelines for Adolescent Preventive Services (GAPS): Recommendation Monograph, 2nd ed. Chicago, IL: American Medical Association; 1995.

The majority of prevention programs have been implemented within school settings.[48–51] Schools provide an excellent setting for implementing prevention interventions in that they reach all adolescents; they provide a captive audience within a learning atmosphere and numerous opportunities for social interactions.[51] Ideally, a school-based intervention includes various components aimed at reaching the general population of adolescents, adolescents at increased risk for eating disorders, and school staff. Interventions should not only aim for changes in levels of personal knowledge and attitudes but also for changes in social norms (e.g., peer norms promoting thinness and dieting), policy changes (e.g., regarding tolerance levels for weight-teasing), and environmental changes (e.g., food availability within the cafeteria). Suggested components for a comprehensive school-based program for preventing eating disorders are shown in Table 15.15.

Interventions aimed at preventing eating disorders may also be implemented within other community-based settings. Neumark-Sztainer and her colleagues utilized the Girl Scout framework to reach fifth- and sixth-grade girls with an intervention aimed at promoting a positive body image and preventing unhealthy dieting.[52] The program, "Free to Be Me," focused on promoting media awareness and advocacy among participants.

Eating Disorders among Adolescents: Summing Things Up

Eating disorders may be viewed on a continuum ranging from body dissatisfaction or dieting behaviors to clinically significant eating disorders including anorexia nervosa, bulimia nervosa, and binge-eating disorder. The high prevalence of eating disorders and their potential harmful physical and psychosocial consequences indicate the need for interventions aimed at their treatment and prevention. The numerous socioenvironmental, personal, and behavioral factors leading to their onset need to be addressed in interventions. Parents, peers, educators, and health care providers play an important role in decreasing the prevalence of eating disorders.

Children and Adolescents with Chronic Health Conditions

Approximately 18% of children and adolescents have a chronic condition or disability. These children and adolescents are at increased risk for nutrition-related health problems because of (1) physical disorders or disabilities that may affect their ability to consume, digest, or absorb nutrients; (2) biochemical imbalances caused by long-term medications or internal metabolic disturbances;

Table 15.15 Suggested components of a comprehensive school-based program for preventing weight-related disorders

Staff training	1. Examination of their own body image and eating behaviors 2. Knowledge about weight-related disorders 3. Skills for working with youth and program implementation
Classroom interventions for the general student body	Modula specifically aimed at preventing weight-related concerns and disorders. Options for approaches include (1) feminist approach, (2) weight control and nutrition, and (3) promotion of life skills such as self-esteem and assertiveness.
Integration of relevant material into existing curricula	Relevant information integrated into existing classes such as science, art, history, health, and physical education
Smaller and more intensive activities for high-risk students	Small-group work or individual counseling for overweight students or students with excessive weight preoccupation and unhealthy dieting behaviors
Referral system within school and between school and community	Training of school staff to be alert to warning signs; referral mechanism to ensure that students don't get overlooked and those in need get help
Opportunities for healthy eating at school	Options for attractive nutrient-dense and low-fat foods at affordable prices in the cafeteria, in vending machines, and at school events
Modifications in physical education and sport activities	Increased time being active in physical education classes; involvement of more students in after-school sports; increased sensitivity to needs of overweight youth in physical education classes
Outreach activities to the community	By students, staff, and parents (e.g., contact with local media)

SOURCE: Adapted from Neumark-Sztainer D. School-based programs for preventing eating disturbances. Journal of School Health 1996;66.

(3) psychological stress from a chronic condition or physical disorder that may affect a child's appetite and food intake; and/or (4) environmental factors, often controlled by parents who may influence the child's access to and acceptance of food.

Nutrition reports of children and adolescents with special health care needs estimate that as many as 40% have nutrition risk factors that warrant a referral to a dietitian. Common nutrition problems in children and adolescents with special health care needs include the following:

- Altered energy and nutrient needs (e.g., inborn errors of metabolism, spasticity of movement, enzyme deficiencies)

- Delayed growth
- Oral-motor dysfunction (e.g., neurological disorders, swallowing disorders)
- Elimination problems
- Drug/nutrient interactions
- Appetite disturbances
- Unusual food habits (e.g., rumination)
- Dental caries, gum disease

Malnutrition has been implicated as a major factor contributing to poor growth and short stature in adolescents with a variety of diseases (e.g., chronic inflammatory bowel disease, cystic fibrosis). Factors such as inadequate nutrient and energy intakes, excessive

---

Table 15.16 Outline of sessions in *Free to Be Me*

Session	Activities
Body Truths I: **Body Development**	• *Let Me Introduce You:* An introduction to the program and an ice-breaker activity based on the game Bingo. • *Stepping Stones:* Question–answer game in which girls learn about stages of body development. Each team advances to the next "stone" if question is answered correctly. • *Feelin' Good:* Take-home activity: Interviews with family members and friends about perceived positive traits.
Body Truths II: **Working with the Ideal Image**	• *Pin the Tail on the Timeline:* Girls attach ideal images of women throughout history to their appropriate spot on the timeline (e.g., Marilyn Monroe in the 1950s, Twiggy in the 1960s, and Tyra Banks in the 1990s). • *Sarah's Story:* Discussion of a story about a girl with poor body image who diets excessively. Discuss reasons for dieting, negative effects of skipping meals, and alternative approaches for Sarah. • *People Watching:* Take-home activity: Girls are asked to look for different body types in their community and then compare them to the body types they see in magazines.
Body Truths III: **What Else Is Out There?**	• *Looking for the Alternative (Stop-n-Go Posters):* Girls look through different teen magazines and make collages of the pictures that promote positive traits versus those that promote negative traits. Pictures are glued to either green poster board ("go") or red poster board ("stop"). • *Take-home activity:* Girls read "girl-friendly" magazine provided such as *New Moon* and discuss magazine with a parent.
Body Myths: **Media Madness**	• *Commercial Crazy:* Girls look at a variety of TV commercials and look for positive and negative media messages (body types shown, messages about healthy eating versus dieting). • *What do you see on TV?* Take-home activity: Girls and their parent(s) watch 15 to 30 minutes of TV and look for positive and negative messages in the commercials.
Take Action I: **What Can You Do to Impact the Media?**	• *Write a Letter:* Each troop writes a letter to a company that has a positive or negative impact on dieting and body image. Letters are posted on website for other advocates to sign. • *Take-home activity:* Girls develop and practice their own "girl-friendly" skit or commercial promoting positive body image and healthy eating.
Take Action II: **Spreading the Good Word**	• *Lights, Camera, Action!* Girls perform their "girl-friendly" commercials or skits in front of their parents and troop. • *Class overview.*

SOURCE: Reprinted with permission. Neumark-Sztainer D, Sherwood N, Coller T, Hannan PJ. Primary Prevention of Disordered Eating Among Pre-Adolescent Girls: Feasibility and Short-Term Impact of a Community-Based Intervention. Journal of American Dietetic Association. In Press.

nutrient losses, malabsorption, and increased nutrient requirements all lead to the chronic malnourished state. Studies have shown that the energy requirements for adolescents with cystic fibrosis or inflammatory bowel disease may be 30–50% higher than the RDA for adequate growth. In addition to the increased energy needs caused by malabsorption (or in the case of adolescents with cystic fibrosis, the increased work of breathing), fever, infection, and inflammation also increase energy requirements. Whereas undernourishment is frequently seen in adolescents with chronic illnesses, obesity is common among youth with gross motor limitations or immobility. Because of limited activity, caloric requirements are lower, and the balance between intake and expenditure is often difficult, resulting in obesity.

Consideration of nutrition needs of children with chronic disabling conditions or illnesses is complex and requires specialized, individualized care by an interdisciplinary team. Assessment of nutrition status followed by nutrition intervention, when necessary, and monitoring will help ensure the health and well-being of adolescents with chronic and disabling conditions. Also, during adolescence issues of personal responsibility and independent living skills related to food purchasing and preparation may need to be addressed.

Key Points

1. The prevalence of overweight is higher among males than females among adolescents.

2. More than half of all adolescents participate in organized sports. The need for adequate hydration and caloric intake are important aspects of sports nutrition for all teen athletes to be aware of.

3. Approximately one-fourth of adolescents have elevated serum lipid levels. Early detection and intervention is important to reduce lifetime risk of cardiovascular disease.

4. While only 1–5% of adolescents have clinically diagnosed eating disorders, 5 to 10-times as many engage in disordered eating practices, including bingeing, purging, diuretic and laxative use and fasting. Also, please change the prepared by line in the CO to match that in Chapter 14.

Resources

American College of Sport Medicine
Website: www.acsm.org

American Diabetes Association
Clinical practice, recommendations, nutrition, exercise, recipes, virtual grocery.
Website: www.diabetes.org

American Heart Association
An office-based practical approach to the child with hypercholesterolemia
Website: www.americanheart.org/Scientific/pubs/hyperchol/toc.html
Nutrition tips and virtual cookbook for heart health
Website: www.deliciousdecisions.org

Diabetes
Betschart J, Thom S. In control—A guide for teens with diabetes. Minneapolis, MN: Chronimed Publishing, 1995. Boland, E. Teens pumping it up! 2nd ed. Sylmar, CA: Minimed Inc.; 1998.
Nutrition in the fast lane: A guide to nutrition and dietary exchange values for fast-food and casual dining. Indianapolis, IN: Franklin Publishing; 2001.
Website: www.FastFoodFacts.com

Family Voices
Health information for children and adolescent with special health needs.
Website: www.familyvoices.org

Food Allergy Network
Educational materials and resources for youth with food allergies.
Website: www.foodallergy.org

Juvenile Diabetes Association, International
Phone: 1-800-533-CURE

National Campaign to Prevent Teen Pregnancy
Website: www.teenpregnancy.org

National Heart Lung and Blood Institute
Health Information for public: cholesterol, obesity.
Health Information for professionals: NCEP ATP III, overweight and obesity.
Website: www.nhlbi.nih.gov

Weight Information Network
Publications and information on child and adolescent obesity treatment and prevention.
Telephone: 1-800-946-8098
Website: www.niddk.nih.gov/NutritionDocs.html

References

1. Nader PR et al. National Institute of Child Health and Human Development Early Child Care Research Network. Identifying risk for obesity in early childhood. Pediatrics. 2006 Sep;118(3):e594–601.

2. Barlow SE, Dietz WH. Obesity evaluation and treatment: Expert Committee recommendations. The Maternal and Child Health Bureau, Health Resources and Services Administration and the Department of Health and Human Services. Pediatrics. 1998;102:E29.

3. Ogden CL et al. Prevalence of overweight and obesity in the United States, 1999–2004. JAMA. 2006 Apr 5;295(13):1549–55.

4. Centers for Disease Control and Prevention. Youth risk behavior surveillance—surveillance summaries 2005. Morb Mortal Wkly Rep. June 9, 2006;55(No SS-5).

5. Zephier E, Himes J. Prevalence of overweight and obesity in American Indian school children and adolescents in the Aberdeen area: a population study. Int J Obes. 1999;23(suppl 2):S28–S30.

6. Deshmukh-Taskar P et al. Tracking of overweight status from childhood to young adulthood: the Bogalusa Heart Study. Eur J Clin Nutr. 2006;60:48–57.

7. Whitaker RC et al. Predicting obesity in young adulthood from childhood and parental obesity. N Eng J Med. 1997;337: 869–73.

8. Gortmaker SL et al. Social and economic consequences of overweight in adolescence and young adulthood. N Eng J Med. 1993; 329:1008–12.

9. Daniels SR et al. Overweight in children and adolescents: pathophysiology, consequences, prevention, and treatment. Circulation. 2005 Apr 19;111(15):1999–2012.

10. Freedman DS et al. The relation of overweight to cardiovascular disease risk factors among children and adolescents: the Bogalusa Heart Study. Pediatr. 1999; 103:1175–82.

11. Pate RR et al. Cardiorespiratory fitness levels among U.S. youth 12 to 19 years of age: findings from the 1999–2002 National Health and Nutrition Examination Survey. Arch Pediatr Adolesc Med. 2006 Oct; 160(10):1005–12.

12. Pinhas-Hamiel O et al. Increased incidence of non-insulin-dependent diabetes mellitus among adolescents. J Pediatr. 1996; 128:608–15.

13. Reinehr T et al. Long-term follow-up of cardiovascular disease risk factors in children after an obesity intervention. Am J Clin Nutr. 2006 Sep;84(3):490–6.

14. Yanovski JA. Intensive therapies for pediatric obesity. Ped Clinics N America. 2001;48:1041–53.

15. Inge TH et al. Bariatric surgery for severely overweight adolescents: concerns and recommendations. Pediatr. 2004;114: 217–23.

16. Lawson ML et al. One year outcomes of Roux-en-Y gastric bypass for morbidly ovese adolescents: a multicenter study from the Pediatric Bariatric Study Group. J Ped Surg. 2006;41:137–43.

17. Balluz LS et al. Vitamin and mineral supplement use in the United States. Results from the third National Health and Nutrition Examination Survey. Arch Fam Med. 2000; 9:258–62.

18. Bell A et al. A look at nutritional supplement use in adolescents. J Adol Health. 2004;34:508–16.

19. Stang J et al. Relationships between vitamin and mineral supplement use, dietary intake, and dietary adequacy among adolescents. J Am Diet Assoc. 2000;100:905–10.

20. Dwyer JT et al. Do adolescent vitamin–mineral supplement users have better nutrient intakes than nonusers? Observations from the CATCH tracking study. J Am Diet Assoc. 2001 Nov;101(11):1340–46.

21. O'Dea J. Consumption of nutritional supplements among adolescents: usage and perceived benefits. Health Ed Res. 2003; 18:98–107.

22. Calfee R, Fadale P. Popular ergogenic drugs and supplements in young athletes. Pediatrics. 2006;117:e577–e589

23. National Collegiate Athletic Association. NCAA study of substance abuse habits of college student athletes. NCAA Committee on Medical Safeguards and Medical Aspects of Sports; 2001.

24. National federation of state high school association, sports medicine: high school drug testing programs. Available at www.nfhs.org, accessed 11/06.

25. Clark JF. Creatine and phosphocreative: a review of their use in exercise and sport. J Athl Train. 1997;32:45–51.

26. Boguslaw W, Bar-Or O. Effect of drink flow and NaCl on voluntary drinking and hydration in boys exercising in the heat. J Appl Physiol. 1996;80:1112–17.

27. Isralowitz RE, Trostler N. Substance use: toward an understanding of its relation to nutrition-related attitudes and behavior among Israeli high school youth. J Adolesc Health. 1996;19:184–89.

28. Neumark-Sztainer D et al. Covariations of eating behaviors with other health-related behaviors among adolescents. J Adolesc Health. 1997;20:450–58.

29. Story M, Holt K, Sofka D, eds. Bright futures in practice: nutrition. Arlington, VA: National Center for Education in Maternal and Child Health; 2000.

30. Centers for Disease Control and Prevention. Recommendations to prevent and control iron deficiency anemia in the United States. Morb Mortal Wkly Rep. 2002; 51(40):897–99.

31. National Heart, Lung, and Blood Institute, National High Blood Pressure Education Working Group on Hypertension Control in Children and Adolescents. 4th Report on the diagnosis, evaluation and treatment of high blood pressure in children and adolescents. Bethesda, MD: National Institutes of Health; 2005.

32. American Heart Association. Dietary Recommendations for Children and Adolescents. A guide for practitioners. Consensus statement from the American Heart Association. Circulation. 2005; 112:2061–2075.

33. Lucas AR et al. 50-year trends in the incidence of anorexia nervosa in Rochester, Minn.: a population-based study. Am J Psychiatry. 1991;148:917–22.

34. Woodside DB. A review of anorexia nervosa and bulimia nervosa. Curr Probl Pediatr. 1995;25:67–89.

35. American Psychiatric Association. American Psychiatric Association diagnostic and statistical manual of mental disorders, 4th ed. Washington, DC: American Psychiatric Association; 1994.

36. Spitzer RL et al. Binge eating disorder: its further validation in a multisite study. Int J Eat Disord. 1993;13:137–53.

37. Neumark-Sztainer D, Hannan PJ. Weight-related behaviors among adolescent girls and boys: results from a national survey. Arch Pediatr Adolesc Med. 2000; 154:569–77.

38. Neumark-Sztainer D et al. Psychosocial concerns and health-compromising behaviors among overweight and nonoverweight adolescents. Obes Res. 1997;5:237–49.

39. Neumark-Sztainer D et al. Weight related concerns and behaviors among overweight and nonoverweight adolescents: implications for preventing weight-related disorders. Arch Ped Adolesc Med. 2002;156(2):171–78.

40. Neumark-Sztainer D et al. Weight-control behaviors among adults and adolescents: associations with dietary intake. Prev Med. 2000;30:381–91.

41. Fisher M et al. Eating disorders in adolescents: a background paper. J Adolesc Health. 1995;16:420–37.

42. Neumark-Sztainer D et al. Obesity, disordered eating, and eating disorders in a longitudinal study of adolescents: how do dieters fare 5 years later? JADA. 2006; 106(4):559–68.

43. Patton GC et al. Abnormal eating attitudes in London schoolgirls—a prospective epidemiological study: outcome at twelve

month follow-up. Psychol Med. 1990; 20:383–94.

44. Neumark-Sztainer D et al. Does body satisfaction matter? Five-year longitudinal associations between body satisfaction and health behaviors in adolescent females and males. J Adolesc Health. 2006;39(2):244–51.

45. Story M et al. Ethnic/racial and socio-economic differences in dieting behaviors and body image perceptions in adolescents. Int J Eat Disord. 1995;18:173–79.

46. Rosen DS, Neumark-Sztainer D. Review of options for primary prevention of eating disturbances among adolescents. J Adolesc Health. 1998;23:354–63.

47. Rock CL, Curran-Celentano J. Nutritional management of eating disorders. Psychiatr Clin North Am. 1996;19:701–13.

48. Neumark-Sztainer D, Butler R, Palti H. Eating disturbances among adolescent girls. Evaluation of a school-based primary prevention program. J Nutr Educ. 1995; 27:24–31.

49. Piran N. On the move from tertiary to secondary and primary prevention: working with an elite dance school. In: Piran N, Levine MP, Steiner-Adair C, eds. Preventing eating disorders: a handbook of interventions and special challenges. Philadelphia: Brunner/Mazel; in press.

50. Smolak L, Levine MP, Schermer F. A controlled evaluation of an elementary school primary prevention program for eating problems. J Psychosom Res. 1998; 44:339–53.

51. Neumark-Sztainer D. School-based programs for preventing eating disturbances. J Sch Health. 1996;66:64–71.

52. Neumark-Sztainer D et al. Primary prevention of disordered eating among pre-adolescent girls: feasibility and short-term impact of a community-based intervention. J Amer Diet Assoc; in press.

"There is no
dress rehearsal for life."
Anonymous

Chapter 16

Adult Nutrition

Chapter Outline

- Introduction
- Healthy People 2010 Objectives
- Physiological Changes of Adulthood
- Maintaining a Healthy Body
- Dietary Recommendations
- Nutrient Recommendations
- Physical Activity Recommendations
- Nutrition Intervention for Risk Reduction

Prepared by **U. Beate Krinke**

Key Nutrition Concepts

1 Enjoyment of good food contributes to quality of life.

2 Food security means consistent access to safe, wholesome, and culturally acceptable food.

3 Good nutritional habits maximize peak physical and mental performance.

4 Hormones regulating reproduction also affect nutritional status.

5 Good nutrition now is an investment in healthy old age.

Introduction

Definition of Adulthood in the Life Cycle

With good luck, good genes, and good habits, adulthood covers a life span of roughly 60 years. There is no standard way to divide up these adult years similar to the common divisions of infancy, childhood, and adolescence in the first 20 years of life. However, it may be helpful to think about the changing nutrient needs of adulthood by using the following arbitrary segmentation:

Early Adulthood: For the most part, individuals have stopped growing by the time they reach their twenties. Some males grow slightly after age 20, men and women continue to develop bone density until roughly age 30, and muscle mass continues to grow as long as muscles are used. The primary tasks of early adulthood include personal and career development and potentially reproduction. Nutritional habits developed now are investments in future health.

Midlife: During their forties and fifties, most adults are reaching the peak of their career achievements. Physiologically, body composition slowly shifts in tandem with hormonal shifts, and probably due to decreased activity. On average, individuals start to gain weight after age 40. It is a good time to reassess earlier nutritional habits.

Old Age: In their sixties and beyond, adults harvest the fruits of earlier health habits. Good food and exercise habits practiced over a lifetime can support continued enjoyment of sports and daily activities.

In this text, nutritional aspects of adulthood from age 20 to 64 will be covered in Chapters 16 and 17. Nutrition specifically for older adults will be covered in Chapters 18 and 19. This arbitrary division segments a natural continuum uniquely experienced by individuals. The focus of Chapter 16 is development and maintenance of the most healthful nutritional habits possible.

Importance of Nutrition

Adult nutrition is about the many roles of food. Food is fuel, enjoyment, comfort, and a symbol of traditions, rituals, and celebrations. Food serves as a connection for socializing. When things go well, we take food for granted. We expect to have sufficient quantities of safe, appetizing food whenever we are hungry. During adulthood, we may be too busy to pay much attention to food.

How does food and nutrition enhance life? Lifestyle factors have a greater influence on long life than genetics, health care systems, and the environment, accounting for approximately half of premature deaths.[1] Nutrition and exercise are at the top of the list of lifestyle factors; therefore, individuals have some control over whether food and nutritional habits will contribute to a long and healthy life. Even though the leading causes of death at age 21 (unintentional injuries, homicide, and suicide) are not nutrition related, good nutrition throughout adulthood will reduce the risk of the leading causes of death of later adulthood, namely heart disease, cancer, stroke, and diabetes mellitus. Nutrition during adulthood supports an active lifestyle, contributes to maintenance of healthy weight, and promotes physical and mental health and well-being.

Healthy People 2010 Objectives

Since the first Healthy People 2000 report[2] was generated under the guidance of the federal government in 1979, several goals have been met. Rates of heart disease, cancer, and stroke deaths have declined; fat intake (as percentage of total calories) has declined. However, rates of obesity and diabetes have increased, sugar intake has risen, and health care disparities still exist. Current national health goals address these issues.[3] Table 16.1 reflects dietary goals for disease prevention and health promotion for adults. The main focus of national public health goals is healthy weight maintenance.

Physiological Changes of Adulthood

Growth and maturation are complete by early adulthood, so nutritional emphasis turns to maintaining physical status, especially muscle strength, and avoiding excess fat deposition, especially around the midriff. It is easy to remember the need for good nutrition in growing children, but we are less likely to appreciate that ongoing cell repair and replacement depends on a reliable supply of nutrients. Overall, approximately 5% of adult body cells are replaced every day. Rejuvenation varies by cell type: cells of the intestinal lining turn over every 1 to 3 days, taste bud cells every 7 days, and red blood cells turn over every 120 days.

Peak capacity for physical performance for most activities is reached during adulthood. Sports examples of athletes at the top of their form at age 20 or older include cycling, running, cross-country skiing, tennis (males), and

Table 16.1 Healthy People 2010: Health promotion goals for adults compared to current levels

	Target Percentage	Population Percentage Baseline, Age 20+	
		Females	Males
Increase the number who are at a healthy weight	60	45	38
Decrease the number who are obese (BMI 30+)	15	25	20
The following baseline information is for age 20–39 (Baselines for ages 40–59 are in parentheses) Increase the proportion of persons who:			
• Eat at least two daily servings of fruit	75	20 (26)	23 (28)
• Eat at least three daily servings of vegetables (at least one-third dark green or deep yellow)	50	4 (4)	3 (4)
• Eat at least six daily servings of grain products (at least three of those as whole grains)	50	4 (4)	10 (10)
• Eat less than 10% of calories from saturated fat[a]	75	41 (42)	32 (33)
• Eat no more than 30% of calories from fat[a]	75	38 (33)	29 (28)
The following baseline information is for age 20–49 Meet dietary recommendations (1000 mg) for calcium (level increases to 1200 mg at age 50)	75	40	64
The following baseline is for all adults, age 20+ Eat 2400 mg or less of sodium daily (Males consume more calories, so they consume more sodium)	65	30	5
Increase the proportion of physician office visits made by patients with a diagnosis of cardiovascular disease, diabetes, or hyperlipidemia that includes counseling or education related to diet and nutrition	75	39	44

SOURCE: Food and Drug Administration and National Institutes of Health. *Healthy People 2010: National Health Promotion and Disease Prevention Objectives*. Washington, DC: U.S. Department of Health and Human Services, 2000.
[a] Updated to recommend 20–35% of calories from fat, minimizing saturated and trans-fatty acid intake. [2002; IOM]

soccer. The peak for endurance sports tends to be later in life than sports based on speed (for example, marathon runners are likely to be older than sprinters). However, for the average individual who participates in sports as a leisure activity, the physiological changes of early and middle adulthood neither enhance nor limit participation in, or enjoyment of, sports.

Sensory Changes

Losses of taste and smell tend to occur later in life. Taste sensations are typically robust until age 60 (see Chapter 19), except when disease diminishes taste-sensing cells. For younger adults, olfactory losses are also associated more with illness than with aging. Approximately 60% of individuals over age 80 have some olfactory impairment, compared to 6% of adults in their fifties.[4] The conditions most associated with impaired olfaction are nasal congestion, upper respiratory tract infections, stroke, epilepsy, and current smoking.[4] Women retain their sense of smell better than men do.[4] One's sense of smell can enhance the enjoyment of eating and drinking, and has been related to improved quality of life.[5] Regardless of the effect that sense of smell has on the enjoyment of food, olfactory impairment

may have a greater impact on food safety, especially detection of spoiled food, than it does on nutritional status.[5]

Hormonal Changes

Physiological changes of adulthood differ for women and men in other ways. Women asked about their health and nutrition concerns are likely to be concerned about "not gaining weight" and about "looking good." Men are likely to say that they do not think about nutrition and health at all, but after probing will relate health concerns on strength, energy, and weight management. However, hormonal shifts related to reproductive capacity have the biggest impact on physiology, and therefore nutritional status and needs. Pregnancy and lactation are discussed in previous chapters. Menopause in women and the climacteric change in men are discussed in the following sections.

Physiological Changes in Males: Climacteric

For men, a gradual decline in testosterone levels begins about age 40 to 50, although sperm is capable of fertilizing human eggs until much later. Decreased sperm production is linked to underweight; malnutrition is linked to declining libido. Alcohol use can result in defective sperm. For the general population, alcohol intake declines with age. Alcohol accounts for 5% of calorie intake between ages 20 to 34, but only 3% between ages 51 and 64.[6]

Based on cross-sectional data, males eat fewer calories as they get older. However, body weight rises slowly beginning around age 40, more likely due to declining activity levels than changing hormones.

Physiological Changes in Females: Menopause

For 40 years females have roughly 13 menstrual cycles per year (minus the ones missed during pregnancy or for other reasons such as extreme loss of body fat), totaling more than 500 menstrual cycles. After cessation of menstrual blood loss, the need for iron decreases from 18 to 8 mg daily.[7] Loss of estrogen leads to atrophy of tissues in the urinary tract and vagina, increased abdominal fat, and greater risk for chronic conditions such as osteoporosis and heart disease. Menopause is associated with, but not the cause of, weight gain and decreases in muscle mass.[5,8] Hormones regulate blood chemicals, that is, hormonal shifts impact nutritional status as shown in Table 16.2.

Oral contraceptives lead to an 8% to 11% increase of total serum cholesterol, with increases in LDL and triglycerides and concurrent decreases in HDL. Even though menopause negates the need for contraceptives, lack of estrogen and progesterone also leads to a rise in total cholesterol. Hormone replacement therapy (HRT), traditionally estrogen combined with progesterone, given as tablets, patch, or cream, is no longer prescribed to maintain bone density, prevent height loss or vertebral fractures. For bone health, the North American Menopause Society's expert consensus group[9] suggests calcium intake at DRI levels, preferably from food, and enough Vitamin D (sun and/or DRI recommendations) to maintain adequate blood levels (serum 25(OH)D of 30 ng/mL).

Phytoestrogen A hormonelike substance found in plants, about 1/1000 to 1/2000 as potent as the human hormone, but strong enough to bind with estrogen receptors and mimic estrogen and anti-estrogen effects.

Nutritional Remedies for Symptoms of Menopause

Menopause does not lead to declining life satisfaction for women nor do most women suffer from menopause.[10,11] Women who do experience menopausal effects such as hot flashes, fatigue, anxiety, sleep disturbances, and memory loss are bombarded with treatment options including hormone replacement therapy, herbals, teas, foods and dietary supplements, exercise regimens, creams, and ointments. A review of randomized, controlled trials of complementary treatments of menopausal symptoms included several nutritional approaches, herbs, dietary *phytoestrogens*, and vitamin E.[12,13] Although most of the trials so far have been small and of short duration, some promising approaches were identified. Black cohosh (*Cimicifuga racemosa*) was found to decrease menopausal symptoms, including hot flashes, but not consistently. None of the other commonly used herbs (chaste tree berry, dong quai, ginseng, evening primrose, motherwort, red clover, and licorice), nor Vitamin E (up to 800 IU daily), were found to decrease menopause symptoms.[12] Potential problems with using nutritional remedies include increased bleeding after taking ginseng (*Panax ginseng* and other *Panax* species) and dong quai (*Angelica sinesis*), and photosensitization from the furocoumarins in dong quai. Evidence for phytoestrogen's effectiveness continues to be mixed.

Benefits of Dietary Phytoestrogens Phytoestrogen effects depend on gut flora and hormonal milieu. Lignans, found in whole grains, beans and peas, and especially in flaxseed, are dietary precursors of phytoestrogens; gut flora convert them to usable forms. Isoflavones are especially concentrated in leguminous plants (plants that fix nitrogen in the soil, such as beans and peas). Soy contains isoflavones and has estrogenic activity. In the presence of high estrogen levels, such as before menopause, phytoestrogens link to estrogen receptors to prevent more potent human estrogens from binding with receptors. When estrogen levels are low, as in post-menopausal women, phytoestrogens bind with estrogen receptors to supplement endogenous estrogen.

This is one reason why soy, especially soy protein and soy isoflavone, is so popular during menopause. An epidemiological view suggests that in Asian cultures, where soy is a dietary staple, women do not report hot flashes.

Table 16.2 Hormonal shifts affect blood chemistries

Hormone	Triglyceride	LDL Cholesterol	HDL Cholesterol	Other Effects
Estrogen	Raises	Lowers	Raises	Slows calcium loss Increases insulin sensitivity Decreases homocysteine
Progesterone	Lowers	Raises	Lowers	. . . . . .

SOURCE: Adapted from I. Contreras and D. Parra, "Estrogen Replacement Therapy and the Prevention of Coronary Heart Disease in Postmenopausal Women." Available at www.medscape.com/ASHP/AJHP/2000/v57.n21/ajhp5721.01.cont-01.html, accessed 1/01.

Fewer hot flashes and night sweats may be a result of early exposure to soy phytoestrogens.[14]

The North American Menopausal Society Recommendations state:

Although the observed health effects in humans cannot be clearly attributed to isoflavones alone, it is clear that foods or supplements that contain isoflavones have some physiological effects.[15]

We do know that soy isoflavones function as antioxidants and may be especially useful to women at increased risk for heart disease, which occurs at menopause.[16] Soy protein, with or without isoflavones, increases HDL cholesterol and decreases LDL and total cholesterol as well as decreasing triglycerides. In fact, the Food and Drug Administration allows a health claim to be made for soy: "Diets low in saturated fat and cholesterol that include 25 grams of soy protein a day may reduce the risk of heart disease." Because one consequence of menopause is an increased risk of heart disease in women, soy may be a healthful dietary addition.

A few ways to eat soy include the following:

- Edamame, young green soybeans, (sweet beans) work in salads or side dishes and make for an attractive and quick snack in their green shells.
- Roasted mature soybeans look like roasted peanuts and can easily be flavored to make sweet or savory snacks (9 grams of soy protein per ¼ cup serving).
- Soy protein meat analogs can substitute for meats (burgers, bacon, sausage, turkey, and chicken) while providing soy nutrients. One 2.5 oz veggie burger provides about 14 grams of protein.
- Soymilks do not taste like dairy milk, but are good protein sources: 1 cup soymilk provides 7 to 10 grams, 6 oz of yogurt about 4 grams of protein. Soymilk production formulas vary; try several to find your favorite.
- Processed cereals, bars, candies, and drinks and other snack foods are convenient sources of isoflavones and soy protein.

Flaxseed has been recommended to reduce menopausal symptoms, but the evidence is inadequate for public health recommendations.[17] The proposed mechanism for flaxseed's effectiveness is thought to result from the intestinal bacteria's conversion of lignan precursors to weak estrogens, enterodiol and enterolactone. Flaxseed (see Table 16.3 for basic nutrition information) is a rich source of omega-3 or alpha-linolenic acid and of antioxidants. The recommendations[7] for alpha-linolenic acid are 1.6 grams for men and 1.1 grams for women daily, roughly the amount in 1 tablespoon of ground flax. Flaxseed must be ground to make the nutrients available for metabolism. Flaxseed can be heated to 300° Fahrenheit without losing nutritional value, but they are most often used in chilled foods such as smoothies or sprinkled over cooked items such as cereal. Toasted flax

Table 16.3 Nutrient content in 2 tablespoons (14 grams) of ground flaxseed	
Kcal	70
Protein	2.8 g
Carbohydrate	4.9 g
Fiber	4.0 g
Fat	4.8 g
Omega-3 fatty acid	2.6 g

Also contains significant amounts of iron, zinc, magnesium, potassium, and some calcium. For fewer calories than in a slice of bread, 2 tablespoons of ground flaxseed provide the protein of half an egg and more fiber than an apple. (A small apple, with skin, contains 2.9 grams of dietary fiber.)

has a nutty flavor; roast seeds until the kernels start to pop. (Keep the pan cover handy; it will remind you of popcorn.)

To Use or Not? Potential Drawbacks of Increasing Dietary Phytoestrogens   Soy has an estrogenic effect.[13] Estrogen increases the risk for breast cancer. It would be unwise to suddenly add several servings of soy to the diet of someone vulnerable to breast cancer development (females with family history of breast cancer).

The taste of soy, as in soy flour and soy powder, is unappealing to many people. Soy oil, which is versatile and mild-flavored, does not contain enough phytoestrogens to impact menopausal symptoms. Tamari, or soy sauce, even in its reduced-sodium form is quite salty (compare 540 mg of sodium in 1 tablespoon light soy sauce, 1230 mg in 1 tablespoon of regular soy sauce, and approximately 2300 mg of sodium in 1 teaspoon of table salt). In the small amounts used, soy sauce does not add much soy nutrition to the daily North American diet. Fermented soy products are good sources of soy protein and isoflavones but may not be widely available. Not many people are familiar with *tempeh* or *miso*.

Tempeh   Fermented soybean cake eaten in stir-fry and veggie burgers.

Miso   Fermented soybean paste used as flavoring; a mainstay in Japanese kitches.

Flax in larger amounts (1.5 oz or more) acts as a laxative. The high fiber content may interfere with medications such as digoxin.

Other Dietary Supplements   Menopausal treatment effects such as increased well-being have been attributed to ginseng, but experimental evidence supporting such activity is lacking.[18] The edible tubers of wild yams (*Dioscorea villosa*) contain steroidal saponins, which are used in progesterone production, leading to speculation about hormonal activity in tubers. Wild yams are primarily used in topical applications and have been shown ineffective as hormone replacement therapy.[12] Kava (*Piper methysticum*), used as a muscle relaxant and anticonvulsant, was tested against placebos and shown to be effective for menopausal symptoms, with only mild side effects.[19] Still, the FDA issued a warning about kava's safety, citing liver toxicity.[20] Alcohol potentiates kava's effect. In Pacific Island cultures (Oceania) where kava is native, the root is made into a bitter ceremonial beverage. The kava beverage first numbs the mouth, then leads to greater sociability, decreased anxiety, less fatigue, and eventually deep sleep. Therapeutic uses include muscle relaxation, sleep induction, and reduction of nervous anxiety. Kava extract contains kavalactones, also called pyrones, accounting for kava's sedative effects. There is controversy about using kava for extended periods; kava functions as medicine, and England has banned kava as a food ingredient. Individuals with liver disease and those with Parkinson's are especially vulnerable to adverse side effects of kava. Table 16.1 contains additional information on herbal and dietary supplements.

Maintaining a Healthy Body

Reproduction, weight management including maintenance of muscle mass while avoiding excess fat, and wellness are primary nutrition concerns of adulthood. Reproductive issues are discussed elsewhere in the text; the following sections focus on energy and other nutrients needed to maintain a healthy body.

Energy for Weight Management

The focus of adulthood is maintenance of physiological status rather than growth, although some males continue to gain height after age 20. Average calorie expenditures peak during late adolescence and early adulthood, and then decline roughly 20% over the course of a lifetime, although the size of this decline has not yet been clearly estimated because of limited longitudinal data.[21] In the mid-1960s, researchers at the Baltimore Longitudinal Study on Aging (BLSA) reported that caloric intake in men decreased 22%, from 2700 to 2100 calories, between age 30 and 80.[22] They suggested that the decrease was due to lowered metabolic rates as well as decreased activity levels. The theory is that activity-related caloric expenditures are lower for women than for men, in part due to lower fat-free body mass. Women's resting metabolic rate (energy expended when the body is not active) declined 2% to 4% after age 50. However, energy expenditures stayed constant as long as fat-free mass stayed constant. The Fels Longitudinal Study[23] examined weight gains over 20 years, starting at age 40, and calculated average annual increases:

- Men gained 0.66 lb per year (0.3 kg)
- Women gained 1.21 lb per year (0.55 kg)

Basal Metabolic Rate (BMR)
Measuring energy expenditure in an individual who has been awake less than 30 mintues and is still at *absolute* rest, fasted for 10 hours or more, and is in a quiet room with normal, comfortable temperatures.

Resting Metabolic Rate (RMR)
Measuring energy expenditure in an individual who has fasted, had no vigorous physical activity prior to the test, has been given time to relax (e.g., rest) for 30 minutes before starting measurement, and is in a quiet room with comfortable temperatures.

Determining Energy Needs

Changes in body weight are due to a complex system of interactions including gender, body size, energy intake, activity levels, health status, hormonal shifts, and individual variation.

- *Gender/body size/muscle mass:* Males in general use 5–10% more calories than females because they have proportionately more lean body mass. Metabolism in muscle burns more oxygen and needs more energy than fat stores. However, persons who are overweight or obese don't necessarily burn fewer calories or have a lower metabolic rate.

- *Activity levels:* Activity requires energy, but more than that, exercise increases subsequent resting energy expenditure.

- *Health status:* Illness affects energy needs. For example, a fever's higher body temperatures burn more energy, 7% more per degree Fahrenheit or 13% per degree Celsius. Starvation slows energy expenditures by 20% to 30%.

- *Hormones:* The thyroid hormone thyroxin accelerates metabolism while lack of thyroid hormones leads to weight gain. Growth hormone is associated with weight gain, as is estrogen loss.

- *Individual variation:* Individuals vary by as much as 20% in their calorie expenditures during light to moderate activity.[24] In addition, the efficiency of metabolism of food (the thermic effect of food) varies from person to person. Each of the following approaches to estimating calorie needs tries to accommodate aspects of individual variation.

The Harris–Benedict Equation The Harris–Benedict equation was derived from 239 subjects, aged 15–74 years old, with a mean age of 27 years and 31 years for men and women, respectively.[25] While Harris and Benedict called their measure a *basal metabolic rate (BMR)*, by today's interpretations, the reported measurements really reflect *resting metabolic rates (RMR)* since subjects had to travel to the testing site. That is, the measurements were not taken at *absolute* rest.

Harris–Benedict Energy Estimation Formula (developed 1919)
Males: RMR = 66.5 + (13.75 × wt) + (5 × ht) − (6.8 × age)
Females: RMR = 655 + (9.6 × wt) + (1.8 × ht) − (4.7 × age)
Weight is in kilograms, height in centimeters, age in years

Healthy adults today have different body compositions and lifestyles than adults in the early 1900s had. Validation studies have have yielded newer estimation formulas.[27–29] Energy estimation formulas that are within ±10% of "true" RMR are considered accurate. The Mifflin–St. Jeor equation accurately reports RMR 82% of the time in nonobese adults (i.e., BMI 18.5–29.9 kg/m²). If the estimation is wrong, the worst underestimation error is 18% and the worst overestimation error is 15%. In contrast, an energy estimate using the Harris–Benedict equation is accurate 69% of the time.[30] Once RMR has been determined, physiological stress factors related to illnesses are added to estimate daily energy expenditure in clinical use. In healthy adults, however, an activity factor is added to arrive at the estimated daily calorie expenditure. Note that metric measurements were used in the original formulas and that neither has a factor for the thermic effect of food.

Mifflin–St. Jeor Energy Estimation Formula (developed 1990)

Males: $RMR = (10 \times wt) + (6.25 \times ht) - (5 \times age) + 5$

Females: $RMR = (10 \times wt) + (6.25 \times ht) - (5 \times age) - 161$

Weight is in kilograms, height in centimeters, age in years

The box below describes the steps for using the Mifflin–St. Jeor equation to calculate the energy expenditure in an individual who is 67 inches tall (170.2 cm), weighs 150 lb (68.1 kg) and is 55 years old.

Example of how to use estimation formulas

Example: Calculate the resting metabolic rate for a 5'7" female, weighing 150 lb who is 55 years old using the Mifflin–St. Jeor equation:

Step 1: Identify which estimation formula will be most reliable for individual
Selected Mifflin–St. Jeor

Step 2: Convert English measurements into metric
67 inches multiplied by 2.54 centimeter/inch = 170.2 centimeters
150 pounds divided by 2.2 kilograms/pounds = 68.2 kg

Step 3: Use formula* based on gender

Step 3a: FEMALE
$RMR = (10 \times 68.2) + (6.25 \times 170.2) - (5 \times 55) - 161$
$RMR = (682) + (1064) - (275) - 161 = \sim 1310$ calories/day

Step 3b: MALE
$RMR = (10 \times wt) + (6.25 \times ht) - (5 \times age) + 5$
$RMR = (10 \times 68.2) + (6.25 \times 170.2) - (5 \times 55) + 5$
$RMR = (682) + (1064) - (275) + 5 = \sim 1476$ calories/day

Step 4: RMR is multiplied by activity or injury and stress factors[29]

Sedentary or weight maintenance	>1.0 to <1.4
Light activity	>1.4 to <1.6
Moderately active	>1.6 to <1.9
Very active	>1.9 to <2.5

*Weight is in kilograms, height in centimeters, age in years.

To explore formula differences, complete the same steps using the Harris–Benedict equation. How different are the results between males and females? Comparing these two equations demonstrates why calorie calculations are called calorie estimates. Calculating caloric needs involves trial and error because energy expenditures involve complex interactions. However, the indirect calorimetry technology for measuring oxygen consumption and carbon dioxide excretion is advancing, offering another choice to identify energy expenditure.

Energy for Weight Change

It takes approximately 3500 calories to gain or lose 1 lb of body weight. To gain or lose 1 lb a week, add or subtract 500 calories daily. These 500 calories can be a combination of intake and activity either added or subtracted. For instance, walking 1 mile uses roughly 100 calories, cycling 5 miles uses 200, and eliminating 16 ounces of cola allows subtraction of 200 calories, summed to total 500 calories. Seven days of burning 300 calories and eating 200 fewer calories leads to a weight loss of approximately 1 lb.

What Is a Desirable Weight?   A Body Mass Index calculator is built into the online MyPramid (see Chapter 1). There is an easy-to-remember method that does not require computer access. The Miller method[28,31] most closely approximates weights at the midpoints of weight ranges in the Metropolitan Life Tables of 1983, which reflect BMI recommendations.

- **Female:** 119 lb for the first 5 feet plus 3 lb for each additional inch
- **Male:** 135 lb for the first 5 feet 3 inches plus 3 lb for each additional inch

The Miller method is not accurate for males taller than 74 inches. Other weight-for-height measures are discussed with obesity in Chapter 17.

Actual Energy Intake

> "We're a culture that eats anywhere and all the time. Food is everywhere. When was the last time you looked for food and couldn't find anything to eat?"

Older adults eat fewer calories than younger ones. Table 16.4 shows findings from data collected by the U.S. Department of Agriculture's Continuing Survey of Food Intake by Individuals (CSFII) and by the Department of Health and Human Services in the National Health and Nutrition Examination Survey (NHANES). Judging by these data, there shouldn't be an obesity epidemic. Nutrition scientists suggest that underreporting by overweight people and over-reporting by underweight individuals skews these numbers. However, they are relatively consistent over time.

Table 16.4 Comparing caloric intakes of adults, by gender, from CSFII (Continuing Survey of Food Intake by Individuals, 1994–1995) and NHANES III (National Health and Nutrition Examination Survey 1988–1994) and NHANES 1999–2000

Age	Daily Caloric Intake, CSFII		Daily Caloric Intake, NHANES III		Daily Caloric from NHANES 99-00	
	Males	Females	Males	Females	Males	Females
20–29	2844	1828	3025	1957	2828	2028
30–39	2702	1676	2872	1883	2828	2028
40–49	2411	1680	2545	1764	2590	1828
50–59	2259	1583	2341	1629	2590	1828
60–69	2100	1496	2110	1578	2123	1596

Another way to examine caloric intake is by reviewing food disappearance data. Food available for consumption has increased by 16% (>500 calories per person per day) since 1970, described by food group in Table 16.5. These per capita figures apply to individuals of all ages and correspond more closely to the increase in weight of the American public.[32]

Fad Diets

Determining adequate energy in the context of maintaining a healthy body is a good place for a brief discussion of fad dieting. Dieting is so pervasive, and weight loss is so difficult to maintain, that weight-loss efforts support a multimillion-dollar industry. Weight cycling (yo-yo dieting) is associated with higher cardiovascular and all-cause mortality,[33] and with weight regains larger than weight lost. Obesity

Table 16.5 Per capita increases in food available for consumption, adjusted for plate waste and spoilage

Commodity group	1970, in lb	2003, in lb	Change in daily calories (Kcal)
Fats and oils	53	86	+216
Grains	136	194	+188
Sugars, sweeteners	119	142	+76
Meat, eggs, nuts	226	242	+24
Vegetables	337	418	+16
Fruits	242	275	+14
Dairy	564	594	−11
TOTAL	1675	1950	+523

Adapted from Farah H and Buzby J. U.S. Food consumption up 16% since 1970. Amber Waves, 11/05. (32) available at www.ers.usda.gov/AmberWaves/November05/Findings/USFoodConsumption/htm, accessed 10/3/2006.

and potential weight-loss interventions are further discussed in Chapter 17. Just how severely can we limit our food intake and stay healthy? Chapter 18 discusses caloric restriction (CR) and longevity, but the focus in CR is on increased longevity and most of the CR data is still from animal studies. Fad diets, however, provide an ongoing test of human resilience.

Fad diets focus on fast and easy weight loss—something easy and "guaranteed to get you into those tight new jeans." Dr. Susan Raatz identifies the following features to flag a fad diet:[34]

- Inadequate nutrient supply (long-term strategies promote nutritious foods)
- Severe energy restriction (eating 800 or fewer calories a day requires needs medical supervision)
- Unusual food restriction (for example, no carbohydrates)
- "Food combinations" (for example: grapefruit with all meals)
- Strict limitations (avoiding certain food groups, such as never eating potatoes)
- Gimmicks (don't eat afer 7 p.m. in the evening, eat only popcorn for lunch, drink caffeine or guarana, or eat hot peppers to speed up your metabolism)

Fad diets are often built on logical weight-loss concepts, but with a twist. Stay in "the zone" or never eat after 7 p.m. are strategies to reduce overall food intake. Caffeine does speed up metabolism for a few hours, as does guarana and exercise. The Grapefruit Diet leads to weight loss, not because the acid in grapefruit neutralizes fat, but because people eat less when they tire of grapefruit. Chewing on celery does not burn fat. Enzymes break down dietary fat, and your body either uses the energy released for activities, or it stores that energy as fat for later use.

Diet scams can be recognized by any one or more of these four features:

1. Defiant claims of a renegade genius (Dr. Bea's soothing peppermint tea diet WILL melt away those pounds and the stress that's troubling you—trust me, I have a doctor's degree!)
2. Discounting calories
3. Enticing testimonials (Elvis used this when he toured; Paris Hilton swears by it)
4. Promise of a quick fix (Wedding this weekend? NO problem! You'll look perfect!)

Dr. Raatz adds that a dead giveaway of a fraudulent diet plan is one suggesting that calories don't matter. Without energy, counted as calories, we die. Excess energy intake is stored, so of course, calories matter. In addition, exercise or physical activity is needed for healthy weight loss. Merely cutting calories will have your body breaking down body tissues to maintain a fuel supply. Body tissues include skeletal muscle, heart, and skin tissues. It takes energy (calories) and exercise to maintain glowing skin, a strong heart, and fit muscles. A popular cliché applies here: "If it sounds too good to be true, it is."

When it comes to weight loss, we all wish for uncomplicated solutions. We know what we should do, but knowledge and action do not necessarily coincide. In describing the development of a 100-point-tool to rate top-selling diet books in Australia (there were 20 books in this study, 100 was the best score), Williams and Williams[35] found four of the lowest scoring books (32–40 points) were *Sugar Busters, Eat Right 4 Your Type, Dr Atkins' New Diet Revolution,* and *Slim Forever.* The highest scoring books (only two scored above 90) were *The Volumetrics Weight Control Plan* and *License to Eat* (97 and 96 points). And do the top-scoring books match the best-seller lists? They do not. Variety and moderation in dieting is excellent advice. Excellent advice—that doesn't sell diet books.

Dietary Recommendations

How would you like your food advice? "Monounsaturated fats contribute to a lower risk for heart disease."

or

"Eating a salad with olive oil dressing makes your blood flow and gives you energy."

To achieve wellness, it is not enough to know how many calories will maintain healthy weight. A more important question is: "What should those calories look like?" Do certain foods contribute to better health? Which foods should I eat? Can some foods cause illness? Dietary guidelines help to answer these questions. To understand dietary recommendations, it is useful to know about nutritional risk factors: How are our dietary habits linked to health and the avoidance of disease?

Much as we try, we cannot always achieve the degree of health we would like. But we can reduce the risk of contracting both short-term and chronic illnesses. Chronic conditions with modifiable risk factors include 5 of the 10 leading causes of death for adults: heart disease, cancer, stroke, diabetes, and hypertension. Short-term illnesses include food poisoning and infections. Table 16.6 summarizes modifiable nutritional risk factors that concern adults.

Governmental and private groups translate nutritional risk factors into dietary recommendations, such as the U.S. government's development of Dietary Guidelines for Americans,[33] the Institute of Medicine's (IOM) Food and Nutrition Board's Recommended Dietary Allowances,

and the American Heart Association's Diet and Lifestyle Recommendations, Revision 2006.[34] Organizations make dietary recommendations according to their mission and goals as well as available science. Three main perspectives underlie dietary guidance tools and advice.

1. *Advocating for reduction of specific disease risk:* For example, the American Heart Association suggests reducing total dietary fat to decrease risk of heart disease. The National Cancer Institute (as well as the fruit and vegetable produce industry) urges us all to eat more fruits and vegetables to reduce risk of certain cancers.

2. *Ensuring adequate population intake of specific nutrients:* For example, Food and Drug Administration (FDA) regulations mandate flour and grain product fortification with B vitamins, folic acid, and iron. Folic acid was added in the late 1990s to prevent neural tube defects. Here, FDA recommendations are more than "advice," they are enacted into law. Enrichment and fortification laws are intended to improve the health status of populations. The Recommended Dietary Allowances established by the National Academy of Sciences are used to determine nutrient standards for food assistance and population feeding programs. For instance, military commissaries plan and serve menus that meet the RDA for their troops.

3. *Offering guidance on what, and how much, to eat:* The U.S. Department of Agriculture and U.S. Department of Health and Human Services together publish the MyPyramid and "Dietary Guidelines for Americans" that include meal patterns for selected caloric levels. On the nongovernmental side, many Dairy Councils have developed dietary guidance tools built on the basic food groups, including the cheese and milk group. Dietary guidance tools such as these are designed to help consumers decide what to eat to be healthy.

Dietary Recommendations to Combat Nutritional Concerns

Dietary recommendations for the public translate and integrate documents such as Healthy People 2010, nutrient guidelines from the National Academy of Sciences Food and Nutrition Board, the U.S. Departments of Agriculture and Health and Human Services as well as the various health advocacy coalition positions, such as those from the American Heart Association, National Cancer Institute, and National Osteoporosis Foundation, into a consensus document. Adults can refer to this document for several reasons: (1) it spells out the national health and nutrition goals and priorities; (2) it makes evident the links between eating, drinking, and health outcomes; and (3) it includes food advice that can guide eating behaviors.

Table 16.6 Illnesses and their modifiable risk factors

Short-Term Illnesses	Nutritional Risk Factors
Food poisoning	Unsafe food handling practices Lack of hand washing, especially after toilet use Cross-contamination, mixing raw meat and vegetable preparation Lack of refrigeration, food at room temperature longer than 2 hours Inadequate temperatures in maintaining hot foods Contaminants such as salmonella, *E. coli,* mercury in some fish, etc.
Lowered immunity, infections	Malnutrition, especially protein and calorie malnutrition Low antioxidant intake Low vitamin/mineral levels Low levels of zinc and excessive levels of zinc Dehydration

Chronic Conditions	Nutritional Risk Factors (Inactivity is associated with most)
Heart disease	Diet high in saturated fat More than 10% of calories from saturated fat (Healthy People 2010) Daily dietary cholesterol above 300 mg, on average
Atherosclerosis	Any trans-fatty acids Low levels of mono-unsaturated fatty acids Low antioxidants, low fruit and vegetable intake Low intake of whole grains Low folic acid intake and high levels of homocysteine No or excess alcohol* Obesity (BMI >30); waist >40″ in men, >35″ in women Elevated plasma Apo B levels High levels of LDL cholesterol in men Low levels of HDL cholesterol in women
Cancer	Low fruit and vegetable intake Low level of antioxidants (especially vitamins A, C) Low intake of whole grains, especially fiber-rich grains High dietary fat intake Nitrosamines, burnt and charred food High intakes of pickled and fermented food Obesity
Hypertension, Stroke	High blood pressure: obesity, waist >40″ in men, >35″ in women High sodium Low potassium Excess alcohol Low levels of antioxidants
Diabetes	Obesity

*In middle-aged adults, moderate alcohol reduces risk of heart disease (for men, after age 45; for women, after age 55).

"Dietary Guidelines for Americans, 2005" addresses current health issues by outlining health goals, describing diet and disease linkages, and including MyPyramid to provide advice on what to eat for good health. Dietary Guidelines 2005 continues a series first issued in 1980 and updated every 5 years. The current version of health-promoting food advice reflects current nutrition science and policy. Food safety was added in the 2000 guide because incidents of food-borne illness had increased. The 1995 statement about sugar changed from "choose a diet moderate in sugars" to "choose beverages and foods to moderate your intake of sugar" in order to reflect the rising intake of nonnutrient sugars, one of the contributors of obesity. Chapter 1 highlights basic components of the Dietary Guidelines and of the latest food guide pyramid, including specific food-based advice.

Food Advice

Food guidance documents have been around at least since 1916, having been spurred by war and economic depression.[35] The current recommendations, "Nutrition and Your Health: Dietary Guidelines for Americans," have grown from collaborations by the U.S. Department of Agriculture

Table 16.7 How to translate 10% of calories from fat to grams of fat

1. Multiply total calories in the diet by 0.1 to determine calories equaling 10%.
2. Divide these 10% of calories by 9 (calories per gram of fat) to calculate the number of grams of saturated fat that represent the "10% of calories from saturated fat" limit suggested for reducing LDL cholesterol in blood.

Amount of fat that represents 10% of calories for selected calorie levels:

Total Calorie Intake	Fat Grams = 10%
1200	13
1500	17
1800	20
2800	31

and U.S. Department of Health and Human Services, who maintain them at www.healthierus.gov/dietaryguidelines.

Deciding what to eat is harder than knowing what to avoid. What makes food advice useful? The message "skip ice cream" tells an audience exactly what to do: avoid eating ice cream. But the message "eat less saturated fat" needs further knowledge to be translated into breakfast, lunch, snacks, or supper. Table 16.7 shows how to translate percentage of calories from fat into gram amount; it is unlikely that consumers would do this type of calculation before planning their meals. In order to help individuals focus on foods to eat, a food guide needs several attributes:

- *Classify foods into relevant groups.* For example, what are the food groups for peanut butter on toast and a glass of juice?
- *Describe what constitutes a serving.* For example, how do you count the lettuce, tomato, pickle, and ketchup on a burger?
- *Identify the number of servings needed from each group to promote health.* For instance, how many grain servings will it take to meet carbohydrate and fiber recommendations? How much milk is optimum? How much fruit is enough?

MyPyramid

One of the benefits of the current online pyramid (available at www.MyPramid.gov) is that adults can key in their height, weight, gender, age, and activity level, and a program will instantly calculate their BMI and suggested energy intake. They can then click on tools to suggest meal patterns, track their intake and activity (MyPyramid Tracker), and find menu ideas that illustrate the suggested food intake for them. Nutrient and energy contributions of the food groups are based on the most nutrient-dense foods of that group,

leaving an additional calories that may be used as discretionary calorie allowance. The discretionary allowance provides flexibility for additional nutrient-dense foods or extras like a larger mocha, chips, wine, extra olives, or other personal favorites.

Future dietary guidance tools will integrate the emerging understanding of *nutrigenomics*. For instance, a female with the Apo E14 allele may be more prone to bone loss after menopause. She may be advised to build maximum bone density by age 30 and to monitor her bone mass density.

This is the first time that the main access to national dietary guidance is online.

Nutrigenomics The science of gene–nutrient interactions.

Food Advice Reflects Cultural Food Patterns

Food group guidance comes in many forms other than the pyramid. For example, "Eating well with Canada's food guide" uses a rainbow (Illustration 16.1). Inclusion of canned and frozen food pictures makes it obvious that processed foods count. The rainbow groups fruits and vegetables, lists metric and English amounts for serving sizes, and does not picture "other" foods.

The United States uses a pyramid, Canada a rainbow, and China uses a pagoda. China's pagoda highlights how food guides reflect their target audience's unique dietary needs and goals (Illustration 16.2). Following pagoda guidance ensures that individuals eat from 1800 to 2800 calories per day. Additional advice to Chinese residents includes distributing food evenly throughout the day, making full use of local food sources, and making good nutrition a lifetime commitment. The gram-weights listed with food groups show that the Chinese diet is cereal-based, includes greater quantities of vegetables than fruit, includes fish daily, and uses small portions of meat. The placement of milk/milk products and bean/bean products together in one group is also unique to the pagoda.

Dietary advice for health promotion, Japanese style, includes the following recommendations to enjoy meals.

Make all activities pertaining to food and eating pleasurable ones:

1. Use the mealtime as an occasion for family communications.
2. Treasure family taste and home cooking.

Alcohol: Food, Drug, and Nutrient

Overall, 61% of adults in the United States drink alcohol, with the highest rates among persons aged 25 to 44 years old (76% of males, 63% of females).[39] Rates decline with age. Alcohol consumption guidance varies by country. For example, Canadian recommendations are to limit alcohol calories to 5% or less of total calories or 2 drinks a day, whichever is less.[40] New Zealand gives guidelines in grams

Illustration 16.1 Canada's Food Guide to Healthy Eating—the rainbow.

SOURCE: Health and Welfare Canada. Nutrition recommendations. Canada: Minister of Supply Services; 1990.

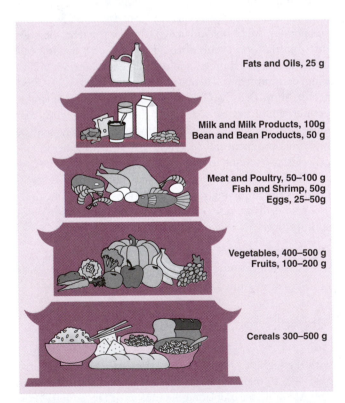

Illustration 16.2 The Food Pagoda Guide for Chinese residents.

Used by permission. The Chinese Nutrition Society, People's Republic of China.

Calculating Calories from Alcohol

- Alcohol is lighter than water, with a specific gravity of 0.79 (0.8)
- The "proof" of alcoholic beverages such as rum, gin, scotch, or vodka represents twice the percentage of alcohol per volume
- Use the Gastineau formula (CF Gastineau. Alcohol and Calories. Mayo Clinic Proceedings 51(2):88,1976):
  - $0.8 \times$ proof $\times$ ounces = calories
  - 0.8×80 proof rum $\times 1.5$ oz = 96 alcohol calories
- For wine and beer where the percentage of alcohol is known, use $1.6 \times \%$ alcohol $\times$ ounces = calories
  - For example, 12 oz of a 10% beer:
  - $1.6 \times 12 \times 10 = 192$ alcohol calories

of alcohol: 20 or fewer grams of alcohol per day for women and 30 grams or less for men. (See Illustration 16.3 for alcohol content of selected beverages.) United States guidelines are "If you drink, do so in moderation," defined as no more than 2 drinks per day for males, and no more than 1 drink per day for females. Saving up drinks to have all 7 to 14 on the weekend is not "moderation." The Canadian guide states that more than 4 drinks at a time is risky.

Amount	Beverage	Alcohol, grams
12 oz	Regular beer	13
12 oz	Light beer	12
1.5 oz	Distilled spirits (gin, whiskey, rum, vodka)	15
5 oz	Red wine	15
5 oz	Dry white wine	14
2 oz	Sherry	9

To calculate grams of alcohol in a beverage, multiply the milliliters of alcoholic beverage by the percent alcohol and the result by 0.79 (the specific gravity of alcohol). For example, to determine the alcohol in 12 oz (or 360 ml) of regular beer: 360 ml times 5% alcohol times 0.79 = 14 grams alcohol (most beers range from 4 to 6% alcohol).

Illustration 16.3 Alcohol content in beverages and how to calculate grams of alcohol in specific beverages.

What is a drink? Equivalent alcohol contents for "a drink" are 12 oz of regular beer, 5 oz of wine, 1.5 oz of 80 proof distilled spirits. A drink contains roughly 13–15 grams of alcohol or 0.5 oz of ethanol (ounces are listed on the blood alcohol charts describing legal limits for drivers).

Alcohol is considered a drug for its central nervous system effects, but it is also a nutrient (although not a required one) because it yields energy; that is, 7 calories per gram. To calculate the caloric value of alcohol in a beverage, see the boxed example on page 418.

Population averages for alcohol consumption under-report individual consumption because nearly 40% of adults abstain from drinking alcohol. Current data reports average alcohol consumption for the alcohol per se. Caloric contributions of alcoholic beverages are higher because alcoholic beverages such as wine and beer also contain some carbohydrates. Mixed drinks may contain carbohydrates, protein, and fat. Based on an analysis of the first (1977–1978) Nationwide Food Consumption Survey of just those individuals who drank alcoholic beverages, males consumed 17% and females 22% of total calories from mixed drinks, wine, and beer.[41]

Water and Other Fluids

Adults can survive roughly 60 days without food but only 3 to 4 days without water.[8,42] Because fluid is such an integral part of metabolism, the body has several regulatory mechanisms to maintain hydration: increased thirst, concentrating urine to reabsorb water, slowing down activities, and shutting down perspiration conserve water. Metabolism of foods and of body tissues also yields water. For instance, breaking down 1 kg of body tissue composed half and half of fat and non-fatty tissue yields nearly 1000 milliliters (ml) of water.[42] Someone who has stopped eating and drinking metabolizes stored tissues and will continue to produce urine.

Methods to determine fluid recommendations relate to caloric intake or body weight. The Food and Nutrition Board recommends 1 ml water per calorie of food ingested. Therefore, 2000 ml of fluid would be recommended for an individual eating the 2000-calorie diet used as reference on food labels. Some calorie-based recommendations also suggest that fluid intake should be at least 1500 ml daily (when calories fall below 1500 calories). So getting a jumpstart on dieting with a 1200-calorie diet means drinking 1500 ml or more rather than drinking 1200 ml or more.

Estimates using body weight vary by activity and age, with the average adult needing 30–40 ml/kg.[42]

Young active adults, aged 16–30 years:	35–40 ml/kg
Adults aged 55–65 years:	30 ml/kg

Using the guideline of 30 ml of fluid per kg body weight, a 143 lb (65 kg) person would need 65 × 30 or 1950 ml fluid daily. A 170 lb (77 kg) person would need 77 × 30 or 2310 ml fluid per day.

A step wise formula[43] that allows adjustment for increasing or decreasing body weight and insures a 1500 ml minimum daily intake was developed for pediatric patients and is useful for smaller, lighter adults:

100 ml/kg body weight for the first 10 kg
50 ml/kg body weight for the next 10 kg
15 ml/kg body weight for the remaining kg

Applying this formula to a 143 lb (65 kg) person leads to a recommendation of just over 2 liters (8 cups), and roughly 10 cups for a 170 lb (77 kg) person.

	Body weight of	
	143 pounds	170 pounds
10 kg × 100 =	1000 ml	1000 ml
10 kg × 50 =	500 ml	500 ml
Remaining ___ kg × 15 =	45 × 15 = 675 ml	57 × 15 = 855 ml
Total =	2175 ml	2355 ml

240 ml = 8 ounes = 1 cup; 4 cups = 1 quart, roughly equivalent to 1 liter

In reviewing available data to establish dietary reference intakes for water, the National Academy of Sciences concluded that average fluid intake as reported in the NHANES III survey data met human needs.[44] Overall, food accounts for 19% of total water intake and beverages supply the remaining 81%.

Even dry foods contain water; water content of selected foods follows:

Butter	16%
Bagel	33% (white bread is approximately 35% water)

Eggs	75%
Banana	76%
Watermelon	91%
Lettuce	96%

Find this and other nutrient data at www.ars.usda.gov nutriendata; look for the nutrient composition tables.

The NHANESIII population intake data were used to establish the fluid DRI at 3.7 liters of total water per day for males (including approximately 3 liters or 13 cups as beverages), and 2.7 liters (including 2.2 liters or 9 cups) of total water per day for females. These recommended levels do not change across the healthy adult life span after age 19. Some special circumstances (extreme heat or cold, sports participation, pregnancy, and lactation) increase water needs. On the other hand, surveys of Canadians showed lower average fluid consumption, so guidelines reflect lower intakes. It seems there is some variation in how much fluid will sustain a human body.

A simple, yet practical guide to adequate fluid intake that does not rely on numbers evaluates the color of urine: when urine is nearly colorless or pale yellow, the individual is drinking enough fluid. However, this color test doesn't always work. Therapeutic amounts of riboflavin supplements as well as some medications and foods (for example, beets and asparagus) affect the color of urine. Food components such as flavins are potent pigments.

Diuretic Effects of Caffeine and Alcohol

What about fluids containing caffeine? Occasionally someone suggests that coffee, tea, or other caffeine-containing beverages not be counted as fluid intake because they have a diuretic effect on the body. Caffeine is a stimulant that relaxes the esophageal sphincter (leads to acid reflux), has a laxative effect, and temporarily increases urine production at high doses.[44] But no evidence connects drinking coffee or tea or other caffeine-containing fluids with dehydration.[45] The DRI committee concluded that caffeine-containing beverages contribute to daily total water intake.

Several countries address caffeine in their dietary guidelines; none address whether caffeine-containing beverages should count as fluids.[38] South Africa suggests restricting caffeine (no amounts given) as part of dietary guidance. Canada is the only country to suggest an amount for caffeine consumption: "The Canadian diet should contain no more caffeine than the equivalent of four regular cups of coffee per day."[40]

Alcohol (ethanol) has a dehydrating effect on cells for the first few hours after consumption; after about 6 hours, an antidiuretic phase begins.[44] In addition, the effects of alcohol seem to change over the course of the day. Although alcoholic beverages are fluids by definition, most health professionals do not include them in their hydration recommendations. With moderate alcohol use, the amounts of fluid in question are necessarily small.

International Guidance Related to Fluids

Here is how several countries address the need for adequate fluid in their dietary guidance documents.[38]

- New Zealand: Drink 1 to 2 liters per day (1000 to 2000 ml, about 8 cups) and choose from tap water and other beverages, including tea and coffee.
- South Africa: Drink at least 1 liter of clean, safe water or fluids.
- Switzerland: More fluids.
- Venezuela: Drink water.
- Hungary, Korea, Philippines, Singapore, and Switzerland suggest "milk every day" or "more milk" as part of their dietary recommendations. This suggestion may have more to do with increasing calcium and other nutrients than fluid intake.

Nutrient Recommendations

"Poison is in everything, and no thing is without poison. The dosage makes it either a poison or a remedy."

Paracelsus, Swiss alchemist, 1493–1541

Nutrient recommendations are constantly evolving. The Institute of Medicine of the National Academies regularly reviews and revises them.[7,24] For instance, pre-2002 public health recommendations, such as those published in Healthy People 2010, were to eat fewer than 30% of total calories as fats; current recommendations are to consume from 20% to 35% of calories in the form of fat. Dietary Reference Intake (DRI) values for all ages as well as the tolerable Upper Intake Levels (UL) are located on the inside covers of this text. DRIs for adults and the recommended nutrient levels used in food labeling as daily values (DV)[46] are included in Table 16.8. The increase of fortified and functional foods means that diets can provide excessive levels of nutrients, which can be as detrimental to health as insufficient intake. Ranges of recommended intake are given where possible. The ULs indicate a potentially toxic limit for vitamins and mineral intake. A diet of unfortified foods alone is unlikely to get anyone to these UL amounts; however, eating fortified foods and taking vitamin–mineral supplements may. Paracelsus was right; too much of a good thing can make you sick.

Nutrients of Public Health Concern

Saturated fat, fiber, folic acid for women, vitamin E, calcium, magnesium, plus iron for women are of public health concern because intake levels do not meet recommendations (Table 16.8).

Fat   In the late 1970s, 41% of calories in U.S. diets came from fat. That level dropped to 33% of calories from fat (11% saturated fats) in 1999–2000.[47,48] Fat intake levels seem to have stabilized. CSFII data from 1994–1995[49]

Table 16.8 Selected nutrient intakes of adults, continuing survey of food intake by individuals, CSFII, 1994–1995 data collection, compared to recommendations

	INTAKE				
	Actual				Recommended[a]
	20–29 years		40–49 years		31–50 years
Nutrient	Males	Females	Males	Females	Males/Females
Protein, g	106	68	96	64	63/50
Total fat, g	104	66	91	63	65 (DV)[b]
Saturated fat, g	36	23	31	21	20 (DV)[b]
Monounsaturated fat, g	40	25	35	24	–
Cholesterol, mg	349	226	339	225	300 or less
Fiber, g	18	12	18	14	25 (DV)
Sodium, mg	4627	2974	4061	2771	Up to 2400
Potassium, mg	3278	2240	3225	2411	3500
Vitamin A, mcg RE	1430	1149	1000	800	900/700
Carotenes, mcg RE	539	506	539	500	–
Vitamin E, mg TE	10.1	7.1	9.4	7.9	15
Vitamin C, mg	123	87	106	93	90/75
Vitamin B_6, mg	2.4	1.6	2.1	1.5	1.3/1.3
Vitamin B_{12}, mcg	6.2	3.8	6.2	4.4	2.4/2.4
Folate, mcg[c]	323	226	288	226	400/400
Calcium, mg	1005	890	717	643	1000/1000
Magnesium, mg	341	229	330	241	420/320[d]
Zinc, mg	15.2	9.7	13.2	9.4	11/8
Iron, mg	19.8	13.5	17.6	13.1	8/18
Copper, mcg	1600	1100	1500	1100	900

SOURCE: Consumption data from Wilson et al. Data Tables: *Combined Results from USDA's 1994 and 1995 Continuing Survey of Food Intakes by Individuals and 1994 and 1995 Diet and Health Knowledge Survey; 1997.*

[a]Recommended intake according to DRI 20015 or to DV nutrient reference amount used on food labels, relevant to a 2000 Kcal diet.[43]
[b]Based on a 2000 Kcal diet; for other caloric levels, recommendation is no more than 30% of calories from fat and 10% from saturated fat.
[c]Synthetic, prior to 1998 grain fortification with folic acid.
[d]Nonfood sources.

reported 33% of total calories from fat and 11% from saturated fatty acids. Total fat intake is now in the acceptable macronutrient distribution range,[50] but the saturated fat intake of adults still exceeds the "no more than 10% of calories" guideline. An additional recommendation is to keep trans-fat consumption as low as possible; trans-fat intake data are changing rapidly because labeling laws led to changes in food manufacturing and restaurant practices.

Fiber Average daily dietary fiber recommendations are 38 grams for men and 25 grams for women, or 14 grams of dietary fiber per 1000 calories.[49] Daily median intakes are 16.5–19.5 grams for adult males and 12.1–13.8 grams for adult females, clearly below the recommendations. Increasing fruit, vegetable, and whole grain intakes could help to meet food guide recommendations. Chapter 18 includes examples for increasing fiber intake.

Folic Acid The evidence for folic acid in the prevention of neural tube defects is so strong that the Food and Drug Administration mandated folic acid fortification of grains and grain products beginning in 1998. Dietary surveys show that fortification is working; mean intake has increased to 327 micrograms for 20 to 39-year-old women and 435 mcg for 20 to 39-year-old men, compared to the 400 mcg DRI.[47] Some evidence also supports folic acid as a way to lower risk of heart disease. (More on folic acid in Chapter 18.)

Vitamin E This versatile antioxidant has been implicated in prostate cancer, heart disease, and cataract risk reduction. The UL for Vitamin E is 1000 mg/day, but a 2005 meta-analysis found that high-dosage (400 IU or more) supplements were associated with increased all-cause mortality.[51] Food sources of Vitamin E include sunflower seeds and oil, walnuts, wheat germ, olive oil and mayonnaise, avocado, and kale. (More in Chapter 18.)

Calcium Adults could meet calcium goals by adding a glass of milk, 8 oz yogurt, or 8 oz calcium-fortified orange juice to their diets daily. Eight ounces of milk contain roughly 300 mg calcium, regardless of butterfat. Some milks and yogurt provide additional calcium (>30% of DV per serving) because additional milk solids have been used to give

Table 16.9 Optimizing calcium intake for osteoporosis prevention

- **Vitamin D:** Adequate intake of Vitamin D (1,25-dihydroxyvitamin D, also called calcitriol) stimulates active transport of calcium in the small intestine and the colon.
- **Sodium:** Lower levels of sodium intake lead to lower levels of urinary calcium; that is, less calcium is excreted while consuming a low-sodium diet.
- **Lower protein intakes:** Excess protein leads to greater urinary excretion of calcium. In the Nurses Health Study, women (aged 35 to 59) who ate more than 95 grams protein per day suffered more osteoporotic forearm (but not hip) fractures over 12 years.[55]
- **Caffeine:** Two to three cups of coffee daily in postmenopausal women, when consumed with less than 800 mg/day of calcium, were associated with bone loss.[53]
- **Timing:** Drink milk and take supplements with a meal because food slows intestinal transit time and allows more calcium to be absorbed from the gut. Calcium carbonate is best taken with meals; calcium citrate may be taken at any time. In addition, stomach acidity enhances absorption; therefore, calcium is better absorbed when not consumed with antacids.
- **Moderation:** Consuming nutrients at recommended levels (enough but not too much) helps to synthesize and develop collagen matrix into which minerals are deposited during bone mineralization.
- **Other vitamins:** Vitamin C plays a role in the development of the protein bone matrix (collagen), but has no defined role in treatment. Vitamin B_6 is a cofactor in stabilizing collagen cross-links and has been linked to preventing hip fractures. Vitamin K is required for the formation of proteins that stimulate osteoblasts to build new bone and attract osteoclasts to initiate bone reabsorption (e.g., glutamic acid residues are carboxylated to form gla-proteins; most studied is osteocalcin).

the product extra body and flavor. See Table 16.9 for strategies that optimize calcium intake.

Calcium is a public health concern for several reasons. Calcium plus vitamin D and magnesium help to develop and maintain bone density, which delays osteoporosis and reduces risk of bone fractures. Adequate calcium is a potential contributor to lowered risk of colon cancer. Milk as part of the DASH diet (see Chapter 19) has been successfully used as part of blood pressure reduction intervention.[52] Calcium's crucial role in metabolism is supported by physiological mechanisms to keep calcium in tight balance including calcium intake, amount of calcium absorbed, and calcium excreted through urine and feces. Of these factors, intake and fractional absorption of calcium account for 25% of the variance in calcium balance, while losses through urine account for another 50%.[53,54]

Because protein and sodium promote urinary calcium excretion, eating a diet high in protein and sodium (like the typical adult diet) contributes to calcium losses through urine.[55] Other factors that decrease calcium absorption are ingestion of glucocorticoids (used to treat inflammatory diseases such as rheumatoid arthritis, asthma, and inflammatory bowel disease). As the body ingests higher levels of calcium, the percentage absorbed decreases. Conversely, as calcium intake drops, the percentage absorbed increases. A larger portion will be absorbed from a 500 mg per day intake than from a 1500 mg intake.

The tolerable upper limit for calcium is 2500 mg. It would be difficult to exceed this from food sources alone. For example, the basic diet without dairy products equals 400 mg calcium. Four 8 oz glasses of milk = 1200 mg calcium; 1 oz cheese = 250 mg. These sources add up to approximately 1850 mg calcium. However, supplement usage adds up more quickly.

Calcium carbonate is the cheapest form of calcium supplement. However, taking it can decrease or interfere with iron absorption. Citrate- or ascorbic-acid-bound calcium supplements, on the other hand, lead to greater iron absorption. Citrate and ascorbic acid calcium supplements are more easily absorbed than calcium carbonate.[57] They usually cost more, too. Calcium and its binder are bulky, so multivitamin and vitamin–mineral supplements tend to contain little calcium. Calcium is best absorbed when taken in divided doses. Take calcium carbonate with food, but take calcium citrate any time.

Magnesium   Poor diet, diabetes, or prolonged illness can cause magnesium deficiency, with adverse effects on bone strength, heart health, and metabolic syndrome. Magnesium deficiency tends to be more problematic in older adults (see Chapter 18). Good magnesium sources are sunflower seeds, almonds, beans, milk, whole grains, spinach, and bananas.

Iron   For women, iron needs drop from 18 to 8 mg per day at menopause. Iron deficiency can be one of the nutritional causes of anemia, although there are non-nutritional causes as well. The benefits of sufficient iron are healthy blood cells that carry oxygen for metabolism and overall energy; drawbacks of excess iron are greater need for antioxidants. High-iron foods include red meats, fortified cereals, strawberries, and dried beans as in baked beans, bean soup, bean dip, and chili. Changing from a high-iron to a low-iron cereal during the menopausal transition can help to adjust for the dramatic shift in iron needs.

Actual Intake of Food

Food Groups and Portion Size   How are Americans doing when it comes to meeting MyPramid guidelines? A pyramid top-heavy in sweets reflects the American plate. Americans are consuming more fruits and vegetables,

Case Study 16.1

Run, Kristen, Run

Photo Disc

Kristen, who was active in competitive sports throughout high school, has decided to run a marathon with some of her college friends. She is 25 years old, 5 ft 8 in tall, and weighs 135 pounds. She eats all sorts of foods, likes fruits and vegetables, but tries to avoid greasy foods. She doesn't like sweets, although she keeps ice cream in her freezer. A family history notes that her mother needed angioplasty to treat occluded arteries shortly after menopause and that her father is not at risk for any chronic conditions. Although she would eventually like to have children, Kristen is not pregnant now.

An analysis of a 24-hour dietary recall shows the following:

2090 calories
377 g carbohydrate (41 g total fiber)
98 g protein
33 g fat (7 g saturated fat, 1 g trans fat,
   1.5 g omega-3 fatty acid, 99 mg cholesterol)
3343 mg sodium
958 g calcium

34 mg iron
158 mg vitamin C
213 IU of vitamin D
35 IU of vitamin E
1548 mcg of folic acid
6 mg of pantothenic acid

Questions

1. How many calories does Kristen need to maintain her weight? Is she eating enough to support daily workouts?
2. Describe three health-promoting aspects of Kristen's diet.
3. Make three suggestions that could improve Kristen's diet.
4. What types of performance-enhancing products might complement this diet?

---

although intake from non-nutrient-dense calories still exceeds recommendations.[3,6] CSFII tracked the percentage of individuals who consume foods from each food group. Among 20- to 29-year-old adults, fewer than half eat fruit, but four of five eat vegetables daily. Two-thirds (67.5% of males) to three-fourths (74.5% of females) consume some milk or milk products every day. On any given day, 7% to 8% eat nuts and seeds, 11% to 13% eat beans and other legumes, 40% to 50% eat sugars and sweets, and 62% to 68% consume carbonated soft drinks.[50] However, the movement to super-size restaurant and take out foods (e.g., fries, drinks, rolls, and cookies) has led to "portion distortion." For example, a review of food intake over 19 years prior to 1997 found that portion sizes increased for salty snacks, desserts, soft drinks, and especially fast foods.[57] Within fast food meals, the largest increase was 133 calories for Mexican foods. Try to estimate your next fast-food meal in pyramid portions. It is not easy to estimate a "normal" portion in a value-conscious environment. Getting a lot more product for just a little extra money may be nice when it comes to soap; it can be detrimental when applied to restaurant portions because we tend to prefer not sharing our restaurant orders.

Using NHANES data, Kant calculated that 27% of the average individual's calories are eaten as energy-dense, nutrient-poor (EDNP) foods, that is, the extras.[6] Add to that the 3–4% of calories alcohol provides, and it turns out that we rely on 69–70% of our calories for the bulk of our vitamins and minerals. This picture has not improved since Kant's analysis. An analysis of subsequent NHANES data from 2001–2002 found that the percentage of solid fat, alcohol, and added sugar (used as a proxy for discretionary calories) ranges from 30% (for women over age 51) to 42% (for men, age 19–30).[58] With this type of eating pattern, 60–70% of our calories must supply 100% of recommended vitamins and minerals. No wonder we are looking for super foods.

> "Now cooks think it is silly to invest 2 hours' work in 2 minutes' enjoyment; but if cooking is evanescent, well, so is the ballet."
>
> Julia Child, TV chef and master cooking teacher

Eating Out Whether for ease and convenience, entertainment, or just the pure pleasure of sharing good food with friends and family, more and more people are eating away from home. For example, 62–72% of 20- to 29-year-olds consume more than a third of their calories in food prepared away from home.[59] The National Restaurant Association estimates that by 2010, 53% of food dollars will be spent away from home, with fast food restaurants as the fastest-growing type of operation.[59] The restaurant industry share of the food dollar has nearly doubled in the last 50 years, from 25% in 1955 to approximately 46% in 2006.[60] Eating greater numbers of fast food meals is associated with higher total and fat calories intake.[61] Eating foods prepared away from home is also associated with eating fewer fruits and vegetables, except for fried potatoes.[62]

Supplements: Vitamins and Minerals

Adults can eat nutritiously without dietary supplements. Circumstances under which dietary supplements may enhance peak functioning are childbearing, high-level athletic performance, illness, or when eating restricted diets—for example, a wheat-free diet due to allergies or a calorie-restricted diet for weight loss. Adults might consider supplementing dietary gaps (potentially calcium, iron, magnesium, vitamin E, and folic acid in women, plus vitamins B_{12} and D for non-meat eaters and vegetarians).

CSFII data reports that 42% of males and 56% of females aged 20 and older consume a vitamin or mineral supplement daily, nearly every day, or "every so often."[50] Supplement use by males gradually rises with age (37–45% of individuals from age 20 to age 70+). Female usage rises from 51% of individuals at age 20 to a peak of 62% between ages 50 to 59 and then dropping to 54% by age 70 and older. The greatest percentage of males who take supplements use a multivitamin. For females, the pattern shifts: a multivitamin is most popular until age 40, after which single vitamins and minerals are used more. At age 70 and older, 23.4% of women use a single vitamin or mineral and another 23.4% of women use a multivitamin. Findings in NHANES are slightly different than CSFII, showing that 40% of the U.S. population (2 months and older) took at least one vitamin or mineral supplement in the previous month (35% of males and 44% of females). The most commonly found nutrients in supplement products are ascorbic acid, vitamin B_{12}, vitamin B_6, niacin, thiamin, riboflavin, vitamin E, beta carotene (vitamin A precursor), vitamin D, and folic acid.[63]

Supplements and CAM (Complementary and Alternative Medicine)

Herbal and other botanical products are often grouped with nutritional supplements as adjuncts of complementary medicine. Table 16.10 lists examples of some of the more popular herbals and suggests potential uses and side effects.

Herbs and foods with special nutritive and medicinal attributes (functional foods) comprise a rapidly growing and largely unregulated market. *When in doubt, throw it out* is a food safety education message that seems appropriate for herbal products as well.

The Food and Drug Administration has banned several herbs, although they occasionally appear on the market (sassafras, yohimbe/yohimbine, comfrey, chaparral, lobelia, ephedra). Others that are unsafe and/or ineffective include apricot pits (cyanide), belladonna, blue cohosh, coltsfoot, dong quai or angelica, ephedra or ma huang, eyebright, garcinia, life root, mistletoe, pokeroot (except the berries), wild yam, willow bark, and wormwood.[18,19] Additional information on the complex topic of herbal supplements can be found via Web sources listed at the end of the chapter.

Cross-Cultural Considerations

The health and nutrition goals outlined in Healthy People 2010 are designed to decrease health disparities among ethnic groups. The traditional diets of diverse cultural groups in North America can serve as models for contemporary eating patterns. For instance, diets of Native Americans in the southwestern part of the United States were based on corn, beans, and squash, the three "sister vegetables." These form a solid nutritional base.

Here are a few examples of dietary patterns from various cultures:

1. Asian diets tend to be high in carbohydrate, low in fat, especially in saturated fats. The diet includes many varieties of fish. Beans, including fermented soy products, supply protein, fiber, and antioxidants. Even Americanized Vietnamese restaurants serve many low-fat, high-vegetable choices. Spring rolls with plum and chili sauce are a nonfried alternative to deep-fried egg rolls. Thai dishes are based on vegetables, noodles, and fish. Fish sauce, although salty, provides flavor and adds protein without fat. Peanut sauce and coconut milk are high in fat and flavor, and usually eaten in small amounts. Peanuts provide protein.

2. Indian diets offer many vegetarian choices, are low in saturated fats and high in fiber. Extensive use of spices provides flavor without added fat.

3. North African cuisine (for example, Amharic, Oromo, Ethiopian, Egyptian) relies heavily on beans and whole grains. Injera or buddeena, the flat breads that feel a bit like a pancake, hold dabs of lentils, split peas, lamb, beef, and yogurt. Sharing is customary. Each person pulls off a piece of the buddeena and uses it to pick up fillings such as peas and lentils (dahl), beef, lamb, or vegetable mixtures. Picture the amount of filling you can pick up with a piece of soft bread. You would be eating pyramid style—mainly grain complemented with a little animal protein.

Table 16.10 Dietary supplement and herbal remedies with proposed claims and potential medication and side effects

Dietary Supplement/Herbs/Other Remedy	Proposed Claims	Potential Side Effects	Potential Interaction with Prescription Medications
Cholestin	Maintains desirable blood cholesterol levels	Safety of some ingredients unknown	
Creatine	Sport supplement (increased performance in short, high-intensity events); inconclusive claims regarding use in congestive heart failure	Kidney disease; side effects possible and include vomiting and diarrhea	
DHEA	Improves physical well-being throughout aging	Increases risk of breast and endometrial cancer in women and prostate cancer in men	
Echinacea	Prevents and treats colds and sore throat; Anti-inflammatory	Allergies to plant components in daisy family; affects oral cavity integrity; depresses immune system if taken longer than 6 weeks	Corticosteroids
Ephedra (Ma Huang)	Promotes short-term weight loss	Insomnia, headaches, nervousness, stomach problems, heart palpitations, seizures, death	Corticosteroids, digoxin, monoamine oxidase inhibitors (MAOI); oral hypoglycemics
Garlic	Lowers blood cholesterol; relieves colds and other infections, asthma	Heartburn, gas, blood thinner	Oral hypoglycemics; blood thinner[a]
Ginger	Calms stomach upset, fights nausea	Central nervous system depression, heart rhythm disturbances if using very large doses	Oral hypoglycemic medications; blood thinners
Ginkgo biloba	Increases mental skills, delays progression of Alzheimer's disease, increases blood flow, decreases depression	Nervousness, headache, stomachache, interacts with blood thinners	Thiazides; blood thinner (e.g., warfarin); MAOI
Ginseng	Increases energy, normalizes blood glucose, stimulates immune function, relieves impotence in males	Insomnia, hypertension, low blood glucose, menstrual dysfunction	Estrogens, insulin, oral hypoglycemics, blood thinner;[a] monamine oxidase inhibitors (MAOI)
Glucosamine and chondroitin	Slows progression of joint space narrowing with long-term use; may relieve joint pain and improve mobility	Gastrointestinal upset (i.e., gas, soft stools, diarrhea, nausea, indigestion, heartburn) and allergic reactions in individual with shellfish allergy	Blood thinner (chondroitin only)
Peppermint	Treats indigestion and flatulence, spasmolytic (relaxes muscles), antibacterial agent	Heartburn allergic reactions	
Saint John's wort	Relieves depression	Dry mouth, dizziness, interacts with many drugs	Cyclosporin, digoxin, iron supplements, oral contraceptives, selective seotonin re-uptake inhibitors; monamine oxidase inhibitors (MAOI)
SAMe	Relieves mild depression, pain relief for arthritis	May trigger manic excitement, nausea	
Saw palmetto	Improves urine flow, reduces urgency of urination in men with prostate enlargement	Nausea, abdominal pain	Estrogens

SOURCE: Table compiled from Shekelle PG, Hardy ML, and Morton SC et al. "Efficacy and Safety of ephedra and ephedrine for weight loss and athletic performance: A Meta-Analysis," *J Am Med Assoc.* 289(12), 2003:1537–45; *AACE Guidelines.* Medical guidelines for the clinical use of dietary supplements and nutraceuticals. *Endocrine Prac.* 2003;9(5):418–470, Table 7; Brown JE, *Nutrition Now.* 3rd ed. Belmont, CA: West/Wadsworth Publishing Company, 1999, Table 24.4; Stupay S, and Sivertsen L. Herbal and nutritional supplement use in the elderly. *Nurs Pract.* 2000;25(9):56–67.
[a]Blood thinners include aspirin, warfarin, coumarin. Blank boxes denote unknown.

4. Caribbean dishes are made with lots of potatoes, vegetables such as spinach, fruit, and small portions of meat, often chicken.

5. Traditionally prepared Mexican food is based on corn, grains, beans, vegetables, and small amounts of meat and fat. Loads of cheese and sour cream on chips with some pickled jalapeno slices are a U.S. invention.

When population groups migrate to another country, nutritional problems may develop as acculturation results in newly adopted lifestyles. For instance, California and Minnesota saw a large influx of Hmong (pronounced "mung") refugees who had been displaced from their farms in Cambodia and Laos after the Vietnam War. Hmong youths (but not their elders) quickly adopted American food habits. However, many families were no longer pursuing their physically active lifestyles. Overweight and obesity have increased dramatically in Hmong youths since coming to America.

Adopting the prevailing cuisine of the new country is part of acculturation. In a study of eating habits of mostly first-generation Korean immigrants (average age 41 years; time in United States 15 years), Korean foods were consumed nearly twice as often as American foods. Consumption of Korean foods declined with adoption of American culture.[64]

Cross-Cultural Dietary Guidance

Several years ago, public health nutritionists Patricia Splett and Karen Zeleznak worked with the Women, Infants, and Children's program (WIC). As they thought about ways to teach health-promoting eating habits to pregnant and breast-feeding women, some of whom did not know English, they decided to ask potential clients about how they perceived food. How did they perceive the role of foods? What might they wish to learn about foods? Their clients wanted to know "What will this food do for me?" So these nutritionists developed a set of posters (see Illustration 16.4) that grouped foods according to function:

1. *Staple foods:* Grains, cereals, and starchy roots provide energy and bulk to the diet. They are the filling basis of the diet and, barring natural disasters or bad luck, are readily available. See Table 16.11 for nutrient profiles of selected staple grains.

2. *Body-building foods:* Beans, lentils, peas, nuts, seeds, eggs, milk, cheese, fish, poultry, and red meats are all high in protein and provide amino acids for tissue building and repair.

3. *Protective foods:* Vegetables and fruits contribute vitamins and minerals, antioxidants, and other phytochemicals to keep the individual healthy.

In their WIC program, grouping food according to function dealt with the milk/dairy dilemma faced by some Asian and African American audiences. Milk, cheese, calcium-fortified soy milk, or yogurt can all be placed in the protein or "body-building" group. The "staple" group depicted grains and energy foods for several local cultures. In cross-cultural nutrition work, it is important to find out what the other person or group thinks and believes. Go ahead. Ask.

Vegetarian Diets

How many people consider themselves to be vegetarian? Among individuals aged 18 and older, 2.3% consider themselves "vegetarian."[65] A little over 1% of the population eats no animal products (that means approximately 40% of vegetarians are vegan), roughly 7% eat no meat

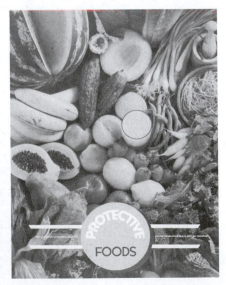

Illustration 16.4 Foods grouped according to function for cross-cultural nutrition counseling.

Photos Courtesy of General Mills, Inc.

Table 16.11 Staple grains around the globe and selected nutrient contributions

Grain, ½ cup cooked unless noted otherwise	Kcal	Carb	Fiber	Protein	Calcium, mg	Iron, mg
Amaranth, 2 Tbsp dry	91	16	2.5	3.5	38	1.8
Barley, ½ cooked	97	22	3	1.8	9	1
Buckwheat, groats	77	17	2.3	2.8	6	0.7
Corn as polenta, grits	73	16	0.3	1.7	2	0.8
Millet	104	21	1	3	5	0.6
Oats, oatmeal	73	13	2	3.0	12	0.8
Quinoa	79	15	1	2.8	15	2.0
Brown rice	108	23	1.8	2.3	10	0.5
White rice	103	22	0.3	2.1	8	1.0
Wild rice	83	18	1.5	3.3	2	0.5
Wheat, bulgur	76	17	4.1	2.8	9	0.9

SOURCES: Nutrient Data Laboratory Food Composition Data. United States Department of Agriculture Agricultural Research Service. Available at: www.ars.usda.gov/main/site_main.htm?modecode=12354500, accessed October 27, 2006; and University of Minnesota NDSR, version 2.6, September 2006.

(with more women abstaining from meat), 9% never eat eggs, 15% never eat seafood, and 6% never eat poultry. One difficult thing about polling people about vegetarianism is defining just exactly what it means to be a vegetarian. Table 16.12 classification gives an overview.

Individuals can achieve a high-quality diet whether or not they eat meat. Often, people eating vegetarian diets are health conscious in other areas of their lives as well. Can adopting a vegetarian diet make one healthier? Yes. A well-chosen vegetarian diet is associated with decreased mortality and morbidity.[66] Potential health benefits of vegetarian diets include:

- Disease specific benefits
  - Decreased risk of mortality and symptoms of heart disease
  - Lower incidence of hypertension
  - Improved risk profile for kidney disease (lower glomerular filtration rates)
  - Decreased risk of prostate and colorectal cancers
- General diet quality
  - Higher intake of vegetables and fruits improves overall nutritional risk picture
  - Diet tends to be lower in saturated fats and may also be lower in calories
- Environmental benefits
  - Vegetable foods are lower on the food chain than meat, fish, and poultry
  - Depending on packaging/transportation, food production uses fewer resources

Nutrients singled out for evaluation in vegetarian diets include protein, vitamins B_{12} and D, calcium, iron, and omega-3 fatty acids.

Table 16.12 Classification of vegetarians by food groups consumed

Type of Vegetarian	Foods Not Eaten	Foods Eaten
Vegan	No animal foods of any kind, no honey	Grains, beans, nuts, seeds, nut butters, fruits, vegetables, sugar, molasses, oils, margarine, soda, alcohol, soy analogs (e.g., textured vegetable bacon, burgers, "meats")
Lacto-vegetarian	No meat, poultry, fish	Above plus milk and other dairy products, cheese, yogurt, butter
Lacto-ovo vegetarian	No meat, poultry, fish	Above plus eggs of any sort
Vegetarian	No beef, pork, lamb, venison, buffalo, other red meat; "No red meat"	Depends on individual interpretation: fish, both fin fish and shellfish; poultry and game birds may be allowed

1. *Protein:* Protein adequacy is based on absolute intake of protein, an adequate caloric intake, and sufficient amounts of amino acids per unit of body weight.[67] Protein absorption rates vary, suggesting that eating a

Table 16.13 Examples of nonmeat protein foods and serving sizes used in MyPyramid

Protein Food Serving	Protein Content grams	Total fat, grams	Kcal
Almonds, 2 Tbsp whole	3.5	9.3	105
Peanuts, 2 Tbsp dry-roasted	4.8	9.0	106
Peanut butter, 2 Tbsp	8.1	16.5	191
Sesame seeds, 2 Tbsp	4.2	8.8	94
Egg, 1 large	6.2	5.0	74
Kidney beans, ½ c cooked	7.7	0.4	112
Soy milk, 1 c	6.6	4.6	79
Lentils, ½ c cooked	8.9	0.4	115
Tofu, firm and soft, ¼ c	5.7	3.4	54
Cheese, 1.5 oz	10.6	14.1	171

variety of proteins throughout the day helps to maintain amino acid balance. The low level of lysine in cereal grains can be complemented by beans (most legumes) and, to some extent, greens. However, eating complementary proteins at each meal is not necessary.[68] Generally, protein intake of vegetarians appears to be adequate, despite lower overall protein quality and intake. See Table 16.13 for nutrient content of several high-protein foods from MyPyramid.

2. *Vitamin B₁₂:* Plant foods are not reliable sources of vitamin B_{12}; most of the B_{12} in sea vegetables and fermented soy is an inactive analog. B_{12} is found in fortified ready-to-eat breakfast cereals, fortified soy milks, nutritional yeast grown on a special medium (like Red Star), and meat analogs. (Check the label; they vary!)

> **Linseed** From the flax plant, also used to make linen, *linum* is another name for flaxseed. Linseed oil is used in paints, varnishes, and inks but is also produced in food form for its rich nutrient content.

3. *Vitamin D:* One good source of D is sunshine (all year in the south; during spring, summer, and fall in northern United States and Canada). Vitamin D formation is blocked by sunscreen. Another source is Vitamin D–fortified milk providing 2.5 mcg (100 IU) per cup of the 10 mcg (400 IU) recommended per day. Fortified cereals, fortified soy and dairy milks, and vitamin tablets are reliable sources of vitamin D.

4. *Calcium:* Absorption varies among foods. The calcium content in vegetables such as spinach and fruits such as rhubarb looks impressive in a food table, but doesn't reach the bones to the extent that calcium from dairy sources does. Some dark leafy greens such as collards are low in oxalic acid so that the calcium can be better absorbed. No general rule fits calcium absorption. Fortunately, vegetarians do not tend to have excessive

protein intakes. If they also keep sodium intakes low, they will excrete somewhat less calcium. The vegetarian food guide pyramid in Illustration 16.5 includes a separate component depicting calcium-rich foods.

5. *Iron:* Enriched flour and breads, beans, and fortified cereals, especially when eaten in the presence of vitamin C and not washed down with coffee or tea, can provide all the iron a healthy adult needs. An example of how to reach the 8 mg of iron needed daily by all adults except premenopausal women (who need 18 mg) is to include two slices whole wheat bread (1.8 g), 6 oz tomato juice (1 g), ½ cup cooked chard (1.9 g), ¼ cup kidney beans (0.75gm), and 2 tablespoons of pumpkin seeds (2.5 g) in the day's intake.

6. *Fats:* For vegetarians who consume eggs and dairy, chosing low-fat dairy products will help to keep the saturated fat within dietary guidelines.[66] Most plant foods do not contain significant amounts of omega-3 fatty acids, as fish and eggs do. However, good plant sources of omega-3 fatty acids are ground flaxseed and *linseed* oil, walnuts, walnut oil, canola oil, and soybean oils.[66]

The American and Canadian Dietetic Associations suggest that a health-promoting vegetarian diet include a variety of foods from each basic food group. Whole, unrefined foods are more nutrient-dense than highly processed ones. They contain phytochemicals that were not lost in processing, and about which we know too little to replace appropriately. For example, red grapes contain resveratrol, a compound implicated in heart health and in reducing cancer risk. Real grapes contain resveratrol; manufactured imitation grape beverages do not.

Physical Activity Recommendations

"I could have danced all night!"

A 2006 review of the health benefits of physical activity concluded that (1) current Canadian physical activity guidelines are sufficient to gain health benefits, especially in previously sedentary individuals, and (2) physical activity can serve as primary and secondary prevention strategy for several chronic diseases including cardiovascular disease, diabetes, and hypertension.[70] Physical activity is so good for us, surely we can assume that everyone is exercising?

Not exactly. See Table 16.14 for the Healthy People recommendations and baseline activity levels. CDC data shows that inactivity in the United States has decreased slightly in the last 10 years.[71] Prevalence of leisure-time inactivity varied by race and ethnicity, and ranged from 18% to 40% for adults age 18 and older in 2004. Compare this with inactivity levels of 25% to 46% in 1994. The trend is in the right direction.

Vegetarian food guide pyramid

Vegetarian food guide rainbow

Illustration 16.5   Vegetarian food guide pyramid, highlighting calcium-rich foods.

SOURCE: A new food guide for North American Vegetarians, reference #69, also on the Web at www.eatright.org.

Table 16.14 Healthy People 2010 physical activity goals and baseline activity levels

	Baseline Levels		Target
	Age 18 and older	Age 25–44	All adults
Engage in no leisure time physical activity	40%	34%	20%
Engage in regular physical activity at least 20 minutes, three times per week	15%	15%	30%
Perform strengthening activities two or more days per week	19%	22%	30%
Perform stretching and flexibility activities	30%	34%	40%

How much activity will yield benefits? CDC, together with the American College of Sports Medicine, suggests "Adults should engage in moderate-intensity physical activities for at least 30 minutes on 5 or more days of the week."[72] Their definitions for moderate-intensity activity are: any activity that burns 3.5–7 Kcal per minute. Warburton and associates cite evidence that the higher end of this caloric expenditure range (1000 calories per week) is linked to health benefits, and suggest that the 500 calories per week expenditure level needs further study to determine if health benefits result.[70]

The Healthy People 2010 goals spell out benefits and guidelines that are similar to the CDC recommendations:[3]

Because the highest risk of death and disability is found among those who do no regular physical activity, engaging in any amount of physical activity is preferable to none. Physical activity should be encouraged as part of a daily routine. While moderate physical activity for at least 30 minutes is preferable, intermittent physical activity also increases caloric expenditure and may be important for those who cannot fit 30 minutes of sustained activity into their daily schedules.

Resistance training (at least 30 minutes per day, two to three times per week) develops muscle mass, strength, and endurance.[73] Benefits of physical activity begin to diminish in 2 weeks of inactivity and disappear after several months, so don't stop exercising.

"Shall we dance?"

Nutrition Intervention for Risk Reduction

A Model Health-Promotion Program

Nutrition intervention occurs through the provision of food as well as by providing counseling and education to help improve nutritional status. Food pantries, community kitchens, and the 5 A Day for Better Health campaign (More Matters) are all examples of nutrition interventions that mobilize individuals and communities to improve nutritional status.

An example of a health promotion program for adults illustrating some of the concepts of nutrition intervention is Health Partner's Better Health Restaurant Challenge.® Health Partners is a large midwestern health maintenance organization (HMO).

- *Health goal:* Improve nutritional health by enabling HMO members to choose tasty, health-promoting, low-fat foods when eating away from home.
- *Target audience:* Health Partners sees its mission as serving the whole community, so the whole metropolitan community was eligible to participate in the program.
- *Theoretical basis:* Health behavior change depends on social support as well as on individual goals and commitments. Health Partners wanted to build social networks that enabled healthy choices.
- *Intervention strategies:* Participating restaurants (sit-down and fast food) were challenged to feature health-promoting items on the menu. Diners voted on favorite items; the HMO compiled and publicized the restaurants and the winning items. The HMO provides additional publicity for restaurants in its publications and through local media advertising, offering discount coupons for members to visit participating restaurants.
- *Evaluation:* Member satisfaction increased program participation from year to year, and a growing list of restaurants wished to join the program. Member surveys addressed the nutritional goal of enabling access to tasty, health-promoting menu items by asking about restaurant ordering patterns and satisfaction with featured menu items. Diner surveys revealed that 90% of diners would order low-fat items again, and 50% thought low-fat items tasted better than expected.

Public Food and Nutrition Programs

One of the aims of national health goals is to minimize health disparities; another is to improve the food security of the population (see Table 16.15). Living in poverty is linked to poor diets and to adverse health outcomes such as increased rates of obesity.[3] The Census Bureau has determined that approximately 11% of adult U.S. residents, age 18 to 64, live in poverty.[74] The poverty threshold is derived from calculating the cost of foods needed for basic dietary requirements (according to the Thrifty Food Plan) and multiplying that cost by 3. This poverty index was developed in the early 1960s by Mollie Orshansky of the Social Security Administration; at that time food costs made up about

Table 16.15 Healthy People 2010 food security goals for adults compared to baseline levels

The following applies to all adults, age 20+	Target	Population Percentage Baseline Females	Males
Increase food security and in so doing, decrease hunger			
All households at or <130% of poverty	94	69*	69*
All households >130% poverty	94	94*	94*

SOURCE: Food and Drug Administration and National Institutes of Health. Healthy People 2010: national health promotion and disease prevention objectives. Washington, DC: U.S. Department of Health and Human Services; 2000.
*Adequate data to distinguish household food security levels for males and females is not available.

one-third of the average household budget. Currently, households spend 9–34% of after-tax income on food, depending on household income. Poor households spend disproportionately higher levels of their income on food.

For 2006, a person was considered to live in poverty if their annual income was $9800 or less.[75] Poverty guidelines are calculated according to household size and adjusted for higher living costs in Alaska and Hawaii. For a four-person household, the poverty guideline is $20,000 in the contiguous states and washington, D.C., $25,000 in Alaska, and $23,000 in Hawaii. The federal poverty guideline is published annually in the Federal Register. It is used to determine eligibility for food and nutrition assistance programs. Benefit levels can be adjusted to meet various household needs. For instance, the Child Nutrition Program (school lunch, breakfast), which can help whole families stretch their food budget, provides free lunches to children living in at 130% and less of poverty and reduced-cost meals when family income is between 130% or 185% of poverty. Families must apply for these programs.

By far the largest part of all nutrition assistance programs is the Food Stamp or Food Support program, as some states call it to reduce the stigma attached. The five largest food assistance programs are all administered by the United States Department of Agriculture (www.fns.usda.gov; in 2004 dollars):

1. Food Stamps or Food Support (over 27,154 million dollars)
2. National School Lunch Program (7627 million dollars)
3. Women, Infants, and Children's Program (4890 million dollars)
4. Child and Adult Care Food Program (2021 million dollars)
5. Nutrition Assitance Program for Puerto Rico (1413 million dollars)

Many other programs work to help hungry individuals and families gain food security. For example:

- Governmental extension programs teach budgeting, shopping, and food handling skills.
- The Second Harvest food bank network keeps food items out of the waste stream, coordinates charitable giving programs, and supplies food shelves and community kitchens.
- Soup kitchens and shelters provide hot meals and snacks for hungry and homeless people.
- Meals-on-Wheels programs serve homebound adults.

Together, governmental and private organizations help individuals gain consistent access to safe, wholesome foods that are culturally acceptable. Such access is the basis of food security.

Nutrition and Health Promotion

What are the most healthful nutritional habits that we can develop? The general message is to follow the principles of variety and moderation in choosing a diet that will achieve and maintain a healthy body weight when combined with activity.

And how does our national diet rate? There is room for improvement. The USDA's Center for Nutrition Policy and Promotion designed the Healthy Eating Index (HEI) in 1995 in order to monitor the dietary status of Americans;

Heart Disease: Good News and Bad News
The good news is that deaths from heart disease have been declining. The bad news is that in the United States the leading cause of morbidity and mortality is still heart disease. The American Heart Association estimated the U.S. financial burden of heart disease at $403.1 billion in 2006, which includes the cost for health professionals, hospitals and nursing homes, medications and lost productivity.[77] Those billions translate to $1343 per person per year. Add to that the personal loss when a family member dies prematurely.

Definition
Heart disease is a common term for conditions affecting the heart, the blood vessels, and the flow of blood (including platelet stickiness, blood clots that block blood flow or lead to a ruptured blood vessel). These are discussed in greater detail later.

Prevalence

But isn't heart disease really a problem for old people? Yes, one-third of adults aged 65 and older die of heart diseases, while only 13% of 25- to 44-year-olds do. You will read more about cardiovascular disease prevalence, risks, effects, and nutritional interventions for younger and older adults in Chapters 17 and 19. But healthy young adults can benefit most from risk reduction strategies because heart disease is largely preventable, and it develops over many years. The plaque buildup that restricts blood flow doesn't happen overnight, and has been found in children as young as 10 years old. Youth is a great time to adopt preventive habits that keep arteries open and clean. The payoff now is a greater ability for the blood to carry oxygen and nutrients to all parts of the body, including the cells of the brain. Good blood flow supports energy for sport and activity, brain health for clear thinking, and healthy organs for optimum hormone balance and metabolic control.

Physiology and Risk Factors

Typically, when describing causes of heart disease, it is the risk factors associated with heart disease that come to mind. For example, analysis of data from the Framingham Study shows that certain factors can predict the relative risk of developing coronary heart disase over the next 10 years. These factors are: age of 55 or older, having total blood cholesterol of 250 mg/dL, HDL cholesterol of 39 mg/dL or less, blood pressure of 146/88 mm, having diabetes, and cigarette smoking. (See Chapter 17 to score your own risk.) However, approximately half of coronary events occur in individuals who have normal lipid patterns. Their blood cholesterol levels appear to be "healthy." Obviously, blood cholesterol is only part of the picture. Willerson and Ridker make the case for *chronic inflammation* (see Chapter 2) as a highly predictive heart disease risk factor.

Briefly, an atheromatous lesion develops when oxidized LDL cholesterol, injury, or perhaps infection damage the structure of the blood vessel's endothelium (lining), stimulating pro-inflammatory cytokines that induce expression of cell adhesion molecules and set in motion an inflammatory cascade response in the body. This cascade leads to the deposition of plaque (cholesthrol, macrophages, tissue factor, and, other cell debris) in the endothelium. Many inflammatory factors are involved in the cascade response, including: interleukin-6, a procoagulant cytokine that increases fibrinogen and C-reactive protein (CRP), interleukin 1, tumor necrosis factor (plays role in coagulation), cell adhesion molecules and CRP. Of these, CRP is widely available as a clinical marker and is highly predictive.[78] The worse (more likely to rupture) the plaques, the higher the levels of high-sensitivity CRP.

Prevention

The basis of preventing heart disease is to reduce LDL cholesterol and inflammation, which function synergistically in lesion development and plaque buildup. Stabilizing or raising HDL cholesterol helps to remove LDL cholesterol; keeping antioxidant levels in the blood high may prevent LDL-C oxidation; statins reduce CRP; and eating omega-3 fish oils, which help to reduce blood platelet stickiness, may help reduce plaque buildup and blood clotting.[79]

And this is where nutritional guidelines come into play. Our general food guidelines were developed to fight the leading killer diseases, especially heart disease. Young adulthood is not too early to start preventing heart disease.

---

a score of 80 or above is considered "good." In a 2003 assessment of older adults, HEI scores ranged from 61 to 67.[76] The ability to evaluate and track the health-promoting quality of our national diet helps to guide our progress toward national health goals.

Knowing does not automatically lead to doing. To make health-promoting behavior changes takes knowledge of desired behaviors, skills to practice them, belief in one's own ability to carry them out, and the intention or commitment to do so.

Key Points

1. Dietary habits during adulthood can raise or lower risk of chronic diseases.

2. Healthy People 2010 spells out the overarching goals for the nation's nutritional health.

3. Adulthood signals a change in nutritional focus from growth and development to maintenance of physiological health.

4. Balancing energy intake and physical activity helps to maintain a healthy body weight

and increase the chances of a long and healthy life.

5. Dietary guidance tools are designed to reduce risks of specific diseases, ensure that the population consumes adequate levels of required nutrients, and help the public to understand what, and how much, to eat.

6. The many different cultures that make up the United States have influenced the population's food

preferences and eating habits; effective dietary guidance accommodates diverse eating patterns.

7. Alcohol contributes calories to a diet and thus may be considered a nutrient. However, alcohol is also considered a drug due to its depressive effect on the central nervous system and potential toxic effects on the liver.

8. Individuals can survive only a few days without water. The recommended level of fluid intake is based on population intake data, and includes water ingested in food form.

9. Adults can be well nourished without vitamin or mineral supplements. However, when certain food groups are eliminated from the diet (such as meats, dairy products, or breads), supplementary vitamins and minerals can help to complete nutrient adequacy.

10. Individuals can achieve a high-quality diet with or without meat. Nutrients of special concern for adults on vegetarian diets are protein, vitamins D and B_{12}, calcium, iron and omega-3 fatty acids.

Resources

American Dietetic Association
Provides information, advice, and links.
Website: www.eatright.org

American Heart Association
Site includes assessments, eating advice, and definitions.
Website: www.americanheart.org

Calorie Control Council
Lists overall diet and fitness information. (Don't look for this site to sponsor the health-at-any-size movement.)
Website: www.caloriecontrol.org

Centers for Disease Control and Prevention
The main site also includes morbidity and mortality data at www.cdc.gov/mmwr.
Website: www.cdc.gov

Consumer Lab
Check out vitamin and mineral supplements. (Also use www.fda.gov.)
Website: www.consumerlab.com

Environmental Protection Agency
Among other things, ratings for local water supplies are available.
Website: www.epa.gov/safewater

Food and Drug Administration
Resource for labeling information, health claims on foods and supplements, fortification rules, and more.
Website: www.nutrition.gov
U.S. government's gateway site for nutrition information.
Website: www.fda.gov

U.S. Census Bureau
The site to go to for who we are and where we live.
Website: www.census.gov

U.S. Department of Agriculture
Wonder about your diet? This site includes a link to MyPyramid Tracker, an interactive online diet and activity assessment tool. Key in what you've eaten and see how your intake stacks up against the MyPyramid recommendations, for your age, gender, height and weight.
Website: www.usda.gov/cnpp

Vegetarian Resource Group
Provides resources for all age groups, recommended books, recipes, and a journal for subscribers.
Website: www.vrg.org

References

1. Mokdad AH et al. Actual causes of death in the United States, 2000. JAMA. 2004;291:1238–45.

2. Healthy People 2000: National health promotion and disease prevention objectives. DHHS Publication No. (PHS) 91–50212. Washington, DC: U.S. Department of Health and Human Services; 1991.

3. Healthy People 2010: National health promotion and disease prevention objectives. Wahsington, DC: U.S. Dept of Health and Human Services; 2000. Available at http://healthypeople.gov, accessed 12/06.

4. Murphy C et al. Prevalence of olfactory impairment in older adults. JAMA. 2002;288:2307–12.

5. MiwaT et al. Impact of olfactory impairment on quality of life and disability. Arch Otolaryngol Head Neck Surg. 2001;127:497–503.

6. Kant AK. Consumption of energy-dense, nutrient-poor foods by adult Americans: nutritional and health implications. The Third National Health and Nutrition Examination Survey, 1988–1994. Am J Clin Nutr. 2000;72:929–36.

7. National Academy of Sciences, Institute of Medicine, Food and Nutrition Board. Dietary Reference Intakes: recommended intakes for individuals. 7-page PDF avaailable at http://foodsafety.nal.usda.gov, accessed 12/06.

8. Forbes GB. Human body composition. Growth, aging, nutrition, and activity. New York: Springer-Verlag; 1987:31.

9. Position Statement. The role of calcium in peri- and postmenopausal women: 2006 position statement of the North American Menopause Society. Menopause. 2006;13(6):862877. Available at www.menopause.org, accessed 12/06.

10. Melby MK, Lock M, Kaufert P. Culture and symptom reporting at menopause. Human Reproduction Update. 2005;11(5):495–512.

11. Dennerstein L et al. Life satisfaction, symptoms, and the menopausal transition. Available from www.medscape.com/Medscape/WomensHealth/journal/2000/v05.n04/wh7254.denn-01.html, accessed 01/01.

12. Kronenberg F, Fugh-Berman A. Complementary and alternative medicine for menopausal symptoms: a review of randomized, controlled trials. Ann Inter Med. 2002;137(10):805–13.

13. Nelson HD et al. Nonhormonal therapies for menopausal hot flashes. JAMA. 2006;295:2057–71.

14. Nagata C et al. Soy product intake and hot flashes in Japanese women: results from

a community-based prospective study. Am J Epidemiol. 2001;153:790–93.

15. The role of isoflavones in menopausal health: consensus opinion of The North American Menopause Society. Menopause 2000;7:215–29. Available from www.menopause.org.

16. Wiseman H et al. Isoflavone phytoestrogens consumed in soy decrease F(2)-isoprostane concentrations and increase resistance of low-density lipoprotein to oxidation in humans. Am J Clin Nutr. 2000;72:395–400.

17. Fragakis AS. The health professional's guide to popular dietary supplements, 2nd ed. Chicago, IL: The American Dietetic Association; 2003.

18. Blumenthal M (senior ed.), Hall T (managing ed.). The ABC clinical guide to herbs. Austin, TX: American Botanical Council; 2003.

19. Schulz V et al. Rational phytotherapy: a reference guide for physicians and pharmacists. Berlin: Springer Verlag; 2004.

20. FDA/CFSAN Consumer Advisory. Kava-containing dietary supplements may be associated with severe liver injury. March 25, 2000. Available at www.cfsan.fda.gov/, accessed 12/06.

21. Harper EJ. Changing perspectives on aging and energy requirements: aging and energy intakes in humans, dogs and cats. J Nutr. 1998;128:2623S–26S.

22. McGandy RB et al. Nutrient intakes and energy expenditure in men of different ages. J Gerontol. 1966;21:581–87.

23. Guo SS et al. Aging, body composition, and lifestyle: the Fels Longitudinal Study. Am J Clin Nutr. 1999;70:405–11.

24. Food and Nutrition Board. National Academy of Sciences (Institute of Medicine) NRC, Subcommittee on the 10th Edition of the RDAs, Commission on Life Sciences. Recommended dietary allowances. Washington, DC: National Academy Press; 1989.

25. Harris JA, Benedict FG. Standard basal metabolism constants for physiologists and clinicians: a biometric study of basal metabolism in man. Washington, DC: Carnegie Institute of Washington; 1919.

26. Frankenfield D, Muth ER, Rowe WA. The Harris-Benedict studies of human basal metabolism: history and limitations. J Am Diet Assoc. 1998;98:439–45.

27. Mifflin MC et al. A new predictive equation for resting energy expenditure in healthy individuals. Am J Clin Nutr. 1990; 51(2):241–47.

28. Owen OE et al. A reappraisal of the caloric requirements in healthy women. Am J Clin Nutr. 1986;44:1–19.

29. Brooks, GA, Butte NF, Rand WM, Flatt JP, Caballero B. Chronicle of the Institute of Medicine recommendation: how a physical activity recommendation came to be dietary recommendations. Am J Clin Nutr. 2004; 79(suppl): 921S–30S.

30. American Dietetic Association. Evidence Analysis Library. Healthy Adults: Individual errors. Available at http://www.adaevidencelibrary.com/topic.cfm?cat=1072. Accessed May 10, 2007.

31. Miller MA. A calculated method for determination of ideal body weight. Nutr Support Serv. 1985;5:31–33.

32. Farah H, Buzby J. U.S. food consumption up 16 percent since 1970. Amber Waves, U.S. Department of Agriculture Economic Research Service, November 2005. Available at www. Ers.usda.gov/AmberWaves/November05/Findings/USFoodConsumption.htm, accessed 10/06.

33. Diaz VA, Mainous AG, Everett CJ. The association between weight fluctuation and mortality: results from a population-based cohort study. Journal of Community Health. 2005;30(3):1532–65.

34. Raatz S. Obesity lecture, Human nutrition and health, School of Public Health, University of Minnesota, October 24, 2006.

35. Williams L, Williams P. Evaluation of a tool for rating popular diet books. Nutr Diet. 2003;60:185–97.

36. Dietary Guidelines for Americans, 2005, available at www.healthierus.gov/dietaryguidelines, accessed 10/06.

37. Lichtenstein AH et al. Diet and lifestyle recommendations Revision 2006. A Scientific statement from the American Heart Association Nutrition committee, available at http://circ.ahajournals.org, accessed 10/06.

38. Welch S. Nutrient standards, dietary guidelines, and food guides. In: Ziegler EE, Filer LJ, eds. Present knowledge in nutrition. Washington, DC: ILSI Press; 1996:630–46.

39. Health behaviors of adults, United States 2002–2004; Chapter 3, "Alcohol use," in CDC Reports Series 10, Number 230:8–9, published September 2006. Available at www.cdc.gov/nchs/data/series/sr_10/sr10_230, accessed 11/06.

40. Health and Welfare Canada. Nutrition recommendations. The report of the Scientific Review Committee. Canada: Minister of Supply Services; 1990:6.

41. Windham CT, Wyse BW, Hansen RG. Alcohol consumption and nutrient density of diets in the Nationwide Food Consumption Survey. J Am Diet Assoc. 1983;82: 364–70,373.

42. Zeman, FJ. Clinical Nutrition and Dietetics. MacMillan Publishing Company; 2nd ed. New York: 1991.

43. Hodgkinson B, Evans D, Wood J. Maintaing oral hydration in older people. A systematic review. NHS Centre for Review and Dissemination; York, UK: 2001:1–66.

44. Food and Nutrition Board, Institute of Medicine of the National Academy of Sciences. Dietary Reference Intakes for water, potassium, sodium, chloride, and sulfate. Washington, DC: The National Academies Press: 2005.

45. Grandjean AC et al. The effect of caffeinated, non-caffeinated, caloric and non-caloric beverages on hydration. J Am Coll Nutr. 2000;19:591–600.

46. Pennington JA, Hubbard VS. Derivation of daily values used for nutrition labeling. J Am Diet Assoc. 1997;97:1407–12.

47. Wright JD, et al. Dietary intake of ten key nutrients for public health, United States: 1999–2000. Advance data from vital and health statistics; no.334. Hyattsville, MD National Center for Health Statistics; 2003.

48. Trends in intake of energy and macronutrients–United States, 1971–2000. MMWR, 2004; 53(4):80–82.

49. Food and Nutrition Board, Institute of Medicine of the National Academy of Sciences. Dietary Reference Intakes for energy, carbohydrate, cholesterol, fat, fatty acids, protein, amino acids, and physical activity. Washington, DC: The National Academies Press; 2002/2005.

50. Wilson JW et al. Data tables: combined results from USDA's 1994 and 1995 Continuing Survey of Food Intakes by Individuals and 1994 and 1995 Diet and Health Knowledge Survey, 1997.

51. Miller ER et al. Meta-analysis: high-dosage vitamin E supplementation may increase all-cause mortality. Ann Intern Med. 2005;142:37–46.

52. Appel LJ et al. Dietary approaches to prevent and treat hypertension: a scientific statement from the American Heart Association. Hypertension. 2006;47:296–308, available at www.hypertensionaha.org, accessed 5/06.

53. Krall EA, Dawson-Hughes B. Osteoporosis. In: Shils ME et al. eds. Modern nutrition in health and disease. Philadelphia, PA: Lippincott, Williams & Wilkins; 1999:1353–64.

54. National Institutes of Health. Optimal calcium intake. Consensus Statement. 1994; 12:1–31.

55. Feskanich D et al. Protein consumption and bone fractures in women. Am J Epidemiol. 1996;143:472–79.

56. Sakhaee K et al. Meta-analysis of calcium bioavailability: a comparison of calcium citrate with calcium carbonate. American Journal of Therapeutics. 1999; 6: 313–21.

57. Nielsen SM, Popkin BM. Patterns and trends in food portion sizes, 1977–1998. J Am Med Assoc. 2003;289(4):450–53.

58. USDA, Center for Nutrition Policy and Promotion. Americans consume too many calories from solid fat, alcohol, and added sugar. Nutrition Insight. June 2006, 33.

59. National Restaurant Association. Restaurant Industry Pocket Factbook; 2000.

60. Frequently Asked Questions, available at www.restaurant.org/aboutus/faqs.cfm, accessed 12/06.

61. Jeffery RW, French SA. Epidemic obesity in the United States: are fast foods and television viewing contributing? Am J Public Health. 1998;88:277–80.

62. Guthrie JF et al. Understanding economic and behavioral influences on fruit and vegetable choices, AmberWaves April 2005, available at www.ers.usda.gov/AmberWaves/April05/Features/FruitandVegChoices.htm, accessed 12/06.

62. Bogden JD et al. Daily micronutrient supplements enhance delayed-hypersensitivity skin test responses in older people. Am J Clin Nutr. 1994;60:437–47.

63. Balluz LS et al. Vitamin and mineral supplement use in the United States. Results from the Third National Health and Nutrition Examination Survey. Arch Fam Med. 2000;9:258–62.

64. Lee S-K, Sobal J, Frongillo EA. Acculturation, food consumption, and diet-related factors among Korean Americans. J Nutr Ed. 1999;31:321–30.

65. Stahler C. How many vegetarians are there? Available at www.vrg.org/nutshell/poll2006.htm; accessed 12/06.

66. Mangels AR, Messina V, Melina V. Position of the American Dietetic Association and Dietitians of Canada: vegetarian diets. J Am Diet Assoc. 2003;103:748–65.

67. Bilsborough S, Mann N. A review of issues of dietary protein intake in humans. International J Sport Nutr & Exercise Metab. 2006;16:129–52.

68. Young VR, Pellett PL. Plant proteins in relation to human protein and amino acid nutrition. Am J Clin Nutr. 1994;59:1203S–12S.

69. Messina V, Melina V, Mangels AR. A new food guide for North American vegetarians. Can J Diet Prac Res. 2003;64:82–86.

70. Warburton DER, Nicol CW, Bredin SSD. Health benefits of physical activity: the evidence. CMAJ. 2006;174(6)801–9.

71. CDC, BRFSS: Prevalence of leisure-time physical inactivity among Americans age 18 and older by race/ethnicity, and sex. MMWR. 2005;54:39.

72. Physical activity for everyone: recommendations. Available at www.cdc.gov/nccdphp/dnpa/physical/recommendations/index.htm, accessed 12/06.

73. Pollock ML et al. AHA Science Advisory. Resistance exercise in individuals with and without cardiovascular disease: benefits, rationale, safety, and prescription: an advisory from the Committee on Exercise, Rehabilitation, and Prevention, Council on Clinical Cardiology, American Heart Association. Position paper endorsed by the American College of Sports Medicine. Circulation. 2000;101:828–33.

74. DeNavas-Walt C, Proctor BD, Lee CH. Income, poverty, and health insurance coverage in the United States: 2005. Current population reports consumer income. Available at http://census.gov/prod/2006pubs/p60–231.pdf, accessed 12/06.

75. U.S. Department of Health and Human Services. The 2006 HHS poverty guidelines. Federal Register. January 24, 2006;71:3848–49. Available from http://aspe.hhs.gov/poverty/06poverty.shtml, accessed 12/06.

76. Finke MS, Huston SJ. Healthy Eating Index scores and the elderly. Family Economics and Nutrition Review. 2003;15(1)67–73.

77. American Heart Association. Heart disease and stroke statistics—2006 update. www.americanheart.org/downloadable/heart, accessed 12/06.

78. Willerson JT, Ridker PM. Inflammation as a cardiovascular risk factor. Circulation. 2004;109[supplII]:II-2–II-10.

79. Zhao G, Etherton TD, Martin K. Dietary alpha-linolenic acid reduces inflammatory and lipid cardiovascular risk factors in hypercholersterolemic men and women. J Nutr. 2004;134:2991–97.

> "Don't tell me what
> I can't eat. Tell me
> about the health
> value of food."
> Margaret Meter

Chapter 17

Adult Nutrition:
Conditions and Interventions

Chapter Outline

- Introduction
- Cancer
- Cardiovascular Diseases: Coronary Heart Disease
- Overweight and Obesity
- Diabetes Mellitus
- HIV/AIDS

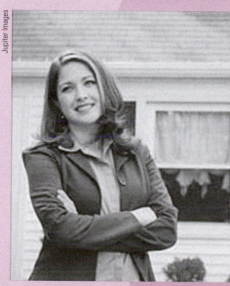

Prepared by U. Beate Krinke

Key Nutrition Concepts

1   Dietary intake, body weight and composition, and physical activity influence changes in health status with age.

2   Optimal nutrition interventions for nutrition-related diseases and disorders rely on good information and a support system for follow-through.

3   Nutritional care practiced in the context of wellness is dynamic; it is evaluated by changing relationships as well as absolute values of health measures.

4   Foods often assume extra signifance when a person is ill.

5   Relevant diet counseling considers the many roles of foods in life.

Introduction

"And this I know, moreover, that to the human body it makes a great difference whether the bread be fine or coarse; of wheat with or without the hull, whether mixed with much or little water, strongly wrought or scarcely at all, baked or raw—and a multitude of similar differences.... Whoever pays no attention to these things, or paying attention, does not comprehend them, how can he understand the diseases which befall a man?"

Hippocrates, 400 BC

Currently, more people in the United States die of lifestyle-related diseases and disorders than from any other cause.[1] Diseases that develop during the adult years may fully or partially result from the cumulative effects of diets high in saturated fats and low in vegetables, fruits, and fiber. Lifestyle factors such as smoking, physical inactivity, and poor diet insidiously influence health on a day-to-day basis. For example, in 2003, cancer, heart disease, cerebrovascular diseases, diabetes, chronic liver disease, and cirrhosis accounted for 61% of all deaths in the United States.[2] Not only do lifestyle-related conditions contribute to premature mortality, they also affect daily living. Heart disease and diabetes are among the chronic conditions that significantly limit the activity of working-age adults.[2] Because these diseases are related to lifestyle, they can be prevented, in part, by healthful changes in diet. One risk factor for lifestyle diseases is obesity, which is also considered to be a disease by some in the health care field. This chapter addresses a number of nutrition-related diseases and disorders that affect adults.

Cancer

Adults are improving some of their lifestyle habits related to cancer: smoking rates are decreasing, alcohol consumption is stable, fruit consumption is up, and dietary fat intake (as a percentage of total calories) has declined, although the percentage of calories eaten from saturated fats is still too high.[3] On the other hand, vegetable consumption is not meeting national guidelines, and obesity rates have skyrocketed (from 13% in 1960 to 31% in 2000 and 2005). Still, death rates for prostate, breast, lung, and colorectal cancers are declining, and there is a large and growing population of individuals who have been free of cancer for 5 or more years. Because cancer prevalence increases with age, additional information on cancers is presented in Chapter 19.

Definition

Cancer is a group of diseases in which genes malfunction so that cells lose normal control over their growth and replication. Cells become self-sufficient, creating new blood vessels and traveling to distant organs from their site of origin (primary site). Epithelial tissue accounts for the majority of all cancers, although any of the four tissue types may be affected: connective (sarcomas and leukemia in blood), muscular (myosarcomas), nervous (gliomas), and epithelial tissue. Epithelial tissues include the skin, gastrointestinal tract (esophagus, stomach, colon, rectum), urogenital system (urinary tract, uterus), some secretory organs (prostate, breast, pancreas) and the respiratory system (lung cancer). These epithelial tissues host the most common cancers.

Prevalence/Etiology

Social, cultural, behavioral, genetic, and environmental factors affect the initiation and progression of cancer; six and seven mutations are needed to produce a cancer cell. It is true that DNA replication is very efficient, but mistakes happen. Most mistakes are repaired, but not all. If enough of the "cancer-risk" mistakes accumulate, a cancerous cell will develop. The more rapidly cells replicate, such as those in epithelial tissue, the greater the chances for mistakes. Populations who migrate tend to develop the cancer rates of their new locales, suggesting that lifestyle and environmental factors impact cancer development more than genetics. Having one of the oncogenes, however, is clearly a risk factor in cancer development. The four most common cancers in adults are prostrate, breast, colorectal, and lung cancers. Blacks have the highest rate of new cancers, and American Indians/Alaska Natives have the lowest.

Risk Factors

While smoking is the most-recognized contributor to cancer occurrence (estimates are that smoking causes about 30% of all U.S. deaths from cancer), obesity and insulin resistance; excess alcohol consumption; and low intakes of fruits, vegetables, and calcium are the main nutrition-related risks. High fat intake is thought to lead to higher levels of bile acids, which irritate the colon lining; salt also irritates the colon lining. Charred foods such as meats cooked to a crisp on an open grill and the N-nitroso compounds in processed meat (such as in hams and smoked sausage) potentially promote mutagenesis. Other related risk factors are intestinal parasites and helicobacter pylori infection. Physical activity is associated with reduced colon and breast cancer, and has been implicated at several other sites.[3]

Nutrition Intervention

Cancer prevention depends on minimizing nutritional risks:[3,4]

- Maintaining or reaching a healthy weight
- Eating 3 to 14 servings of fruits and vegetables daily
- Eating a low-fat diet (most likely it is due to the saturated fats)
- Consuming alcohol in moderation, if at all
- Exercise (decreases fecal transit time, helps maintain healthy weight)

Research on how fats and types of fat alter cancer risk continues. What is needed is a reliable and precise test that accurately measures the influences of fat on cancer risk in the body. Another promising research area is coming from animal models, which have shown that energy restriction has reduced cancer incidence. The main prevention lesson for humans? Don't get fat.

After a diagnosis of cancer, good nutrition supports treatment strategies (surgery, chemotherapy, radiation) or palliative care in terminal patients. Health care professionals provide guidance on dealing with weight loss, taste aversions, types of foods likely to be tolerated for dry or sore mouth and throat, treating constipation and diarrhea, and nutritional therapy to maintain overall stamina. The American Cancer Society has synthesized evidence-based guidance on nutrition during and after cancer, such as suggesting nutrient-dense supplements to provide adequate calories and protein for recovery.[5] The National Cancer Institute maintains a Physicians Database Query (PDQ), with information for patients and for health professionals available at http://cancer.gov/cancertopics/pdq. Find additional cancer information in Chapter 19, and find specific details about treatment options for adult cancers at http://cancer-gov/cancertopics/pdq/adulttreatment.

Alternative Medicine and Cancer Treatment

In some cases, complementary treatments are available to enhance traditional cancer therapies. The NCI database provides up-to-date information regarding complementary and alternative medicine for many purported anti-cancer

Illustration 17.1 Example of PDQ summary CARTILAGE (Bovine and Shark)

Mechanism: Believed to contain an *angiogenesis inhibitor* because cartilage contains no blood vessels.

How sold, and number of brands sold in the U.S.: As a dietary supplement; there are 40 different brand names. No FDA approval is needed for dietary supplements unless specific disease prevention or treatment claims are made.

Number of human studies using cartilage products since the 1970s: Over 12 clinical studies; 5 studies have been published in peer-reviewed, scientific journals, including 1 randomized controlled trial.

Doses and cartilage treatment duration: Varies because of the many cartilage products available and the wide array of administration methods (i.e., topical, oral, enema or as subcutaneous injection).

Adverse effects: *Dyspepsia*, taste changes, nausea, vomiting, fever, fatigue, dizziness; make physician aware of usage due to the very high calcium content of cartilage.

Overall level of evidence for cartilage: Reported data from controlled and uncontrolled research trials are inconclusive.

Recommendation: Use of cartilage (bovine or shark) as a treatment for cancer cannot be recommended outside the context of well-designed clinical trials.

SOURCE: National Cancer Institute website: www.Cancer.gov. Link to Cancer Topics and then PDQ. Final website location is: http://cancer.gov/cancertopics/pdq/cam/cartilage/healthprofessional, accessed November 25, 2006.

remedies. An example of the types of information included in a PDQ summary is given for Cartilage (Bovine or Shark) in Illustration 17.1. Good information helps cancer patients and their families to balance the benefits and risks of various therapies.

Cardiovascular Disease: Coronary Heart Disease

Cardiovascular diseases (CVD) consist of coronary heart disease (CHD, often called "heart disease"), *stroke*, and other diseases and disorders related to the heart, blood vessels, and circulation. CHD is the number one cause of death in men and women in the United States. Table 17.1 lists nutrition-related, Health Objectives for the Nation for decreasing rates of heart disease as well as stroke.

Angiogenesis Inhibitor Angiogenesis is the formation of new blood vessels. An angiogenesis inhibitor slows or stops vessel formation. Tumors cannot grow or expand without additional blood vessels to carry oxygen and other nutrients.

Dyspepsia "Upset stomach"; impaired gastric function may include nausea, pain, and burning sensation.

Stroke The event that occurs when a blood vessel in the brain becomes occluded due to a clot or ruptures, cutting off blood supply to a portion of the brain. Also called a cerebral vascular accident.

Table 17.1 Healthy People 2010 nutrition-related health objectives to reduce heart disease and stroke among adults

	Percentage of All Adults	
	Target	Baseline
Increase the proportion of adults with high blood pressure whose blood pressure is under control.	50%	18%
Increase the proportion of adults with high blood pressure who are taking action (for example, losing weight or reducing sodium intake) to help control their blood pressure.	95%	72%
Reduce the mean total blood cholesterol levels among adults.	199 mg/dL	206 mg/dL
Reduce the proportion of adults with high total blood cholesterol levels (240 mg/dL or greater).	17%	21%
Eat 2400 mg or less of sodium daily.	65%	21%

SOURCE: *Healthy People 2010: National Health Promotion and Disease Prevention Objectives.* Washington, DC: U.S. Department of Health and Human Services; 2000: Sections 12 and 19.

Table 17.2 Deaths from heart disease and stroke compared to national goals

National Goals to Reduce Mortality

	Per 100,000 Population	
	Target	Baseline, 1997
Reduce coronary heart disease deaths	166	216
American Indian or Alaska Native	166	134
Asian or Pacific Islander	166	125
Black or African American	166	257
Hispanic or Latino	166	151
White	166	214
Reduce stroke deaths	48	60
American Indian or Alaska Native	48	39
Asian or Pacific Islander	48	55
Black or African American	48	82
Hispanic or Latino	48	40
White	48	60

SOURCE: *Healthy People 2010: National Health Promotion and Disease Prevention Objectives.* Washington, DC: U.S. Department of Health and Human Services; 2000: Tables 12–15 and 12–18.

Definition

CHD and stroke are characterized by atherosclerosis, or hardening of the *arteries* due to plaque buildup in the walls of arteries. Atherosclerosis occuring in blood vessels of the heart is related to CHD, whereas atherosclerosis in the cerebral artery of the brain is related to stroke. Atherosclerosis narrows arteries, increasing the risk that a blood clot will form, shut off blood flow, and cause a heart attack (myocardial infarction) or stroke.

Arteries Blood vessels that carry oxygenated blood to cells.

Arteriosclerosis Age-related thickening and hardening of the artery walls, much like an old rubber hose that becomes brittle or hard.

Prevalence

Mortality rates from CHD and stroke fell by 24% and 10%, respectively, in 2000.[1] Still, the American Heart Association (AHA) estimates that close to 1 million deaths occur annually due to heart disease.[6] Heart disease is the third (13% of all deaths) leading cause of death in 25- to 44-year-olds, the second (23%) in 45- to 64-year-olds, and the first in adults aged 65 years and older, accounting for 31% of all their deaths. Table 17.2 shows the annual number of deaths from heart disease by ethnic group. The prevalence of CVD in American adults increases with age from 30%

for 45- to 54-year-olds to about 50% in 55- to 64-year-olds.

Etiology

Atherosclerosis begins when nutrient substances such as cholesterol, fatty acids, and calcium become part of tissues that form over injured arterial wall cells. Blood lipids, such as LDL cholesterol and fatty acids, become incorporated into plaque over time, increasing the extent of atherosclerosis.

Atherosclerosis is a multifactorial disease that can develop from:

- Fatty streaks in the smooth muscle of the artery wall
- Chronic inflammation leading to plaque
- Vessel thickening as the result of endothelial cell damage
- Formation of fibrous plaques (deposits of fats, cholesterol, collagen, muscle, and other cells and metabolites) inside the artery walls at the point of injury or lipid accumulation
- Calcification of fibrous plaques into lesions that eventually deteriorate, become infected, or in general weaken the artery wall

High blood levels of homocysteine, inflammatory diseases, abnormal blood clotting factors, and other conditions also influence the development of atherosclerosis.[7–10] The progression of atherosclerosis can be slowed, neutralized, or partially reversed by dietary and lifestyle modifications.

Effects of CHD

The effects of CHD can range from shortness of breath after exertion to chest pain (angina) and death from a heart attack. Atherosclerosis decreases blood circulation to the heart, resulting in decreased energy, decline in organ function, and the inability to perform activities of daily living. Buildup of plaque and lesions inside the blood vessels leave less room for blood flow. Consequently, the heart has to work harder to pump blood through this narrower space to reach all parts of the body, leading to higher blood pressure levels. An analysis of data from more than 12,000 middle-aged men in the Seven Countries Study found that increases in blood pressure (diastolic and systolic) led to increased heart disease, just as higher levels of blood cholesterol can lead to atherosclerosis.[11]

Risk Factors

High total blood cholesterol, high blood pressure, and smoking are three risk factors for CHD that have declined since 1960,[6] but diagnosed diabetes and obesity rates have increased substantially. The following factors increase risk of CHD:[10,12–18]

- Excess weight for height (e.g., BMI >30kg/m^2)
- Excess abdominal fat: waist circumference of 35 inches or more for females; 40 inches or more for males
- Blood pressure above 140/90 mm Hg
- High-density lipoprotein (HDL) cholesterol levels less than 40 mg/dL, especially in women
- Low-density lipoprotein (LDL) levels of 130–159 mg/dL (borderline high), 160–189 (high), 190+ (very high)
- Total cholesterol levels of 200–240 mg/dL (moderate risk) or 240 mg/dL or more (high risk)
- Elevated blood triglyceride (TG) of >150 mg/dL
- Elevated levels of high-sensitivity C-reactive protein
- Elevated plasma apolipoprotein B (an atherogenic lipoprotein)
- Diabetes, especially if uncontrolled; elevated fasting plasma insulin levels
- High saturated fats and trans-fatty acid intake
- Consumption of few vegetables, fruits, and whole grains
- Inadequate folate intake; high blood homocysteine levels
- Infrequent intake of fish (low omega-3 fatty acid intake)
- Smoking cigarettes and cigars; chewing tobacco
- Lack of physical activity
- Unresolved emotional stress, hostility, angry personality

Other risk factors for CHD include:

- Family history of, genetic disposition toward CHD
- Male gender, females after menopause
- Old age

A Web-based self-assessment that lets you calculate your risk of CHD is available at http://hp2010.nhlbiin.net/atpiii/evalData.asp. It is based on predictive risk factors indentified in the Framingham Heart Health study.[19] An example of the scoring sheet is shown in Illustration 17.2.

Nutrition Interventions

Nutrition interventions for heart disease include risk reduction and management of symptoms related to atherosclerosis, and are best begun early in life. Abnormal blood lipid levels are the strongest risk factors for heart disease, and the strongest evidence for atherosclerotic plaque reduction supports limiting saturated and trans-fatty acids in the diet. However, no single dietary change or other action guarantees freedom from heart disease.

The National Heart, Lung, and Blood Institute initiated the National Cholesterol Education Program (NCEP) in 1985 in order to decrease the population's average blood cholesterol, which was considered to be the most modifiable nutrition-related risk factor for heart disease. The NCEP Expert Panel on Detection, Evaluation, and Treatment of High Blood Cholesterol developed population-based strategies to reduce heart disease[13,14] through identification of cut-points for diagnosis of risk-associated blood lipid levels, development of corresponding dietary and pharmaceutical treatment suggestions, and implementation of health promotion campaigns. Population levels of cholesterol have decreased.

Originally, NCEP's dietary guidelines were made in the form of the Step I and Step II diets.[13] Since then, both NCEP[14] and the American Heart Association (AHA)[12] have released new guidelines. The Nutrition Committee of AHA based its report on three underlying philosophies: (1) diet and lifestyle practices can be safely followed throughout life; (2) individual intake is to be evaluated over extended time rather than a single meal; and (3) guidelines are a population framework into which individual needs are to be integrated. The AHA dietary guidelines underscore the potential benefits of eating a diet that emphasizes vegetables, fruits, whole grain products and other high-fiber foods, nonfat dairy products, nuts, fish, and lean meats. (The AHA diet is similar to the DASH diet plan found in MyPyramid guidelines and also described in Chapter 19.) The following NCEP guidelines result in reduction of LDL-cholesterol levels and reduced risk of CHD. They are called therapeutic lifestyle Changes (TLC):[14]

- Saturated fat intake should be less than 7% of total calories, monounsaturated fatty acids should contribute up to 20% of calories, and polyunsaturated fatty acids not more than 10% of calories.

Heart Health Assessment

This is a risk assessment tool that uses data from The Framingham Heart Health Study to estimate the 10-year risk of coronary heart disease. This self-assessment is not meant to be a replacement for medical advice.

Men
Estimate of 10-Year Risk for Men

AGE	POINTS
20–34	−9
35–39	−4
40–44	0
45–49	3
50–54	6
55–59	8
60–64	10
65–69	11
70–74	12
75–79	13

Total Cholesterol	POINTS				
	Age 20-39	Age 40-49	Age 50-59	Age 60-69	Age 70-79
<160	0	0	0	0	0
160–199	4	3	2	1	0
200–239	7	5	3	1	0
240–279	9	6	4	2	1
≥280	11	8	5	3	1

	POINTS				
	Age 20-39	Age 40-49	Age 50-59	Age 60-69	Age 70-79
Nonsmoker	0	0	0	0	0
Smoker	8	5	3	1	1

HDL (mg/dL)	POINTS
≥60	−1
50–59	0
40–49	1
<40	2

Systolic BP (mmHg)	If Untreated	If Treated
<120	0	0
120–129	0	1
130–139	1	2
140–159	1	2
≥160	2	3

| < less than |
| ≥ greater than |

Your Points

Age Points	
Cholesterol Points	
Smoking Points	
HDL Points	
BP Points	
Total Points	

Women
Estimate of 10-Year Risk for Women

AGE	POINTS
20–34	−7
35–39	−3
40–44	0
45–49	3
50–54	6
55–59	8
60–64	10
65–69	12
70–74	14
75–79	16

Total Cholesterol	POINTS				
	Age 20-39	Age 40-49	Age 50-59	Age 60-69	Age 70-79
<160	0	0	0	0	0
160–199	4	3	2	1	1
200–239	8	6	4	2	1
240–279	11	8	5	3	2
≥280	13	10	7	4	2

	POINTS				
	Age 20-39	Age 40-49	Age 50-59	Age 60-69	Age 70-79
Nonsmoker	0	0	0	0	0
Smoker	9	7	4	2	1

HDL (mg/dL)	POINTS
≥60	−1
50–59	0
40–49	1
<40	2

Systolic BP (mmHg)	If Untreated	If Treated
<120	0	0
120–129	1	3
130–139	2	4
140–159	3	5
≥160	4	6

Your 10-Year Risk _____ %

Discuss the results of this survey with your physician.

Point Total	<0	0	1	2	3	4	5	6	7	8	9	10	11	12	13	14	15	16	≥17
10-Year Risk %	<1	1	1	1	1	1	1	2	2	3	4	5	6	8	10	12	16	20	25	≥30

Point Total	<9	9	10	11	12	13	14	15	16	17	18	19	20	21	22	23	24	≥25
10-Year Risk %	<1	1	1	1	1	2	2	3	4	5	6	8	11	14	17	22	27	≥30

Illustration 17.2 Framingham heart health assessment.

- Total fat intake should range between 25–35% of calories [similar to the Acceptable Macronutrient Distribution Range of 20 to 35% of calories].
- Dietary cholesterol intake should be less than 200 mg per day.
- Carbohydrates should consist of 50–60% of total calories.
- Dietary fiber intake should be 20–30 grams per day with 10–25 grams coming from soluble fibers.
- Plant stanols or sterols (2 grams per day) should be consumed (e.g., from spreads).
- Weight reduction if overweight or obese should be a goal.
- Expenditure of at least 200 calories per day should be achieved through physical activity.

Further guidance about fat intake comes in the form of Dietary Reference Intakes (DRIs, see macronutrient distribution guidelines in Chapter 1). The DRIs recommend that saturated fatty acid intake be kept as low as possible while consuming a nutritionally adequate diet and that trans-fatty acid intake be as low as possible. The DRIs further recommend that 5–10% of calories come from linoleic acid (an omega-6 fatty acid) and that 0.6–1.2% of calories come from alpha-linolenic acid (an omega-3 fatty acid). These fat intake levels are meant to promote health in general, and especially heart health.[18]

Fatty fish (lake whitefish, salmon, herring, trout, flaxseed, and walnuts) are good sources of the omega-3 fatty acids. Purslane, treated as a garden weed by some, is a good plant source; brussel sprouts, broccoli, and kale also contain some omega-3 fatty acids.

Eating as few trans fats as possible means limiting processed foods, especially those made with hydrogenated oils. French fries, pot pies, breaded and fried fish or chicken, doughnuts, cakes, cookies, biscuits, and crackers are examples of foods containing trans-fatty acids, although manufacturers removed trans fats from many snack foods when trans-fat labeling became mandatory. Foods that naturally contain very few trans-fatty acids are vegetables, cereal grains such as rice, corn, and oats, oils, and fruits. A bowl of salad greens tossed with herbs, onion, olive oil, and a splash of vinegar and topped with grilled chicken, poached salmon, or hard-cooked eggs is an example of a meal that qualifies as trans-fatty-acid-free.

Dietary Supplements and Heart Disease

Many modern medicines are based on plant-derived chemicals. Aspirin or acetylsalicylic acid, derived from willow bark (Latin plant name: *salix*) and also from spiraea, is one example; cocaine from coca leaves as local anesthetic is another. It's no surprise that herbs and plants are rich sources of complementary approaches for prevention and/or treatment of heart disease.[20-23] Because these *botanicals* are considered dietary supplements, efficacy does not have to be demonstrated. Examples of herbs used for heart disease include:

> **Botanicals** Herbs and other plant-based products (e.g., chamomile tea to soothe dyspepsia).
>
> **Myopathy** From the Greek, *mys* = muscle and *pathos* = suffering; any abnormal condition of muscle tissue.

- *Hawthorn.* The leaves, flowers, and fruits contain antioxidants (flavonoids and procyanidins), which are used in the early stages of congestive heart failure and arrhythmias.
- *Garlic.* Cloves of garlic contain allicin and antioxidants. Garlic has antibiotic properties, decreases blood cholesterol, and inhibits platelet aggregation or blood clotting. Recommended doses range from 1 clove (4 grams) to 5–20 cloves of fresh garlic per day; chopping and drying release the active compound allicin, although special processing can result in active dried garlic for supplement use. The German Commission E approves garlic to treat hyperlipoproteinemia and to slow arteriosclerosis.
- *Green tea.* Green tea and its extract contain antioxidants (polyphenols and proanthocyanidins). Regular consumption of green tea is associated with decreased LDL-cholesterol and triglycerides and increased HDL-cholesterol levels.
- *Red yeast rice (Monascus).* Chinese records from AD 800 report that red yeast rice is useful for treating diarrhea, indigestion, and poor stomach health and blood circulation. Red yeast is a fungus used in rice wine fermentation and is the food coloring in Peking duck. Red yeast rice is used to treat hyperlipoproteinemia due to its content of monacolin K, also known as the cholesterol-lowering drug lovastatin (with the same side effects). Red yeast contains several potentially active monacolins; the monacolin K in red yeast blocks cholesterol formation. While red yeast has a long tradition of use, it also has the potential to interfere with statin drugs.
- *Coenzyme Q$_{10}$ or ubiquinone.* It is not an essential nutrient. Present in most cells (it is ubiquitous, with the highest levels present in heart, liver, kidney, and pancreas), Q$_{10}$ facilitates electron transport in mitochondria. Low levels have been associated with aging, congestive heart failure, and various cancers. Coenzyme Q$_{10}$ is used to combat *myopathy* and liver dysfunction associated with statin use and some cancer treatments. Recommendations are to take Q$_{10}$ with a meal containing fat because it is fat soluble.

- *Carnitine*. Derived from amino acids (primarily meat and dairy), carnitine functions in long-chain fatty acid metabolism. Some evidence supports its use for cardiac performance and reduction of hyperhomocysteinemia.[20]

Overweight and Obesity

"The second day of a diet is always easier than the first. By the second day you're off it."

Jackie Gleason

Bigger portions, bigger people? Obesity has become an "epidemic." Three states now have one-third of their population weigh in "obese," while only four states had obesity prevalence rates less than 20% in 2005, see Illustration 17.3.[24] It seems as if every other person is dieting, and everyone "knows" how to lose weight, yet the number of obese individuals has doubled since 1980. To see a time line of how the U.S. population has gotten so big since 1980, go to cdc.gov and download the obesity trends maps, which are depicted as a PowerPoint® slide show, www.cdc.gov/nccdphp/dnpa/obesity/trend/maps/index.htm.

Overweight and obesity are often examined together because obesity is a degree of overweight. In the 1980s, health professionals used to use the term *ideal body weight*.[25] Ideal weights were based on survivorship of healthy adults applying for life insurance, published as the 1983 Metropolitan Life Insurance Company actuarial tables. These weights were also called "normal," "suggested," and "acceptable" weight. In the mid-1980s, the term *ideal body weight* was replaced with the term *desirable weight*, which was subsequently replaced with the term *healthy body weight* in 1995 and subsequent Dietary Guidelines. In general, a healthy weight is one that can be maintained through a health-promoting lifestyle. Weight becomes "overweight" when it reaches a level associated with higher risk for disease, disability, and death. The word "obesity" was not introduced into the Dietary Guidelines until 2000.[25, 26]

Definition

The Obesity Education Initiative of the National Institute of Health classifies overweight and obesity by body mass index (BMI, kg body weight divided by height in meters, squared).[27] BMI can be found in an interactive online calculator at cdc.gov or calculated by following these steps:

1. Multiply body weight in pounds by 703.
2. Divide that number by height in inches.
3. Divide that number by height once more.

BMI ≥ 30, or ~ 30 lb. overweight for 5'4" person

2005

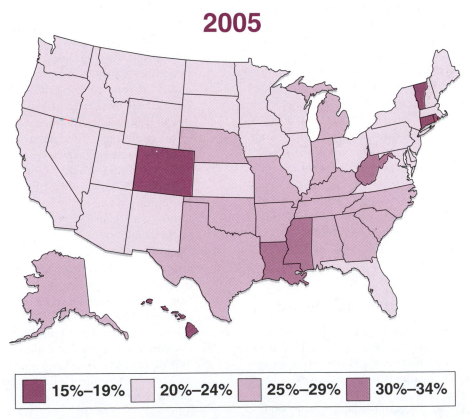

| ■ 15%–19% | □ 20%–24% | ■ 25%–29% | ■ 30%–34% |

Illustration 17.3 Overweight and Obesity: U.S. Obesity Trends: 2005.

For example, calculating the BMI for a 5'7" individual weighing 150 lb:

1. 150 lb × 703 = 105,450
2. 105,450 divided by 67 inches = 1573.880
3. 1573.880 divided by 67 inches = 23.49 BMI

Based on this single number expressing the relationship between weight-to-height, overweight in adults is defined by BMIs of 25.0–29.9 and obesity by BMIs of 30 or greater.[27] A BMI of 30 or greater is roughly equivalent to being 30 or more pounds overweight.

Body fat content is a more important indicator of health than is weight-for-height. Although BMI approximates body fat for most healthy individuals, a BMI value is not the same as a measurement of body fat. For instance, a heavily muscled football player who is 6 feet tall and weighs 200 lb has a BMI of 27.2 but may have a low body fat content. BMI measures don't accurately represent healthy weights for:

- Athletes or others with greater-than-average percentages of muscle mass
- Individuals with little muscle mass
- Individuals with dense, large bones
- Dehydrated and overhydrated individuals

Prevalence

Obesity rates have increased rapidly in the United States in the last two decades. Nearly one-third of U.S. adults have a BMI or 30 or greater, (see Illustration 17.3) with black females having the highest rate (49.6%) and males aged 75 and older the lowest (18.0%).[2,28] No adult population group meets the Healthy People 2010 target of 15% obesity prevalence. Using the criterion of BMI between 18.5–24.9, only 33% of U.S. adults are at a healthy weight, down from 51% in 1960.[2] For gender differences in overweight, obesity, and healthy weight, see Table 17.3.

Etiology

Between 1970 and 1990, average caloric intake of adults in the United States increased by as much as 300 calories daily.[28] Mean caloric intake ranges from 2028 to 2825 for 20- to 39-year-old females and males, respectively.[29] Small

Table 17.3 Percentage of the adult population (age 20–74) who is at healthy weight, overweight, or obese, by gender

	Overweight	Obese	Healthy Weight
Males	69%	28%	32%
Females	62%	34%	36%

SOURCE: Data from National Center for Health Statistics. Health. United States, 2005. Hyattsville, Maryland: 2005.

changes can add up: eating or not eating 100 calories per day adds up to approximately 10 lb per year gained or lost. Thirty-eight percent of U.S. adults do not participate in leisure time physical activity, and 33% participate in regular physical exercise (defined as 20 minutes of vigorous activity, three times per week or more time spent in moderate activity).[2] An abundant food supply (in the United States, we now have approximately 500 more calories available for consumption, per person per day, than we did in 1970) coupled with sedentary lifestyles contribute to the population's rising weight.

Overweight and obesity may appear to be simply a matter of intake exceeding output. But if it were so simple, the weight-loss industry would hardly be so huge. Overweight and obesity are complex and chronic conditions, stemming from multiple causes. For instance, scientists are beginning to learn more about the balance of hormones involved in energy metabolism and appetite regulation:

- Thyroid hormones control metabolism and thermogenesis; in the case of hypothyroidism, the body responds by conserving energy and slowed metabolism.
- Insulin acts in the central nervous system to inhibit food intake and in the peripheral nervous system to synthesize and store fat.
- Leptin is secreted by adipose tissue and increases feelings of satiety.
- Ghrelin, primarily secreted in the stomach, and therefore suppressed after gastric bypass surgery, stimulates food intake and slows metabolism and fat oxidation. Levels rise during dieting.

Other factors contributing to the obesity epidemic such as food advertising, the easy availability of highly palatable snack foods and beverages, and larger portions served in fast food and sit-down restaurants will need societal efforts to develop effective interventions. Environmental, genetic, physiological, psychological, socioeconomic, and cultural factors all play a role in the development of obesity.

Effects of Obesity

Excess body weight is associated with increased morbidity, and at a BMI of 35 and above, with increased mortality.[30] Obesity poses a health risk serious enough that the Surgeon General convened a Conference on Obesity to plan a nationwide anti-obesity campaign.[31] Overweight and obesity are associated with an increased risk of hypertension, dyslipidemia, coronary heart disease, type 2 diabetes, stroke, gallbladder disease, osteoarthritis, sleep apnea, back problems, and endometrial, breast, prostate, and colon cancers.[27] As anyone who has ever been obese can tell you, social stigma also accompanies obesity. Weight loss in overweight and obese persons decreases the presence and severity of obesity-related health problems.

Case Study 17.1

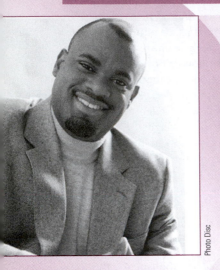

Photo Disc

Maintaining a Healthy Weight

Adam is 5'11" tall and weighs 190 pounds. He lives alone. The commute to his software development job takes about 90 minutes each day. He likes his coworkers and the work environment and is happy that his workplace provides a cafeteria so he doesn't have to bring a lunch. But he is trying to work from home occasionally to cut car expenses. His main hobby is golf; the course he plays encourages carts. He's an avid football fan. In his spare time, he is restoring an old car.

Questions

1. Calculate Adam's current BMI. How would you classify his weight status based on the NIH classifications?
2. What would it take for Adam to achieve a BMI of 24? Calculate an energy level and estimate the number of weeks it would take at that level.
3. What would you consider a "healthy weight" for Adam?
4. What are some suggestions you would discuss with Adam in order to decrease his BMI?

Nutrition Interventions

See Case Study 17.1 for a few issues related to maintaining a healthy weight. A successful weight-loss program includes physical activity, a diet that allows for lifelong personal adherence, and an acceptable diet that is safe and that results in long-term maintenance of weight loss.[32] Compared to normal-weight individuals or those who regain lost weight, individuals who successfully maintain weight loss use more behavioral strategies to support weight loss and maintenance. These behaviors include consistently controlling caloric intake, exercising more often and more strenuously, tracking weight, and eating breakfast (See Table 17.4 for details).[32–34]

Keeping excess weight off over time is often harder than losing it in the first place. This is illustrated by the widely varying results of a meta-analysis of weight-loss treatments.[35] After 1 year, average weight change with diet therapy and/or physical activity ranged from a 1.9 kg weight gain to 8.8 kg weight loss.[35] Counseling for low-calorie diets (1000–1200 calories per day) reduced body weight by an average of 8% over 3 to 12 months and decreased abdominal fat. Counseling for physical activity led to weight losses of 2% to 3% and also reduced abdominal fat.

Components of a successful weight-management program include:

1. *Realistic goals:* Identifies a healthy weight and a feasible rate of loss (0.5–1.0 lb/week), with the ability to self-monitor progress.

Table 17.4 Weight-loss strategies of individuals in the National Weight Control Registry by gender

	Women	Men
Number sampled	629	155
Age (years)	44.4 ± 11.5	49.1 ± 11.9
Weight change (kg)	28.7 ± 13.6	35.4 ± 20.7
Weight loss duration (years)	5.5 ± 6.8	5.8 ± 6.9
Maintenance Diet and Exercise		
Calories/day	1296 ± 454	1725 ± 647
Fat (%)	24 ± 9	23 ± 8
Protein (%)	19 ± 4	18 ± 4
Carbohydrate (%)	55 ± 10	56 ± 10
Eating frequency/day	4.9 ± 3.2	4.5 ± 1.6
Weight-Loss Dietary Strategies		
Restricted intake of certain type or classes of food	88%	87%
Ate all foods but limited quantity	47%	32%
Counted calories	45%	3%
Limited percentage of daily energy from fat	31%	37%

SOURCE: Adapted from Klem ML, Wing RR, McGuire MT, Seagle HM, Hill J. A descriptive study of individuals successful at long-term maintenance of substantial weight loss. Am J Clin Nutr. 1997;66:239–46. To join the National Weight Loss Registry, an individual must be 18 or older, have lost at least 30 pounds, and maintained a weight loss of >30 pounds for one year or more. www.lifespan.org

2. *Caloric deficit:* Develops an individualized meal plan with sufficient calories to lose weight gradually.

3. *The meal plan:* Builds meal plans around a variety of foods that can be readily obtained and enjoyed by the entire household.

4. *Long-term effectiveness:* The weight management plan is built around learning and practicing behaviors that can be maintained for a lifetime.

5. *Practice problem-solving techniques:* Develops strategies to anticipate and solve potential weight-management problems.

6. *Stress-management:* Teaches strategies other than eating to deal with stressful situations.

7. *Maintenance:* Makes available support for weight loss and maintence of the loss.

8. *Regular exercise:* Includes strength training and aerobic exercise.

9. *Cultivates self-image:* Builds self-confidence while guarding against disordered eating.

Although weight loss can benefit health, the way in which it is lost can undo that benefit. Inappropriate eating habits, potentially hazardous herbal remedies (such as ephedra), and diet drugs are harmful to health. Using the preceding checklist can provide some assurance that the proposed weight-loss approach will be helpful rather than harmful. If weight loss were easy, we would not be paying more than $33 billion per year to the weight-loss industry.

Bariatric Surgery and Role of the RD

Individuals diagnosed as morbidly obese (BMI ≥ 40 kg/m²) or overweight (BMI ≥ 25 kg/m²) with several comorbidities may be considered appropriate candidates for gastric bypass surgery (GBS). Candidates have been unsuccessful in diet and exercise attempts, are motivated to learn about complicated aftercare, and present acceptable operative risks. For GBS candidates, the medical complications associated with GBS, particularly nutritional deficiencies, are expected to be outweighed by the potential health benefits (e.g., lower blood glucose, better lipid profile, lower blood pressure, and increased mobility).

Dietitians assist physicians and psychologists in both preoperative and postoperative assessment and education. The role of the bariatric surgery dietitian is to promote adherence to the strict eating behavior guidelines and supplement prescriptions necessary to reach desired weight goals and minimize complications. Counseling can help to prevent the onset of acute (nausea, vomiting, dehydration, dumping syndrome) and long-term nutritional complications (micronutrient deficiencies including iron, folate, vitamin B$_{12}$, calcium, and vitamin D). The most important postoperative eating guidelines include reducing portion sizes, chewing food completely, eating slowly, consuming foods and beverages separately, avoiding poorly tolerated foods, and compliance with dietary supplement intake.[36]

Diabetes Mellitus

"What AIDS was in the last 20 years of the twentieth century, diabetes is to be in the first 20 years of this century."

Paul Zimmet, International Diabetes Institute[37]

The major forms of diabetes are *type 1*, *type 2*, and gestational diabetes. Type 1 diabetes results from an autoimmune process, where the body's immune system destroys pancreatic beta cells that produce insulin; peak incidence is during puberty. Daily insulin injections are needed to prevent cell starvation and death. The other major type is due to insulin resistance, related to obesity. Type 2 diabetes is the most common by far and is fueling the diabetes epidemic. Rates of type 2 diabetes are escalating worldwide due, in part, to rising rates of obesity. Although most often diagnosed in people over the age of 40, type 2 diabetes is becoming increasingly common in children and adolescents.[38]

Diabetes exists when blood glucose levels are above normal. Type 1 and 2 diabetes are diagnosed when fasting levels of blood glucose are 126 mg/dl and higher.

> **Type 1 Diabetes** A disease characterized by high blood glucose levels resulting from destruction of the insulin-producing cells of the pancreas. This type of diabetes was called juvenile-onset diabetes and insulin-dependent diabetes in the past.
>
> **Type 2 Diabetes** A disease characterized by high blood glucose levels due to the body's inability to use insulin normally, or to produce enough insulin. This type of diabetes was called adult-onset diabetes and non-insulin-dependent diabetes in the past.
>
> **Prediabetes** A condition in which blood glucose levels are higher than normal but not high enough for the diagnosis of diabetes. It is characterized by impaired glucose tolerance, or fasting blood glucose levels between 100 and 126 mg/dl.

Prevalence

Diabetes affects approximately 200 million individuals worldwide, including 21 million people in the United States.[39] Less than 1% of the U.S. population was diagnosed with diabetes in 1960. The figure has grown to 9.6% of adults (aged 20 years and over) having diabetes in 2005, although approximately one-third of these cases are undiagnosed.[38] Prevalence increases with age and depends on ethnic background. The age-adjusted prevalence of diagnosed diabetes was higher among non-Hispanic blacks, American Indians, Alaska Natives, and Hispanic/Latino Americans than among non-Hispanic whites.[39] The rise in diabetes correlates with a rise in overweight adults. In a review of data on overweight and obese adults from the Third National Health and Nutrition Examination Survey (NHANES III; 1988–1994, 45- to 74-year-olds), 10.8% had undiagnosed diabetes, 22.6% had *prediabetes*, and only 54% had normal glucose metabolism.[40] If these percentages seem impersonal, think about a city of 1,500,000 inhabitants: that is how many new cases of diabetes were diagnosed last year—1.5 million individuals.

Etiology

Type 2 diabetes is caused by *insulin resistance* combined with insulin deficiency. Insulin is a hormone produced by the beta cells of the pancreas. It performs many functions, one of which is to moderate the postprandial (post-meal) increases in blood glucose by facilitating the passage of glucose into cells. If insulin is produced in insufficient amounts, or if insulin receptors on cell membranes are too few or not sensitive to the action of insulin, cells become starved for glucose. As a result, the functional levels of tissues and organs decline. Adverse side-effects of high levels of blood glucose include elevated blood levels of triglycerides, increased blood pressure, and hardening of the arteries.

> **Insulin Resistance**   A condition is which cells "resist" the action of insulin in facilitating the passage of glucose into cells.
>
> **Metabolic Syndrome**   A cluster of metabolic abnormalities including excessive central body fat, blood insulin, and fasting blood gludose, that increses the risk of type 2 diabetes and heart disease.

People who develop type 2 diabetes often have impaired fasting glucose (IFG) and/or impaired glucose tolerance (IGT) ("prediabetes") years before type 2 diabetes is diagnosed.[40] Having IGT is associated with the metabolic syndrome and indicates a relatively high risk of developing diabetes.

Insulin Resistance

Obesity, low levels of physical activity, and a genetic predisposition are common contributors to insulin resistance. Abdominal obesity (an apple-shaped body or waist circumference >40 inches for males, >35 inches for females) is a particularly potent risk factor for insulin resistance. Insulin resistance reduces glucose passage into cells. This in turn prompts the beta cells of the pancreas to produce and secrete additional insulin. Higher than normal blood levels of insulin are generally sufficient to keep blood glucose transfer into cells and blood levels of glucose under control for a number of years. Cells in the pancreas may become exhausted, however, from the years of overwork. In such cases, production of insulin slows down or stops, and as a result, glucose accumulates in blood.[36]

Insulin resistance is related to the development of a spectrum of metabolic abnormalities besides diabetes that have far-reaching effects. Collectively, the adverse effects of insulin resistance are included in a disorder called *metabolic syndrome*. (See Chapter 3, In Focus, and subsequent section, also Case Study 17.2.)

Effects of Diabetes

Health effects of diabetes vary depending on how well blood glucose levels are controlled and on the presence of other health problems such as hypertension or heart disease. In the short run, untreated or poorly controlled diabetes produces blurred vision, frequent urination, weight loss, increased susceptibility to infection, delayed wound healing, and extreme hunger and thirst. In the long run, diabetes may contribute to heart disease, hypertension, blindness, kidney failure, stroke, and the loss of limbs due to poor circulation. The number one cause of death among people with diabetes is heart disease. Many of the side effects of diabetes can be prevented or delayed if blood glucose levels are maintained within the normal range.

Risk Factors

Type 2 diabetes, insulin resistance, and metabolic syndrome share a set of risk factors that include obesity, high levels of central body fat, physical inactivity, and eating patterns providing few whole grains and little fiber. There are genetic components to these disorders, as evidenced by the fact that they track in families and are more likely to occur in certain groups (Hispanic American, African Americans, Asian and Pacific Islanders, and Native Americans) than others.[35]

Effects of weight loss and exercise on the prevention of type 2 diabetes can be quite dramatic.[41] In one large study that took place over a 3-year period, people with prediabetes reduced their risk of developing type 2 diabetes by over 50% with loses in body weight of around 7% and 150 minutes a week of exercise.[42] Diets rich in whole grain and high-fiber foods are protective against the development of type 2 diabetes and appear to aid weight loss.[43] High-fiber, whole grain foods raise blood glucose levels to a lesser extent that do refined grain products and appear to provide nutrients and other biologically active substances that lessen diabetes risk.

Nutrition Interventions for Type 2 Diabetes

Diet and exercise are the cornerstones of the treatment of type 2 diabetes.[44,45] Modest weight loss alone (5–10% of body weight) has been repeatedly shown to significantly improve blood glucose control in overweight and obese people with type 2 diabetes.[46]

In general, diets developed for diabetes emphasize:

- Whole grain breads and cereals and other high-fiber foods, vegetables, fruits, non- and low-fat milk, lean meats, and fish
- Unsaturated fats
- Regular meals and snacks[45,47]
- Carbohydrate counting as a strategy used to quantify the amount and monitor effects of dietary carbohydrate on blood glucose levels (See Illustration 17.4).

Chromium supplements (500–1000 micrograms per day) are sometimes recommended. It appears that the essential mineral chromium improves blood glucose and lipid levels in many people with type 2 diabetes.[48]

Case Study 17.2

Managing Metabolic Syndrome in Adults: Dan Goes Dancing

Dan Beek is 59 years old, semiretired, and lives with his wife in a midtown apartment complex. Dan was diagnosed with metabolic syndrome 10 years ago, and has since gained 15 lb. He attributes his weight gain to lazy afternoons in front of the television and frequent suppers at a local buffet restaurant. Though he plans to take his wife ballroom dancing on the evening of their wedding anniversary, Dan fears he will be out of shape and uncomfortable in the tight confines of his old suit. His wife suggests that the couple speak with a health professional regarding the management of his metabolic syndrome before attempting to lose weight. The following information is obtained at a recent medical follow-up.

Height: 5'9"
BMI: 32 kg/m²
Waist Circumference: 42"
Weight History (in lb):
   Current: 225
   Highest: 225
   Lowest: 200
Healthy Body Weight: 155 to 165 lb

TCHOL: 218 mg/dL
HDL: 33 mg/dL
LDL: 154 mg/dL
TRIG: 155 mg/dL
FBS: 125 mg/dL
TSH: Normal
HgbA1C: 7.1% (4.0%–6.0%)

Questions

1. From the information gathered at his medical visit, how well do you think Dan is managing his metabolic syndrome? Why?

2. What are the desired goals for metabolic syndrome factors (i.e., anthropometric and laboratory indicators)?

3. List the primary sequelae of poorly managed metabolic syndrome.

4. What sort of lifestyle modifications would you discuss with Dan in order to improve the management of his condition?

5. During your visit with the couple, Dan's wife asks you explain "glycemic load" and "GI," wondering if eating foods with a low glycemic index will help her husband's condition. What do you say?

Dietary management of type 2 diabetes[47] should focus on heart disease risk reduction as well as blood glucose control. Food sources of monounsaturated fats, such as vegetable oils, nuts, seeds, lean meats, and seafoods are recommended over foods high in saturated or trans fats. Monounsaturated fats tend to reduce LDL-cholesterol levels while not also reducing blood levels of HDL-cholesterol; they improve insulin resistance and lower blood glucose levels somewhat.[45] Diet and weight-loss interventions may be supplemented by oral medications that decrease insulin resistance and blood lipids, and by insulin if needed.[49] Dietary recommendations for type 2 diabetes are currently not consistent across developed countries, indicating that scientific consensus is yet to be reached on several important issues related to diet and diabetes.[49,50]

Glucophage and Statins

Dietary and exercise interventions may be supplemented by oral medications that decrease insulin resistance and blood lipids, and by insulin if needed. Metformin (brand names: Glucophage®, Glucophage XR®, Glumetza®) is one commonly prescribed oral medication used to treat type 2 diabetes as well as to prevent the onset of diabetes in people at risk and those who have polycystic ovarian syndrome. Metformin acts by increasing the insulin sensitivity of liver, muscle, adipose, and other tissues. It does not increase the

Although carbohydrates have the greatest effect on blood sugar, it is the total diet that contributes to health. Counting calories to maintain a healthy weight, as well as monitoring saturated fat and cholesterol intake, will help reduce the risk of developing heart disease and stroke.[47]

These two lunches provide the same amount of carbohydrate. However, Lunch No. 2 is higher in calories, fat, and cholesterol.

Foods	Carbs (grams)	Calories (grams)	Fat (mg)	Cholesterol
Lunch No. 1	70	480	14	21
1 c vegetable soup with 4–6 crackers				
1 turkey sandwich				
2 slices whole wheat bread				
1 oz turkey				
1 oz low-fat cheese				
1 tsp mayonnaise				
1 small apple				
8 fl oz unsweetened iced tea				
Lunch No. 2	70	823	45	120
Cheeseburger				
1 bun				
1 hamburger patty (not lean)				
2 oz cheddar cheese				
1 tsp catsup				
1 tsp mustard				
3 slices dill pickle				
1 leaf lettuce				
1 slice tomato				
1 cup French fries (fried in oil and lard)				
8 fl oz 2% milk				

Illustration 17.4 Look Beyond the Carbs.

HMG-CoA Reductase The primary enzyme in the metabolic pathway that produces cholesterol. Statins lower blood cholesterol because they slow the action of HMG-CoA.

amount of insulin in the blood. Increased insulin sensitivity allows for improved glucose passage into cellular tissues, thereby reducing blood glucose levels. Statins are a class of drug used to lower blood cholesterol levels (brand names: Lipitor®, Lescol®, Crestor®). Statins work by blocking the enzyme (*HMG-CoA*) responsible for making cholesterol in the liver, resulting in reduced formation of new plaques and reduced size of existing plaques lining arterial walls. Statins also stabilize plaques, making them less prone to rupturing and forming clots that can block arteries. A review paper by Yoshida[51] concluded that statins also reduce arterial inflammation, which contributes to atherosclerosis.

Metabolic Syndrome

The more difficult it is to measure specific health outcomes from a set of complex diagnostic criteria, the more we rely on interpretation and consensus to define certain conditions. The metabolic syndrome is one such instance. Criteria to diagnose metabolic syndrome are not universally accepted. There are at least three sets of definitions for the metabolic syndrome: World Health Organization, International Diabetes Federation Consensus Conference

2005 (IDF), and the National Cholesterol Education Program ATPIII panel with a 2004 update. However, there is clear agreement that reduction of metabolic syndrome will lead to less atherosclerotic disease and less diabetes.

Depending on the definition, it is estimated that 20–30% of adults in the United States and 15% in Canada have metabolic syndrome, with the bulk of cases being made up of overweight and obese inactive adults.[52] With greater levels of visceral fat, there is decreasing ability to dispose of circulating glucose. Metabolic syndrome includes a cluster of metabolic abnormalities that increase the risk of type 2 diabetes and heart disease and include:

- High levels of central body fat
  - IDF: 94 cm for male/80 cm for female Europids (it's ethnicity-specific)
  - ATPIII: greater than 102 cm for males/88 cm for females (40 inches/35 inches)[14]
- High blood insulin levels (IDF, not ATPIII)
- High blood pressure (130/85 mm Hg or higher)
- Elevated blood triglyceride levels (150 mg/dl or higher)
- Low levels of protective HDL cholesterol (less than 50 mg/dL in women, and 40 mg/dL in men)
- Elevated fasting glucose levels (IDF: 100 mg/dL or higher, ATPIII: 110 mg/dL or higher)

The diagnosis of metabolic syndrome is made when three or more abnormalities are identified.[14] Weight loss and physical activity are key components of prevention and management of metabolic syndrome. The ATPIII panel recommends treatment of the underlying causes of the metabolic syndrome and reaching the LDL cholesterol goal, beginning with TLC (therapeutic lifestyle changes). TLC features are to limit saturated fat to less than 7% of calories, limit cholesterol to less than 200 mg per day, and adding 15–25 gram of soluble (viscous) fiber and 2 gram of plant stanols/sterols to the diet each day. This is in addition to intensified weight management and increased physical activity.[14]

HIV/AIDS

Advances in drug therapy have shifted HIV/AIDS in the United States away from a terminal disease characterized by malnutrition and severe weight loss toward being a chronic condition that can be managed.

Definition

AIDS, or acquired immune deficiency syndrome, is caused by the human immunodeficiency virus (HIV). This virus invades and kills specific immune cells, causing increased susceptibility to infection and cancer. Individuals infected with the virus have HIV, whereas people with symptoms of disease have AIDS.

Prevalence

The United Nations, which track the worldwide epidemic of HIV/AIDS, reports that 40 million people live with HIV/AIDS.[53] Of these, 37 million are adults and 2.5 million are children under 15. Worldwide, AIDS has caused the deaths of an estimated 3 million people.

In the United States, the estimated number of diagnosed AIDS cases among 20- to 59-year-olds was nearly 900,000 (reported through 2004).[54] Rates of new cases of AIDS are declining. According to the U.S. Centers for Disease Control and Prevention, AIDS affects nearly seven times more African Americans and three times more Hispanics than whites.[55]

Etiology

HIV is transmitted through blood and body fluid exchange. Unprotected sex and sharing of contaminated needles are routes of viral transmission among adults. Women can transmit HIV to their babies during pregnancy, at birth, or by breastfeeding. Approximately one-quarter to one-third of all untreated pregnant women infected with HIV will pass the infection to their babies. HIV also can be spread to babies through the breast milk of mothers infected with the virus. If the mother takes the drug AZT during pregnancy, she can significantly reduce the chances that her baby will get infected with HIV.[55]

Effects of HIV/AIDS

HIV is able to make its own DNA and replicate itself by using genetic material from the host's cells. It penetrates the body's immune cells and eventually destroys the cells. Decreased functional level of the immune system increases the risk that people with HIV will develop infections and cancer, although not all individuals with HIV go on to develop AIDS. CDC is responsible for tracking the spread of AIDS in the United States.

Nutrition-related symptoms common in people with AIDS affect dietary intake and include:

- Coughing and shortness of breath
- Seizures and lack of coordination
- Difficult or painful swallowing
- Mental disturbances such as confusion and forgetfulness
- Severe and persistent diarrhea
- Fever
- Nausea, abdominal cramps, and vomiting
- Weight loss and extreme fatigue

AIDS progression often begins when HIV mutates and drugs used to treat the virus become ineffective. AIDS progression is marked by decreased appetite, increased nutrient needs, nutrient deficiencies, weight loss, and tissue wasting.[56] HIV/AIDS raises nutrient requirements as long as the infection is present, and may cause other nutrition-related conditions, such as nutrient malabsorption due to infection in the gastrointestinal tract.[56]

Drug therapy for HIV/AIDS causes changes in body shape due to the redistribution of fat stores. Fat stores shift from the arms and legs to the central part of the body. Increased central body fat stores elevate blood triglyceride and cholesterol levels, and may increase the risk of heart disease and insulin resistance in the long term.[57]

Nutrition Interventions

Nutrition interventions for people with HIV/AIDS center on maintaining weight and nutritional adequacy, and reducing blood lipid levels and insulin resistance related to drug treatment. During the early phase of the disease, adequate nutrient intakes improve immune function and decrease susceptibility to infection. Currently, even the best nutritional advice and self-care cannot prevent the eventual progression of AIDS and restore immune function, however. Weight maintenance and nutritionally adequate diets can help people with the disease increase their level of control and sense of well-being.[56]

Key Points

1. Habitual consumption of high-calorie, high-saturated fat diets that are low in fruits, vegetables, and whole grains contributes to disability and death from chronic disease.

2. Cancer is a group of diseases; risks of developing cancer are reduced in persons who maintain a healthy weight, eat 3 to 14 servings of fruits and vegetables per day, and consume alcohol in moderation, if at all.

3. The population of cancer survivors is growing rapidly, highlighting the need for greater research into the aftereffects of aggressive cancer treatment.

4. The plaque that occludes arteries in cardiovascular disease develops over time, which means prevention strategies are best begun early. Key nutritional risk-reduction strategies are similar to those for reducing risk of diabetes and are reflected in the Dietary Guidelines for Americans.

5. Obesity is a major public health problem. Over half of U.S. adults were at a healthy weight in 1960; now about one-third of adults are at a healthy weight.

6. The Body Mass Index is used as a proxy for body fat and as a descriptor of weight status, but it is not an adequate indicator of body fat for athletes, for those with little muscle mass, for dehydrated individuals, or for those with large, dense bones.

7. Maintaining weight loss is more difficult than achieving weight loss. Components of successful weight-management programs include long-term commitment, setting realistic goals, adopting individualized eating plans, having problem-solving techniques to deal with backsliding, and an exercise component to avoid losing muscle mass.

8. The treatment cornerstones for type 2 diabetes are diet and exercise; similarly, maintaining normal body weight through good nutrition and exercise is a strategy to reduce risk of type 2 diabetes.

Resources

Diabetes and Metabolic Syndrome Resources

Go to the "Conditions Center" on the WedMD gateway site and select diabetes and hit "go." Get the latest information on the treatment and a risk assessment for diabetes. Site leads to reliable information on insulin resistance and metabolic syndrome as well.
Website: **www.webmd.com**

Search insulin resistance, diabetes, metabolic syndrome, and hypoglycemia to find reliable reports.
Website: **www.healthfinder.gov**

AIDS/HIV Resources

Pan American Health Organization (PAHO). For current statistics on the number of reported AIDS cases in North, Central, and South America, contact the regional World Health Organization office for the Americas at 525 23rd Street, N.W., Washington, D.C. 20037, telephone: 202–861–4346.

AIDSinfo provides referrals and current information on federally and privately funded clinical trials for AIDS patients and others infected with HIV. AIDSinfo is primarily Web-based and can be found at http://aidsinfo.nih.gov. AIDSinfo also operates a telephone service from 12:00 p.m. to 5:00 p.m. Eastern Time, Monday through Friday. English and Spanish-speaking health information specialists are available to answer questions about HIV/AIDS, treatment options, and navigating the website.

Telephone: 800-HIV-0440 (1–800–448–0440)
International: 301–519–0459
TTY/TDD: 888–480–3739
Email: ContactUs@aidsinfo.nih.gov

References

1. Healthy People 2010: national health promotion and disease prevention objectives. Washington, DC: U.S. Dept of Health and Human Services; 2000.

2. National Center for Health Statistics. Health. United States, 2005 Chartbook. Hyattsville, MD: 2005.

3. Cancer Trends Progress Report—2005 Update, National Cancer Institute, NIH, DHHS, Bethesda, MD, December 2005, http://progressreport.cancer.gov, accessed 11/06.

4. American Cancer Society. *Cancer Facts and Figures 2006.* Atlanta: American Cancer Society, 2006.

5. National Cancer Institute. Physical Activity and Cancer: Fact Sheet, available at http://cancer.gov/cancertopics, accessed 11/06.

6. The American Heart Association. American Heart Association heart and stroke statistics. 2005 update. Available at www.americanheart.org, accessed 11/06.

7. Gauthier GM, Keevil JG, McBride PE. The association of homocysteine and coronary artery disease. Clin Cardiol. 2003; 16(12):563–68.

8. Stanger O et al. On behalf of the DACH-LIGA Homosystein e.V. DACH-LIGA Homocystein (German, Austrian and Swiss Homocysteine Society): Consensus paper on the rational clinical use of homocysteine, folic acid and B-vitamins in cardiovascular and thrombotic diseases: guidelines and recommendations.

Clin Chem Lab Med. 2003;41(11): 1392–1403.

9. Ridker PM et al. C-reactive protein and other markers of inflammation in the prediction of cardiovascular disease in women. N Eng J Med. 2000;342:836–43.

10. Willerson JT, Ridker PM. Inflammation as a cardiovascular risk factor. Circulation. 2004;109:2–10.

11. Van den Hoogen PC et al. The relation between blood pressure and mortality due to coronary heart disease among men in different parts of the world. Seven Countries Study Research Group. N Eng J Med. 2000; 342:1–8.

12. Krauss RM et al. AHA dietary guidelines: revision 2000: a statement for healthcare professionals from the Nutrition Committee of the American Heart Association. Circulation. 2000;102:2284–99.

13. Report of the National Cholesterol Education Program Expert Panel on Detection, Evaluation, and Treatment of High Blood Cholesterol in Adults. The Expert Panel. Arch Intern Med. 1988;148:36–69.

14. Third report of the National Cholesterol Education Program (NCEP) Expert Panel on Detection, Evaluation, and Treatment of High Blood Cholesterol in Adults (Adult Treatment Panel III). J Am Med Assoc. 2001;285:2486–97; Available at www.nhlbi .nih.gov/guidelines/cholesterol/atp_iii.htm, accessed 11/06.

15. Grundy SM et al. Definition of metabolic syndrome: report of the National Heart, Lung, and Blood Institute/American Heart Association Conference on Scientific Issues Related to Definition. *Circulation.* 2004;109: 433–438.

16. Lamarche B et al. Fasting insulin and apolipoprotein B levels and low-density lipoprotein particle size as risk factors for ischemic heart disease. JAMA. 1998;279:1955–61.

17. Giles WH et al. Association between total homocyst(e)ine and the likelihood for a history of acute myocardial infarction by race and ethnicity: results from the Third National Health and Nutrition Examination Survey. Am Heart J. 2000;139: 446–53.

18. National Academy of Sciences. Food and Nutrition Board. Dietary Reference Intakes for energy, carbohydrates, fiber, fat, fatty acids, cholesterol, protein and amino acids. Institute of Medicine, National Academy of Sciences, Washington, DC:National Academies Press; 2002. http://books.nap.edu/ books/0306085373/html, accessed 05/03.

19. Wilson PWF et al. Prediction of coronary heart disease using risk factor categories. Circulation. 1998;97:1837–47.

20. American Association of Clinical Endocrinologists. Medical guidelines for the clinical use of dietary supplements and nutraceuticals. Endocrine Prac. 2003;9(5): 418–70. Available at aace.com/clin/ guidelines/nutraceuticals, accessed 11/03.

21. Schulz V et al. Rational phytotherapy. A reference guide for physicians and pharmacists. Berlin: Springer Verlag; 2004.

22. Blumenthal M, ed. The ABC clinical guide to herbs. Austin, TX: American Botanical Council; 2003.

23. Malesky G. Therapies: statins, muscle damage, and coenzyme Q10. Nutrition in Complementary Care, A Dietetic Practice Group of the American Dietetic Association. Summer 2005; 8(1):1,6–8.

24. Overweight and Obesity. Obesity trends: U.S. obesity trends 1985–2005. Available at www.cdc.gov/nccdphp/dnpa/obesity/ trend/maps/, accessed 11/06.

25. Flegal KM, Troiano RP, Ballard-Barbash R. Aim for a healthy weight: what is the target?" J Nutr. 2001;131:440S–450S.

26. U.S. Department of Agriculture, U.S. Department of Health and Human Services. Nutrition and your health: dietary guidelines for Americans, 5th ed. 2000.

27. NHLBI Obesity Education Initiative Expert Panel on the Identification Evaluation, and Treatment of Overweight and Obesity in Adults. Clinical guidelines on the identification, evaluation, and treatment of overweight and obesity in adults: the evidence report. Bethesda, MD: National Institutes of Health; 1998. Available at www.nhlbi.nih .gov/guidelines/obesity/ob_home.htm.

28. Flegal KM et al. Prevalence and trends in obesity among US adults, 1900–2000. J Am Med Assoc. 2002;288(14):1723–27.

29. Wright JD et al. Dietary intake of ten key nutrients for public health, United States: 1999–2000. Advance data from vital and health statistics; no.334. Hyattsville, MD: National Center for Health Statistics; 2003.

30. Flegal KM et al. Excess deaths associated with underweight, overweight, and obesity. JAMA. 2005;293:1861–67.

31. Satcher D. Surgeon General's Conference on Obesity, a year-long process to develop a weight reduction plan, co-sponsored by National Institutes of Health, Centers for Disease Control and Prevention, and the Office of Public Health and Science; January 8, 2001.

32. Position of the American Dietetic Association: weight management. J Am Diet Assoc. 2002;102(8):1145–55.

33. McGuire MT et al. Behavioral strategies of individuals who have maintained long-term weight losses. Obes Res. 1999;7:334–41.

34. Wyatt HR et al. Long-term weight loss and breakfast in subjects in the National Weight Control Registry. Obes Res. 2002;10:78–82.

35. McTigue KM et al. Screening and interventions for obesity in adults: summary of the evidence for the U.S. Preventive Services Task Force. Ann Intern Med. 2003;139:933–49.

36. Parkes E. Nutritional management of patients after bariatric surgery. Am J Med Sci. 2006;331(4):207–13.

37. World seen facing diabetes catastrophe, Impact may outpace AIDS. International Diabetes Federation Conference, Paris 2003. Reported in www.medscape.com, accessed 9/03.

38. Cowie CC. MMWR, CDC Surveillance System, 2003:52:833–37 and Diabetes rose slightly in the 1990s. Centers for Disease Control Press Release, 9/12/03.

39. National Diabetes Fact Sheet 2005. Available at www.diabetes.org/ucdocuments/ National DiabtesFactSheetRev.pdf, accessed 11/06.

40. American Diabetes Association. Diagnosis and classification of diabetes mellitus. Diabetes Care. 2006;29(1):S43–S48.

41. Sheard NF. Moderate changes in weight and physical activity can prevent or delay the development of type 2 diabetes mellitus in susceptible individuals. Nutr Rev. 2003;61: 76–79.

42. Diabetes Prevention Program Research Group. Reduction in the incidence of type 2 diabetes with lifestyle intervention or metformin. N Engl J Med. 2002;346; 393–403.

43. McKeown NM et al. Whole grain is favorably associated with metabolic risk factors for type 2 diabetes and cardiovascular disease in the Framingham Offspring Study. M J Clin Nutr. 2002;76:390–98.

44. Brand-Miller J et al. Low glycemic index foods and glycemic control in diabetics. Diabetes Care. 2003;26:2261–67.

45. Solomon CG. Reducing cardiovascular risk in type 2 diabetes. N Engl J Med. 2003; 248:457–60.

46. Heymsfield S et al. Effects of weight loss with Orlistat on glucose tolerance and progression to type 2 diabetes in obese adults. Arch Inter Med. 2000;160:1321–26.

47. Franz MJ et al. Evidence-based nutrition principles and recommendations for the treatment and prevention of diabetes and related complications. Diabetes Care. 2002; 25:148–66.

48. Wong Z et al. Chromium supplements and glucose sensitivity in type 2 diabetes. 18th IDF Congress, abstract 154, 756, 762, Aug. 28, 2003.

49. Nathan DM. Initial management of glycemia in type 2 diabetes mellitus. N Engl J Med. 2002;347:1242–50.

50. Foster-Powell K, Holt SH, Brand-Miller JC. International table of glycemic index and glycemic load values: 2002. Am J Clin Nutr. 2002;76:5–56.

51. Yoshida M. Potential role of statins in inflammation and atherosclerosis. J Atheroscler Thromb. 2003;10(3):140–54.

52. Park Y-W et al. The metabolic syndrome: prevalence and associated risk factors. Arch Int Med. 2003;163:427–36.

53. Joint United Nations Programme on HIV/AIDS (UNAIDS)/World Health Organization (WHO) Annual Report. AIDS

epidemic update 2003. December 2003, Geneva, Switzerland. Available at www.unaids.org

54. Centers for Disease Control and Prevention. Division of HIV/AIDS Prevention-Surveillance and Epidemiology, National Center for HIV, STD, and TB Prevention. Atlanta, GA. Available at www.cdc.gov/hiv/topics/, accessed 11/06.

55. Gayle HP, Janssen RS. Successful implementation of perinatal HIV prevention guidelines. A multistate surveillance evaluation. Morbid & Mortal Week Rep. 2001;50; (RR-6):15–28.

56. Kline DA. AIDS in the 21st century. Today's Dietitian. July 2003;12–16.

57. Kuritzkes DR, Currier J. Cardiovascular risk factors and antiretroviral therapy. N Engl J Med. 2003;348:679–80.

"Nutrition is one of the major determinants of successful aging."

Chapter 18

Nutrition and Older Adults

Chapter Outline

- Introduction
- A Picture of the Aging Population: Vital Statistics
- Theories of Aging
- Physiological Changes
- Nutritional Risk Factors
- Dietary Recommendations
- Nutrient Recommendations
- Food Safety Recommendations
- Physical Activity Recommendations
- Nutrition Policy and Intervention for Risk Reduction
- Community Food and Nutrition Programs

Masterfile

Photo Disc

Photo Disc

Prepared by U. Beate Krinke

Key Nutrition Concepts

1 Eating and enjoying a varied diet contributes to mental and physical well-being.

2 Generalizations relative to health status changes with aging are unwise because "older adults" are a heterogeneous population.

3 Diseases and disabilities are *not* inevitable consequences of aging.

4 Functional status is more indicative of health in older adults than chronological age.

5 Body composition changes that occur with aging can alter lifestyle; these have the greatest impact on nutritional needs.

Introduction

"An ounce of prevention is worth a pound of cure."

Traditional

In "normal" aging, inevitable and irreversible physical changes occur over time, and some diseases are more prevalent. The leading causes of death are heart disease, cancer, and stroke; these are linked with nutrition in complex ways.[1] Most older people consider themselves to be healthy. More than anything, they want to remain healthy and independent and certainly do not want to be a burden to others.[2] Older adults feel that good nutrition and exercise are the most important health habits they can maintain in order to avoid losing autonomy and independence.[2]

Just exactly what is good nutrition for older adults? They can meet their decreasing energy requirements by choosing more nutrient-dense foods. They can drink more water to stay hydrated, even when not thirsty. By eating adequate vegetables, fruits, and whole grains, keeping fats in balance, and drinking alcohol only in moderation, older adults can reduce their risks of getting chronic diseases. Diet quality is linked to the longevity of older men and women.[3,4] Older adults can live longer and better (i.e., postponing disability and shortening the period of decreased functional capabilities at the end of life) with good health habits.[5] This chapter defines aging and provides information about the nutrient requirements, dietary recommendations, and food and nutrition programs designed to support healthy aging.

Longevity Length of life; it is a measure of life's duration in years.

What Counts As Old?

Many chronological ages have been used as cut-points to mark the beginning of "old age." While no biological benchmark signals a person's becoming old, there are societal and governmental definitions for *old*. Societal definitions label someone as "old" at a fairly young age. Turning 50 makes you eligible to join the American Association for Retired Persons (AARP). Views on "normal" retirement ages are changing, however. At age 60, many cafeterias, movie theaters, and shops give senior discounts. What counts as "old" depends on who is counting.

Governmental definitions of old age include the following categories suggested by the U.S. Census Bureau:

- 65 to 74 years is "young old"
- 75 to 84 years is "aged"
- 85 and older is "oldest old"

The Elderly Nutrition Program, first funded under the Older Americans Act in 1972, calls "older adults" those aged 60 years and older. The Social Security program identifies people 65 years old as being eligible for retirement benefits (62 years is "early retirement"). These various age cut-points are important because services and programs that support nutrition for the elderly do not use the same age limits. The arbitrarily set retirement age of 65 years is most commonly used in textbooks, and we will use that label here. Readers will also learn that many factors besides chronological age affect nutrition in the elderly.

Chronological age is a marker of where we are on life's path. Our functional status, defined as how well we accomplish the desired tasks of daily living, is more indicative of health than chronological age. Rather than ask, "How old are you?" we should ask, "What can you do?" Nonetheless, chronological age is an easy measure, so we continue to ask that question. Our perception is that age is a proxy for predicting health status and functional abilities.

Food Matters: Nutrition Contributes to a Long and Healthy Life

"We found that tomatoes were the primary food associated with higher functional status in centenarians … even when controlled for other factors such as illness, depression, and gastrointestinal problems."

Dr. Mary Ann Johnson, commenting on findings that elderly nuns with higher functional status also had higher blood lycopene levels[6]

In our search for magic bullets, tomatoes will probably not turn out to be nature's perfect food. The cumulative effects of lifelong dietary habits determine nutritional status in old age. Good nutrition throughout life contributes to optimal growth, to appropriate weight, and to nutrient levels in blood and other tissues that provide disease resistance. In trying to assess the contribution good nutrition can make to longer life, the Centers for Disease Control and Prevention (CDC) suggest that *longevity* depends 19% on genetics, 10% on access to high-quality health care, 20% on environmental factors such as pollution, and 51% on lifestyle factors.

Besides not smoking, diet and exercise are estimated to be the lifestyle factors contributing most to decreased mortality, or longer life.[7] In a longitudinal study including diet monitoring, older women who ate the healthiest diets were 30% less likely to die during a 6-year study period than the women who ate few whole grains, fruits, vegetables, low-fat dairy products, and lean meats.[4]

The role of food and nutrition often changes during aging. Besides reducing risk of disease and delaying death, diet plays a role in health and longevity by contributing to wellness. Wellness means having the energy and ability to do the things one wants to do and to feel in control of one's life. Being able to choose, purchase or prepare, and eat a satisfying diet every day; enjoying traditional foods at holidays, birthdays, and other special occasions; and having the resources to purchase desired foods on a regular basis all contribute to independence and a higher quality of life. Good nutrition, as defined by dietary guidelines covered later in this chapter, can help to "add life to years" as well as add years to life.

A Picture of the Aging Population: Vital Statistics

More and more of us are growing old and older. During Roman times, fewer than 1% of the population reached age 65; but today, 12.4% of the North American population is aged 65 and older. Of 300 million people in the United States in 2006, 37.2 million are at least 65 years old. By 2050, one of five people is expected to be age 65 or older.[7]

Persons aged 85 years and older are the fastest growing segment of the population.[7] A White House conference on aging called this age wave "a demographic revolution" and predicted that it will change the twenty-first century much as information technologies revolutionized the twentieth century, and the industrial revolution affected the nineteenth century.[8]

Global Population Trends: Life Expectancy and Life Span

Today's life expectancy of 78 years is quite different from 1900, when average life expectancy was 47 years. The National Center for Health Statistics (NCHS) reports that women born in 2001 in the United States can expect to live 80 years, whereas men can expect to live 74 years (see Table 18.1). For babies born in 2050, life expectancy is expected to increase to 84 years for women and 80 years for men.[9] Life expectancy in the United States lags behind that of many other countries, partially because infants in the United States have higher mortality rates than those in other nations. Other contributors to life expectancy estimates include rates of childhood mortality, infectious and chronic diseases, and death from violence and accidents.

Since the early 1900s, immunizations and other risk-reduction measures, treatment of disease, decreased infant and childhood mortality rates, and clean water and safe

Table 18.1 Range of life expectancy for 15 of the 37 countries reported in Health, United States 2005, for 2001, according to gender

Country	Female Life Expectancy at Birth, years	Country	Male Life Expectancy at Birth, years
Japan	84.9	Japan	78.1
Switzerland	83.0	Sweden	77.6
France	82.9	Switzerland	77.4
Spain	82.9	Canada	77.1
Italy	82.8	Israel	77.1
Australia	82.4	Australia	77.0
Canada	82.2	Italy	76.7
Sweden	82.1	Singapore	76.5
Israel	81.6	Norway	76.2
Costa Rica	79.9	Finland	74.6
United States	79.8	United States	74.4
Ireland	79.7	Portugal	73.5
Bulgaria	75.4	Hungary	68.1
Romania	75.0	Romania	67.7
Russian Federation	72.3	Russian Federation	59.1

SOURCE: Health, United States 2005, with Chartbook in Trends in the Health of Americans. December 2005. Internet version at www.cdc.gov/nchs, accessed October 16, 2006.

food have increased average *life expectancies*, which are getting closer to the potential human *life span*. For instance, from 1980 to 2000 life expectancy increased by more than 4 years for men and 2 years for women, largely due to fewer deaths from heart disease, stroke, and accidents among older adults.[9] Overall, the United States has the second highest number of people 80 years and over (China ranks first).

Although life expectancy is rising and the population is aging, human life span remains stable at around 110 to 120 years.[10] Few people currently survive to age 120. However, Jeanne Calment, the oldest known person, who lived to age 122, is an exception. Nonetheless, the number of centenarians (persons living to age 100 and beyond) in the United States has increased markedly and may add to our understanding of current limits on the human life span. Demographers are predicting that by 2030, there will be 381,000 centenarians in the United States—up from 50,364 in the year 2002.

Life Expectancy   Average number of years of life remaining for persons in a population cohort or group; most commonly reported as life expectancy from birth.

Life Span   Maximum number of years someone might live; human life span is projected to range from 110 to 120 years.

Nutrition: A Component of Health Objectives for the Older Adult Population

Behaviors that can enhance the health of an aging population are reflected in national health objectives. Table 18.2 identifies dietary goals related to disease prevention and to health promotion for older adults. The current set of health objectives emphasizes public education about health consequences of overweight and obesity. The greatest dietary improvement in the older adult population would be to eat more vegetables and grains, especially whole grain products.

Theories of Aging

What triggers aging? Theories explaining aging grow from human desire to understand the biological processes that determine how long and how well we live. Aging theory tries to explain the mechanisms behind loss of physical resilience, decreased resistance to disease, and other physical and mental changes that accompany aging.

Biological systems are much too complex to have one theory that is robust enough to explain the mechanisms of aging.[11] Basic biological processes involved in aging are largely determined by genetics. However, environmental factors influence expression of the genetic code by exacerbating certain traits. For example, height and weight are genetically programmed, but diet and other environmental exposures can moderate these outcomes.[12] Persons with a family history of high levels of LDL cholesterol

Table 18.2 Healthy People 2010, food and nutrition goals for older adults

| | | Percentage of Population | |
| | | Current, Age 60+ | |
	Target	Females	Males
Increase the number who are at a healthy weight	60	37	33
Decrease the number who are obese (BMI 30+)	15	26	21
Increase the proportion of persons who:			
• Eat at least 2 daily servings of fruit	75	35	40
• Eat at least 3 daily servings of vegetables (at least one-third dark green or deep yellow)	50	6	5
• Eat at least 6 daily servings of grain products (at least 3 of those as whole grains)	50	4	11
• Eat less than 10% of calories from saturated fat	75	47	42
• Eat no more than 30% of calories from fat	75	40	34
Baseline information is for age 50 and older			
Meet dietary recommendations (1200 mg) for calcium	75	27	35
Baseline information is for age 65+			
Increase the proportion of physician office visits made by patients with a diagnosis of cardiovascular disease, diabetes, or hyperlipidemia that include counseling or education related to diet and nutrition	75	33[a]	33[a]

SOURCE: *Healthy People 2010.*[10]

[a]Adequate data to distinguish levels for males and females is not available.

(i.e., bad cholesterol) and early death from cardiovascular disease can moderate the outcomes suggested by their genetic programming through weight loss, a diet low in saturated fat, adequate exercise, and no smoking. Someone with a family history of type 2 diabetes can reduce his or her risk of developing the disease with a healthy diet, exercise, and maintenance of normal weight.

Theories of aging fall into three major categories: (1) programmed aging theories, (2) "wear-and-tear" theories, and (3) caloric restriction. Categories of aging theories overlap to some extent because it is clear that neither genetic programming nor environmental exposures fully account for changes related to aging.

Programmed Aging

Hayflick's Theory of Limited Cell Replication

Hayflick proposed that all cells contain a genetic code that directs them to divide a certain number of times during their life span.[10] After cells divide according to their programmed limit, and barring disease or accident, cells begin to die. For example, if individual cells of a fly have a 3-day life span and replicate 15 times, a fly can live 45 days. Using this theory, Hayflick calculated the potential human life span to be in the range of 110 to 120 years, estimating that human cells replicate from 40 to 60 times. Although most human cells can regenerate (e.g., blood, liver, kidney, and skin cells reproduce themselves), not all cells have that capacity (e.g., spinal cord, nerves, and brain cells). Hayflick's theory is difficult to prove in humans because we die from age-associated chronic disease more often than from old age itself.

Molecular Clock Theory

Another theory of programmed aging is that of the molecular clock. *Telomeres* that cap the ends of chromosomes act as clocks, becoming a bit shorter with each cell division. Eventually loss of telomeres stops the ability of chromosomes to replicate, and they become *senescent*. Loss of chromosomal replication may produce signs of aging because new cells cannot be formed, and the function of existing cells declines with time. A major thrust of current research is to identify ways to limit loss of telomeres and thus prolong cell replication.[13]

Wear-and-Tear Theories of Aging

Wear-and-tear theories are built on the concept that things wear out with use. Mistakes in the replication of cells or buildup of damaging by-products from biological processes eventually destroy the organism. Cytotoxicity (poisoning of the cell) results when damaged cell components accumulate and become toxic to healthy cells. According to this theory, the accumulation of damaged cells and waste by-products leads to aging.

Oxidative Stress Theory

Oxygen is an integral and versatile part of metabolic processes; it can both accept and donate electrons during chemical reactions. One cause of aging is thought to be oxidative stress due to the buildup of reactive (unstable) oxygen compounds. Unstable oxygen, formed normally during metabolism (e.g., hydroxyl radicals) can also damage cells by initiating reactions that break down cell membranes and modify the normal metabolic processes that protect people from disease. Exposure to oxidizing agents is increased by smoking, ozone, solar radiation, and environmental pollutants. Unstable oxygen compounds are neutralized, however, when they combine with an antioxidant. This prevents them from interfering with normal cell functions. The body produces antioxidant enzymes (such as catalases, glutathione, peroxidase reductases, and superoxide dismutase), but part of our need for antioxidants is met from the diet. Dietary antioxidants include selenium, vitamins E and C, and other phytochemicals such as beta-carotene, lycopene, flavonoids, lutein, zeaxanthin, resveratrol, and isoflavones.

Rate of Living Theory

The rate of living theory is similar to the oxidative stress theory in that it suggests that "faster" living results in faster aging.[14] For example, higher metabolic rate and energy expenditure leads to greater turnover of all body tissues. Theoretically, fast-paced living shortens life span, whereas living more slowly leads to a longer life. Scientists have not adequately examined old people, including centenarians, to fully understand how this theory works.

Calorie Restriction and Longevity

Animal studies (e.g., of fruit flies, water fleas, spiders, guppies, mice, rats, and other rodents) show that an energy-restricted diet that meets micronutrient needs can prolong healthy life.[15] For example, laboratory mice and rats fed calorie-restricted diets live longer and have fewer age-associated diseases than their counterparts whose diets are unrestricted. In the 1930s, McCay and colleagues suggested that delays in aging result after food restriction, due to slowed growth and development.[16] But since then, rodent studies have shown that instituting *caloric restrictions* in midlife, after growth and development were completed, results in longer life spans.[15,17]

Caloric restriction is also being examined in primates.[18] Because nonhuman primate life expectancy is 30 to 40 years, it will be years until we learn whether life extension through caloric restriction might be applied to primates. Stay tuned.

Could calorie-restricted diets also extend human life? Experimental findings in small animals have led some individuals, such as Dr. Roy Walford of Biosphere 2, to personally adopt very low-calorie diets. Walford coordinated the calorie-restricted diets of eight normal-weight people living in Biosphere 2.[19] However, such a small study lasting 2 years is too limited to determine human life span extension results from caloric restriction. Luigi Fontana and colleagues tested the theory that caloric reduction also reduces a thyroid hormone (T3) that controls cell respiration and free radical production.[20] They compared hormone levels of healthy lean, weight-stable adults eating an 1800 kcal diet for 3–15 years with those of two matched groups eating typical western diets of 2400 (sedentary group) or 2800 kcals (exercise group). Fontana's group concluded that reducing metabolic rate and oxidative stress may slow the rate of aging.

> **Telomere**   A caplike structure that protects the end of chromosomes; it erodes during replication.
>
> **Senescent**   Old to the point of nonfunctional.
>
> **Calorie Restriction**   Decreasing the energy level of one's diet by 25 to 30% while meeting protein, vitamin, and mineral needs.

From an ecological view, people in France and Japan have lower caloric intake than do people in the United States, and people in both of those countries have longer life expectancies.[21] Physiologically, we know that nutrition affects human longevity by moderating risks of developing chronic diseases, ameliorating certain chronic conditions, and contributing to healing in a variety of acute conditions.[9,22] Severe caloric restriction during famine leads to malnutrition and starvation, with poor outcomes in human reproduction, growth, development, immune status, and healing. Obesity leads to early death, and so does starvation. An example of reducing calories might be adding vegetables to a diet in order to decrease chronic disease risk. Eating nutrient-dense foods and avoiding obesity enhances prospects for longevity. Scientists search for the perfect range of energy and nutrient intake; one that will maintain a weight range associated with optimal health, longevity, and quality of life.

> "Everything in moderation. Including moderation."
>
> Joyce Hendley, food writer, citing Julia Child

Physiological Changes

Normal aging is associated with shifts in body composition and subsequent loss of physical *resilience*. Physiological system changes commonly associated with healthy aging are described in Table 18.3. As scientists come to understand the human aging process, they will learn to sort through age-associated physiological changes to be able to distinguish exactly which changes are due to genetic factors and which are due to poor diets, inactivity, or other lifestyle-related factors. Yet aging is not all loss or decline. Rather, healthy aging is associated with continuing psychosocial, personal, moral, cognitive, and spiritual development. Diseases that affect older persons' health are discussed further in Chapter 19.

Body Composition Changes

Lean Body Mass (LBM) and Fat   Individual shifts in body composition are common but neither inevitable nor irreversible.[23] Of all physiologic changes that do occur during aging, the biggest effect on nutritional status is due to the shifts in the musculoskeletal system (Table 18.3). Many individuals experience a decline in *lean body mass* with aging (see Table 18.4). On average, fat-free mass decreases by 15% in the 50 years from the mid-twenties to the mid-seventies.[24] and is known as sarcopenia. These composition changes are associated with lower levels of physical activity, food intake, and hormonal changes in women.[25] Loss of mineral and muscle mass is also accompanied by loss of body water.

Resilience   Ability to bounce back, to deal with stress and recover from injury or illness.

Lean Body Mass   Sum of fat-free body tissues: muscle, mineral as in bone, and water.

Table 18.3 Age-associated physiological system changes that affect nutritional health*

Cardiovascular System
- Reduced blood vessel elasticity, stroke volume output
- Increased blood pressure

Endocrine System
- Reduced levels of estrogen, testosterone
- Decreased secretion of growth hormone
- Reduced glucose tolerance
- Decreased ability to convert provitamin D to previtamin D in skin

Gastrointestinal System
- Reduced secretion of saliva and of mucus
- Missing or poorly fitting teeth
- Dysphagia or difficulty in swallowing
- Reduced secretion of hydrochloric acid and digestive enzymes
- Slower peristalsis
- Reduced vitamin B_{12} absorption

Musculoskeletal System
- Reduced lean body mass (bone mass, muscle, water)
- Increased fat mass
- Decreased resting metabolic rate
- Reduced work capacity (strength)

Nervous System
- Blunted appetite regulation
- Blunted thirst regulation
- Reduced nerve conduction velocity affecting sense of smell, taste, touch, cognition
- Changed sleep as the wake cycle becomes shorter

Renal System
- Reduced number of nephrons
- Less blood flow
- Slowed glomerular filtration rate

Respiratory System
- Reduced breathing capacity
- Reduced work capacity (endurance)

*Some of these age-associated changes, such as the increase in blood pressure, are usual but not normal.

Overall, older people have lower mineral, muscle, and water reserves to call upon when needed.

At the same time, many older adults gain body fat. Compared to males in their twenties, males in their seventies have roughly 24 pounds less muscle (a decrease from 53 to 29 lb on average) and 22 pounds more fat (an increase from 33 to 55 lb on average). Think about losing 24 lb of muscle: It is roughly equivalent to the muscle mass that girls or boys gain throughout the puberty growth spurt. Over 50 years,

Table 18.4 Comparison of body composition of a young and an old adult		
	20 to 25 Years	70 to 75 Years
Protein/cell solids	19%	12%
Water	61%	53%
Mineral mass	6%	5%
Fat	14%	30%

SOURCE: Data from Chernoff R (ed). Geriatric Nutrition The Health Professional's Handbook, Sudbury, MA: Jones and Bartlett; 2006, p. 435; based on NW Shock, *Biological Aspects of Aging*, New York, NY: Columbia University Press, 1962.

24 lb of muscle slowly goes away to be replaced by 22 lb of fat. Of course, the extra fat does provide a reserve of energy for periods of low food intake, recovery from illness or surgery, acts an insulator in cold weather, and cushions falls. These shifts in body composition tend to occur even when weight is stable.[25]

Muscles: Use It or Lose It   Many older people expect decreases in physiologic function with increasing age. However, physical activity contributes to staying strong, no matter the age.[24] For example, the HEalth, RIsk Factors, Training, And GEnetics (HERITAGE) Family Study compared the effects of a 20-week strength-training program on older and younger men and women, finding that the training response differed by gender and by race, but *not* by age.[26] Training exercises led to increases in fat-free mass; decreases in total, subcutaneous, and visceral fat mass; and to weight loss. Weight-bearing and resistance exercise increase lean muscle mass and bone density.[24,27] Because muscle tissue contains more water than fat tissue, building muscle also results in more stored water. Regular physical activity, including strengthening and flexibility exercise, contributes to maintenance of *functional status.*

> "I would not wish to imagine a world in which there were no games to play and no chance to satisfy the natural human impulse to run, to jump, to throw, to swim, to dance."
>
> Sir Roger Bannister, 1989

Weight Gain   Weight gain, although not inevitable, tends to accompany aging. Large cross-sectional studies show that mean body weight increases gradually during adulthood. Weight and BMI peak between 50 and 59 years.[28] Reasons for age-associated gains are uncertain, but longitudinal studies are showing that lack of exercise could be a factor.[23] For example, men in the Baltimore Longitudinal Study on Aging decreased their energy expenditure by 17 to 24 calories per day after age 55, but gained weight.[29] In the Fels Longitudinal Study, men gained 0.7 pounds per year and women gained an average of 1.2 pounds as they aged.[25]

Subjects were 40 to 66 years old when entering the study and were followed for up to 20 years. Weight gains were concurrent with decreases in lean body mass and increases in body fat. This overall weight and body composition shift was moderated by physical activity. For example, the two groups with moderate or high physical activity levels (as opposed to the least active group) increased lean body mass and decreased total and percentage of body fat as age increased. Physical activity effects differed by gender. In women, higher levels of physical activity were associated with higher levels of lean body mass. However, lack of estrogen seems to promote fat accumulation, and total weight increased regardless of the group's activity level. (The menopausal transition was covered in Chapter 16.) Men in the highest physical activity groups in the Fels Longitudinal Study slowed their total body weight and body fat gains.

Changing Sensual Awareness: Taste and Smell, Chewing and Swallowing, Appetite and Thirst

Taste and Smell   Although there is some argument about the extent to which aging affects the sense of taste, there is general agreement that taste and smell senses eventually decline.[30,31] Eating is a sensuous activity, involving taste buds and olfactory nerves. In healthy adults, aging is associated with a decline in ability to identify smells beginning sometime around age 55 for men and age 60 or later for women.[32] Some smells are perceived more easily than others. A blunted sense of smell can lead to a blunted sense of enjoyment of food as well as decreased ability to detect spoiled or overcooked foods. Women's abilities to identify smells remain higher than men's throughout the life span.

Functional Status   Ability to carry out the activities of daily living, including telephoning, grocery shopping, food handling and preparation, and eating.

The controversy around taste is that the number and structure of taste buds are not significantly altered during aging. In addition, taste perception for sucrose does not decline with age. Bartoshuk[33] argues that taste is so important to biological survival that the body has developed "redundancy in the mediation of taste," meaning that several pathways (nerves and receptors) control taste mechanisms. All of the pathways would have to be damaged before the ability to identify sweet, bitter, salty, sour, or savory tastes is lost.

Disease and medications affect taste and smell more than does age itself. Roughly three of four adults have had a temporary smell loss at some time, most often due to colds, flu, or allergies. Yet, age makes a difference. For example, during illness or with the use of medications, younger individuals maintained greater ability to detect salty, bitter, sour, sweet, and savory tastes than older ones.[34]

Oral Health: Chewing and Swallowing   Being able to chew and swallow contributes to eating enjoyment.

Oral health depends on several organ systems working together: gastrointestinal secretions (saliva), the skeletal system (teeth and jaw), mucus membranes, muscles (tongue, jaw), taste buds, and olfactory nerves for smelling and tasting. Disturbances in oral health are associated with, but not necessarily caused by aging. Poor dietary habits are a modifiable risk factor that may contribute to caries and potential tooth loss. The Healthy People 2010 oral health objective for adults[9] aged 65–74 years, is to reduce the percentage of individuals who have lost all their teeth from 26% to 20%. The prevalence of missing teeth varies by population and has been found to be as high as 42% in Native American elders.[35] Reduction of periodontal disease and early cancer detection are other public health goals relating to older adults.[9]

Missing teeth and ill-fitting dentures can make chewing difficult and have a negative impact on eating.[36] An oral health assessment asks questions about soft tissues, teeth, and other factors that may affect dietary intake.[37] One such factor is functional status, which may affect the ability to brush and floss, potentially resulting in periodontal disease.

What and when we eat affects oral health. Oral health, in turn, affects what and how we eat. Healthy teeth are protected by an enamel layer, but bacterial action on the breakdown products of food slowly erodes tooth enamel. For about 15 minutes after we ingest food or drink, oral bacteria feast on the food breakdown products, especially those of sucrose. Foods such as caramels or raisins stick around longer, especially when they lodge between teeth. Frequent eating and drinking of sugary beverages provides a continuous substrate for bacteria. The acid in carbonated beverages adds to the corrosive potential of food.

Saliva, which lubricates the mouth and begins the digestive process (amylase in "spit" begins starch breakdown), also helps to keep tooth enamel clean. However, saliva seems to become thicker and more viscous with age. Lack of saliva slows nutrient absorption. Lack of sufficient and effective saliva, especially in the presence of gingivitis and periodontal disease, also makes the oral cavity more sensitive to temperature extremes and coarse textures, resulting in pain while eating.

Pain and discomfort with chewing foods can result in eating fewer fruits, vegetables, and whole grains. A loss of self-esteem associated with missing teeth and worry regarding how to pay for dental care can affect quality of life. Edentulous older people are less likely to visit the dentist for oral care (denture adjustment, periodontal disease management) than are those individuals having their natural teeth.[38]

Appetite and Thirst   Hunger and satiety cues are weaker in older than in younger adults. Roberts and colleagues examined the ability of 17 young men (mean age 24) and 18 older men (mean age 70) to adjust caloric intake after periods of overeating and of undereating.[39] All men were healthy and not taking medications. Food intake and weight were monitored for 10 days, and then men were overfed by roughly 1000 calories, or underfed by about 800 calories, for 21 days. Periods of over- or underfeeding were followed by 46 days of "ad-lib" intake during which all men were free to eat as much or as little as desired. After the periods of over- and underfeeding, young men adjusted their caloric intake to get back to their initial calorie intake level and weight. Older men kept overeating if they had been in the overfed group, and undereating if they had been in the underfed group. The authors suggest that older adults may need to be more conscious of food intake levels because their appetite-regulating mechanism may be blunted. Whereas healthy young people adjust to cycles of more and less food intake, healthy older people's inability to adapt to these changes may lead to overweight or anorexia.

Elderly people don't seem to notice thirst as clearly as younger people do. A small study demonstrated that the thirst-regulating mechanism of older adults was less effective than that of younger individuals.[40–42] Researchers compared thirst response to fluid deprivation in a group of seven 20- to 31-year-old men, and seven 67- to 75-year-old men. Subjects lost 1.8% to 1.9% of body weight during 24 hours without fluids. Both groups were asked about feeling thirsty, mouth dryness, and how pleasant it would be to drink something. After fluid deprivation, the younger group reported being thirsty and having dry mouth. The older group, however, reported no change in thirst or mouth dryness. Both the older and younger groups thought that it would be pleasant to drink something after fluid deprivation. Blood measures showed that older men lost more blood volume than younger men did, indicated by their plasma concentrations of sodium. Researchers also measured how much water the men drank in the hour after their 24-hour period of fluid deprivation. Older men drank less water than their younger counterparts did. Younger people made up for fluid loss in 24 hours; older people did not drink enough to achieve their prior state of hydration. It appears that dehydration occurs more quickly after fluid deprivation and that rehydration is less effective in older men.

Nutritional Risk Factors

Identifying nutritional risk factors before chronic illness occurs is basic to health promotion. Decreasing risk factors forms the basis of dietary guidance. For instance, the leading causes of death for older adults are heart disease, cancer, and stroke. In all adults, dietary risk factors that increase the likelihood of developing these diseases are consuming a high-fat diet that is also high in saturated fat; low intake of vegetables, fruits, and whole grain products; and excessive calorie intake leading to obesity. Healthy People 2010, MyPyramid, and the Dietary Guidelines for Americans emphasize eating patterns that reduce the risk of the leading killer diseases.

Another approach to nutritional risk factor identification is to compare adequacy of current dietary intake to dietary intake recommendations such as the Recommended Dietary Allowances and the Dietary Reference Intakes. According to national intake data[43] older adults are not consuming enough protein, vitamin E, folic acid, magnesium, and calcium to meet recommended nutrient levels.

A third approach is to examine a population and determine how environmental factors combine with dietary factors to predict nutritional health. This approach was used by a consortium of care providers, policy makers, and researchers to develop the Nutrition Screening Initiative (NSI).[44] The American Academy of Family Physicians, the American Dietetic Association (ADA), and the National Council on the Aging, Inc., sponsored the development of this health-promotion campaign. The NSI consortium used literature review, expert discussion, and a consensus process to generate a list of warning signs of poor nutritional health in older adults. See Table 18.5 for a condensed version of the acronym (DETERMINE).[45,46]

These warning signs were integrated into a screening tool, the NSI DETERMINE checklist (Illustration 18.1 on the next page), and pilot-tested for use in community settings. Ten risk factors remained after testing a longer list.[47] This tool is used by community agencies, educators, and providers to screen program participants and the aged public for risk of malnutrition.

The nutritional risk factors identified during the NSI process[45] are reflected in the list of risk factors identified in the ADA position on nutrition and aging.[1] Presence of any of the following conditions places older adults at greater nutritional risk:

- Hunger
- Poverty
- Inadequate food and nutrient intake
- Functional disability
- Social isolation
- Living alone
- Urban and rural demographic areas
- Depression
- Dementia
- Dependency
- Poor dentition and oral health; chewing and swallowing problems
- Presence of diet-related acute or chronic diseases or conditions
- Polypharmacy (use of multiple medications)
- Minority status
- Advanced age

Why are factors such as poverty and minority status included in a list of nutritional risk factors? Economic security contributes to food security, one of the health goals for

---

Table 18.5 DETERMINE: Warning signs of poor nutritional health

Disease. Any disease, illness, or chronic condition (i.e., confusion, feeling sad or depressed, acute infections) that causes changes in the way you eat, or makes it hard for you to eat, puts your nutritional health at risk.

Eating poorly. Eating too little, too much, or the same foods day after day, or not eating fruits, vegetables, and milk products daily will cause poor nutritional health.

Tooth loss/mouth pain. It is hard to eat well with missing, loose, or rotten teeth, or dentures that do not fit well or cause mouth sores.

Economic hardship. Having less or choosing to spend less than $37.20 (female) to $41.80 (male) weekly for groceries makes it hard to get the foods needed to stay healthy. [These costs are calculated for individuals living in 4-person households; add 20% to adjust for living alone.]

Reduced social contact. Being with people has a positive effect on morale, well-being, and eating.

Multiple medicines. The more medicines you take, the greater the chance for side effects such as change in taste, increased or decreased appetite and thirst, constipation, weakness, drowsiness, diarrhea, nausea, and others. Vitamins or minerals taken in large doses can act like drugs and can cause harm.

Involuntary weight loss or gain. Losing or gaining a lot of weight when you are not trying to do so is a warning sign to discuss with your health care provider.

Needs assistance in self-care. Older people who have trouble walking, shopping, and buying and cooking food are at risk for malnutrition.

Elder years above 80. As age increases, risk of frailty and health problems also rises.

SOURCE: Warning signs adapted from Nutrition Screening Initiative, a project of the American Academy of Family Physicians, the American Dietetic Association, and the National Council on the Aging, Inc., and funded by a grant from Ross Products Division, Abbott Laboratories, Inc.; dollar amounts under economic hardship inserted by author using the August 2006, USDA low-cost food plan.

---

the nation (see Table 18.6). Lack of food security leads to food intakes below 67% of the RDA, especially in older persons, who eat fewer calories and less magnesium, calcium, zinc, and vitamins E, C, and B_6 when food insecure.[48] Although older people on average are less likely to live in poverty than are children, they are a heterogeneous population, and several groups are at high risk of poverty. Minority status is related to economic status. In 2001 nearly 60% of the black and Hispanic population aged 65 and over were either poor or near poor (199% of the poverty threshold; see Chapter 16).[7] Furthermore, minority

DETERMINE YOUR NUTRITIONAL HEALTH

The Warning Signs of poor nutritional health are often overlooked. Use this checklist to find out if you or someone you know is at nutritional risk.

Read the statements below. Circle the number in the yes column for those that apply to you or someone you know. For each yes answer, score the number in the box. Total your nutritional score.

	Yes
I have an illness or condition that made me change the kind and/or amount of food I eat.	2
I eat fewer than 2 meals per day.	3
I eat few fruits or vegetables, or milk products.	2
I have 3 or more drinks of beer, liquor, or wine almost every day.	2
I have tooth or mouth problems that make it hard for me to eat.	2
I don't always have enough money to buy the food I need.	4
I eat alone most of the time.	1
I take 3 or more different prescribed or over-the-counter drugs a day.	1
Without wanting to, I have lost or gained 10 pounds in the last 6 months.	2
I am not always physically able to shop, cook, and/or feed myself.	2
TOTAL	

Total your nutrition score. It it's . . .

0–2   Good! Recheck your nutritional score in 6 months.

3–5   You are at moderate nutritional risk.
See what you can do to improve your eating habits and lifestyle. Your office on aging, senior nutrition program, senior citizens center, or health department can help. Recheck your nutritional score in 3 months.

6 +   You are at high nutritional risk.
Bring this checklist the next time you see your doctor, dietitian, or other qualified health care or social service professional. Talk with them about any problems you may have. Ask for help to improve your nutritional health.

Remember: warning signs suggest risk, but do not represent diagnosis of any condition.

These materials developed and distributed by the Nutrition Screening Initiative, a project of: AMERICAN ACADEMY OF FAMILY PHYSICIANS, THE AMERICAN DIETETIC ASSOCIATION, and NATIONAL COUNCIL ON AGING, INC.

Illustration 18.1 Determine your nutritional health checklist.

SOURCE: Reprinted by permission by the Nutrition Screening Initiative, a project of the American Academy of Family Physicians, The American Dietetic Association and the National Council on the Aging, Inc., and funded by a grant from Ross Products Division, Abbott Laboratories, Inc.

agents, diuretics, and anti-hypertensives.

Taken individually, the risk factors identified in the DETERMINE acronym and in the ADA position statement on aging are not unique to older adults. But each is more likely to lead to nutritional problems in a frail, vulnerable population. For instance, functional disability can affect dietary intake at any age, but very old people are more likely to live alone and to have fewer resources to compensate for any type of lost function. Consequently, diet quality may decline. Fewer 65-year-olds live alone than 75- and 85-year-olds, and more women than men live alone. Race and ethnicity affect living situations; Hispanic men and women are least likely to live alone, whereas white women over age 85 are most likely to live alone; 65% of older white women live alone, compared to 47% of African American and 35% of Hispanic women.

A common perception is that older adults living alone eat poorly. Although living alone is a nutritional risk factor, it's not clear whether the issue is eating alone or living alone. On average, meals eaten with other people last longer and supply more calories than do meals eaten alone.[50] The effect of living alone starts at a younger age for men and affects nutrition more extensively than for women.[51] Women aged 75 and older eat less protein and also less sodium than those living with others; men aged 65 and older eat less protein, beta-carotene, vitamin E, phosphorus, calcium, zinc, and fiber. Men and women living alone did not consume greater amounts of any of the 23 nutrients (alcohol was not included) examined than did men and women living with someone.

The purpose of a screening tool is to identify aging individuals at risk for diseases and provide interventions that reduce or eliminate risk. The intensive push for elderly persons to receive a flu shot is an example of this type of tool.

populations are more likely to be in fair or poor health while non-Hispanic whites are most likely to be in excellent or good health. High health care costs may contribute to food insecurity. All of us can think of a story about seniors having to choose between drugs and food.

Poverty is one risk factor for malnutrition; polypharmacy is another. Reported medication use increases with age. The Slone Survey at Boston University tracks medication use. Women aged 65 are older use the most medications: 63% use at least 5, and 17% use 10 or more.[49] The most commonly used drugs by men and women are identified in Table 18.7. Common prescription drugs include cholesterol-lowering

Table 18.6 Healthy People 2010, food security goals for adults aged 65 and older

	Target	Current Goal Males	Females
Increase food security and in so doing, decrease hunger			
Households <130% of poverty, with elderly persons	94	85[a]	85[a]
Households >130% poverty, with elderly persons	94	98[a]	98[a]

SOURCE: *Healthy People 2010.*

[a]Adequate data to distinguish levels for males and females is not available.

Table 18.7 Top six most commonly used medications by men and women aged 65 and older

Men	Women
Aspirin (40%)	Aspirin (36%)
Acetaminophen (17%)	Acetaminophen (25%)
Lisinopril (15%)	Levothyroxine (18%)
Atorvastatin (14%)	Atorvastatin (14%)
Metoprolol (13%)	Hydrochlorothiazide (14%)
Simvastatin (13%)	Metoprolol (12%)

SOURCE: Patterns of medication use in the United States 2005. A report from the Slone Survey. Compiled from Table 2: Thirty Most Commonly Used Prescription and Over-the-Counter Drugs Taken by Adults in 2005, According to Sex and Age. www.bu.edu/slone/SloneSurvey/AnnualRpt/slonesurveyreport2005.pdf, accessed October 31, 2006.

In secondary prevention, such as cholesterol screening or obtaining a bone scan to evaluate bone health, early symptoms of the disease are identified and treated to prevent symptoms or worsening of the condition. Thus, the question remains, "Does the nutrition screening initiative tool prevent and/or identify those older persons at nutritional risk?" The results are mixed. The cumulative score resulting from using the NSI tool weakly predicted mortality in an aging population of mostly white, educated adults.[52] In another study, the NSI checklist did not identify all those individuals who had poor health or low nutrient consumption.[47] The two-stage Mini Nutritional Assessment (MNA, accessible from www.mna-elderly.com) combines 6 screening questions in stage 1 with 12 assessment questions in stage 2. It includes midarm circumference and calf circumference measurements and questions related to lifestyle, medication and mobility, dietary intake, and self-perception of health.[53] A large study compared the NSI and the MNA using dietary intake, anthropometrics, and blood biochemistries and found that each tool was of limited value.[54]

Use of the NSI checklist to screen frail individuals who had been newly admitted into a nursing home identified problem areas, linked problems to care protocols, and resulted in expedited treatment.[55]

Dietary Recommendations

Sometimes, in all the discussions about nutritional effects on health and disease status of older adults, we forget that old age is not a disease. However, recommendations for specific nutrients do change with age, so the food on one's plate should also change over time.

Food-Based Guidance: MyPyramid

MyPyramid is easily recognized as dietary advice. Pyramid adaptations for older adults have been developed by several groups, including the American Dietetic Association,[56] the Senior Nutrition Awareness Project (SNAP) of Connecticut and Rhode Island, and Tufts University researchers (see Illustration 18.2). The American Dietetic Association pyramid for persons aged 50 and over depicts fluids, but does not include vitamin–mineral supplements. In the Tufts pyramid for older adults, authors portray the need for

Illustration 18.2 Tufts University modified food pyramid for 70+ adults.

SOURCE: Used by permission of Tufts University.

Table 18.8 Using MyPyramid recommendations to show the daily amount of food needed to meet nutritional needs for older adults at various activity levels (See Chapter 1 for food measure equivalents)

	Female, age 72, sedentary	Female, age 60 and older, active	Male, age 65, moderately active
Calorie Level	1600	2000	2400
Fruits	1.5 c	2 c	2 c
Vegetables	2 c	2.5 c	3.5 c
Grains	5 oz equivalent	6 oz-eq	8 oz-eq
Meat and Beans	5 oz-eq	5.5 oz-eq	6.5 oz-eq
Milk	3 c	3 c	3 c
Oils	5 tsp	6 tsp	7 tsp
Discretionary Calorie Allowance	132	267	362

SOURCE: www.mypyramid.gov, accessed 10/29/2006.

fewer calories with a narrower pyramid base.[57] They also emphasize adequate fluid intake by depicting 8 glasses of water at the pyramid base, but this depiction is confusing for some older audiences who interpret the water glasses to mean that water is the only fluid that counts (personal communication with Olga Monell, nutrition provider with the Senior Nutrition Program, October 1999). Potentially higher need for supplemental calcium and vitamins D and B_{12} is flagged at the top of the pyramid. Symbols depicting fat, sugar, and fiber in foods are intended to help older adults recognize sources of those food constituents and subsequently choose a more nutrient-dense diet.

The pyramid shows food groups and corresponding serving sizes to be used for meal plan development. Table 18.8 lists food-group suggestions for common calorie levels, which can be modified by choosing the more or less nutrient-dense foods from each group. For instance, citrus and berries are among the most nutrient-dense fruits. A baked potato has more vitamins and minerals per bite than french fries. Among grain-based products, whole grain breads and cereals provide fiber as well as the usual nutrients found in grain. Consistently choosing nutrient-dense foods decreases caloric intake. Fortification and enrichment, while enhancing nutrient intake, complicate nutrient density calculations. For instance, unless whole grain products are also fortified with folic acid, they may actually be considered to be less nutrient-dense than their plain white, but enriched, counterparts. That is one reason variety remains a part of dietary advice. Regularly choosing a variety of foods from each group enhances overall diet quality.

Actual Food Group Intake

A picture of the eating habits of older Americans can be developed from records of consumption according to food group, by analyzing occasions for eating (e.g., snacks) and by nutrients consumed (see Tables 18.9 and 18.10).

Eating Occasions

Eating Out   Older adults eat away from home at a lower rate than younger persons. Whereas 62–72% of individuals in their twenties consume more than one-third of their calories away from home, 39–50% of individuals in their sixties consume one-fifth of their calories away from home, and approximately 28% of people aged 70 and older consume one-eighth of their calories away from home. Food

Table 18.9 Daily consumption of named food groups by adults aged 70 years and older using Continuing Survey of Food Intake by Individuals (CSFII), 1996 data collection

Number of servings eaten per day

Food Group	70+ Years Males	70+ Years Females
Grain products	6.0	4.8
Whole-grain	1.2	1.0
Vegetables	2.9	2.8
Potatoes	0.7	0.7
Tomatoes	0.4	0.4
Fruits	2.1	1.7
Citrus fruits, melons, berries	0.9	0.8
Milk and milk products	1.2	1.1
Meat, poultry, fish, (Ounces or lean meat equivalents)	4.1	3.4
Eggs	0.4	0.3
Legumes (dried peas and beans)	0.2	0.1
Nuts and seeds	0.2	0.1

SOURCE: Actual intakes amounts are taken from 1996 Continuing Survey of Food Intakes by Individuals, 2-day average. USDA, ARS, Beltsville Human Nutrition Research Center, Food Surveys Research Group.

Table 18.10 Caloric intake comparison of younger and older adults, by gender, from NHANES, 1999–2000 data collection

Age	Actual Daily Calorie Intake		Recommended Calorie Intake	
	Males	Females	Males	Females
20–29	2828	2028	2900	2200
60–74	2123	1596	2300	1900

SOURCE: Wright JD, Kennedy-Stephenson J, Wang CY, McDowell MA, Johnson CL. Trends in intake of energy and macronutrients—United States, 1971–2000. MMWR, February 6, 2004;53(04):80–82. Available from www.cdc.gov/mmwr/preview/mmwrhtml/mm5304a3.htm, accessed 10/16/06.

eaten away from home supplies relatively more protein and fat than carbohydrate. For example, the population aged 70 years and older obtains the following percentages of the day's calories with foods eaten away from home:[43]

- 13–14% of protein
- 14–15% of fat
- 11–12% of carbohydrate

Snacking   Older adults snack is less than other age groups. Snack foods do not appear to be nutrient dense. Two-thirds of people aged 70 and older (65% of women, 67% of men) eat snacks that supply 12% of their food energy but only 6–10% of vitamins, minerals, and potassium consumed. Here are the energy–nutrient contributions of snacks for this group:

- 7% of protein consumed
- 10–11% of fat consumed
- 13–14% of carbohydrate consumed

Information about nutrient intake of older adults (see Tables 18.11 and 18.12, on the following page) helps to answer questions such as, "How does Uncle Al's diet compare to that of other men his age?" and "Are there unusual patterns in Al's diet that might reflect some potential problem?" Although change is good when it helps people to stay flexible, unexplained changes in diet and nutritional status need to be evaluated.

Nutrient Recommendations

Nutrient recommendations change as scientists learn more about the effect of foods on human function. Specific nutrient levels for population groups above age 51 were first established in 1997 as the Dietary Reference Intakes (DRIs).[60] The complete tables, including Tolerable Upper Intake (UL), are found on the inside covers of this text. This section first addresses the macronutrient recommendations (carbohydrates, protein, and fat) and then the micronutrient recommendations relevant to older adults.

Estimating Energy Needs

The main goal for energy calculations is to maintain a healthy body weight. Decrease of physical activity and basal metabolic rate from early to late adulthood results in 20% fewer calories needed for weight maintenance.[61] Total energy expenditure is primarily determined from basal metabolic rate (see Chapter 16), diet-induced thermogenesis, and physical activity energy expenditure. Genetics, gender, and body composition also influence metabolic rates, resulting in broad ranges of energy needs for older populations. Basal metabolic rates (BMRs), using doubly-labeled water measurements, have ranged from 1004 kcal/d to 2060 kcal/d in older individuals with normal body mass index (BMI).[62] A small international study of the oldest old (i.e., >90 years) who were living at home demonstrated variability in physical activity and energy expenditures. Mean resting metabolic rates were 1523 kcal for women and 1944 kcal for men with physical activity ratios (i.e., the measured total energy expenditure/resting metabolic rate) of 1.2, ranging up to 1.7.[63]

Traditionally, the Harris-Benedict equation has been used to estimate individual calorie needs (see Chapter 16). However, validation was done with mostly young adults. Newer energy estimation formulas are being developed to account for greater individual variation. The Mifflin–St. Jeor formula most closely predicts energy requirements of healthy adults in general (see Chapter 16), and validation did include individuals up to age 80, but it was not specifically developed for older adults.

Arcerio and colleagues developed formulas for use with older adults.[64–66] The equation used with older adult females includes a factor for hormonal status (e.g., menopause). Males completed the Minnesota Leisure Time Activity survey as well as a set of three chest skinfold measurements.[66] Using these formulas requires training; they are not in common use.

Table 18.10 shows that on average, adults do eat fewer calories as they age. Women aged 60 and older eat the least and are especially vulnerable to malnutrition, even though approximately one in four is obese (BMI = 30 or more). Calorie intake and weight status is an area that illustrates the dangers of generalizing from populations to individuals, or the reverse. For older adults, it is difficult to meet vitamin and mineral needs at caloric levels below 1600, but adding activity can maintain caloric needs. This is often easier to say than to do.

Nutrient Recommendations: Macro- and Micronutrients of Concern

Potentially problematic nutrients for older adults are presented in Tables 18.11 and 18.12, listing both actual and recommended intake. Each listed nutrient plays a role in normal, healthy aging and is eaten in amounts different from the recommended level (except for dietary cholesterol

Table 18.11   Selected macronutrient intakes of older adults, NHANES 1999–2000[a] and Continuing Survey of Food Intake by Individuals, CSFII, 1994–1996, 1999–2000 data collection,[b] with relevant recommendations

| | Actual Intake[a] | | Actual Intake[b] | | | | Recommended Intake | |
| | All ethnicities | | Caucasian | | Black | | All ethnicities | |
Nutrient	Males	Females	Males	Females	Males	Females	Males	Females
Protein, g	NA	NA	80	59	72	54	56	46
Total fat, g	77	57	75	52	66	50	65	65 (DV)
Saturated fat, g	25	19	25	17	21	16	20	20 (DV)
Monounsaturated fat, g	30	21	29	20	26	19	Up to 20% calories	
Polyunsaturated fat, g	16	13	16	11	13	11	Up to 10% calories	
Cholesterol, mg	282	229	285	198	335	228	300 or less	

[a]Actual intake amounts from Wright JD, Kennedy-Stephenson J, Wang CY, McDowell MA, Johnson CL. Trends in intake of energy and macronutrients—United States, 1971–2000. MMWR, February 6, 2004, 53(04):80–82. Available from www.cdc.gov/mmwr/preview/mmwrhtml/mm5304a3.htm, accessed 10/16/06.

[b]Actual intakes amounts are taken from 1994–1996 Continuing Survey of Food Intakes by Individuals, 2-day average. USDA, ARS, Beltsville Human Nutrition Research Center, Food Surveys Research Group with the exception for cholesterol, which was taken from 1999–2000 data in Wright JD, Wang CY, Kennedy-Stephenson J, Erving RB. Dietary intake of ten key nutrients for public health, United States: 1999–2000. *Advance data from vital and health statistics*; no. 334. Hyattsville, Maryland: National Center for Health Statistics. 2003. Recommended intake data are based on Dietary Reference Intake (DRI) Institute of Medicine of the National Academies. *Dietary Reference Intakes: Energy, carbohydrate, fiber, fat, fatty acids, cholesterol, protein, and amino acids*. National Academies Press, Washington, DC, 2002, and the nutrient reference amounts used on the Nutrition Facts Panel food labels for sodium; Daily Values (DV) used on the Nutrition Facts Panel food labels for fat and saturated fat.

and zinc). The reasons that each nutrient is of concern follow next in the text.

Carbohydrate and Fiber

Adequate carbohydrate intake of 45% to 65% of calories is generally not a problem.[62] Of individuals aged 70 and over, males eat 51% of their calories from carbohydrate and women eat 54%.[43]

Table 18.12   Selected micronutrient intakes of older adults, CSFII, 1994–1996, 1999–2000 with relevant recommendations for adults aged 60 and older

| | Actual Intake | | Recommended Intake | |
	Males	Females	Males	Females
Sodium, mg	3447	2532	2400 (DV)	2400 (DV)
Vitamin A, mcg RE	1282	939	900	700
Vitamin E, mg	9.4	7.2	15	15
Vitamin B₁₂, mcg	6.0	4.5	2.4	2.4
Folate, mcg	387	312	400	400
Calcium, mg	754	587	1200	1200
Magnesium, mg	286	224	420	320
Zinc, mg	12.2	9.3	11	8
Iron, mg	17.3	12.8	8	8

SOURCE: Actual intake amounts of A, E, B₁₂ taken from 1994–1996 Continuing Survey of Food Intakes by Individuals (CFSII) conducted by the Agricultural Research Service of the U.S. Department of Agriculture and available online; the remaining micronutrient actual intakes are based on 1999–2000 data in J. D. Wright, C. Y. Wang, J. Kennedy-Stephenson, and R. B. Erving, "Dietary Intake of Ten Key Nutrients for Public Health, United States: 1999–2000," *Advance Data from Vital and Health Statistics*; no. 334. Hyattsville, MD: National Center for Health Statistics, 2003. Recommended intake data are based on Dietary Reference Intake (DRI) Institute of Medicine of the National Academies. *Dietary Reference Intakes: Energy, Carbohydrate, Fiber, Fat, Fatty Acids, Cholesterol, Protein, and Amino Acids* (Washington, DC: National Academies Press, 2002) and the nutrient reference amounts used on the Nutrition Facts Panel food labels for sodium.

Following the MyPyramid recommendations can ensure that the carbohydrate quantity and quality is adequate and that fiber guidelines are met. For example, men who need 2300 calories/day would need to eat between 288 to 345 grams of carbohydrate per day to meet the recommended 50% to 60% of calories from carbohydrate sources; women would need 188 to 285 grams. A listing of foods that provides at least 50% carbohydrates of a 1500-calorie diet is presented in Table 18.13.

Depending on caloric intake, a range of 21–30 grams dietary fiber (14 g/1000 cal) daily is recommended for adults aged 51 and older.[62] Males eat 18 grams and females eat 14 grams of fiber daily, which is less than the recommended levels (see Chapter 1).

Table 18.14 shows how common portions of foods eaten by older adults can supply adequate fiber. Reasons to eat adequate fiber include the strong associations between fiber intake and incidence of diverticular disease in men, non-insulin-dependent diabetes in women, coronary heart disease in men, and hypertension in women and men.[67] The role of fiber in gastrointestinal problems is discussed in Chapter 19.

When increasing fiber intake, additional fluids are needed to process the fiber. Slowly adding both fiber and fluids allows the intestinal system to adapt to the additional bacterial substrate.

Protein

Although most adults in North America eat sufficient or even excess amounts

Table 18.13 Using the MyPyramid groups to estimate adequate carbohydrate and fiber for daily intake

Basic Food Groups

Number of Servings/ Grams of Carbohydrate per Serving	Carbohydrate, grams	Approximate Total Fiber Content, grams
6 servings of grain at 15 g each	90	12
2 servings of fruit at 15 g each	30	6
3 servings of milk at 12 g each	36	0
3 servings vegetable at 5 g each	15	9
Total from basic groups	**171**	**27**

Other Carbohydrate-Containing Foods

1 Tbsp sugar for coffee or tea	12	0
2 Fig Newton cookies	20	1
Total, including "other" group	**203**	**28**

NOTE: Mixed dishes such as soups, sandwiches, and salads count as partial servings from their contributing food groups.

of protein, older adults (especially those over 75 years old, living alone, inactive) may be eating too little.[51] Inadequate protein intake contributes to muscle wasting (sarcopenia), a weakened immune status, and delayed wound healing. Protein guidelines for individuals aged 18 years and older are currently set at 56 grams per day for males and 46 grams for females.[62] Some researchers found that older persons need higher levels of protein intake to maintain nitrogen balance.[24,68] Nitrogen-balance studies used to determine current recommendations of 0.8 gram per kilogram (or 0.36 gram per pound) were done primarily with young adults who have proportionately more muscle mass than older adults, male or female, and are more efficient in maintaining nitrogen balance.[69]

Nitrogen balance is also easier to achieve when the protein eaten is of high quality (such as meat, milk, and eggs;

Table 18.14 An example of fiber-containing foods that might comprise one day's intake

Food Item	Grams of Fiber
Oatmeal (½ cup) with wheat germ (¼ cup)	8
Banana	2
Peanut butter (2 Tbsp)/whole wheat sandwich (2 slices)	6
Orange	3
Baked potato with skin	4
Green beans (½ cup)	3
Bran muffin (1 med)	2
Pear	4
Total dietary fiber	**32**

NOTE: Meat, poultry, fish, eggs, milk, sugar, and oils do not contain dietary fiber.

see Table 18.15), is ingested with adequate calories, and when an individual is doing resistance exercises. During resistance training, less nitrogen is excreted and thus nitrogen balance improves. Consuming a low-calorie diet, as many older adults do, leads to a greater need for protein in order to maintain nitrogen balance. Evans and Cyr-Campbell recalculated nitrogen-balance study data and did new laboratory tests with older adults.[24] They found that older adults on average maintain nitrogen balance eating 0.91 gram of high-quality protein per kilogram body weight per day (see Table 18.15 for quality scores of common proteins). They also calculated that 1.25 gram protein/kg body weight/day would be a safe intake for 97.5% of the older adult population, recommending that older adults eat 1 to 1.25 gram of protein/kg body weight daily. This translates to 0.45 to 0.57 gram protein/lb body weight. While Rand and associates' meta-analysis of nitrogen balance studies[69] concluded that the RDA for protein should be 0.83 gram of good-quality protein per day, a review of protein needs of older adults led Morais and associates to propose intakes of 1.0 to 1.3 gram protein per kg per day.[68] Proportionately less lean body mass in an older person does not lead to lower protein requirements. In fact, someone who is losing muscle due to inactivity requires extra protein.

Leading protein sources for all adults, based on NHANES II surveys,[70] are (1) beef steak, roast; (2) hamburgers; (3) white bread, rolls; (4) whole milk; and (5) pork. The chemical protein scores of four of the five leading protein sources are 100 (see Table 18.15), but the

Table 18.15 Protein sources and protein quality measures: examples of protein scores used by the Institute of Nutrition of Central America and Panama (INCAP)

Protein Source	Chemical Quality Score*	True Digestibility in Humans
Meat, eggs, milk	100	95
Beans	80	78
Soy protein isolate	97	94
Rice	73	88
Oats, oatmeal	63	86
Lentils	60	Not available
Corn	50	85
Wheat	44	Refined, 96; whole wheat, 86

SOURCE: Data from B. Torun, M. T. Menchu, and L. G. Elias, *Recomendaciones dieteticas diarias del INCAP*. Guatemala: INCAP, 1994 [INCAP Publicacion ME/0571]; and M. E. Shils, J. A. Olson, and M. Shike et al., *Modern Nutrition in Health and Disease*. Philadelphia: Lippincott, Williams & Wilkin, 1999.

*Amino acid score relative to amino acid score of egg, the reference protein.

digestibility scores of the five leading protein sources are all in the 90s (see the same table).

People eat mixed diets, so individual protein scores tell only part of the story. For example, using the human protein digestibility scale, the average American and Chinese mixed diets score 96, the rural Mexican diet scores 80, and an Indian rice-and-bean diet scores 78. However, for older adults who are following modified diets, who are cutting back on meat because they can't afford it or lack energy to prepare it, or who may eat too few calories, protein quality can make the difference between a good and a poor diet.

On average, older adults eat slightly more protein as percentage of total calories than adults in general (16.1% and 16.0% of calories from protein for males and females, compared to 14.9% and 14.6% percent in younger persons.)[59] But when the population is heterogeneous, average data are not enough. The following questions address protein adequacy for older individuals:

- Based on height and weight, how much protein will meet the individual's need?
- Are enough calories eaten so that protein does not have to be used for energy?
- If marginal amounts of protein are eaten, is the protein of high quality?
- Are there additional needs: wound healing, tissue repair, surgery, fracture, infection?
- Is the individual exercising? It is harder to achieve nitrogen balance while sedentary.

To minimize the risks of sarcopenia in aging adults, it seems wise to keep within the range of 1–1.3 g/kg daily recommended by Morais and associates.[68] The reference older woman weighs 143 lb, and the reference older man weighs 170 lb. Eating 1 gram of protein per kilogram of body weight would mean eating 65 grams/day for women and 77 grams/day for men, significantly more than the 50 grams recommended as the basis for the Daily Value food-product nutrition labels.[71] Fifty grams of protein would be adequate for an older adult weighing 50 kg or 110 lb.

Fats and Cholesterol   The role of fat does not seem to change with age; high saturated fat and trans-fatty acid intake continues to be a risk factor for chronic disease. Minimizing the amount of saturated fat in the diet and keeping total fat between 20 and 35% of calories is a reasonable goal for older adults to maintain a beneficial blood cholesterol ratio. Cholesterol intake for older adults is already below recommendations for the nation (see Table 18.11). Eggs, which have high cholesterol content, are a nutrient-dense, convenient, and safe food for most people (those without lipid disorders; e.g., high triglyceride and high serum cholesterol). McNamara, a professor of nutrition sciences and director of the Lipid Metabolism Laboratory at the University of Arizona,

calculated that eating 7 eggs per week would raise total serum cholesterol by 1.3 mg/dL.[72] Most older people can enjoy lean high-cholesterol foods such as shrimp and eggs for their high-quality protein, antioxidants, and other nutrients, without added risk of CVD. Even individuals who are hyperresponders to dietary cholesterol have been shown, albeit in a small study, to maintain their LDL:HDL ratio after high dietary cholesterol intake.[73]

Recommendations for Fluid

Water as percentage of total body weight decreases with age, resulting in a smaller water reservoir and leaving a smaller safety margin for staying hydrated. Drinking 6 or more glasses of fluid per day prevents dehydration (and subsequent confusion, weakness, and altered drug metabolism) in individuals whose thirst mechanism may no longer be very sensitive.

To individualize fluid recommendations, provide 1 milliliter (mL) of fluid per calorie eaten, with a minimum of 1500 mL. For a 2000-calorie diet, that would be 2000 mL or 2 liters of fluid, roughly 8 cups. Foods such as stews, puddings, fruits, and vegetables contribute significant amounts of fluid to the diet but are not counted as part of the fluid allowance in healthy individuals. The Tufts (see Illustration 18.2) Food Guide Pyramid for healthy older adults show 8 glasses of water, which is adequate for a 2000-calorie diet. Individuals who need additional calories can use milk, juice, shakes, and soups as nutrient-dense fluids.

Age-Associated Changes in Metabolism: Nutrients of Concern

The nutritional health of older adults depends on modifying dietary habits that address age-associated changes in absorption and metabolism. The nutrients discussed in the following sections are of special concern for older adults because of age-associated metabolic changes or low dietary intake.

Vitamin D, Calciferol   Age-related metabolic changes affect vitamin D status, independent of dietary intake, primarily due to decreased ability of the skin to synthesize previtamin D_3 from its precursor, 7-dehydrocholesterol.[74,75] Declining photochemical production may be compounded by limited exposure to sunlight due to use of sunscreen or pollution, institutionalization, or being homebound. Furthermore, in northern regions (above 42° north, the latitude of Boston and Chicago) between November and February, ultraviolet (UV) light is not powerful enough to synthesize vitamin D in exposed skin.[76] The sun's rays are even weaker in Edmonton, Canada (at 52° N latitude), so that previtamin D_3 is not synthesized between mid-October and mid-April. Older individuals who live in these northern latitudes are at greater risk for vitamin D deficiency because they need more vitamin D than younger people do. How far south does one have to go for the winter sun to convert vitamin D precursors? Tests in Los Angeles, at 34° N, showed vitamin D production in skin even in January. The sun's rays in Puerto Rico (latitude of 18° N) were even more effective.[76] Fortunately for people living in northern regions, the body stores vitamin D, so that getting summer sun builds a winter reserve.

Another reason that vitamin D is of concern for older adults (who use more drugs than younger persons) is that commonly used medications interfere with vitamin D metabolism. Examples include barbiturates, cholestyramine, phenytoin (Dilantin), and laxatives.

Recommendations for vitamin D intake are 10 mcg (400 international units—IU) for persons aged 51 to 70 and 15 mcg (600 IU) for individuals older than 70 years. Potentially toxic doses are those above 50 mcg (2000 IU), the Tolerable Upper Intake Level (UL) set by the National Academy of Sciences. Symptoms of toxicity are hypercalcemia (high blood calcium levels), anorexia, nausea, vomiting, general disorientation, muscular weakness, joint pains, bone demineralization, and calcification (calcium deposits) of soft tissues. Toxicity is rare from food sources. The relatively few good food sources of vitamin D are fortified cereal, milk (but not cheese), eggs, liver, salmon, tuna, catfish, and herring. Mushrooms also contain a small amount of vitamin D. Cod liver and other fish oils contain medicinal levels of Vitamin D (about 21 mcg, or 840 IU per teaspoon). Although vitamin D intake has not been tracked in national food-monitoring surveys (because technically, humans can derive vitamin D from nonfood sources; i.e., UV light), researchers have assessed individual intake. For instance, Dawson-Hughes and associates[77] found that individuals 65 years old and older consumed 4.5 to 5.0 mcg (180–200 IU) on average.

For healthy aging, the critical function of vitamin D is maintenance of blood calcium levels through intestinal absorption or through bone resorption when the diet is insufficient in calcium.[67] Osteoporosis and osteomalacia (adult rickets) prevention are covered in Chapter 19.

Vitamin B$_{12}$ Although population intake of vitamin B$_{12}$ or serum cobalamine is higher than the DRI level of 2.4 mcg (see Table 18.12), vitamin B$_{12}$ status suggests that many older adults are unable to use B$_{12}$ efficiently. Vitamin B$_{12}$ blood levels decrease with age even in healthy adults. Compared to 18% of healthy 22- to 63-year-olds in the Framingham Heart Study, 40% of 65- to 99-year-olds had low serum B$_{12}$ levels. Despite adequate food intake, an estimated 30% of older adults suffer from atrophic gastritis. In atrophic gastritis, a bacterial overgrowth in the stomach leads to inflammation, and decreased secretion of hydrochloric acid and pepsin with or without decreased secretion of intrinsic factor and subsequent inability to split Vitamin B$_{12}$ from its food protein carrier. It takes years to develop a B$_{12}$ deficiency; but once developed, the neurological

symptoms are irreversible. Symptoms include deterioration of mental function, change in personality, and loss of physical coordination.

"Food first" is usually sound advice regarding nutritional needs, but B$_{12}$ is one of the two vitamins better absorbed in synthetic or purified form. Folic acid is the other one. Synthetic, not protein-bound, vitamin B$_{12}$ is found in fortified foods such as cereals and soy products. Protein-bound B$_{12}$ is found in all animal products, although poultry is a surprisingly poor source (2.4% of B$_{12}$ in adult diets is derived from poultry, compared to 27.3% from beef). The leading food sources of vitamin B$_{12}$ among U.S. adults are beef, milk, fish (excluding canned tuna), and shellfish.[70]

Vitamin A Older adults are more likely to overdose with vitamin A than to be deficient in the nutrient. They eat more than the DRI of 900 mcg for males and 700 mcg RE for females (see Table 18.12). Plasma levels and liver stores of vitamin A increase with age. This may be due to increased absorption but is more likely due to decreased clearance of vitamin A metabolites (retinyl esters) from the blood.[78] Kidney disease further elevates serum vitamin A levels because retinol-binding protein, another vitamin A metabolite, can no longer be cleared from blood. Thus older adults are more vulnerable to vitamin A toxicity and possible liver damage than younger individuals are. The UL for vitamin A is 3000 mcg (3 mg).[79] Older adults have a smaller margin of safety for vitamin A than do younger adults.

The plant precursor of vitamin A, beta-carotene, will not damage the liver, although supplements used as antioxidants to prevent cardiovascular disease have been linked to higher all-cause mortality.[80] Excess dietary beta-carotene, because it is water soluble, may give old skin a yellow-orange tint; but it will not lead to hair loss, dry skin, nausea, irritability, blurred vision, or weakness as excess vitamin A would. Leading food sources of vitamin A and beta-carotene among U.S. adults are carrots, ready-to-eat cereals, and milk.[70] Rich vitamin A sources are liver, milk, cheese, eggs, and fortified cereals.

Iron The need for iron decreases with aging for women after menopause. Like vitamin A, iron is stored more readily in the old than in the young; high intakes of vitamin C enhance absorption. On average, older adults eat more iron than the DRI of 8 mg (males and females; see Table 18.12). Excess iron contributes to oxidative stress, increasing the need for antioxidants to deal with oxidant overload. The UL for iron is 45 mg per day.[79]

Older adults are a heterogeneous population, and not every older person has adequate iron stores. Reasons for inadequate iron status include blood loss from disease or medication (e.g., aspirin), poor absorption due to antacid interference or decreased stomach acid secretion, and low caloric intake. Leading sources of iron in the American diet are ready-to-eat cereals, yeast bread, and beef.[70]

Low Dietary Intake: Nutrients of Concern

Vitamin E Also known as tocopherol, vitamin E is a potent antioxidant. It is a problematic nutrient because dietary intake is well below the recommended 15 mg or 15 IU alpha-tocopherol equivalents (TE; see Table 18.12). Vitamin E plays a special role in the health of older adults due to its antioxidant functions, such as hindering development of heart disease and cataracts.[81] Vitamin E is associated with enhanced immune function[82] and cognitive status, although not with reduced cardiovascular disease risk.[80,83] The UL is 1000 mg (or IU, seen on supplement labels) alpha-TE. Even though vitamin E is fat soluble and can be stored in the body, it seems safe in doses up to 400 IU.[84] At higher doses, vitamin E may increase all-cause mortality and is linked to longer blood-clotting times and increased bleeding. Aspirin, anticoagulants, and fish oil supplements also increase blood-clotting time and are incompatible with high doses of vitamin E intake. Vitamin E is thought to be more effective in its "d" (rather than "dl") form.

Leading food sources of vitamin E are salad dressings/mayonnaise, margarine, and ready-to-eat cereals. Other good sources of vitamin E are oils (especially sunflower and safflower oils), fats, whole grains, wheat germ, leafy green vegetables, tomatoes, nuts, seeds, and eggs.

Folate, Folic Acid For persons with low serum folate levels, dietary increases of folic acid (100 to 400 mcg) can lower serum homocysteine levels and subsequent risk of heart disease.[85] The DRI for males and females is 400 mcg folate (the food source) per day, with a UL of 1000 mcg of folic acid (the synthetic form). On average, older adults did not meet recommended levels prior to 1998 (see Table 18.12). But mandatory fortification of grain products has added 215–240 µg folic acid to our daily diets.[86] Leading food sources for U.S. adults are ready-to-eat cereals, yeast bread, and orange/grapefruit juice.[79]

Despite potentially low intakes levels, folate deficiency as measured by serum levels is uncommon in healthy older adults. Depletion is measured as serum folate <3 ng/mL and red blood cell folate <160 ng/mL.[77] Absorption of folate, like vitamin B_{12}, may be impaired by atrophic gastritis. Moreover, alcoholism is associated with folate deficiency and subsequent pernicious anemia. Folic-acid deficiency can also be secondary to vitamin B_{12} deficiency, which is more common in the elderly than folic-acid deficiency. Folic-acid deficiency should be treated only with folic acid, or treatment may mask potential B_{12} deficiency, which will continue to cause irreversible neurological damage. Medications commonly used by older adults—such as antacids, diuretics, phenytoin (Dilantin), sulfonamides, and anti-inflammatory drugs—affect folate metabolism.

Calcium A 1994 consensus conference sponsored by the National Institutes of Health[87] recommended calcium intake for women depending on estrogen status; namely, 1500 mg for women aged 50 to 64 not taking supplementary estrogen and 1000 mg calcium per day for postmenopausal women taking estrogen. After age 65, it was recommended that all men and women consume 1500 mg calcium per day. This recommendation of 1500 mg was lowered in 1997, when the National Academy of Sciences set the DRI for adults aged 51 and older at 1200 mg daily, independent of gender or hormone status. The Canadian 2002 Clinical Practice Guidelines outline the highest level of evidence to support these DR recommendations. In addition, they state that calcium and vitamin D should not be used as the sole treatment of osteoporosis; rather, they are adjuncts.[88] Recommended levels are met by a small portion of the population; on average, calcium intake of older men and women ranges between 600 and 800 mg per day, with the lowest intake among older black women.[43,58]

Low calcium intake has been linked to colon cancer, overweight, and hypertension. Study results are strongest for the protective effects of calcium intake in the development of hypertension.[89] For example, blood pressure decreases in a subgroup of hypertensive individuals with higher calcium intake. On average, participants who adhered to the Dietary Approaches to Stop Hypertension (DASH) diet reduced their blood pressure. The diet includes 2 or more servings of low-fat dairy products (1265 mg calcium from food in the experimental group); 10 servings of fruits and vegetables; and reduced fat, saturated fat, and cholesterol.

The National Academy of Sciences has set the UL for calcium at 2500 mg per day. Individuals should not consume supplements in excess of this amount. Adverse effects of excess calcium (reported at 4 grams per day) include high blood calcium levels, kidney damage, and calcium deposits in soft tissues and outside the bone matrix, such as bone spurs on the spine.[79] High calcium intake may interfere with zinc, iron, and magnesium absorbtion, and it may result in elevated urinary excretion of calcium, leading to new kidney stones in individuals with a history of kidney stones.

Magnesium Of adults aged 70 and older, approximately one in four meet 100% or more of their magnesium DRI. Adequate magnesium intake (see Table 18.12) is needed for bone and tooth formation, nerve activity, glucose utilization, and synthesis of fat and proteins. Magnesium deficiency can result not only from low intake but also from malabsorption due to gastrointestinal disorders, chronic alcoholism, and diabetes. Signs of deficiency include personality changes (irritability, aggressiveness), vertigo, muscle spasms, weakness, and seizures. An indicator of the wide-ranging functions of magnesium is that it plays a part in over 300 enzyme systems.[90]

Age does not seem to affect magnesium metabolism. However, drugs commonly used by older adults, such as magnesium-based antacids and laxatives, may lead to magnesium overdose. The DRI is 420 mg for males and 320 mg daily for females, with a UL of 350 mg from non-food (supplementary) sources. Signs of magnesium toxicity are diarrhea, dehydration, and impaired nerve activity. Magnesium from food sources does not result in toxicity. Leading food sources for adults are milk, yeast breads, coffee, ready-to-eat cereals, beef, and potatoes.

Zinc   The role of zinc is changing from problematic nutrient because of insufficient intake to potentially problematic because of too much zinc in the diet. Few diets supplied adequate zinc before the requirement was adjusted downward to 8 mg daily for females and 11 mg for males by the 2001 DRI. Fewer than 20% of adults aged 70 and older ingested 100% of the zinc recommended in 1989 (12–15 mg for females and males). However, most people meet the current DRI (see Table 18.12). The UL is 40 mg; excess dietary zinc results in decreased immune response, decreases in HDL cholesterol, and copper deficiency. Leading food sources of zinc among U.S. adults are beef, ready-to-eat cereals, milk, and poultry.[70]

Healthy older individuals maintain zinc balance as well as younger ones do, even though they may absorb less zinc.[91] Zinc absorption is affected by conditions of the intestinal tract such as infections, surgery, muscle-wasting diseases, pancreatic insufficiency, and alcoholism, which all increase the need for zinc. Medications including diuretics, antacids, laxatives, and iron supplements also increase the need for zinc. High intakes of calcium, copper, and iron interfere with zinc absorption.

Of particular interest to older adults are the relationships among zinc deficiency and delayed wound healing, decreased taste acuity and immune response, and increased susceptibility to dermatitis. Zinc supplementation seems to be beneficial only when zinc status is depleted; supplementation in the absence of zinc deficiency does not improve wound healing, taste acuity, or dermatitis. Studies dealing with improved immune status following zinc supplementation are inconsistent.[91]

Nutrient Supplements: Why, When, Who, What, and How Much?

When to Consider Supplements   Can older adults benefit from taking nutrient supplements? They may. Vitamin and mineral supplements may decrease infection-related illness.[92] In some elderly persons, they may provide nutrients often lacking in diets.[92,93] Population surveys show that diets of many older persons fall short of meeting recommended nutrient levels (see Table 18.12). A boost from appropriately used supplements completes the nutritional balance of an inadequate regular diet.[94] Writing in

the Clinician's Corner of the *Journal of the American Medical Association,* Fletcher and Fairfield recommend that all adults take one multivitamin daily.[95]

Any discussion about supplements is based on the assumption that whole foods are the ideal source of nutrients, and supplements are meant to boost a marginal diet. Sometimes it turns out that, as is the case with beta-carotene, the pill form of a nutrient is harmful while the food form promotes health. Furthermore, the interactions among nutrients and the composition of plants and animals making up our food supply are much too complex to replicate in supplements. However, vitamin B_{12} and folic acid are two nutrients better absorbed in a synthetic form than in their protein-bound food form; but this becomes important only when normal metabolic processes fail. It is possible for healthy older adults to live well without dietary supplements.

Age-associated nutritional risk factors such as poverty, isolation, chronic illness, or lack of appetite highlight circumstances when dietary intake might be inadequate and further assessment and intervention, potentially including a supplement, may be useful. Such factors include:

- Lack of appetite resulting from illness, loss of taste or smell, or depression
- Diseases or bacterial overgrowths in the gastrointestinal tract that prevent absorption
- Poor diet due to food insecurity, loss of function, dieting, or disinterest in food
- Avoidance of specific food groups such as meats, milk, or vegetables
- Contact with substances that affect absorption or metabolism: smoke, alcohol, drugs

Who Takes Supplements?   The most likely individuals to take supplements are non-Hispanic white females, as are individuals with more education and higher incomes.[96] Women are more likely than men to use supplements.

What to Take?   The question: "What should I take?" may not have a simple answer. A multivitamin was the most commonly used vitamin/mineral preparation, and around 30% of elderly men and women use multivitamins.[96] Vitamin E was the second highest vitamin/mineral supplement taken by men (i.e., 14%), whereas calcium was next for women (23%). Finally, the third most often used supplement was vitamin C, with 12% and 19% in men and women, respectively. Most of the nearly 2500 products reported in NHANES III were vitamin–mineral combinations or vitamin–single nutrient combinations.[96] The most common supplement ingredients in this survey were vitamin C, vitamin B_{12}, vitamin B_6, niacin, thiamin, riboflavin, vitamin E, beta-carotene,

USP (United States Pharmacopeia) A nongovernmental, nonprofit organization (since 1820); establishes and maintains standards of identity, strength, quality, purity, processing, and labeling for health care products.

NF (National Formulary)   A uniformity standard for herbs and botanicals.

cholecalciferol or vitamin D, and folic acid. In cases where older adults feel that a supplement will prevent cancer or cardiovascular disease, the U.S. Preventive Services Task Force (USP-STF) concluded that "the evidence is insufficient to recommend for or against the use of supplements of vitamins A, C, or E; multivitamins with folic acid; or antioxidant combinations for the prevention of cancer or cardiovascular disease."[97] This was echoed by the National Institutes of Health.[98]

Considerations that guide supplement choices are the same for older as for younger adults, and address the following five questions:

1. Does the supplement contain a balance of vitamins and minerals?
2. When all supplements and fortified foods are combined, is the dose still safe?
3. Does the supplement contain the missing nutrients?
4. Does the supplement carry a *USP* (U.S. Pharmacopeia, a mark that indicates that the manufacturer followed recognized standards when making the

product) or *NF* (National Formulary) code to assure potency and purity?
5. Is the supplement safe? ("Natural" does not mean safe.)

How Much to Take?   In general, multivitamin/mineral supplements should be used in physiologic rather than high-potency doses to maintain a balance of vitamins and minerals. Physiologic dose formulas unique for older adults are available with little or no iron and additional vitamins B_{12} and E.

Generously fortified breakfast cereals, "power" bars, and drinks count as vitamin or mineral supplements. As a group, older adults need the vitamins B_{12}, E, D, and folic acid. The others are superfluous and possibly unsafe (e.g., beta-carotene supplements). Balluz and associates[96] report that adults use more than 300 nonvitamin and nonmineral products, including some that have toxic effects at normal doses (e.g., blue cohosh, chaparral). Because older adults eat fewer calories and are less resilient physically, they also have less tolerance for mistakes in dietary supplement use.

Why? Older people are motivated by wellness and want to take responsibility for their own health, wishing to increase their energy, improve memory, and reduce susceptibility to illness and stress.[49,96,99] Table 18.16 is a summary of some of the vitamins, minerals, and other dietary supplements of special interest to older adults.

Table 18.16 Dietary supplements potentially used by older adults for selected conditions

Condition or Health Status	Supplement
Poor appetite or dieting, leading to intake below 1200–1600 calories	Multivitamin/mineral
Weight loss, chronic underweight	Add high-calorie/protein foods/fats as oils
Vegetarian or vegan	Vitamins B_{12}, D, calcium, zinc, iron
Arthritis*	Antioxidants potentially useful (vitamins C, E), lactobacillus, fiber
Age-related macular degeneration; eye health and cataracts	AREDS formulation (high levels of C, E, A, Zn, Cu) for AMD; Vitamins E, C, B_2, B_6, selenium to reduce cataract risk
Constipation	Fiber (cellulose, bran, psyllium); Fluid to accompany fiber
Diarrhea	Fluid, multivitamin/mineral
Energy boosters	Evaluate total nutrient intake, adequate calories, iron if blood levels are low
Immune status enhancement	Multivitamin/mineral (DRI dosage)
Memory aids, dementia	Ginkgo leaf extracts as part of comprehensive therapy; (no effect on cognitive decline from B_6 and folic acid supplements)[98,100]
Osteoporosis	Vitamins D, K, calcium, fluoride, magnesium (avoid excess K if on blood thinners)
Sleep aids	Milk and a sweet at bedtime; valerian, lavender; avoid guarana, caffeine, alcohol
Stress reduction	Eat well and play in the sunshine (not available in a pill or tonic)

SOURCE: Schulz V et al. Rational Phytotherapy A reference guide for physicians and pharmacists. Fifth Edition. Berlin Springer Verlag, 2004.[100]

*Hanninen K, Rauma AL, Nenonen M et al. Antioxidants in vegan diet and rheumatic disorders. Toxicology. 2000;155:45–53.

Dietary Supplements, Functional Foods, Nutraceuticals: Special Interest for Older Adults

"Eat leeks in March and wild garlic in May,
And all the year after, physicians may play."

Welsh rhyme

The growing availability of functional foods can be a boon for older adults, who benefit from the convenience of nutrient-dense foods such as calcium-fortified orange juice, yogurt–juice beverages, fortified high-fiber cereals, yogurt with live cultures, soy milks, breakfast powders, and various fortified chews, bars, and drinks. However, regular users of functional foods, no matter what age, need to add up all intake in order to avoid potentially toxic nutrient levels (see Table 18.17 for recommended and potentially toxic intake levels). Advice that "you can't overdose on nutrients from foods" does not apply to fortified foods. High nutrient doses can act like drugs, and should be treated as such. Just like drugs, nutrient supplements interact with each other (and with medications); side effects may outweigh the desired benefits.

Several other "ingestibles" besides vitamins and minerals affect older adults' health status. Examples include quasi-functional foods, herbs, stimulants, and other nonvitamin or mineral food components used by older adults to promote health and ward off chronic disease. *Herbal therapy* is defined as the use of plants for medicinal purposes rather than for food consumption, and it comes in many forms: capsules, pills, infusions, tea; tinctures or extracts; and oils and salves. Different parts of the plant are used to produce supplements.[100] Herbal therapies used alongside traditional medical prescriptions are called complementary, and those used alone are called alternative.

Here is a guide for common botanicals and nutraceuticals:

1. *Herbs and spices.* Rosemary, sage, thyme, cinnamon, clove, and ginger enhance the taste of food and protect against oxidative stress.

2. *Caffeine and guarana.* Stimulants to which older people may become increasingly sensitive.

3. *Black, white, and green teas (Camellia sinensis).* Contribute fluid *and* phytochemicals, especially the antioxidant catechins and flavonols.

4. *Garlic.* Contains allicin, which has antibacterial activity, and ajoene, which enhances blood thinning. Eating garlic has been linked to colon cancer protection and better heart health. Individuals taking aspirin or blood thinners shouldn't take garlic pills or eat large amounts of garlic (one or two garlic cloves are considered a small amount).

5. *Herbal treatments.* Widely used and often effective, but herbs used as medicinal treatment can also be dangerous. Herbs should help, not hinder health! Rule #1 in using herbs is also the first rule of medicine: "Do no harm." A list of unsafe herbs is found in Chapter 16.

6. *Phytochemicals.* These plant-based compounds (such as resveratrol, flavonoids, carotenoids, indoles, isoflavones, lignans, and salicylates) are of special interest in aging because they have been linked to reduction of chronic disease. Eating fruits, vegetables, and whole grains automatically increases phytochemical intake.

7. *Pre- and probiotics.* Prebiotics are nondigestible food ingredients that feed health-promoting colon bacteria, and probiotics are live health-promoting bacterial cultures such as lactobacillus acidophilus and bifidobacterium in yogurt. After a course of antibiotic treatment, biotics can reestablish intestinal bacterial life.

8. *Plant stanols.* Plant products that compete with cholesterol in the small intestine because they are similar in structure to cholesterol. Found in corn, wheat, oats, rye, and some other foods; also processed from wood. Eating margarine spreads containing stanols results in decreased levels of LDL but stable HDL cholesterol.

9. *Hormones.* DHEA (dehydroepiandrosterone) is taken to increase muscle mass and immune function, pregnenolone to enhance memory, and melatonin to enhance sleep. There is no evidence that pregnenolone enhances human memory or improves concentration, but there is some evidence that the other two can work. Despite equivocal evidence for melatonin, it is popularly used to induce sleep and reduce jet lag.[101] Secretion of this biorhythm regulator normally decreases with age.

Nutrient Recommendations: Using the Food Label

The Nutrition Facts Panel on food packages is structured to provide nutrient content information in relation to nutrient needs. The Food and Nutrition Board of the National Academies of Sciences, National Research Council, worked with scientists from Canada to establish an integrated set of reference values for specific nutrients.

The recommended nutrient amounts for older adults (aged 70 years and older) are slightly different from the amounts used as references for the nutrition label and that are recommended for "all" adults (see Table 18.17). For example, older adults need more

Table 18.17 Vitamin and mineral recommendations for nutrition labeling (Daily Reference Values or DV) compared with Dietary Reference Intakes (DRI) and Tolerable Upper Intake Levels (UL) for older adults

Mandatory Vitamin and Mineral Components of the Nutrition Label

Nutrient	Daily Values (1993, 1995) All Adults	Dietary Reference Intake (2001) Adults over 70 Males	Females	Upper Limit (UL)
Vitamin A, IU or RE (5 IU = 1 RE = 1 RAE, Retinol Activity Equivalent)	5000 IU	900 RE	700 RE	3000 RE
Vitamin C, mg	60	90	75	2000
Calcium, mg	1000	1200	1200	2500
Iron, mg	18	8	8	45

Voluntary Vitamin and Mineral Components of the Nutrition Label

Nutrient	Daily Values (1993, 1995) All Adults	Males	Females	Upper Limit (UL)
Vitamin D, mcg (1 mcg = 40 IU)	400 IU	15 mcg	15 mcg	50 mcg
Vitamin E, mg (1 mg = 1 TE = 1 IU)	30 IU	15 mg	15 mg	1000 mg
Vitamin K, mcg	80	120	90	–
Thiamin, mg	1.5	1.2	1.1	–
Riboflavin, mg	1.7	1.3	1.1	–
Niacin, mg	20	16	14	35
Vitamin B_6, mg	2.0	1.7	1.5	100
Folic acid, mcg	400	400	400	1000[a]
Vitamin B_{12}, mcg	6.0	2.4	2.4	–
Biotin, mcg	300	30	30	–
Pantothenic acid, mg	10	5.0	5.0	–
Phosphorus, mg	1000	700	700	3000
Iodine, mcg	150	150	150	1100
Magnesium, mg	400	420	320	350[b]
Zinc, mg	15	11	8	40
Selenium, mg	70	55	55	400
Copper, mcg	2000	900	900	10,000
Manganese, mg	2.0	2.3	1.8	11
Chromium, mcg	120	30	20	–
Molybdenum, mcg	75	45	45	2000
Chloride, mg	3400	2300	2300	–
Potassium, mg	3500	4700	4700	–
Choline, mg	–	550	425	3500
Fluoride, mg	–	4.0	3.0	10

SOURCE: J. A. Pennington and V. S. Hubbard, "Derivation of Daily Values Used for Nutrition Labeling," *J Am Diet Assoc* 97 (1999):1407–12; P. Trumbo et al., "Dietary Reference Intakes: Vitamin A, Vitamin K, Arsenic, Boron, Chromium, Copper, Iodine, Iron, Manganese, Molybdenum, Nickel, Silicon, Vanadium, and Zinc," *J Am Diet Assoc* 101 (2001):294–301; 2004 DRIs for Macronutrients.

[a] synthetic
[b] nonfood sources

vitamin C and calcium than the reference amounts listed in the mandatory components of the label, and they need less iron and zinc than the label value. Values also differ for vitamins D, E, and B_{12}. Table 18.17 lists the levels used for food labels and the current DRI values for older adults. The UL is listed because with today's food supply, it is easy to overdose vitamins and minerals.

Older adults can still use the percentages on food labels for dietary guidance, as long as they adjust them to get more than 100% calcium and vitamins D and C and less than 100% for iron and zinc. The vitamin B_{12} label recommendation is higher than the DRI, but poor absorption in older adults makes unsafe intake from foods unlikely. In nutrition labeling and dietary guidance, "one size does not fit all."

Cross-Cultural Considerations in Making Dietary Recommendations

Food habits develop in cultural contexts, and we can learn about them through various ways. Travel through North America would allow us to observe cultures that make up our society. Visiting ethnic restaurants, stores, and farmer's markets can be another way to get a glimpse of cultural food diversity. Cookbooks, films, talking with individuals about their food history, and participating in ethnic celebrations are other sources of insight into food patterns of various cultures. Each new immigrant wave adds unique food traditions to the country's mix. Older adults may be stronger advocates for upholding traditional food patterns than young people are. In working with them on food issues, it is useful (and interesting) to determine whether food and lifestyle habits are patterned on specific cultural considerations.

National food-monitoring programs survey the population with proportionately larger samples drawn from minority groups in order to develop a balanced picture of the whole population. North America is home to rapidly growing Hispanic, Asian, Russian, and African immigrant groups. The U.S. Census tracks minority groups, but it is completed only once every 10 years. The local Area Agencies on Aging and Senior Nutrition Programs also track population trends, and are likely to have greater insight about some of the smaller ethnic population groups in their unique communities and regions.

Cultural differences are reflected in approaches to dietary guidance. For example, Chile has separate guidelines for older and younger adults. In New Zealand, older adults are encouraged to socialize at mealtimes to improve appetite. In France, South Korea, and Japan, people of all ages are encouraged to enjoy mealtimes and take pleasure in eating. Other unique guidelines are those of China, which suggest people eat 20 to 25 grams (nearly an ounce) of fish daily. Guatemala uses a bean pot as a nutritional icon. India has developed one set of guidelines for the rich (i.e., overall energy intake should be restricted to levels commensurate with the sedentary occupations of the affluent, so that obesity is avoided; total fat intake is not to exceed 20% of total energy; and use of clarified butter, a prized Indian culinary ingredient, should be restricted to special occasions) and one for the poor (addressing the fact that at least one-third of the households in India are not able to afford even the minimum nutritional requirements, even though they are spending 80% of their income on food, and so recommendations identify food combinations that are most likely to meet recommended dietary intakes).

Communicating effectively and avoiding misinterpretation in intercultural settings is probably the most important thing a nutritionist can learn to do when working with older adults from various cultures; it requires skill in transferring information, developing and maintaining relationships, and gaining compliance.[102] Developing individual skills takes time, commitment, and practice. On the other hand, nutrition education and guidance tools have been developed in many languages and for diverse cultures, although not typically for elders of ethnic groups. Culturally appropriate resources can be found through local extension services, cross-cultural education centers, diabetes education programs, public health agencies, and some commodity groups such as the Dairy Council. Multilingual versions of the MyPyramid can be downloaded from the Nutrition Education for New Americans Project in the Department of Anthropology and Geography at Georgia State University (http://multiculturalhealth.org). Food pyramids in nearly 40 languages are available, including Amharic, Russian, Somali, and Vietnamese.

Food Safety Recommendations

Foodborne illness risk defines a biological, chemical, or physical property of food (a hazard) that may cause illness or injury under a given set of conditions. Older adults are vulnerable to foodborne illness because they often have a compromised immune status. How widespread this problem is is basically unknown, because many foodborne illnesses are not reported when individuals think it is "the flu."

Poor storage, thawing, and food-handling practices leading to microorganism growth are generally to blame for these illnesses. Bacteria and viruses, especially campylobacter, salmonella, and Norwalk-like viruses, are among the most prevalent causes of foodborne illnesses. Signs and symptoms may appear within half an hour of eating a contaminated food or may not develop for up to 3 weeks. They include gastrointestinal distress, diarrhea, vomiting, and fever.

Leading practices that put an older person at risk are as follows:

- Improper holding temperatures of foods
- Poor personal hygiene
- Contaminated food preparation equipment (cutting boards, knives)
- Inadequate cooking time

The new Dietary Guidelines for Americans provide the following suggestions to keep food safe:

- Wash hands and surfaces often.
- Separate raw, cooked, and ready-to-eat foods while shopping, preparing, or storing.
- Cook foods—especially raw meat, poultry, fish, and eggs—to a safe temperature.
- Refrigerate or freeze perishable or prepared foods within 2 hours.

- Follow the label for food-safety preparation and storage instructions.
- Serve hot foods hot (140°F or above) and cold foods cold (40°F or below).
- When it doubt, throw it out!

Physical Activity Recommendations

"There is no segment of the population that can benefit more from exercise than the elderly."

William J. Evans[103]

Exercise is a true fountain of youth. Physical activity builds lean body mass, helps to maintain balance and flexibility, contributes to aerobic capacity and to overall fitness,[24,104] and boosts immune status.[105] Only resistance exercises (such as weight lifting) build muscle and bone. All physical activities count toward energy expenditure and health maintenance.

National health goals encourage increased physical activity levels for older adults (see Table 18.18). On average, older adults are less active than younger ones. Lower activity levels as well as deteriorating strength, endurance, and sense of balance are associated with, but not caused by, increasing age.

Older people benefit from exercise even more than younger people do because strength training is the only way to maintain and build muscle mass. When evaluating strength in men and women, muscle mass (not function) is the major determinant of age- and gender-related differences in strength, independent of where muscle is located.[24] Besides making a person stronger, increased muscle mass also leads to higher caloric turnover and less likelihood of weight gain.

Age does not hinder training effects. For example, in a comparison of changes in body composition between young and old individuals during 20 weeks of an endurance training program, all individuals gained total body density and fat-free mass while losing fat mass and abdominal fat. Responses to training differed by gender and by race, but *not* by age.[26]

Exercise Guidelines

How can one predict whether physical activity will exacerbate existing medical conditions? Physician screening or assessment by completing a questionnaire like the one in Table 18.19 can identify potential problem areas.[24] Kligman and colleagues identify a more detailed assessment and recommend ways to evaluate cardiovascular fitness, strength, function, balance, flexibility, body composition, bone density, and lipid levels.[106] Overall, an evaluation by an individual's physician is needed to identify contraindications to exercise. Individuals cleared for participation

Table 18.18 Healthy People 2010 physical activity goals and baseline activity levels comparing older and younger adults

Type of Activity	Baseline Levels Target		Target for All Adults
	Baseline for Age 75+	Age 18–24	
Engage in no leisure-time physical activity	65%	31%	20%
Engage in regular physical activity at least 20 min, 3 times/wk	23%	36%	30%
Perform strengthening activities 2 or more days/wk	8%	30%	30%
Perform stretching and flexibility activities	21%	39%	40%

SOURCE: *Healthy People 2010: National Health Promotion and Disease Prevention Objectives,* Washington, D.C.: U.S. Department of Health and Human Services, 2000.

find that even small increases in exercise add up. Even patients with heart disease can be compliant with exercise recommendations.[107]

Many types of physical activity are good. Muscle mass is built through resistance or weight-bearing activities such as walking, running, and jumping, which move the body's own weight against gravity. Playing tennis and jumping rope are examples of games or training activities that incorporate weight-bearing exercise. Lifting weights is another example. Water exercises can be done either as resistance training, by pushing webbed gloves or empty plastic containers against the water, or as aerobic activity.

Aerobic exercise builds endurance and contributes to cardiac fitness, even when mobility is limited. Nonimpact aerobics and chair aerobics classes can help people with arthritis or hip replacements, or those who are wheelchair bound, to get and stay strong. An explicit integrative review on types of interventions to promote physical activity in older adults found that some older adults are able to increase activity levels. However, there is no single intervention component that dramatically or consistently produces successful outcomes.[108] Elaine Souza, RD, MPH, has developed the following guide for planning effective exercise sessions:

- Decide on frequency: 2–3 times per week is effective for strength training, using 8–10 different exercises with 8–12 repetitions each, the whole routine to be done in 20–30 minutes.
- For general health, exercise for 30 minutes on most days of the week.

Table 18.19 Keep moving—fitness after 50 chart

A. Do I get chest pains while at rest and/or during exercise?

B. If the answer to question A is "yes": Is it true that I have not had a physician diagnose these pains yet?

C. Have I ever had a heart attack?

D. If the answer to question C is "yes": Was my heart attack within the last year?

E. Do I have high blood pressure?

F. If you do not know the answer to question E, answer this: Was my last blood pressure reading more than 150/100?

G. Am I short of breath after extremely mild exertion and sometimes even at rest or at night in bed?

H. Do I have any ulcerated wounds or cuts on my feet that do not seem to heal?

I. Have I lost 10 lb or more in the past 6 months without trying and to my surprise?

J. Do I get pain in my buttocks or the back of my legs—my thighs and calves—when I walk? (This question is an attempt to identify persons who suffer from intermittent claudication. Exercise training may be extremely painful;

however, it may also provide relief from pain experienced when performing lower-intensity exercise.)

K. When at rest, do I frequently experience fast irregular heartbeats or, at the other extreme, very slow beats? (Although a low heart rate can be a sign of an efficient and well-conditioned heart, a very low rate can also indicate a nearly complete heart block.)

L. Am I currently being treated for any heart or circulatory condition, such as vascular disease, stroke, angina, hypertension, congestive heart failure, poor circulation to the legs, valvular heart disease, blood clots, or pulmonary disease?

M. As an adult, have I ever had a fracture of the hip, spine, or wrist?

N. Did I have a fall more than twice in the past year (no matter what the reason)? (Many older persons have balance problems and at the initiation of a walking program will have a high chance of falling. These persons may benefit from balance training and resistance exercise before beginning a walking program.)

O. Do I have diabetes?

SOURCE: From W. J. Evans, D. Cyr-Campbell, "Nutrition, Exercise, and Healthy Aging," JADA. 1997;97:632–8. Copyright the American Dietetic Association. Reprinted by permission from the author.

- Drink water when exercising (before, during, and after exercise).
- Do warm-up and cooldown activities of 5 to 10 minutes each.

Individuals can take charge of their own healthy aging by developing appropriate and effective exercise habits. Simple and, occasionally challenging, fitness advice: Eat good food, drink lots of water, and play hard! See Case Study 18.1

Nutrition Policy and Intervention for Risk Reduction

Nutrition policy promotes health by combining nutrition education for individuals and population interventions. The ultimate goal of nutrition intervention is to produce a better health outcome.

Nutrition Education

> "The human mind, once stretched to a new idea, never goes back to its original dimension."
>
> Oliver Wendell Holmes

Contrary to some beliefs, older people do learn and change. Someone born in 1920 has seen the invention of fast food restaurants, microwaves, television and TV dinners, and a whole host of computer-controlled kitchen appliances. To age is to grow and adapt. Learning new nutrition habits is part of aging. Nutrition education is different from education in general because its goal is changed dietary behaviors. Nutrition education consists of a set of learning experiences to facilitate voluntary adoption of nutrition-related behaviors that are conducive to health and well-being. Several requirements must be met for nutrition education—that is, behavior change—to occur (see Illustration 18.3). Think of it as the 4 C's of nutrition education.[109]

1. *Commitment.* Commitment means being motivated to adopt health-promoting behaviors and intending to adopt and maintain a new food behavior.

2. *Cognitive processing.* Understanding how a food behavior contributes to health and planning how it will fit into your life constitutes cognitive processing.

3. *Capability.* Acquiring the skills to practice new food behaviors is part of nutrition education. An example is learning to identify whole grain breads or to prepare vegetables when intending to adopt a high-fiber diet.

4. *Confidence.* "Nothing breeds success like success!" The best predictor that someone will practice new dietary habits is their personal confidence in being able to do so.

Case Study 18.1

Masterfile

JT—Spiraling Out of Control?

JT, a retired computer company executive, eats out four times a week since his wife died last year. Meals at home consist of microwave dinners or supreme pizza. He belongs to a health club, which he visits three times a week. After working out and socializing, JT and his friends go for beer. He developed type 2 diabetes five years ago. Last week, he visited the clinic for his annual check-up. He was measured at 5 feet 9 inches and weighed 235 pounds.

Questions

1. If you were his nutritionist, what nutrition remedies would you prescribe for JT?
2. What sort of advice would you give JT about weight management?
3. What would you ask JT about his food and fitness routine, and how would you convince him that he needs an aggressive nutrition and fitness program?
4. What sort of fluid recommendation would you make?

Educational sessions for older learners are best designed around their potential limitations, such as declines in visual acuity and hearing loss.[109] Adaptation for written material includes:

- Larger type size
- Serif lettering (Helvetica is a typeface with no serifs; Times Roman is a typeface with serifs.)
- Bold type
- High contrast (black on white)
- Nonglossy paper to decrease glare

- Avoid blue, green, and violet (paper and print color) due to decreased ability to discriminate among these colors
- Reading level of fifth to eighth grade

Educational strategies can result in better diets for individuals, but these alone may not be sufficient to enhance the dietary patterns of populations. Cultural environments support, ignore, or punish desired behavior change. An example is a peer group that values health and fitness; group members will support each other's health-promoting behaviors.

Food and nutrition policies arise from public values, beliefs, and opinions that define the cultural context in which dietary behaviors exist. Policies can be overt or unspoken. Public policies supporting the health of older adults are evident in the Social Security program, which provides financial support for postretirement living, and the Food Stamp program, which makes grocery money available when living resources are inadequate.

Model Programs Exemplify Intervention Goals

Nutrition Partners   Nutrition Partners is a model nutrition intervention program for free-living older adults in a large midwestern metropolitan community. Nadine Reiser, RD, project director of the Senior Dining Program, applied for and received a grant to pilot a home-visiting program to assist elders with their nutritional problems. Dietetic

Illustration 18.3 Four essential elements to achieve and maintain individual dietary behavior change.

SOURCE: Adapted from U. B. Krinke, "Effective Nutrition Education Strategies to Reach Older Adults," in R. R. Watson (ed.), *Handbook of Nutrition in the Aged.* Boca Raton, FL: CRC Press, 2001.

staff of the nutrition program make home visits in response to referrals from community agencies, physicians, and program site staff; staff members do dietary assessments, develop and prioritize treatment plans together with the client, and make return visits periodically to monitor the intervention. Clients make changes appropriate to their specific lifestyles. For instance, to avoid dehydration and side effects from medications, one client decided to keep a large bottle of water in the refrigerator, refilling it each morning and drinking until it was empty at the end of the day. A daily mark on the calendar tracking refills helps to ensure that lack of thirst does not lead to dehydration.

Store-to-Door After "retiring" in the 1980s, Dr. Dave Berger surveyed his community to see where he might do some good. Many older adults told him that getting to the grocery store was impossible, and that even when they could get to the store, bringing the bags home was difficult. Winter ice and snow made things worse, for people feared falling. There had been a grocery delivery service, but it closed because profits were not meeting shareholder expectations. So Dr. Berger joined forces with his wife, Fran, and with friends, colleagues, and a lot of volunteers to start up the nonprofit Store-to-Door, a home-delivered grocery program for older people and those who are disabled. For a small fee, volunteers will shop for you. They buy items that discount grocers offer: food, of course, but also greeting cards, medicines, paper goods, and cleaning supplies, although no alcohol or cigarettes. Customers can get credit for coupons. Volunteers deliver the groceries throughout the year. After starting Store-to-Door in the Midwest, Dr. Berger started another in Portland, Oregon. When last heard from, he was in Ventura, California, running Shop Ahoy, his latest home-delivered grocery program.

Community Food and Nutrition Programs

Elderly Nutrition Programs

Governmental programs for older adults include the USDA's Food Stamp and extension programs, Adult Day Services food programs, Nutrition Assistance Programs for Seniors (NAPS), Meals-on-Wheels and other home-delivered meal programs, and the Senior Nutrition Program of the Older Americans Act.

Nongovernmental home health programs provide food and nutrition services as part of a broader range of screening and assessment, nursing, and other support services. For instance, home health aides will shop for and prepare food, and clean up the kitchen afterward. Home care services allow individuals to receive the necessary support to stay in their homes for as long as they wish. Remaining in one's home indefinitely is sometimes referred to as "aging in place." A broader definition of this concept is found in the Position of the American Dietetic Association.[1]

Aging in place has many definitions and does not necessarily mean living in one setting or one home for a lifetime. Ideally, aging in place offers choices from a spectrum of living options and medical and supportive services customized to accommodate those who are fully active and have no impairments, those who require limited assistance, and those with more severe impairments who require care in long-term care facilities.

Other food and nutrition programs that contribute to the continuum of nutrition services include commodity foods, food pantries and soup kitchens, cooperative buying groups such as Fare for All, and screening and referral services.

Senior Nutrition Program

Congress first appropriated funds under Title VII of the Older Americans Act of 1965 to begin the Senior Nutrition Program, also called the Elderly Nutrition Program (ENP). The Senior Nutrition Program was created to alleviate poor nutritional intake and reduce social isolation among older adults. It was based on evidence that older adults do not eat adequately because of the following:

1. Lack of income limits ability to purchase food.
2. Lack of skills limits ability to select and prepare nourishing meals.
3. Limited mobility affects shopping and meal preparation.
4. Feelings of isolation and loneliness decrease the incentive to eat well.

Senior Nutrition, now Title IIIC of the Older Americans Act, is a community-based nutrition program that provides meals (congregate and home-delivered), increased social contact, nutrition screening and education, and information and linkages to other support programs and services, as well as volunteer opportunities. Anyone who is 60 or more years of age (and spouse regardless of age) is eligible to participate in the congregate dining program; home-delivered meal clients must be homebound and unable to prepare their own meals. Typically, $1.00 of Title III funds spent on congregate services is supplemented by an additional $1.70 from other sources; the average cost of an ENP meal is $5.17, and a home-delivered one is $5.31.[110] Title VI grant programs are similar to Title III funds and were established to help deliver social and nutrition services to older American Indians, Alaskan Natives, and Native Hawaiians. About 25% of participants are minorities, almost twice the national percentage of minority adults over age 60.[110]

Today there is less poverty among younger seniors; but there are greater nutritional and social needs among frail elderly and individuals with low incomes, chronic health conditions, limited mobility, and limited English-speaking ability, and among minority and isolated elders. About one-third of Title III congregate meal participants and more than one-half of Title VI meal participants have incomes at or below the Department of Health and Human Services (DHHS) poverty threshold.[111]

Senior dining sites are targeted to neighborhoods where older, frail, impoverished seniors live. Dining sites are located in community centers, senior centers, civic buildings, subsidized housing units, schools, and other accessible locations. Meals are delivered to the homes of individuals who are 60 years of age or older, homebound by reason of illness or disability, and unable to prepare meals. Services are adapted to meet each unique community's setting. For instance, meal vouchers for use at local cafes or diners are available in some small communities.

Other services to meet the needs of frail older seniors include multiple meals, weekend meals, take-home snacks, liquid supplements, nutrition screening and education, and one-to-one nutrition counseling. Also available are special diets for medical reasons and special meals for Jewish and ethnic elders.

Senior or elderly nutrition programs have successfully brought together millions of people to socialize and enjoy nutritious meals. In 2002, about 250 million congregate and home-delivered meals were served to approximately 2.6 million older adults.[110] The Older Americans Act 2000 Amendment, Section 339 (Nutrition) (H.R. 782) states that nutrition projects shall use a dietitian (or person with compatible expertise) to provide meals that comply with the Dietary Guidelines for Americans. Nutrition program meals were found to exceed the one-third RDA standard. Compared to nonparticipants, participants had up to 31% higher intake of recommended nutrients. In other words, the program is working.

Surveys show that targeting those who need services most is successful, and a national evaluation of the senior nutrition program found that the program is targeted to those most in need.[111] Increasing socialization was one of the original program goals and continues to be one of the outcomes. Participants have 16–18% more social contacts per month than nonparticipants do. Relative to the general older population, participants are older and more likely to be female, to belong to an ethnic minority, to live alone, and to have incomes well below poverty level.

Grocery shopping assistance is an important service for frail seniors. A variety of models are used, including volunteer escorts to the supermarket, bus rides, and grocery delivery to the door. However, Title IIIC funds may not currently pay for grocery shopping assistance.

The Promise of Prevention: Health Promotion

> "Grow old along with me, the best is yet to be!"
>
> Robert Browning

Although good nutritional habits make a greater impact when started early in life,[21] sometimes individuals are not motivated to pursue these risk-reduction strategies until later in life or after experiencing a health problem. Successful strategies to reach an older audience reflect their specific needs and interests. It is time for the belief that an 80-year-old is too old to learn and practice health-promotion strategies to become an outdated myth.

Key Points

1. Functional ability (the demonstrated ability to carry out activities of daily living) is more important than chronological age in assessing the health status of older adults.

2. Good nutrition, good health habits, environment, access to health care, and genetics contribute to human life expectancy, which is still significantly shorter than the potential human life span. Theories of aging, such as wear-and-tear theories, help to explore which factors contribute most to a longer, disease-free life.

3. Of all the physiological changes associated with aging, loss of lean body mass and the concomitant gains of body fat may well be the most important in determining functional age.

4. "Use it or lose it" applies to the body and the spirit: keep learning to maintain acute brain function; stay active to build muscle and bone; eat well to maintain and repair tissue; and cultivate a positive approach to aging for enhanced quality of life as well as to live longer.

5. The DETERMINE acronym is a reasonable summary of warning signs associated with poor nutritional health.

6. While adults in general consume more than enough calories and protein, populations of older adults may be lacking in adequate dietary protein and energy.

7. The thirst mechanism of older adults is not as sensitive as that of younger adults, placing them at higher risk of dehydration.

8. Physiological changes that lead to malnutrition in older adults are decreased absorption of Vitamins D and B_{12} and increased storage of Vitamin A and iron.

9. In general, older adults eat better than younger adults, but they do not consume enough Vitamin E, folic acid, calcium, or magnesium to meet recommended intake levels.

10. Vitamin and mineral supplements can be helpful for older adults who have lost their appetite, avoid certain food groups, have poor diets due to food insecurity, loss of function, dieting or depression, or who have gastrointestinal bacterial overgrowth that prevents nutrient absorption.

11. Excellent food safety practices are especially important for older adults who may be more vulner-

able to infection for many reasons, such as a higher prevalence of chronic diseases, sensory and functional losses, and decreased resilience in healing and recovery from illness.

12. Older adults are often more interested in nutrition education and health promotion than younger adults. The stereotype that older adults will not change is just that: an old stereotype.

Resources

AARP

Information related to growing old in America: finances, politics, health issues, travel, and population statistics.
Website: www.aarp.org/visit/brpromo.htm

American Dietetics Association

Position papers, educational materials.
Website: www.eatright.org

Centers for Disease Control and Prevention

Mortality and morbidity data.
Website: www.cdc.gov/mmwr

Florida National Policy and Resource Center on Nutrition and Aging

Aging Policy and education center.
Website: www.nutritionandaging.fiu.edu

National Heart, Lung, and Blood Institute

Gateway site for consumers and health professionals; links and information materials about heart health and more.
Website: www.nhlbi.nih.gov

Oral Health in America

A report of the Surgeon General.
Website: www.nidr.nih.gov/sgr/sgrohweb/home.htm

Tufts University Center on Aging

Popular gateway to information on aging.
Website: www.navigator.tufts.edu

U.S. Administration on Aging

Fact sheet, news, links to other resources.
Website: www.aoa.gov

References

1. Position Paper of the American Dietetic Association: Nutrition across the spectrum of aging. J Am Diet Assoc. 2005;105: 616–33.

2. Maloney SK, Fallon B, Wittenberg CK. Executive summary. Aging and health promotion: market research for public education. Washington, DC: Office of Disease Prevention and Health Promotion, U.S. Department of Health and Human Services, PHS; 1984.

3. Nube M et al. Scoring of prudent dietary habits and its relation to 25-year survival. J Am Diet Assoc. 1987;87:171–75.

4. Kant AK et al. A prospective study of diet quality and mortality in women. JAMA. 2000;283:2109–15.

5. Vita AJ et al. Aging, health risks, and cumulative disability. N Engl J Med. 1998; 338:1035–41.

6. Johnson MA. Achieving 100 candles: the Georgia Centenarian Study lights the way. USDA 2000 Millenium Lecture Symposium. September 28, 2000.

7. Health, United States 2005 chartbook, trends in the health of Americans 017–022–01546–3. Government Printing Office; December 2005. Internet version at www.cdc.gov/nchs/data/hus/hus05.pdf (550pp), accessed 10/06.

8. 1995 White House Conference on Aging. The road to aging policy for the 21st Century. Washington, DC: U.S. National Commission on Libraries; 1996.

9. Healthy People 2010: National health promotion and disease prevention objectives. Washington, DC: U.S. Dept. of Health and Human Services; 2000.

10. Hayflick L. How and why we age. Exp Gerontol. 1998;33:639–53.

11. Bernarducci MP, Owens NJ. Is there a fountain of youth? A review of current life extension strategies. Pharmacotherapy. 1996;16:183–200.

12. Forbes GB. Human body composition. Growth, aging, nutrition, and activity. New York: Springer-Verlag; 1987:31.

13. Buys CH. Telomeres, telomerase, and cancer. N Engl J Med. 2000;342:1282–83.

14. Parsons PA. The limit to human longevity: an approach through a stress theory of ageing. Mech Ageing Dev. 1996; 87:211–18.

15. Weindruch R, Sohal RS. Seminars in medicine of the Beth Israel Deaconess Medical Center. Caloric intake and aging. N Engl J Med. 1997;337:986–94.

16. McCay C, Crowell M, Maynard L. The effect of retarded growth upon the length of

life and upon ultimate size. J Nutr. 1935; 10:63–79.

17. Masoro EJ. Nutrition and aging in animal models. In: Munro HN, Danford DE, eds. Nutrition, aging, and the elderly. New York: Plenum Press; 1989:25–41.

18. Weindruch R. The retardation of aging by caloric restriction: studies in rodents and primates. Toxicol Pathol. 1996;24:742–45.

19. Walford RL et al. Calorie restriction in biosphere 2: alterations in physiologic, hematologic, hormonal, and biochemical parameters in humans restricted for a 2-year period. J Gerontol A Biol Sci Med Sci. 2002;57: B211–24.

20. Fontana L et al. Effect of long-term calorie restriction with adequate protein and micronutrients on thyroid hormones. J Clin Endocrine Metab, in press. First published online May 23, 2006 as doi10.1210/jc.2006–0328.

21. Weisburger JH. Approaches for chronic disease prevention based on current understanding of underlying mechanisms. Am J Clin Nutr. 2000;71:1710S–4S; discussion 1715S–9S.

22. Walker AR, Walker BF. Nutritional and non-nutritional factors for 'healthy' longevity. J R Soc Health. 1993;113:75–80.

23. Forbes GB. Longitudinal changes in adult fat-free mass: influence of body weight. Am J Clin Nutr. 1999;70:1025–31.

24. Evans WJ, Cyr-Campbell D. Nutrition, exercise, and healthy aging. J Am Diet Assoc. 1997;97:632–38.

25. Guo SS et al. Aging, body composition, and lifestyle: the Fels Longitudinal Study. Am J Clin Nutr. 1999;70:405–11.

26. Wilmore JH et al. Alterations in body weight and composition consequent to 20 weeks of endurance training: the HERITAGE Family Study. Am J Clin Nutr. 1999;70:346–52.

27. Fiatarone MA et al. High-intensity strength training in nonagenarians. Effects on skeletal muscle. JAMA. 1990;263:3029–34.

28. Villareal DT et al. Obesity in older adults: technical review and position statement of the American Society for Nutrition and NAASO, The Obesity Society. Obes Res. 2005;13:1849–63.

29. Elahi VK et al. A longitudinal study of nutritional intake in men. J Gerontol. 1983;38:162–80.

30. Kaneda H et al. Decline in taste and odor discrimination abilities with age, and relationship between gustation and olfaction. Chem Senses. 2000;25:331–37.

31. Schiffman SS. Taste and smell losses in normal aging and disease. JAMA. 1997;278:1357–62.

32. Ship JA et al. Longitudinal changes in smell identification. J Gerontol A Biol Sci Med Sci. 1996; 51:M86–91.

33. Bartoshuk LM. Taste. Robust across the age span? Ann N Y Acad Sci. 1989;561:65–75.

34. Schiffman S. Changes in taste and smell: drug interactions and food preferences. Nutr Rev. 1994;52(II):S11–14.

35. Jones DB, Niendorff WJ, Broderick EB. A review of the oral health of American Indian and Alaska Native elders. J Public Health Dent. 2000;60 Suppl 1:256–60.

36. Appollonio I et al. Influence of dental status on dietary intake and survival in community-dwelling elderly subjects. Age Ageing. 1997;26:445–56.

37. Mobley C, Saunders MJ. Oral health screening guidelines for non-dental health care providers. J Am Diet Assoc. 1997;97 (suppl 2):S123-26.

38. Vargas CM, Kramarow EA, Yellowitz JA. The oral health of older Americans. Aging Trends; No. 3. Hyattsville, MD: National Center for Health Statistics; 2001.

39. Roberts SB et al. Control of food intake in older men. JAMA. 1994;272:1601–6.

40. Rolls BJ. Regulation of food and fluid intake in the elderly. Ann N Y Acad Sci. 1989;561:217–25.

41. Phillips PA, Johnston CI, Gray L. Disturbed fluid and electrolyte homoeostasis following dehydration in elderly people. Age Ageing. 1993;22:S26–33.

42. Phillips PA et al. Reduced thirst after water deprivation in healthy elderly men. N Engl J Med. 1984;311:753–59.

43. Wilson JW et al. Data tables: combined results from USDA's 1994 and 1995 Continuing Survey of Food Intakes by Individuals and 1994 and 1995 Diet and Health Knowledge Survey, 1997.

44. White JV et al. Nutrition screening initiative: development and implementation of the public awareness checklist and screening tools. J Am Diet Assoc. 1992;92:163–7.

45. A Consensus Conference sponsored by the Nutrition Screening Initiative. Report of nutrition screening I: toward a common view. Washington, DC, April 8–10, 1991. The Nutrition Screening Initiative.

46. White JV et al. Consensus of the Nutrition Screening Initiative: risk factors and indicators of poor nutritional status in older Americans. J Am Diet Assoc. 1991;91:783–87.

47. Posner BM et al. Nutrition and health risks in the elderly: the nutrition screening initiative. Am J Public Health. 1993;83:972–78.

48. Rose D, Oliveira V. Nutrient intakes of individuals from food-insufficient households in the United States. Am J Public Health. 1997;87:1956–61.

49. Mitchell AA, Kaufman DW, Rosenberg L. Patterns of medication use in the United States 2004. A report from the Slone Survey. www.bu.edu/slone/SloneSurvey/AnnualRpt/slonesurveyreport2004.pdf, accessed 10/06.

50. De Castro JM. Social facilitation of food intake: people eat more with other people. Food Nutr News. 1994;66:29–30.

51. Gerrior SA et al. How does living alone affect dietary quality? U.S. Department of Agriculture, ARS, Home Economics Research Report No. 51. Fam Econ Nutr Rev. 1995;8:44–46.

52. Sahyoun NR et al. Nutrition Screening Initiative checklist may be a better awareness/educational tool than a screening one. J Am Diet Assoc. 1997;97:760–64.

53. Garry PJ, Owen GM, Eldride TO. The New Mexico Aging in Process Study. 1980–1997. Albuquerque: University of New Mexico Health Sciences Center; 1997.

54. deGroot LC et al. Evaluating the DETERMINE Your Nutritional Health Checklist and the Mini-Nutritional Assessment as tools to identify nutritional problems in elderly Europeans. Eur J Clin Nutr. 1998;52(12):877–83.

55. Crogan NL, Corbett CF, Short RA. The minimum data set: predicting malnutrition in newly-admitted nursing home residents. Clin Nurs Res. 2002;11(3):341–53.

56. American Dietetic Association Foundation. Nutrition and health for older Americans toolkit, 1998. Content available from www.eatright.org/olderamericans.

57. Russell RM, Rasmussen H, Lichtenstein AH. Modified Food Guide Pyramid for people over seventy years of age. J Nutr. 1999;129:751–53.

58. 1994–1996 Continuing Survey of Food Intakes by Individuals, 2-day average. USDA, ARS, Beltsville Human Nutrition Research Center, Food Surveys Research Group.

59. Wright JD et al. Dietary intake of ten key nutrients for public health, United States: 1999–2000. Advance data from vital and health statistics; no. 334. Hyattsville, MD: National Center for Health Statistics; 2003.

60. Yates AA, Schlicker SA, Suitor CW. Dietary Reference Intakes: the new basis for recommendations for calcium and related nutrients, B vitamins, and choline. J Am Diet Assoc. 1998;98:699–706.

61. Harper EJ. Changing perspectives on aging and energy requirements: aging and energy intakes in humans, dogs and cats. J Nutr. 1998;128:2623S–26S.

62. Institute of Medicine of the National Academies. Food and Nutrition Board Dietary Reference Intakes: energy, carbohydrate, fiber, fat, fatty acids, cholesterol, protein, and amino acids (macro nutrients). The National Academies Press; Washington, DC: 2005. www.nap.edu/openbook/0309085373/html/index.html.

63. Rothenberg EM et al. Resting energy expenditure, activity energy expenditure and total energy expenditure at age 91–96 years. Br J Nutr. 2000;319–24.

64. Arcerio PJ et al. A practical equation to predict resting metabolic rate in older men. Metabolism. 1993;42(8):950–57.

65. Arcerio PJ et al. A practical equation to predict resting metabolic rate in older females. J Am Geriatric Soc. 1994;41:389–95.

66. Taylor HL et al. A questionnaire for the assessment of leasure time physical activities. J Chronic Dis. 1978;31:741–55.

67. Fuchs CS et al. Dietary fiber and the risk of colorectal cancer and adenoma in women. N Engl J Med. 1999;340:169–76.

68. Morais JA, Chevalier S, Gougeon R. Protein turnover and requirements in the healthy and frail elderly. The Journal of Nutrition, Health, & Aging. 2006;10(4):272–83.

69. Rand WM, Pellett PL, Young VR. Meta-analysis of nitrogen balance studies for estimating protein requirements in healthy adults. Am J Clin Nutri. 2003;77:109–27.

70. Subar AF et al. Dietary sources of nutrients among US adults, 1989 to 1991. J Am Diet Assoc. 1998;98:537–47.

71. Pennington JA, Hubbard VS. Derivation of daily values used for nutrition labeling. J Am Diet Assoc. 1997;97:1407–12.

72. McNamara DJ. Increase in egg intake minimally affects blood cholesterol levels (editorial). Washington, DC: Nutrition Close-up, Egg Nutrition Center; 1994.

73. Heron KL et al. Pre-menopausal women, classified as hypo- or hyper-responders, do not alter their LDL/HDL ratio following a

high dietary cholesterol challenge. J Am Coll Nutr. 2002;21:250–58.

74. Dawson-Hughes B, Harris SS, Dallal GE. Plasma calcidiol, season, and serum parathyroid hormone concentrations in healthy elderly men and women. Am J Clin Nutr. 1997;65:67–71.

75. Holick MF. High prevalence of vitamin D inadequacy and implications for health. Mayo Clin Proc. 2006;81(3):353–73.

76. Webb AR, Kline L, Holick MF. Influence of season and latitude on the cutaneous synthesis of vitamin D3: exposure to winter sunlight in Boston and Edmonton will not promote vitamin D3 synthesis in human skin. J Clin Endocrinol Metab. 1988; 67:373–78.

77. Dawson-Hughes B et al. Effect of calcium and vitamin D supplementation on bone density in men and women 65 years of age or older. N Engl J Med. 1997; 337:670–76.

78. Krasinski SD et al. Relationship of vitamin A and vitamin E intake to fasting plasma retinol, retinol-binding protein, retinyl esters, carotene, alpha-tocopherol, and cholesterol among eldery people and young adults: increased plasma retinyl esters among vitamin A-supplement users. Am J Clin Nutr. 1989;49:112–20.

79. Trumbo P et al. Dietary reference intakes: vitamin A, vitamin K, arsenic, boron, chromium, copper, iodine, iron, manganese, molybdenum, nickel, silicon, vanadium, and zinc. J Am Diet Assoc. 2001;101:294–301.

80. Vivekananthan DP et al. Use of antioxidant vitamins for the prevention of cardiovascular disease: meta-analysis of randomised trials. Lancet. 2003;361:2017–23.

81. Jacques PF. The potential preventive effects of vitamins for cataract and age-related macular degeneration. Int J Vitam Nutr Res. 1999;69:198–205.

82. Meydani M. Effect of functional food ingredients: vitamin E modulation of cardiovascular diseases and immune status in the elderly. Am J Clin Nutr. 2000;71:1665S–8S; discussion 1674S–5S.

83. Sano M et al. A controlled trial of selegiline, alpha-tocopherol, or both as treatment for Alzheimer's disease. The Alzheimer's Disease Cooperative Study. N Engl J Med. 1997;336:1216–22.

84. Miller ER et al. Meta-analysis: high-dosage vitamin E supplementation may increase all-cause mortality. Ann Intern Med. 2005;142:37–46.

85. Malinow MR et al. Reduction of plasma homocyst(e)ine levels by breakfast cereal fortified with folic acid in patients with coronary heart disease. N Engl J Med. 1998;338:1009–15.

86. Quinlivan EP, Gregory JF. Effect of food fortification on folic acid intake in the United States. Am J Clin Nutr. 2003;77:221–25.

87. Optimal calcium intake. National Institutes of Health Consensus Statement. 1994;12:1–31.

88. Brown JP et al. For the Scientific Advisory Council of the Osteoporosis Society of Canada. 2002 clinical practice guidelines for the diagnosis and management of osteoporosis in Canada. Can Med Assoc J. 2002;167(10 suppl):S1–S33.

89. Appel LJ et al. A clinical trial of the effects of dietary patterns on blood pressure. DASH Collaborative Research Group. N Engl J Med. 1997;336:1117–24.

90. Shils ME. Magnesium. In: Ziegler EE, Filer LJ, eds. Present knowledge in nutrition. Washington, DC: ILSI Press; 1996:256–64.

91. Fosmire G. Trace metal requirements. In: Chernoff R, ed. Geriatric nutrition: the health professional's handbook. Sudbury, MA: Jones & Bartlett; 2006.

92. Girodon F et al. Effect of micronutrient supplementation on infection in institutionalized elderly subjects: a controlled trial. Ann Nutr Metab. 1997;41:98–107.

93. El-Kadiki A, Sutton AJ. Role of multivitamins and mineral supplements in preventing infection in elderly people: systematic review and meta-analysis of randomized controlled trials. BMJ, doi:10.1136/bmj.38399. 495648.8F (published 31 March 2005)

94. Fairfield KM, Fletcher RH. Vitamins for chronic disease prevention in adults: scientific review. J Am Med Assn. 2002; 287:3116–26.

95. Fletcher RH, Fairfield KM. Vitamins for chronic disease prevention in adults: clinical applications. J Am Med Assn. 2002;287: 3127–29.

96. Balluz LS et al. Vitamin and mineral supplement use in the United States. Results from the Third National Health and Nutrition Examination Survey. Arch Fam Med. 2000;9:258–62.

97. U.S. Preventive Services Task Force. Routine vitamin supplementation to prevent cancer and cardiovascular disease: recommendations and rationale. Ann Intern Med. 2003;139:51–55. www.preventiveservices.ahrq.gov

98. National Institutes of Health State-of-the-Science conference statement: multivitamin/mineral supplements and chronic disease prevention. Ann Intern Med. 2006; 145.

99. National Council for Reliable Health Information. "Prevention magazine assesses dietary supplement use." Newsletter 2000; 23:cites BA Johnson in Herbalgram 2000; 48:65 and Natural Health Line, 3/10/00.

100. Schulz V et al. Rational phytotherapy: a reference guide for physicians and pharmacists, 5th ed. Berlin: Springer Verlag; 2004.

101. Fragakis AS. The health professional's guide to popular dietary supplements. 2nd ed. Chicago: American Dietetic Association; 2003.

102. Kittler PG, Sucher K. Food and culture in America. Belmont, CA: Wadsworth; 1998.

103. Evans WJ. Exercise training guidelines for the elderly. Med Sci Sports Exerc; 1999; 31:12–7.

104. Pollock ML et al. AHA Science Advisory. Resistance exercise in individuals with and without cardiovascular disease: benefits, rationale, safety, and prescription: an advisory from the Committee on Exercise, Rehabilitation, and Prevention, Council on Clinical Cardiology, American Heart Association; Position paper endorsed by the American College of Sports Medicine. Circulation. 2000;101:828–33.

105. Ventrakaman JT, Fernandes G. Exercise, immunity and aging. Aging Clin Exp Res. 1997;9:42–56.

106. Kligman EW, Hewitt MJ, Crowell DL. Recommending exercise to healthy older adults. The Physician Sportsmed. 1999; 27(11):1–11.

107. Papadopoulou SK et al. Health status and socioeconomic factors as determinants of physical activity level in the elderly. Med Sci Monit. 2003;9(2):79–83.

108. Conn VS et al. Integrative review of physical activity intervention research with aging adults. J Am Geriatr Soc. 2003;51: 1159–68.

109. Krinke UB. Effective nutrition education strategies to reach older adults. In: Watson RR, ed. Handbook of nutrition in the aged. Boca Raton, FL: CRC Press; 2001:319–31.

110. Wellman NS, Rosenzweig L, Lloyd JL. Thirty years of the Older Americans Nutrition Program. J Am Diet Assoc. 2002; 102(3):348–50.

111. Ponza M, Ohls J, Millen B. Serving elders at risk. The Older Americans Act Nutrition Programs—National Evaluation of the Elderly Nutrition Program, 1993–1995. Washington, DC: Mathematica Policy Research, Inc. and Administration on Aging; 1996.

"Knowing is not enough;
we must apply.
Willing is not enough;
we must do."
Goethe

Chapter 19

Nutrition and Older Adults:
Conditions and Interventions

Chapter Outline

- Introduction: The Importance of Nutrition
- Nutrition and Health
- Heart Disease/Cardiovascular Disease
- Stroke
- Hypertension
- Cancer
- Diabetes
- Obesity
- Osteoporosis
- Oral Health
- Gastrointestinal Diseases
- Changes in Nutrient Availability: Vitamin B$_{12}$ Deficiency
- Inflammatory Diseases: Osteoarthritis
- Cognitive Disorders: Alzheimer's Disease
- Medications and Polypharmacy
- Low Body Weight/Underweight
- Dehydration
- Bereavement

Prepared by U. Beate Krinke
with Lori Roth-Yousey

Key Nutrition Concepts

1 Multiple health problems put older adults at higher nutritional risk; interventions with the greatest potential for benefit are targeted to treat specific conditions.

2 Nutrient thresholds depend on individual circumstances: dietary nutrients that are beneficial in treating a deficiency have little effect when the diet is adequate, and they can be dangerous when consumed in excess of recommendations.

3 Successful nutrition interventions complement other lifestyle choices to enhance physical and mental resilience and to stabilize physiological changes.

Introduction: The Importance of Nutrition

Aging adults want to stay healthy until death. But too often, chronic illnesses[1] interfere with their *health* goals (see Table 19.1). Good nutrition can ameliorate effects of illness and contribute to *quality of life*. Medical nutrition therapy becomes part of a comprehensive treatment plan that encourages health-promoting food choices once diseases have occurred. Outcome data regarding the effectiveness of *medical nutrition therapy (MNT)* by registered dietitians (RD) has resulted in reimbursement for MNT provided to Medicare Part B beneficiaries with diabetes mellitus and kidney disease,[2] using nationally recognized nutrition protocols and evidence-based practice guidelines.[3]

Numerous studies support the link between nutritional status and health as measured by health care utilization of older adults.[4–7] The Lewin Group researchers estimated

Health More than the absence of disease, health is a sense of well-being. Even individuals with a chronic condition may properly consider themselves to be healthy. For instance, a person with diabetes mellitus whose blood sugar is under control can be considered healthy.

Quality of Life A measure of life satisfaction that is difficult to define, especially in an aging heterogeneous population. Quality of life measures include factors such as social contacts, economic security, and functional status.

Medical Nutrition Therapy (MNT) Comprehensive nutrition services by registered dietitians to treat the nutritional aspects of acute and chronic diseases.

that covering medical nutrition therapy for older managed care clients who had cardiovascular disease, diabetes, or renal disease would recover Medicare costs after 3 years and would begin to save system dollars by the fourth year.[5] Malnourished older patients have higher postoperative complication rates and longer hospital stays, therefore incurring greater health care costs.[8] In free-living older adults, nutritional risk status was found to be the most important predictor of total number of physician visits, visits to physicians in the emergency room, and hospitalization rates.[9] Aging adults use proportionately more health care services and products than younger persons do; therefore, nutrition interventions are particularly important.

Nutrition and Health

When asked, older people generally say that they feel good about their health. Only 26% say their health is fair or poor, while 74% consider their health to be good, very good, or excellent.[1] Race and Hispanic origin affect this perception, which ranges from 60% in black groups and 62% in Latino groups to 76% in whites. This perception holds even when many are troubled with a chronic health problem. In contrast to older adults' perceptions, public health professionals objectively monitor health by measuring leading causes of death (mortality) and leading diagnoses of health conditions (morbidity), as seen in Table 19.1.

Heart disease and cancer are the leading causes of death for all persons aged 65 and older,[10] followed by cerebrovascular disease and diabetes, as listed below:

Heart disease	31%
Cancer	22%
Cerebrovascular disease	8%
Diabetes	3%

The majority (87%) of older Americans have one or more chronic conditions, such as high blood cholesterol, hypertension, osteoporosis, chronic respiratory disease, and obesity.[11] Eating patterns of older adults contribute to the incidence and course of diseases such as hypertension, heart disease, and cancer, and this can, in turn, impact functional ability. For example, an overweight individual with heart disease may continue to overeat and become obese, further complicating arthritis management. In turn, arthritis limits functional abilities. Weight loss of even 1 pound will result

Table 19.1 Percentage of people age 65 and older who reported selected chronic conditions, by sex, 2003–2004

Sex	Heart Disease	Hypertension	Stroke	Any Cancer	Diabetes	Arthritis
Total	31.7	51.9	9.2	20.6	17.0	49.9
Men	37.2	48.1	10.1	23.8	19.5	42.9
Women	27.7	54.7	8.5	18.2	15.1	55.0

SOURCE: Federal Interagency Forum on Aging-Related Statistics. Older Americans Update 2006: Key Indicators of Well-Being. Federal Interagency Forum on Aging-Related Statistics. Washington, DC: U.S. Government Printing Office. May 2006. Accessed Indicator 15, 11/5/2006.

in significant reduction of stress on the knees during daily activities and may lead to enhanced management of heart disease as well.[12] Overall, nutritional status is a major factor in disease prevention, treatment, and recovery of health.

Heart Disease/ Cardiovascular Disease

Heart disease (cardiovascular disease, CVD) is the leading cause of death in older adults and is potentially reversible by adopting a healthy lifestyle. The adult risk factors and course of heart diseases have been discussed in Chapter 17. Specifics for older adults are highlighted in this section, including stroke and hypertension.

Prevalence

Heart disease prevalence varies by race and gender (see Table 19.2).[1,10] Approximately one-third of adults aged 65 and older have some form of heart disease. CVD is the primary diagnosis of 10% of individuals being admitted to nursing homes and 11% of home health clients.[13]

Risk Factors

Risk factors for cardiovascular disease in old age remain the same as in younger adults, except that the factors have less predictive value in old age.[14] Of adults aged 65 years and older examined in the NHANES III survey, 86% had one or more modifiable cardiovascular risk factors. These include hypertension (140 mm Hg/90 mm Hg), elevated LDL cholesterol (at least 130 mg/dL), and/or diabetes mellitus (physician diagnosis or fasting plasma glucose greater than 126 mg/dL).[15,16] Race is associated with risk, and older African Americans are especially vulnerable. They are nearly three times more likely to have one of the three cardiovascular risk factors than the average population.

Obesity increases the risk of CVD in aging adults. For example, 93% of those with a BMI over 30 (i.e., obesity) had one or more nutrition-related CVD risk factors (see Chapter 17). In the overweight category (BMI of 25 to 29.9), 87% had one or more risk factors.

Table 19.2 Prevalence of heart diseases in adults aged 65 and older by gender and race

	Total (%)	Gender (%) Male	Gender (%) Female	Race (%) White	Race (%) Black
Heart disease	30.8	36.2	26.9	31.5	26.1
Hypertension	40.3	34.9	44.2	39.5	53.3
Cerebrovascular diseases	7.1	7.9	6.5	7.1	*
Atherosclerosis	4.1	4.5	3.9	4.5	*

*Information not available.

Nutritional Remedies for Cardiovascular Diseases

Assertive treatment can modify the course of heart disease at any age. Assessment of personal motivation regarding diet and health, functional status, and lifestyle habits is among the special considerations for older adults contemplating dietary changes. Is the individual interested in making needed diet changes? If willing to change, does the individual have the knowledge and skill to adopt new dietary behaviors? What motivated this individual to seek nutrition advice? What is the individual's physical and mental functional level? Is assistance needed to shop for groceries or to prepare meals? Does the individual eat alone, with family and friends, at home, or in restaurants? Questions such as these impact nutritional intake and help to predict whether lifestyle changes will be adopted and maintained.

Dietary guidelines to fight cardiovascular disease in adults of all ages include those from the National Cholesterol Education Program (NCEP)[17] and the American Heart Association.[18] Cardiac rehabilitation programs adapting these guidelines were described in Chapter 17 and age-related considerations are described in Table 19.3.

Stroke
Definition

The American Heart Association defines stroke as "a cardiovascular disease that affects the blood vessels supplying blood to the brain". A stroke can occur in several forms, including cerebral thrombosis (a blood clot, or thrombus, blocks blood flow to the brain), cerebral embolism (an embolus, or wandering blood clot, lodges in an artery and blocks blood flow in or to the brain), or as hemorrhage (a blood vessel bursts and part of the brain goes without oxygen as a result).

Prevalence

Of adults aged 65 and older, 9% of females and 10% of males have had a stroke.[1] Stroke is one of leading causes of death of older adults.

Etiology

Factors that can lead to a stroke include blocked arteries, easily clotting blood cells, and weak heartbeats that are unable to keep blood circulating through the body, allowing pools of blood to form and clot. Hypertension is another stroke contributor because the force of blood may break weak vessels.

Effects of Stroke

Strokes deprive the brain of needed oxygen and other nutrients. Brain and nerve cells deprived of oxygen for only a few

Table 19.3 Treatment factors for older adults with heart disease

Target Area	Adults (≥ 65 years)
Decrease amount and type of fat	Focus on 1–2 items to decrease fat intake in individual's regular diet rather than change all things
• Use lean meats	Ensure adequate protein intake
• Substitute saturated fatty acids with PUFA & MUFA	Focus on oils currently using and suggest one to change, if appropriate
• Decrease trans-fatty acids	Decrease trans-fatty acids
	Consider giving a brief description of trans-fatty acids and sources: margarine, vegetable shortening, cookies, pastries, and other processed fats based on mental awareness and readiness to change
Reduce cholesterol intake	Focus on 1–2 food items; research is conflicting on the role of cholesterol in older adults; liver makes less
Increase fiber, fruits, and vegetables	Work with the fruits and vegetables that the individual can chew (e.g., if dentures, do they fit?)
Healthy cooking	May not be controllable if Meals-on-Wheels; overall goal is adequate nutrient intake
Limit salt	Focus on "no added salt," and no salt shaker on the table
Label reading	May be difficult if eyesight is poor; consider financial limits; know bargain strategies
Exercise	Obtain doctor's approval prior to starting; emphasize health benefits that include mobility, agility, and strength; emphasize that walking is exercise
Maintain healthy weight	Strongly influenced by functional status of indvidiual; emphasize adequate nutrient intake
Reduce stress	Exercise, relaxation, and socialization with friends
Quit smoking	Refer to smoking cessation program; continue the no smoking followed while in hospital; discuss potential for weight gain

SOURCE: Adapted from Gerlach, Anne F. Principles in a cardiac rehabiliation program. Guest lecturer for Nutrition for Adults and the Elderly, University of Minnesota, Minneapolis, MN, 2002.

minutes can die. Brain cells do not regenerate. As a result, stroke leads to loss of function (speaking, walking, talking, and eating) for parts of the body controlled by the oxygen-deprived nerves. Quick recognition of stroke results in faster treatment and better recovery. Although brain cells do not regenerate, new nerve pathways can develop in the gray matter reservoirs of the brain. This provides hope for successful rehabilitation therapies. Relearning how to feed oneself, to chew, and to swallow may well be a part of the slow, arduous process of rehabilitation.

Carotid Artery Disease   The arteries that supply blood to the brain and neck becoming damaged.

Atrial Fibrillation   Degeneration of the heart muscle causing irregular contractions.

Transient Ischemic Attacks (TIAs)   Temporary and insufficient blood supply to the brain.

Risk Factors

Gender is not a risk factor for stroke, although more women die from a stroke than men do. The following factors place an individual at higher risk for stroke:

- Family history
- African American, Asian, and Hispanic ethnic groups
- Having had a prior stroke

- Long-term high blood pressure (either systolic or diastolic)
- Cigarette smoking
- Diabetes mellitus
- *Carotid artery disease, atrial fibrillation, transient ischemic attacks (TIAs)*
- High red blood cell count
- Sickle cell anemia
- Living in poverty
- Excessive use of alcohol; use of cocaine and illicit intravenous drugs

The role of alcohol is complex and controversial. Moderate amounts of any type of alcohol can be protective against stroke while excessive amounts increase risks significantly. In Japan, where fewer than 5% of women drink alcohol, a 10.5 year prospective study with nearly 3000 men aged 40 to 69 years found the least risk at 42 grams of alcohol per day (2 to 3 drinks, see Chapter 16).[19] Men who consumed more than 70 grams of alcohol per day were 2.5 times more likely to have a stroke than the low-risk group. In a Framingham study of alcohol use and ischemic strokes in older adults, up to 2 drinks (or 24–30 grams of alcohol) per day counted as "moderate."[20] Moderate intake was protective. Stroke rates were triple the lowest rate for individuals drinking 7 drinks per day.

Nutritional Remedies

The focus of dietary advice in stroke prevention is to normalize blood pressure.[21] Other dietary goals are to reduce overweight and obesity (particularly abdominal fat) and to moderate alcohol intake.

The Dietary Approaches to Stop Hypertension, or DASH, diet (see Table 19.4) has been a promising strategy to decrease blood pressure[21] and risk of stroke in adults under age 65. It has been shown to enhance perceptions of quality of life.[22] A study of the effects of lifestyle modifications on blood pressure concluded that established lifestyle recommendations plus the DASH intervention "should especially be the focus of implementation efforts in individuals over age 50 years."[23] Researchers found that older people benefitted more than their under-50 counterparts. Other nondrug interventions have successfully lowered the blood pressure of older adults (e.g., using weight reduction

and/or sodium restriction of 1500 to 1800 mg per day over 30 months).[24,25]

Hypertension

Definition

In Western societies, blood pressure increases with age and stabilizes around age 50 to 60 for men. A blood pressure greater than or equal to 140/90 mm Hg is defined as stage 1 hypertension. Systolic blood pressure of 120–139 mm Hg or diastolic pressure of 80–89 is "prehypertension." Although it has been suggested that older adults can tolerate higher blood pressure and may even benefit from increased blood flow to the brain, old age does not change the diagnosis criteria for high blood pressure. Higher blood pressure puts more force on potential vessel blockages and increases chances of blood vessel breakage. An individual who controls high blood pressure with medication is still considered to have hypertension.

Table 19.4 The DASH eating plan for blood pressure control

	Servings per Day	Serving Sizes of Foods within the Food Group
Grains and Grain Products Especially whole grain[a]	7–8	Breads: 1 slice or 1 oz Cereal: ½ c cooked or dry Rice, pasta: ½ c cooked
Vegetables Fresh, frozen, no-salt-added canned	4–5	Raw, 1 c; cooked, ½ c
Fruits Fresh, frozen, or canned in juice	4–5	Juice: 6 oz Fresh: 1 med piece Mixed or cut: ½ c Juice: 6 oz Dried: ¼ c
Dairy Foods Skim or 1% milk, fat-free dairy products	2–3	Milk: 8 oz Yogurt: 1 c Cheese: 1½ oz
Meats, Poultry, and Fish	Up to 2	3 oz, cooked
Nuts, Seeds, Dry Beans	4–5 per week	⅓ c or 1½ oz nuts 2 Tbsp or ½ oz seeds ½ c cooked beans (legumes)
Fats and Oils[b] Select olive, canola, corn, and safflower oils	2–3	1 tspn soft margarine, oils, mayonnaise 1 Tbsp low-fat mayonnaise 2 Tbsp light salad dressing
Sweets	Up to 5 per week	1 Tbsp jam, jelly, syrup, or sugar ½ oz fat-free candy, jelly beans, or 12-oz sweetened beverage

SOURCE: Adapted from a 2000-calorie eating plan, the National High Blood Pressure Education Program's HeartFile, Winter 1999, National Heart, Lung, and Blood Institute and the National Dairy Council's "DASH TO THE DIET," 2000.

[a]Whole grain is the entire edible part of wheat, corn, rice, oats, barley, and other grains. Whole grain bread has the words "whole grain" before the type of flour is listed; whole grain breakfast cereals include the word "whole" or "whole grain" before the grain name (e.g., whole grain wheat).
[b]One serving is equivalent to 5 grams of fat.

Prevalence

Hypertension is even more prevalent than arthritis. Of adults aged 65 and older, 48% of men and 55% of women have hypertension.[1] Prevalence increases with age and differs by gender. In adults aged 75 years old and older, only one-third of men and one-fourth of women do not have hypertension. Over half of individuals with diagnosed hypertension control it with prescribed medications. Uncontrolled hypertension is a real public health challenge: it has serious consequences and is manageable, yet only 34% of individuals with hypertension have it under control.

Etiology

Family history and ethnic background increase the risk of hypertension; African Americans are most likely to have hypertension. Salt intake can also contribute to hypertension. Researchers with the Intersalt Study calculated that over time, 20% of hypertension in Western societies is attributable to salt intake.

Effects of Hypertension

Prolonged high blood pressure puts extra tension on blood vessels and organs in the body, wearing them out before the natural aging process. Damaged kidneys are a common sign of uncontrolled hypertension.

Risk Factors

Nutritional risk factors are drinking alcohol to excess, high-saturated-fat diets leading to dyslipidemia and atherosclerosis, lifestyles resulting in overweight and obesity, and a diet low in calcium.[26]

Nutritional Remedies

The DASH diet described in the earlier section on stroke is an important nutrition intervention for persons with a history of hypertension. Two strategies that achieved the greatest reduction in systolic blood pressure were achieving and maintaining a healthy weight, and reducing sodium intake to no more than 2400 mg/day.[26] The Dietary Guidelines for Americans also suggest limiting dietary sodium to 2400 mg per day. This is markedly less than older adults eat. CSFII data estimated the sodium intake of 70-year-olds (not counting salt added at the table) at 3122 mg daily for males and 2376 mg for females.[27] Daily sodium intake data collected for the NHANES survey counted all salt ingested, and used ages 60 and older for reporting, but is similar: 3447 mg sodium for males, 2532 mg sodium for females. On average, older adults could reduce their blood pressure by decreasing their sodium intake. In the DASH-sodium study, the greatest overall blood pressure reduction occurred in the subjects with the strictest sodium intake limit (1500 mg a day). Blood pressure reduction occurred

whether individuals were normo- or hypertensive. About 75% of the sodium in our daily foods comes from various manufacturing and preservation processes. Overall, salt at the table contributes about one-fourth of sodium intake. Choosing less processed foods can limit sodium intake. For instance, the sodium in a serving of baked potato is 10–15 mg of sodium while that in a serving of potato chips is 150–200 mg.

Other recommendations to reduce blood pressure are moderation in drinking alcohol and maintaining adequate potassium, magnesium, and calcium intakes. The MyPyramid meal plans include the DASH diet patterns because it is considered to be health-promoting for the general public as well as for people with hypertension.

Cancer

Cancer is a name for a group of diseases resulting from uncontrolled growth of abnormal cells. It is a complex progression through several stages: activation, initiation (injury or insult by a carcinogen), promotion (damaged DNA divides during a lag period, potentially over 10 to 30 years), progression (growth and spread), and possible remission (successful treatment or reversal). The focus of nutrition intervention depends on the stage of the cancer continuum: prevention, treatment, or care during survivorship. Cancer survivorship is a growing practice area; in 2002, there were 10.1 million Americans with a history of cancer, either still under care or posttreatment.[28]

Prevalence of Cancer

Cancer can occur at any age, but is most common in older adults. The American Cancer Society reports that 76% of all cancers are diagnosed in individuals aged 55 and older.[28] In women, the most common sites are breast, lung, and colon/rectum while in men the most common sites are prostate, lung, and colon/rectum.[28] For the most common male cancer (prostate), 80% of all cases occur in men aged 65 and older.[29]

Cancer is the second leading cause of death after heart disease. Of people aged 65 and older, 563,390 died of heart disease in 2003, compared to 388,911 who died of cancers (listed as malignant neoplasms in some charts).[11] To put those causes of death into perspective, about 1.8 million persons aged 65 and older died in 2003.

Etiology and Effects of Cancer

Certain genetic patterns predispose an individual to contracting cancer. However, environmental factors largely determine whether or not an individual develops cancer. As humans live longer, it becomes more likely that some insult or error will damage RNA or adversely affect the DNA replication process and ultimately cause cancer. Infectious agents such as helicobacter in stomach cancer or the human immunodeficiency virus in HIV/AIDS are preventable. And what about the role of diet? While older texts suggested that 35% of all cancers are due to poor dietary habits, current reporting by the American Cancer Society combines nutrition, overweight/obesity, and lack of physical activity, estimating that these factors account for about one-third of all cancer deaths.[28]

Cancer development is age-associated but not age-dependent; development is neither consistent nor linear. In a healthy, resilient individual, initiation may be repaired and subsequent cancer avoided or delayed. In someone with impaired immunity or suffering major physiological stress, initiation may proceed through promotion or progression. For older adults, cancer treatment presents a greater nutritional challenge than cancer prevention because of the lag times for development.

Risk Factors

Diet can refocus the genetic role in cancer development. Although dietary patterns make the biggest impact when begun early in life, older adults show greater interest in risk reduction. Vegetable and fruit consumption is the primary diet strategy for cancer risk reduction. Vegetable, fruit, and antioxidant intake are related to reductions in cancer initiation and progression. Above-average vegetable and fruit consumption (see Table 18.9) is associated with decreased risk of cancer including oral, esophageal, stomach, colorectal, and lung cancers. The American Cancer Society Expert Committee concluded that evidence points to a possible benefit from increasing vegetable and fruit intake to prevent breast, colorectal, and prostate cancer and for overall survival in breast, colorectal, and lung cancer.[30] This Expert Committee also concluded that increasing fiber intake may benefit overall survival in colorectal cancer, and omega-3 fatty acids aid treatment of lung cancer. However, insufficient evidence exists for increasing fiber, omega-3 fatty acids, and soy intake to improve overall survival in breast, lung, and prostate cancer. Research is currently underway to examine how the antioxidant components in green tea hinder the body's abilities to develop cancer-promoting chemicals.

The studies supporting information in Table 19.5 do not deal specifically with older adults but with adults of all ages. Particularly promising research is dealing with the role of lycopene (a flavonoid found in tomatoes and other vegetables) and other antioxidants as dietary strategies to decrease prostate cancer risk and lengthening survival.

Reducing risk of secondary cancers during the growing survivorship period is a relatively new area of research. Healthy eating patterns are encouraged by the National Cancer Institute: "It is generally accepted that individuals posttreatment should follow cancer prevention dietary recommendations established for the general population."[31]

Table 19.5 Nutritional risk factors for the most common cancers

Type of Cancer	Nutritional Risk Factors with Convincing or Probable Evidence
Lung	Low vegetable and fruit intake
Stomach	Low vegetable and fruit intake
	High salt intake
	High use of smoked, cured, and pickled foods (salt and nitrites)
Breast	Low vegetable and fruit intake
	Increased obesity (especially in postmenopausal women)
	Increased alcohol
Colon/rectum	Low vegetable and fruit intake
	High meat consumption (especially fat from red meats)
	Excess alcohol (folate may help offset detrimental effects)
Mouth/pharynx/ nasopharynx	Low vegetable and fruit intake
	Excess alcohol, together with tobacco (smoke or chew)
	High salted fish intake (nasopharynx)
Liver	Excess alcohol
	Eating contaminated (especially aflatoxins[a]) foods
Cervix	Low vegetable and fruit intake
Esophagus	Low vegetable and fruit intake
	Diets generally deficient in nutrients
	High alcohol intake
Prostate	High meat or meat fat or dairy fat

SOURCE: Adapted from Word Cancer Research Fund and American Institute for Cancer Research, Food, Nutrition, and Prevention of Cancer, p. 542.

[a]Aflatoxins are toxins produced by molds in some foods such as peanuts.

Nutritional Remedies for Cancer

Cancer survival rates are increasing; the current 5-year relative survival rate for all cancers is 65%.[28] The focus of cancer treatment is to maximize nutritional status, to maintain strength and energy, and to ensure an intake that helps to preserve lean body mass and avoids dehydration. This can be a challenging task. Anticancer medications and radiation treatments are associated with nausea, vomiting, diarrhea or constipation, fatigue, and weight loss. Finding food items that are tolerated, especially if they are calorie-dense, is one way family and friends can help medical care providers support a cancer patient's treatment. The National Cancer Institute maintains a compilation of the most current treatment guidelines, with versions for health professionals and also for patients at www.cancer.gov/cancertopics/pdq.[31]

Cancer treatment affects all aspects of nutrition. Side effects include mouth sores and altered sense of taste and smell, which affects appetite and eating. This may include changes in intake, digestion, and absorption related to chemotherapy, surgery, and radiation. For example, treatment may leave the cancer patient tired and lacking immunity to fight common diseases. In addition, ability to socialize over meals or to grocery shop may be curtailed. Medications also affect nutrient absorption. Overall, cancer treatment effects cause significant amounts of stress and nutritional insult in older adults.

Treating cancer is a process that tends to generate quests for alternative methods. The hope for remission and cure is a powerful motivator to try herbs, new nutrient combinations, and complementary healing practices. Herbal products have potentially useful roles in cancer treatment to ameliorate nausea and common symptoms; for example, chamomile tea for gastrointestinal discomfort and peppermint tea as a digestive aid. Other herbs may be considered:

- *Ginger:* possible antinausea properties but may not be as effective during chemotherapy treatment because it acts on the stomach, and chemotherapy nausea is triggered in the central nervous system; yet the taste and smell of small amounts of ginger may be calming.
- *Garlic:* possible antibacterial or antifungal effects of garlic with possible cancer prevention in moderate amounts; excess consumption via garlic supplements can cause stomach pain and gas and may interfere with bleeding time with aspirin and warfarin.

The first rule of treatment, "Do no harm," is especially relevant during the vulnerable period of cancer treatment. A product that can be enjoyed or perhaps tolerated by a healthy person may exacerbate illness in someone with health problems. Discussing all contemplated treatments with a physician, registered dietitian, and pharmacist will help to prevent dangerous interactions. A registered dietitian can also assist with providing individualized recommendations. Aging, food preferences, dietary intakes, the effects of the tumor, and anticancer treatments and side effects (that include pain) are all considered when planning therapy for elderly cancer patients.[32]

Diabetes

In the National Health Interview Survey 2003–2004, nearly 1 in 5 (17%) of all adults aged 65 years and older reported having diabetes, (see Table 19.1). Most older adults (90–95%) have type 2 diabetes. Prevalence is higher in Mexican American, African American, Native American, and Alaska Native older adults than in Caucasians. CDC has reported that Hispanic black and Hispanic individuals are twice as likely to have diabetes as whites.[33] Diabetes is the seventh leading cause of death in this age cohort.[10]

Diabetes diagnosis criteria for older adults are the same as for younger adults, fasting blood glucose of 126 mg/dl or higher,[16] see Chapter 17. Hypoglycemia in older adults may lead to weakness, confusion, and possible falls and fractures. Alcohol and drugs such as salicylates (aspirin) also contribute to drops in blood sugar. Older adults using oral hypoglycemic drugs (for example, sulfonylureas that stimulate insulin secretion) are at increased risk for hypoglycemia.

Effects of Diabetes

Diabetes affects all organs. Older adults who are experiencing declines in organ function are likely to be more affected by diabetes. For instance, blood glucose levels above a threshold of roughly 120 mg/dL exceed the kidney's capacity for reabsorption, and subsequently glucose starts spilling into the urine. With aging, there are fewer nephrons, less blood flow, and a slowed glomerular filtration rate, further slowing glucose reabsorption.

In four of five older people, diabetes occurs as one of several coexisting conditions that includes heart disease, hypertension, elevated blood lipids, and obesity. Individuals with diabetes are at greater risk for heart disease and its complications; diabetes itself is an independent risk factor for atherosclerosis. Furthermore, individuals with diabetes are more likely to die after suffering a myocardial infarction (MI) than those without diabetes.[34]

Diabetes leads to a tenfold greater risk of amputations, macular degeneration, visual loss, cataracts, glaucoma, and neuropathies (nerve damage, pain, or tingling) of the hands and feet. Other effects of hyperglycemia may include sodium depletion and dehydration, trace mineral depletion (zinc, chromium, magnesium), insomnia, nocturia, blurred vision, increased platelet adhesiveness related to atherosclerosis, increased infection and decreased wound healing, and aggravated peripheral vascular disease.

Risk Factors

Risk factors for type 1 diabetes are primarily genetic and mediated by viral and other diseases. Obesity is unlikely. Type 1 diabetes is less common in old age. Risk factors for type 2 diabetes are more lifestyle related (e.g., obesity) and also include:

- *Body weight and abdominal fat:* Higher body fat levels (BMI over 25, but even more so over 30) are associated with higher incidence of type 2 diabetes, especially when fat is stored in the abdominal area (waist circumference greater than 35 inches or more in females, 40 inches or more in males).
- *Race/ethnicity:* Blacks have a higher risk of developing type 2 diabetes than whites at normal or low

body weight. Risks of developing diabetes are similar for blacks and whites at higher body weights.[35] NHANES data (30- to 74-year-olds) collected from 1971 to 1992 yielded risk calculations for developing type 2 diabetes relative to BMI (weight for height). Cumulative risk increases with BMI in all race and sex groups. At a BMI of 22 (i.e., desirable weight), the probability of developing diabetes was 1.87 for blacks compared to the probability of 1.76 (p <0.01) for whites.[35] At BMI levels of 32 (obesity), the probability of developing diabetes was not significantly different for blacks and whites. The difference in risk may be associated with visceral adiposity (fat storage in midsection). Greater visceral adiposity impacts blacks more than whites at normal weight (BMI levels of 22) but not at levels defined as overweight or obese (BMI levels 25 or greater).

- *Hyperhomocysteinemia*, which is linked to increased risk of heart disease, is a stronger risk factor for overall mortality in individuals with type 2 diabetes, independent of CVD and other risk factors, than in individuals without diabetes.[36]

Nutritional Interventions

The goal of nutritional interventions is to achieve the best possible glycemic (blood sugar) control and reduction of cardiovascular disease risk factors while maintaining the quality of life, health, and safety of the individual. Diabetes care providers tend to agree that the following components comprise the nutritional management of both types of diabetes.[37–39] See Case Study 19.1.

1. Maintain as close to normal blood glucose levels as possible by balancing food intake, exercise, and insulin/medications; avoid hypoglycemia.

2. Choose health goals similar to the USDA/HHS nutritional guidelines for the public, with an emphasis on regular physical activity.

3. Individualize dietary recommendations according to the individual's personal needs and preferences.

4. Where nephropathy (chronic kidney disease) is present, limit protein intake to 0.8 g/kg/day.

5. Weight loss in overweight or obese individuals improves blood glucose control. However, many practitioners suggest that a stable weight (even if overweight) is better than weight cycling, since weight cycling is associated with greater risk for complications of diabetes (such as cardiovascular disease).

6. Assess dietary adequacy and supplement with vitamins and minerals where needed.

7. Monitor physical and mental abilities. Factors that affect glycemic control in older adults include altered senses; decreased mobility; difficulty in buying, preparing, or eating food; stress, loss, social changes, and depression; and comorbid conditions. Attend to the psychosocial needs of an aging individual.

Constant attention to meal planning in diabetes management has led many individuals to search for foods that will not affect blood sugar. One example is a vegetable called Jerusalem artichoke, not related to the globe artichoke or artichoke hearts. Jerusalem artichokes contain inulin, a fructose polymer, which does *not* turn to insulin. Inulin is absorbed more slowly than other starches and serves as prebiotic because it provides substrate for bacteria in the gut. Although Jerusalem artichokes are a perfectly fine vegetable, eating them raises blood sugars as do other vegetables. Inulin does not lower blood sugar. However, noncaloric sweeteners are a class of foods that do not raise blood sugar; FDA-approved examples are saccharin, aspartame, acesulfame potassium, and sucralose. Sugar alcohols (polyols such as xylitol or sorbitol) are used to sweeten sugar-free candies and gums. These substances are much sweeter than sucrose and fructose, and provide very little energy. Older adults may need to be more cautious in eating large amounts of polyols because high doses can lead to diarrhea.

Carbohydrate and fiber recommendations do not change with age. A high-carbohydrate diet is also likely to elevate fiber intake. Eating for blood sugar control and decreased risk of cardiovascular disease means eating high-fiber foods, oats, rye, and other whole grains, fruits, and vegetables. Fibers such as pectins and gums in citrus, apples, oats, barley, and dried beans have been shown to slow glucose absorption and lower serum cholesterol.[40] Evidence is inconclusive about whether fiber is most effective as soluble or insoluble fiber and whether it can be supplemental or must be eaten as part of a mixed diet. Older individuals can increase their fiber intake to the recommended 14 grams of fiber per 1000 kcal by eating high fiber cereals once a day.

Glycosylated Hemoglobin   A laboratory test that measures how well the blood sugar level has been maintained over a prolonged period of time, also called Hemoglobin A_1C.

Moderate increases in carbohydrate (raising percent of energy from 45 to 55%) have been well tolerated in 42- to 79-year-old adults with diabetes when given as breakfast cereal over a 6-month period.[41] However, individuals with elevated blood triglyceride levels may be sensitive to a high-carbohydrate diet and modify carbohydrate intake while maintaining other heart-healthy features of the meal plan.

Overall, the American Diabetes Association recommends that carbohydrate and monosaturated fat together should provide 60–70% of energy intake, making adaptations according to the individual's metabolic profile.[38] Because older adults are more likely to be undernourished than overnourished, consuming needed nutrients from a mixed diet is generally solid nutrition advice for older adults with type 1 or type 2 diabetes.

The American Diabetes Association suggests a daily multivitamin supplement may be appropriate for older adults in order to complement inadequate intake or unique metabolic needs.[38] Chromium picolinate supplements have shown some promise to reduce insulin needs and oral hyperglycemic medications when chromium deficiency exists.[38,42] Chromium-rich foods include brewer's yeast, oysters, meats (especially liver), whole grains, skin-on potatoes, and apples.

Magnesium deficiency has been linked with insulin resistance, carbohydrate intolerance, and abnormal blood lipid metabolism. Magnesium may be lost through the urine when excess glucose is being excreted, or intake may be insufficient or intestinal absorption limited. Magnesium supplementation should be monitored closely to avoid magnesium toxicity, especially if kidney function is impaired.

Occasionally, complementary medicine is used to treat diabetes. This is especially true in Mexican American and Native American populations. Evening primrose oil, milk thistle, fenugreek seeds, and prickly pear cactus (nopales) are some examples of botanical treatments. Safe complementary nutritional remedies can enhance the standard nutritional therapies.

Individualized medical nutrition therapy for older adults takes into account food availability, food preferences, and the client's treatment goals. Diabetes educators provide ongoing self-management education and care that supports quality of life. Cost-benefit analyses have shown that tight control of blood sugars (maintaining *glycosylated hemoglobin* below 7%) can lead to better quality of life for the older patient with diabetes and can result in fewer complications from diabetes.[43–45]

Obesity

Definition

The National Heart, Lung, and Blood Institute[46] and the World Health Organization[47] define obesity as a BMI of 30.0 or higher and extreme obesity as a BMI of 40 or higher.

Prevalence

In the United States, nearly one-third of older adults are not just overweight, but obese. Obesity rates have risen noticably since the 1980s, (see Table 19.6). The highest prevalence of obesity (by gender) among adults 60 years of age or older occurs in non-Hispanic black women and non-Hispanic white men. The lowest prevalence of obesity occurs in non-Hispanic, black women and non-Hispanic white men, and even in these groups, the obesity prevalence is above 25%.[48]

Table 19.6 Rising obesity rates in older adults, as percentage of 65-year-old men and women

Sex	1988–1994	2003–2004
Men	20.3%	29.7%
Women	23.6%	30.4%

Etiology/Effects/Risk Factors of Obesity

Analyses from the National Center for Health Statistics at CDC supports the notion that in adults, the body mass index associated with the lowest mortality falls within the range of 18.5–30.0.[49] Flegal's most recent analysis of BMI data and mortality found that "the majority of deaths associated with obesity were associated with BMI 35 and above." However, the associations among health, BMI, and aging are controversial. Based on Ernsberger's analysis,[50] morbidity and mortality is not any higher, and sometimes is lower, in older people who are at the high end of the BMI continuum. A study of older, community-dwelling Canadians, found that an increased BMI was associated with lower mortality.[51] However, studies consistently associate obesity, poorer health-related quality of life, and functional decline in older adults.[52]

For older adults, extra weight during illness episodes, especially hospitalizations, seems to be protective. Materials developed in 1999 by the American Dietetic Association Long Term Care Task Force in conjunction with Health Care Finance Administration (HCFA, now the CMS or Centers for Medicare and Medicaid Services) suggest a BMI range of 19–27 as an acceptable and health-promoting weight range for older adults.[53] This range is similar, yet broader, than that used by the Nutrition Screening Initiative, where BMI cut-points are 22–27. The point is that health care providers are more flexible when it comes to BMI recommendations for older than for younger adults.

Nutritional Remedies

Older obese individuals should undertake a healthy eating program based on enough nutrient-dense calories to support a slow weight loss. Age is not a factor in formulas to calculate calorie levels supporting desired activity and weight loss, although comorbidities are. Ensuring adequate nutrient intake by eating and drinking a balance of servings from basic food groups, for instance as outlined in the MyPyramid, or the DASH eating pattern promotes health in older as well as younger adults.

All adults benefit from physical activities designed to ensure functional independence. The Healthy People 2010 goal is for 40% of the population to perform physical activities that enhance or maintain flexibility. Approximately 22% of adults 65 years or older achieved this goal.

Physical activity is the only way to prevent muscle loss; this is especially important to help older people maintain function and avoid becoming frail. Due to functional limitations of older adults, health professionals will need to acquire special skills and to coordinate activities between health care and community or senior center programs. The U.S. Assisstant Surgeon General summarized the issue: "Public health interventions to decrease obesity prevalence must apply the same kind of multifaceted and coordinated approach that reduced tobacco use in order to change individual behavior patterns and effectively address the environmental barriers to physical activity and healthful food choices."[54]

Osteoporosis

> "Watch it when you hug Grandma."
>
> Mary Nelsestuen, afraid that a strong hug might break her frail mother's osteoporotic ribs

Definition

Osteoporosis means "porous bone," which results from reduction in bone mass and disruption of bone architecture. Osteoporosis can develop rapidly or slowly, depending on the homeostatic mechanism involved. An accelerated phase of bone loss occurs due to estrogen or testosterone loss. Bone mass loss is greater for women, who are most vulnerable to bone loss in the 3 to 5 years past menopause due to estrogen decline.[55] Men develop osteoporosis later than women. They start out with more bone, and their testosterone levels decline slowly over a long period, typically between ages 40 to 70. Men's bone mass losses double after surgical or hormonal castration (a form of treatment for prostate cancer). Several years after menopause, womens' bone loss slows. Osteoporosis is an age-associated disease because effects of slow bone loss are usually not seen until late adulthood.

Diagnosis of osteoporosis can be aided by assessment of clinical risk and biochemical markers and by bone density measurement, for example, dual-energy X-ray absorptiometry (DEXA or DXA).[56]

Just as bones grow strong over time, they become brittle slowly. In addition, bone tissue is not all the same. Up to 50% of trabecular or spongy bone (wrist, vertebrae, and ends of long bones) may be lost, and up to 35% of cortical or compact bone (shafts of long bones) will be lost during a lifetime.[57] Average bone loss is 0.5–1% per year after approximately age 50; when bone density is below the "young normal" mean, osteoporosis is diagnosed.[58]

Prevalence

Osteoporosis is four times more common in women than men (80% compared to 20%). Men have larger bodies and denser bones than women, and thus have greater peak and total bone mass. Blacks have denser bones and less osteoporosis than whites.[59]

Osteoporosis is dissimilar in men and women. Women tend to break hips and backbones, while wrist fractures are less common. Men with osteoporosis are also likely to have hip fractures but suffer fractures of or near the wrist more often than women. One out of every two women and one out of every eight men over fifty will have an osteoporosis-related fracture in their lifetime.[60]

Etiology

Bone mass is gained primarily during growth periods, with peak bone density reached between ages 18 and 30.[59] Subsequently, bone mass remains stable until about age 40 to 50 for women and about age 60 for men. Inadequate building of peak bone mass coupled with significant bone loss leads to a low bone density and increased risk for fractures.

Inadequate Bone Mass

Although osteoporosis is seen most often in the elderly, the risk for developing osteoporosis in later years begins during childhood and adolescence. Development of osteoporosis is delayed when an individual develops bigger, denser bones during youth.[61] For example, an epidemiological study in Yugoslavia showed that higher calcium intake in youth led to higher peak bone mass, independent of exercise and other factors. Higher bone mass has also been associated with slower decline in later life.[62] Studies in the United States have also shown that getting enough calcium during growth spurts (between ages 11 and 17) reduces the risk of osteoporosis. In an intervention study, girls aged 11 to 12 years who received calcium supplements (500 mg calcium citrate-malate) gained an additional 1.3% bone mineral density per year compared to controls.[63] In a cross-sectional study of 18- to 31-year-old women, positive correlations were found between adolescent milk intake in childhood and bone mineral density (total skeleton, spine, femoral neck, radius).[64] While there are many actions that build bones in older adults, preventing brittle bones is best done early in life.

Lack of exercise or inactivity leads to osteoporosis. Bedrest, hospitalization, and sedentary lifestyles make bones weak! Weight-bearing or resistance exercises are needed to develop bone mass because bone grows in response to pressure on the bone tissue. The more often and the harder you push on the bone (not enough to break it, of course), the more the body will respond by depositing minerals (calcium, magnesium, phosphorous, fluoride, and boron) into the bone matrix. For instance, tennis players have significantly higher bone mass in their playing arm than nonathletic controls.[65] However, even the controls had more bone mass in their dominant arm than the less-used one. Exercise also stimulates growth hormone, which in turn stimulates bone development. "Use it or lose it" applies here. Lack of exercise leads to loss of bone and eventually a fragile skeleton. Fear of falling and breaking a bone keeps some older adults from getting much-needed exercise, contributing to a vicious circle.

Increased Bone Loss

The skeleton acts as structural support and as a calcium reservoir for the body. Bone tissue includes jawbones and teeth. Bones and teeth contain about 99% of the calcium in an adult, roughly 2.2 to 3.3 lb.[65] The remaining 1% of calcium is found linked with protein in blood, soft tissues, and extracellular fluids. This reservoir is needed for nerve transmission, muscle contraction, and enzyme systems such as those controlling blood clotting. Maintaining nerve transmissions takes physiologic priority over maintaining bone structure. In order to be consistently available to perform many functions, calcium is tightly regulated by hormone systems. When calcium levels in the blood fall, the body responds by secreting more parathyroid hormone (PTH). PTH acts on bone to release calcium and thus raise the blood calcium levels. Too much calcium in the blood stimulates calcitonin secretion. The hormone calcitonin slows release of stored calcium.[65] Bone mineral reserves are dissolved (resorption) and rebuilt constantly to maintain adequate calcium levels for these messenger functions of calcium.

A consistent dietary supply of bone-building minerals (i.e., calcium, magnesium, phosphorus, fluoride, boron) and vitamins (primarily D and K) coupled with regular weight-bearing exercise helps maintain the skeleton reserves and supply calcium when needed. When some portion of this build-dissolve-rebuild cycle is malfunctioning, the body's first priority is to maintain blood calcium levels for nerve, muscle, and enzyme functions. Bone loss results from inadequate nutrient levels or calcium absorption, or from excess calcium excretion.

Osteoporosis can also develop from a shortage of the mineral phosphate during bone mineralization. A balanced calcium–phosphorus ratio as provided in a varied diet allows both nutrients to be used by the bone-building cells.

Lack of sufficient phosphorus promotes release of calcium from the skeleton. Although phosphorus is abundant in the food supply, some antacids bind with phosphorus, making it unusable by the body. In the absence of phosphorus, bone mineral formation is delayed until more phosphate is available. Shortage of vitamin D also delays bone mineral formation. Finally, the skin's ability to make vitamin D from the sun is less efficient with aging.

Effects of Osteoporosis

Falls and Fractures   Avoiding fractures is an important goal for older adults because fractures contribute to earlier death. Ten to 20% of older persons who break a hip die within a year.[66] Death is not due to the fracture itself but to complications resulting from the break. One of these complications is impaired mobility, complicating all the activities of daily living (including eating and exercising). If an older adult has also had a stroke, impaired mobility becomes the leading cause of institutionalization in the United States. Furthermore, 50% of older individuals who fracture a hip have permanent functional disabilities.

Shrinking Height, Kyphosis   In contrast to hip fractures, most vertebral fractures (67%) are asymptomatic. Postmenopausal women with compression and/or a bone fracture in the spinal column have a condition known as "shrinking height," leading to dowager's hump (also known as *kyphosis*, meaning a bent upper spine). Shrinking in height is slow and usually not painful. For example, a woman who was 5 feet 6 inches tall at age 30, may measure 5 feet even at age 83. The individual may not notice the gradual height loss until someone else comments or until they notice that clothes no longer fit.

Risk Factors

A typical osteoporosis patient is a petite elderly white female. Brittle bones develop from a complex array of physiological factors, including nutrition and exercise. Major risk indicators are male and female glucocortcoid treatment and past history of fractures. Table 19.7 lists risk factors related to osteoporosis.

Nutritional Remedies

The first remedy is optimal diet, including intake of the recommended daily allowance of calcium for older adults, which is 1200 mg per day, not to exceed the tolerable upper limit of 2500 mg (or 1500 mg calcium per day in Canada).[67] The dietary intake goal is to provide enough available calcium and vitamin D through diet or supplements so that despite declining absorption rates, bone loss is minimized. Calcium retention increases with increasing calcium intake up to each individual's threshold. Beyond that threshold, greater intake of calcium does not lead to higher levels of calcium retention. A sample meal plan that provides the DRI for calcium in older adults is shown in Table 19.8.

In older persons, taking calcium supplements without other bone-building activity has not been shown to increase bone density. In fact, an individual at bed rest or otherwise immobilized loses bone mass rapidly. Thus, Canadian practice guidelines for nonpharmacologic interventions in osteoporosis recommends physical activity ≥30 minutes per day three times per week.[67] However, compliance may be an even larger issue. In a 5-year, double-blind, placebo-controlled study with nearly 1500 women, mean age 75 years, the authors concluded that supplementation with 1200 mg of calcium carbonate per day "is effective in those patients who are compliant."[68] The problem in

Table 19.7 Risk factors associated with osteoporosis in older adults

Not Modifiable

Female, number of pregnancies, length of time between pregnancies

Age, age at which pregnancy(ies) occurred, breastfeeding

Caucasian, Asian

Thin, small-boned rather than large-boned

Family history of osteoporosis

Inadequate bone mass achieved during youth

Low body weight

Premature menopause

History of amenorrhea

Hypogonadism

Glucocorticoid fracture

Maternal history of hip fracture

Potentially Modifiable

Lack of weight-bearing exercise

Cigarette smoking

Long-term dietary phosphorus deficiency (e.g., use of phosphorus-binding antacid)

Heavy alcohol consumption

Underweight

Malnourished

Inadequate dietary calcium (<1200 mg) and vitamin D (<400/600 IU) intake

Still Controversial or Not Yet Clear

Diet high in phosphorus while low in calcium (mixed evidence)

Inadequate fluoride, boron, and magnesium in diet

Consistently high protein and/or sodium intake

Eating soy products (soymilk, textured protein, etc.) for its estrogen-like activity

Table 19.8 An example of 1 day's food that provides at least 1200 mg calcium

Food	Amount of Calcium (mg)
Oatmeal made with milk, 1 c total	266
Banana, one medium	6
Coffee, 10 oz with 1 oz (2 Tbsp) evaporated milk	87
Turkey sandwich on whole wheat bread, lettuce, mayonnaise	54
Cheese added to sandwich, 1 oz cheddar	148
Canned fruit cocktail, ½ c	9
Iced tea, plain	0
Orange juice, calcium-fortified, 8 oz	289
Roasted almonds, 2 Tbsp	33
Pasta with chicken, 1½ c	54
Tomato slices, 2	2
Milk, 1%, 8 oz	279
Sugar cookie, 1 medium	5
Chocolate milk, 8 oz	287

Table 19.9 Examples of dietary contributors to calcium loss

	Calcium lost in urine
Salt: for 1 gram salt (2300 mg sodium)	~ 26 mg
Protein: for 1 gram dietary protein	~ 1 mg
Protein amount in 1 oz of meat, fish, poultry, or egg	~ 7 mg
Quarter-pound hamburger	~ 25–30 mg
Caffeine: 6 oz of regular coffee	~ 40 mg

recommending calcium supplementation to prevent clinical fractures was "poor long-term compliance."

Several dietary components (protein, sodium, caffeine, vitamins) are closely linked to calcium metabolism and their intake can interfere with appropriate supplementation. (See Table 19.9) High protein intakes result in less calcium being available because protein leads to greater excretion of calcium into the urine. In the Nurses Health Study, women (aged 35 to 59, followed for 12 years) who ate more than 95 grams protein per day suffered more osteoporotic forearm (but not hip) fractures.[69] High sodium intake leads to higher levels of urinary calcium, that is, more calcium is excreted when higher levels of sodium are eaten. Caffeine to equal 2 to 3 cups of coffee daily (in postmenopausal women), consumed with a low level of calcium, has been associated with bone loss.

Along with supplementation, a number of nutritional remedies can improve calcium intake and absorption:

- Drink milk and take supplements with a meal since food slows intestinal transit time and allows more calcium to be absorbed from the gut.

- Divide calcium supplements to be taken at a different time than antacids, because stomach acidity enhances absorption; for older adults who have decreased acidity, calcium *citrate* is considered more soluble.

- Consume foods that are rich in bone-building vitamins (C, D,

Osteoblasts Bone cells involved with bone formation; bone-building cells.

Osteoclasts A bone cell that absorbs and removes unwanted tissue.

B_6, K) at recommended levels. This will help to synthesize bone and develop the collagen matrix, into which minerals are deposited during bone mineralization.[70]

- Vitamin C plays a role in the development of the protein bone matrix (collagen), but has no defined role in treatment.

- Vitamin D (1,25-dihydroxyvitamin D, also called calcitriol) stimulates active transport of calcium in the small intestine and the colon; for those states where elderly are unlikely to get adquate sun exposure, ingested forms of vitamin D are encouraged to meet adequate intake of 10–15 mcg of vitamin D_3 (see inside front cover).

- Vitamin K is required for the formation of proteins that stimulate *osteoblasts* to build new bone and attract *osteoclasts* to initiate bone resorption. This has the potential to slow bone loss, and further research is needed to understand associations with osteoporosis.

Other Issues Impacting Nutritional Remedies

Vitamin K Caution   Vitamin K contributes to building a stronger bone matrix, so some calcium supplements now also include vitamin K. However, vitamin K also plays an important role in the blood clotting process. When older adults with a history of strokes are placed on an anticlotting medication like warfarin, nutrition counselors can advise them to maintain a stable vitamin K intake.

Two forms of vitamin K are found in foods. Phylloquinone, also known as vitamin K_1 is naturally occuring in plants, and menatetrenone, also known as vitamin K_2 is found in meat, cheese, and fermented products. Big portions of broccoli and greens for a few weeks in summer could increase chances of blood clotting, and this can increase the chances of causing another stroke. In addition, when a vitamin K–containing supplement is added to the diet (e.g., taking 2–3 Viactiv® calcium chews as per recommendation), 150% of Vitamin K is instantly provided. One can easily get too much of a good thing.

How much is too much? As yet, no tolerable upper levels are set for vitamin K. But for fat-soluble vitamins, a general rule (although one that does not apply to vitamin E) is to stay under 500% of the RDA or DRI.

Xerostomia   Dry mouth, or xerostomia, can be a side effect of medications (especially antidepressants), of head and neck cancer treatments, of diabetes, and also a symptom of Sjogren's syndrome, which is an autoimmune disorder for which no cure is known.

Hormones   Estrogen, testosterone, growth hormone, insulin-like growth factor-I, and parathyroid hormone increase calcium absorption rate. Hormone replacement, with or without additional calcium, has been shown to increase bone mineral density in early postmenopausal women (4.5% and 1.5% respectively).[71] Pines and associates also found that the women not receiving hormone therapy, either with or without calcium supplementation, lost bone mineral density (1.4% and 3.7% respectively). However, in 2002, the Women's Health Initiative reported increased rates of breast cancer, coronary heart disease, stroke, and venous thromboembolism in addition to decreased rates of hip fracture and colorectal disease.[72] This information was quickly disseminated to health professionals and women, the risks and benefits for use of hormone replacement therapy must be carefully considered on a case by case basis.

Individuals choosing not taking hormone replacements may try soy (e.g., milk, textured protein, bars, soy nuts) for its estrogen-like activity (e.g., phytoestrogens). Naturally occuring phytoestrogens are weak estrogen-like chemicals. Evidence is not definitive that soy will mimic the effects of estrogen in bone development; to obtain a definitive answer, studies will need to ensure that there was adequate intake of calcium and vitamin D.

Exercise   While there is a relationship between calcium intake, exercise, and bone mass, it is not known whether exercise increases absorption or whether it enhances bone mineralization.

In summary, the best osteoporosis prevention strategy is exercise and adequate diet in young people when bones are first growing. For older individuals who have brittle bones, exercise supplemented with calcium and vitamin D strengthens bone mass. As much as possible, stay active, enjoy sunny days, and drink milk!

Oral Health

"We can't have a healthy mouth, a great smile, or a good conversation without it. Saliva or 'spit' lubricates living."

Nelson Rhodus, professor, Division of Oral Medicine and Diagnosis, University of Minnesota

It is possible to be well nourished without a full set of teeth, or even without any teeth at all. However, the ability to bite and chew increases fruit and vegetable intake, which is linked to better health. Fortunately, more people are keeping their teeth as a result of better dental care. At the turn of the century, about one-third of all adults had no natural teeth; now that number has dropped to about one-fourth of older adults who have no natural teeth, see Table 19.10, which shows the prevalence of being edentulous by sex and poverty. Older adults who are poor are most likely to have no natural teeth.

Changes in oral health are most likely to be a result of disease, medical treatment, or medications rather than aging itself. Periodontal disease (PD) and *xerostomia*

Table 19.10 Percentage of people age 65 and older who reported having no natural teeth, by sex and poverty, 2004[1]

Sex	Age and poverty status	No natural teeth
Both sexes	65 and over	25.7
	Below poverty	41.6
	Above poverty	24.8
Men	65 and over	24.2
Women	65 and over	26.8

SOURCE: Centers for Disease Control and Prevention, National Center for Health Statistics, National Health Interview Survey.

(known as dry mouth) are two conditions that can interfere with tolerance and enjoyment of food. PD results from bacterial infections of the **gingiva**, with destruction of the ligaments attaching teeth to the jawbone and gums that reced from the teeth. Additional plaque builds up in the resultant pockets, contributing to further infection and eventual tooth loss. Persons whose overall health and immune system are compromised are at greater risk for periodontal disease. Prevention of PD emphasizes oral hygiene to remove plaque, enhancing immune status, and ensuring optimal nutrition. Besides deficiencies of vitamin C, folic acid, and zinc that are associated with PD, correcting potential deficiencies of calcium, vitamin D, and magnesium will help postmenopausal women keep their bones, including the jaw, strong. Ensuring optimal nutrition includes management of diabetes. High blood sugars can accelerate periodontal disease and also make the mouth more susceptible to yeast infection (candidiasis).

Except for Sjogren's syndrome, where xerostomia is a side effect, medications and other treatments are the likely causes of dry mouth, and thus prevention may be impossible. For example, diuretic treatment for hypertension leads to decreased salivary secretion; other medications with this side effect include antianxiety drugs, antidepressants, sedatives, and antihistamines. Head and neck cancer treatment can also lead to xerostomia when the salivary glands are involved. Lack of saliva for any reason gives bacteria a better environment to build plaque. Further interfering with enjoyment of food with xerostomia is the tendency to have loss of taste (dysgeusia) and pain of the tongue (glossodynia). The key treatment of xerostomia is good oral hygiene, especially after meals, plus stimulating saliva with sugar-free candy or frozen fruit bits, chewing xylitol-flavored gums, and sipping water liberally. Artificial saliva may also help to keep the oral cavity moist.

Good oral health for older adults also includes caries prevention. Brushing, flossing, and dietary recommendations for older adults match those of younger ones. Tea sippers (black tea, no sugar) may even have an advantage here; the polyphenols in black tea seem to interfere with the bacteria's ability to stick to the plaque. Yes, we all still need to brush and floss to remove plaque.

Gastrointestinal Diseases

The gastrointestinal (GI) system is roughly the length of a football field and serves so many functions that it should not be surprising to learn that, by late adulthood, it occasionally malfunctions. It seems miraculous that we so consistently eat what we like, without much thought, and our body converts that food to energy for daily living. Parts of the GI system most likely to malfunction in old age are:

1. The esophageal-stomach juncture: weakened muscle results in *gastroesophageal reflux disease* (GERD)
2. The stomach: decreased acidity leading to *changes in nutrient absorption* or increased acidity causes ulcers
3. The intestines: resulting in *constipation*, diarrhea, and some food intolerance

Often these problems are secondary to other diseases. No matter what the cause, older adults are at higher risk for some of the GI conditions discussed next, any of which may impair older adults' activities.

Gastroesophageal Reflux Disease (GERD)

Definition   Gastroesophageal reflux disease occurs when stomach contents flow back into the esophagus.

Prevalence   Approximately 19 million Americans or one of five older adults have GERD.[73]

Etiology and Effects   It is not clear if acid in the esophagus leads to a weakened lower esophageal sphincter (*LES*) or if a weakened sphincter leads to GERD.[74] The main symptoms of GERD are heartburn and acid regurgitation. Stomach contents, which are highly acidic, spill back into the esophagus, resulting in irritation, belching, hoarseness and substernal pain. Ulceration and swallowing disorders are symptoms of severe cases of GERD.

> **Gingiva**   Gum tissue.
>
> **LES**   Lower esophageal sphincter, which is the muscle enabling closure of the junction between the esophagus and stomach.

Nutritional Risk Factors   Alcohol in excess of seven drinks per week, obesity, and smoking are consistently linked to GERD episodes.[73] In addition, regular and decaffeinated coffee are associated with heartburn.

Nutritional Remedies   The primary dietary remedy is to omit foods that are chemically or mechanically irritating. There is little consistency about which foods these are for an individual. However, general guidelines are to choose a low-fat diet and nonspicy foods; high-fat meals and alcohol both lower LES pressure. Fermented beverages and caffeine stimulate gastric acid, so reducing them

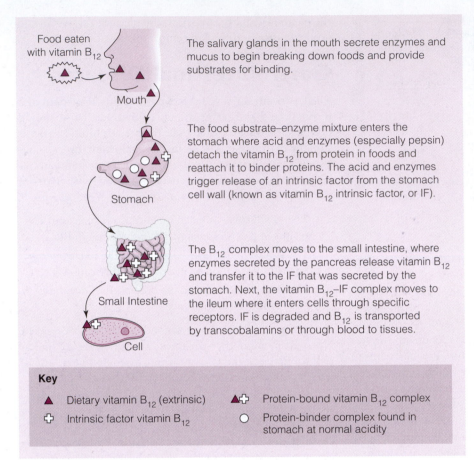

Food eaten with vitamin B$_{12}$

Mouth

The salivary glands in the mouth secrete enzymes and mucus to begin breaking down foods and provide substrates for binding.

Stomach

The food substrate–enzyme mixture enters the stomach where acid and enzymes (especially pepsin) detach the vitamin B$_{12}$ from protein in foods and reattach it to binder proteins. The acid and enzymes trigger release of an intrinsic factor from the stomach cell wall (known as vitamin B$_{12}$ intrinsic factor, or IF).

Small Intestine

The B$_{12}$ complex moves to the small intestine, where enzymes secreted by the pancreas release vitamin B$_{12}$ and transfer it to the IF that was secreted by the stomach. Next, the vitamin B$_{12}$–IF complex moves to the ileum where it enters cells through specific receptors. IF is degraded and B$_{12}$ is transported by transcobalamins or through blood to tissues.

Cell

Key

▲ Dietary vitamin B$_{12}$ (extrinsic)

✚ Intrinsic factor vitamin B$_{12}$

▲✚ Protein-bound vitamin B$_{12}$ complex

○ Protein-binder complex found in stomach at normal acidity

Illustration 19.1 Overview of vitmain B$_{12}$ absorption.

may give some relief. Chew thoroughly, eat slowly, and remain upright after eating to take advantage of gravity and make it difficult for stomach acids to reflux upward.

Changes in Nutrient Availability: Vitamin B$_{12}$ Deficiency

The normal process by which the body absorbs vitamin B$_{12}$ from foods requires several conditions. The stomach environment needs to be acidic and producing enzymes (especially pepsin), and the stomach cell lining needs to be secreting an intrinsic factor (IF). Stomach acids and enzymes split off and transfer vitamin B$_{12}$ from foods to carrier proteins (mostly secreted by the salivary gland in the mouth). Then, the vitamin B$_{12}$-carrier protein complex moves into the small intestine where the vitamin B$_{12}$ will again be broken off and bound to the IF (produced in the stomach and migrated to the small intestine). The vitamin B$_{12}$—IF linked complex then binds to a specific site in the lining of the small intestine, is transported across, and released into blood serum to be taken to tissue cells. (See Illustration 19.1).

The two primary types of vitamin B$_{12}$ deficiency in older adults are: (1) pernicious anemia due to lack of intrinsic factor (IF) being released from the stomach cell wall, and (2) food-bound malabsorption of vitamin B$_{12}$, often due to hypochlorhydria.

Pernicious Anemia

Definition and Effects Pernicious anemia results when a lack of intrinsic factor (IF) prevents absorption of vitamin B$_{12}$. Physical signs of pernicious anemia (macrocytic megaloblastic anemia) include large, undeveloped red blood cells, increased redness and swelling in the mouth (glossitis) and tongue fissures, and irreversible neurological symptoms. Individuals with pernicious anemia also tend to have shrunken stomach glands and mucosa leading to a decreased enzyme and hydrochloric acid secretion (hypochlorhydria), which, in turn, leads to impaired iron absorption.

Prevalence Pernicious anemia is uncommon, even in older adults.[75] Typically, it takes 5 to 6 years to develop; it is rare before age 35 and is most often seen in older women.

Risk Factors Risk factors are a history of *Helicobacter pylori* (*H. pylori*) infections, decreased stomach acid production, and a family history of pernicious anemia.

Nutritional Remedies Vitamin B$_{12}$ may be given orally or by injection. Injections are administered daily until lab values stabilize, then six to eight times per year throughout life. In Scandinavian countries, injections are prescribed until pernicious anemia is in remission. Then, maintenance doses of oral vitamin B$_{12}$ are used. One reason that oral maintenance therapy works is that approximately 1–2% of synthetic vitamin B$_{12}$ (not food-bound) is absorbed passively, removing the need for either intrinsic factor or carrier protein.

Foods high in vitamin B$_{12}$ will not maintain adequate blood levels in pernicious anemia. However, foods fortified with high levels of crystalline or synthetic vitamin B$_{12}$ increase the likelihood of passive absorption. Other dietary recommendations include eating proteins of high biological value and, if there is an iron deficiency, iron.

Food-Bound Vitmain B_{12} Malabsorption

Definition and Etiology   The most common cause of vitamin B_{12} deficiency in older adults is related to abnormal stomach function. For instance, after prolonged inflammation, the stomach mucosa shrinks and secretes less acid and enzymes, a condition known as atrophic gastritis. Older adults with atrophic gastritis still absorb some dietary vitamin B_{12}, but they will become deficient over time. Bacterial overgrowth (usually related to infection with *H. pylori*) will also use vitamin B_{12} and diminish the amount available for absorption. Finally, antacid treatment in individuals neutralizes stomach acid (i.e., pH increases), and this rise in stomach pH does not allow all of the available vitamin B_{12} to be split from the protein carrier, even though the intrinsic vitamin B_{12} factor is present. Symptoms of vitamin B_{12} deficiency begin to appear after 3–6 years of poor absorption.[76,77]

Prevalence   Nilsson-Ehle found prevalence rates for Vitamin B_{12} deficiency of up to 80% in the papers he reviewed, but found that typical estimates were much lower. Protein-bound vitamin B_{12} deficiency is estimated to occur in 20% of people over 69 years old (known or undiagnosed).[75,77]

Serum *cobalamin* levels are used to detect deficiency, although the cut-off limit may have been set too low for older adults.[77] It is more effective to measure the blood levels of intermediate breakdown products in pathways where vitamin B_{12} is needed. For example, vitamin B_{12} is needed to change *methylmalonic acid (MMA)* into a component used in the *Krebs cycle*. When adequate amounts of vitamin B_{12} are unavailable, blood levels of MMA rise. Since B_{12} is the only coenzyme to catalyze this reaction, a test for levels of MMA is specific to vitamin B_{12} deficiency.

Another test is to measure blood levels of the coenzyme *homocysteine*. If vitamin B_{12} isn't available to complete the pathway that forms DNA, then homocysteine (intermediate metabolite) levels become elevated. Since folate is also needed in this metabolic pathway, lack of either folate or vitamin B_{12} can lead to high levels of homocysteine.

Therefore, elevated MMA and homocysteine levels confirm a vitamin B_{12} deficiency. However, if MMA levels are normal but homocysteine levels are high, folic acid deficiency or another cause of increased homocysteine levels should be investigated.

Effects   Vitamin B_{12} deficiency leads to irreversible neurological damage, walking and balance disturbances, and cognitive impairment (including confusion and mood changes). High levels of homocysteine are known to be a risk factor for heart and peripheral vascular disease.[78]

Risk Factors   Risk factors include advanced age, gastrointestinal disorders, genetic family patterns, medications, and (to some extent) inadequate food intake. An example of aging as a risk factor comes from Nilsson-Ehle, who reported that 1.5% of 50-year-olds have achlorhydria and B_{12} deficiency as compared to 18% of 80-year-olds.[77] Gastrointestinal disorders that affect vitamin B_{12} absorption include atrophic gastritis (with higher risks in diabetes mellitus and autoimmune thyroid disorders), partial stomach removal, and *H. pylori* infection. Medications that suppress stomach acid secretion or impair absorption are associated with the risk of deficiency. Examples of these types of medicines are oral biguanides (e.g., metformin used to treat type 2 diabetes mellitus), modified-release potassium preparations, anesthesia, hydrogen-receptor antagonists (e.g., Cimetidine or Tagamet) and proton pump inhibitors (e.g., omeprazole/Prilosac given for gastroesophogeal reflux disease). In genetically linked cases, individuals often have antibodies for the intrinsic factor.

Vegetarians not taking a B_{12} supplement are at risk of vitamin B_{12} deficiency. For example, a study of 245 Australian Seventh Day Adventist ministers found most (70%) had a vitamin B_{12} deficiency due to a low dietary intake.[79]

Even with intrinsic vitamin B_{12} factor present, 53% of the group had serum cobalamin levels below the reference level for this study (171–850 picomoles per liter). Some ethnic and gender differences may exist for developing a vitamin B_{12} deficiency. Elderly white men had the highest prevalence rates, whereas black and Asian women had the lowest rates among community-dwelling seniors (aged 60 or older) in Los Angeles.[80] In contrast, Zeitlin and associates.[81] concluded that there were no age- or gender-related differences in the mean levels of serum B_{12} after studying approximately 440 subjects in the Bronx Aging Study (mean age = 79 years).

Nutritional Remedies   The DRI of vitamin B_{12} for men and women over 50 years is 2.4 mcg/day,[82] and the tolerable upper intake level is not determined.[83] Good food sources include meats, shellfish (shrimp, crab, mussels), and milk and milk products. Oral pharmacological doses for vitamin B_{12}-deficient patients are 0.2–1 mg (200–1000 mcg) and are above the DRI because roughly 1–2% is absorbed through passive diffusion in the small intestine.[84]

Cobalamin   Another name for vitamin B_{12}. Important roles of cobalamin are fatty acid metabolism, synthesis of nucleic acid (i.e., DNA, a complex protein that controls the formation of healthy new cells), and formation of the myelin sheath that protects nerve cells.

Methylmalonic Acid (MMA)   An intermediate product that needs vitamin B_{12} as a coenzyme to complete the metabolic pathway for fatty acid metabolism. Vitamin B_{12} is the only coenzyme in this reaction; when it is absent, the blood concentration of MMA rises.

Krebs Cycle   A series of metabolic reactions that produce energy from the proteins, fats, and carbohydrates that constitute food.

Homocysteine   Another intermediate product that depends on vitamin B_{12} for complete metabolism. However, both vitamin B_{12} and folate (another B vitamin) are coenzymes in the breakdown of certain protein components in this pathway. Thus, elevated homocysteine levels can result from vitamin B_{12}, folate, or pyridoxine deficiencies.

Diverticulitis   Infected "pockets" within the large intestine.

> Power Pudding, also known as Behm's Special Recipe
> 1 cup applesauce
> 1 cup unprocessed bran, all-bran, or bran buds
> ½ cup prune juice
>
> Mix all ingredients and refrigerate of freeze.
> Serve 2 tablespoons at supper, follow by extra fluid such as 1 cup of water.
>
> You can even find Power Pudding at Cooks.com (accessed 11/18/2006):
>
> 1½ cups pitted prunes
> 1 cup applesauce
> ½ cup All-Bran
> ¾ cup prune juice
> Put all ingredients into blender and puree until smooth. Serve ¼ cup daily, followed by a glass of water.

Illustration 19.2 High-fiber recipes as alternative to laxatives.

A dose-finding study tested vitamin B_{12} doses from 2.5mcg to 1000 mcg with healthy, free-living adults over age 70 and found that mild deficiencies can be normalized with doses of 500 to 1000 mcg of cyanocobalamin, depending on the blood indicator measured.[85] Foods high in folate and iron are also necessary in cases where folate deficiency and/or iron deficiency has been diagnosed. In 1999, forti-fied flour and breakfast cereals supplied almost two-thirds of the total folate in the food supply.[86] Thus, for older adults, a bowl of vitamin B_{12}- and folate-fortified whole grain cereal is a "power food."

Constipation

Definition   There are many definitions of "normal bowel pattern." Individuals can consistently experience from two or three bowel movements per day to two or three bowel movements per week.

Prevalence   When "two or fewer bowel movements per week" is used as the definition of constipation, the preva-lence of constipation was 3.8% in older adults aged 60–69 years and 6.3% in older adults aged 80 years and older.[87] In a midwestern community of people ages 65 years and older, as many as 40.1% reported some type of constipation.[88]

Etiology   The physiological causes of constipation in older adults are emerging.[89] Animal studies indicate that aging intestinal muscles respond less to triggers and that the brain-muscle transmitters are either inadequate or less responsive. Older adult's thirst mechanisms decline. Lower levels of stomach secretions and potentially less muscle strength affect peristalsis. Due to chewing problems, there is a potential for eating fewer fiber-rich foods. Declining cognitive function may become so severe that the individ-ual may not recognize the urge to defecate; this can also lead to constipation.

Effects   Persons with laxation difficulties are prone to be more anxious and focused on bowel movements as an aspect of health. Lower-fiber diets can also exacerbate *diverticulitis*.

Risk Factors   Nutritional risk factors for constipation include:

- Low intake of water and other fluids
- Eating small amounts of food
- Medications, such as nonsteroidal anti-inflammatory drugs
- High-iron mineral supplement

Nutritional Remedies   Remedies are to encourage increased dietary fiber in tandem with an increased fluid intake. Bran in breads and cereals is one way to add to fecal bulk that stimulates persistalsis to move wastes through the colon. The fiber section of Chapter 18 offers suggestions for incorporating more fiber into a diet. Long-term care nutrition services have devised special recipes to wean residents off laxatives, such as incorporating psyllium fiber, apple sauce, and juice into fruit flavored gelatin. One particularly popular one is called Behm's Special Recipe or Power Pudding (see illustration 19.2). There is one caution before increasing fiber: find out if there is potential for fe-cal impaction due to disease such as colon cancer. Increased exercise will improve peristalsis and alleviate constipation; it may also relieve individuals who worry about bloating and flatulence from eating additonal high-fiber foods.

Inflammatory Diseases: Osteoarthritis

Definition

Inflammatory diseases include osteoarthritis, rheumatoid arthritis, atrophic gastritis (typically due to *Helicobacter pylori* infection), celiac disease (***gluten*** intolerance), irritable bowel disease (IBS), diverticulitis (an infection in the large intestine), and asthma. Of these conditions, arthritis affects the greatest number of older individuals.

Osteoarthritis is the most common form of arthritis, affecting approximately 7% of the population (over 20 million people) in the United States and is the focus of this section.[90] Prevalence of osteoarthritis at the knee is twice as common as at the hip; the knee and hip are the two most common sites for persons aged 30 and older. Prevalence increases with age and peaks between 70 and 79 years. More men than women have osteoarthritis before age 50; after age 50, women are more often affected. (Study results are consistent about the potential benefits of estrogen treatment in osteoarthritis.)

Etiology

Cartilage loss, bone changes such as bone outgrowth, hardening of soft tissues, and inflammation all lead to tissue damage. Some wear and tear of aging also contributes to the complex disease of osteoarthritis. Variability in prevalence and progression of knee, hip, and hand osteoarthritis suggests that the disease may be a group of different and unique conditions. Rheumatoid arthritis is an autoimmune collagen disease characterized by inflammation and increased protein turnover.

Effects of Osteoarthritis

Pain occurring with joint movement is common, but the cause of pain is still unexplained.[90] Osteoarthritis is among the most disabling conditions of older adults. In a review of the disease, Felson and colleagues concluded that:[90]

> The risk for disability (defined as needing help walking or climbing stairs) attributable to osteoarthritis of the knee is as great as that attributable to cardiovascular disease and greater than that due to any other medical condition in elderly persons.

Individuals with osteoarthritis tend to have pain, depressive symptoms, muscle weakness, and poor aerobic capacity.

Risk Factors

Obesity, continuous overexposure to oxidants, and possibly low vitamin D levels are risk factors for developing osteoarthritis. Low intake of vitamins C and D are risk factors for its progression. Risk factors may turn out to be different for knee or hip and hand osteoarthritis.

Nutritional Remedies

Weight loss is the first remedy advised.[11,12,90] Felson and associates summary report[90] found that women in the Framingham study who lost 11 lb cut their risk for knee osteoarthritis in half. Messier and associates[12] found that weight loss as small as 1 lb reduces load exerted on the knee; the beneficial effect adds up as steps accumulate during the day. Other nutritional approaches to symptom reduction follow:

Gluten   A protein found in wheat, spelt, kamut, dinkle/dinkel and triticale (genus *Triticum*), barley and and rye. Spelt, dinkel, and kamut are ancestral forms of today's wheat. Oats appear on some "gluten" lists; oats are inherently gluten-free, but may be contaminated by gluten-containing grains during processing.

Hyaluronans   Derivatives of hyaluronic acid, a lubricant found naturally in the joints.

Antioxidants   Individuals with the highest levels of vitamin C intake had significantly slower (threefold) disease progression and less knee pain than individuals with the lowest intakes. Results of increased vitamin E and beta-carotene intake were inconsistent.

Vitamin D   Progression at higher intakes and serum levels is roughly one-third slower than at the lowest levels. The DRI for older adults is 400 to 600 IU, with a tolerable upper intake level of 2000 IU.

Flavonoids   This large group of phytochemicals (found in vegetables, fruits, tea, whole grains) with antioxidant and anti-inflammatory properties that act to maintain cell membranes. Higher levels of antioxidants are required to scavenge oxidized metabolites and free radicals in inflammatory diseases.

Chondroitin and Glucosamine   These two substances naturally occur in the body as substrate for cartilage repair. Several clinical trials of their combined use demonstrate favorable effects, and the results in osteoarthritis seem to be best when knee pain is moderate to severe. A caution: product tests have shown that quality is surprisingly varied. Not all products contain the specified levels of active compound. Check with the FDA or go to www.consumerlab.com to find out how specific brands rate.

SAM-e   S-adenosylmethionine may reduce pain and stiffness and possible depression; in large doses it may also cause nausea.

Capsaicin   The compound that provides the heat in peppers is used in a topical cream for pain relief.

Other Treatments   Fatty acids and oils are other anti-inflammatory therapies with potential to lessen signs and symptoms in a variety of conditions including osteo- as well as rheumatic arthritis. Borage seed and evening primrose oils contain GLA or gamma-linolenic acid, which plays a role in prostaglandin synthesis (GLA competes with other fatty acids, limiting the production of inflammatory omega-6 fatty acids) and can decrease pain of arthritis after several months use.[91–93] Reductions in pain,

Case Study 19.1

Bridget Doyle Remembers Laura

Photo Disc

Just because she lives at Lenoir Manor, a continuing care retirement facility, Laura, a petite (4 ft. 8 in., 97 pounds) widow of the local college dean, does not consider herself as old. She is 87. She has had no major nutritional or health problems and her appetite is good. She had been a good cook and entertained graciously, but in the residential care facility, meals are prepared for her. Because she has had slight fluid retention over the past year, she no longer adds salt to her meals. She tells Bridget Doyle, her nutritionist, that yes, occasionally she does not like her meals and misses cooking for herself.

One Monday morning, Laura is found in bed with her left side paralyzed. The diagnosis is a right-sided stroke, resulting in 3 weeks of hospitalization. Back at the skilled care wing of Lenoir, Laura needs a nasogastric tube for feeding. She is alert and knows people, but is limited in speech. Overnight, Laura's care has changed from an individual needing routine nutritional monitoring to someone with many interrelated problems:

- Inability to communicate her overall medical and nutritional concerns clearly
- Weight loss of 9 pounds during the 3-week hospital stay
- Intense dislike of the nasal tube, as demonstrated by repeated attempts to pull it out, leading to restraints of her hands

Questions

1. What nutritional parameters should be assessed and monitored now that Laura is back at Lenoir Manor?
2. What disciplines should be involved in Laura's care plan, and why?
3. The interdisciplinary care team wants to meet Laura's needs in a dignified and respectful manner. How can the care team address both clinical and ethical concerns?
4. How could Bridget ensure that Laura's nutritional needs are met?
5. What are strategies young adults can adopt to reduce their risk of stroke?

---

morning stiffness, and swelling of rheumatoid arthritis follow intake of omega-3 fatty acids (fish oil, marine fatty acids) for at least 12 weeks.[93] Flaxseed and purslane also contain a type of omega-3 fatty acid. Traditional Native American medical therapy used echinacea (three varieties) for pain relief, rheumatism, and arthritis; ginseng for asthma and rheumatism; garlic for asthma; and evening primrose oil for obesity.[94] Today, echinacea and ginseng are marketed to boost immune function, and evening primrose oil is marketed as an antioxidant.

Vegetarian Diets Plants are rich in antioxidants that may play a role in managing inflammatory diseases in general. A review of the roles of vegetarian diets[95] found that it is difficult to design good experiments to answer the question: "Do vegetarians have less inflammatory disease?"

Food Allergies Gluten has a clear role in celiac disease, which is an inflammatory disease. Food allergies have also been suggested as a cause for inflammatory diseases such as irritable bowel syndrome, Crohn's disease, and rheumatoid arthritis, but evidence from well-done studies that can point to allergens as a mechanism for osteoarthritis are not yet available.

Cognitive Disorders: Alzheimer's Disease

Often considered an expected (and dreaded) aspect of "getting old," dementia is a condition of progressive cognitive decline, typically characterized by impaired thinking, memory, decision-making, and linguistic ability. Dementia is not a disease itself, but rather a group of symptoms that are often associated with particular degenerative neurological conditions:[96]

Alzheimer's disease
Vascular dementia

Table 19.11 Percentage of persons aged 65 years and older with moderate to severe cognitive disorders[22]

Age	Males	Females
65–74 years	15.4	9.7
75–84 years	39.0	30.6
85+ years	37.3	35.0

Dementia with Lewy bodies
Frontal-temporal lobe dementia
Parkinson's disease
Alcohol-related dementia
AIDS-related dementia

Dementia is not a part of normal aging, but rather the manifestation of various forms of physiologic damage.

Prevalence of Dementia

According to the World Health Organization, approximately 25–29 million people in the world suffer from dementia[97] and Alzheimer's disease (AD) is estimated to contribute to at least 70% of dementia cases among Western populations. AD now ranks fifth in leading causes of death for adults aged 65 and older in the United States.[10] The prevalence of AD is thought to double every 5 years after the age of 65,[98] and Table 19.11 reflects the rising prevalence of *memory impairment* with age.

Etiology of Cognitive Disorders

Dementia can be caused by a variety of conditions resulting in traumatic physiologic changes, including vascular changes ("mini strokes" or obstructions), degenerative diseases (Alzheimer's, Parkinson's, diabetes), physical trauma and infection, depression, chronic drug use, and malnutrition, including deficiency of vitamin B_{12} and dehydration. In addition, decreased learning ability and depression accompany deficiencies in niacin, thiamin, folic acid, biotin, iron, and selenium.

The symptoms of dementia may be completely or partially reversible in certain conditions. These include chronic drug abuse, metabolic disorders such as a vitamin B_{12} deficiency and hypoglycemia, and medical interventions such as surgery or medications.[99] Examples of nontreatable forms of dementia include degenerative diseases such as Alzheimer's, Huntington's, Parkinson's, Lewy body, and multi-infarct dementia.

Aluminum, copper, carnitine, and choline have been examined as possible causes of AD due to their role in neurological function, but they have not been demonstrated to be causal. Current research is focused on the B vitamins and the role of homocysteine as potential contributors to

the development of AD; see Illustration 19.3 for a graphic depiction of how these nutrients are metabolized.

Deficiencies of vitamin B_{12} and folate are related to high concentrations of homocysteine, an amino acid associated with the promotion of poor vascular health and cognitive decline.[99,100] In order to prevent the build-up of homocysteine in the blood and neural tissue, vitamin B_{12} and folate and are needed to convert it to methionine, a harmless amino acid. Methionine contributes to the synthesis of S-adenosylmethionine, which is widely distributed throughout the central nervous system for use in methylation reactions. Primary examples of methylation reactions include vitamin B_{12} in the production of myelin (the insulation cover on nerves) and folate in the DNA cycle (cell replication). Excess homocysteine in brain tissue is thought to contribute to the development of Alzheimer's disease either through vascular mechanisms or as a neurotoxin.[102] While vitamin B_{12} deficiency leads to neurological damage and cognitive decline, it is not an underlying factor in Alzheimer's disease.

Memory Impairment Moderate or severe impairment is when 4 or fewer words can be recalled from a list of 20.

Effects of Cognitive Disorders

Confusion, anxiety, agitation, loss of oral muscular control, impairment of hunger and appetite regulation, changes in smell and taste, chewing and swallowing difficulties, and dental problems are all aspects of Alzheimer's disease that make it difficult to maintain good nutritional habits. As the disease advances, individuals with AD will require more and more assistance with meal preparation and eating. In later stages of the disease, wandering and restless movements expend energy and increase caloric need. Behavioral, physical, or neurologic problems may impede adequate food intake. Consequently, individuals with late-stage AD suffer from unintentional weight loss (this is discussed in the section Low Body Weight/Underweight).

Nutrition Interventions for Cognitive Disorders

Many dangers lurk in the kitchens of persons with dementia who are living at home. Ensuring food safety and safe use of kitchen tools and equipment are primary considerations for good nutrition. When a caretaker is present, concerns turn to ensuring adequate intake rather than preventing injury. Additional calories may be needed for increased energy expended by individuals who pace and wander. Persons with dementia can benefit from meals offered in a calm dining environment, free of loud noises and confusion (no highly patterned tableware, nothing extraneous on the table). Maintaining the focus on eating,

MTHF = 5-methyltetrahydrofolate
Vit B$_{12}$ MET synthetase = vitamin B$_{12}$–dependent methionine synthetase
Vit B$_6$ CBS = vitamin B$_6$–dependent cystathione-β-synthase
MAT = methionine adenosine transferase
SAM-e = S-adenosylmethionine

Illustration 19.3 Folate, vitamin B$_{12}$ and homocysteine metabolism.

SOURCE: Adapted from Mattson, M, Chan, S, Duan, W. Modification of brain neurodegenerative disorders by genes, diet, and behavior. Physiology Review. 2002;82:637–72.

providing plenty of time to eat, serving finger foods, and encouraging regular drinks between bites are additional strategies to promote food and fluid intake. In cases of decreased physical coordination, adaptive eating utensils, such as slip-resistant placemats, mugs without spoons, and silverware with built-up handles may promote independent eating. Caretakers often use visual and verbal cues to facilitate the eating process.

Medications and Polypharmacy

In Health 2005, prescription costs of older adults were reported at $1807 per person per year.[1] They now represent more than 10% of Medicare costs. Prescription medicines are those ordered by a physician or other health care provider within the scope of their license, and do not include vitamins, herbal medicines, or over-the-counter (OTC) medicines. OTC medications are any pills, liquids, salves, creams, and supplements that are purchased at a pharmacy, discount, or food store without prescription. Most complementary medicines such as botanicals and herbs are sold as OTCs.

Prevalence of medication use for community-residing older adults was discussed in Chapter 18. Polypharmacy (taking several different drugs each day) is common in older adults, and was included as a nutritional risk factor in the Nutrition Screening Initiative. Comorbid conditions lead older adults to take more medications, both OTC and prescription, than younger adults, and thereby increase the potential for drug interactions and side effects. Disease progression, drug effects, and functional limitations such as poor eyesight or impaired memory, all increase risks for using a drug incorrectly. The potential for error also increases with the number of drugs.

Effects of Medications Medications may require dietary restrictions and can interfere with appetite, digestion, and metabolism and general alertness. For example, the blood-thinning drug warfarin requires a stable vitamin K

Table 19.12 Amounts of vitamin K in selected fruits and vegetables (DRI for ages 51 and older is 120 mcg/day for males, 90 mcg/day for females)

Less than 5 mcg vitamin K:

1 c servings of corn, mushrooms, onions, baked and navy beans, potatoes, applesauce, cherries, pineapple, strawberries

1 apple, banana, nectarine, peach, pear, orange, tangerine, 5 dried apricots, 5 dates

5 to 10 mcg vitamin K:

1 c canned apricots, fresh raspberries, chickpeas, lima beans, all varieties of squash, stewed tomatoes, sweet potatoes, fruit cocktail, papaya

2 figs

>10 and <50 mcg vitamin K:

1 c yellow and green string beans, blueberries, red cabbage, carrots, cauliflower, celery, cucumber, with and without peel, iceberg lettuce, roasted peppers, canned plums, grapes (red or green)

1 kiwi, 2 sprigs parsley

>50 mcg vitamin K (1 c cooked unless specified):

Beet greens, turnip greens (529–851)

Broccoli (183–220)

Cabbage, any (27–58)

Collards (836–1059)

Kale (1062–1147)

Romaine lettuce (57)

Green leaf lettuce (97)

Mustard greens (419)

Okra (64)

Raw green onion (207)

Prunes or dried plums (65)

Rhubarb (71)

Raw spinach (145)

Cooked spinach (1027)

SOURCE: www.nal.usda.gov/fnic/foodcomp/Data/SR18/nutrlist/sr18a430.pdf, accessed 8/24/06.

intake (see Table 19.12). The cost of drugs may also negatively affect a person's food budget. Inability to eat or forgetting to eat lead to unintentional weight loss, while some medications lead to undesired weight gain. Table 19.13 describes nutritional implications associated with medications used to treat diseases that are prevalent in older adults.

Low Body Weight/ Underweight

There is no consensus or universal definition for underweight in the frail elderly. The most common methods that measure changes in nutrition status are body mass index (BMI) and unplanned weight loss within the past 3 to 6 months. A BMI or weight found in the lowest percentiles of a reference standard in a comparable population is a starting point. Because of the National Health and Nutrition Examination Survey III, we now have weight percentiles for older adults by decade (beginning at 50 years), gender, and ethnic-racial groups and are able to recognize underweight in older adults (see Table 19.14).[103] Individuals with weights falling at the 5th or lower percentile of this population can be defined as "underweight."

Another approach is to compare an individual's current weight to "usual" body weight. Terms such as sarcopenia (muscle loss), anorexia (no appetite), and cachexia (another way to say "no appetite") are related to undernutrition in older adults and are associated with becoming frail.

The National Heart, Lung, and Blood Institute (NHLBI) defined underweight as a BMI <18.5 kg/meter squared for all adults.[104] The World Health Organization (WHO) further defines levels of underweight as grades of "thinness."[105]

BMI 17.0–18.49 indicates grade 1 thinness
BMI 16.0–16.99 indicates grade 2 thinness
BMI <16.00 indicates grade 3 thinness

Approximately one-third of older adults are underweight.

Table 19.13 Medications associated with chronic conditions in older adults: nutritional implications

Coronary Heart Disease
- Cardiac glycosides (e.g., digitalis) may result in anorexia and/or nausea.
- Statins may result in elevated enzymes.
- High doses of niacin may be associated with flushing, hyperglycemia, hypotension, hypoalbuminemia, upper GI distress and liver enzyme elevation.

Hypertension
- Diuretics may result in depletion of sodium, calcium, magnesium, and/or potassium.
- Centrally acting antihypertensives may result in a decline in food intake due to sedation, confusion, and depression.
- Medications such as beta-blockers may cause constipation and delayed gastric emptying.
- Potential related side effects are dizziness, dry mouth, or mouth pain.

Diabetes Mellitus
- Medications to treat diabetes may cause hypoglycemia, especially if nutritional intake is erratic and/or if increased or decreased appetite or diarrhea.
- Oral hypoglycemics (e.g., sulfonylureas such as glipizide) may cause heartburn, nausea, hypoglycemia, decreased appetite; biguanides (such as Glucophage) may cause decreased appetite, diarrhea, and vomiting.

Dementia
- Cholinesterase inhibitors (e.g., rivastigmine or galantamine) may cause nausea, diarrhea or weight loss.
- Antipsychotic/antidepressants with anticholinergic side effects may lead to dry mouth, delayed stomach emphtying, and constipation.
- Antidepressants may enhance appetite in depressed patients, but SSRIs may cause a decrease in appetite.

Congestive Heart Failure
- Diuretics (e.g., Diuril) may lead to electrolyte abnormalities, especially sodium and potassium and/or thiamine deficiency (e.g., furosemide).

Cancer
- Radiation, chemotherapy, and/or surgery can negatively affect nutritional status and metabolism.

Chronic Obstructive Pulmonary Disease (COPD)
- Xanthine derivatives (e.g., theophylline) may result in anorexia and nausea.

Lower Respiratory Infections
- Antibiotics (e.g., amoxicillan) may result in nausea, anorexia, diarrhea, and stomatitis.
- Radiation, chemotherapy, and/or surgery can negatively affect nutritional status and metabolism.

SOURCE: American Academy of Family Physicians, Nutrition Screening Initiative, American Dietetic Association. A physicians guide to nutrition in chronic disease management for older adults. Waldorf, MD, 2002. Can be ordered at e-mail address nsi@gmmb.com.

Table 19.14 NHANES III, 1988–1994 reported data of 5th percentile mean weight in pounds by age and gender in older adults residing in the United States

Age	Male	Female
60–69	134.5	109.0
70–79	128.7	100.5
≥ 80	114.3	92.0

SOURCE: Third National Health and Nutrition Examination Survey (1988–1994). www.cdc.gov/nchs/about/major/nhanes/Anthropometric%20Measures.htm, accessed July 9, 2003.

Etiology

Underweight is not considered problematic when the individual has had a lifelong low weight. However, weight cycling is problematic. In addition, unintentional weight loss is likely due to disease. A loss of 10% or more of total body weight in a 6-month period is associated with increased mortality. Intended weight loss should be consistent with reasonable weight-for-height standards.

For older adults, underweight is much more serious than overweight. Being thin has been related to increased incidence of diseases, but it is impossible to tell from the data whether thin precedes or follows incidence of disease. Overall, the effects of malnutrition impact immune response, muscle and respiratory function, and wound healing; malnutrition is associated with, but not caused by, aging.

Protein–calorie malnutrition leads to underweight. Underlying causes may be illness, poverty, or functional decline. See Case Study 19.2.

Nutrition Interventions

Avoiding weight loss is desirable but not always possible. Weight loss in the elderly is intertwined with

Ms. Wetter: A Senior Suffering through a Bad Stretch

Ms. Wetter: About to turn 81, Elizabeth Wetter is 5 feet 6 inches tall and weighs 106 pounds. She has had Parkinson's disease for 5 years, but that is not what concerns her. Her problem is pain from arthritis and lack of energy. She saw an ad on television for a vitamin–mineral supplement with ginseng that promises "more energy." Her son also told her to take a liquid dietary supplement to "feel better." Eighteen months ago, she had successful surgery for colon cancer, which was followed by chemotherapy treatments. She is now free of cancer. After the cancer treatment she fell and broke a hip. This healed well, but serious leg pains started shortly afterward. There seems to be no cause for the pain, and a cure is unavailable. She is no longer able to take her walks through the neighborhood or tend her prize-winning garden. She would like to weigh 118 pounds again (her "usual") and is seeking nutritional counseling to try to regain some of her energy.

Questions

1. What are some of the nutritional issues faced by Ms. Wetter? (Hint: Calculate her current weight to usual body weight as a percentage.)
2. How would you prioritize these in a nutritional care plan?
3. Calculate her energy needs and suggest strategies she might use to regain some energy.
4. What other information would you want to know in order to counsel Ms. Wetter?

---

disease-related biochemical and physiologic mechanisms, which, in turn, affect functional status and appetite. In a systematic review of studies to evaluate the effect of nutrition treatment in protein–energy malnutrition in connection with multiple disorders in the elderly, 20 studies noted an improvement in anthropometric or biochemical measures and 10 studies reported an improvement in function. In contrast, there was insufficient evidence to determine how nutrition treatment should be formulated, due to inconsistent or uncertain treatment adherence, and inability to separate treatment effects further complicates nutrition monitoring.[106] In a systematic review of 22 studies reported by the Cochrane Library, protein and energy supplementation appears to produce a small but consistent weight gain (2–4%), reduced mortality, and shorter length of hospital stays.[107] Supplementation was associated with nausea, diarrhea, and other gastrointestinal disturbances. Overall, medical nutrition therapy for a frail, elderly, malnourished person should occur in consultation with an experienced registered dietitian. Refeeding and rehydration are done gradually:

- *Calories:* eat and exercise to build muscle mass, strength.
- *Protein:* 1 to 1.5 grams of protein per kilogram body weight is adequate; 1.5–2 g/kg/day is

recommended for severe depletion by the American Dietetic Association's Consultant Dietitians in Health Care Facilities Practice Group.[108] Exceptions are patients with renal or liver failure, who may need a protein restriction.

- *Water:* 1 mL per Kcal; rehydrate slowly (see section Dehydration).

Dehydration

Dehydration is the physiological state in which cells lose water to the point of interfering with the metabolic processes. Normal urination does not cause dehydration. Phillips and colleagues[109] defined dehydration as losing nearly 2% of initial body weight; this can occur after avoiding all fluid and eating only dry foods for 24 hours.

There are three types of dehyration (isotonic, hypotonic, and hypertonic), and they are related to the balance of proportional sodium and water losses. Abnormally high serum sodium levels (>150 millequivalents per liter) or a high ratio of blood urea nitrogen to creatinine (>25) can also be used to diagnose "significant dehydration."[110] Hypertonic dehydration (i.e., serum sodium is >145 mmol/L) is the type seen in iatrogenic cases (i.e., fluid deprivation with possible neglect). It can also be seen in individuals with fever.

Connection: Fluid

8 ounces = 1 cup = 240 ml (milliliters) = 240 cc (cubic centimeters)

A super-sized soda (32 ounces) equals approximately 4 cups (or 4 × 240 = 960 ml). A 2-liter bottle of soda provides a little more than 8 cups of fluid. Some foods also count as fluid. For example, soup, jello, and sherbet are considered fluids.

Table 19.15 Percent of initial weight lost due to dehydration and physiological signs

Percent Lost	Physiological Signs
1%	Thirst (true for young people, not necessarily in older men or women)
4–6%	Economy of movement, flushed skin, sleepiness, apathy, nausea, tingling in arms, hands, feet, headache, heat exhaustion in fit men, increases in body temperature, pulse rate, respiratory rate
8%	Dizziness, slurred speech, weakness, confusion
12%	Cognitive signs: wakefulness, delirium
20%	Bare survival limit

For someone who normally weighs 160 lb:

1% loss means weight down to 158.4 lb

4% loss means weight down to 153.6 lb

6% loss means weight down to 150.4 lb

20% loss means weight down to 128.0 lb

SOURCE: Adapted from Briggs and Calloway, originally NASA, 1967.[103]

Dehydration can be measured as percentage of body weight lost when normal body weight is known. In a continuum of dehydration shown by Briggs and Calloway[111] (Table 19.15), originally designed for the National Aeronautics Space Administration (NASA), the indicators are somewhat different from the indicators that Gross found. The NASA continuum shows how the human body responds to water losses.

Weight loss of 4% normal weight would be hard to ignore unless the individual had a compromised mental or cognitive status. Flushed skin, nausea, and apathy or lack of energy occur when 4% body weight is lost due to dehydration.

Although the human body is quite resilient in that it can lose up to 10% of its water weight and survive, dehydration in smaller degree is common for older people. Prevalence data is hard to gather because dehydration is usually temporary, and there is no simple or commonly used standard definition for diagnosing dehydration in the elderly. For example, patients coming to a nursing home from a hospital may have IV tubes, and nursing notes may simply say, "The patient looks well-hydrated."

Etiology

Aging itself does not cause dehydration, even though the percentage of total body water shrinks from infancy to old age. Dehydration occurs more often in the elderly as a result of illness or other problems. Older people, are less sensitive in detecting thirst than younger people, and therefore may not think to drink.[109] Once fluids are consumed, aging kidneys may lose the ability to concentrate urine and *antidiuretic hormone* may become less effective. Swallowing problems, depression, or dementia may cause individuals to forget to eat or drink. Decreased mobility impairs older adults' access to water, and subsequent mobility to the bathroom. Fear of incontinence, in general, is another reason leading to decreased fluid intake and subsequent dehydration.

Effects of Dehydration

In older adults, thirst and skin turgor (pinch a skin fold on the forearm, forehead, or over the breastbone, and observe it fall back) are not good indicators of dehydration.[110] In the same review of 38 potential indicators of dehydration in older adults (61 to 98 years old, median age 82; $n = 55$; half were free living; half were admitted to the hospital emergency department from extended care), seven signs and symptoms were strongly related to dehydration (p <0.01 or p <0.001), and *not* to patient age:

1. Upper body muscle weakness
2. Speech difficulty
3. Confusion
4. Dry mucous membranes in nose and mouth
5. Longitudinal tongue furrows
6. Dry tongue
7. Sunken appearance of eyes in their sockets

These signs were also confirmed in a systematic review of maintaining oral hydration in older people.[112] Although regulation of body temperature is one of the functions of the body's water compartment, fever or elevated temperature did not identify dehydration status in Gross's study of individuals admitted to emergency rooms. However, fever or elevated temperature may be an indicator of impending dehydration.

Dehydration increases the resting heart rate, susceptibility of developing urinary tract infection, pneumonia, and pressure ulcers; it also leads to confusion, disorientation, and dementia.[113,114] Because confusion and delirium are signs of—as well as risk factors for—dehydration, getting enough fluids can become a vicious circle for someone at risk for cognitive decline.

Nutritional Interventions

The Institute of Medicine's DRIs for water does not change as adults get older:[114] "The AI for total water (drinking water, beverages, and foods) for the elderly is set based on median total water intake of young adults." Some health professionals suggest fluid levels to be 1 mL per calorie eaten, with a minimum of 1500 mL per day (approximately 6 cups). The ultimate beverage is *water*—tap or flavored, *not* sugared. Water is generally accessible, adds no calories to the diet, provides traces of minerals needed for metabolism, and is very low in sodium, even when softened. When a water beverage provides calories, sugar or some other carbohydrate was added. Pure water doesn't provide energy, but lacking water to the point of dehydration can dramatically reduce one's energy.

Many beverages contribute nutrients in addition to providing fluid:

1. Tea, especially green tea, has been reported as "promising, but requiring future studies" in relation to cardiovascular and cancer risk reduction; the flavonoids in tea act as antioxidants.[115]

2. Milk provides calcium, protein, riboflavin, and vitamin D and plays a role in treating hypertension and weight maintenance when it is also low in fat.

3. Regular use of cranberry juice reduces urinary tract infection in older women.[116]

4. Fruit and vegetable juices can be counted as part of the recommended fruit and vegetable servings.

Rehydrate Slowly

To treat dehydration in older adults, replace fluids slowly. Guidelines are to provide roughly one-fourth to one-third of the overall fluid deficit each day in the form of water or a 5% glucose solution (when the individual's blood values are stable).[112] For individuals with swallowing problems, or dysphagia, thickened liquids count as fluid. When bedridden older adults are offered fluids hourly and also with medication, they achieve higher levels of of hydration.[112]

Dehydration at End-of-Life

Lack of hydration can be an issue for people with a terminal disease or who are near death. Some individuals stop eating and/or drinking hours, days, or even weeks before death. Laboratory values for blood and urine are likely to become abnormal. Treatment for dehydration at the end of life may differ from treatment of dehydration during an acute disease episode. Suggestions for treating dehydration in a dying person are to integrate four Cs:[117]

1. *Common sense:* Use approaches that benefit the whole patient; ask, "What does the patient want?"

2. *Communication:* Respect and acknowledge the sadness felt by friends and family.

3. *Collaboration:* With the patient's permission, bring in experienced people such as hospice workers to guide treatment.

4. *Caring:* Listen; respond with love and compassion.

> **Antidiuretic Hormone** Hormone that causes the kidneys to dilute urine by absorbing more water.

Dehydration at the end of life contributes to an overall slowing down of body systems, including production of body fluids, resulting in less congestion, less edema and ascites or water retention, and less gastrointestinal action. A person experiencing dehydration at the end of life may experience slight thirst, although many dying patients are not thirsty. They may experience dry mouth that can be alleviated by sucking on ice chips or using artificial saliva. Decreased urine output can be a benefit because there is less need to go to the toilet. The most commonly reported symptoms in the last week of life of an individual with advanced progressive disease are loss of appetite, asthenia (loss of strength), dry mouth, confusion, and constipation.[118] Dehydration may also lead to increasing levels of confusion and drowsiness, which can reduce fear and anxiety related to dying.

Bereavement

Bereavement is the loss felt when someone who is personally significant dies. Losses of friends and family members happen more often in the lives of older persons. Grief, a very powerful emotion, is a natural response to bereavement. The grieving process, with its stages of shock and denial, disorganization, volatile reactions, guilt, loss and loneliness, relief, and reestablishment,[119] diverts attention from normal activities. Shopping and food preparation, eating, and drinking may get lost in the grieving process. Any loss of long-shared relationships through death, dementia, or moving brings about lack of interest in activities surrounding meal planning, preparation, shopping, and eating. People who are in mourning are vulnerable to malnutrition.

Widowhood has been shown to trigger disorganization and changes in daily routine, especially related to food preparation and eating.[120] Widowed persons who are able to enjoy mealtimes, have good appetites, have higher-quality diets, and receive social support work through the grieving process with fewer health consequences. They demonstrate healthy aging, which is, in the words of Dr. Tamara Harris, chief of Geriatric Epidemiology at the National Institutes of Health, "the ability of the individual to be resilient, to be adaptive, to be flexible, and to mobilize compensatory areas as they face adversities in all areas associated with health, disease, and decline in old age."

Key Points

1. Nearly three of four older adults consider their health to be good, very good, or excellent, despite the high prevalence of chronic disease in older adults.

2. Hypertension, which affects nearly half of older adults, can be moderated with the DASH diet; further blood pressure reductions are achieved by restricting sodium intake in addition to following the DASH eating pattern.

3. Most cancers develop over such long periods that, for older adults, cancer treatment presents a greater nutritional challenge than cancer prevention. However, the growing numbers of cancer survivors are encouraged to follow a cancer-prevention diet when their treatment is complete.

4. Of adults over age 65, nearly one in five reports having diabetes, primarily type 2, suggesting that meals planned for older adult nutrition programs automatically promote heart health and support diabetes management.

5. Body Mass Index should be combined with other measures such as waist circumference to assess weight status in older adults; a BMI range of 19 to 27 seems reasonable for older adults.

6. Bed rest and inactivity is detrimental to strong bones. Exercise combined with good nutrition—including adequate calcium, vitamins D, C, B_6, and K—help to maintain bone mass density. Older adults on blood thinners need to monitor vitamin K intake, which is included in some bone strengthening supplements.

7. Polypharmacy is a risk factor for malnutrition and may contribute to poor oral health; dry mouth and the associated gingival disease can result from drug use (such as antihistamines, antidepressants, sedatives and antianxiety drugs).

8. Despite adequate dietary intake of vitamin B_{12}, older adults may still have low blood levels due to malabsorption. Monitoring B_{12} blood levels in vulnerable populations can help to prevent the irreversible nerve damage caused by B_{12} deficiency. vitamin B_{12} is one instance when the synthetic version of a vitamin is better absorbed than the food-bound version.

9. Osteoarthritis is the most common inflammatory disease of older adults. Weight loss helps to reduce the load exerted on knees and also reduces risk of developing osteoarthritis (as do adequate intake of vitamins C and D).

10. Dementia associated with old age can be due to vascular and degenerative diseases, physical trauma and infection, depression, and malnutrition. However, most of the dementia in older adults in the United States is associated with Alzheimer's disease. Nutritional efforts focus on treatment and maintaining quality of life.

11. Weight loss in an older adult signals potential illness and malnutrition; an individual who is losing weight without trying needs a thorough assessment.

12. Confusion and muscle weakness related to dehydration is avoidable. Adequate fluid intake prevents dehydration and can also contribute needed nutrients to the diet of an older adult, such as calcium, vitamin D, protein and riboflavin from milk, antioxidants from tea and coffee, vitamins, minerals, and fiber from juices and nectars. Socializing over a beverage may add to quality of life. Some long-term care institutions have implemented "happy hour" programs to encourage fluid intake.

Resources

Alzheimer's Disease, Education and Referral Center
Site provides information about Alzheimer's disease and related disorders. It is a service of the National Institute of Aging (NIA).
Website: www.alzheimers.org

American Cancer Society
Cancer statistics and cancer detection guide.
Website: www.cancer.org

American Diabetes Association
Site with areas for professionals and for the public; includes links.
Website: www.diabetes.org

American Dietetic Association
Site offers food and nutrition tips, fact sheets, and search to find a registered dietitian.
Website: www.eatright.org

Choices in Dying
1035 30th Street NE, Washington, DC 2007
1-800-989-9455.
Information about hydration and artifical nutrition during end-of-life decision making.
Website: www.choices.org

Food and Drug Administration and American Association of Retired Persons (AARP)
Provides information for seniors about safe handling of foods.
Website: http://vm.cfsan.fda.gov/~dms/seniorsd.html

Fowkes WC. Prolonging Death—An American Tragedy
Long Beach, CA: The Archstone Foundation. 562-590-8655
Debates the use of life support.

National Center for Complementary and Alternative Medicine

Site provides fact sheets for complementary medicine associated with dietary supplements, cancer prevention and treatment, and other dietary components. The website is supported by the National Institutes of Health (NIH). *Website:* www.nccam.nih.gov

National Institute of Diabetes and Digestive and Kidney Diseases

Site that offers health information for diabetes, metabolic illnesses, and kidney disease. Research and clinical trial information is also available at this site. The site provides online interactive diet planning interview to develop client-personalized food recommendations. A written handout worksheet can be printed. *Website:* www.niddk.nih.gov

National Institute of Health, Office of Dietary Supplements

Website: http://ods.od.nih.gov/showp

National Institute of Mental Health

Major depression is most prevalent in developed nations. This site provides a link for the public with a specific topic discussing depression. A further link, titled older adults, provides comprehensive information on definitions of depression and the treatment options available. *Website:* www.nimh.nih.gov

National Osteoporosis Foundation

Website: www.nof.org

Partnership for Caring

Website: www.partnershipforcaring.org/homepage/index.html

Surveillance, Epidemiology and End Results (SEER) Registry

Cancer surveillance data. *Website:* www.seer.cancer.gov

References

1. Federal Interagency Forum on Aging-Related Statistics. Older Americans update 2006: key indicators of well-being. Washington, DC: U.S. Government Printing Office; May 2006.

2. Federal Register, 42 CFR, Part 405 et al. Vol. 66, No. 212, November 1, 2001; Medicare Program; Revisions to Payment Policies and Five-Year Review of and Adjustments to the Relative Values Units Under the Physician Fee Schedule for Calendar Year 2002: Final Rule, pages 55275–81.

3. Lacey K, Pritchett E. Nutrition care process and model: ADA adopts road map to quality care and outcomes management. J Am Diet Assoc. 2003;103(8):1061–72.

4. Sheils JF, Rubin R, Stapleton DC. The estimated costs and savings of medical nutrition therapy: the Medicare population. J Am Diet Assoc. 1999;99:428–35.

5. Chima CS et al. Relationship of nutritional status to length of stay, hospital costs, and discharge status of patients hospitalized in the medicine service. J Am Diet Assoc. 1997;97:975–78; quiz 979–80.

6. Barents Group. The clinical and cost-effectiveness of medical nutrition therapy: evidence and estimates of potential Medicare savings from the use of selected nutrition interventions. Washington, DC: KMPG Peat Marwick LLP; 1996.

7. American Dietetic Association Position Paper. Cost effectiveness of medical nutrition therapy. J Am Diet Assoc. 1995;95:88–91.

8. Sullivan DH, Sun S, Walls RC. Protein-energy undernutrition among elderly hospitalized patients: a prospective study. J Am Med Assoc. 1999;281:2013–19.

9. Wolinsky FD et al. Health services utilization among the noninstitutionalized elderly. J Health Soc Behav. 1983;24:325–37.

10. National Center for Health Statistics. Health, United States, 2005 with chartbook in trends in the health of Americans.

Washington, DC: U.S. Dept of Health and Human Services; 2005.

11. Institute of Medicine. Role of nutrition in maintaining health in the nation's elderly: evaluating coverage of nutrition servce for the Medicare population. Washington DC: National Academy Press; 2002. Available at http://nap.edu.books accessed.

12. Messier SP et al. Weight loss reduces knee-joint loads in overweight and obese older adults with knee osteoarthritis. Arthritis & Rheumatism. 2005;52:2026–32.

13. National Center for Health Statistics. Fast stats from A to Z, heart disease. www.cdc.gov/nchs/faststats/heart.htm, accessed 11/06.

14. Erlinger TP, Pollack H, Appel LJ. Nutrition-related cardiovascular risk factors in older people: results from the Third National Health and Nutrition Examination Survey. J Am Geriatr Soc. 2000;48:1486–89.

15. American Diabetes Association. Report of the Expert Committee on the Diagnosis and Classification of Diabetes Mellitus. Diabetes Care. 2001;24:S5–S20.

16. American Diabetes Association. Diagnosis and classificaton of diabetes mellitus. Diabetes Care. 2006;29(1):S43–S48.

17. Grundy SM et al. Implications of recent clinical trials for the National Cholesterol Education Program Adult Treatment Panel III Guidelines. Circulation. 2004;110:227–39.

18. Krauss RM et al. American Heart Association dietary guidelines: revision 2000: a statement for healthcare professionals from the Nutrition Committee of the American Heart Association. Circulation. 2000;102:2284–99.

19. Hiroyasu et al. Alcohol intake and the risk of cardiovascular disease in middle-aged Japanese men. Stroke. 1995;26(5):767–73.

20. Sacco RL et al. The protective effects of moderate alcohol consumption on ischemic stroke. JAMA. 1999;281:53–60.

21. Appel LJ et al. Dietary approaches to prevent and treat hypertension. A scientific statement from the American Heart Association. Hypertension. 2006;47;296–308.

22. Plaisted CS et al. The effects of dietary patterns on quality of life: a substudy of the Dietary Approaches to Stop Hypertension trial. J Am Diet Assoc. 1999;99:S84–89S.

23. Svetkey LP et al. Effect of lifestyle modifications on blood pressure by race, sex, hypertension status, and age. Journal of Human Hypertension. 2005;19:21–31.

24. Whelton PK et al. Sodium reduction and weight loss in the treatment of hypertension in older persons: a randomized controlled trial of nonpharmacologic interventions in the elderly (TONE). TONE Collaborative Research Group. J Am Med Assoc. 1998;279:839–46.

25. Sacks FM et al. Effects on blood pressure of reduced dietary sodium and the Dietary Approaches to Stop Hypertension (DASH) diet. DASH-Sodium Collaborative Research Group. N Engl J Med. 2001; 344:3–10.

26. The seventh report of the Joint National Committee on Prevention, Detection, Evaluation, and Treatment of High Blood Pressure. (JNC 7). NIH Pub No. 03–5233. May 2003. Available at www.nhlbi.nih.gov/guidelines/, accessed 11/03.

27. Wilson JW et al. Data tables: combined results from USDA's 1994 and 1995 Continuing Survey of Food Intakes by Individuals and 1994 and 1995 Diet and Health Knowledge Survey, 1997.

28. American Cancer Society. Cancer facts and figures 2006. Atlanta: American Cancer Society; 2006.

29. World Cancer Research Fund & American Institute for Cancer Research. Food, nutrition, and the prevention of cancer: a global perspective. Washington, DC, 1997. (New edition due 2007, see www.wcrf.org)

30. National Cancer Institute, U.S. National Institutes of Health, available at www.cancer.gov/cancertopics/pdq/supportivecare/nutrition/healthprofessionalversion/, accessed 10/06.

31. Brown JK et al. Nutrition and physical activity during and after cancer treatment: an American Cancer Society Guide for Informed Choices. CA Cancer J Clin. 2003;53:28–291.

32. deCicco M et al. Supportive therapy of elderly cancer patients. Crit Rev Oncol Hematol. 2002;42(2):189–211.

33. Centers for Disease Control and Prevention.Unrealized prevention opportunities; reducing the health and economic burden of chronic disease.Centers for Disease Control and Prevention, National Center for Chronic Disease Prevention and Health Promotion; 1997.

34. Haffner SM et al. Mortality from coronary heart disease in subjects with type 2 diabetes and in nondiabetic subjects with and without prior myocardial infarction. N Eng J Med. 1998;339:229–34.

35. Resnick HE et al. Differential effects of BMI on diabetes risk among black and white Americans. Diabetes Care. 1998;21:1821–35.

36. Hoogeveen EK et al. Hyperhomocysteinemia increases risk of death, especially in type 2 diabetes: 5-year follow-up of the Hoorn Study. Circulation. 2000;101: 1506–11.

37. Haffner SM. Management of dyslipidemia in adults with diabetes. Diabetes Care. 1998;21:160–78.

38. Franz MJ et al. Evidence-based nutrition principles and recommendations for the treatment and prevention of diabetes and related complications. Diabetes Care. 2002; 25:148–98.

39. Professional Practice Committee. Summary of revisions for the 2006 Clinical Practice Recommendations. Diabetes Care; 29 January 2006, S3.

40. Chandalia M et al. Beneficial effects of high dietary fiber intake in patients with type 2 diabetes mellitus. N Eng J Med. 2000; 342:1392–98.

41. Tsihlias EB et al. Comparison of high- and low-glycemic-index breakfast cereals with monounsaturated fat in the long-term dietary management of type 2 diabetes. Am J Clin Nutr 2000;72:439–49.

42. Preuss HG, Anderson RA. Chromium update: examining recent literature 1997–1998. Curr Opin Clin Nutr Metab Care. 1998;1:509–12.

43. The effect of intensive treatment of diabetes on the development and progression of long-term complications in insulin-dependent diabetes mellitus. The Diabetes Control and Complications Trial Research Group. N Eng J Med. 1993;329:977–86.

44. Effect of intensive blood-glucose control with metformin on complications in overweight patients with type 2 diabetes (UKPDS 34). UK Prospective Diabetes Study (UKPDS) Group. Lancet. 1998;352:854–65.

45. Intensive blood-glucose control with sulphonylureas or insulin compared with conventional treatment and risk of complications in patients with type 2 diabetes (UKPDS 33). UK Prospective Diabetes Study (UKPDS) Group. Lancet. 1998; 352:837–53.

46. National Heart, Lung, and Blood Institute Expert Panel on the Identification, Evaluation, and Treatment of Overweight and Obesity in Adults. Clinical guidelines on the idenification, evaluation, and treatment of overweight and obesity in adults: the evidence report. Obes Res. 1998;6(suppl 2): 51S–209S.

47. World Health Organization Consultation on Obesity. Obesity: preventing and managing the global epidemic. Geneva, Switzerland: World Health Organization; 2000;WHO Technical Report Series 894.

48. Flegal KM et al. Prevalence and trends in obesity among US Adults, 1999–2000. J Am Med Assoc. 2002;288(14):1723–27.

49. Flegal KM et al. Excess deaths associated with underweight, overweight, and obesity. JAMA. 2005;293:1861–67.

50. Ernsberger P, Koletsky RJ. Biomedical rationale for a wellness approach to obesity: an alternative to a focus on weight loss. J Soc Issues. 1999;55:221–59.

51. Ostbye T et al. Predictors of five-year mortality in older Canadians: the Canadian Study of Health and Aging. J Am Geriatr Soc. 1999;47:1249–54.

52. Arterburn DE, Crane PK, Sullivan SD. The coming epidemic fo obesity in elderly Americans. J Am Geriatr Soc. 2004;52: 1907–12.

53. The American Dietetic Association Long Term Care Task Force. Nutrition risk assessment form, guides, strategies and interventions. Chicago, IL: American Dietetic Association; 1999.

54. Blumenthal SJ, Hendi JM, Marsillo L. A public health approach to decreasing obesity. J Am Med Assoc. 288(17):2178.

55. Looker AC et al. Prevalence of low femoral bone density in older U.S. adults from NHANES III. J Bone Miner Res. 1997;12: 1761–68.

56. Madhok R, Allison T. Bone mineral density measurement in the management of osteoporosis: a public health perspective. In: Fordham JN, ed. Manual of bone densitometry measurements. An aid to the interpretation of bone densitometry measurements in a clinical setting. New York: Springer; 2000:1–16.

57. Faine MP. Dietary factors related to preservation of oral and skeletal bone mass in women. J Prosthet Dent. 1995; 73:65–72.

58. Nordin BE et al. Nutrition, osteoporosis, and aging. Ann NY Acad Sci. 1998;854: 336–51.

59. Chumlea WC, Guo SS. Body mass and bone mineral quality. Curr Opin Rheumatol. 1999;11:307–11.

60. McGowan J. Osteoporosis research, prevention and treatment. Senate Special Committee on Aging; 1999.

61. Rubin LA et al. Determinants of peak bone mass: clinical and genetic analyses in a young female Canadian cohort. J Bone Miner Res. 1999;14:633–43.

62. Matkovic V et al. Bone status and fracture rates in two regions of Yugoslavia. Am J Clin Nutr. 1979;32:540–49.

63. Lloyd T et al. Calcium supplementation and bone mineral density in adolescent girls. J Am Med Assoc. 1993;270:841–44.

64. Teegarden D et al. Dietary calcium, protein, and phosphorus are related to bone mineral density and content in young women. Am J Clin Nutr. 1998;68: 749–54.

65. Heaney RP. Bone biology in health and disease. In: Shils ME et al., eds. Modern nutrition in health and disease. Philadelphia: Lippincott, Williams & Wilkins; 1999: 1327–38.

66. Bischoff-Ferrari HA et al. Fracture prevention with vitamin D supplementation: a meta-analysis of randomized controlled trials. JAMA. 2005;293:2257–64.

67. Brown JP, Josse RG, for the Scientific Advisory Council of the Osteoporosis Society of Canada. 2002 clinical practice guidelines for the diagnosis and management of osteoporosis in Canada. CMAJ. 2002;167 (10 suppl):S1–S34.

68. Prince RL et al. Effects of calcium supplementation on clinical fracture and bone structure. Arch Intern Med. 2006;166: 869–75.

69. Feskanich D et al. Protein consumption and bone fractures in women. Am J Epidemiol. 1996;143:472–79.

70. Weber P. The role of vitamins in the prevention of osteoporosis—a brief status report. Int J Vitam Nutr Res. 1999;69:194–97.

71. Pines A et al. The effect of various hormonal preparations and calcium supplementation on bone mass in early menopause. Is there a predictive value for the initial bone density and body weight? J Intern Med. 1999;246:357–61.

72. Writing Group for the Women's Health Initiative Investigators. Risks and benefits of estrogen plus progestin in healthy postmenopausal women. Principal results from the Women's Health Initiative randomzied controlled trial. J Am Med Assoc. 2002; 288:321–33.

73. Locke GR et al. Risk factors associated with symptoms of gastroesophageal reflux. Am J Med. 1999;106:642–49.

74. Castell DO. Gastroesophageal reflux and abnormal esophageal pressures: cause or effect? J Clin Gastroenterol. 2000;30:3.

75. Ho C, Kauwell GP, Bailey L. Practitioner's guide to meeting the vitamin B-12 Recommended Dietary Allowance for people aged 51 years and older. J Am Diet Assoc. 1999;99:725–27.

76. Kasper H. Vitamin absorption in the elderly. Int J Vitam Nutr Res. 1999; 69:169–72.

77. Nilsson-Ehle H. Age-related changes in cobalamin (vitamin B_{12}) handling. Implications for therapy. Drugs Aging. 1998;12:277–92.

78. Lindenbaum J et al. Prevalence of cobalamin deficiency in the Framingham elderly population. Am J Clin Nutr. 1994; 60:2–11.

79. Hokin BD, Butler T. Cyanocobalamin (vitamin B-12) status in Seventh-Day Adventist ministers in Australia. Am J Clin Nutr. 1999;70:576S–78S.

80. Carmel R et al. Serum cobalamin, homocysteine, and methylmalonic acid concentrations in a multiethnic elderly population: ethnic and sex differences in cobalamin and metabolite abnormalities. Am J Clin Nutr. 1999;70:904–10.

81. Zeitlin A, Frishman WH, Chang CJ. The association of vitamin B_{12} and folate blood levels with mortality and cardiovascular morbidity incidence in the old old: the Bronx aging study. Am J Ther. 1997; 4:275–81.

82. Yates AA, Schlicker SA, Suitor CW. Dietary Reference Intakes: the new basis for recommendations for calcium and related nutrients, B vitamins, and choline. J Am Diet Assoc. 1998;98:699–706.

83. Trumbo P et al. Dietary Reference Intakes: vitamin A, vitamin K, arsenic, boron, chromium, copper, iodine, iron, manganese, molybdenum, nickel, silicon, vanadium, and zinc. J Am Diet Assoc. 2001;101:294–301.

84. Norberg B. Turn of tide for oral vitamin B_{12} treatment. J Intern Med. 1999; 246:237–38.

85. Eussen SJPM et al. Oral Cyanocobalamin supplementation in older people with vitamin B_{12} deficiency a dose-finding trial. Arch Intern Med. 2005;165:1167–72.

86. Bente LB, Gerrior SA. Selected food and nutrient highlights of the 20th century: U.S. food supply series. Family Econ Nutr Rev. 2002;14(1):43–51.

87. Harari D et al. Bowel habit in relation to age and gender. Findings from the National Health Interview Survey and clinical implications. Arch Intern Med. 1996;156:315–20.

88. Talley NJ et al. Constipation in an elderly community: a study of prevalence and potential risk factors. Am J Gastroenterol. 1996;91:19–25.

89. Camilleri M et al. Insights into the pathophysiology and mechanisms of constipation, irritable bowel syndrome, and diverticulosis in older people. J Am Geriatr Soc. 2000;48:1142–50.

90. Felson DT et al. Osteoarthritis: new insights. Part 1: the disease and its risk factors. Ann Intern Med. 2000;133: 635–46.

91. Galperin C, German BJ, Gershwin ME. Nutrition and diet in rheumatic diseases. In: Shils ME et al., eds. Modern nutrition in health and disease. Philadelphia, PA: Lippincott, Williams & Wilkins; 1999:1339–51.

92. Felson DT et al. Osteoarthritis: new insights. Part 2: treatment approaches. Ann Intern Med. 2000;133:726–37.

93. Kremer JM. n-3 fatty acid supplements in rheumatoid arthritis. Am J Clin Nutr. 2000;71:349S–51S.

94. Borchers AT et al. Inflammation and Native American medicine: the role of botanicals. Am J Clin Nutr. 2000;72:339–47.

95. Johnston PK. Nutritional implications of vegetarian diets. In: Shils ME et al., eds. Modern nutrition in health and disease. Philadelphia, PA: Lippincott, Williams & Wilkins; 1999:1755–67.

96. American Psychiatric Association. Diagnostic and statistical manual of mental disorders 4th edition, text revision (DSM-IV-TR). Washington, DC: American Psychiatric Press, Inc. 2000.

97. World Health Organization. World health organization will focus on aging. Bulletin of World Health Organization 1999;77:293–94.

98. Brookmeyer R, Gray S, Kawas C. Projections of Alzheimer disease in the United States and the public health impact of delaying disease onset. Am J Public Health. 1998;88:1337–42.

99. Cleveland Clinic Health System. 2004, for your health. Types of dementia. www.cchs.net/health/main.asp, accessed 09/06.

100. Hofman A, Ott A, Breteler MMB. Atherosclerosis apolipoprotein E and the prevalence of dementia and Alzheimer's disease in the Rotterdam Study. The Lancet. 1997;349:151–54.

101. Breteler MM. Vascular risk factors for Alzheimer's disease: an epidemiologic prospective. Neurobiol Aging; 2000;21: 153–60.

102. Seshadri S, Beiser A, Selhub J. Plasma homocysteine as a risk factor for dementia and Alzheimer disease. N Eng J Med. 2002; 346:476–83.

103. Kuczmarski MF, Kuczmarski RJ, Najjar M. Descriptive anthropometric reference data for older Americans. J Am Diet Assoc. 2000;100:59–66.

104. National Institutes of Health. Clinical guidelines on the identification, evaluation, and treatment of overweight and obesity in adults—the evidence report. Obes Res. 1998;6(Suppl 2):51S–209S.

105. The World Health Organization/Food and Agricultural Organization/United Nations University. Physical status: the use and interpretation of anthropometry. Report of a WHO Expert Committee. Geneva: World Health Organization; 1995.

106. Akner G, Cederholm. Treatment of protein-energy malnutrition in chronic nonmalignant disorders. Am J Clin Nutr. 2001;74:6–24.

107. Milne AC, Potter J, Avenell A. Protein and energy supplementation in elderly people at risk from malnutrition. The Cochrane Database of Systematic Reviews. 2002; (Issue 4):1–71. Online access via Ovid on November 14, 2002.

108. Consultant Dietitians in Health Care Facilities, Niedert K, Dorner B. Nutrition care of the older adult. 2nd edition. Chicago: The American Dietetic Association. 2004.

109. Phillips PA et al. Reduced thirst after water deprivation in healthy elderly men. N Eng J Med. 1984;311:753–59.

110. Gross CR et al. Clinical indicators of dehydration severity in elderly patients. J Emerg Med. 1992;10:267–74.

111. Briggs GM, Calloway DH. Bogert's nutrition and physical fitness. New York: Holt, Rinehart and Winston; 1984.

112. Hodgkinson B, Evans D, Wood J. Maintaining oral hydration in older people. The Joanna Briggs Institute for Evidence Based Nursing and Midwifery; Systematic Review. 2001;12:1–66.

113. Chidester JC, Spangler AA. Fluid intake in the institutionalized elderly. J Am Diet Assoc. 1997;97:23–38; quiz 29–30.

114. Institute of Medicine of the National Academies. Dietary Reference Intakes for water, potassium, sodium, chloride, and sulfate. Washington, DC: The National Academies Press; 2005.

115. Cabrera C, Artacho R, Gimenez R. Beneficial effects of green tea—a review. J Am Coll Nutri. 2006;25:79–99.

116. Fleet JC. New support for a folk remedy: cranberry juice reduces bacteriuria and pyuria in elderly women. Nutr Rev. 1994; 52:168–78.

117. Fordyce M. Dehydration near the end of life. Ann Long-Term Care 2000;8. www.mmhc.com/nhm/articles/NHM0005/fordyce.html, accessed 5/2000.

118. Conill C et al. Symptom prevalence in the last week of life. J Pain Symptom Manage. 1997;14:328–31.

119. Leming MR, Dickinson GE. Understanding dying, death, and bereavement. Fort Worth, TX: Holt, Rinehart, and Winston; 1990.

120. Rosenbloom CA, Whittington FJ. The effects of bereavement on eating behaviors and nutrient intakes in elderly widowed persons. J Gerontol. 1993; 48:S223–29.

CDC Growth Charts

Weight-for-length percentiles: Girls, birth to 36 months

Published May 30, 2000 (modified 6/8/00).
SOURCE: Developed by the National Center for Health Statistics in collaboration with
the National Center for Chronic Disease Prevention and Health Promotion (2000).

Head circumference-for-age percentiles: Girls, birth to 36 months

Published May 30, 2000.
SOURCE: Developed by the National Center for Health Statistics in collaboration with
the National Center for Chronic Disease Prevention and Health Promotion (2000).

CDC Growth Charts: United States

Length-for-age percentiles: Boys, birth to 36 months

Published May 30, 2000.
SOURCE: Developed by the National Center for Health Statistics in collaboration with
the National Center for Chronic Disease Prevention and Health Promotion (2000).

CDC Growth Charts: United States

Weight-for-age percentiles: Boys, birth to 36 months

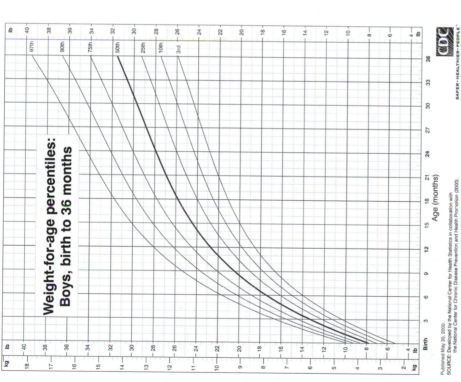

Published May 30, 2000.
SOURCE: Developed by the National Center for Health Statistics in collaboration with
the National Center for Chronic Disease Prevention and Health Promotion (2000).

CDC Growth Charts: United States

Weight-for-length percentiles: Boys, birth to 36 months

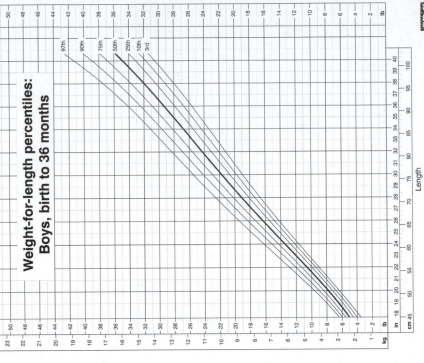

Published May 30, 2000 (modified 6/8/00).
SOURCE: Developed by the National Center for Health Statistics in collaboration with the National Center for Chronic Disease Prevention and Health Promotion (2000).

CDC Growth Charts: United States

Head circumference-for-age percentiles: Boys, birth to 36 months

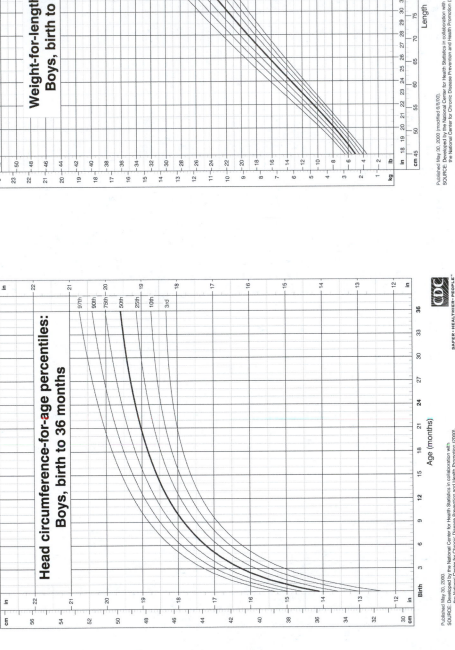

Published May 30, 2000.
SOURCE: Developed by the National Center for Health Statistics in collaboration with the National Center for Chronic Disease Prevention and Health Promotion (2000).

CDC Growth Charts: United States

Stature-for-age percentiles:
Girls, 2 to 20 years

Published: May 30, 2000.
SOURCE: Developed by the National Center for Health Statistics in collaboration with
the National Center for Chronic Disease Prevention and Health Promotion (2000).

CDC Growth Charts: United States

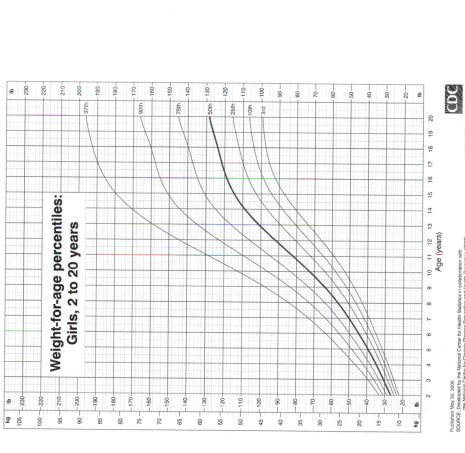

Weight-for-age percentiles:
Girls, 2 to 20 years

Published: May 30, 2000.
SOURCE: Developed by the National Center for Health Statistics in collaboration with
the National Center for Chronic Disease Prevention and Health Promotion (2000).

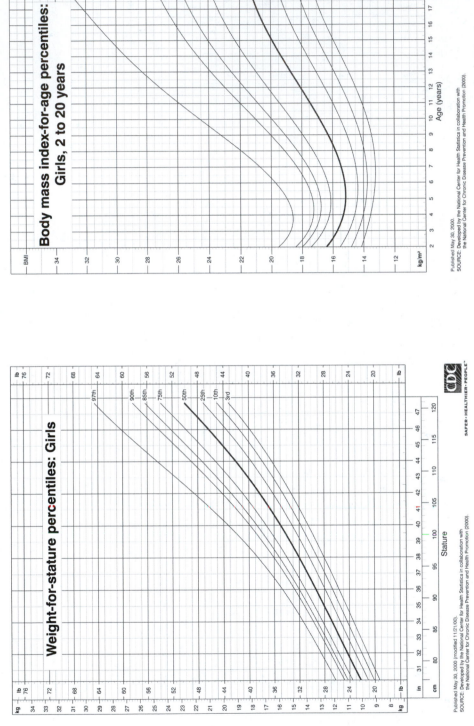

CDC Growth Charts: United States

Body mass index-for-age percentiles:
Girls, 2 to 20 years

Published May 30, 2000.
SOURCE: Developed by the National Center for Health Statistics in collaboration with
the National Center for Chronic Disease Prevention and Health Promotion (2000).

CDC Growth Charts: United States

Weight-for-stature percentiles: Girls

Published May 30, 2000 (modified 11/21/00).
SOURCE: Developed by the National Center for Health Statistics in collaboration with
the National Center for Chronic Disease Prevention and Health Promotion (2000).

CDC Growth Charts: United States

Stature-for-age percentiles: Boys, 2 to 20 years

Published May 30, 2000.
SOURCE: Developed by the National Center for Health Statistics in collaboration with
the National Center for Chronic Disease Prevention and Health Promotion (2000).

CDC Growth Charts: United States

Weight-for-age percentiles: Boys, 2 to 20 years

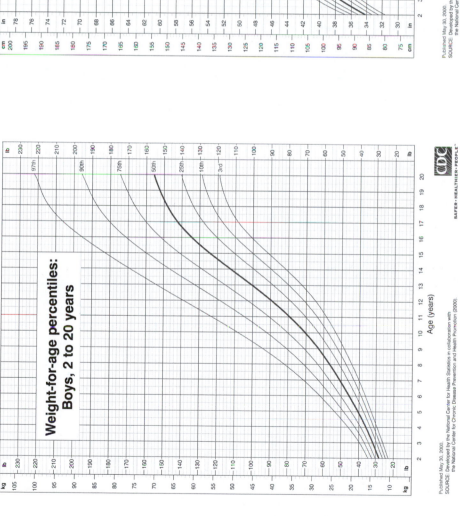

Published May 30, 2000.
SOURCE: Developed by the National Center for Health Statistics in collaboration with
the National Center for Chronic Disease Prevention and Health Promotion (2000).

CDC Growth Charts: United States

Body mass index-for-age percentiles: Boys, 2 to 20 years

Published May 30, 2000.
SOURCE: Developed by the National Center for Health Statistics in collaboration with
the National Center for Chronic Disease Prevention and Health Promotion (2000).

SAFER • HEALTHIER • PEOPLE™

CDC Growth Charts: United States

Weight-for-stature percentiles: Boys

Published May 30, 2000 (modified 11/21/00).
SOURCE: Developed by the National Center for Health Statistics in collaboration with
the National Center for Chronic Disease Prevention and Health Promotion (2000).

SAFER • HEALTHIER • PEOPLE™

Nutrient Intakes of Adults Aged 70 and Older

From CSFII[48]

Nutrient	Males	Females
Energy, kcal	1854	1377
Protein, gm	74	57
Total fat, gm	69	48
Saturated fatty acids, gm	23	16
Monounsaturated fatty acids, gm	27	18
Polyunsaturated fatty acids, gm	14	10
Cholesterol, mg	274	185
Total carbohydrate, gm	233	184
Dietary Fiber, gm	18	14
Vitamin A, mcg RE	1430.0	1149.0
Carotenes, mcg RE	648.0	602.0
Vitamin E, mg alpha TE	9.0	6.0
Vitamin C, mg	99.0	96.0
Thiamin, mg	1.6	1.2
Riboflavin, mg	2.0	1.5
Niacin, mg	22.2	17.5
Vitamin B_6, mg	2.0	1.5
Folate, mcg	283.0	234.0
Calcium, mg	754.0	587.0
Phosphorus, mg	1211.0	912.0
Magnesium, mg	286.0	224.0
Iron, mg	16.7	12.3
Zinc, mg	11.8	8.4
Copper, mg	1.3	1.0
Sodium, mg	3122.0	2376.0
Potassium, mg	2825.0	2824.0

Measurement Abbreviations and Equivalents

Metric or SI Unit

Unit	Abbreviation
kilogram	kg
gram	g
milligram	mg
microgram	μg, mcg
nanogram	ng
meter	m
centimeter	cm
millimeter	mm
liter	L
deciliters	dL
milliliter	mL
millimole	mmol
micromole	μmol
picomole	pmol

Nonmetric

Unit	Abbreviation
ounce	oz
pound	lb
tablespoon	Tabl, tb
teaspoon	tsp
cup	c
pint	pt
quart	qt
gallon	gal
inch	in.
foot	ft
yard	yd

Equivalents

Weight: Metric

1 kilogram	=	2.2 pounds, 1000 grams
1 gram	=	0.035 ounce; 1000 milligrams
1 milligram	=	1000 micrograms
1 microgram	=	1000 nanograms

Weight: Nonmetric

1 ounce	=	28.35 grams
1 pound	=	0.45 kilograms; 454 grams

Linear

1 millimeter	=	0.039 inch
1 centimeter	=	0.01 meter; 0.39 inch
1 meter	=	100 centimeters; 39.4 inches; 3.28 feet
1 inch	=	2.54 centimeters; 0.025 meter
1 foot	=	30.5 centimeters; 0.31 meter
1 yard	=	3 feet; 0.91 meters

Fluid Volume

1 teaspoon	=	5 mL
1 Tablespoon	=	15 mL; 0.5 ounce; 3 teaspoons
1 ounce	=	30 mL; 29.57 grams; 6 teaspoons; 2 tablespoons
1 cup	=	240 mL; 8 ounces; 48 teaspoons; 16 tablespoons
1 pint	=	480 mL; 2 cups; 16 ounces; 1 pound ("A pint is a pound the whole world round.")
1 quart	=	0.95 liter; 2 pints; 4 cups; 32 ounces; 2 pounds
1 liter	=	1.06 quarts; 1000 mL
1 gallon	=	3.79 liters; 4 quarts; 8 pints; 16 cups; 8 pounds ("Two cups in a pint, two pints in a quart, four quarts in a gallon.")

Conventional Units and SI Units

a. To convert conventional units to SI Units, multiply by the conversion factor.

b. To convert from SI units to conventional units, divide by the conversion factor.

	Conventional Units	Conversion Factor	SI Unit
Calcium	mg/dL	0.25	mmol/L
Cholesterol	mg/dL	0.0259	mmol/L
HDL cholesterol	mg/dL	0.0259	mmol/L
Folate	mg/mL	2.266	nmol/L
Glucose	mg/dL	0.0555	mmol/L
Hematocrit	%	0.01	Proportion of 1.0
Hemoglobin	g/dL	10.0	g/L
Homocysteine	mg/dL	7.397	μmol/L
Insulin	μIU/mL	6.945	pmol/L
Iron	μg/dL	0.179	μmol/L
LDL cholesterol	mg/dL	0.0259	mmol/L
Lipoprotein (a)	mg/dL	0.0357	μmol/L
Triglycerides	mg/dL	0.0113	mmol/L

Body Mass Index (BMI)

Height	18	19	20	21	22	23	24	25	26	27	28	29	30	31	32	33	34	35	36	37	38	39	40
										Body Weight (pounds)													
4'10"	86	91	96	100	105	110	115	119	124	129	134	138	143	148	153	158	162	167	172	177	181	186	191
4'11"	89	94	99	104	109	114	119	124	128	133	138	143	148	153	158	163	168	173	178	183	188	193	198
5'0"	92	97	102	107	112	118	123	128	133	138	143	148	153	158	163	168	174	179	184	189	194	199	204
5'1"	95	100	106	111	116	122	127	132	137	143	148	153	158	164	169	174	180	185	190	195	201	206	211
5'2"	98	104	109	115	120	126	131	136	142	147	153	158	164	169	175	180	186	191	196	202	207	213	218
5'3"	102	107	113	118	124	130	135	141	146	152	158	163	169	175	180	186	191	197	203	208	214	220	225
5'4"	105	110	116	122	128	134	140	145	151	157	163	169	174	180	186	192	197	204	209	215	221	227	232
5'5"	108	114	120	126	132	138	144	150	156	162	168	174	180	186	192	198	204	210	216	222	228	234	240
5'6"	112	118	124	130	136	142	148	155	161	167	173	179	186	192	198	204	210	216	223	229	235	241	247
5'7"	115	121	127	134	140	146	153	159	166	172	178	185	191	198	204	211	217	223	230	236	242	249	255
5'8"	118	125	131	138	144	151	158	164	171	177	184	190	197	203	210	216	223	230	236	243	249	256	262
5'9"	122	128	135	142	149	155	162	169	176	182	189	196	203	209	216	223	230	236	243	250	257	263	270
5'10"	126	132	139	146	153	160	167	174	181	188	195	202	209	216	222	229	236	243	250	257	264	271	278
5'11"	129	136	143	150	157	165	172	179	186	193	200	208	215	222	229	236	243	250	257	265	272	279	286
6'0"	132	140	147	154	162	169	177	184	191	199	206	213	221	228	235	242	250	258	265	272	279	287	294
6'1"	136	144	151	159	166	174	182	189	197	204	212	219	227	235	242	250	257	265	272	280	288	295	302
6'2"	141	148	155	163	171	179	186	194	202	210	218	225	233	241	249	256	264	272	280	287	295	303	311
6'3"	144	152	160	168	176	184	192	200	208	216	224	232	240	248	256	264	272	279	287	295	303	311	319
6'4"	148	156	164	172	180	189	197	205	213	221	230	238	246	254	263	271	279	287	295	304	312	320	328
6'5"	151	160	168	176	185	193	202	210	218	227	235	244	252	261	269	277	286	294	303	311	319	328	336
6'6"	155	164	172	181	190	198	207	216	224	233	241	250	259	267	276	284	293	302	310	319	328	336	345

Under-weight (<18.5)	Healthy Weight (18.5–24.9)	Overweight (25–29.9)	Obese (≥30)

Find your height along the left-hand column and look across the row until you find the number that is closest to your weight. The number at the top of that column identifies your BMI. The area shaded in green represents healthy weight ranges. The figure below presents silhouettes of various BMI.

Women
17 18 20 22.5 24 32 35

Men
18 21 23.5 24.5 26.5 31.5 37

SOURCE: Reprinted from material of the Dietitians of Canada.

Glossary

Adiposity or BMI Rebound   A normal increase in body mass index that occurs after BMI declines and reaches its lowest point at 4 to 6 years of age.

AIDS = Acquired Immunodeficiency Syndrome.

Allergy   Hypersensitivity to a physical or chemical agent.

Alveoli   Rounded or oblong cavities present in the breast (singular = alveolus).

Amenorrhea   Absence of menstrual cycle.

Amino Acids   The "building blocks" of protein. Unlike carbohydrates and fats, amino acids contain nitrogen.

Amniotic Fluid   The fluid contained in the amniotic sac that surrounds the fetus in the uterus.

Amylophagia   Compulsive consumption of laundry starch or cornstarch.

Anaphylaxis   Sudden onset of a reaction with mild to severe symptoms, including a decrease in ability to breathe, which may be severe enough to cause a coma.

Androgens   Types of steroid hormones produced in the testes, ovaries, and adrenal cortex from cholesterol. Some androgens (testosterone, dihydrotestosterone) stimulate development and functioning of male sex organs.

Anemia   A reduction below normal in the number of red blood cells per cubic mm in the quantity of hemoglobin, or in the volume of packed red cells per 100 ml of blood (hematocrit). This reduction occurs when the balance between blood loss and blood production is disturbed.

Anencephaly   Condition initiated early in gestation of the central nervous system in which the brain is not formed correctly, resulting in neonatal death.

Angiogenesis Inhibitor   Angiogenesis is the formation of new blood vessels. An angiogenesis inhibitor slows or stops vessel formation. Tumors cannot grow or expand without additional blood vessels to carry oxygen and other nutrients.

Anorexia Nervosa   A disorder characterized by extreme underweight, malnutrition, amenorrhea, low bone density, irrational fear of weight gain, restricted food intake, hyperactivity, and disturbances in body image.

Anovulatory Cycles   Menstrual cycles in which ovulation does not occur.

Anthropometry   The science of measuring the human body and its parts.

Antidiuretic Hormone   Hormone that causes the kidneys to dilute urine by absorbing more water.

Antioxidants   Chemical substances that prevent or repair damage to cells caused by exposure to oxidizing agents such as oxygen, ozone, and smoke and to other oxidizing agents normally produced in the body. Many different antioxidants are found in foods; some are made by the body.

Appropriate for Gestational Age (AGA)   Weight, length, and head circumference are between the 10th and 90th percentiles for gestational age.

Arteries   Blood vessels that carry oxygenated blood to cells.

Arteriosclerosis   Age-related thickening and hardening of the artery walls, much like an old rubber hose that becomes brittle or hard.

Assisted Reproductive Technology (ART)   An umbrella term for fertility treatments such as *in vitro* fertilization (IVF, a technique in which egg cells are fertilized by sperm outside the woman's body), artificial insemination, and hormone treatments.

Asthma   Condition in which the lungs are unable to exchange air due to lack of expansion of air sacs. It can result in a chronic illness and sometimes unconsciousness and death if not treated.

Atherosclerosis   A type of hardening of the arteries in which cholesterol is deposited in the arteries. These deposits narrow the coronary arteries and may reduce blood flow to the heart.

Athetosis   Uncontrolled movements of the large muscle groups as a result of damage to the central nervous system.

Atrial Fibrillation   Degeneration of the heart muscle, causing irregular contractions.

Attention Deficit Hyperactivity Disorder (ADHD)   Condition characterized by low impulse control and short attention span, with and without a high level of overall activity.

Autism   Condition of deficits in communication and social interaction with onset generally before age 3 years, in which mealtime behavior and eating problems occur along with other behavioral and sensory problems.

B-lymphocytes   White blood cells that are responsible for producing immunoglobulins.

Baby-Bottle Tooth Decay   Dental caries in young children caused by being put to bed with a bottle or allowed to suck from a bottle for extended periods of time. Also called baby- or nursing-bottle dental caries.

Basal Metabolic Rate (BMR)   Measuring energy expenditure in an individual who has been awake less than 30 minutes and is still at absolute rest, has fasted for 10 hours or more, and is in a quiet room with normal, comfortable temperatures.

Binge-Eating Disorder (BED)   An eating disorder characterized by periodic binge eating, which normally is not followed by vomiting or the use of laxatives. People must experience eating binges twice a week on average for over 6 months to qualify for this diagnosis.

Body Mass Index   An index that correlates with total body fat content or percent body fat, and is an acceptable measure of adiposity or body fatness in children and adults. It is calculated by dividing weight in kilograms by the square of height in meters (kg/m^2).

Bone Age   Bone maturation; correlates well with stage of pubertal development.

Bronchopulmonary Dysplasia (BPD)   Condition in which the underdeveloped lungs in a preterm infant are damaged so that breathing requires extra effort.

Bulimia Nervosa   A disorder characterized by repeated bouts of uncontrolled, rapid ingestion of large quantities of food (binge eating) followed by self-induced vomiting, laxatives or diuretic use, fasting, or vigorous exercise in order to prevent weight gain. Binge eating is often followed by feelings of disgust and guilt. Menstrual cycle abnormalities may accompany this disorder.

Calorie   A unit of measure of the amount of energy supplied by food. Also known as the "kilocalorie," or the "large Calorie."

Carotenemia   A condition, caused by ingestion of high amounts of carotenoids (or carotenes) from plant foods, in which the skin turns yellowish orange.

Carotid Artery Disease   Condition in which the arteries that supply blood to the brain and neck become damaged.

Catch-Up Growth   Period of time shortly after a slow growth period when the rate of weight and height gains is likely to be faster than expected for age and gender.

Celiac Disease Celiac disease is an autoimmune disease that occurs in people with a genetic susceptibility to the protein gliadin found in the gluten component of wheat, rye, and barley. Oats do not contain gluten, but commercial oats and oat products are often cross-contaminated by grains that do. Celiac disease is also called *celiac sprue* and *nontropical sprue*.

Cerebral Palsy A group of disorders characterized by impaired muscle activity and coordination present at birth or developed during early childhood.

Cerebral Spinal Atrophy Condition in which muscle control declines over time as a result of nerve loss, causing death in childhood.

Children with Special Health Care Needs A federal category of services for infants, children, and adolescents with, or at risk for physical or developmental disability, or with a chronic medical condition caused by or associated with genetic/metabolic disorders, birth defects, prematurity, trauma, infection, or perinatal exposure to drugs.

Cholesterol A fat-soluble, colorless liquid found in animals but not plants.

Chronic Condition Disorder of health or development that is the usual state for an individual and unlikely to change, although secondary conditions may result over time.

Chronic Inflammation Low-grade inflammation that lasts weeks, months, or years. Inflammation is the first response of the body's immune system to infection or irritation. Inflammation triggers the release of biologically active substances that promote oxidation and other potentially harmful reactions in the body.

Cleft Lip and Palate Condition in which the upper lip and roof of the mouth are not formed completely and are surgically corrected, resulting in feeding, speaking, and hearing difficulties in childhood.

Cobalamin Another name for vitamin B$_{12}$. Important roles of cobalamin are fatty acid metabolism, synthesis of nucleic acid (i.e., DNA, a complex protein that controls the formation of healthy new cells), and formation of the myelin sheath that protects nerve cells.

Coenzymes Chemical substances that activate enzymes.

Cognitive Function The process of thinking.

Colic A condition marked by a sudden onset of irritability, fussiness, or crying in a young infant between 2 weeks and 3 months of age who is otherwise growing and healthy.

Colostrum The milk produced in the first 2–3 days after the baby is born.

Colostrum is higher in protein and lower in lactose than milk produced after the milk supply is established.

Commodity Program A USDA program in which food products are sent to schools for use in the child nutrition programs. Commodities are usually acquired for farm price support and surplus removal reasons.

Competitive Foods Foods sold to children, in food service areas during mealtimes, that compete with the federal meal programs.

Congenital Abnormality A structural, functional, or metabolic abnormality present at birth. Also called congenital anomalies. These may be caused by environmental or genetic factors, or by a combination of the two. Structural abnormalities are generally referred to as congenital malformations, and metabolic abnormalities as inborn errors of metabolism.

Congenital Anomaly Condition evident in a newborn that is diagnosed at or near birth, usually as a genetic or chronic condition, such as spina bifida or cleft lip and palate.

Corpus Luteum (*corpus* = body, *luteum* = yellow) A tissue about 12 mm in diameter formed from the follicle that contained the ovum prior to its release. It produces estrogen and progesterone. The "yellow body" derivation comes from the accumulation of lipid precursors of these hormones in the corpus luteum.

Critical Periods Preprogrammed time periods during embryonic and fetal development when specific cells, organs, and tissues are formed and integrated, or functional levels established. Also called sensitive periods.

Cystic Fibrosis Condition in which a genetically changed chromosome 7 interferes with all the exocrine functions in the body, but particularly pulmonary complications, causing chronic illness.

Daily Values (DVs) Scientifically agreed-upon standards for daily intakes of nutrients from the diet developed for use on nutrition labels.

Development Progression of the physical and mental capabilities of an organism through growth and differentiation of organs and tissues, and integration of functions.

Developmental Delay Conditions represented by at least a 25% delay by standard evaluation in one or more areas of development, such as gross or fine motor, cognitive, communication, social, or emotional development.

Developmental Disabilities General term used to group specific diagnoses together that limit daily living and functioning and occur before age 21.

Diaphragmatic Hernia Displacement of the intestines up into the lung area due to incomplete formation of the diaphragm in utero.

Dietary Fiber Complex carbohydrates and lignins naturally occurring and found mainly in the plant cell wall. Dietary fiber cannot be broken down by human digestive enzymes.

Dietary Reference Intakes (DRIs) Quantitative estimates of nutrient intakes, used as reference values for assessing the diets of healthy people. DRIs include Recommended Dietary Allowances (RDAs), Adequate Intakes (AI), Tolerable Upper Intake Level (UL), and Estimated Average Requirement (EAR).

Dietary Supplements Any product intended to supplement the diet, including vitamin and mineral supplements, proteins, enzymes, amino acids, fish oils, fatty acids, hormones and hormone precursors, and herbs and other plant extracts. In the United States, such products must be labeled "Dietary Supplement."

Differentiation Cellular acquisition of one or more characteristics or functions different from that of the original cells.

DiGeorge Syndrome Condition in which chromosome 22 has a small deletion, resulting in a wide range of heart, speech, and learning difficulties.

Diplegia Condition in which the part of the brain controlling movement of the legs is damaged, interfering with muscle control and ambulation.

Disproportionately Small for Gestational Age (dSGA) Newborn weight is ≤ 10th percentile of weight for gestational age; length and head circumference are normal. Also called asymmetrical SGA.

Diverticulitis Infected "pockets" within the large intestine.

Doula An individual who gives psychological encouragement and physical assistance to a mother during pregnancy, birth, and lactation; the doula may be a relative, friend, or neighbor and is usually but not necessarily female.

Down Syndrome Condition in which three copies of chromosome 21 occur, resulting in lower muscle strength, lower intelligence, and greater risk for overweight.

Dysmenorrhea Painful menstruation due to abdominal cramps, back pain, headache, and/or other symptoms.

Early Intervention Services Federally mandated evaluation and therapy services for children in the age range from birth to 3 years under the Individuals with Disabilities Education Act.

Edema   Swelling (usually of the legs and feet, but can also extend throughout the body) due to an accumulation of extracellular fluid.

Embryo   The developing organism from conception through 8 weeks.

Empty-Calorie Foods   Foods that provide an excess of calories relative to their nutrient content.

Endocrine   A system of ductless glands, such as the thyroid, adrenal glands, ovaries, and testes, that produces secretions that affect body functions.

Endometriosis   A disease characterized by the presence of endometrial tissue in abnormal locations, such as deep within the uterine wall, in the ovary, or in other sites within the body. The condition is quite painful and is associated with abnormal menstrual cycles and infertility in 30–40% of affected women.

Endothelium   The layer of cells lining the inside of blood.

Enrichment   The replacement of thiamin, riboflavin, niacin, and iron that are lost when grains are refined.

Enteral Feeding   Fluid or food being delivered directly into the gastrointestinal system. The delivery can be by mouth or through a tube that is placed into the stomach or intestines.

Epididymis   Tissues on top of the testes that store sperm.

Epithelial Cells   Cells that line the surface of the body.

EPSDT   The Early Periodic Screening, Detection, and Treatment Program is a part of Medicaid and provides routine checkups for low-income families.

Essential Fatty Acids   Components of fat that are a required part of the diet (i.e., linoleic and alpha-linolenic acids). Both contain unsaturated fatty acids.

Essential Nutrients   Substances required for growth and health that cannot be produced, or produced in sufficient amounts, by the body. They must be obtained from the diet.

Exposure Index   The average infant milk intake per kilogram body weight per day × (the milk to plasma ratio divided by the rate of drug clearance) × 100. It is indicative of the amount of the drug in the breast milk that the infant ingests and is expressed as a percentage of the therapeutic (or equivalent) dose for the infant.

Extremely Low-Birth-Weight Infant (ELBW)   An infant weighing 1000 g or 2 lb 3 oz at birth.

Failure to Thrive (FTT)   Condition of inadequate weight or height gain thought to result from a caloric deficit, whether or not the cause can be identified as a health problem.

Familial Hyperlipidemia   A condition that runs in families and results in high levels of serum cholesterol and other lipids.

Fatty Acids   The fat-soluble components of fats in foods.

Fecundity   Biological ability to bear children.

Fertility   Actual production of children. The word best applies to specific vital statistic rates, but is commonly taken to mean the ability to bear children.

Fetal-Origins Hypothesis   The theory that exposures to adverse nutritional and other conditions during critical or sensitive periods of growth and development can permanently affect body structures and functions. Such changes may predispose individuals to cardiovascular diseases, type 2 diabetes, hypertension, and other disorders later in life. Also called *metabolic programming* and the *Barker Hypothesis*.

Fetus   The developing organism from 8 weeks after conception to the moment of birth.

Fine Motor Skills   Development and use of smaller muscle groups demonstrated by stacking objects, scribbling, and copying a circle or square.

Fluorosis   Permanent white or brownish staining of the enamel of teeth caused by excessive ingestion of fluoride before teeth have erupted.

Food Allergy (Hypersensitivity)   Abnormal or exaggerated immunologic response, usually immunoglobulin E (IgE) mediated, to a specific food protein.

Food Insecurity   Limited or uncertain availability of safe, nutritious foods.

Food Intolerance   An adverse reaction involving digestion or metabolism but not the immune system.

Food Security   Access at all times to a sufficient supply of safe, nutritious foods.

Fortification   The addition of one or more vitamins or minerals to a food product.

Full-Term Infants   Infants born between 37 and 42 weeks of gestation.

Functional Fiber   Nondigestible carbohydrates including plant, animal, or commercially produced sources, that have beneficial effects in humans.

Functional Foods   Generally taken to mean food, fortified foods, and enhanced food products that may have health benefits beyond the effects of essential nutrients they contain.

Functional Status   Ability to carry out the activities of daily living, including telephoning, grocery shopping, food handling and preparation, and eating.

Galactosemia   A rare genetic condition of carbohydrate metabolism in which a blocked or inactive enzyme does not allow breakdown of galactose. It can cause serious illness if not identified and treated soon after birth.

Gastroesophageal Reflux (GER)   Movement of the stomach contents backward into the esophagus, due to stomach muscle contractions. The condition may require treatment depending on its duration and degree. Also known as *gastro-esophageal reflux disease (GERD)*.

Gastrostomy Feeding   Form of enteral nutrition support for delivering nutrition by tube placement directly into the stomach, bypassing the mouth through a surgical procedure that creates an opening through the abdominal wall and stomach.

Geophagia   Compulsive consumption of clay or dirt.

Gestational Diabetes   Carbohydrate intolerance with onset or first recognition in pregnancy.

Glucogenic Amino Acids   Amino acids such as alanine and glutamate that can be converted to glucose.

Gluten   A protein found in wheat, oats, barley, rye and triticale (all in the genus Triticum); gliadin is the toxic fraction of gluten.

Glycemic Index (GI)   A measure of the extent to which blood glucose levels are raised by a specific amount of carbohydrate-containing food compared to the same amount of glucose or white bread.

Glycemic Load (GL)   A measure of the extent to which blood glucose levels are raised by a specific amount of carbohydrate-containing food. It is calculated by multiplying the carbohydrate content of an amount of food consumed by the glycemic index of the

Glycerol   A component of fats that is soluble in water. It is converted to glucose in the body.

Glycosylated Hemoglobin   A laboratory test that measures how well the blood sugar level has been maintained over a prolonged period of time; also called Hemoglobin A_1C.

Gravida   Number of pregnancies a woman has experienced.

Gross Motor Skills   Development and use of large muscle groups as exhibited by walking alone, running, walking up stairs, riding a tricycle, hopping, and skipping.

Growth   Increase in an organism's size through cell multiplication (hyperplasia)

and enlargement of cell size (hypertrophy).

Growth Velocity The rate of growth over time.

Gynecological Age Defined as chronological age minus age at menarch. For example, a female with the chronological age of 14 years minus age at first menstrual cycle of 12 equals a gynecological age of 2.

Health More than the absence of disease, health is a sense of well-being. Even individuals with a chronic condition may properly consider themselves to be healthy. For instance, a person with diabetes mellitus whose blood sugar is under control can be considered healthy.

Heart Disease The leading cause of death and a common cause of illness and disability in the United States. Coronary heart disease, the principal form of heart disease, is caused by buildup of cholesterol deposits in the coronary arteries that feed the heart.

Hematocrit An indicator of the proportion of whole blood occupied by red blood cells. A decrease in hematocrit is a late indicator of iron deficiency.

Heme Iron Iron contained within a protein portion of hemoglobin that is in the ferrous state.

Hemoglobin A protein that is the oxygen-carrying component of red blood cells. A decrease in hemoglobin concentration in red blood cells is a late indicator of iron deficiency.

Hemolytic Anemia Anemia caused by shortened survival of mature red blood cells and inability of the bone marrow to compensate for the decreased life span.

Hemolytic Uremic Syndrome (HUS) A serious, sometimes fatal complication associated with illness caused by *E. coli* O157:H7, which occurs primarily in children under the age of 10 years. HUS is characterized by renal failure, *hemolytic anemia*, and a severe decrease in platelet count.

High-Poverty Neighborhoods Neighborhoods where 40% or more of the people are living in poverty.

HIV = human immunodeficiency virus.

Homeostasis Constancy of the internal environment. The balance of fluids, nutrients, gases, temperature, and other conditions needed to ensure ongoing, proper functioning of cells and, therefore, all parts of the body.

Homocysteine Another intermediate product that depends on vitamin B_{12} for complete metabolism. However, both vitamin B_{12} and folate (another B vitamin)

are coenzymes in the breakdown of certain protein components in this pathway. Thus, elevated homocysteine levels can result from vitamin B_{12}, folate, or pyridoxine deficiencies.

Hydrolyzed Protein Formula Formula that contains enzymatically digested protein, or single amino acids, rather than protein as it naturally occurs in foods.

Hyperbilirubinemia Elevated blood levels of bilirubin, a yellow pigment that is a by-product of the breakdown of fetal hemoglobin.

Hypertonia Condition characterized by high muscle tone, stiffness, or spasticity.

Hypoallergenic Foods or products that have a low risk of promoting food or other allergies.

Hypocalcemia Condition in which body pools of calcium are unbalanced, and low levels are measured in blood as a part of a generalized reaction to illnesses.

Hypogonadism Atrophy or reduced development of testes or ovaries. Results in immature development of secondary sexual characteristics.

Hypothyroidism A condition characterized by growth impairment and mental retardation and deafness when caused by inadequate maternal intake of iodine during pregnancy. Used to be called cretinism.

Hypotonia Condition characterized by low muscle tone, floppiness, or muscle weakness.

Iatrogenic Used in reference to disease, it is a condition induced by a medical treatment.

Immunoglobulin A specific protein that is produced by blood cells to fight infection.

Immunological Having to do with the immune system and its functions in protecting the body from bacterial, viral, fungal, or other infections and from foreign proteins (i.e., those proteins that differ from proteins normally found in the body).

Indirect Calorimetry Measurement of energy requirements based on oxygen consumption and carbon dioxide production.

Infant Health and Development Program (IHDP) Growth charts with percentiles for VLBW (,1500-g birthweight) and LBW (,2500-g birthweight).

Infant Mortality Death that occurs within the first year of life.

Infant Mortality Attributable to Birth Defects (IMBD) Category used in tracking infant deaths in which specific diagnoses have a high mortality.

Infecundity Biological inability to bear children after 1 year of unprotected intercourse.

Infertility Commonly used to mean a biological inability to bear children.

Innocenti Declaration The Innocenti Declaration on the Protection, Promotion, and Support of Breastfeeding was produced and adopted by participants at the WHO/UNICEF policymakers' meeting on "Breastfeeding in the 1990s: A Global Initiative," held at the Spedale degli Innocenti, in Florence, Italy, on August 1, 1990. The Declaration established exclusive breastfeeding from birth to 4–6 months of age as a global goal for optimal maternal and child health.

Insulin Hormone usually produced in the pancreas to regulate movement of glucose from the bloodstream into cells within organs and muscles.

Insulin Resistance A condition in which cells "resist" the action of insulin in facilitating the passage of glucose into cells.

Intrauterine Growth Retardation (IUGR) Fetal undergrowth from any cause, resulting in a disproportionality in weight, length, or weight-for-length percentiles for gestational age. Sometimes called intrauterine growth restriction.

Iron Deficiency A condition marked by depleted iron stores. It is characterized by weakness, fatigue, short attention span, poor appetite, increased susceptibility to infection, and irritability.

Iron-Deficiency Anemia A condition often marked by low hemoglobin level. It is characterized by the signs of iron deficiency plus paleness, exhaustion, and a rapid heart rate.

Jejunostomy Feeding Form of enteral nutrition support for delivering nutrition by tube placement directly into the upper part of the small intestine.

Juvenile Rheumatoid Arthritis Condition in which joints become enlarged and painful as a result of the immune system; generally occurs in children or teens.

Kernicterus or Bilirubin Encephalopathy The end result of very high untreated bilirubin levels. Excessive bilirubin in the system is deposited in the brain, causing toxicity to the basal ganglia and various brain-stem nuclei.

Ketogenic Diet High-fat, low-carbohydrate meal plan in which ketones are made from metabolic pathways used in converting fat as a source of energy.

Ketones Metabolic by-products of the breakdown of fatty acids in energy formation. b-hydroxybutyric acid, acetoacetic acid, and acetone are the major ketones, or "ketone bodies."

Klinefelter's Syndrome A congenital abnormality in which testes are small and firm, legs abnormally long, and intelligence generally subnormal.

Krebs Cycle   A series of metabolic reactions that produce energy from the proteins, fats, and carbohydrates that constitute food.

Kwashiorkor   A disease syndrome in children, primarily caused by protein deficiency. It is generally characterized by edema (or swelling), loss of muscle mass, fatty liver, rough skin, discoloration of the hair, growth retardation, and apathy.

L. Monocytogenes, or Listeria   A foodborne bacterial infection that can lead to preterm delivery and stillbirth in pregnant women. Listeria infection is commonly associated with the ingestion of soft cheeses, unpasteurized milk, ready-to-cat deli meats, and hot dogs.

Lactation Consultant   A health care professional who provides education and management to prevent and solve breastfeeding problems and to encourage a social environment that effectively supports the breastfeeding mother-infant dyad. Those who successfully complete the International Board of Lactation Consultant Examiners (IBLCE) certification process are entitled to use IBCLC (International Board Certified Lactation Consultant) after their names (www.iblce.org/).

Lactiferous Sinuses   Larger ducts for storage of milk behind the nipple.

Lactogenesis   Another term for human milk production.

Lactose   A form of sugar or carbohydrate composed of galactose and glucose.

Large for Gestational Age (LGA) Weight for gestational age exceeds the 90th percentile for gestational age. Also defined as birthweight greater than 4500 g (≥ 10 lb) and referred to as excessively sized for gestational age, or macrosomic.

LDL Cholesterol   Low-density lipoprotein cholesterol, the lipid most associated with atherosclerotic disease. Diets high in saturated fat, trans fatty acids, and dietary cholesterol have been shown to increase LDL-cholesterol levels.

Le Leche League   An international, nonprofit, nonsectarian organization dedicated to providing education, information, support, and encouragement to women who want to breastfeed. It was founded in 1956 by seven women who had learned about successful breastfeeding while nursing their own babies. (www.lelecheleague.org).

Lean Body Mass   Sum of fat-free body tissues: muscle, mineral as in bone, and water.

Leptin   A protein secreted by fat cells that, by binding to specific receptor sites in the hypothalamus, decreases appetite, increases energy expenditure, and stimulates gonadotropin secretion. Leptin levels are elevated by high, and reduced by low, levels of body fat.

Life Expectancy   Average number of years of life remaining for persons in a population cohort or group; most commonly reported as life expectancy from birth.

Life Span   Maximum number of years someone might live; human life span is projected to range from 110 to 120 years.

Lignin   Noncarbohydrate polymer that contributes to dietary fiber.

Linseed   From the flax plant, linum; linseed is another name for flaxseed. Linseed oil is used in paints, varnishes, and inks but is also produced in food form for its rich nutrient content.

Liveborn Infant   The World Health Organization developed a standard definition of liveborn to be used by all countries when assessing an infant's status at birth. By this definition, a liveborn infant is the outcome of delivery when a completely expelled or extracted fetus breathes, or shows any sign of life such as beating of the heart, pulsation of the umbilical cord, or definite movement of voluntary muscles, whether or not the cord has been cut or the placenta is still attached.

Lobes   Rounded structures of the mammary gland.

Long-Chain Fats   Carbon molecules that provide fatty acids with 12 or more carbons, which are commonly found in foods.

Longevity   Length of life; it is a measure of life's duration in years.

Low-Birth-Weight Infant (LBW)   An infants weighing, 2500 g or ,5 lb 8 oz at birth.

Lower Esophageal Sphincter (LES)   The muscle enabling closure of the junction between the esophagus and stomach.

Macrobiotic Diet   This diet falls between semivegetarian and vegan diets and includes foods such as brown rice and other grains, vegetables, fish, dried beans, spices, and fruits.

Macrocephaly   Large head size for age and gender as measured by centimeters (or inches) of head circumference.

Macrophages   A white blood cell that acts mainly through phagocytosis.

Malnutrition   Poor nutrition resulting from an excess or lack of calories or nutrients.

Mammary Gland   The source of milk for offspring, also commonly called the breast. The presence of mammary glands is a characteristic of mammals.

Maple Syrup Urine Disease   Rare genetic condition of protein metabolism in which breakdown by-products build up in blood and urine, causing coma and death if untreated.

MCT Oil   A liquid form of dietary fat used to boost calories; composed of medium-chain triglycerides.

Meconium   Dark green mucilaginous material in the intestine of the full-term fetus.

Medical Neglect   Failure of parent or caretaker to seek, obtain, and follow through with a complete diagnostic study or medical, dental, or mental health treatment for a health problem, symptom, or condition that, if untreated, could become severe enough to present a danger to the child.

Medical Nutrition Therapy (MNT) Comprehensive nutrition services by registered dietitians to treat the nutritional aspects of acute and chronic diseases.

Medicinal Herbs   Plants used to prevent or remedy illness.

Medium-Chain Fats   Carbon molecules that provide fatty acids with 6–10 carbons, again not typically found in foods.

Memory Impairment   Moderate or severe impairment is present when four or fewer words can be recalled from a list of 20.

Menarche   The occurrence of the first menstrual cycle.

Meningitis   Viral or bacterial infection in the central nervous system that is likely to cause a range of long-term consequences in infancy, such as mental retardation, blindness, and hearing loss.

Menopause   Cessation of the menstrual cycle and reproductive capacity in females.

Menses   The process of menstruation.

Menstrual Cycle   An approximately 4-week interval in which hormones direct a buildup of blood and nutrient stores within the wall of the uterus and ovum maturation and release. If the ovum is fertilized by a sperm, the stored blood and nutrients are used to support the growth of the fertilized ovum. If fertilization does not occur, they are released from the uterine wall over a period of 3 to 7 days. The period of blood flow is called the menses, or the menstrual period.

Mental Retardation   Substantially below average intelligence and problems in adapting to the environment, which emerge before age 18 years.

Metabolic Syndrome   A constellation of metabolic abnormalities that increase the risk of type 2 diabetes and heart disease. It is characterized by insulin resistance, abdominal obesity, high blood pressure and triglyceride levels, low levels of HDL cholesterol, and impaired glucose tolerance. Also called Syndrome X and insulin-resistance syndrome.

Metabolism The chemical changes that take place in the body. The conversion of glucose to energy or body fat is an example of a metabolic process.

Methylmalonic Acid (MMA) An intermediate product that needs vitamin B_{12} as a coenzyme to complete the metabolic pathway for fatty acid metabolism. Vitamin B_{12} is the only coenzyme in this reaction; when it is absent, the blood concentration of MMA rises.

Microcephaly Small head size for age and gender as measured by centimeters (or inches) of head circumference.

Middle Childhood Children between the ages of 5 and 10 years; also referred to as school-age.

Milk/Plasma Drug Concentration Ratio (M/P Ratio) The ratio of the concentration of drug in milk to the concentration of drug in maternal plasma. Since the ratio varies over time, a time-averaged ratio provides more meaningful information than data obtained at a single time point. It is helpful in understanding the mechanisms of drug transfer and should not be viewed as a predictor of risk to the infant as it is the concentration of the drug in milk, and not the M/P ratio, is critical to the calculation of infant dose and assessment of risk.

Miscarriage Generally defined as the loss of a conceptus in the first 20 weeks of pregnancy. Also called spontaneous abortion.

Mitochondria Intracellular unit in which fatty acid breakdown takes place and many enzyme systems for energy production inside cells are regulated.

Monounsaturated Fats Fats in which only one pair of adjacent carbons in one or more of its fatty acids is linked by a double bond (e.g., –C–C=C–C–).

Monovalent Ion An atom with an electrical charge of +1 or −1.

Morbidity The rate of illnesses in a population.

Mortality Rate The rate of death.

Myoepithelial Cells Specialized cells that line the alveoli and can contract to cause milk to be secreted into the duct.

Necrotizing Enterocolitis (NEC) Condition with inflammation or damage to a section of the intestine, with a grading from mild to severe.

Neonatal Death Death that occurs in the period from the day of birth through the first 28 days of life.

Neural Tube Defects (NTDs) Spina bifida and other malformations of the neural tube. Defects result from incomplete formation of the neural tube during the first month after conception.

Neurobehavioral Pertains to control of behavior by the nervous system.

Neuromuscular Term pertaining to the central nervous system's control of muscle coordination and movement.

Neuromuscular Disorders Conditions of the nervous system characterized by difficulty with voluntary or involuntary control of muscle movement.

Neutrophils A class of white blood cells that are involved in protecting against infection.

NF (National Formulary) A uniformity standard for herbs and botanicals.

Nonessential Nutrients Nutrients required for growth and health that can be produced by the body from other components of the diet.

Nonheme Iron Iron contained within a protein of hemoglobin that is in the ferric state.

Nonorganic Failure to Thrive Inadequate weight or height gain without an identifiable biological cause, so that an environmental cause is suspected.

Nutrient-Dense Foods Foods that contain relatively high amounts of nutrients compared to their caloric value.

Nutrients Chemical substances in foods that are used by the body for growth and health.

Nutrigenomics The science of gene-nutrient interactions.

Nutrition Programming The process by which exposure of the fetus to certain levels of energy and nutrients modify the function of genes in ways that affect metabolism and the development of diseases later in life.

Nutrition Support Provision of nutrients by methods other than eating regular foods or drinking regular beverages, such as directly accessing the stomach by tube or placing nutrients into the bloodstream.

Obesity BMI-for-age greater than the 95th percentile with excess fat stores as evidenced by increased triceps skinfold measurements above the 85th percentile.

Oral-Gastric (OG) Feeding A form of enteral nutrition support for delivering nutrition by tube placement from the mouth to the stomach.

Organic Failure to Thrive Inadequate weight or height gain resulting from a health problem, such as iron-deficiency anemia or a cardiac or genetic disease.

Osmolarity Measure of the number of particles in a solution, which predicts the tendency of the particles to move from high to low concentration. Osmolarity is a factor in many systems, such as in fluid and electrolyte balance.

Osteoblasts Bone cells involved with bone formation; bone-building cells.

Osteoclasts Bone cells that absorb and remove unwanted tissue.

Osteoporosis Condition in which low bone density or weak bone structure leads to an increased risk of bone fracture.

Ova Eggs of the female produced and stored within the ovaries (*singular* = ovum).

Overweight Body mass index at or above the 95th percentile.

Oxytocin A hormone produced during letdown that causes milk to be ejected into the ducts.

Oxidative Stress A condition that occurs in the endothelium when it is exposed to a surplus of oxidizing agents relative to its supply of antioxidents.

Pagophagia Compulsive consumption of ice or freezer frost.

Parenteral Feeding Delivery of nutrients directly to the bloodstream.

Palpebral Fissure The space between the top and bottom eyelid when the eye is open. This opening is small in children with FAS.

Parity The number of previous deliveries experienced by a woman; *nulliparous* = no previous deliveries, *primiparous* = one previous delivery, *multiparous* = two or more previous deliveries. Women who have delivered infants are considered to be "parous."

Pediatric AIDS Acquired immunodeficiency syndrome in which infection-fighting abilities of the body are destroyed by a virus.

Pelvic Inflammatory Disease (PID) A general term applied to infections of the cervix, uterus, fallopian tubes, or ovaries. Occurs predominantly in young women and is generally caused by infection with a sexually transmitted disease, such as gonorrhea or chlamydia, or with intrauterine device (IUD) use.

Periconceptional Period Around the time of conception, generally defined as the month before and the month after conception.

Perinatal Death Death occurring at or after 20 weeks of gestation and through the first 28 days of life.

Philtrum The vertical groove between the bottom of the nose and the upper lip. The philtrum is smooth, or flat, when there is no groove.

Phytochemicals (*Phyto* = plants) Chemical substances in plants, some of which affect body processes in humans that may benefit health.

Phytoestrogen   A hormone-like substance found in plants, about 1/1000 to 1/2000 as potent as the human hormone, but strong enough to bind with estrogen receptors and mimic estrogen and anti-estrogen effects.

PICA   An eating disorder characterized by the compulsion to eat substances that are not food.

PKU (Phenylketonuria)   An inherited error in phenylalanine metabolism most commonly caused by a deficiency of phenylalanine hydroxylase, which converts the essential amino acid phenylalanine to the nonessential amino acid tyrosine.

Placenta   A disk-shaped organ of nutrient and gas interchange between mother and fetus. At term, the placenta weighs about 15% of the weight of the fetus.

Platelets   A component of the blood that plays an important role in blood coagulation.

Polycystic Ovary Syndrome (PCOS) (*polycysts* = many cysts; i.e., abnormal sacs with membranous linings) A condition in females characterized by insulin resistance, high blood insulin and testosterone levels, obesity, menstrual dysfunction, amenorrhea, infertility, hirsutism (excess body hair), and acne.

Polyunsaturated Fats   Fats in which more than one pair of adjacent carbons in one or more of its fatty acids are linked by two or more double bonds (e.g., –C–C=C–C=C–).

Postictal State   Time after a seizure of altered consciousness; appears like a deep sleep.

Pouring Rights   Contracts between schools and soft-drink companies whereby the schools receive a percentage of the profits of soft-drink sales in exchange for the school offering only that soft-drink company's products on the school campus.

Prader-Willi Syndrome   Condition in which partial deletion of chromosome 15 interferes with control of appetite, muscle development, and cognition.

Preadolescence   The stage of development immediately preceding adolescence; 9 to 11 years of age for girls and 10 to 12 years of age for boys.

Prebiotics   Certain fiberlike forms of indigestible carbohydrates that support the growth of beneficial bacteria in the lower intestine. Nicknamed "intestinal fertilizer."

Preeclampsia   A pregnancy-specific condition that usually occurs after 20 weeks of pregnancy (but may occur earlier). It is characterized by increased blood pressure and protein in the urine and is associated with decreased blood flow to maternal organs and through the placenta.

Prediabetes   A condition in which blood glucose levels are higher than normal but not high enough for the diagnosis of diabetes. It is characterized by impaired glucose tolerance, or fasting blood glucose levels between 110 and 126 mg/dl.

Preloads   Beverages or foods such as yogurt in which the energy/macronutrient content has been varied by the use of various carbohydrate and fat sources. The preload is given before a meal or snack and subsequent intake is monitored. This study design has been employed by Birch and Fisher in their studies of appetite, satiety, and food preferences in young children.

Premenstrual Syndrome   (*premenstrual* = the period of time preceding menstrual bleeding; *syndrome* = a constellation of symptoms) A condition occurring among women of reproductive age that includes a group of physical, psychological, and behavioral symptoms with onset in the luteal phase and subsiding with menstrual bleeding. Also called premenstrual dysphoric disorder (PMDD).

Preschool-Age Children   Children between the ages of 3 and 5 years, who are not yet attending kindergarten.

Preterm Infants   Infants born at or before 37 weeks of gestation.

Primary Malnutrition   Malnutrition that results directly from inadequate or excessive dietary intake of energy or nutrients.

Probiotics   Strains of Lactobacillus and bifidobacteria that have beneficial effects on the body. Also called "friendly bacteria."

Programming   The process by which exposure to adverse nutritional or other conditions during sensitive periods of growth and development produces long-term effects on body structures, functions, and disease risk.

Prolactin   A hormone that stimulates milk production.

Proportionately Small for Gestational Age (pSGA)   Newborn weight, length, and head circumference are ≤10th percentile for gestational age. Also called symmetrical SGA.

Prostacyclins   Biologically active substances produced by blood vessel walls that inhibit platelet aggregation (and therefore blood clotting), dilate blood vessels, and reduce blood pressure.

Prostaglandins   A group of physiologically active substances derived from the essential fatty acids. They are present in many tissues and perform such functions as the constriction or dilation of blood vessels, and stimulation of smooth muscles and the uterus.

Psychostimulant   Classification of medication that acts on the brain to improve mental or emotional behavior.

Puberty   The time frame during which the body matures from that of a child to that of a young adult, and becomes biologically capable of reproduction.

Pulmonary   Related to the lungs and their movement of air for exchange of carbon dioxide and oxygen.

Quality of Life   A measure of life satisfaction that is difficult to define, especially in heterogeneous aging population. Quality of life measures include factors such as social contacts, economic security, and functional status.

Recommended Dietary Allowances (RDAs)   The average daily dietary intake levels sufficient to meet the nutrient requirements of nearly all (97% to 98%) healthy individuals in a population group. RDAs serve as goals for individuals.

Recumbent Length   Measurement of length while the child is lying down. Recumbent length is used to measure toddlers less than 24 months of age, and those between 24 and 36 months who are unable to stand unassisted.

Reflex   An automatic (unlearned) response that is triggered by a specific stimulus.

Registered Dietitian   An individual who has acquired knowledge and skills necessary to pass a national registration examination and who participates in continuing professional education.

Resilience   Ability to bounce back, to deal with stress and recover from injury or illness.

Resting Energy Expenditure   The amount of energy needed by the body in a state of rest.

Resting Metabolic Rate (RMR)   Measuring energy expenditure in an individual who has fasted, had no vigorous physical activity prior to the test, has been given time to relax (e.g., rest) for 30 minutes before starting measurement, and is in a quiet, private room with privacy and normal, comfortable temperatures.

Rett Syndrome   Condition in which a genetic change on the X chromosome results in severe neurological delays, causing children to be short, thin appearing, and unable to talk.

Rooting Reflex   Action that occurs if one cheek is touched, resulting in the infant's head turning toward that cheek and the infant opening his mouth.

Saturated Fats   Fats in which adjacent carbons in the fatty acid component are linked by single bonds only (e.g., –C–C–C–C–).

Scoliosis   Condition in which the vertebral bones in the back show a side-to-side curve, resulting in a shorter stature than expected if the back were straight.

Scrotum A muscular sac containing the testes.

Secondary Condition Common consequence of a condition, which may or may not be preventable over time.

Secondary Malnutrition Malnutrition that results from a condition (e.g., disease, surgical procedure, medication use) rather than primarily from dietary intake.

Secondary Sexual Characteristics Physiological changes that signal puberty, including enlargement of the testes, penis, and breasts and the development of pubic and facial hair.

Secretory Cells Cells in the acinus (milk gland) that are responsible for secreting milk components into the ducts.

Secretory Immunoglobulin A One of the proteins found in secretions that protect the body's mucosal surfaces from infections. It may act by reducing the binding of a microorganism with cells lining the digestive tract. It is present in human colostrum but not transferred across the placenta.

Seizures Condition in which electrical nerve transmission in the brain is disrupted, resulting in periods of loss of function that vary in severity.

Self-Efficacy The ability to make effective decisions and to take responsible action based upon one's own needs and desires.

Semen The penile ejaculate containing a mixture of sperm and secretions from the testes, prostate, and other glands. It is rich in zinc, fructose, and other nutrients. Also called seminal fluid.

Senescent Old to the point of nonfunctional.

Sensorimotor An early learning system in which the infant's senses and motor skills provide input to the central nervous system.

Serum Iron, Plasma Ferritin, and Transferring Saturation Measures of iron status obtained from blood plasma or serum samples.

Sex Hormone-Binding Globulin (SHBG) A protein that binds with the sex hormones testosterone and estrogen. Also called steroid hormone binding globulin, because testosterone and estrogen are produced from cholesterol and are thus considered to be steroid hormones. These hormones are inactive when bound to SHBG, but are available for use when needed. Low levels of SHBG are related to increased availability of testosterone and estrogen in the body.

Short-Chain Fats Carbon molecules that provide fatty acids less than 6 carbons long, as products of energy generation from fat breakdown inside cells. Short-chain fatty acids are not usually found in foods.

Shoulder Dystocia Blockage or difficulty of delivery due to obstruction of the birth canal by the infant's shoulders.

Small for Gestational Age (SGA) Newborn weight is ≤10th percentile for gestational age. Also called small for date (SFD).

Social Marketing A marketing effort that combines the principles of commercial marketing with health education to promote a socially beneficial idea, practice, or product.

Spastic Quadriplegia A form of cerebral palsy in which brain damage interferes with voluntary muscle control in both arms and legs.

Stature Standing height.

Steroid Hormones Hormones such as progesterone, estrogen, and testosterone produced primarily from cholesterol.

Stroke The event that occurs when a blood vessel in the brain becomes occluded due to a clot or ruptures cutting off blood supply to a portion of the brain. Also called a cerebral vascular accident.

Subfertility Reduced level of fertility characterized by unusually long time to conception (over 12 months) or repeated, early pregnancy losses.

Subscapular Skinfold Thickness A skinfold measurement that can be used with other skinfold measurements to estimate percent body fat; the measurement is taken with skinfold calipers just below the inner angle of the scapula, or shoulder blade.

Suckle A reflexive movement of the tongue moving forward and backward; earliest feeding skill.

T-lymphocytes White blood cells that are active in fighting infection may also be called T-cells; the T in T-cell stands for thymus). These cells coordinate the immune system by secreting hormones that act on other cells.

T. Gondii, or Loxoplasmosis A parasitic infection that can impair fetal brain development. The source of the infection is often hands contaminated with soil or the contents of a cat litter box; or raw or partially cooked pork, lamb, or venison.

Telomere A cap-like structure that protects the end of chromosomes; it erodes during replication.

Teratogenic Exposures that produce malformations in embryos or fetuses.

Testes Male reproductive glands located in the scrotum. Also called testicles.

Thromboxanes Biologically active substances produced in platelets that increase platelet aggregation (and therefore promote blood clotting), constrict blood vessels, and increase blood pressure.

Toddlers Children between the ages of 1 and 3 years.

Tolerable Upper Intake Levels Highest level of daily nutrient intake that is likely to pose no risk of adverse health effects to almost all individuals in the general population; gives levels of intake that may result in adverse effects if exceeded on a regular basis.

Total Fiber Sum of dietary fiber and functional fiber.

Tracheoesophageal Atresia Incomplete connection between the esophagus and the stomach in utero, resulting in a shortened esophagus.

Trans Fatty Acids Fatty acids that have unusual shapes resulting from the hydrogenation of polyunsaturated fatty acids. Trans fatty acids also occur naturally in small amounts in foods such as dairy products and beef.

Transient Ischemic Attack (TIA) Temporary and insufficient blood supply to the brain.

Transpyloric Feeding (TP) Form of enteral nutrition support for delivering nutrition by tube placement from the nose or mouth into the upper part of the small intestine.

Triceps Skinfold A measurement of a double layer of skin and fat tissue on the back of the upper arm. It is an index of body fatness and measured by skinfold calipers. The measurement is taken on the back of the arm midway between the shoulder and the elbow.

Type 1 Diabetes A disease characterized by high blood glucose levels resulting from destruction of the insulin-producing cells of the pancreas. In the past, this type of diabetes was called juvenile-onset diabetes and insulin-dependent diabetes.

Type 2 Diabetes A disease characterized by high blood glucose levels due to the body's inability to use insulin normally, or to produce enough insulin. In the past this type of diabetes was called adult-onset diabetes and non-insulin-dependent diabetes.

Unsaturated Fats Fats in which adjacent carbons in one or more fatty acids are linked by one or more double bonds (e.g., –C–C=C–C=C–).

USP (United States Pharmacopeia) A nongovernmental, nonprofit organization (since 1820); establishes and maintains standards of identity, strength, quality, purity, processing, and labeling for health care products.

Vegan Diet The most restrictive of vegetarian diets, allowing only plant foods.

Venous Thromboembolism   A blood clot in a vein.

Vermillion Border   The exposed pink or reddish margin of a lip. A thin vermillion border in FAS denotes a thin upper lip.

Very-Low-Birth-Weight Infant (VLBW)   An infant weighing <1500 g or <3 lb 5 oz at birth.

Waist-to-hip Ratio   The ratio of the waist circumference, measured at its narrowest, and the hip circumference, measured where it is widest. This ratio is an easy way to measure body fat distribution, with a higher ratio indicative of an abdominal fat pattern. A high waist-to-hip ratio is associated with a high risk of chronic disease.

Weaning   Discontinuation of breastfeeding or bottle-feeding and substitution of food for breast milk or infant formula.

Work of Breathing (WOB)   A common term used to express extra respiratory effort in a variety of pulmonary conditions.

Working-Poor Families   Families where at least one parent worked 50 or more weeks a year and the family income was below the poverty level.

Xerostomia   Dry mouth, or xerostomia, can be a side effect of medications (especially antidepressants), of head and neck cancer treatments, and of diabetes, and is also a symptom of Sjogren's syndrome, which is an autoimmune disorder for which no cure is known.

Index

A

Academy of Breastfeeding Mothers, 177
Adequate Intakes (AIs), 4
Adiposity rebound, 279–280, 310, 315
Adolescent nutrition
 for athletes, 391–392
 cardiovascular disease and, 393–395
 chronic conditions and, 402–404
 eating disorders and, 393–402
 energy/nutrient needs, 362–369
 ergogenic supplement use and, 390–391
 health/eating-related behaviors and, 358–362
 iron-deficiency anemia and, 392–393
 normal growth/development and, 355–357
 normal psychosocial growth and, 357–358
 nutritional needs of, 354–355
 nutrition screening/assessment/intervention, 369–375
 overweight/obesity and, 384–389
 physical activity/sports and, 375–377
 pregnancy nutrition and, 149–150
 and promotion of healthy behaviors, 377–380
 resources, 380–381
 substance abuse and, 392
 vitamin/mineral supplement use and, 387, 390
Adolescent pregnancies, 149–150
Adult nutrition
 and actual intake of food, 422–424
 cancer and, 438–439
 cardiovascular disease and, 439–444
 cultural considerations and, 417–418, 424, 426
 diabetes and, 447–450
 dietary recommendations for, 415–420
 exercise recommendations for, 428, 430
 HIV/AIDS and, 450, 452
 interventions for risk reduction in, 430–432
 and maintenance of healthy body, 412–415
 and nutrients of public health concern, 420–422
 overweight/obesity and, 444–447
 and physiological changes of adulthood, 408–412
 resources, 433, 453
 supplementation/alternative medicine and, 424–425
 vegetarian diets and, 426–428
Adult-onset diabetes. See Type 2 diabetes
Aging adults. See Older adult nutrition
Aging theories, 458–460
AIDS. See HIV/AIDS
Alcohol
 adolescent nutrition, 374, 392
 adult nutrition and, 417–420

 fertility and, 56–57
 kava and, 412
 lactation and, 198–199
 male factors of fertility and, 57
 osteoporosis and, 499
 pregnancy outcome and, 105
Allergic diseases, 207
Allergies
 infant nutrition and, 207–208, 239–240
 inflammatory diseases and, 507–508
 in school settings, 349
 toddler/preschooler nutrition and, 303
Alpha-linolenic acid
 adolescent nutrition and, 365
 children/preadolescents and, 318–319
 as essential nutrient, 3
 as essential nutrients, 8
 fish and, 443
Alternative medicine, 424–425. See also Herbal remedies/supplements
Alveoli, 157
Alzheimer's disease, 508–509
Amenorrhea, fertility and, 54–55
American Academy of Pediatric Dentistry, 277
American Academy of Pediatrics (AAP)
 breastfeeding and, 213, 228
 Committee on Drugs, 192
 fluoride and, 277
 school lunches and, 325
American College of Obstetricians and Gynecologists (ACOG), 194
American Dietetic Association (ADA), 228, 277, 465–466
American Heart Association, 415, 441
Amino acids, 5
Amniotic fluid, 93
Amylophagia, 117
Anabolic phase of pregnancy, 88
Anaphylaxis, toddler/preschooler nutrition and, 303
Androgens, 52
Anemia, 113–115. See also Iron
 cow's milk and, 228–229
 hemolytic, 279
 toddler/preschooler nutrition and, 275–276
Anencephaly, 110, 257
Angiogenesis inhibitor, 439
Anorexia nervosa, 70–71, 395–396
Anovulatory cycles, 55
Anthropometric assessments, 38
Antibiotics, prebiotics/probiotics and, 34
Antidiuretic hormone, 513
Antioxidants, 13
 fertility and, 79
 inflammatory diseases and, 506
 male factors of fertility and, 57
 pregnancy outcome and, 116

Appetite/satiety
 aging and, 462, 473
 toddler/preschooler nutrition and, 274–275
Appropriate for gestational age (AGA), 96–97
Arteries, 440
Arteriosclerosis, 440
Artificial sweeteners, intake during pregnancy of, 105
Assessments
 anthropometric, 38
 biochemical, 38
 clinical/physical level assessment, 36
 community-level assessment, 36
 dietary assessment, 36–37
 dietary history, 37
 food frequency questionnaires, 37
 individual-level assessment, 36
 USDA Automated Multiple-Pass Method for assessment, 37–38
 Web dietary assessment resources, 37
Assisted reproductive technology (ART), 141
Asthma, 302
Atherosclerosis, 281
 adult nutrition and, 440
 nutritional risk factors for, 416
Athetosis, 345
Atrial fibrillation, 490
At-risk infants and nutrition, 241, 246–247
Attention deficit hyperactivity disorder (ADHD), 297
 children/preadolescents and, 347
Autism, 253, 297, 300
Autoimmune disease, 73
Autoimmune response, 78

B

Babinski reflex, 222
Baby-bottle tooth decay, 277. See also Dental caries
Baby foods, commercial, 234. See also Infant nutrition
Baby Friendly Hospital Initiative, 177, 179
Bariatric surgery, 447
Basal metabolic rate (BMR)
 adolescent nutrition and, 365
 adult nutrition and, 412
Bee propolis, fertility and, 79
Behavioral Risk Factor Surveillance System, 38
Bereavement, 514–515
Berger, Dave, 481
Best Start Social Marketing, 175–176, 181–182
Bilirubin encephalopathy, 204
Bilirubin metabolism, 202–203
Binge eating, 396–398

Biochemical assessment, 38
Biotin
    function/deficiency of, 14
    overdose consequences/food source of, 15
Birthweight. *See also* Weight
    growth charts and, 250
    as measure of health, 220
    of single/multifetal newborns, 143
Black cohosh, 410
Blink reflex, 222
Blood pressure control. *See* Dietary
        Approaches to Stop Hypertension
        (DASH) Eating Plan
B-lymphocyte, 165
Body dissatisfaction, 396, 399
Body image/dieting, 313, 399
Body mass index (BMI). *See also* Weight
    adiposity rebound and, 280, 310, 315
    adolescents and, 369–370
    children/preadolescents and, 309–311
    definition, 268
    fertility and, 54
    obesity and, 445, 496
    older adult nutrition and, 510–511
    overweight classification and, 279
Body water changes during pregnancy, 88
Bone ages, 315
Botanicals, 443
Bottle caries, 239
Breast augmentation after lactation, 168
*Breastfeeding: A Guide for the Medical
        Professional*, 160
Breastfeeding and Human Lactation Study
        Center, 194
Breastfeeding of multiples, 206–207
Breastfeeding Promotion in Physicians
        Office Practices (BPOPIII), 175, 213
Breast-milk/breastfeeding. *See also*
        Lactation nutrition
    human milk composition, 160–164
    maternal energy balance and milk
        composition, 173–174
    preterm delivery and, 97
Breast-milk jaundice syndrome, 205
Breast non-feeding jaundice, 204–205
*Bright Futures Guidelines for Health
        Supervision of Infants, Children, and
        Adolescents*, 287–288
Bright Futures in Practice: Nutrition,
        287–288
Bronchopulmonary dysplasia (BPD), 302
Bulbourethral gland, 50
Bulimia nervosa, 70–71, 396–397

C
Cabbage leaves, 198
Caffeine
    adult nutrition and, 420
    fertility and, 56
    lactation and, 200
    older adult nutrition and, 475
    PMS and, 68
    pregnancy outcome and, 116–117
Calciferol, 471–472
Calcium
    adolescent nutrition and, 366–367
    adolescents and, 377

adult nutrition and, 421–422
children/preadolescents and, 320
fertility and, 78
food sources of, 26
function/deficiency of, 22
hypocalcemia in infants, 249
menopause and, 410
older adult nutrition and, 472
osteoporosis and, 500
overdose consequences/food source of, 23
pregnancy outcome and, 112–113
toddler/preschooler nutrition and, 276,
        284–285
Calorie, 2
    children/preadolescents and, 337
    dementia and, 509
    gestational diabetes and, 139
    infant nutrition and, 224, 247
    intake during pregnancy of, 105
    male/female use of, 412
    sample 2000-cal. menu, 43–44
    spina bifida and, 340
    toddler/preschooler nutrition and, 295
    translation to grams, 417
Calorie reduction, 459
CAM. *See* Complementary and Alternative
        Medicine (CAM)
Campylobacter sp., 279
Cancer
    adult nutrition and, 438–439
    medications and nutritional
        implications, 511
    older adult nutrition and, 492–496
CANfit Program, 380
Capsaicin, 507
Carbohydrates
    adolescent nutrition and, 365–366
    breastmilk and, 162–163
    children/preadolescents and, 321
    description/DRIs of, 4–5, 7
    as essential nutrients, 3
    galactosemia, 241
    gestational diabetes and, 139
    infant nutrition and, 229
    intake during pregnancy of, 105
    maternal metabolism of, 90
    older adult nutrition and, 468–469
    toddler/preschooler nutrition and, 285
Cardiovascular disease
    adolescent nutrition, 393–395
    adult nutrition and, 431–432, 439–444
    children/preadolescents and, 319
    green tea and, 514
    medications and nutritional
        implications, 511
    nutritional risk factors for, 416
    older adult nutrition and, 489
    prevention in toddlers/preschoolers of,
        280–281
Caries
    bottle, 239
    children/preadolescents and, 314
    toddler/preschooler nutrition and, 277
Carnitine, 444
Carotenemia, fertility and, 56
Carotid artery disease, 490
Casein, 162

Catabolic phase of pregnancy, 88
Catch-up growth, 248
CDC Growth Charts, 309–310
CDC Guide to Breastfeeding Interventions,
        177
Celiac disease, 77–78
    single-gene defects and, 31
    toddler/preschooler nutrition and, 299
Celiac sprue, 78
Cell function, 21, 29
Centers for Disease Control and Prevention
        (CDC)
    HIV/AIDS and, 450
    iron deficiency and, 277
    lead poisoning and, 278
    preconception nutrition and, 62
Cerebral palsy (CP), 97, 300–301, 344–345
    children/preadolescents and, 336
Cerebral spinal atrophy, 338
Cesarean section, preeclampsia and, 132
Channel One, 379
Chaste tree berry extracts, fertility and,
        78–79
Child and Adult Care Food Program, 431
Child and Adult Care Food Program
        (CACFP), 40
Child care, 286–287, 303
Child/preadolescent nutrition
    body image/dieting and, 313
    and development of feeding skills,
        311–313
    dietary recommendations for, 319–322
    energy/nutrient needs and, 313–314
    exercise recommendations for, 322–324
    growth and, 338–339
    herbal remedies/supplements and,
        347–348
    normal growth/development and,
        309–310
    nutrition problems and, 314
    nutrition recommendations for, 339–342
    nutrition services and, 348–350
    physiological/cognitive development of,
        310–311
    prevention of disorders related to,
        314–319
    public food/nutrition programs and,
        324–329
    resources, 330–331, 350
    risk reduction programs for, 324–327
    special health care needs and, 336–338,
        340, 342–347
    tracking of, 308–309
Children with special health care needs,
        246, 294
Chloride
    function/deficiency of, 24
    overdose consequences/food source of, 25
Cholesterol
    adult hormonal shifts and, 410–411
    breastmilk and, 162
    children/preadolescents and, 321
    coronary heart disease (CHD) and, 441
    definition, 9
    HMG-CoA Reductase and, 450
    LDL cholesterol, 280
    toddler/preschooler nutrition and, 285

Cholestin, 425
Chondroitin, 425, 506–507
Chromium
    function/deficiency of, 24
    overdose consequences/food source of, 25
Chronic health conditions
    adolescent nutrition and, 402–404
    definition, 31
    nutritional risk factors for, 416
    toddler/preschooler nutrition and, 294
Chronic hypertension, 130–131
Chronic inflammation, 70
Chronic Obstructive Pulmonary Disease
    (COPD), 511
Cimicifuga racemosa, 410
Cleft lip/palate, 257
Clitoris, 50
Cobalamin, 503
Coenzyme Q$_{10}$, fertility and, 79
Coenzymes, 13, 443
Cognitive development
    children/preadolescents and, 310–311
    infants and, 223
    preschoolers and, 271–272
    toddlers and, 269
Cognitive function, breastfeeding and, 166
Cognitive maturation (average age), 354
Colic, 238
Colostrum, 160
Community outreach
    adolescent nutrition and, 380
    lactation and, 180, 183
Competitive foods, 325
Complementary and Alternative Medicine
    (CAM), 424
Complex carbohydrates, 7. *See also*
    Carbohydrates
Congenital abnormality
    definition, 72
    folate and, 110–111
    infant nutrition and, 256–258
Congenital anomalies
    gestational diabetes and, 135
    infant nutrition and, 256–258
Constipation
    infant nutrition and, 239
    older adult nutrition and, 504–505
    pregnancy and, 122
    toddler/preschooler nutrition and, 278
Consumers Union Education Services, 379
Continuing Survey of Food Intakes by
    Individuals (CSFII), 466–468
    toddler/preschooler nutrition and, 285
Contraceptives, preconception nutrition
    and, 58–60
Contraindications to breastfeeding, 210–212
Copper
    function/deficiency of, 24
    overdose consequences/food source of, 25
Coronary heart disease (CHD)
    adult nutrition and, 439–444
    medications and nutritional
        implications, 511
    older adult nutrition and, 489–491
Corpus luteum (CL), 49
Counseling, preconception nutrition and, 49
Cow's milk, 228–229

Creatine, 425
Critical periods in embryonic growth, 93
Critical periods in oral feeding
    development, 223
Cross-sectional study, 173
Cultural considerations
    adult nutrition and, 417–418, 424, 426
    child/preadolescent nutrition and, 322
    infant nutrition and, 240
    older adult nutrition and, 476
    pregnancy diets and, 118–119
    toddler/preschooler nutrition and, 286
Cup drinking in infants, 232–233
Cystic fibrosis
    children/preadolescents and, 336
    definition, 295
    eating/feeding problems in children
        with, 342–343

D

Daily Values (DVs), 4–5
Deficiencies, 29–30. *See also specific*
    nutrient, i.e. Selenium
Dehydration, older adult nutrition and,
    513–514
Dementia. *See also Alzheimer's disease*
    medications and nutritional
        implications, 511
Dental caries
    children/preadolescents and, 314
    infant nutrition and, 239
    toddler/preschooler nutrition and, 277
Department of Health and Human Services
    (DHHS), 482
Depo-Provera, 59
Depot medroxyprogesterone acetate. *See*
    Depo-Provera
Development, embryonic/fetal. *See*
    Embryonic/fetal growth and
    development
Developmental delay, toddler/preschooler
    nutrition and, 297, 302
Developmental disabilities, 237
Development delay, 252–253
DHEA (dehydroepiandrosterone), 425, 475
Diabetes. *See also Gestational diabetes*
    adult nutrition and, 447–450
    carbohydrate counting/glycemic
        index/glycemic load and, 451
    children/preadolescents and, 336
    eating/feeding problems in children
        with, 343
    medications and nutritional
        implications, 511
    nutritional risk factors for, 416
    obese children/preadolescents and, 315
    older adult nutrition and, 494–496
    pregnancy nutrition and, 134–140
    type 1, 72–73
    type 1 during pregnancy, 140
    type 2, 5, 31, 72–75
Diabetes Control and Complications Trial
    (DCCT), 451
Diabetes mellitus, 72–75
Diabetogenic effect of pregnancy, 90
Diagnostic and Statistical Manual of
    Mental Disorders (DSM IV), 68

Diaphragmatic hernia, 257
Diarrhea
    infant nutrition and, 239
    toddler/preschooler nutrition and, 299
Dibromochloropropane, male factors of
    fertility and, 57
Dichlorodiphenyltrichloroethane (DDT),
    male factors of fertility and, 58
Dietary Approaches to Stop Hypertension
    (DASH) Eating Plan, 40, 42, 44,
    491–492
Dietary assessment, adolescents and,
    369–374
Dietary fiber, 319
Dietary folate equivalents (DFE), 111
Dietary Guidelines
    definition, 40, 42
    older adult nutrition and, 482
    toddler/preschooler nutrition and, 282
    weight management and, 280
Dietary Guidelines for Americans, 415
Dietary history for assessment, 37
Dietary intake
    adolescent nutrition and, 362
    in twin pregnancy, 144–146
Dietary recommendations for
    child/preadolescent nutrition, 319–322
Dietary Reference Intakes (DRIs)
    adolescent nutrition and, 363–364
    children/preadolescents and, 313–314,
        337–338
    definition, 3–4
    obesity and, 318–319
    school breakfasts and, 328–329
    toddler/preschooler nutrition and, 275
    Tolerable Upper Intake Levels, 281
Dietary supplements, 32–33. *See also*
    Herbal remedies/supplements
Dieting
    adolescents and, 396, 398–399
    children/preadolescents and, 313
    fads, 414–415
Differentiation of cells, 93
DiGeorge syndrome, 258
Digestive system development in infants, 223
Dioscorea villosa, 411
Diplegia, 295
Disabilites Education Act (IDEA). *See*
    Individuals with Disabilities Education
    Act (IDEA)
Disordered eating behaviors, 396, 398
Disproportionately Small for Gestational
    Age (dSGA), 95–96
Diverticulitis, 504
Dizygotic (DZ) twins, 142
Docasahexaenioic acid (DHA)
    breastmilk and, 162
    pregnancy outcome and, 107–109
Double Talk, 206
Doula, 176
Down syndrome, 246–247, 339
Dr. Atkins' New Diet Revolution, 415
Drinking from cup, 232–233
Drug abuse. *See specific substance*;
    Substance abuse
Dutch Hunger Winter (1943-1944),
    pregnancy outcome and, 103

Dysmenorrhea, 68
Dyspepsia, 439

E
Ear infections in infants, 239
Early Head Start programs, 288, 303
    nutrition services and, 260–261
Early intervention program, 256
Early intervention services,
    toddler/preschooler nutrition and, 294
Ear Right 4 Your Type, 415
Eating disorders
    anorexia nervosa, 70–71, 395–396
    binge eating, 396–398
    body dissatisfaction and, 396, 399
    bulimia nervosa, 70–71, 396–397
    dieting and, 396, 398–399
    discussion, 72
    disordered eating behaviors, 396, 398
    etiology of, 400
    pregnancy and, 147–148
    pregnancy nutrition and, 147–148
    prevention of, 401–402
    treatment of, 400
Eating out and older adults, 466–467
Eating-related behaviors, adolescent
    nutrition and, 358–362
Echinacea, 197, 425
E. coli, 279
Edema, 88
Education
    adolescent nutrition and, 374–380
    older adult nutrition and, 479–480
    prevention of weight-related disorders,
        402
Eicosapentaenoic acid (EPA), pregnancy
    outcome and, 107–109
Ejaculatory duct, 50
Elderly. See Older adult nutrition
Elderly Nutrition Program (ENP), 481
Embryo, 60
Embryonic/fetal growth and development
    critical periods of, 93–94
    metabolic disorders affecting, 76
    normal, 94
    nutrition/miscarriages/preterm delivery, 97
    variations in, 95–97
Empty-calorie foods, 32
Encephalocele, preconception nutrition
    and, 110
Endocrine, 48
Endometriosis
    bee propolis and, 79
    definition, 52
Endometrium, 50
Endothelium, 131
Energy
    adolescent nutrition, 362–369
    adult nutrition and, 413–414
    aging and, 459
    breastmilk and, 160–161
    child/preadolescent nutrition, 313–314
    children/preadolescents and, 337
    children's ability to control intake of,
        272–273
    infant nutrition and, 224–226, 247–249
    lactation and, 172

older adult nutrition and, 467
    requirements in pregnancy, 104–105
    toddler/preschooler nutrition and,
        275–276
Engorgement, 191, 198
Enrichment/fortification, labeling and, 33
Enteral feeding, 253–254
Environmental factors
    as determinant of nutritional health, 463
    exposure during lactation of, 200–202
    genotype interactions with, 31
    lead exposure and, 226
    and medical contraindications to
        breastfeeding, 211
    that increase oxidative stress, 131
Ephedra (Ma Huang), 197, 425
Epididymis, 50, 52
Epithelial cells, 165
Ergogenic supplement use, adolescent
    nutrition and, 390–391
Essential fatty acids, 8–9
Essential/nonessential nutrients, 3
Estimated Average Requirements (EARs), 4
Estrogen, 52, 410
Ethnicity, dietary needs based on, 35
Evening primrose oil, fertility and, 79
Exercise
    adolescent nutrition and, 391–392
    adolescents and, 373, 375–377
    adult nutrition and, 428, 430
    breastfeeding and, 173–174
    children/preadolescents and, 322–324
    fertility and, 55
    gestational diabetes and, 137
    infants and, 236
    infertility and, 71–72
    older adult nutrition and, 478–479
    osteoporosis and, 498–499, 501
    PMS and, 68–69
    pregnancy and, 120–121
Exposure index, 192
Extension programs, 431, 481
Extremely low birth-weight infant (ELBW),
    246, 251, 254–255, 258

F
Failure to thrive (FTT), 237–238, 258
    toddler/preschooler nutrition and,
        299–300
Failure to thrive, breastfeeding and,
    170–172
Familial hyperlipidemia, 280
Famine and pregnancy outcome
    Dutch Hunger Winter (1943–1944), 103
    Japanese food shortages (WWII),
        103–104
    Siege of Leningrad (1942), 103
Farmer's Market Nutritionprogram, 288
Fats (lipids), 248
    adolescent nutrition and, 365–366
    adult nutrition and, 420–421
    body fat changes in pregnancy, 101
    breastmilk and, 162
    calories translation to grams of, 417
    children/preadolescents and, 320–321
    description, 6
    food sources of, 10–12

infant nutrition and, 224–225, 229
    maternal metabolism of, 91
    older adult nutrition and, 468, 470
    preterm delivery and, 97
    recommended intake of, 9–10
    toddler/preschooler nutrition and,
        284–285
Fatty acids
    definition, 6
    inflammatory diseases and, 507
Fecundity, 48
Feeding Infants and Toddlers Study of 2002
    (FITS), 285
Feeding skills development/behavior
    child/preadolescent nutrition, 311–313
    preschoolers and, 272
    toddlers and, 269–271
Female athlete triad, 71–72. See also Exercise
Fenugreek, 197–198
Fertility
    breastfeeding and, 164
    celiac disease and, 77–78
    definition, 48
    diabetes mellitus and, 72–75
    diet and, 55–57
    eating disorders, 72
    effect of obesity on, 69–70
    female athletes and, 71–72
    hypothalamic amenorrhea and, 70–71
    impaired, 53
    male factors of, 57–58
    phenylketonuria (PKU) and, 76
    polycystic ovary syndrome (PCOS) and,
        75–76
    premenstrual syndrome (PMS) and,
        68–69
Fertility Blend, 79
Fetal alcohol spectrum, pregnancy nutrition
    and, 148–149
Fetal alcohol syndrome, 31, 148–149
Fetal-origins hypothesis of later disease
    risk, and pregnancy nutrition, 97–99
Fetus, 60
Fiber. See also Carbohydrates
    adolescent nutrition and, 364–366
    adult nutrition and, 421
    children/preadolescents and, 319–320
    constipation and, 504–505
    definition, 7
    infant nutrition and, 226
    older adult nutrition and, 468–469
    toddler/preschooler nutrition and,
        284–285
FightBAC™, 279
Fine motor skills, 266, 270
First foods for infants, 233–235
Fish
    adult nutrition and, 443
    and medical contraindications to
        breastfeeding, 211
    mercury in, 109, 121, 201–202
    toddler/preschooler nutrition and, 282
5 A Day for Better Health, 325
504 Accommodation, 348
Fluids. See also Water
    adult nutrition and, 419–420
    children/preadolescents and, 320

older adult nutrition and, 470
toddler/preschooler nutrition and,
284–285
Fluoride
function/deficiency of, 22
infant nutrition and, 225–226, 236
older adult nutrition and, 513–514
overdose consequences/food source of, 23
pregnancy outcome and, 113
toddler/preschooler nutrition and, 277
Fluorosis, 277
Focus group interviews, 182
Folate (folacin/folic acid)
adolescent nutrition and, 368–369
function/deficiency of, 14
older adult nutrition and, 472
overdose consequences/food source of, 15
Folate and neural tube defects, 60–61,
109–111
Folic acid
adult nutrition and, 421
breastmilk and, 163
older adult nutrition and, 472
Follicle stimulating hormone (FSH),
49–51, 52
Food allergy (hypersensitivity), 207, 239–240
Food and Drug Administration (FDA),
herbal remedies and, 120
Food frequency questionnaires for
assessment, 37
Food guide pyramid. *See* MyPyramid
Food guide
Food insecurity, 2
Food intolerance in infants, 207–208
Food preferences
infant nutrition and, 236
toddler/preschooler nutrition and, 273
Food safety/security
definition, 2
nutritional risk factors and, 416
older adult nutrition and, 477–478
pregnancy outcome and, 121
toddler/preschooler nutrition and,
278–279, 298
Foods for infants. *See* Infant nutrition
Food Stamp Program, 40, 288–289, 431, 481
Food texture and infants, 233
Formula. *See also* Infant formulas
children/preadolescents and, 341
Fortification, labeling and, 33
Framingham Heart Health Study, 442
Free to Be Me, 403
Friendly bacteria, 34
Full-term infants, 220
Functional fiber, 319
Functional foods, labeling and, 34
Functional status, 461

G

Galactosemia, 241, 342
Garlic, 443, 494
older adult nutrition and, 475
Gastric bypass surgery (GBS), 447
Gastroesophageal reflux (GER), 502
infant nutrition and, 223, 256, 261
Gastrointestinal diseases, older adult
nutrition and, 502

Gastrostomy feedin, 299
Gastrostomy feeding, 254, 341
Genes
nutrigenomics and, 30–31
nutrition programming and, 98–99
Genetic disorders, infant nutrition and,
257–259
Geophagia, 117
Gestational diabetes, 73, 134–140
Gestational hypertension, 130–131
Ginger, nausea/vomiting during pregnancy
and, 122
Ginger and nausea, 120, 425, 494
Gingiva, 501
Ginseng, 120
Ginseng root, 197, 425
Glucogenic amino acid, 91
Glucophage, 449–450
Glucosamine, 425, 506–507
Glucosamine/chondroitin Arthritis
Intervention Trial (GAIT), 506–507
Glucose, 74–75. *See also* Diabetes;
Gestational diabetes
Gluten, 505
Glycemic index (GI) of carbohydrates, 74
Glycemic index of carbohydrates, 5–6, 105
gestational diabetes and, 139–140
preeclampsia and, 133
Glycemic load (GL), 74
Glycerol
definition, 9
structure of, 8
Glycogen, preterm delivery and, 97
Glycols, male factors of fertility and, 58
Glycosylated hemoglobin, 496
Gonadotropin-releasing hormone (GnRH),
49–50, 52, 72
Gravida, 100
Green tea, 31, 443, 514
older adult nutrition and, 475
Gross motor skills, 266, 270
Growth
2000 CDC Growth Charts (children),
267–268
adolescent nutrition and, 372
adolescent psychosocial, 357–358
adolescents and, 355–357
catch-up, 248
children/preadolescents, 309–310
children/preadolescents and,
338–339
intrauterine growth as predictor,
250–251
normal toddler/preschooler, 266–267
and physiological changes of adulthood,
408–412
preterm infants and, 249–250
specialty charts for, 339
Growth, embryonic/fetal. *See*
Embryonic/fetal growth and
development
Growth assessment of infants, 226–228
Growth velocity, 266
Guidelines for Adolescent Preventive
Services (GAPS), 369
*Guidelines for School and Community
Programs to Promote Lifelong*

*Physical Activity Among Young
People*, 375, 377
*Guidelines for School Health Programs to
Promote Lifelong Healthy Eating*, 377

H

Halal, 36
Halogens, male factors of fertility and, 57
Handbook of Milk Composition, The, 160
Harris, Tammy, 515
Harris-Benedict equation, 412–413
Hawthorn, 443
Hayflick's theory of limited cell replication,
459
Head Start Program, 40, 288, 303
Health, definition, 488
Health/eating-related behaviors, adolescent
nutrition and, 358–362
Health Maintenance organization (HMO),
430
Health Partner's Better Health Restaurant
Challenge®, 430
Healthy Eating Index (HEI), 321
Healthy People 2010 objectives,
toddler/preschooler nutrition and, 267
Heartburn, pregnancy and, 122
Heart disease, as cause of death, 280
Heart Health Study, 442
Heavy metal exposure, male factors of
fertility and, 57
Hematocrit, 276
Heme iron, 368
Hemochromatosis, single-gene defects
and, 31
Hemoglobin, 276
Hemoglobin Alc, gestational diabetes
and, 135
Hemolytic anemia, 279
Hemolytic uremic syndrome (HUS), 279
Herbal remedies, 33
and inappropriateness during
pregnancy/lactation, 195–196
pregnancy and, 120
used in U.S., 195–196
Herbal remedies/supplements
child/preadolescent nutrition, 347–348
children/preadolescents and, 347–348
for infertility, 78–79
menopause and, 411–412
older adult nutrition and, 475
toddler/preschooler nutrition and,
281–282, 303
Herbal teas, 197
High 5 Alabama, 326
HIV/AIDS
adult nutrition and, 450, 452
breastfeeding and, 210–212
children/preadolescents and, 347
Innocenti Declaration and, 177
pediatric AIDS, 295
pregnancy nutrition and, 145–147
St. John's wort and, 197
HMG-CoA Reductase, 450
Homeostasis, 29
Homocysteine, 503–504
Hormonal changes during pregnancy, 88–89
Hormone replacement therapy (HRT), 410

Hormones
    breast development/lactation and,
        158–159
    female reproductive system and, 49–52
    male factors of fertility and, 58
    male reproductive system and, 50, 52
    older adult nutrition and, 475
    osteoporosis and, 500–501
    and physiological changes of adulthood,
        409–412
    thyroxin, 412
Hospitals and breastfeeding, 177–179
Human milk and preterm infants, 208–210
Human Milk Banking Association of North
        American (HMBANA), 212–213
Human milk collection/storage, 212–213
Hyaluronans, 506
Hydrogenation, trans FA and, 9
Hydrolyzed protein formula, 240
Hyperactive letdown, 191
Hyperbilirubinemia, 201–203
Hyperplasia, embryonic
        growth/development and, 93–94
Hypertension
    medications and nutritional
            implications, 511
    milk and, 514
    nutritional risk factors for, 416
    older adult nutrition and, 491–492
Hypertensive disorders of pregnancy,
        130–134
Hypertonia, 299, 339
Hypertrophy, embryonic
        growth/development and, 94–95
Hypoallergenic, 233
Hypocalcemia, 249
Hypogonadism, male factors of fertility
        and, 57
Hypothalamic amenorrhea, 70–71
Hypothyroidism, 116, 241
Hypotonia, 299, 339

I
IDEA, 303
Immunoglobin, 165
Immunological, 48
Impaired glucose tolerance, 75
Inappropriate/unsafe foods, 235
Indirect calorimetry, 247
Individuals with Disabilities Education Act
        (IDEA), 336, 348–349
    nutrition services and, 260
Infant allergies, lactation nutrition and,
        207–208
Infant feeding position, 232
Infant feeding skill development, 229–231
Infant formulas, 229, 255
Infant Health and Development Program
        (IHDP), 250
Infant mortality, 166
    birth weight/preterm delivery and, 86–87
    congenital anomalies/chronic illness
            and, 257
    definition, 220
    reduction of, 87
    as reflection of general population's
            health, 85–86

Infant mortality attributable to birth
        defects (IMBD), 257
Infant nutrition
    at-risk infants and, 246–247
    common problems, 237–240
    congenital anomalies/chronic illness
            and, 256–258
    cross-cultural considerations, 240
    drinking from cup, 232–233
    in early infancy, 228–229
    energy/nutrient needs, 224–226, 247–249
    feeding problems and, 258, 260
    first foods, 233–235
    food texture, 233
    growth and, 249–251
    inappropriate/unsafe foods, 235
    infant development and, 221–224
    infant feeding position, 232
    infant feeding skill development,
            229–231
    interventions to, 260
    introduction of solid food, 231–232
    learning of food preferences, 236
    newborn health assessment and, 220–221
    nutrition guide for, 236–237
    and nutrition intervention for risk
            reduction, 241
    nutrition services, 260–261
    physical growth assessment and,
            226–228
    quantity of food for, 235–236
    reluctance to eat, 233
    resources, 241–242, 262
    severe preterm birth and, 253–256
    special health care needs and, 251–253
    vegetarian diets and, 240
    water and, 235
Infants, breastfeeding and, 168–172.
        See also Lactation nutrition
Infection of the breast, 191–192
Infecundity, 48
Infertility. See also Fertility
    definition, 48
    herbal remedies for, 78–79
Inflammatory diseases and older adults,
        505–508. See also Chronic
        inflammation.
Innocenti Declaration, 177
Innocenti Declaration on the Protection,
        Promotion, and Support of
        Breastfeeding, 181
Institute of Medicine's (IOM) Food and
        Nutrition Boards Recommended
        Dietary Allowances, 415
Institute of Nutrition of Central American
        and Panama (INCAP), 469
Insulin, 72, 343
Insulin-dependent diabetes. See Type 1
        diabetes
Insulin resistance, 5, 69, 72–75, 132, 448.
        See also Type 2 diabetes
Intestinal fertilizer, 34
Intrauterine growth retardation (IUGR), 221
Iodine
    function/deficiency of, 22
    overdose consequences/food source of, 23
    pregnancy outcome and, 116

Iron
    adolescent nutrition, 392–393
    adolescent nutrition and, 367–368
    adult nutrition and, 422
    anemia due to deficiency of, 238–239
    children/preadolescents and, 314, 319
    deficiency, 113
    feeding in early infancy and, 228
    food sources of, 27
    function/deficiency of, 22
    infant nutrition and, 236
    older adult nutrition and, 471
    overdose consequences/food source of, 23
    preconception nutrition and, 49, 62
    pregnancy outcome and, 113–115
    toddler/preschooler nutrition and,
            275–277, 283
Iron deficiency anemia, 113
Isoflavones, 410–411

J
Japanese food shortages (WWII), pregnancy
        outcome and, 103–104
Jaundice/kernicterus, 201–206
Jejunostomy feeding, 254
Joint Commission on Accreditation
        of Healthcare Organizations
        (JCAHO), 202
Juvenile-onset diabetes. See Type 1 diabetes
Juvenile rheumatoid arthritis, 339

K
Kava, 411–412
Kernicterus, 201–206
Ketogenic diet, 342
Ketones, 91, 343
Kilocalorie, 2
Klinefelter's syndrome, fertility and, 57
Krebs cycle, 503
Kwashiorkor, 6

L
Labeling
    dietary supplements, 32–33
    enrichment/fortification and, 33
    functional foods and, 34
    ingredient label, 33
    nutrition facts panel and, 32–33
Labels, Daily Values (DVs) and, 4–5
Labium major, 50
Labium minor, 50
Lactation consultant, 179
Lactation nutrition
    alcohol/drug exposures and, 198–201
    breastfeeding benefits for
            mothers/infants, 164–167
    breastfeeding infant and, 168–172
    breastfeeding of multiples, 206–207
    breastfeeding promotion/support,
            175–180
    breastmilk supply/demand, 167–168
    common breastfeeding conditions,
            190–192
    factors of breastfeeding
            initiation/duration, 174–175, 178
    feeding in early infancy, 228–229
    herbal remedies and, 195–198

human/cow milk composition compared, 161
human milk and preterm infants, 208–210
human milk collection/storage, 212–213
human milk composition, 160–164
infant allergies and, 207–208
maternal diet and, 172
maternal energy balance and milk composition, 173–174
maternal medications and, 192–195
medical contraindications to breastfeeding, 210–212
model breastfeeding promotion programs, 181–183, 213–214
near-term infants and, 208
neonatal jaundice/kernicterus and, 201–206
physiology and, 157–160
public food/nutrition programs and, 180–181
resources, 184, 215
United States goals for, 156–157
Lactation support, 175–180
Lactiferous sinuses, 157
Lactobacillus, 34
Lactogenesis, 158–159
Lacto-ovovegetarians. *See also* Vegetarian diets
adolescent nutrition and, 361
child/preadolescent nutrition and, 323
toddler/preschooler nutrition and, 287
Lactose, intolerance in infants, 240
Lactose intolerance, single-gene defects and, 31
La Leche League, 179–180
Lamaze, 180
Large for gestational age (LGA), 96–97
LDL cholesterol, 280, 443
adult hormonal shifts and, 410–411
Lead
infant nutrition and, 226
toddler/preschooler nutrition and, 278
Lean body mass (LBM), 460
Leptin, 54
Letdown failure, 190–191
Letdown reflex of milk, 159–160
Liability/legal issues, accuracy of *Physician's Desk Reference (PDR)*, 194
License to Eat, 415
Licorice and oral contraceptives, 59
Life expectancy/span, 457–458
Lignins, 319
Linoleic acid
adolescent nutrition and, 365
children/preadolescents and, 318–319
as essential nutrient, 3
as essential nutrients, 8
Linseed, 428
Listeria, 121
Listeria monocytogenes, 279
Liveborn infants, 86
L. Monocytogenes, 121
Lobes of mammary gland, 158
Long-chain fats, 225
Longevity, 456, 459–460

Loving Support Makes Breastfeeding Work, 182
Low-birth-weight infant (LBW), 246, 303
Lower esophageal sphincter (LES), 502
Luteinizing hormone (LH), 49–51, 52

M
Macrobiotic diet, 286
Macrocephaly, 251
Macrophages, 165
Magnesium
adult nutrition and, 422
fertility and, 78
food sources of, 26
function/deficiency of, 22
older adult nutrition and, 472–473
overdose consequences/food source of, 23
PMS and, 69
Ma Huang, 197, 425
Malnutrition, 29–30
breastfeeding and, 170–172
Mammary gland, 157–158
Manganese
function/deficiency of, 24
overdose consequences/food source of, 25
Maple syrup urine disease, 258, 261
Marijuana, lactation and, 200
Maslow's hierarchy, 3
Mastitis, 191–193
Maternal and Child Health Block Program of HHS, 348
Maternal and Child Health Bureau, 175, 180–181
Maternal and Child Health Bureau (MCHB), 213, 260–261, 349–350
Maternal medications, lactation nutrition and, 192–195
MCT oil, 248, 254
Meals-on-Wheels, 431, 481
Meconium, 206
Media, influence on toddler/preschooler nutrition and, 273–274
Medical contraindications to breastfeeding, 210–212
Medical neglect, 299
Medical Nutrition Therapy, 488
Medications and older adults, 465
older adult nutrition and, 509–510
Medications/polypharmacy, older adult nutrition and, 509–510
Medicinal herbs, 195–196
Medium-chain fats, 225
Memory impairment, 508
Menarche, 356
Meningitis, 296
Menopause, 49, 410–412
Menstrual cycle
in athletes/sedentary women, 55
definition, 49
Menstrual flow, 51, 55
Mental retardation, toddler/preschooler nutrition and, 302
Mercury contamination, 109, 121, 201–202
and medical contraindications to breastfeeding, 211
toddler/preschooler nutrition and, 282

Metabolic changes, older adult nutrition and, 470–475
Metabolic syndrome, 70, 73, 448–450
Metabolism, 13
Metformin, 449
Methylmalonic acid (MMA), 503
Microcephaly, 251
Micronutrients, older adult nutrition and, 467–468
Middle childhood, 308
Mifflin-St. Jeor energy estimation equation, 413
Milk. *See* Lactation nutrition
Milk banking, 212–213
Mil to plasma drug concentration ratio (M/P ratio), 192
Minerals, 248
adolescent nutrition and, 364–365, 387, 390
adolescents and, 374
adult nutrition and, 421–422, 424
bariatric surgery and, 447
breastmilk and, 163–164, 208
children/preadolescents and, 314, 319
description, 13, 21
as essential nutrients, 3
food sources of, 26–28
maternal metabolism of, 91
multifetal pregnancies and, 145–146
older adult nutrition and, 470–476
pregnancy outcome and, 112–113, 119–120
summary of individual, 22–25
toddler/preschooler nutrition and, 275–276, 285
Miscarriage
definition, 48
nutrition and, 97
Miso, 411
Mitochondria, 89, 225
Model breastfeeding promotion programs, 213–214
Molecular clock theory of aging, 459
Molybdenum
function/deficiency of, 24
overdose consequences/food source of, 25
Monounsaturated fats, description, 8–9
Monovalent ion, 163
Monozygotic (MZ) twins, 142
Montreal Diet Dispensary (MDD), 123
Morbidity, breastfeeding and, 166
Moro reflex, 222
Mortality
birthweight and, 246
breastfeeding and, 166
Mother-Baby Friendly Hospital, 183
Mothers of Super Twins (MOST), 206
Motor development
infants and, 221–222, 266
toddlers/preschoolers and, 266, 270
Multifetal pregnancies, pregnancy nutrition and, 140–145
Multiparous, 100, 206–207
Muscle coordination problems, toddler/preschooler nutrition and, 300–301
Myoepithelial cells, 157
Myometrium, 50

Myopathy, 443
MyPyramid Food guide, 40–44
    adolescents and, 377
    adult nutrition and, 415, 417
    aging and, 465–466
    children/preadolescents and, 314, 336
    lactation and, 172
    preconception nutrition and, 61–62
    pregnancy outcome and, 118–119
    toddler/preschooler nutrition and,
        282–284

N
Natality statistics, 84–85
National Academy of Sciences Food and
        Nutrition Board, 415
National Alliance of Breastfeeding
        Advocacy (NABA), 180
National Center for Chronic Disease
        Prevention and Health Promotion
        (CDC), infant growth and, 227–228
National Center for Complementary and
        Alternative Medicine (NCCAM), 282
National Cholesterol Education Program
        (NCEP), 441
National Formulary (NF), 474
National Health and Nutrition Examination
        Survey (NHANES), 38, 280, 309, 314,
        316, 362, 419–420, 423
National Health Examination Survey
        (NHES), 316
National Health Interview Survey, 494
National Heart, Lung, and Blood
        Institute, 496
National Heart, Lung, and Blood Institute
        (NHLBI), 511
National Restaurant Association, 424
National School Lunch Program,
        326–329, 431
National School Lunch Program (NSLP), 378
National Training and Technical Support
        Center for Breastfeeding, 183
National Weight control Registry, 446
Nationwide Food Consumption Survey
        (NFCS), 38
Nausea during pregnancy, 121–122, 132
Near-term infants, lactation nutrition
        and, 208
Necrotizing enterocolitis (NEC), 254
Nelson's Textbook of Pediatrics, 269
Neonatal death, 246
Neonatal jaundice/kernicterus, lactation
        nutrition and, 201–206
Neural tube defects (NTDs)
    definition, 60–61
    preconception nutrition and, 110–111
Neurobehavioral, 347
Neuromuscular disorders
    children/preadolescents and, 338–339
    toddler/preschooler feeding problems
        and, 298–299
Neutraceuticals, 34
Neutrophils, 165
Niacin (vitamin B$_3$)
    function/deficiency of, 14
    overdose consequences/food source of, 15
Nicotine, lactation and, 199–200

Nonheme iron, 368
Non-insulin-dependent diabetes. See Type 2
    diabetes
Nonorganic failure to thrive, 237
Nonprotein nitrogen, breastmilk and, 162
Nontropical sprue, 78
Norplant, 59
North American Human Milk Banking
        Association, 212–213
North American Menopausal Society, 411
Nulliparous, 100
Nurses Health Study, 312
Nutrient-dense foods, 32
Nutrients
    adolescent nutrition and, 362–369
    adult nutrition and, 420–428
    categories of, 3
    child/preadolescent nutrition, 313–314
    definition, 2
    inadequate/excessive, 29–30
    infant nutrition and, 224–226, 247–249
    maternal metabolism of, 89–91
    placental transport of, 91–93
    preeclampsia and, 133
    of term/preterm formulas, 256
Nutrigenomics, 30–31
Nutrition, and outcome of multifetal
        pregnancy, 144, 146
Nutritional risk of aging, 462–465
"Nutrition and Your Health: Dietary
        Guidelines for Americans", 416–417
Nutrition Assistance Program for Puerto
        Rico, 431
Nutrition Assistance Program for Seniors
        (NAPS), 481
Nutrition basics
    for adolescent pregnancies, 149
    anthropometric assessments, 38
    biochemical assessment, 38
    chronic diseases and, 31–32
    clinical/physical level assessment, 36
    community-level assessment, 36
    contemporary prenatal research results,
        104–105
    dietary assessment, 36–37
    dietary history for assessment, 37
    dietary needs based on ethnicity, 35
    dietary needs based on religion, 35–36
    food frequency questionnaires for
        assessment, 37
    good/bad foods, 32
    high-risk malnutrition groups and, 31
    importance of adequacy/balance, 32
    individual-level assessment, 36
    infant nutrition and, 220, 246
    key concepts, 2
    monitoring U.S, 38
    priorities for improvements, 39, 41
    public food/nutrition programs, 39
    resources, 45
    USDA Automated Multiple-Pass
        Method for assessment, 37–38
    U.S. guidelines, 40–44
    Web dietary assessment resources for
        assessment, 37
    for women with gestational diabetes,
        138–140

Nutrition Facts Panel, 475–476
Nutrition facts panel, labeling and, 32–33
Nutrition guide, infant nutrition and,
        236–237
Nutrition monitoring, 38
Nutrition Partners, 480–481
Nutrition problems, child/preadolescent
        nutrition, 314
Nutrition programming, 98–99
Nutrition recommendations, for
        child/preadolescent nutrition, 339–342
Nutrition screening/assessment/intervention,
        adolescent nutrition, 369–375
Nutrition Screening Initiative (NSI), 463
Nutrition services
    child/preadolescent nutrition, 348–350
    toddler/preschooler nutrition and,
        303–304
Nutrition support, 253
Nutrition surveillance, 38

O
Obesity
    adolescent nutrition and, 384–389
    adolescent pregnancies and, 149
    adult nutrition and, 444–447
    children/preadolescents and, 314–318
    coronary heart disease (CHD) and, 441
    definition, 279
    fertility and, 54
    gestational diabetes and, 136
    infertility and, 69–70
    metabolic syndrome and, 70
    older adult nutrition and, 496–497
    oral contraceptives and, 59
    preconception nutrition and, 49
    in toddlers/preschoolers, 279–280
Obesity Education Initiative of the
        National Institute of Health, 444–445
Older adult nutrition
    and aging theories, 458–460
    Alzheimer's disease and, 508–509
    bereavement and, 514–515
    cancer and, 492–496
    carbohydrate needs for, 468–469
    cardiovascular disease and, 489
    community food/nutrition programs
        and, 481–482
    cultural considerations and, 477
    dehydration and, 513–514
    dietary recommendations for, 465–467
    energy needs for, 467
    exercise recommendations for,
        478–479
    fat needs for, 468, 470
    fiber needs for, 468–469
    fluid needs for, 470
    food safety and, 477–478
    gastrointestinal diseases and, 502
    health and, 488–489
    hypertension and, 491–492
    importance of, 488
    low body weight/underweight, 510–512
    medications/polypharmacy and, 509–510
    metabolic changes and, 470–475
    micronutrients and, 467–468
    nutrient recommendations for, 475–476

nutritional risk factors and, 462–465
nutrition as contributor to health, 456–457
   obesity and, 496–497
   oral health and, 501
   osteoarthritis and, 505–508
   osteoporosis and, 497–501
   and physiological changes, 460–462
   protein needs for, 468–470
   resources, 483, 515–516
   risk reduction programs for, 479–481
   stroke and, 489–491
   supplementation and, 473–475
   vital statistics of aging population, 457–458
   vitamin B_{12} deficiency and, 502–505
   vitamins/minerals and, 470–476
Older American Act, 481–482
Oligosaccharides, breastmilk and, 163
Omega-3 fatty acids
   pregnancy outcome and, 107–109
   toddler/preschooler nutrition and, 282
Omega-6 to omega-3 fatty acid ratio, 8
Oral contraceptives, 59, 194
Oral-gastric (OG) feeding, 254
Oral health, older adult nutrition and, 501
Organic failure to thrive, 237
Osmolality, 163
Osmolarity, 223
Osteoarthritis, 505–508
   older adult nutrition and, 505–508
Osteoblasts, 500
Osteoclasts, 500
Osteoporosis, 342, 422, 474
   older adult nutrition and, 497–501
Ova, 49
Ovary, 50
Over-the-counter drugs, and medical contraindications to breastfeeding, 211
Overweight, 279. See also Obesity; Weight
Oviduct, 50
Oxidative stress, 131
Oxidative stress theory of aging, 459
Oxytocin, 157

P
Pagophagia, 117
Palmar reflex, 222
Palpebral fissure, 149
Pantothenic acid (pantothenate)
   function/deficiency of, 16
   overdose consequences/food source of, 17
Parenteral feeding, 253
Parenting and infant nutrition, 223–224
Parents of Multiple Births Association (POMBA), 206
Parity, 100
Partnership for Fod Safety Education, 279
Pathologic jaundice, 203–204
Pediatric AIDS, 295
   children/preadolescents and, 347
Pelvic Inflammatory Disease (PID), 52
Penis, 50
Peppermint, 425
Periconceptional period, 68
Perinatal death, 246
Persistent organic pollutants (POPs), 201

Phenylketonuria (PKU), 76
   children/preadolescents and, 336, 345–346
   in infants, 241
   ketogenic diet and, 342
   single-gene defects and, 31
Philtrum, 149
Phosphorus
   food sources of, 28
   function/deficiency of, 22
   overdose consequences/food source of, 23
Physicians Database Query (PDQ), 439
Physician's Desk Reference (PDR), 194
Physiologic jaundice, 203–204, 258
Physiology of adulthood
   growth/maturation changes, 408–412
Physiology of aging, 460–462
Physiology of growth/development
   of child/preadolescent nutrition, 310–311
   in preschoolers, 271–273
   in toddlers, 268–271
Physiology of lactation, 157–160
Physiology of pregnancy
   maternal, 87–88
   normal changes, 88–91
Phytochemicals, 13, 475
Phytoestrogen, 410–411
PICA, 117
Pima Indians, gestational diabetes and, 135
Piper methysticum, 411
Pituitary gland
   definition, 49
   lactation and, 160
Placenta, 87
   hormones of, 90
   nutrient transport across, 91–93
Plant stanols, 475
Plasma ferritin, 367–368
Platelets, 279
Plugged duct, 191–192
Polychlorinated biphenyls (PCBs), male factors of fertility and, 58
Polycystic ovary syndrome (PCOS), 73, 75–76
Polypharmacy and older adults, 509–510
Polyunsaturated fats, description, 8–9
Ponderal index (PI), 97
Postictal state, 343
Postpartum weight retention, 101
Potassium
   food sources of, 28
   function/deficiency of, 24
   overdose consequences/food source of, 25
Pouring rights, 325
Prader-Willi syndrome, 295
Prader-Willi Syndrome, 337
Preadolescence, 308. See also Child/preadolescent nutrition
Preadolescent nutrition. See Child/preadolescent nutrition
Prebiotics, 34–35, 475
Preconception nutrition
   2010 objectives for, 48–49
   celiac disease and, 77–78
   contraceptives and, 58–60
   diabetes mellitus and, 72–75

   eating disorders, 72
   effect of obesity on fertility, 69–70
   female athletes and, 71–72
   female infertility factors, 52–57
   female reproductive system and, 49–52
   hypothalamic amenorrhea and, 70–71
   infertility herbal remedies, 78–79
   male infertility factors, 57–58
   male reproductive system and, 50, 52
   model programs for, 61–62
   nutrition services and, 260–261
   overview, 48–49
   phenylketonuria (PKU) and, 76
   polycystic ovary syndrome (PCOS) and, 75–76
   premenstrual syndrome (PMS) and, 68–69
   RDIs for, 61
   resources, 63, 79–80
   very early pregnancy nutrition exposures, 60–61
Prediabetes, 447
Preeclampsia, 109
Preeclampsia-eclampsia, 130–134
Pregnancy nutrition. *See also* Lactation nutrition; Preconception nutrition
   adolescents and, 149–150
   common pregnancy health problems, 121–122
   course/outcome of pregnancy and, 103–117
   diabetes and, 134–140
   eating disorders and, 147–148
   and embryonic/fetal growth and development, 93–97
   exercise and, 120–121
   fetal alcohol spectrum and, 148–149
   and fetal-origins hypothesis of later disease risk, 97–99
   food safety and, 121
   health diets, 117–120
   HIV/AIDS and, 145–147, 450
   and hypertensive disorders of pregnancy, 130–134
   and multifetal pregnancies, 140–145
   and physiology of pregnancy, 87–93
   pregnancy weight gain and, 99–102
   reduction of infant mortality/morbidity and, 87
   resources, 124–125, 151
   risk reduction programs for, 123–124
   and status of outcomes, 84–86
Preloads, 274
Premenstrual dysphoric disorder (PMDD), 68
Premenstrual syndrome (PMS), 68–69
Preschool age children, 266
Preschooler nutrition. See Toddler/preschooler nutrition
Preterm infants
   definition, 220
   delivery, 97
   feeding problems and, 258, 260
   growth and, 249–250
   human milk and preterm infants, 208–210
   nutrition and severe VLBW, 253–256

Primary malnutrition, 30
Primiparous, 100
Probiotics, 34–35, 475
Produce for Better Health Program, 325
Progesterone, 52, 410
Prolactin, 159
Promotion of healthy behaviors, adolescent nutrition and, 377–380
Proportionately Small for Gestational Age (pSGA), 95–97
Prostacyclins, 8–9, 131–132
Prostaglandins, 8–9, 51
Prostate gland, 50
Protein
   adolescent nutrition and, 365–366
   breastmilk and, 162, 208
   children/preadolescents and, 314, 321, 337
   description/DRIs of, 5–6
   hydrolyzed protein formula, 239–240
   infant nutrition and, 224–225, 229, 248
   intake during pregnancy of, 105–106
   maternal metabolism of, 91
   older adult nutrition and, 468–470
   toddler/preschooler nutrition and, 275–276, 285
Psychosocial growth, adolescents and, 357–358
Psychosocial maturation (average age), 354
Psychostimulant, 347
Puberty, 49, 354–356. See also Child/preadolescent nutrition
Public food programs, 39–40
   child/preadolescent nutrition, 324–329
Pulmonary problems, toddler/preschooler nutrition and, 294, 301–302
Pumping/expressing breastmilk, 167–168, 254

Q
Quality of life, 488
Quantity of food, infant nutrition and, 235–236

R
Raatz, Susan, 414–415
Rate of living theory, 459
Recommended Dietary Allowances (RDAs), 4
Recommended Dietary Intakes (DRIs)
   multifetal pregnancies and, 145
   preconception nutrition and, 61, 63
   toddler/preschooler nutrition and, 275
Recumbent length, 267, 269
Red yeast rice, 443
Reflex, 221–222
Registered dietitian, 36
   celiac disease and, 77–78
Reiser, Nadine, 480
Religion, dietary needs based on, 35–36
Reluctance to eat, 233
Reproductive anatomy/physiology
   female, 49–52
   male, 50, 52
Resilience, 460
Resources, 123–124
   adolescent nutrition and, 380–381
   adult nutrition and, 433, 453

child/preadolescent nutrition and, 330–331, 350
   infant nutrition and, 241–242, 262
   lactation and, 184, 215
   nutrition basics, 45
   older adult nutrition, 483
   older adult nutrition and, 515–516
   preconception nutrition and, 79–80
   pregnancy nutrition and, 124–125, 151
   toddler/preschooler nutrition and, 289–290, 304
Respiratory distress syndrome, preeclampsia and, 132
Resting energy expenditure, 316
Resting metabolic rate (RMR), adult nutrition and, 412–413
Rett syndrome, 296
Riboflavin (vitamin B$_2$)
   function/deficiency of, 14
   overdose consequences/food source of, 15
Risk groups for malnourishment, 31
Risk reduction programs, for child/preadolescent nutrition, 324–327
Rooting reflex, 221–222
Rotavirus, 78
Rush Mothers' Milk Club, 213–214

S
Salmonella sp., 279
SAMe, 425, 507
Saturated fats
   adult nutrition and, 443
   children/preadolescents and, 321
   description, 8–9
   toddler/preschooler nutrition and, 285
Saw palmetto, 425
School Breakfast and Lunch Programs, 40, 326–329
School Breakfast Program (SBP), 378
School Health Index (SHI), 325–326
School Health Policies and Programs Study (SHPPS), 378–379
School health programs, adolescents and, 374–380
School lunch. See National School Lunch Program
School programs, 324–325
Science of nutrition, principles of, 2
Scoliosis, 338
Scrotum, 58
Secondary condition, 338
Secondary malnutrition, 30
Secondary sexual characteristics, 355
Second Harvest, 431
Secretory cells, 157
Secretory Immunoglobin A, 165
Seizures, 246–247, 338
   eating/feeding problems in children with, 343–344
Selenium
   food sources of, 26
   function/deficiency of, 24
   overdose consequences/food source of, 25
Self-efficacy, 358
Semen, definition, 52
Seminal vesicle, 50

Senescent, 459
Senior Nutrition Awareness Project (SNAP), 465
Senior Nutrition Program, 481–482
Senses and older adults, 461–462
Sensorimotor development in infants, 223–224
Serum iron, 367–368
Sex hormone binding globulin (SHBG), 69
Sexuality, adolescent maturity ratings and, 355
Sexual Maturation Rating (SMR), 355, 376
Short-chain fats, 225
Shoulder dystocia, 97
Siege of Leningrad (1942), pregnancy outcome and, 103
Silicone breast implants, 168
Simple sugars, 7. See also Carbohydrates
SlimForever, 415
Small for gestational age (SGA), 95, 221
Smoking. See Tobbaco
Social marketing, 182
Socioeconomics, breastfeeding and, 166–167, 174–175
Sodium
   food sources of, 26–27
   function/deficiency of, 24
   infant nutrition and, 226
   overdose consequences/food source of, 25
   pregnancy outcome and, 116
Soft drinks, 379
   children/preadolescents and, 320–321
Solid food introduction to infants, 231–232
Sore nipples, 190
Soup kitchens, 431
Soy, 410–411, 499
Spastic quadriplegia, 300–301
Special health care needs
   child/preadolescent nutrition, 340, 342–347
   child/preadolescent nutrition and, 336–338
   fluids and, 342
   nutrition and, 251–253
Special Supplemental Nutrition Program for Women, Infants, and Children (WIC). See WIC program for nutrition
Spina bifida, 49, 110–111, 340
Splett, Patricia, 426
Sports. See also Exercise
   adolescent nutrition and, 391–392
   adolescents and, 375–377
   children/preadolescents and, 324
Statins, 450
Stature, 267, 269
Stepping reflex, 222
Steroid hormones, 89
Steroids, male factors of fertility and, 58
St. John's wort, 197, 425
Storage of human milk, 212–213
Store-to-Door, 481
Streptococcus mutans, 277
Stroke, 439–444
   older adult nutrition and, 489–491
Subfertility, 58
   definition, 48
Subscapular skinfold thickness, 316

Substance abuse. *See also specific substance*
adolescent nutrition, 392
adolescents and, 374
dementia and, 508
infant seizures and, 246–247
lactation and, 198–200
and medical contraindications to
breastfeeding, 211
Suckle, 221–222
Sudden infant death syndrome (SIDS), 199
Sugar Busters, 415
Summer Food Service Program, 40, 329
Supplemental Nutrition Program for
Women, Infants, and Children
(WIC), 61
Supplementation, older adult nutrition
and, 473–475

T
Tanner Stages, 355
Tea, 31, 443, 514
older adult nutrition and, 475
Team Nutrition, 40, 329
Teenagers, adolescent pregnancies, 149
Television, overweight/obesity and,
316–317
Telomeres, 459
Tempeh, 411
Temperament, toddler/preschooler nutrition
and, 273
Teratogenic, 72
Testes, 356
definition, 52
Testis, 50
Testosterone, 52, 409–410
T. Gondii, 121
Therapeutic Lifestyle Changes (TLC),
441, 443
Thiamin (vitamin B$_1$)
function/deficiency of, 14
overdose consequences/food source of, 15
Thirst and aging, 462
Thromboxanes, 8–9, 131–132
Thyroxin, 412
Title IIIC, 481–482
Title V Maternal and Child Health
Directors, 180
T-lymphocyte, 165
Tobacco
adolescent nutrition, 392
adolescents and, 374
coronary heart disease (CHD)
and, 441
lactation and, 199–200
Toddler/preschooler nutrition
autism and, 297, 300
chronic conditions and, 295
common problems, 275–279
developmental delay and, 297, 302
diarrhea/celiac disease and, 299
dietary recommendations for, 282–287
energy/nutrient needs, 275
exercise recommendations for, 287
failure to thrive (FTT) and, 299–300
feeding problems and, 296–299
food allergies/intolerances and, 303
food preferences and, 273

growth assessment and, 295–296
herbal remedies/supplements and, 303
media influence and, 273–275
muscle coordination problems/cerebral
palsy and, 300–301
normal growth/development and,
266–268
nutrition services and, 303–304
physiological/cognitive development of
preschooler and, 271–273
physiological/cognitive development of
toddlers and, 268–271
prevention of disorders related to,
279–282
public food/nutrition programs and,
288–289
pulmonary problems and, 294,
301–302
resources, 289–290, 304
risk reduction programs for, 287–288
special health care needs and, 294–295
temperament differences and, 273
tracking of, 266
Toddlers, 266
Tolerable Upper Intake Levels (ULs), 4
Tolerable Upper Intake Levels,
toddler/preschooler nutrition and, 281
Tooth decay, breastfeeding and, 171
Total Diet Study, 38
Total fiber, 319
Toxicity, 29–30. *See also specific nutrient,
i.e. Selenium*
Toxoplasma gondii, 121
Toxoplasmosis, 121
Tracheoesophageal atresia, 257
Trans fatty acids, 9
Trans fatty acids, 280
adult nutrition and, 443
breastmilk and, 162
Transferrin saturation, 367–368
Transient ischemic attacks (TIAs), 490
Transpyloric feeding (TP), 254
Triceps skinfold, 280
Triglyceride
adult hormonal shifts and, 410–411
green tea and, 443
structure of, 8
Triplets. *See* Multifetal pregnancies
24-hour dietary recalls/records, 36–37
Twins. *See* Multifetal pregnancies
Type 1 diabetes, 72–73, 140, 447
Type 2 diabetes, 5, 31, 72–75, 315, 447–449

U
Ubiquinone, 443
Undernutrition, 53-54 *See also* Weight
United States Pharmacopeia (USP), 474
Unsaturated fats, 9
Urethra, 50
Urinary ketone testing, 139–140
U.S. Breast Milk Monitoring Program, 201
USDA Automated Multiple-Pass Method
for assessment, 37–38
USDA Child Nutrition Program, 348
USDA Food and Nutrition Information
Center, 181
USDA Food and Nutrition Service, 289

U.S. Department of Health and Human
Services (DHHS), 180–181
Uterus, 50

V
Vagina, 50
Vanishing twin phenomenon, 142–143
Vas deferens, 50
Vegan diet, 286, 323, 361
Vegetarian diets
adolescent nutrition and, 360–362
adult nutrition and, 426–429
child/preadolescent nutrition and, 322
infant nutrition and, 240
inflammatory diseases and, 507
during lactation, 174
during pregnancy, 106-108
older adult nutrition and, 474
toddler/preschooler nutrition and,
286–287
Vitamin B$_{12}$ (cyanocobalamin) and, 504
Venous thromboembolism, 59
Vermillion border, 149
Very low birth-weight infant (VLBW), 246,
251, 254–256, 258, 261
Vitamin A
adolescent nutrition and, 369
breastmilk and, 163
function/deficiency of, 16
infant nutrition and, 236
older adult nutrition and, 471
overdose consequences/food source of, 17
pregnancy outcome and, 111
toddler/preschooler nutrition and, 281
Vitamin B$_{12}$ (cyanocobalamin)
breastmilk and, 163
dementia and, 508
function/deficiency of, 14
infant nutrition and, 236
older adult nutrition and, 471, 502–504
oral contraceptives and, 59
overdose consequences/food source of, 15
toddler/preschooler nutrition and, 286
Vitamin B$_{12}$, older adult nutrition and,
502–505
Vitamin B$_6$ (pyridoxine)
fertility and, 78
function/deficiency of, 14
nausea/vomiting during pregnancy
and, 122
overdose consequences/food source of, 15
PMS and, 69
Vitamin C (ascorbic acid)
adolescent nutrition and, 369
function/deficiency of, 16
inflammatory diseases and, 506
overdose consequences/food source of, 17
Vitamin D (1,25 dihydroxycholecalciferol)
adolescent nutrition and, 369
adult nutrition and, 422
breastmilk and, 163
fertility and, 78
function/deficiency of, 16
infant nutrition and, 226, 236
inflammatory diseases and, 506
male factors of fertility and, 57
menopause and, 410

older adult nutrition and, 470
overdose consequences/food source of, 17
PMS and, 69
pregnancy outcome and, 111–112
toddler/preschooler nutrition and, 281, 286

Vitamin E (alpha-tocopherol)
adult nutrition and, 421
breastmilk and, 163
fertility and, 78
function/deficiency of, 16
menopause and, 410
older adult nutrition and, 472
overdose consequences/food source of, 17

Vitamin K (phylloquinone, menaquinone)
breastmilk and, 163, 169
function/deficiency of, 18
older adults' medications and, 510
osteoporosis and, 500
overdose consequences/food source of, 19

Vitamin K-deficiency bleeding (VKDB), 169–170

Vitamins, 248
adolescent nutrition and, 364–365, 369, 387, 390
adolescents and, 374
adult nutrition and, 421, 424
bariatric surgery and, 447
breastmilk and, 163, 208
children/preadolescents and, 314, 319
description, 12–13
as essential nutrients, 3
fertility and, 78–79
food sources of, 18–21
infant nutrition and, 226, 236
male factors of fertility and, 57
menopause and, 410
multifetal pregnancies and, 145–146
older adult nutrition and, 470–476, 506
osteoporosis and, 498–500
PMS and, 69
pregnancy outcome and, 111–112, 119–120

summary of individual, 14–18
toddler/preschooler nutrition and, 275, 285

Volumetrics Weight Control Plan, The, 415
Vomiting during pregnancy, 121–122

W
Waist-to-hip ratio, 316
Water
adult nutrition and, 419–420
breastmilk and, 160
dehydration and, 513–514
as essential nutrient, 3
as essential nutrients, 21
infant nutrition and, 235
older adult nutrition and, 513–514
during pregnancy, 109

Weaning, infant feeding position, 232
Web dietary assessment resources for assessment, 37
Weight. See also Obesity
adolescent nutrition and, 356–357, 372, 384–389
adult maintenance of healthy, 412–415
adult nutrition and, 444–447
adult promotion goals of, 409
children/preadolescents and, 309–310, 314–318
eating disorders and, 393–402
fertility and, 54–55, 70
infant growth and, 227–228
loss during breastfeeding, 173
milk and, 514
multifetal pregnancy gain of, 144, 146
older adult nutrition and, 510–512
osteoporosis and, 499
pregnancy gain of, 99–102
toddlers/preschoolers and, 269
type 2 diabetes and, 74–75

Wellstart International, 182–183
Western Ontario MacMaster University Osteoarthritis Index (WOMAC score), 506

Whey proteins, breastmilk and, 162
WIC program for nutrition, 39–40, 431
adolescent nutrition and, 379–380
adult nutrition and, 426
lactation and, 181–182
pregnancy and, 123–124, 260
toddler/preschooler nutrition and, 288, 303

Widowhood. See Bereavement
Withdrawl reflex, 222
Women, Infants, and Children's Program (WIC). See WIC program for nutrition
Work of Breathing (WOB), toddler/preschooler nutrition and, 302
Workplace, breastfeeding and, 179–180
World Health Organization (WHO)
on AIDS and breastfeeding, 211–212
breast milk substitutes and, 176
infant mortality and, 166
lactation and, 201
obesity and, 496
oral contraceptives and, 194
and U.S. life expectancy, 39

X
Xerostomia, 501

Y
Yams, 411

Z
Zeleznak, Karen, 426
Zinc
adolescent nutrition and, 368
breastmilk and, 164
food sources of, 27–28
function/deficiency of, 22
male factors of fertility and, 57
older adult nutrition and, 473
overdose consequences/food source of, 23
pregnancy outcome and, 115–116

om

Library Materials

To:

Interlibrary Loan Department
Parks Library
Iowa State University
Ames, Iowa 50011-2140

Explore nutrition through each stage of life and succeed in your course with this best-selling text

As one of the most respected nutrition life cycle texts, *Nutrition Through the Life Cycle*, Third Edition, helps you understand the nutritional foundations necessary for the growth, development, and normal functioning of individuals in each stage of the life span. Plus, this text is written by an author team of nutrition researchers who will give you a comprehensive and accessible introduction to all the major concepts!

Filled with resources to guide your study:

» *Key Points* at the end of the chapter are great for review. They give you brief facts that address key content or concepts covered in the chapter.

» **Examples, illustrations, and study questions** offer a current look at the topics you're interested in—from eating disorders to heart disease—to bring the text's concepts to life in a way that reflects *your* life and your needs.

» *In Focus* boxes explore issues relevant to your life, providing a forum for discussing topics of current interest in more detail. Topics include metabolic syndrome, hypertension, and diabetes.

» **Margin definitions** allow you to quickly focus on primary topics and important terminology used in the course.

» **Case studies**, featured throughout the text, illustrate theory in practice, making complex normal and clinical nutritional concepts both memorable and easy for you to grasp.

» Each chapter begins with a *Chapter Outline* and list of *Key Nutrition Concepts* to better prepare you for what you will be learning.

GO ONLINE TO PRACTICE FOR YOUR NEXT EXAM!

Book Companion Website
www.thomsonedu.com/nutrition/brow

Make the grade with this online resource! When you use *Nutrition Through the Life Cycle*, Third Edition, you have access to a rich array of learning resources, including lecture notebooks and key terms practice exercises that will help you prepare for your next exam.

InfoTrac® College Edition

New copies of this text include four months of access to **InfoTrac® College Edition**, a fully searchable online database offering anytime, anywhere access to reliable resources. You'll find more than 25 years' worth of full-text articles (not abstracts) from almost thousands of diverse sources, including *American Journal of Health Studies, Better Nutrition, Nutrition Health Review, Nutrition Today, Time, Newsweek,* and *USA Today.*

Visit Thomson Wadsworth at **www.thomsonedu.com**

ISBN-13: 978-0-495-11637-0
ISBN-10: 0-495-11637-8

9 780495 116370